Advances in Carbohydrate Chemistry and Biochemistry

Volume 64

Advances in Carbohydrate Chemistry and Biochemistry

Editor
DEREK HORTON

Volume 64

Amsterdam • Boston • Heidelberg • London • New York • Oxford
Paris • San Diego • San Francisco • Singapore • Sydney • Tokyo
Academic Press is an imprint of Elsevier

Academic Press is an imprint of Elsevier
Radarweg 29, PO BOX 211, 1000 AE Amsterdam, The Netherlands
Linacre House, Jordan Hill, Oxford OX2 8DP, UK
32 Jamestown Road, London NW1 7BY, UK
30 Corporate Drive, Suite 400, Burlington, MA 01803, USA
525 B Street, Suite 1900, San Diego, CA 92101-4495, USA

First edition 2010

ISBN: 978-0-12-380854-7
ISSN: 0065-2318

British Library Cataloguing in Publication Data
A catalogue record for this book is available from the British Library

Library of Congress Cataloging-in-Publication Data
A catalog record for this book is available from the Library of Congress

For information on all Academic Press publications
visit our website at books.elsevierdirect.com

Printed and bound in the USA
10 11 12 13 10 9 8 7 6 5 4 3 2 1

CONTENTS

Contributors ix

Preface xi

Roy Lester Whistler 1912–2010
James N. BeMiller

Per Johan Garegg 1933–2008
Stefan Oscarson and Olle Larm

Structure and Engineering of Celluloses
Serge Pérez and Daniel Samain

I. Introduction 26
II. Cellulose in Its Cell-Wall Environment 28
III. Conformations and Crystalline Structures of Cellulose 33
1. Chemical Structure of the Cellulose Macromolecule 33
2. Crystallinity and Polymorphism of Cellulose 40
3. Crystalline Structures of Native Celluloses 41
4. Cellulose II. 48
5. Cellulose III 52
6. Cellulose IV 53
7. Alkali Cellulose and Other Solvent Complexes 53
8. Cellulose Acetate and Cellulose Derivatives 55
IV. Morphologies of Celluloses 55
1. Polarity of Cellulose Crystals. 56
2. Crystalline Morphology of Native Celluloses 57
3. Whiskers and Cellulose Microfibrils. 58
4. Surface Features of Cellulose. 62
5. Microfibril Organization 65
V. Chemistry and Topochemistry of Cellulose 66
1. Conditions for the Reactions of Cellulose. 66
2. Main Reactions Sites in Cellulose 69
3. Etherification Reactions of Cellulose 72
4. Acylation Reactions of Cellulose 73

5. Deoxygenation of Cellulose . . . 74
6. Topochemistry of Cellulose . . . 75
7. Enzymatic Alterations and Modifications of Cellulose Fibers . . . 81
VI. Tomorrow's Goal: An Economy Based on Cellulose . . . 84
1. Cellulose–Synthetic Polymer Composites . . . 85
2. Protective Films . . . 89
3. Cellulose–Cellulose Composites . . . 91
VII. Conclusions . . . 98
Acknowledgments . . . 100
References . . . 100

Chemical Structure Analysis of Starch and Cellulose Derivatives
PETRA MISCHNICK AND DANE MOMCILOVIC

I. Introduction . . . 118
1. History . . . 118
2. Polysaccharide Derivatives as Functional Polymers of Renewable Resources . 119
II. Chemical Modification of (1→4)-Glucans . . . 121
1. Cellulose . . . 121
2. Starch . . . 125
3. Kinetically Controlled Reactions . . . 127
4. Thermodynamically Controlled Reactions . . . 131
5. Structural Complexity of Glucan Derivatives . . . 132
III. Structure Analysis of (1→4)-Glucan Derivatives . . . 144
1. Average Degree of Substitution . . . 144
2. Monomer Composition: Substituent Distribution in the Glucose Residue . . . 150
3. Distribution of Monomer Residues along the Polymer Chain . . . 159
4. Heterogeneities in the Bulk Material . . . 180
IV. Summary and Perspectives for the Future . . . 183
Acknowledgment . . . 184
References . . . 184

Glyconanoparticles: Polyvalent Tools to Study Carbohydrate-based Interactions
MARCO MARRADI, MANUEL MARTÍN-LOMAS, AND SOLEDAD PENADÉS

I. Introduction . . . 212
1. Background . . . 212
2. Glyconanoparticles and Glyconanotechnology . . . 217
II. Glyconanoparticles: Synthesis, Characterization, and Types . . . 218
1. Noble-Metal Glyconanoparticles . . . 219
2. Magnetic GNPs . . . 239
3. Glyco Quantum Dots . . . 245
III. Applications of GNPs . . . 250
1. GNPs in Carbohydrate–Carbohydrate Interaction Studies . . . 251
2. GNPs in Carbohydrate–Protein Interactions . . . 254
3. Other Interaction Studies . . . 258
4. Glyconanoparticles as Antiadhesion Agents . . . 260

5. GNPs in Cellular and Molecular Imaging 264
6. Other Applications of GNPs 269
IV. Future and Perspectives 269
References 270

1-Amino-1-deoxy-D-fructose ("Fructosamine") and its Derivatives
VALERI V. MOSSINE AND THOMAS P. MAWHINNEY

I. Introduction 292
1. Historical Perspective 292
2. Nomenclature 293
II. Methods of Formation 294
1. Non-Enzymatic 295
2. Enzymatic 313
III. Structure and Reactivity 316
1. Tautomers in the Solid State and Solutions 316
2. Analytical Methods 322
3. Acid/Base- and Metal-Promoted Reactions 327
4. Enzyme-Catalyzed Reactions 338
IV. Fructosamines as Intermediates and Scaffolds 343
1. The Maillard Reaction in Foods and *in vivo* 344
2. Neoglycoconjugates 349
3. Synthons 351
V. Biological Functions and Therapeutic Potential 354
1. Endogenous Fructosamines in Plants and Animals 354
2. Fructosamines as Potential Pharma- and Nutra-ceuticals 358
VI. Concluding Remarks 366
References 367

Sialidases in Vertebrates: A Family of Enzymes Tailored for Several Cell Functions
EUGENIO MONTI, ERIK BONTEN, ALESSANDRA D'AZZO, ROBERTO BRESCIANI, BRUNO VENERANDO, GIUSEPPE BORSANI, ROLAND SCHAUER, AND GUIDO TETTAMANTI

I. Introduction 405
II. The Lysosomal Sialidase NEU1 406
1. Background 406
2. Lysosomal Routing and Formation of the Multienzyme Complex 409
3. Human NEU1 Deficiency 411
4. New Functions of NEU1 in Tissue Remodeling and Homeostasis 413
5. Mouse Model of NEU1 Deficiency and Study of Disease Pathogenesis 418
III. The Cytosolic Sialidase NEU2 422
1. General Properties of NEU2 422
2. Crystal Structure of Human Sialidase NEU2 423
3. Functional Implication of NEU2 425
IV. The Plasma Membrane-Associated Sialidase NEU3 429
1. General Properties of NEU3 429
2. Functional Implication of NEU3 431

V. The Particulate Sialidase NEU4 435
1. General Properties of NEU4 435
2. Functional Implication of NEU4 436
VI. Sialidases and Cancer 438
VII. Sialidases and Immunity 439
VIII. Further Evidence for Possible Functional Implications of Sialidases 442
IX. *in silico* Analysis of Sialidase Gene Expression Patterns 444
X. Amino Acid Sequence Variants in Human Sialidases 448
XI. Sialidases in Teleosts 449
XII. Trans-Sialidases: What Distinguishes Them from Sialidases? 452
XIII. Final Remarks 459
References 460

AUTHOR INDEX 481
SUBJECT INDEX 529

LIST OF CONTRIBUTORS

James N. BeMiller, Department of Food Science, Purdue University, West Lafayette, Indiana, 47907-2009, USA

Erik Bonten, Department of Genetics, St. Jude Children's Research Hospital, Memphis, Tennessee, 28105-2794, USA

Giuseppe Borsani, Department of Biomedical Science and Biotechnology, University of Brescia, 25123, Italy

Roberto Bresciani, Department of Biomedical Science and Biotechnology, University of Brescia, 25123, Italy

Alessandra D'Azzo, Department of Genetics, St. Jude Children's Research Hospital, Memphis, Tennessee, 28105-2794, USA

Olle Larm, ExThera AB, Karolinska Science Park, Stockholm, Sweden

Marco Marradi, Laboratory of Glyconanotechnology, Biofunctional Nanomaterials Unit, CIC biomaGUNE/CIBER-BBN, San Sebastian, 20009, Spain

Manuel Martín-Lomas, Laboratory of Glyconanotechnology, Biofunctional Nanomaterials Unit, CIC biomaGUNE/CIBER-BBN, San Sebastian, 20009, Spain

Thomas P. Mawhinney, Department of Biochemistry, University of Missouri, Columbia, Missouri, 65211, USA

Petra Mischnick, Technische Universität Braunschweig, Institute of Food Chemistry, Schleinitzstr. 20, D-38106 Braunschweig, Germany

Dane Momcilovic, Department of Fibre and Polymertechnology, Royal Institute of Technology, Teknikringen 56-68, SE-100 44 Stockholm, Sweden

Eugenio Monti, Department of Biomedical Science and Biotechnology, University of Brescia, 25123, Italy

Valeri V. Mossine, Department of Biochemistry, University of Missouri, Columbia, Missouri, 65211, USA

Stefan Oscarson, Centre for Synthesis and Chemical Biology, UCD School of Chemistry and Chemical Biology, University College Dublin, Belfield, Dublin 4, Ireland

Soledad Penadés, Laboratory of Glyconanotechnology, Biofunctional Nanomaterials Unit, CIC biomaGUNE/CIBER-BBN, San Sebastian, 20009, Spain

Serge Pérez, European Synchrotron Radiation Facility, 6 rue Jules Horowitz, BP 220, F-38043 Grenoble Cedex, France; CERMAV-CNRS, BP 53, F-38041 Grenoble Cedex 9, France

Daniel Samain, CERMAV-CNRS, BP 53, F-38041 Grenoble Cedex 9, France

Roland Schauer, Institute of Biochemistry, University of Kiel, Kiel, D-24098, Germany

Guido Tettamanti, Laboratory of Stem Cells for Tissue Engineering, IRCCS Policlinico San Donato, San Donato Milanese, 20097, Italy

Bruno Venerando, Department of Medical Chemistry, Biochemistry and Biotechnology, University of Milan, Segrate, 20090, Italy

PREFACE

The two most abundant organic substances on Earth, cellulose and starch, have been known since antiquity and have been put to many uses, although their chemical identity as polymers of D-glucose were not recognized until the early part of the 20th century. Remarkably, while our detailed knowledge of the structural and functional complexity of a great many carbohydrates in biological systems has burgeoned in recent years, many aspects of these two biopolymers have remained poorly understood. This current Volume 64 of *Advances* features two complementary chapters: Pérez and Samain (Grenoble) focus on the structure, morphology, and solubilization of cellulose, while Mischnick (Braunschweig) and Momcilovic (Stockholm) examine the chemical modification of cellulose and starch with the powerful newer analytical tools now available.

Fiber X-ray crystallography is the key tool used by the Grenoble group to focus on such organizational levels of cellulose structure as chain conformation, chain polarity, chain association, crystal polarity, and microfibril structure. Sundararajan and Marchessault, in Volume 36 of *Advances,* reviewed knowledge earlier gleaned by this technique, but long-standing questions have remained concerning the packing of chains in native cellulose in relation to cellulose modified by chemical treatment. Pérez and Samain now offer resolution of many of these questions and also provide a glimpse of promising new applications for cellulose based on the design of materials having unique physicochemical properties.

The chemical derivatization of cellulose and starch at the three available hydroxyl groups is a reaction of inherent high complexity on account of differences in the reactivity of the individual groups, questions of accessibility in heterogeneous systems, and their behavior of the groups under kinetic or thermodynamic conditions. Mischnick and Momcilovic take advantage of advanced spectroscopic and chromatographic methods to elevate the classic Spurlin model to a higher level of predictability for understanding the structure–properties relationships in chemically modified cellulose and starch. Such interpretations point the way for improving reproducibility and the rational design of products having desired properties. Much of the older work found in the patent literature on starch chemistry, as detailed in by Tomasik and coworkers in Volume 59 of this series, employed a largely empirical approach.

A key driving force behind much current research on carbohydrates has been the potential for applications in medicine and biology, as exemplified in the extensive chapter on neoglycoconjugates by Chabre and Roy in the previous volume (63) of this series. In the current volume, Marradi, Martín-Lomas, and Penadés (San Sebastian), working at the forefront of carbohydrate nanotechnology, detail the preparation and properties of nanoparticles functionalized with carbohydrates, with particular focus on gold glyconanoparticles wherein neoglycoconjugates bearing a thiol functional group are clustered around a gold nanoparticle to present a glycocalix surface. These carbohydrate-based nanoparticles have been used as tools in interaction studies with proteins in biological systems. The gold-based approach is further extended to the use of a variety of other materials, enabling access to multifunctional glyconanoparticles having a range of optical, electronic, mechanical, and magnetic properties.

The interaction of simple sugars with amines to initiate the series of transformations termed the Maillard reaction is well known in food chemistry, with its initial reaction to form a glycosylamine being succeeded by the Amadori "rearrangement" to afford a 1-amino-1-deoxy ketose derivative. The Amadori rearrangement reaction between glucose and such biologically significant amines as proteins leads to an N-substituted 1-amino-1-deoxy-D-fructose (fructosamine) structure by the process termed "glycation." Since the 1980s, fructosamine has witnessed a boom in biomedical research, mainly due to its relevance to pathologies in diabetes, Alzheimer's disease, and related aging processes. In an extensive survey, Mossine and Mawhinney (Columbia, MO) detail important new knowledge on, and applications of, fructosamine-related molecules in basic chemistry, and in the food and health sciences.

Again reflecting medical and glycobiological aspects, research development on vertebrate sialidase biology during the past decade is addressed here by Monti (Brescia) along with coauthors Bonten (Memphis, TN), d'Azzo (Memphis, TN), Bresciani (Brescia), Venerando (Milan), Borsani (Brescia), Schauer (Kiel), and Tettamanti (Milan). Since Schauer's original chapter on sialic acids in Volume 54 of *Advances,* there have been many developments in our understanding of the important roles played by sialic acid-containing compounds in numerous physiological processes, including cell proliferation, apoptosis and differentiation, control of cell adhesion, immune surveillance, and clearance of plasma proteins. Within this context, sialidases, the glycohydrolases that remove the terminal sialic acid at the nonreducing end of various glycoconjugates, perform an equally pivotal function. The authors detail the subcellular distribution of these enzymes in mammalian tissues and their particular function, and also highlight the trans-sialidases, enzymes abundant in trypanosomes and expressing pathogenicity in humans.

Finally, this volume pays tribute to the life and work of two recently deceased carbohydrate scientists of note, Roy L. Whistler at the age of almost 98 and, younger by a whole generation, Per J. Garegg at the age of 75.

In a uniquely personal account, Whistler's student, James BeMiller, traces the threads of his mentor's long career in industry, government, and academia. Whistler's complex and ambitious character cast a powerful influence over the many who worked with him on the structure and applications of polysaccharides, where he was able to bridge effectively the academia–industry divide in the practical exploitation of numerous polysaccharides. Over the course of many decades, he was the key mover in the establishment and development of the international carbohydrate meetings, as detailed by Angyal in Volume 61 of *Advances*. He was also the author and editor of several books on polysaccharides and their industrial applications, as well as being a contributor to *Advances* and a member of its Board of Advisors.

In contrast, Per Garegg left his mark as a superbly talented synthetic carbohydrate chemist leading an enthusiastic and devoted group of coworkers in the construction of remarkably complex oligosaccharide sequences of relevance in medicine and biology. Working in Stockholm with his mentor Bengt Lindberg and later as Lindberg's successor, his synthetic targets focused on the carbohydrate sequences of bacterial polysaccharides whose structures had been elucidated by the Lindberg group. The tribute here by Oscarson (Dublin) and Larm (Stockholm), together with input from other friends and colleagues, paints a picture of a genial man much admired by his student coworkers, an important contributor to the *Advances*, and a familiar figure at the major carbohydrate meetings.

DEREK HORTON

Washington, DC
July, 2010

ROY LESTER WHISTLER

1912–2010

There are relatively very few persons who will be remembered for an extended period of time. Roy Lester Whistler will be one of those because of what he accomplished. In a 1979 issue of *Carbohydrate Research* dedicated to him on the occasion of his formal retirement from Purdue University, but not from active research, I wrote that he was an internationally recognized scientist "primarily for his investigations of polysaccharide structure and function and, more especially, for his efforts towards the industrial utilization of polysaccharides, although he has been active in several areas of carbohydrate chemistry and biochemistry. Because of his interest in, and knowledge of, the industrial modification and uses of polysaccharides, and because of his business sense, Professor Whistler is known and respected by both academicians and industrialists. He has been sought as a consultant to many corporations, more in fact than he could serve because of conflicts of interest and limitations of time, and he has served on the Boards of Directors of several companies. This association with industry has kept industrialists supplied with ideas and has brought their questions and problems into his academic research laboratory."[1]

Those of us who knew Professor Whistler a little better than others did probably saw him as a driven person, but we may never know what was the origin of that drive. We do know that Roy Lester Whistler was born[2] on March 31, 1912 in Morgantown, West Virginia. He was raised by his paternal grandparents in Tiffin, Ohio, from a quite young age following the death of his mother,[3] according to what he told me. He also told me that his father lived in Fostoria, Ohio. He spoke highly of his grandfather, William Harris Whistler, whom he described as being "very smart, but self-taught," having had to quit school after the 8th grade. He also described him as a tough, honest, upright, and

[1] J. N. BeMiller, Professor Roy L. Whistler, *Carbohydr. Res.*, 70 (1979) 179–184.

[2] There is some discrepancy about this date. His Purdue University records list his date of birth as March 21, but government records give it as March 31.

[3] From time to time, he told me a little, but only a little, about his youth. What he told me is congruent with what he told his son William, which was also sketchy.

ISBN: 978-0-12-380854-7
DOI: 10.1016/S0065-2318(10)64001-2

kind person. As a young man after the U.S. Civil War, his grandfather spent time in the "Wild West" with General Miles. Dr. Whistler also described his grandfather as one who could handle a gun (with a quick draw) and could hold his own in a fist fight. Not only the young Roy, but also the mature Dr. Whistler admired his grandfather's ruggedness. In Tiffin, his grandfather lived with the "poor people" on Water Street across the railroad track. His backyard sloped down to the Sandusky River. He was a plasterer by trade. Dr. Whistler also told me that his grandfather died when he was "about a freshman in high school" (9th grade). We presume that he must have continued living with his grandmother, although he never recounted anything about her.

Dr. Whistler attended the Minerva Street Elementary School in what he called a "rough, working-class neighborhood." He then attended Columbian High School. He told me that he worked in a grocery store during his high school years (6–8 am before school, 30 minutes at noon so that other workers could have lunch, 4–6 pm after school, and an additional 6–8 pm on Saturdays). In the summer, he worked in Tiffin's playgrounds doing maintenance work. At the end of one summer, he worked in the frit kiln of a pottery.

Dr. Whistler said that he became attracted to chemistry in Junior High School. As a high school senior, he won The Ohio State Garvan Award (a state-wide essay contest sponsored by the American Chemical Society) with an essay on "The Relation of Chemistry to National Defense." He was most proud of that accomplishment, the first of many awards he was to receive.

After high school, he attended Heidelberg College in Tiffin, majoring in chemistry, physics, and mathematics and worked part time in the chemical storeroom. Upon earning a B.S. degree (1934), he matriculated in the graduate program in chemistry at The Ohio State University. As a graduate student, he took a course in carbohydrates taught by the department head, William Lloyd Evans, who had been a student of J. U. Nef and who was known for his work on the effects of alkali and oxidants on simple sugars. Professor Evans, whom he called an inspiring teacher, wrote on his term paper A^+ and a note saying, "This is a splendid piece of work. I hope that you will continue to do good carbohydrate research in your experimental efforts, June 5th, 1935." This recognition must have also meant much to him because he kept the paper for the rest of his life. (I now have the paper, which is reproduced as an appendix to this article because it exemplifies his love of writing, even as a beginning graduate student, and the state of carbohydrate science when he entered the field.) Earlier, young Roy Whistler had had discussions with each of the faculty members in the department and selected Melville L. Wolfrom, who had joined the chemistry faculty in 1929, as his major professor. Although Dr. Whistler was a graduate student of Dr. Wolfrom for only one year, they became lifelong friends.

Dr. Whistler was a graduate student in the middle of the Great Depression. In the spring of 1935, Governor Davy of Ohio cut the university's budget and no assistantships were awarded. Dr. Whistler told Dr. Wolfrom that he could not continue school without a source of income. According to him, Dr. Wolfrom was just leaving for the spring meeting of the American Chemical Society and, on his return, called him into his office to tell him that, although it was late in the school year, Professor Ralph Hixon at Iowa State College (now Iowa State University) had said that he would offer him an assistantship. So after completing requirements for the M.S. degree in the summer of 1935, Roy Whistler was off to Ames, Iowa.

Before leaving Columbus, he and Leila Anna Barbara Kaufman, who had graduated from The Ohio State University in 1934, were married (September 6). Lee, as she was known to him and their friends, preceded him in death. They adopted a son and named him William Harris Whistler, after his great grandfather. Earlier, I wrote, "Carbohydrate chemists the world over know Lee as a charming, intelligent person and hostess."[1] At the end of her life, Dr. Whistler said of her, "During all the years I've known her, I never heard her say a bad word about anyone." That impressed him. Dr. Whistler also had social graces. The two of them traveled abroad extensively—Dr. Whistler acquainting himself with the major carbohydrate chemists of the time and spreading his influence throughout the world. He was an invited lecturer and/or advisor to universities in nine countries outside the United States [Australia (three times), Egypt (twice), New Zealand (three times), South Africa (six times), and once each to Argentina, Brazil, The Peoples Republic of China, Taiwan, and the USSR] and to Academies of Sciences in four countries [France (once), Hungary (twice), Poland (four times), and the USSR (once)]. Many of those he met in his international travels were also guests of his in West Lafayette, where they were received cordially and entertained in their home. In the 1950s, he invited Professor Ziro Nikuni (Japan) to come to his laboratory as a visiting scientist and several younger people from Japan and Germany to work in his laboratory as postdoctoral researchers. Altogether, some 102 postdoctoral research associates and visiting scientists from 25 countries worked in his laboratory.

His laboratory work while a Ph.D. student of Professor Hixon's resulted in five papers, two of which were not part of his thesis research. One of his non-thesis accomplishments was the development of a commercial process for producing β-D-glucopyranose, a substance desired, and rapidly put to use, by the soft drink industry as a sweetener because of its cold water, solubility. This development probably marked the beginning of his long and extensive association with industry.

Dr. Whistler told me that, after completing the requirements for the Ph.D. degree in 1938, he was offered a position as an Instructor in the Chemistry Department at Iowa State College (where he had done some teaching as a graduate student), a postdoctoral position at Rockefeller University (with an associate of P. A. Levene), and a fellowship at the National Bureau of Standards, an offer that was made upon the recommendation of Dr. Wolfrom with the backing of Dr. Hixon. He told me that he felt very fortunate to have a choice, because there were only about 100 postdoctoral positions available in the entire United States at the time. He chose the National Bureau of Standards (NBS) fellowship, which was in a group headed by Dr. Milton Harris, whom he called "a truly inventive and energetic scientist and an astute business administrator." At the NBS, he was able to rub shoulders with Dr. Horace S. Isbell, Dr. W. Ward Pigman, and Dr. Claude S. Hudson, who was an advisor to the Harris laboratory. In the Harris laboratory, he developed a method to measure the uronic acid contents of polysaccharides (published papers in 1940 and 1941). It is not clear whether this was his first association with polysaccharides or if it was in the laboratory of Professor Hixon, who was beginning to change his focus to starch while Dr. Whistler was a student there.

After two years in Washington, DC, Dr. Whistler was hired to head the Starch Structure Section at the newly established Northern Regional Laboratory (NRL) of the U.S. Department of Agriculture (USDA), Peoria, Illinois, a position he held from 1940 through 1945. There and ever since, he followed a theme of basic research related to practical applications, driven by a strong belief that carbohydrates could be important commercial commodities.

In 1946, after World War II had ended, Dr. Whistler accepted a position as an Assistant Professor in the Department of Agricultural Chemistry (now the Department of Biochemistry) at Purdue University, West Lafayette, Indiana. There his career blossomed. It was a time of the famous carbohydrate laboratories, such as those of J. E. Courtois, K. Heyns, F. Micheel, M. Stacey, M. L. Wolfrom, and others, and the establishment of the laboratories of D. French, J. K. N. Jones, N. K. Kochetkov, R. U. Lemieux, B. Lindberg, H. Paulsen, and others. He made himself one of the established group of carbohydrate chemists in the world, something I think he could do because he stood out even at that age. For example, in 1951, Professor Wolfrom, who chaired the U.S. carbohydrate nomenclature committee, asked Dr. Whistler to join the committee, upon which he immediately developed an interest in systematic rules for polysaccharide nomenclature, which he developed and incorporated into his book on *Polysaccharide Chemistry* (1953) and promoted strongly. He also chaired the Polysaccharide Nomenclature subcommittee of the National Research Council Committee on Nomenclature.

As best I could determine, during his career, he and those in his laboratory published 276 research papers in the peer-reviewed literature. In addition, he authored or co-authored 106 review articles, book chapters, encyclopedia articles, and methods; 26 book reviews; and 16 non-technical or less-technical articles about realized and potential industrial uses of polysaccharides and agricultural crops and residues.

He was an author of 3 books and an editor of 15. Patents issued to, or applied for by, him and his colleagues numbered 24.

Immediately upon arriving at Purdue University, he championed the use of agricultural crops, and especially crop residues, in industrial applications, and initiated a research program on the polysaccharides of annual agricultural crops, particularly the hemicelluloses. He and his students made holocellulose from corn (maize) stalks and cobs and extracted, fractionated, purified, and structurally characterized hemicelluloses from a variety of sources, but mostly from tissues of corn, in a plethora of papers beginning with one in 1948 and ending with one published in 2004 and a paper presented at a meeting of The American Chemical Society in 2005. In the process, he developed the method of fragmentation analysis of hemicelluloses. To separate the numerous oligosaccharides produced by this method, he developed the method of carbon column chromatography (1950), which was short-lived, but during its lifetime was widely used for producing clean separations of rather large amounts of material and was a citation classic.

Beginning with his days at the NRL of the USDA, he perceived of and championed the industrial potential of amylose. First, he explored and found new low-cost agents that would complex with amylose and allow it to be separated from amylopectin by crystallization (1945, 1950) and later used the technique to isolate an intermediate fraction (1961). Then, he made amylose triacetate and films and fibers from it (1943, 1944, 1958). Later, he did the same with xylan (1949) and guaran (1949). He convinced Professor H. H. Kramer, a corn geneticist at Purdue University, to join him in developing high-amylose corn. Together they discovered how the amylose content of starch is controlled by a gene (to become known as the *ae* gene) and simultaneously with another group (that of Detherage *et al.*) raised the amylose content of corn starch (1949–1957). Soon thereafter, high-amylose corn (amylomaize) starch was commercialized.

He appears to be the first to study retrogradation by investigating the effect of acid-catalyzed hydrolysis on the retrogradation of amylose (1948) and applied the method thus developed to examine the rates of retrogradation of a variety of native starches (1954).

Beginning with his first days at Purdue University (1946), he foresaw the industrial potential of guar gum and began its structural analysis. The public articles he wrote about guar seemingly intrigued agricultural and industrial gum interests, because not long thereafter, guar was cultivated in Texas and Oklahoma and guar gum was being produced and used. (Guar is now grown almost entirely in India and Pakistan.) In a series of papers (1948–1952), he worked out the structure of guaran. He also called to the attention of the United Nations the potential value of the protein of the guar seed, present to the extent of ~40%. He sought out Professor T. Hymowitz, an agronomist at the University of Illinois, and together they produced the book *Guar: Agronomy, Production, Industrial Use and Nutrition* (1979), which stimulated work on the nutritional value of guar protein to aid world health and expand use of the crop.

His laboratory produced a number of papers on the oxidation of polysaccharides, with a particular focus on the products of the oxidation of starch with alkaline solutions of chlorine (hypochlorite) (1951–1968). Starches oxidized with hypochlorite were commercial products, but until his work, the actual products of oxidation were unknown.

He also contributed extensively to the understanding of the alkaline degradation of polysaccharides (1955–1966), beginning with the addition to his laboratory of several postdoctoral associates from the laboratory of James Kenner of the British Rayon Research Association.

He promoted the expanded use of modified starches and cellulose and investigated new derivatives of the two polysaccharides (1961–1969), with an emphasis on starch derivatives. He developed procedures for making sulfate half-esters of starch and other polysaccharides (1961–1963) and showed that 6-*O*-sulfated D-glucosyl units of starch could be converted into 3,6-anhydro-D-glucosyl units (1961).

For many years (1962–1988), he investigated the replacement of the ring oxygen atom of sugars and compounds containing them with another hetero atom (S, Se, NH, P) and, after those in his laboratory had synthesized the analogues, examined their chemical properties and biological activities. 5-Thio-D-glucose turned out to be similar to D-glucose, having the same ring conformation and most of the same chemical and physical properties. Biochemically, however, it was quite different and potentially useful, since it had very low toxicity. Its effect of inhibiting D-glucose uptake by cells led to its investigation as an inhibitor of sperm development, and in this application it was found effective. When withdrawn, sperm development resumed.

He also found that 5-thio-D-glucose was sweeter than D-glucose and reasoned that the similar analogue of D-fructofuranose would be sweeter than the natural sugar. He made it and found that indeed it was, while producing no adverse effects on the animal consuming it. The concept of sweetness in relation to molecular structure stimulated him to develop a mathematical relation between aspects of carbohydrate structure and sweetness. From this activity, he designed and made 1,6-di-S-methyl-1,6-dithio-D-fructose, a substance 15–20 times sweeter than sucrose. The research on sweetness was published in the period 1982–1990.

His interest in high-intensity sweeteners coincided with a national interest in reducing calorie intake. Incorporation of a high-intensity sweetener into a food product, other than a beverage, requires simultaneous use of a bulking agent to replace the physical functionalities imparted by sugar, so his attention was turned to non-caloric food bulking agents, which brought him back to hemicelluloses.

He also investigated the synthesis of monosaccharide derivatives containing polymerizable constituent groups and their polymerization (1961–1970), and grafted carbohydrate units onto polysaccharides (1962–1970).

Among other ideas he researched, such as isolation and characterization of oligosaccharides from starch (1954–1955), he looked into the biological and physiological effects of polysaccharides on animals (1967–2004).

In conducting his research, he was always on the lookout for new instrumental techniques. He was among the first to explore the use of infrared (IR) spectrophotometry for analysis of carbohydrates, although IR turned out not to be of much use. He obtained one of the very first transmission electron microscopes made by RCA, and used it in pioneering work on investigations of the fine structure of starch granules (1955–1960), a technique that continues to be extremely useful. In conducting his research he was aggressive and, like other professors of his era, more of a competitor than a collaborator with others in the field.

There were many different aspects to the life, work, and personality of Roy Whistler. In addition to being a scientist, he was a creator and leader in local, national, and international scientific and professional organizations; a bridger of academia and industry; a communicator, especially a compiler of the literature; a promoter; an educator and advisor; an outdoorsman; and above all an achiever. He was quite good at reading the characters and personalities of others and was at heart a very kind person, and both of these traits helped him to accomplish what he did.

He was deeply involved in the creation and leadership of the International Carbohydrate Organization. There is some confusion/disagreement about which "International Carbohydrate Symposium" should be called the first one.[4] Professor Whistler was an honorary co-chairman of a meeting organized by the Centre National de la Recherche Scientifique (CNRS) with participation of the Societé de Chimie Biologique, held in Gif-sur-Yvette in 1960, chaired by Professor J. E. Courtois and entitled the International Colloquium on the Biochemistry of Sugars—Structure and Specificity. Professor Courtois said in his closing remarks, "Surely it is the wish of all that programs such as this will be repeated often in the years ahead." That meeting may thus be regarded as the first of the International Carbohydrate Symposia.

Dr. Whistler recounted later that, on the close of the symposium, he had a discussion with Professors Courtois, B. Lindberg, and F. Micheel, during which he again

[4] S. J. Angyal, International Carbohydrate Symposia—A history, *Adv. Carbohydr. Chem. Biochem.*, 61 (2008) 29–58.

suggested that there should be more international meetings on carbohydrates. It was reported[4] that most of those present said that they would organize a symposium at their university. Several symposia were subsequently organized, but without an international organization. According to Angyal,[4] in 1965 Professor Whistler wrote to eight carbohydrate chemists, asking them to join a proposed Steering Committee for International Carbohydrate Symposia. The eight potential members, each of whom accepted, were S. J. Angyal (Australia), J. E. Courtois (France), J. K. N. Jones (Canada), N. K. Kochetkov (USSR), B. Lindberg (Sweden), F. Micheel (Germany), Z. Nikuni (Japan), and M. Stacey (Great Britain). V. Bauer (Czechoslovakia) and R. Bognár (Hungary) joined the group soon thereafter. The purpose of the steering committee was to plan a regular succession of future symposia at 2-year intervals. Professor Whistler was not only the instigator in the organization of this steering committee, he also acted as its Secretary (1960–1971) and saw to it that the group established a constitution and arrangements to have the 11 members made official national representatives of their respective countries by having them selected by an official body, such as an Academy of Sciences or Chemical Society.

According to him, in 1967 the President of the American Association of Cereal Chemists (AACC) asked Whistler to explore the possibility of creating a Carbohydrate Division. Thereupon, he formed a committee, which quickly developed and presented to the AACC Board of Directors a statement of purpose and operating procedures for the Division, and received authority from it to form a Division. He served as its first Chair.

In the week preceding the 1980 International Carbohydrate Symposium (ICS) held in Sydney, Australia, Professor Bruce A. Stone organized a satellite conference in Melbourne for discussion of the chemistry and enzymology of crop carbohydrates. Professor Whistler and Robert D. Hill attended that conference and concluded that it was valuable enough, and not in conflict with the ICS, that such satellite conferences should be continued. In 1982, Robert Hill organized a similar conference in Winnipeg, Canada, prior to the ICS in Vancouver. In 1984, Professor Lars Munck, with the assistance of Dr. Hill and Professor Whistler (who served as the conference chair), held another such conference in Copenhagen prior to the ICS in Utrecht. These meetings, variously named, but most commonly termed Plant Polysaccharide Symposia (or Workshops), became regular biennial satellite conferences to the ICS. In all of these ways, Roy Whistler was a unifier of the community of carbohydrate scientists.

He was a master at conducting meetings and provided leadership to other scientific and professional organizations, beginning with being President of the Peoria Academy of Sciences in 1945. He held the following offices in the American Chemical Society: Division of Carbohydrate Chemistry Executive Committee (many years) and Chair

(1951), Executive Committee (many years) and Chair of the Division of Cellulose, Wood, and Fiber Chemistry (1962), Chair of the Purdue University Section (1949–1950), National Councilor (many years) and a member of Several Council Committees, and a member of the Board of Directors (1955–1958) and several Board Committees. In the American Association of Cereal Chemists, he chaired the Carbohydrate Division (1967–1968), was President (1973), and chaired the Board of Directors (1974). With the International Steering Committee on Carbohydrate Chemistry (the forerunner of the International Carbohydrate Organization), he was the National Representative of the United States (1960–1978), Secretary (1960–1971), President-elect (1971–1972), and President (1972–1974). He was President of the Purdue University Section of the Society of Sigma Xi (1957–1959) and a member of the National Executive Committee (1958–1962). He was President of the American Institute of Chemists (1980–1981) and chaired its Board of Directors (1982–1983). In addition, he was a member of the International Standards Organization (1973–1983), the CHEM Study Steering Committee (a committee organized to revise high school chemistry curricula in the United States) (1959–1963), the Scientific Advisory Committee of the American Sugar Association (1987–1995; Chairman, 1987–1992), and Chairman at the Wheat Utilization Committee of the National Association of Wheat Growers Foundation (1988–1992).

He was equally at home in the academy and in the world of business. In the business arena, Professor Whistler was a consultant to 20 companies on a long-term basis; was a director of 11 companies (both carbohydrate-producing companies and others); served as the board chair of several, including USAir; and was an advisor to the Caribbean Development Bank and Instituto Mexicanos de Investigaciones Technologicas A.C. of the Bank of Mexico. Perhaps of special note here is that he was a consultant to a corn wet-milling company and several companies producing industrial gums during the major developmental periods of the two industries following World War II.

A major contribution he made was in his efforts to compile the carbohydrate literature, especially in seeing a need to organize the primary literature for those in industry. Shortly after arriving at Purdue University, Professor Whistler began writing the book *Polysaccharide Chemistry* with the help of his student Charles Louis ("Lou") Smart. This book, published in 1953, was the first compilation of the chemistry of all known polysaccharides. It was described by one reviewer as "not only a work of science but of literature." In this book, he established the beginning of systematic polysaccharide nomenclature, using the rules he and his committee had proposed, including changing the name carrageenin to carrageenan. He specifically proposed that all polysaccharides, excepting only those whose names had become highly ingrained from common usage, should have names ending in -an. As the chair of a new ad hoc subcommittee of the U.S. Carbohydrate Nomenclature Committee, he submitted a set of proposed rules to the parent committee, which submitted them to the British Nomenclature Committee, and eventually (in a joint submission) to IUPAC. The rules established the word *glycan* as being synonymous with polysaccharide and introduced the terms *homoglycan* and *heteroglycan*, with the monosaccharide units in a heteroglycan being listed in alphabetical order, as in galactomannan and arabinoxylan, and that polysaccharides could be named for their source, such as guaran, or indicating a geographical source, as in carrageenan.[5]

Next, with another of his students, he turned to editing the book *Industrial Gums*. Now in its third edition, *Industrial Gums* is a practical book that describes the chemistry, properties, and applications of water-soluble or water-dispersible polysaccharides and polysaccharide derivatives with commercial value or potential. In the introductory chapter of this book, Professor Whistler presented important concepts about the relationships of structure to properties of polysaccharides, an area he championed after it had been introduced by D. A. Rees.

He initiated the *Methods in Carbohydrate Chemistry* series, which grew out of a "Laboratory Procedure Book" that was being used by those working in his laboratory, and came from a recognition by him that research in carbohydrate chemistry and biochemistry, and a number of related fields, depended on establishing a collection of reliable experimental procedures, and their dissemination and use. The *Methods* series was described by reviewers as "a must for sugar chemists" and "a monumental work of permanent value."

The treatise *Starch: Chemistry and Technology*, which he initiated and co-edited, is also in its third edition. Few writers in any field have been responsible for such significant, clearly written, and useful contributions to the secondary literature. In addition, he co-authored with me the first edition of the textbook *Carbohydrate Chemistry for Food Scientists* (1996).

In addition to his writing and editing, he was on the Board of Advisors of *Advances in Carbohydrate Chemistry and Biochemistry* (1950–2003) and the Editorial Boards of *Carbohydrate Research* (1965–1991), *Starch/Stärke* (1970–2010), *Organic Preparations and Procedures International* (1965–1997), and *Journal of Carbohydrates-Nucleosides-Nucleotides* (1970–1976).

He was always a strong promoter of new uses for agricultural crops and products from them: guar gum, hemicelluloses, and, in later years, banana starch, along with compounds such as amylose acetate and 5-thio-D-glucose.

[5] Nomenclature of Carbohydrates, *Adv. Carbohydr. Chem. Biochem.*, 52 (1997) 43–177.

For his research and other professional contributions, Professor Whistler was recognized with the Annual Research Award of the Purdue University Chapter of The Society of the Sigma Xi (1953), the Claude S. Hudson Award of the Division of Carbohydrate Chemistry of the American Chemical Society (1960), the Anselme Payen Award (1967) and a Honorary Fellow Award (1983) of the Cellulose, Wood, and Fiber Division of the American Chemical Society, the Medal of Merit of the Japanese Society of Starch Science (1967), the Alsberg–Schoch Award (1970) and the Osborne Medal (1974) of the American Association of Cereal Chemists, the Saare Medal of the Association for Cereal Research (Arbeitsgemeinschaft Getreide-forschung) (1974), the Spencer Award of the Kansas City Section of the American Chemical Society (1975), the Sterling B. Hendricks Memorial Lecture Award and medal of the U.S. Department of Agriculture (1991), the Gold Medal of the American Institute of Chemists (1992), the Fred W. Tanner Lectureship of the Chicago Section of the Institute of Food Technologists (IFT) (1994), and the Nicholas Appert Award of the Institute of Food Technologists (1994), plus other awards and honors. He was awarded four honorary doctorate degrees, including one from Purdue University (1985) and one from Iowa State University (2003), and an Outstanding Achievement Citation from the Iowa State University Alumni Association (1997). At Purdue University, the Roy L. Whistler Hall of Agriculture Research (1997) and the Whistler Center for Carbohydrate Research, when it was initiated in 1986, were named to honor him.

As a university Professor, he was, of course, an educator and an advisor, but more broadly he had a profound influence on the lives, thoughts, attitudes, and ethics of many, not only his students and postdoctoral research associates, to whom he was affectionately known as "Doc," but also those he worked with in companies. He enjoyed and took pride in his teaching, especially in his classes for beginning students. His teaching career (other than that as a graduate assistant) began when he taught an evening course in carbohydrate chemistry at Bradley University in Peoria, Illinois, during the time he lived and worked there. At Purdue University, in addition to teaching carbohydrate and polymer chemistry, he taught biochemistry for non-majors. For more than 20 years, he and I co-taught short courses on industrial gums/hydro-colloids for professional organizations, companies, and a university.

The example he set for his graduate students, of whom there were 74, was one of accomplishments that resulted from determination and the hard work required to transform possibilities into achievements. He could not tolerate mediocrity, either in himself or in others. His former students and postdoctoral associates include many who went on to successful careers in carbohydrate research and other areas of science, in both academia and industry. I, like many others, am most grateful for what he did for me and my own career. In his relationships with his students, postdoctoral associates, and others, he was usually gracious, but there was a side of him that could be demanding and dogmatic. However, these aspects of his character should not detract from what he accomplished, nor cause anyone to forget the kindliness or the friendship and help he freely gave to those like myself whom he took under his wing. He was very loyal to friends and former students and postdoctoral associates. He often spoke of a person as one of "my guys" or "our guys," implying that people could be divided into two groups—"my guys" and all others.

Dr. Whistler had a great love for nature, with a special interest in birds. He traveled the world extensively to visit wilderness areas. When it was acceptable to do so, he spent several weeks each summer hunting big game in Africa and returned home with trophies. He also hunted game birds in the United States. Hunting appeared to be his only recreation. After hunting became no longer acceptable, he traveled to observe wildlife, such as polar bears north of Hudson Bay and birds in the foothills of the Himalayas. He also visited Antarctica. At the age of 86, he took a trip into the Amazon rain forest to observe birds. The next year (1999), he traveled to South Africa (again to observe birds) and to the People's Republic of China (to go down the Yangtze River before the lake filled up). He was planning a trip to the Seychelles when he became unable to travel. Beginning with the influence of his grandfather, I think, he developed a strong interest in preserving wildlife and, in 1997, established the Roy Whistler Foundation, Inc. "to promote and support the preservation of natural land and wildlife, including both plant and animal." A man of tall stature, he was a believer in fitness and kept fit. Until his last days, if he was ever ill, he hid it very effectively.

I believe that strength in every aspect of his life was most important to him. I think that he thought that any display of weakness would invite challenges to his position, which he had worked so hard to achieve, and that any criticism, or even a difference of opinion, would be seen as a sign of weakness, and thus a sign that he had lost the control that was so important to him.

Dr. Whistler served Purdue University in many capacities. He rapidly moved up the professional ranks, becoming an Associate Professor in 1948 and a Full Professor in 1950. Also in 1948, he was appointed Executive Officer of the Department of Agricultural Chemistry (later to become the Department of Biochemistry), although the strained relationship between him and the Department Head was no secret. Also in 1948, he was named chairman of an interdepartmental committee concerned with industrial uses of agricultural products. The latter activity grew, and in 1958, Purdue University established an Institute for Agriculture Utilization Research, with Professor Whistler as its Head (1958–1976). In 1975, he was named Hillenbrand Distinguished Professor of Biochemistry.

He served on a wide range of university committees and chaired many of them. He was a member of the committee established to organize a University Senate and served as a Senator for four years. For two years, he chaired the Faculty Affairs Committee of the senate. He served for more than 20 years on the Patent Advisory Committee of the Purdue Research Foundation. He served almost as long on the Purdue Athletic Committee and was its Executive Secretary. From 1966 until his formal retirement from the university faculty in 1977, he was Purdue's Faculty Representative to the Intercollegiate Conference of Faculty Representatives (the Big Ten), into which position he threw all his energy, as he did with all activities in which he was engaged. For several years, he headed its Rules and Agenda Committee. He also chaired a university committee to suggest ways and means for creating a Division of Women's Intercollegiate Athletics.

Dr. Whistler died on February 7, 2010, just weeks short of his 98th birthday. To me, he was a somewhat complicated, but rather understandable person. My reading of him is that the things he desired were fame, wealth, being respected and admired, influence, even control, over those with whom he interacted, being liked, and posthumous remembrance, all of which he achieved.

In conclusion, I am confident that few have ever exerted so deep an influence over a discipline for such a long period. We may never see his like again. He was a complicated, but readable person. He was dedicated to his career—determined to create an image about himself and one who, I am certain, planned every stage of his ascent. He was extremely self-confident. He had a ferocious work ethic and, at times, exhibited a ruthless competitive streak, although never admitting to it. These traits allowed him to overcome obstacles and to create a lifetime of accomplishments, beginning with rising above his humble beginnings and difficult early years during the Great Depression. His outlook was also shaped by the two World Wars, although he was directly involved in neither. During his career, he had a significant impact on carbohydrate science and a variety of businesses. Many are grateful for the opportunity he gave us to be associated with him, and for what he did to advance polysaccharide chemistry while teaching us in the process. In 1992, I said of him, while introducing him as the recipient of the Gold Medal of the American Institute of Chemists, "Professor Whistler is truly a man and a scientist of unique stature. For 50 years, he has been both a pioneer and a leader in carbohydrate research and applications. He promoted cooperation and friendship among carbohydrate researchers around the world. He bridged academic science and industry and is held with esteem and admiration by all who know him."[6] I stand by what I said almost 20 years ago.

James N. BeMiller

[6] R. L. Whistler—1992 Gold Medalist, *The Chemist* (December 1991) 9.

APPENDIX

Contributions of The Carbohydrates to The Field of Organic Chemistry

BY ROY L. WHISTLER

Ohio State University
Spring, 1935

(Whistler's 1935 term paper, reproduced verbatim)

The field of the carbohydrates stretches vast and fertile. Their propagation is as far flung as the vital organisms to which they are inherent. The carbohydrates serve as fuel for the burning fires of metabolism and as structure for the organisms that burn them. They are inherent to life and anything so vital must of necessity be complex. Only compounds whose reaction balance is finely poised can possibly fulfill the multifarious and intricate demands of that marvelous unit of life, the living cell. Hence it is of little wonder that early chemists avoided these inobedient compounds. Neither is it of wonder that this branch of chemistry has developed with such seeming slowness. Nor has this retarded elucidation been due to lack of workers. On the contrary, some of the most brilliant minds of science have toiled with carbohydrate problems. In this field labored the greatest chemist of all time, Emil Fischer. The beauty of mastering and the grandeur of its secrets more and more turns curious minds to its problems.

Though it seems that the simple sugar, glucose, was recognized as far back as the pre-Christian era, its chemical history did not begin until Lowitz (1) isolated the pure crystalline substance from honey in 1792. In 1802 Proust (2) obtained the same substance from grape juice. From then on it was found in numerous and divers places. Chevreul (3) recovered glucose from the urine of persons suffering from diabetes. By prolonged treatment with sulfuric acid it was obtained from starch by Kirchoff (4) and from cellulose by Braconnot (5). The name glucose was suggested by Dumas (6) in 1843, deriving it from the Greek word *glukos* meaning sweet.

The 19th century saw many workers enter the field of the sugars. Yet in these early days the sugars offered the greatest difficulties to investigators. No clear ideas regarding chemical structure had been evolved. Purely empirical and descriptive results were all that could be hoped for. Soon, however, came the great days of structural chemistry evolved by Kekuli and fresh light was thrown upon the whole field. Then appeared Van't Hoff's extension of chemical formulae into three dimensions, and with these advances the hitherto incomprehensible complexity of the carbohydrate problem began to fade out.

Nevertheless the disentangling of these closely related compounds offered practical stumbling blocks which even the finest experimental skill could hardly surmount. It was Fischer's discovery of phenylhydrazones and osazones that opened afresh the pathway toward complete understanding. The new tool supplied by phenylhydrazine was most powerful and produced a complete change in one aspect of the carbohydrate chemistry. In an almost incredibly short time, Fischer carried to a successful conclusion a vast research which ended with the determination of the structures and spatial configurations of all the known pentoses and hexoses.

With the close of Fischer's activity in the region of chemistry, a time elapsed in which no comparable advances were made. During this time the action of enzymes on various sugars was studied and some further attempts were made in constitutional examination.

It was not until the beginning of the present century that another great impetus came to the study of the carbohydrates. This new surge was started by Purdie and his collaborators and the tool they used was methylation. As phenylhydrazone formation clarified the constitution of the straight chains in the sugar molecule, so methylation threw new light on ring formation and opened a powerful attack on the constitution of the more complex carbohydrates.

The methods of methylation were first launched by the preparation of crystalline tetramethyl-glucose (7). Galactose (8), sucrose, and maltose (9) were subjected to methylation, and in 1906 the method was extended to determine the structure of the natural glucoside, salicin (10). Denham and Woodhouse (11) in 1910 were able to introduce methyl groups into cellulose. A modification of the methylation method enabled it to be applied to the preparation of glucosides and methylated glucosides (12); and the simplification of work in such cases facilitated the constitutional study of sucrose, lactose, maltose, and cellobiose (13, 14, 15).

Methylation is still a major tool in the elucidation of structural configurations, but we are now entering a new period of carbohydrate chemistry. This new period is on a much vaster scale than at any time previous. Our perspective is widened and our views more consolidated. We are now able to set forth tentatively certain broad generalizations. Many of our rules are incorrect yet they lead to a general perspective and intimacy which certainly could not be felt only a few years ago.

PER JOHAN GAREGG

1933–2008

Per Garegg passed away on September 23, 2008. He was born on July 11, 1933, in a small village, Berkåk, close to Trondheim on the west coast of Norway. His father, Syver Garegg, was a medical doctor and his mother, Hjördis, was a housewife. He had two younger sisters, Inger and Tori, who later worked closely together, one a nurse and the other a medical doctor. Per was married twice, first to Sheila Hamilton, with whom he had two daughters, Inez, born 1962, and Irene, born 1964, and subsequently in 1991 to Margaret Clarke.

Per was early regarded as very brilliant, especially in music and mathematics, and he started in the local country school at the age of six, one year early. During his beginning school years in the local village school, Norway was under German occupation. Per's parents were staunch anti-Nazis, and there was a prison camp close to the village where Per grew up. Per's mother sent the preteen on rather hazardous missions to throw packages of food over the camp fence where members of the Norwegian resistance had been imprisoned by the Germans.

When Per was 11 years old, the headmaster contacted his parents and recommended that the gifted boy should continue his education in Oslo, the capital of Norway. He went there alone, but after a short time the family followed. His father had obtained a position as a lung specialist in Oslo. Per continued to excel at school, jumped over another class, and finished school when he was only sixteen years old. He was the youngest ever to do that in Norway. As he was too young to start at university in Norway, he worked in industry for one year. At the age of seventeen he moved to Edinburgh to start studying chemistry at Heriot-Watt University. There he met a Scottish lass, Sheila Hamilton, whom he would later marry. He graduated with a B.Sc. degree in Chemistry in 1953 when he was 20 years old, and in 1955 he started to work in the research and development department of a company in Birmingham making plastics.

In 1958 he received a stipend from Norway that made it possible for him to go to Stockholm and start Ph.D. studies at the Swedish Institute for Wood Research. His

ISBN: 978-0-12-380854-7
DOI: 10.1016/S0065-2318(10)64002-4

supervisor was Professor Bengt Lindberg, a carbohydrate chemist with a growing international reputation. This turned out to be the most important step in his career, since he was destined to settle in Sweden and collaborate with Bengt Lindberg for the rest of his life. He received a licentiate degree in 1960, and in 1965 he moved with Bengt Lindberg to Stockholm University, where Bengt had been called to the chair in Organic Chemistry. Per was awarded his Ph.D. in 1965, obtaining the highest mark, and was entitled to call himself Docent in Organic Chemistry. The title of his thesis was *Studies on some partially substituted glycosides and polysaccharides.* He then worked as a lecturer in Organic Chemistry at Stockholm University before going to Edmonton in Canada in 1968 for a postdoctoral sojourn with Professor Ray Lemieux of the University of Alberta. His wife Sheila and their two daughters, Inez and Irene, accompanied him.

In 1969 Per returned to Stockholm University and the Department of Organic Chemistry, working as an Associate Professor. In 1984 he was appointed full Professor of Organic Pharmaceutical Chemistry at the Pharmaceutical Faculty of Uppsala University, but held this position for only one year before being appointed as Professor of Organic Chemistry at Stockholm University, where he succeeded Bengt Lindberg. He held this position until his retirement in the year 2000. He continued to work as Professor Emeritus at the same department until his death, by a stroke, in 2008.

He published more than 250 scientific papers, including notable articles “Thioglycosides as Glycosyl Donors in Oligosaccharide Synthesis” in Volume 52 of this series, and “Synthesis and Reactions of Glycosides” in Volume 59. He received the Lindblom Award from the Royal Swedish Academy in 1974 and in 1979 the Norblad—Ekstrand Medal of the Swedish Chemical Society. In 1991 the American Chemical Society recognized his outstanding contributions with the Claude S. Hudson Award.

Per Garegg jointly with Bengt Lindberg formed a research environment in carbohydrate chemistry that was recognized throughout the world. Within the department they built up frontline expertise in both analysis and chemical synthesis, a rare combination that brought with it a great deal of synergism. Per Garegg was responsible for the synthetic part of the collaboration, and his contributions to synthetic carbohydrate chemistry and oligosaccharide synthesis were an important component of advancing the broad field and making it come of age. His efforts made possible the synthesis of complex carbohydrate structures that had earlier been considered out of reach. He was active both in the development of methodology and protective-group manipulation as well as glycosylation techniques, and in the total synthesis of natural saccharide structures, mainly the monosaccharide sequences of bacterial polysaccharides, whose structures had been elucidated by the Lindberg group. Within the protecting-group area, methodologies were developed for alternative formation of acetals and selective protection of polyols, using, for example, tin activation, phase-transfer alkylations and acylations, and, perhaps the most recognized, reductive opening of acetals. Improved halogenation and oxidation methodologies were also developed. Within the glycosylation field, the Garegg group worked with most promoter systems. He had a special interest in the formation of β-mannosides, employing both indirect methods (epimerization by an oxidation–reduction sequence at the 2-position in β-glucosides) and direct methods (use of heterogeneous silver catalysts and the Mitsunobu reaction on α-hemiacetals) for their construction. Another major contribution was the development of thioglycosides as efficient glycosyl donors. A number of chemoselective thiophilic promoters were investigated and found to activate effectively the otherwise stable thioglycosides. This allowed the construction of large building blocks as well as their eventual activation and use as glycosyl donors in convergent oligosaccharide synthesis. All of these methods were then imaginatively employed to achieve the total synthesis of complex oligosaccharide structures. Numerous bacterial oligosaccharide structures were synthesized and conjugated to proteins to produce candidates for the development of conjugate vaccines. In addition, oligosaccharide sequences from parasitic, plant, and human sources were produced for use in biological experiments. Per was also instrumental in securing funding and organizing the setup of a nucleotide group at the department, where the H-phosphonate methodology was investigated and advanced into a standard method for construction of nucleotide linkages. His input advanced the field of synthetic carbohydrate chemistry in a major way and proved that nothing is impossible, not even in oligosaccharide synthesis.

Per Garegg was a tall, slim, fair-haired man, familiar to the carbohydrate community for his regular participation in the international carbohydrate meetings. His oral presentations in English were characterized by traces of a Scandinavian/Scottish accent and delivered with a slight hesitancy. In his second marriage to Margaret Clarke (who in 1998 predeceased him) he became an inveterate world traveller, spending time at his home in Stockholm as well as in Louisiana where Margaret was based. At meetings in the summer, he was instantly recognizable by his "trademark" pale-blue seersucker suit.

Besides his main obsession, chemistry, Per took an interest in literature, yoga, and philosophy. He was also a good bridge player and when competing against other teams his regular partner was Bengt Lindberg. These two were always in agreement about chemistry and at work. However, when losing at the bridge table, hard words could be exchanged.

Per was very interested in music, especially jazz. He was an excellent pianist. On the piano he mostly played ragtime, and his family relates that, after dinner, he regularly played Scott Joplin. The daughters, Inez and Irene, each brought him two grandchildren, together one boy and three girls. He adored these children and never tired of playing with them.

Many discussions, even at work, involved questions concerning aspects of life other than chemistry. He was generous, not only with his research funding, which he shared with younger people at the department to help them in their career development, but also with his own time. As a supervisor Per gave a lot of freedom to everyone involved, making his research group a dynamic, human, and friendly place, which attracted creative people who prospered and felt at home in this environment. Members of his group benefitted greatly from their time in this constructive milieu, which provided an excellent start for a continuing career. Fondly remembered by his students are exciting times in a research group working in an area undergoing rapid development and with major innovative results. Per is remembered not only as an excellent chemist and supervisor, but also as a person who cared about his fellow human beings and manifested a deep engagement with other values in life.

STEFAN OSCARSON
OLLE LARM

ADVANCES IN CARBOHYDRATE CHEMISTRY AND BIOCHEMISTRY, VOL. 64

STRUCTURE AND ENGINEERING OF CELLULOSES

BY SERGE PÉREZ[a,b] and DANIEL SAMAIN[b]

[a] European Synchrotron Radiation Facility, 6 rue Jules Horowitz, BP 220, F-38043 Grenoble cedex, France

[b] CERMAV-CNRS, BP 53, F-38041 Grenoble Cedex 9, France

ISBN: 978-0-12-380854-7
DOI: 10.1016/S0065-2318(10)64003-6

Abbreviations

AFM, atomic force microscope/microscopy; AGP, arabinogalactan–protein; CAN, ceric ammonium nitrate; CBH, cellobiohydrolase; CMC, *O*-(carboxymethyl)cellulose; CP/MAS, cross-polarization/magic-angle spinning; CPK:, Corey–Pauling–Koltun (space-filling model); CTA, cellulose triacetate; DCC, *N,N*-dicyclohexylcarbodiimide; DMAC, *N,N*-dimethylacetamide; DMAP, 4-dimethylaminopyridine; DMI, 1,3-dimethyl-2-imidazolidinone; DP, degree of polymerization; EDA, ethylenediamine; FT-IR, Fourier-transform infrared; GRP, glycine-rich protein(s); HEC, *O*-(2-hydroxyethyl)cellulose; HPC, *O*-(3-hydroxypropyl)cellulose; NMNO, *N*-methylmorpholine *N*-oxide; NMR, nuclear magnetic resonance; PP, 4-pyrrolidinopyridine; PRP, proline-rich protein(s); PVA, polyvinyl alcohol; TEMPO, 2,2,6,6-tetramethylpiperidine-1-oxyl.

I. Introduction

Photosynthetic organisms such as plants, algae, and some bacteria produce more than 180 billion tonnes of organic matter each year from the fixation of carbon dioxide. Half of this biomass is made up of the biopolymer cellulose, which is thus the most abundant organic molecule on the planet. This carbohydrate macromolecule is the principal structural component of the cell-wall of most plants. Cellulose is also a major component of wood, as well as of cotton and other textile fibers such as linen, hemp, and jute (ramie). For this reason, cellulose has always played an important role in the life of humans, and its applications could even constitute a landmark in the understanding of human evolution. Fine lingerie and rough cottons have both been recovered from the tombs of the Egyptian pharaohs. Methods for the fabrication of cellulose substrates for writing and printing go back to the early Chinese dynasties. Exploration, trade, and battles relied for many centuries on man's ability to build wooden ships, and making cotton sails and hemp ropes. Cellulose and its derivatives are principal materials used for much industrial exploitation (as paper, cellulose nitrate, cellulose acetate, methyl cellulose, carboxymethyl cellulose, and others) and they represent a considerable economic investment. Up to the beginning of the 20th century, cellulose as well as other biomacromolecules extracted from renewable resources constituted the raw materials for fuel, chemical, and material production. They were gradually replaced by petroleum-based derivatives. The depletion of petroleum resources, along with the current concerns about anthropogenic global warming, has motivated a shift in dependence away from fossil resources to renewable biomass resources, both in terms of energy production and commodity products. This can be performed based on the "intelligent" use of the unique biomolecular and biomacromolecular architectures that are derived from biosynthetic pathways, and that are not attainable throughout thermodynamically driven processes. Detailed knowledge about the different levels of structural organizations is needed to provide rational ways of conducting chemical modifications while maintaining the biodegradable and recyclable features of the starting raw material.

This chapter collates the developments and conclusions of many of the extensive studies that have been conducted on cellulose, with particular emphasis on the structural and morphological features, while not ignoring the most recent results derived from the elucidation of unique biosynthetic pathways. The presentation of structural and morphological data gathered together in this chapter follows the historical development of our knowledge of the different structural levels of cellulose and its various organizational levels. These levels concern features such as chain conformation, chain polarity, chain association, crystal polarity, and microfibril structure and organization. Several reviews have been published on cellulose research.[1–9] A recently published book summarizes the three-dimensional structures of crystalline cellulose and cellulose derivatives.[10] These articles provide some historical landmarks related to the evolution of concepts in the field of biopolymer science, which parallel the developments of novel methods for characterization of complex macromolecular structures. The elucidation of the different structural levels of organization opens the way to relating structure to function and properties. The chemical and biochemical methods that have been developed to dissolve and further modify cellulose chains are briefly covered. Particular emphasis is given to the facets of topochemistry and topoenzymology where the morphological features play a key role in determining unique physicochemical properties. A final section addresses what might be considered tomorrow's goal in amplifying the economic importance of cellulose in the context of sustainable development. Selected examples illustrate the types of result that can be obtained when cellulose fibers are no longer viewed as inert substrates, and when the polyhydroxyl nature of their surfaces, as well as their entire structural complexity, is taken into account.

II. Cellulose in Its Cell-Wall Environment

Tissues of higher plants such as trees, cotton, flax, sugar beet residues, ramie, cereal straw, and the like constitute the main sources of cellulose. The name cellulose, as coined by Anselme Payen[11] in 1838, indicates that it is the sugar (the "ose") from cells. This sugar (the "glucose unit") is the repeating unit of a long polymeric chain. Cellulose is also found in prokaryotic organisms (*Acetobacter xylinum, Agrobacterium tumefaciens, Rhizobium* spp.,[12] *Escherichia coli, Klebsiella pneumoniae, Salmonella typhimurium,*[13] *Sarcina ventriculi*) and also eukaryotes (Fig. 1), including certain animals (tunicates), algae, fungi, such vascular plants as mosses and ferns, gymnosperms, and angiosperms,[14] and the cellular slime mold *Dictyostelium discoideum.*[15] Proof of cellulose biosynthesis in cyanobacteria, which were probably among the earliest forms of life on earth, has also been demonstrated.[16] Cellulose is also synthesized by such bacteria as *Acetobacter*. It is also found in a highly crystalline form in the cell-walls of such algae as *Valonia* and *Microdicyon*. The animal kingdom also provides examples of several types of cellulose, the best studied of which is the membrane of marine animals belonging to the *Ascite* family, commonly termed tunicates.[1]

With about 10^{11} tons of cellulose growing and disappearing annually, cellulose is the most abundant renewable organic material on earth. In plant sources such as the wood of mature trees, the content of cellulose is in the order of 35–50%. The secondary walls of cotton fibers are almost pure (94%) cellulose. Throughout their lifetime, the cells of living plants continue to divide with the production of certain cells having the unusual property of being able to grow indefinitely while retaining the quality of young plants. These meristematic cells and those deriving from them grow and then differentiate into specialized cells for various functions, such as support, protection, flow of sap, and others. A collection of cells specialized for a single function constitutes a tissue.

Plant cell-walls are distinguished from animal cells by the presence, around the plasmalemma, of a wall within which complex physicochemical and enzymatic phenomena take place. In the course of cell growth, the dimensions of the cell-wall vary according to the types of macromolecules of which it is composed. The first wall deposited after cell division is called the "middle lamella" and is composed essentially of pectic material. The cell then lays down a wall composed of pecto-cellulosic material to supplant the middle lamella of the "primary" cell-wall (Fig. 2).

The primary cell-wall is actually a glycoproteinaceous layer composed of pectin, cellulose, hemicellulose, and proteins. As the cell ages and differentiates, it secretes new materials, which form a mixture with the constituents of the primary cell-wall and

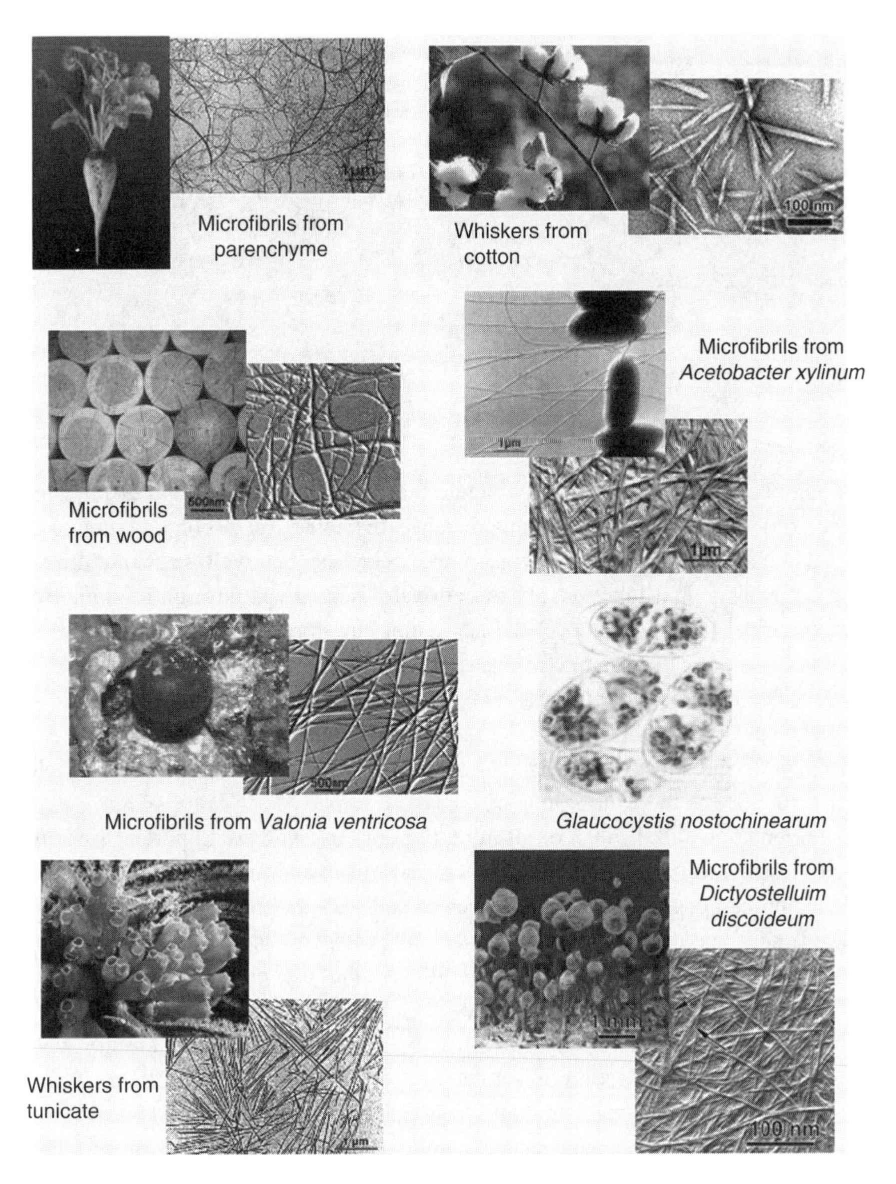

FIG. 1. The ubiquitous occurrence of cellulose in nature. (See Color Plate 1.)

thereby leads to the formation of a "secondary" cell-wall. The nature of the constituents of the secondary cell-wall depends on the cell type and the tissue to which the cell belongs. In general, totally differentiated cells cease expanding and cannot divide further.

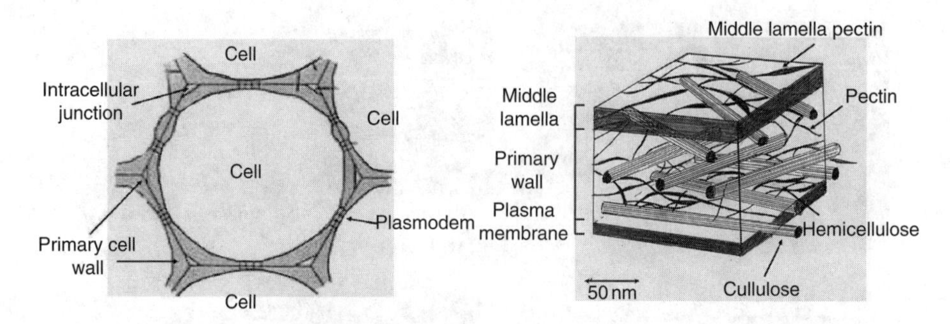

FIG. 2. Schematic representation of the plant cell-wall, and the location of the main polysaccharide components (from Ref. 17).

Young plant cell-walls feature a structure that is simultaneously rigid and dynamic. Indeed, rigidity is required to counterbalance the effect of turgor pressure on the plasmalemma. To allow cell extension to occur, the cell-wall structure must be deformable. This dual functionality of cell-walls is achieved through the mixture of polysaccharides and proteins. Cellulose chains are formed into microfibrils* that constitute the basic framework of the cell, conveying a great resistance to tensile forces.[18] The cellulose microfibrils comprise about 20–30% of dry weight cell-wall material, occupying about 15% of cell-wall volume. In cell-walls that have differentiated and have synthesized a secondary cell, the proportion of cellulose reaches 40–90% of the wall biomass.[19]

The orientation and disposition of microfibrils in the wall are important since they more or less control the capacity of the wall to deform and the direction in which the deformation can occur. In the final stages of cell-wall differentiation, notably in the middle lamella and primary cell-wall, other wall polymers ("lignins") are incorporated into the spaces around the polysaccharide fibrillar components to form lignin–polysaccharides. Lignins arise by free-radical polymerization of alcohols of *p*-hydroxycinnamic acid, and they constitute between 10 and 30% of the dry weight of wood, placing them second only to cellulose. They contribute to the mechanical strength of the plant cell-wall and confer resistance to pathogens. Their hydrophobicity confers resistance to water and controls solute transport and water content. In the course of differentiation, cellulose microfibrils associated with smaller molecules

* The historical term "microfibril" can be a source of confusion as it has been used by different authors along with closely related terms such as "protofibril," "elementary fibril," and "nanofibers." As pointed out by Nishiyama,[4] the term "cellulose microfibril" addresses features that are not limited to its size. For this reason, it is used throughout this article.

(hemicelluloses and lignins) provide a type of "liquid-crystalline" matrix in which microfibrils are able to slide past one another or else generate a disordered arrangement that resists further cell-wall extension[18] (Fig. 3).

The hemicelluloses constitute a large number of different polysaccharide molecules, which actually form a matrix for the cellulose microfibrils engaged in molecular interactions such as hydrogen bonds and van der Waals forces. In addition to structural properties, hemicellulose may also have other functions, such as cell signaling or as

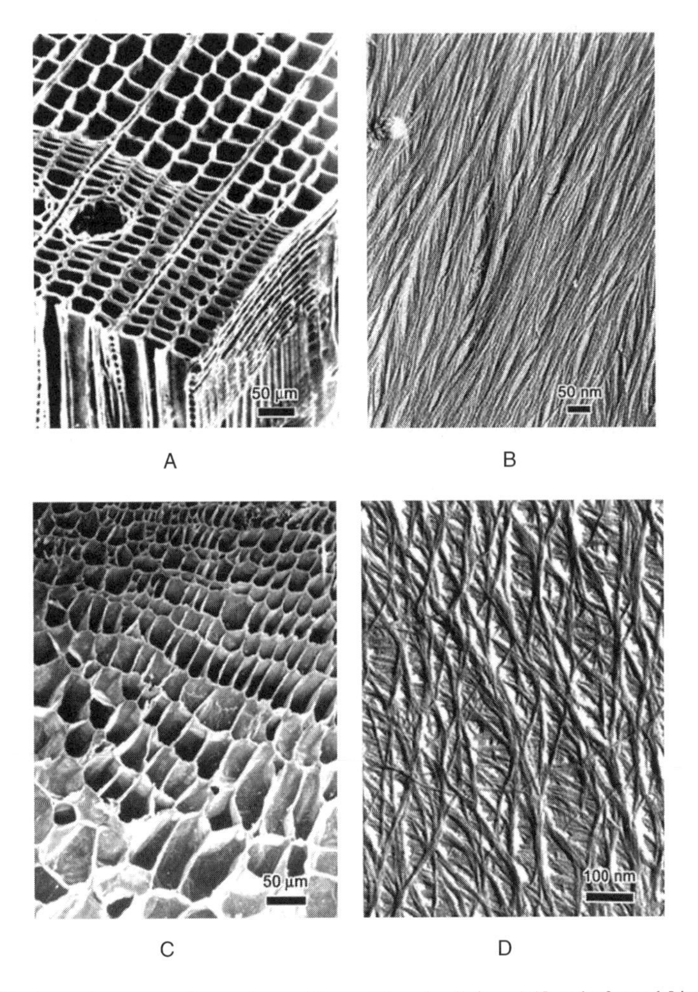

FIG. 3. Electron microscopy of parenchyma [tissue (A) and cellulose (B)] and of wood [tissue (C) and cellulose (D)] (courtesy of H. Chanzy).

precursors of signaling molecules or as reserve substances. Xyloglucans are major components of the hemicelluloses of higher plant dicotyledons and comprise 20% of the dry weight primary cell-wall material.[1] In monocotyledons, xyloglucan constitutes only 2% of the dry material mass. In this case, xylans and β-(1→3)- β-(1→4)-glucans comprise the major hemicellulose components with about 15–20% of dry weight cell-wall mass. Xyloglucans, like the xylans, are closely associated with cellulose microfibrils through intermediary hydrogen bonds.

Pectins constitute a major component of dicotyledonous higher plants, about 35% of the cell-wall dry weight. In monocotyledons their proportion is less, and their type is different. Pectins constitute a complex range of carbohydrate molecules whose backbone is composed chiefly of chains of (1→4)-α-D-galacturonan interrupted by segments of (1→2)-α-L-rhamnan. The rhamnose-rich regions are frequently branched, with side chains composed of neutral sugars of the arabinan–arabinogalactan type. These segments constitute the so-called "hairy" regions, in contrast to unsubstituted galacturonan segments or "smooth" regions. In addition to structural and developmental functions, pectins are responsible for the ion-exchange capacity of the cell-wall and control of the ionic environment and pH of the cell interior.

The plant cell-wall contains range of proteins, which are implicated in the organization and metabolism of the cell-wall. The structural proteins can be gathered into five main families: extensins (rich in hydroxyproline), proteins rich in glycine (GRP), proteins rich in proline (PRP), lectins, and proteins associated with arabinogalactans (AGP). Cell-wall enzymes may also be grouped into families according to function: (i) peroxidases, which participate in the lignification processes of the cell-wall; (ii) transglycosylases, which catalyze the breaking and making of glycosidic bonds in the cell-wall; (iii) many hydrolases (glycosidases, glucanases, cellulases, galacturonanases) and, just as important, esterases, a group of enzymes that constitute the machinery for efficient degradation of the cell-wall; and (iv) "expansions," proteins capable of rupturing the hydrogen bonds between cellulose microfibrils and xyloglucans.[19]

Available evidence suggests that cellulose is formed at or outside the plasma membrane. Groups of rosette particles or terminal complexes are seen in the plasma membrane. These groups of rosette particles can be seen to be associated with the ends of microfibrils (collections of cellulose chains) and are thought to be cellulose synthase complexes, involved in the elongation of whole cellulose microfibrils. The catalytic subunit is a transmembrane protein having a transmembrane region. At the initiation of polymerization, two uridine diphosphate glucose (UDP)-glucose molecules are present in the substrate binding-pocket. As the chain elongates, glucose is added to the nonreducing end. The globular region of the protein is thought to be located in the cytoplasm, the UDP-glucose being in the cytosol. A general model has been set up to explain the molecular organization of the cellulose synthase molecules from the molecular level of organization to the level of the rosette terminal complex.[20] Current estimates of the diameter of the cellulose synthase complex on the endoplasmic region of the plasma membranes are in the range of 40–60 nm, which makes it one of the largest protein complexes thus far observed. This complex is responsible for the synthesis of a microfibril that has 36 cellulose chains. Each of the six subunits of the rosette must consist of six glucan synthase molecules. The hydrophobic regions coordinate the insertion of hydrophilic domains on the cytoplasmic side of the plasma membrane; they facilitate the aggregation and association to form the rosette subunit particles. These particles are believed to synthesize glucan chain sheets,[21] which have been shown to be the first products of the crystallization phase. Sheets of glucan chains are then assembled to form the native cellulose. The movement of fluorescently labeled cellulose synthase complexes has been studied in *Arabidopsis* hypocotyl cells. These cellulose synthase complexes migrate in the plasma membrane along linear trajectories that appears to be defined by co-labeled microtubules. The cellulose synthase complexes are propelled through the plasma membrane by the forces generated by polymerization of the glucan chains and their crystallization into the microfibrils. The cortical microtubules somehow control the direction of movement, perhaps through direct interactions between the microtubules and the cellulose synthase complex (Fig. 4). While the lifetime of the cellulose synthase complex is estimated to be in the order of 30 minutes, the velocity is in the range of 300 nm/min.[23–26] The rates of movement correspond to the addition of about 300–1000 glucose per glucan chain per minute.

III. Conformations and Crystalline Structures of Cellulose

1. Chemical Structure of the Cellulose Macromolecule

Even though the early work of Braconnot[27] concerning the acid hydrolysis of the substance constituting plant cell-walls goes back to the 19th century,[27] it is with Anselme Payen[11] that the honor lies of establishing the fact that the fibrous component of all plant cells has a unique chemical structure.[11] It is also from the studies of Payen that the word *cellulose* was first coined. However, it required another 50 years for the basic cellulose formula to be established by Willstätter and Zechmeister,[28] and for the volume of the crystalline mesh to be evaluated by Polyani in 1921. The concept of cellulose as a macromolecule gave rise to a lively debate, since the generally accepted

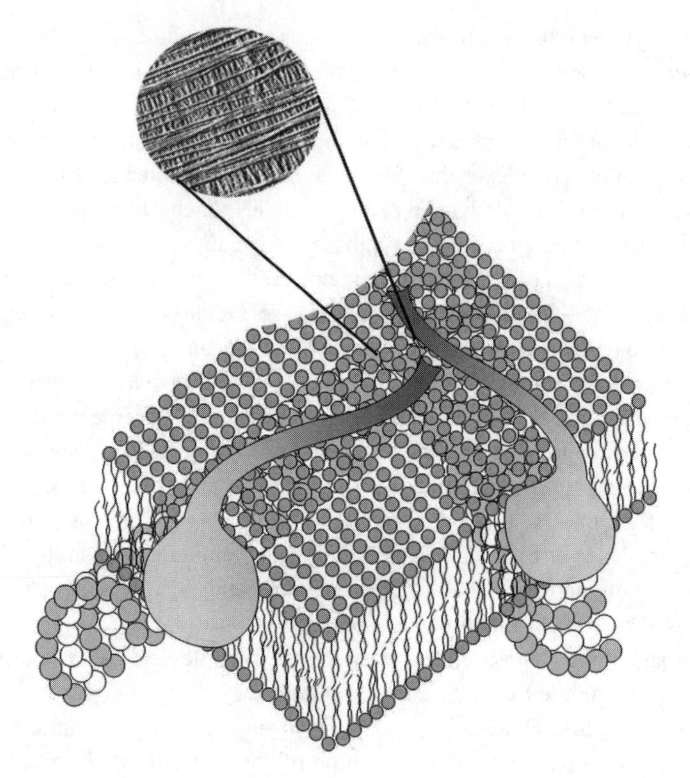

FIG. 4. Cartoon representation of a cellulose synthase complex moving inside the plasma membrane, leaving a cellulose microfibril as one component of the primary cell-wall (inset: transmission electron microscopy). The cellulose synthase complex becomes active while unattached to a microtubule.[22] (See Color Plate 2.)

idea was that the crystalline mesh corresponded exactly to the volume occupied by either one molecule or a restricted number of molecules. It was due to the contribution of Staudinger[29] that the macromolecular nature of cellulose was finally recognized and accepted.

Following this determination, Irvine and Hirst[30] and then Freudenberg and Braun[31] showed that 2,3,6-tri-*O*-methyl glucose was the sole product resulting from methylation and subsequent hydrolysis of cellulose. This work showed that, in cellulose, carbon atoms C-2, C-3, and C-6 bear free hydroxyl groups available for reaction. Complementing these investigations were those in which the structures of glucose and cellobiose[32–34] had been established and others in which it was determined that cellulose was a homopolymer of β-(1→4)-linked D-glucopyranose residues.

Crystallographic investigations of D-glucose and cellobiose[35] established unambiguously that the D-glucose residues had the $^4C_1$ chair conformation.

All of these investigations led to the establishment of the primary structure of cellulose as a linear homopolymer of glucose residues having the D configuration and connected by β-(1→4) glycosidic linkages (Fig. 5). The two chain ends are chemically different. One end has a D-glucopyranosyl group in which the anomeric carbon atom is involved in a glycosidic linkage whereas the other end has a D-glucopyranose residue in which the anomeric carbon atom is free. This latter cyclic hemiacetal function is in an α,β anomeric equilibrium via a small proportion of the free aldehyde form: this gives rise to the observed properties at this end of the chain. Determination of the relative orientation of cellulose chains in the three-dimensional structure has been one of the major problems in the study of cellulose.

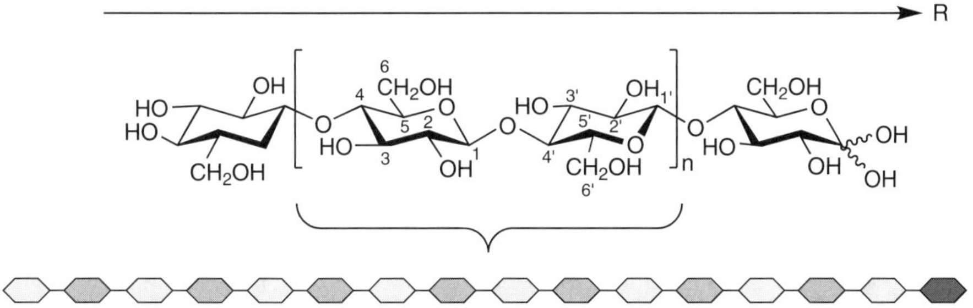

FIG. 5. Molecular representation of the chain structure of cellulose. Labeling of the cellobiose repeat unit is given in accordance with IUPAC nomenclature; *n* indicates the number of repeating disaccharide units in a given cellulose chain. The reducing end of the chain is indicated; a schematic representation of the chemical polarity of the chain is shown underneath; this representation is used throughout this chapter. (See Color Plate 3.)

The degree of polymerization (DP) of native celluloses depends on the source, and is not well established. Indeed, the combination of procedures required to isolate, purify, and solubilize cellulose generally causes scission of the chains. The DP values obtained are therefore minimal and depend on the methods used.[36,37] Values of DP ranging from hundreds to several tens of thousands have been reported.[36] For the same reasons the distribution of chain lengths of cellulose is not well established. Nonetheless, some authors suggest that the molecular mass distribution must be homogeneous for a cellulose of a given source.[38]

The first known X-ray fiber diffraction diagram of cellulose was recorded in 1913,[39] from which a qualitative interpretation was made regarding the presence of oriented micelles parallel to the long axis of the fiber. X-Ray diffraction patterns obtained from fibrillar samples do not provide sufficient experimental information to resolve the crystallographic structure unambiguously. Indeed a fiber is composed of an assembly of crystallites having a common axis, but random orientation. Added to this source of disorder, there is also that arising from the disorientation of the chains in the interior of the crystallite domains, as well as their small dimensions. These various sources of disorder are the origin of the low number of reflections found in the fiber diagrams. On these diffractograms, the reflections are distributed in horizontal rows, the spacing of which corresponds to the fiber repeat when the polymer axis is parallel to the fiber axis (Fig. 6). Thus, this periodicity is a geometric parameter that can be determined unambiguously from a fiber diagram, and this usually corresponds to the c dimension of the unit cell. Systematic absences of (0, 0, l) reflections also provide information about the helical symmetry of the polymer chain. The possibility of unambiguous determination of the other unit-cell parameters, as well as systematic absences in all the reciprocal space, depends on the ability to index the observed reflections, and this in turn depends on the quality of the samples.

The conformation of the cellulose chain can be determined by means of molecular modeling, taking into account experimental data such as the helical parameters derived from the X-ray fiber diffraction diagrams. In the case of the cellulose chain, the conformational variations depend principally on the rotations around the glycosidic linkage. The first step involves construction of a map of the energies corresponding to the variations of the angles (Φ, Ψ) that make up the glycosidic linkage. In the same way, it is

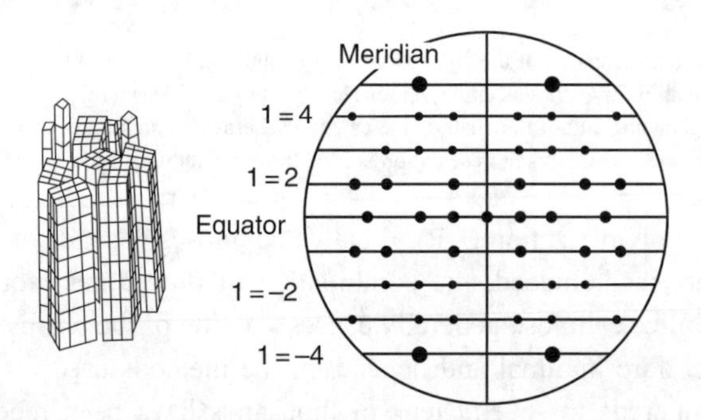

FIG. 6. Idealized fiber diffractogram from X-ray or neutron scattering. An assembly of partially ordered blocks of microcrystallites diffracts to produce diffraction spots that are recorded on a flat film or an area detector. The periodicity of the macromolecular constituents within the microcrystallites is indicated by a series of diffracting lines having a regular spacing. The equator corresponds to the layer line 0, intersecting the nondiffracted central beam. The meridian is perpendicular to the equator and it lies parallel to the fiber axis. The spacing along the meridian provides information about the periodicity of the macromolecule and its helical symmetry. The so-called helical parameters, n and h, are directly related to the symmetry of the macromolecular chain: n is the number if residues per turn, and h the projection of the residue on the helical axis.

possible to superimpose values of helical parameters on the iso-energy maps and permit the construction of a stable model (Fig. 7).

The representation of the three-dimensional structure of the cellulose chain shows some key structural characteristics. As a consequence of the $^4C_1$ chair conformation and the (1→4) glycosidic linkage of the β-D-glucopyranose residues, the structure is much extended and corresponds to a two-fold helix having a periodicity of 10.36 Å, which corresponds to the lowest-energy conformation (Fig. 8).

This conformation is situated in the low-energy zone in which van der Waals interaction and the anomeric effect are optimized. An intramolecular hydrogen-bond between O-3 and the ring O-5 of another residue provides additional stabilization (O-5…O-3: 2.75 Å). This linkage is standard in cellulose chains with two-fold symmetry,

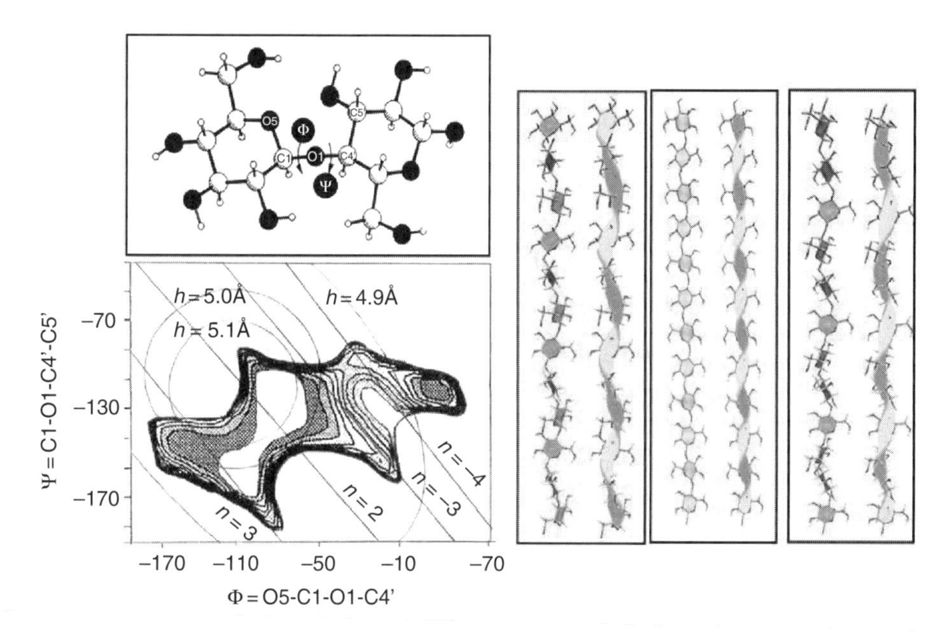

FIG. 7. Selected helical parameters, n and h, computed for a regular cellulose chain as a function of glycosidic torsion angles Φ and Ψ. The iso-n and iso-h contours are superimposed on the potential-energy surface for cellobiose. Arbitrarily, positive values of n and h designate a right-handed helix, and opposite signs correspond to left-handed chirality. The screw sense of the helix changes to the opposite sign whenever the values $n = 2$ are interchanged. In practice the regular parameters are readily derived from the observed fiber diffraction pattern ($n = 2$, $h = 5.18$ Å). Values of the torsional angles consistent with the observed parameters are found at the intersection of the corresponding iso-h and iso-n contours. Discrimination between possible solutions is based on the magnitude of the potential energy. The 3-D representations of a cellulose chain having, respectively, a right-handed three-fold, a two-fold, and a left-handed three-fold helical symmetry are shown.

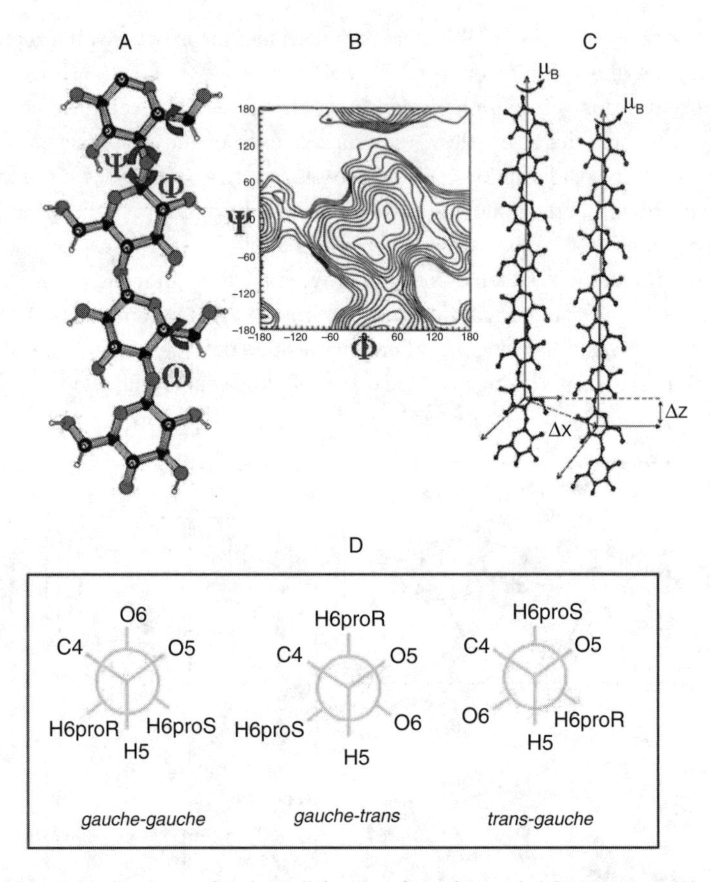

FIG. 8. Structural descriptors for the cellulose conformation and 3-D structure. (A) The main conformational descriptors of the cellulose chains are the glycosidic torsional angles Φ=O-5–C-5–C-1–O-1, Ψ =C-5–C-1–O-1–C-4′, and the orientation of the primary hydroxyl group about the C-5–C-6 bond: *ω*. The low-energy conformations of the chains are derived from the potential energy surface that shows conformational energy with respect to the Φ and Ψ torsion angles (B). The primary hydroxyl group of a pyranose is locked to the C-6 carbon atom, whereas the secondary hydroxyl groups are linked to the ring carbon atoms. The orientation of the primary hydroxyl group is referred to two different torsion angles. The first is O-5–C-5–C-6–O-6 and the second is C-4–C-5–C-6–O-6. These torsion angles are said to be in a *gauche* conformation when they have a ±60° angle, and in a *trans* conformation at 180°. The three significant conformations are thus *gauche–gauche* (*gg*), *gauche–trans* (*gt*), and *trans–gauche* (*tg*). Usually, one of the three possible conformers, *gg/gt/tg*, is discarded because of the unfavorable 1-3-*syn*-axial interactions between the O-6 and O-4 atoms. The prochiral hydrogen atoms (as pro*R* and pro*S*) are noted.[40] The relative orientation of two cellulose chains is dictated by the interhelical parameters μA, μB: rotation of chains A and B; Δ*Z*: translation of chain B with respect to chain A; Δ*X*: distance between the helical axis of chains A and B. (See Color Plate 4.)

but is absent when other less-stable conformations are adopted under different external environments. Other conformations generating such helical conformations having three-fold symmetry can also occur. The exocyclic primary hydroxyl groups (O-6) can adopt three low-energy conformations [*gauche–gauche* (*gg*), *gauche–trans* (*gt*), and *trans–gauche* (*tg*)] as a consequence of a gauche stereoelectronic effect[40] (Table I).

TABLE I

Glycosidic Torsion Angles and O-3...O-5′ Distances for Cellobiose Fragments in Crystal Structures

Compound	Φ (°)	Ψ (°)	O-3...O-5′ Distance (Å)	H-bond	O-6′ Conf.	O-6 Conf.	Reference
α-Cellobiose 2NaI, $2H_2O$	−77.7	−105.3	3.31	Yes	*gg*	*gg*	41
β-Cellobiose	−76.3	132.3	2.77	Yes	*gt*	*gt*	35
Me β-cellotrioside (average of 8)	−94.4	−146.6	2.864	Yes	*gt*	*gt*	42
Cellotetraose (average of 6)	−94.4	−146.8	2.875	Yes	*gt*	*gt*	43
Me 4-*O*-Me cellobioside (monoclinic form)	−88.1	−151.3	2.81	Yes	*gt*	*gg*	44
Me 4-*O*-Me cellobioside (triclinic form)	−90.0	−159.2	2.76	Yes	*gt*	*gg*	44
Me β-cellobioside·MeOH	−91.1	−160.7	2.76	Yes	*gt*	*gg*	45
Cellulose II mercerized	−96.8	−143.5	2.79	Yes	*gt*	*gt*	46
	−93.3	−150.8	2.75	Yes	*gt*	*gt*	46
Cellulose II regenerated	−95.4	−147.7	2.78	Yes	*gt*	*gt*	47
	−95.1	−150.6	2.76	Yes	*gt*	*gt*	47
Cellulose Iβ	−98.8	−141.9	2.77	Yes	*tg*	*tg*	48
	−88.7	−147.1	2.70	Yes	*tg*	*tg*	48
Cellulose Iα	−98.0	−138.0	2.47	Yes	*tg*	*tg*	49
	−99.0	−140.0	2.93	Yes	*tg*	*tg*	49

Although the *trans–gauche* orientation is rarely observed in the crystalline structures of oligosaccharides,[50] this conformation would yield a second hydrogen bond between chains (O-2–H...O-6 = 2.87 Å) that brings an extra stabilizing factor to the chain conformation of cellulose. Nevertheless it should be mentioned that cellulose can adopt other low-energy conformations, in particular at the interface of crystalline and amorphous zones where stacking constraints are weaker. Obviously, the possibilities for the formation of intra- and inter-chain hydrogen bonds can give rise to various possibilities for the formation of stable three-dimensional structures. These possibilities are also reflected in differences of reactivity of the different functional groups, in particular in etherification reactions, since it has been shown that the O-3 and O-6 hydroxyl groups are much less reactive than O-2.

2. Crystallinity and Polymorphism of Cellulose

The free hydroxyl groups present in the cellulose macromolecule can be involved in a number of intra- and inter-molecular hydrogen bonds, and consequently give rise to various ordered crystalline arrangements. These crystalline arrangements in the case of cellulose are usually imperfect to the extent that, in terms of crystal dimensions, even the chain orientation and the purity of the crystalline form must be taken into consideration. The crystal density can be gauged from the crystallographic data, as can the importance of the amorphous components generally present. The density of the crystalline phase is 1.59 g/cm^3 but, when determined for natural samples, is of the order of 1.55 g/cm^3, which corresponds to a value of ~70% for the crystalline component.[51] The degree of crystallinity can also be estimated by infrared spectroscopy as a function of the relative intensity of certain bands.[52] Four principal allomorphs have been identified[53] for cellulose: I, II, III, and IV. Each of these forms can be identified by its characteristic X-ray diffraction pattern. Progress achieved in characterization of the ultrastructure of cellulose has shown that subgroups exist within these four allomorphic families. The relationships among the various allomorphs are shown schematically in Fig. 9.

The natural form of cellulose, termed cellulose I or native cellulose, is apparently the most abundant form. Its three-dimensional structure is highly complex and may comprise two distinct crystalline forms, cellulose Iα and Iβ. This was a major discovery and led to a revival of interest in the study of cellulose structure.[54] Cellulose I can be made to undergo an irreversible transition to a stable crystalline form, cellulose II, by two distinct processes: regeneration and mercerization. The cellulose II allomorph is known by the term "regenerated" cellulose. Regeneration involves either preparing a solution of cellulose in an appropriate solvent or of an intermediate derivative followed by coagulation and recrystallization. This process is used to produce rayon fibers. Mercerization involves intracrystalline swelling of cellulose in concentrated aqueous sodium hydroxide (NaOH), followed by washing and recrystallization. This process is used to improve the properties of natural yarns and fabrics. The transition from cellulose I to cellulose II is not reversible, and this implies that cellulose II is a stable form as compared with the metastable cellulose I.

Treatment with liquid ammonia or with certain amines such as 1,2-diaminoethane (ethylenediamine, EDA) allows the preparation of cellulose III from either cellulose I (which leads to the form cellulose III_I) or cellulose II (which leads to the form III_{II}). Cellulose II_I treated at high temperature in glycerol is transformed into cellulose IV. Here again two types exist: cellulose IV_I and IV_{II}, respectively, obtained from cellulose III_I and III_{II}. It is generally accepted that cellulose IV_I is a disordered form

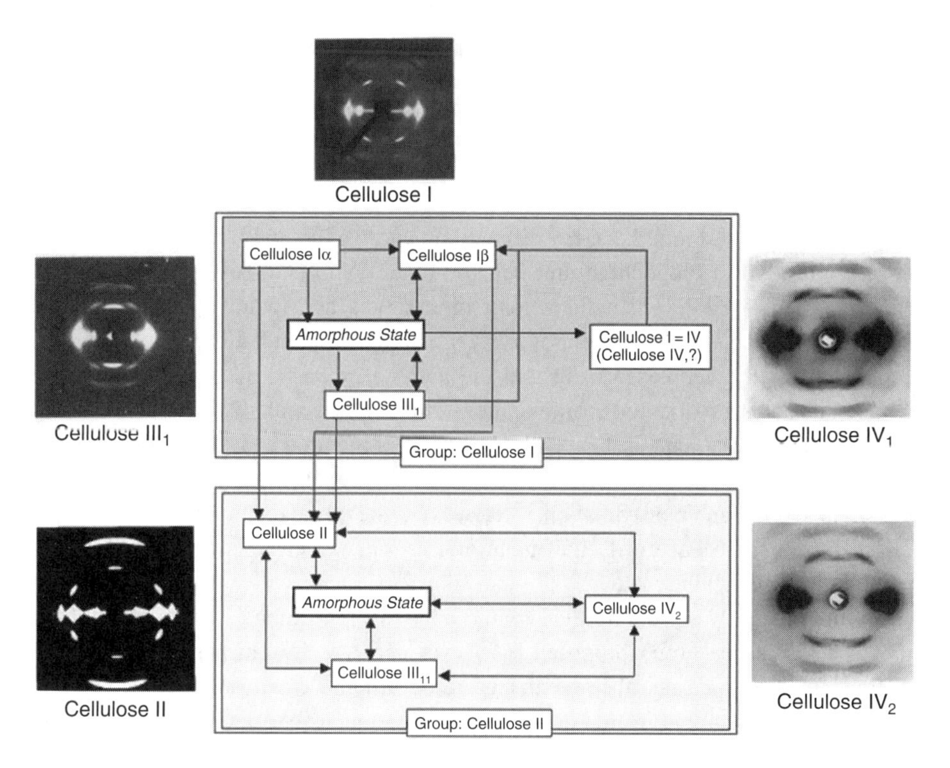

FIG. 9. Relationships among the different cellulose allomorphs, together with some prototypical fiber diffractograms.

of cellulose I. This could explain the reported occurrence of this form in the native state in some plants, as determined by X-ray diffraction.[55,56]

3. Crystalline Structures of Native Celluloses

X-ray investigations of native cellulose samples were made in the 1990s following early observations by optical microscopy that suggested the existence of submicroscopic birefringent and oriented domains.[57,58] The analysis of X-ray diffraction patterns has played and continues to play a major role in structural studies of cellulose.[59] Prior to the discovery of the crystalline dimorphism of cellulose, most crystallographic studies focused on determination of a basic unit cell. The controversy concerning the cellulose I unit-cell dimensions and space group (considered to be unique) has persisted for many years, in spite of observations by various workers[60,61] who reported experimental data showing that diffraction intensities and spacings varied greatly according to the origin of the sample. For this reason the literature is especially confusing on these points, and is overloaded with conflicting experimental data and structural models.

a. Cellulose I.—From the X-ray diffraction pattern of cellulose from ramie, Meyer *et al.*[62] proposed a monoclinic unit cell [a = 8.35 Å, b = 7.0 Å, and c = 10.3 Å (fiber axis), γ = 84°], which for a long time served as a point of reference. The symmetry elements in the space group $P2_1$ are compatible with a two-fold helicoidal symmetry for the cellulose chain, and the authors proposed a structural model in which the chains are oriented in antiparallel fashion. Later, more-elaborate studies that took advantage of methods for resolving crystal structures and taking the packing energies into account showed that the original proposal of Meyer and Misch constitutes only an approximation.[62] However, the principal modification to the original proposal concerned the chain orientation, which was concluded to be parallel in the crystalline lattice.[63,64]

Studies on highly crystalline algal cellulose led to a reopening of the question of the unit cell and space group proposed by Meyer and Misch.[62] In particular, electron diffraction studies, made at low temperature on *Valonia* cellulose, produced results that were incompatible with both the unit-cell dimensions and the space-group symmetry proposed previously. The results, confirmed by independent studies, contradicted the two-fold symmetry of the chain, and suggested that *Valonia* cellulose had the space group P_1 and a triclinic unit cell.[64,65]

b. Celluloses Iα and Iβ.—Away from the main controversy, other work suggested that celluloses from *Valonia* and bacterial sources had the same crystalline unit cell. Native celluloses of different origins might, in the same way, crystallize in different arrangements with different dimensions.[66,67] It was ten more years before the existence of two families of native cellulose was confirmed by the application of solid-state nuclear magnetic resonance (NMR) [^{13}C cross-polarization/magic-angle spinning (CP/MAS)] to a range of cellulose samples of different origins. From a detailed analysis of the carbon atom couplings observed in the solid-state NMR spectrum, Vanderhaart and Atalla[68,69] established (Fig. 10) that native cellulose was a composite of two distinct crystalline phases, named Iα and Iβ.

The crystalline phases Iα and Iβ can occur in variable proportions according to the source of the cellulose. The celluloses produced by primitive organisms (bacteria, algae, and the like) are enriched in the Iα phase, whereas the cellulose of higher plants (for instance, woody tissues, cotton, and ramie) consists mainly of the Iβ phase. Study

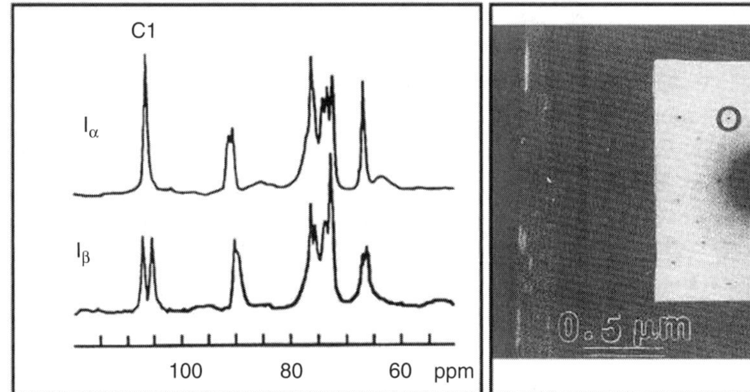

FIG. 10. Solid-state NMR spectrum of cellulose Iα and Iβ, together with transmission electron microscopic evidence of the simultaneous occurrence of these two allomorphs within the same microfibril.

of the cellulose of the outer membrane of marine animals showed that this is uniquely composed of the Iβ phase. Hence, this cellulose may be considered to be the standard for the Iβ phase.[70] Cellulose from *Glaucocystis* has been shown to consist of essentially cellulose Iα. The discovery of the crystalline dimorphism of cellulose was the starting point for a number of research projects the aim of which was to evaluate the properties of each allomorph and procedures for their interconversion.[69,71–74] The observed reflections could be indexed to a monoclinic unit-cell having space group $P2_1$ and dimensions $a = 8.01$ Å, $b = 8.17$ Å, $c = 10.36$ Å, $\gamma = 97.3°$. This unit cell is close to that proposed originally by Meyer and Misch from their work on cellulose from ramie, now known to be enriched in phase Iβ.[62] Phase Iα corresponds to a triclinic symmetry with space group P_1 and dimensions $a = 6.74$ Å, $b = 5.93$ Å, $c = 10.36$Å, $\alpha = 117°$, $\beta = 113°$, and $\gamma = 97.3°$.

The discovery of the crystalline dimorphism of cellulose and the existence of two families of native cellulose explained the number of inconsistencies that have characterized fifty years of crystallographic studies of cellulose. Thus the eight-chain unit cell[75] can be explained as an artifact arising from the superposition of thc diffraction diagrams of the phases Iα and Iβ, which are both present in *Valonia* cellulose.

The occurrence of this dimorphism in native cellulose has been confirmed by systematic investigations by X-ray and neutron diffraction of a wide range of samples. The dimorphic concept has also permitted elucidation of several features of the spectra reported in infrared[76,77] and Raman spectroscopic studies.[78] In subsequent X-ray and electron-diffraction studies, the space group and the chain packing of the Iα and Iβ phases have been further characterized.[79,80] Cellulose Iα has a triclinic unit cell containing one chain, whereas cellulose Iβ has a monoclinic unit cell containing two parallel chains, similar to the approximate unit cell proposed previously for cellulose I.[63,64] The "parallel-up" chain-packing organization favored by Sarko and Muggli has been confirmed by an electron microscopy study.[81] These results have allowed a number of molecular descriptions for Iα and Iβ to be produced by molecular modeling methodology.[82–85] There has also been a reexamination of the cellulose Iβ structure as determined from X-ray patterns of *Valonia* cellulose.[86]

The experimental revision of the structure of cellulose I, in light of this dimorphism, awaited the development of new structural tools, such as those provided by synchrotron and neutron techniques. To achieve this, methods have been developed for deuteration of the intracrystalline regions of native cellulose without affecting the overall structural integrity.[87,88] The neutron-diffraction diagrams obtained in these studies are presented in Fig. 11 for cellulose I and in Fig. 12 for cellulose II. These fiber-diffraction diagrams are recorded at a resolution of 0.9 Å, and several hundred independent diffraction spots can be measured, opening the road for the establishment of unambiguous three-dimensional structures.

The deuterated fibers give high-resolution neutron-diffraction patterns with intensities that are substantially different from the intensities observed on neutron-diffraction patterns obtained from hydrogenated fibers.

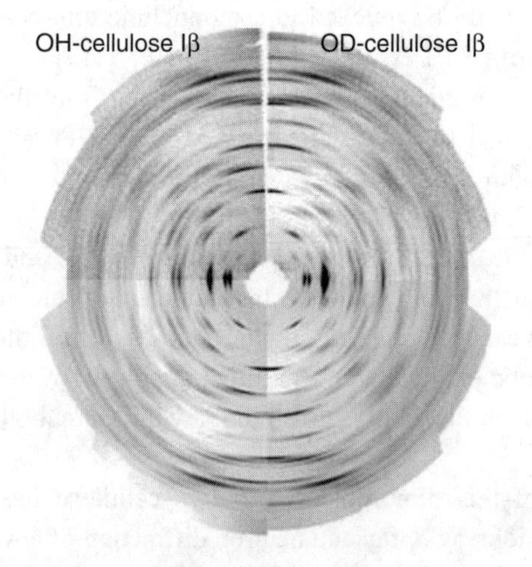

FIG. 11. Neutron-diffraction pattern recorded on a normal (OH, left) and on a deuterated (OD) sample of cellulose Iβ.[48]

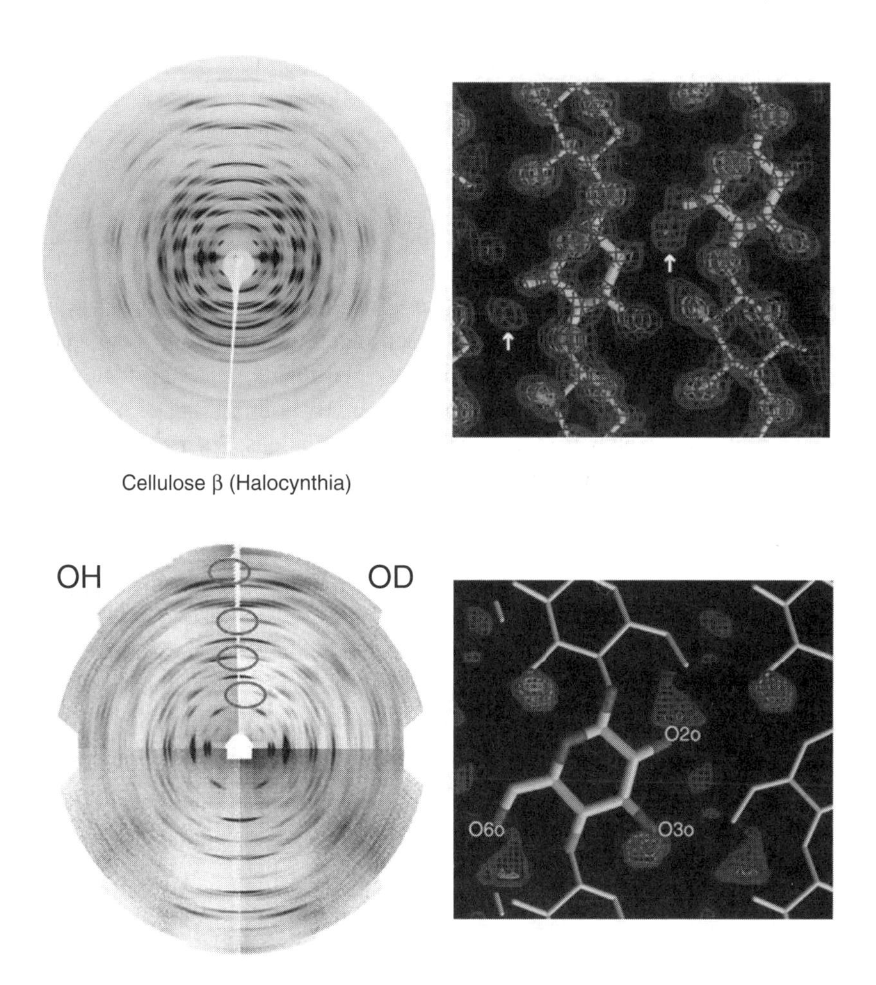

FIG. 12. X-Ray and neutron fiber-diffraction patterns recorded from native cellulose (OH) and fully deuterated (OD) cellulose samples. The difference in the scattering properties of hydrogen and deuterium (indicated by the red circles) allows the precise location of the electron density (shown in blue), which corresponds to the position of hydrogen atoms in the crystalline lattice. (See Color Plate 5.)

The crystal structure and hydrogen-bonding system in cellulose Iβ was elucidated by the combined use of synchrotron X-ray and neutron fiber diffraction.[48] Oriented fibrous samples were prepared by aligning cellulose microcrystals from tunicin, reconstituted into oriented films. These samples diffracted both synchrotron X-rays and neutrons to a resolution better than 1 Å, yielding more than 300 unique reflections and an unambiguous assignment of the monoclinic unit-cell dimensions (a = 7.784 Å,

$b = 8.201$ Å, $c = 10.380$ Å, $\gamma = 96.5°$) in the space group $P2_1$. The X-ray data were used to determine the positions of the carbon and oxygen atoms. The positions of hydrogen atoms involved in hydrogen bonding were determined from Fourier difference analysis, using neutron-diffraction data collected from hydrogenated and deuterated samples.

The chains are located on the 2_1 axes of the monoclinic cell; therefore they are not linked by any symmetry operation. The resulting structure consists of two parallel chains having slightly different conformations, both in terms of backbone and glucose conformations. All of the hydroxymethyl groups adopt the *trans–gauche* orientation, which allows the formation of intrachain hydrogen bonds involving O-2 and O-6 groups interacting throughout multiple possibilities. By contrast, the O-3...O-5 intramolecular hydrogen bond is well localized and unambiguous. Such a multiple hydrogen-bonding scheme explains the complex O–C stretching bands observed in the infrared spectra of cellulose Iβ.[89] The cellulose chains are organized in sheets packed in a "parallel-up" fashion. There are no intersheet O–H...O hydrogen bonds in cellulose Iβ, and therefore, the cellulose sheets are held together only by hydrophobic interactions and weak C–H...O bonds (Fig. 13).

The occurrence of nonequivalent chains may explain the fine details displayed by the ^{13}C CP-MAS spectra of cellulose Iβ.[89] The resonances assigned to the C-1, C-4, and C-6 atoms exhibit distinct splitting. The different conformations of the glycosidic linkages and at the primary hydroxyl groups for the nonequivalent chains provide a structural explanation for these splittings.

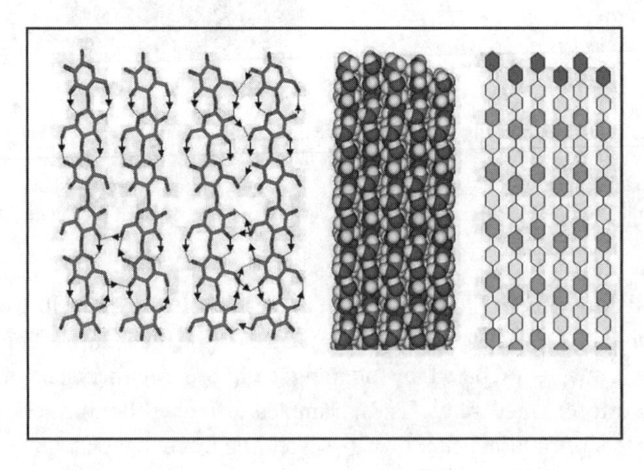

FIG. 13. Structural details of cellulose Iβ.[81] (See Color Plate 6.)

The crystal and molecular structures of the cellulose Iα allomorph have been established using synchrotron- and neutron-diffraction data recorded from oriented fibrous samples prepared by aligning cellulose microcrystals from the cell-wall of the freshwater alga *Glaucocystis nostochinearum.*[49] The X-ray data recorded at 1 Å resolution were used to determine positions of the C and O atoms. The positions of hydrogen atoms involved in hydrogen bonding were determined from a Fourier difference analysis, using neutron-diffraction data collected from hydrogenated and deuterated samples. The resulting structure is a one-chain triclinic unit cell of dimensions: $a = 6.717$ Å, $b = 5.962$ Å, $c = 10.400$ Å, $\alpha = 118.08°$, $\beta = 114.80°$, $\gamma = 80.37°$, space group P_1. The structure consists of a parallel chain arrangement of the "parallel-up" type packed in a very efficient way, the density being 1.61 (Fig. 14). Contiguous residues along the chain axis adopt a conformation remarkably close to a two-fold screw, which is not required by the space-group symmetry, all of the hydroxymethyl groups being in a *trans–gauche* conformation. The occurrence of the intrachain hydrogen bond O-3...O-5 is found throughout the structure, with an alternation of two slightly different geometries. The hydrogen bonds associated with O-2 and O-6 are distributed between a number of partially occupied, but still well-defined positions. As with cellulose Iβ, these partially occupied positions can be described by two mutually exclusive hydrogen-bonding networks, and there is no hint of intersheet O–H...O hydrogen bonds.

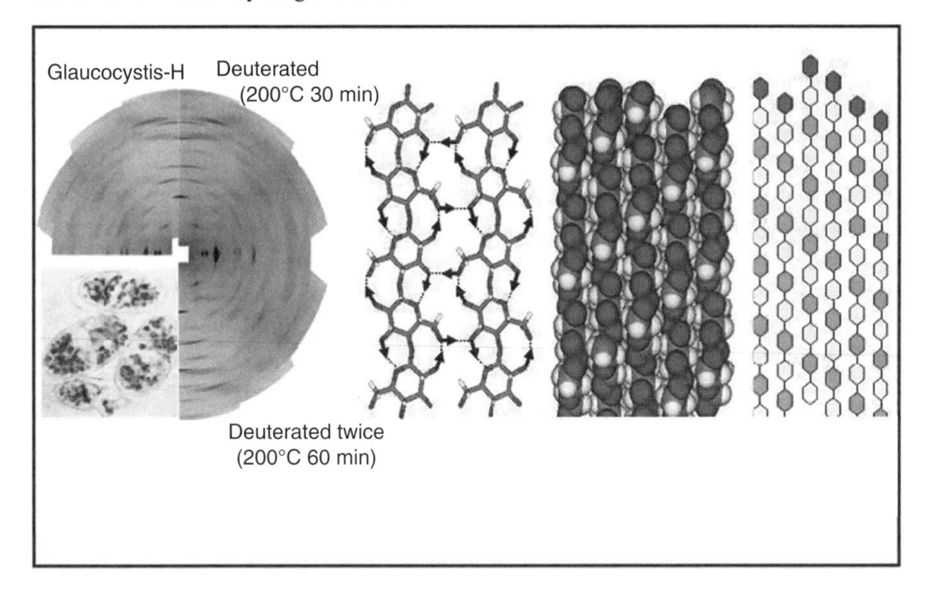

FIG. 14. Structural details of cellulose Iα.[84] (See Color Plate 7.)

Given the relationship between monoclinic and triclinic unit cells, as well as the Iα→Iβ transformation effected by annealing in the solid state, it is likely that cellulose Iα also packs in a "parallel-up" fashion. The projections of the crystal structures of cellulose Iα and Iβ down the chain axis are remarkably similar (see Fig. 18). As is shown by the projection perpendicular to the chain axis in the plane of the hydrogen-bonded sheets, the main difference is the relative displacement of the sheets in the chain direction. In both Iα and Iβ, there is a relative shift of about $c/4$ in the "up" direction between neighboring sheets. The most probable route for the solid-state conversion of cellulose Iα→Iβ is the relative slippage of the cellulose chains past one another. Such a movement does not require the disruption of the hydrogen-bonded sheets (along the 010 planes for cellulose Iβ and 110 planes for cellulose Iα), but slippage by $c/2$ at the interface of the sheets. The exact location of the Iα and Iβ phases along the crystalline cellulose microfibrils is another subject of interest. The respective components could be identified as alternating along the microfibril of the highly crystalline algal cellulose in the cell-wall of *Microdiction*.[90]

Modeling studies have established that the two crystalline arrangements correspond to the two low-energy structures that could arise from parallel associations of cellulose chains. Within the framework of these studies, three-dimensional models have also been proposed that allow comparison of the similarities and the differences that characterize the two allomorphs of native cellulose.[85]

Several hypotheses have been proposed to account for the occurrence of two phases in native cellulose. In general, samples that are rich in Iα are biosynthesized by linear terminal complexes containing a number of cellulose synthases, assembled in biological spinnerets at the cell membranes. Those rich in Iβ are organized in a rosette fashion.[91] However, a notable exception is tunicin, where linear terminal complexes produce almost pure Iβ.[92] Obviously, a comparison between the morphologies of Iβ tunicin microfibrils with those of Iα-rich seaweeds would be instructive. The former have a parallelogram shape, whereas the latter have a square shape (Fig. 15). Therefore, despite their common linear geometry, the terminal complexes of tunicates and those of seaweeds produce microfibrils of different shapes and crystalline polymorphism. Other factors may play a key role in inducing crystalline structures of cellulose.

4. Cellulose II

Early work on the solid-state structure of cellulose dates from 1929,[93] when it was proposed that the unit cell has dimensions: $a = 8.14$ Å, $b = 14$ Å, $c = 10.3$ Å, $\gamma = 62°$, and contained two cellulose chains. This proposal has generated little controversy,

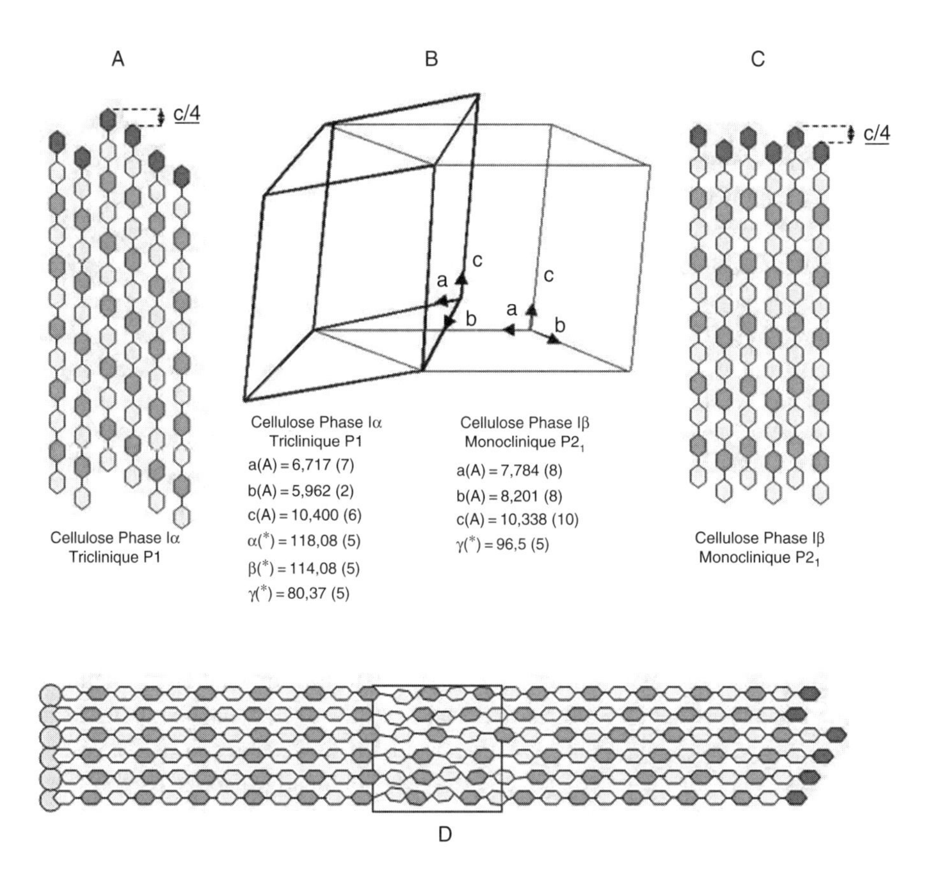

FIG. 15. The relationships between the unit cell of cellulose Iα and Iβ, and a schematic representation of the simultaneous occurrence of the two allomorphs within the same microfibril. Such an event (boxed area) is likely to be the site of an amorphous moiety within the microfibril. (See Color Plate 8.)

despite the difficulty in precisely indexing the X-ray diffraction reflections. However, a larger unit cell ($a = 15.92$ Å, $b = 18.22$ Å, $c = 18.22$ Å, $\gamma = 117°$) was proposed on the basis of a neutron-diffraction study, which called into question the previous assignment of the monoclinic space group $P2_1$.[94] These variations could arise from the use of neutron diffraction, which is sensitive to structural defects and disorder arising from various factors affecting the orientation of the hydrogen atoms in the hydroxyl groups. Furthermore, it could be argued that the methods of preparation of this allomorph might account for some of the differences. There are indeed two main routes to cellulose II: mercerization, which involves treatment with alkali and solubilization, followed by regeneration (recrystallization). In spite of the similarities in unit-cell dimensions, there are some differences that seem to be significant. For example, the value of the a dimension in a cellulose regenerated from ramie is 8.662 Å, but is 8.588 Å in cellulose obtained by mercerization. Similarly, the value of the angle γ is always more significant for mercerized celluloses than for regenerated celluloses. It also seems probable that the sample purity has a bearing on the quality of the crystalline domains and unit-cell parameters leading, as in the case of regeneration, to elevated rates of conversion.[95] There are few reports of the occurrence of the type II allomorph in native celluloses.[96] However, a structure corresponding to cellulose II has been proposed for cellulose from a mutant strain of *A. xylinum.*[97]

Several structural determinations have used a combined approach of X-ray diffraction data and modeling methods for minimizing the packing energies of cellulose chains in the unit cell.[64,65,98] In spite of some minor differences, the results agree sufficiently well to propose a model in which the cellulose chains have almost perfect two-fold symmetry and is compatible with the occurrence of two intermolecular hydrogen bonds between consecutive residues [OH-3′…O-5 (2.70 Å) and OH-2… O-6′ (2.70 Å)]. Within the crystalline mesh, a network of hydrogen bonds ensures the formation of layers composed of cellulose chains. A notable feature of this three-dimensional arrangement is the antiparallel orientation of the cellulose chains.

The similarities that exist between X-ray powder diffraction diagrams of cellulosic oligomers and that of cellulose II have excited the curiosity of crystallographers, since it seemed likely that high-quality structural data from single crystals could be used to construct a model for the polymer. However, in spite of early success in crystallizing cellotetraose[99] and attempts at simulation, it was not until 1995 that the structure was resolved by two independent research groups.[43,100] Cellotetraose crystallizes in a triclinic unit cell (a = 8.023 Å, b = 8.951 Å, c = 22.445 Å, α = 89.26°, β = 85.07°, γ = 63.93°) having space group P_1 containing two independent molecules. A major conclusion of these studies concerned the significant differences in the geometry of the two cellotetraose molecules, which are oriented in antiparallel fashion in the unit cell. Application of these new data to the resolution of cellulose II[43,100] has, in the main, confirmed the conclusions of these studies with regard to the relative chain orientation, the network of hydrogen bonds, the chain conformation, and unambiguous assignment of the *gauche–trans* conformation of the primary hydroxyl groups, all in accord with spectroscopic data.

Progress with methods of intracrystalline deuteration[87,88] has also made an important contribution to elucidation of the cellulose II structure. Indeed, the combination of X-ray and neutron-diffraction data has allowed precise analysis of the complex network of inter- and intra-molecular hydrogen bonding in cellulose II obtained by regeneration. This is the best model available for cellulose II[101] (Fig. 16). In this

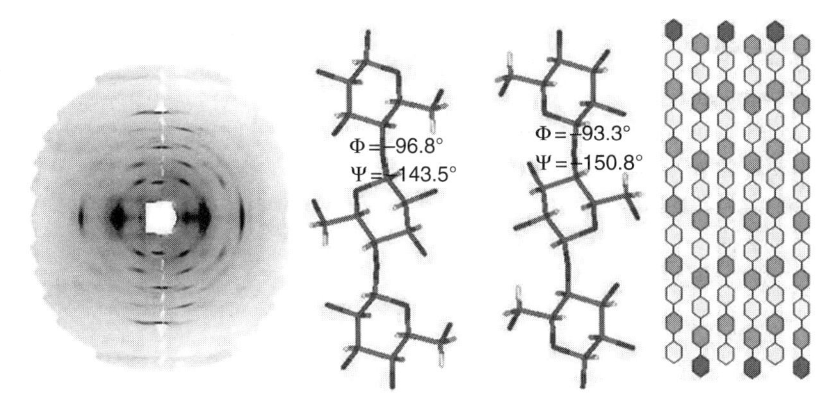

FIG. 16. Structural details of cellulose II.[46] (See Color Plate 9.)

model, the structure of cellulose II is based on a two-chain unit cell of dimensions: $a = 8.10$ Å, $b = 9.04$ Å, $c = 10.36$ Å, $\gamma = 117.1°$. The chains are located on the 2_1 axes of the monoclinic cell and they are antiparallel. The two chains have different backbone and glucose conformations. The glucose residues of the central chain are strained, and the chains are displaced relative to each other by about ¼ of the fiber repeat. The hydroxymethyl groups of the central chains are disordered, and occupy both *trans–gauche* and *gauche–trans* positions. The precise location of hydrogen atoms provides a detailed description of the hydrogen-bonding system. A systematic three-center intrachain hydrogen-bond pattern is observed in both chains. This bond has a major component between O-3 and O-5, with O-3 as a donor. A similar three-center hydrogen-bond interaction occurs in the β-cellotetraose structure. The intermolecular hydrogen bonding differs substantially from that observed in β-cellotetraose. One consequence of the difference is that O-6 of the origin chain can donate a hydrogen bond to three possible acceptors, the major component being to O-6 of the center chain. These three acceptors already interact with each other through a three-center hydrogen bond. It is not clear to what extent disorder of the O-6 group of the center chain is responsible for this intricate hydrogen-bonding arrangement.

The use of synchrotron X-ray data collected from ramie fibers after ad hoc treatment in NaOH provided a revised crystal-structure determination of mercerized cellulose II at 1 Å resolution.[46] The unit-cell dimensions of the $P2_1$ monoclinic space group are $a = 8.10$ Å, $b = 9.04$ Å, $c = 10.36$ Å, $\gamma = 117.1°$. As with the regenerated cellulose, the chains are located on the 2_1 axes of the cell. This indicates that the different ways of preparing cellulose II result in similar crystal and molecular structures. The crystal structure consists of antiparallel chains having different conformations, but with the hydroxymethyl groups of both chains near the *gauche–trans* orientation. There are nevertheless some significant differences between the conformations of the hydroxymethyl group of the center chain as compared to that found in regenerated cellulose. This may be related to the difference observed in the amount of hydroxymethyl group disorder: 30% for regenerated cellulose and 10% for mercerized cellulose. Whether this disorder is confined to the surface of the crystallites or is pervasive is not yet known.

5. Cellulose III

The crystalline forms of cellulose III (III_I and III_{II}) are reversible; they can be prepared by the same treatment in dry liquid ammonia, starting from native or regenerated cellulose. This suggests that, as with allomorphs I and II, the chain orientation is the same as in the starting material.[102,103] From unit-cell dimensions $a = 10.25$ Å, $b = 7.78$ Å, $c = 10.34$ Å, $\gamma = 122.4°$, a structural model in which the chains did not have strict two-fold symmetry was proposed. Several research investigations have focused on the reversible transformations between cellulose I and cellulose III, using techniques such as electron microscopy,[104] solid-state NMR,[72,105] X-ray diffraction,[103] and molecular modeling.[106]

From one study of the cellulose I→II transformation involving an intermediate cellulose I–EDA complex, it was concluded that a liquid-crystalline phase was involved.[106] In *Valonia* cellulose, the conversion from form I to II was accompanied by an important decrystallization and fragmentation of the cellulose crystal. This reverse transition resulted in partial recrystallization, but it did not allow complete restoration of the damage done to the morphological surface. Characterization by electron diffraction revealed that the uniplanar–uniaxial orientation of the crystalline cellulose microfibrils was lost completely during the stage of swelling and washing necessary for the conversion into cellulose III_I. Washing with methanol resulted in the formation of irregularities into which were inserted crystalline domains of small dimensions. The final material that crystallized in the cellulose I form was obtained by treatment with hot water, and characteristically displayed an increase in the accessible surface and consequently reactivity.

Solid-state NMR studies have shown a significant decrease in the lateral crystallite dimensions during the cellulose I→III_I transition. At the same time, the cellulose chains show conformational changes arising from the primary hydroxyl groups, which change from a *trans–gauche* arrangement in cellulose I (65.7 ppm) to *gauche–trans* in the cellulose I–EDA (62.2 ppm) in the allomorph III_I. Thus the regenerated cellulose

I provides a spectrum that differs from that of the native form. Electron microscopy shows that cellulose I complexed with EDA is composed of nonuniform crystalline domains, whereas the III_I allomorph is characterized by well-defined crystalline zones. The conformational changes observed for the primary hydroxyl groups are of interest, as they provide possible markers for study of the various conformational transitions associated with cellulosic systems.

Whereas the molecular and crystal structure of cellulose III_{II} has not yet been solved, that of cellulose III_I was established from synchrotron and neutron fiber diffraction.[107] It consists of a one-chain unit cell in a monoclinic $P2_1$ space group, which implies a parallel arrangement of the cellulose chains. The chain conformation resembles the center chain of the cellulose II structure, as established previously.[46] The primary hydroxyl groups are in the *gauche–trans* conformation, and there is one bifurcated hydrogen bond: O-3…O-5 and O-3…O-6.

6. Cellulose IV

Heating cellulose III in glycerol for 20 minutes at 260 °C yields cellulose IV[108] and the two allomorphs IV_I and IV_{II} originate from cellulose I and II, respectively. The conversions are never totally complete, which explains the difficulties in the production of good-quality X-ray diffraction patterns. However, unit-cell dimensions have been obtained for the two allomorphs of which IV_I has a = 8.03 Å, b = 8.13 Å, c = 10.34 Å, which is close to those found for form IV_{II} a = 7.99 Å, b = 8.10 Å, c = 10.34 Å.[109] In both instances the poor quality of the diffraction patterns does not allow determination of the space group. The authors suggest space group P_1, but this is not compatible with the proposed unit-cell dimensions.

7. Alkali Cellulose and Other Solvent Complexes

Existing research reports have tended to focus on the relative arrangement of cellulose chains in the cellulose I and II allomorphs. Whereas in native architectures the chains are parallel, regenerated or mercerized celluloses have antiparallel arrangements. Elucidation of the detailed events taking place during the transformation of cellulose I and II is of great interest, especially as the process of mercerization does not appear to require solubilization of the cellulose chains. It would seem, therefore, that the cellulose structure should be preserved. To understand the mechanisms that come into play, a great number of investigations have been dedicated to the study of intermediate structures.[92,110] By X-ray diffraction characterization, Okano and Sarko[111] put forward evidence for the occurrence of five different types of structure that could be classified as a function of the cellulose-chain conformation. Soda celluloses of type I, III, and IV were characterized by a repeat of 10.3 Å, whereas types IIA and IIB had a repeat of 15.0 Å, corresponding to a helicoidal repeat of 3, a value not seen in previous studies, but in agreement with the prediction from molecular modeling studies. In addition, all of the "soda celluloses" showed a reasonable degree of crystallinity and orientation. It thus seems difficult to reconcile a change of orientation with the mercerization process. These workers proposed that, since soda cellulose I could not be converted back into cellulose I, the chains must be arranged in antiparallel fashion, namely, as in cellulose II. Despite the failure to identify the mechanisms coming into play during the transformation from a parallel arrangement (cellulose I) to an antiparallel arrangement (cellulose II), these reports nevertheless have the merit of identifying the intermediate stage (Na-cellulose), from which the structural rearrangement could arise without chain solubilization.

An extension of this research can be found in the studies of Hayashi *et al.*,[92] in which nine types of Na-cellulose (which could be formed from allomorphs I, II, III, and IV) were identified.[112,113] From these studies, it was concluded that the irreversibility probably depended more on conformational changes of the cellulose chain than chain rearrangements. This argument seems unconvincing, since the energy differences between two-fold and three-fold helices are too small to account for the irreversibility observed for type I and II structures. Another tentative explanation was made from the occurrence of two types of microfibril in samples of *Valonia* cellulose, one of which had the chains oriented along the O*z* crystal axis, the other having chains oriented in the opposite sense.[114] Most of the investigations dealing with mercerization of cellulose have focused on global measurements recorded on whole fibers, namely, an assembly of large number of organized microfibrils. The structural and morphological changes accompanying mercerization of isolated cellulose microfibrils have been monitored by transmission electron microscopy, X-ray, Fourier-transform infrared (FT-IR), and ^{13}C CP-MAS NMR. The changes in morphology when going from cellulose I to cellulose II were spectacular, as all the microfibrillar morphology disappeared during the treatment. The conclusion of this investigation is that it is impossible for isolated cellulose microfibrils to become mercerized while maintaining their initial morphology.[115]

A few crystal structures with solvent molecules incorporated within the crystal lattice have been investigated: cellulose II–hydrazine,[116] cellulose II–hydrate,[117] and the cellulose–EDA complexes.[118] A large number of cellulose complexes have been characterized by X-ray diffractograms.

8. Cellulose Acetate and Cellulose Derivatives

Acetylation of all hydroxyl groups of the cellulose chain yields cellulose triacetate (CTA); it is soluble in chloroform but insoluble in acetone. The cellulose acetate most commonly used in industrial processes is obtained by a partial deacetylation of CTA to a degree of substitution of about 2.6, and this product is soluble in acetone. As early as 1929, Hess and Trogus[119] identified two types of crystalline structures, namely, CTA I and CTA II. CTA I is obtained by heterogeneous acetylation of native cellulose without ever dissolving the starting material. By contrast, CTA II can be obtained either by homogeneous acetylation, the cellulose acetate being completely dissolved, or by a heterogeneous reaction of cellulose II. The two polymorphs return to their original polymorph, cellulose I and cellulose II, upon saponification.[120]

The first attempt to establish the crystal and molecular structure of CTA I was made by Stipanovic and Sarko,[121] who suggested the occurrence of parallel-arranged cellulosic chains within an orthorhombic unit-cell. Later, the structural analysis of the diffraction pattern from an acetylated material derived from the green alga *Cladophora* sp. established it as a one-chain unit cell crystallizing in the $P2_1$ space group ($a = 5.94$ Å, $b = 11.43$ Å, $c = 10.46$ Å, $\gamma = 95.4°$, density = 1.375).

The first structural model for CTA II was proposed by Dulmage in 1957,[122] and it was not until 1978 that a reliable structure was obtained, from a single-crystal structure elucidation by electron diffraction.[123] An orthorhombic unit-cell having dimensions of $a = 24.68$ Å, $b = 11.52$ Å, $c = 10.54$ Å, density = 1.29, space group $P2_12_121$ indicated an antiparallel arrangement of the chains (Fig. 17). Successive attempts to further refine this crystal structure have not yet modified significantly the structural features established through electron crystallography.

IV. Morphologies of Celluloses

Much research has been devoted to experimental and theoretical studies of the crystalline component of native cellulose, often in a context in which knowledge of the molecular and crystallographic structure of native cellulose was lacking.

It was during the 19th century that Nägeli developed a theory to deal with the birefringent materials in plant cells and starch grains. This theory introduced the concept of crystalline micelles having submicroscopic dimensions.[59] This, in turn, led to the proposal that crystalline micelles were separate, well-defined entities, stacked like bricks whose length coincided with the axis of the constituent cellulose molecules. In order to take account of the amorphous content of cellulose, the idea of

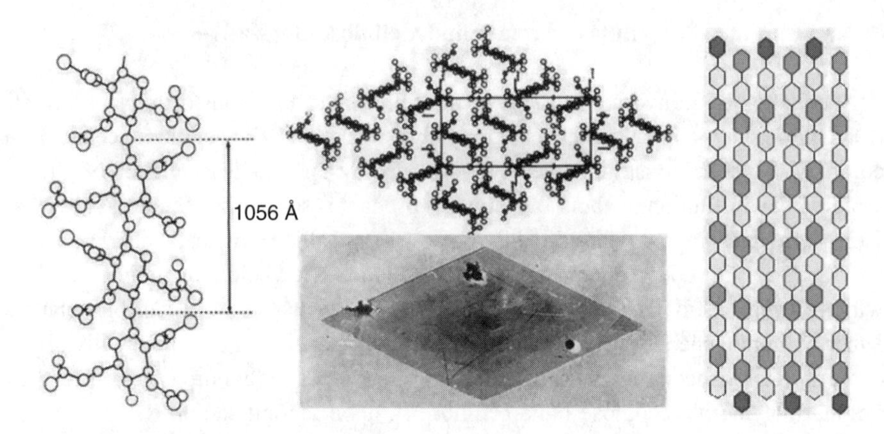

FIG. 17. Cellulose triacetate (CTA) II. (A) Transmission electron microscopy of a polymeric single crystal of CTA II. (B) Projections of the chains in the *a–b* plane. (C) Electron diffraction pattern of a tip of a single crystal of CTA II in the *a–b* plane. [123] (See Color Plate 10.)

individual micelles evolved into the hypothesis of fringed micelles.[124] In this model, the micelles are considered to be ordered regions statistically distributed in a mass of chains that are more or less parallel. The interface between crystalline zones and amorphous zones is blurred, and the micelle length need not necessarily correspond to the constituent chain length and a single chain may even pass through several micelles. The microfibrillar structure of cellulose has been established beyond doubt through the application of electron microscopy[8,125] and great variations in dimensions, depending on origin of the sample, have been reported.[55,56,126] The question of whether or not intermediate structural components termed elementary fibrils exist has been a topic of great controversy. However, the application of transmission electron microscopy[127,128] has established with certainty that the microfibril is the basic crystalline component of native cellulose.[1,127–129] It appears that the different levels of structural organization of cellulose are now well characterized.

1. Polarity of Cellulose Crystals

The cellulose chain possesses a "polarity" that arises from the chemical difference of the two ends of the molecule, and this confers particularly interesting properties to the crystalline architecture. In effect, two types of arrangement can be envisaged depending on whether the reducing groups are all located at the same end of the chain assembly (parallel arrangement) or whether the reducing and nonreducing ends are arranged in alternating fashion within the assembly (antiparallel arrangement). The answer to this question has been the aim of numerous investigations, but has equally given rise to a number of controversies. In their original model, Meyer and Misch[62] had proposed an antiparallel arrangement that was supported by the observations of Colvin on the production of bacterial cellulose[130] in which the reducing ends of cellulose had been stained with silver nitrate.[131] However, other attempts to identify reducing-chain ends using conditions similar to those of Colvin were interpreted as supporting a parallel chain arrangement. The use of exocellulases provided final experimental proof of the parallel arrangement[132–134] in the family of native celluloses.[135] Investigations using complementary enzymatic and chemical staining of reducing ends supported this model and, at the same time, produced precise descriptions of the orientation of the chains relative to the crystal axes.[81] Hence, the crystalline microfibrils possess the same polarity as the chains of which they are composed. These conclusions are in agreement with the body of crystallographic and molecular modeling studies and reflect the constraints imposed by the biosynthetic requirements of native cellulose.

2. Crystalline Morphology of Native Celluloses

The availability of an accurate description of the crystalline structure of cellulose Iβ, along with the predicted features of cellulose Iα, provides new insights into the crystalline morphology of native celluloses. These models can be used to generate different ordered atomic surfaces, and evaluate their occurrence along with their respective features. The schematic representation of the crystalline arrangements of cellulose Iα and Iβ in relation to their respective unit cells is shown in Fig. 18. Irrespective of the fine structural differences, the same gross features are exhibited by the two polymorphs, indicating that the same morphological features are expected to occur in the native celluloses. From such structural arrangements, distinct types of crystalline surfaces can be readily identified.

The type 1 surface represents the faces that run through the diagonal of the of the a–b plane of the Iβ monoclinic unit-cell, or through the a and b axis of the Iα triclinic unit cell. These surfaces are tortuous, displaying grooves extending parallel to the c axis. They are created by free spaces between the chains. Hydroxyl groups point outward, emphasizing the hydrophilic character of these surfaces. The type 2 surface represents the faces that run through either the b axis of the Iβ crystal or the first diagonal of the a–b plane of the Iα crystals. The cellulosic chains display C–H groups at the surface, and this surface is flat and hydrophobic. The type 3 surface represents

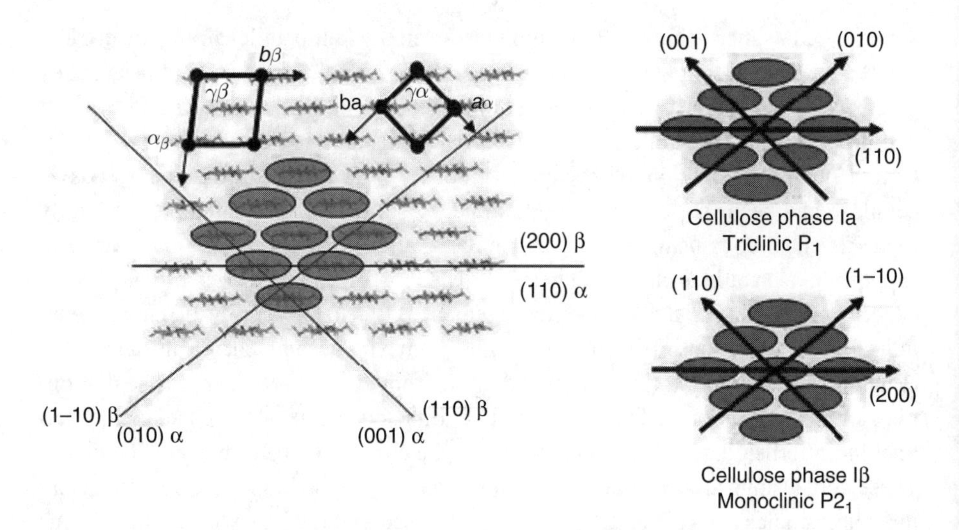

FIG. 18. Schematic representation of the crystalline arrangement of cellulose Iα and Iβ in relation with their unit cells. The triclinic and monoclinic unit-cells are shown, along with the main crystallographic directions relevant for the crystalline morphologies. (See Color Plate 11.)

the faces that are parallel to the *a* axis of the Iβ unit cell or to the second diagonal of the *a*–*b* plane of the Iα crystals (Fig. 19).

3. Whiskers and Cellulose Microfibrils

Depending on their origin, cellulose microfibrils have diameters from 20 to 200 Å, while their lengths can attain several tens of microns[1] (Fig. 20).

These characteristics confer very interesting mechanical properties on microfibrils. Transmission electron-diffraction methods have made a contribution to the quantification of the degree of crystallinity. Thus, using the technique of "image reconstruction" it was shown that, in the microfibril of *Valonia* cellulose, which has a diameter of about 200 Å, there could be more than 1000 cellulose chains, all aligned parallel in an almost perfect crystalline array.

Some imperfections arise from dislocations at the interface of microcrystalline domains along the microfibril length.[127,128] These imperfections were exploited to advantage by treatment of the sample with acid to produce nanocrystals called "whiskers," having the same diameter as the starting microfibrils but of much shorter length.

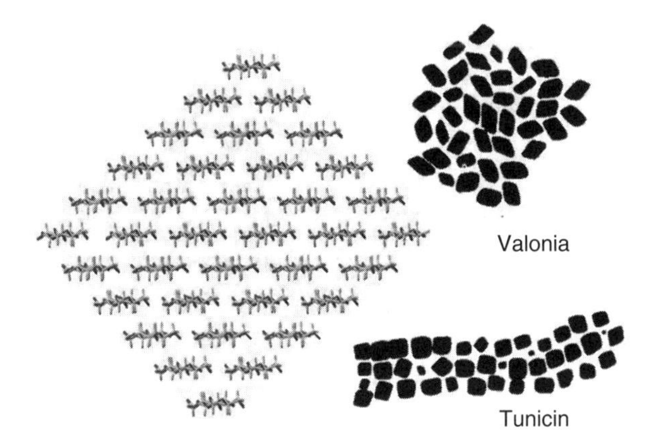

FIG. 19. Molecular model of a microfibril of cellulose, projected along the fibril axes compared with the typical morphologies observed for *Valonia* cellulose and tunicin, along with the CPK (Corey–Pauling–Koltun) representation of the main crystalline faces for cellulose I. (See Color Plate 12.)

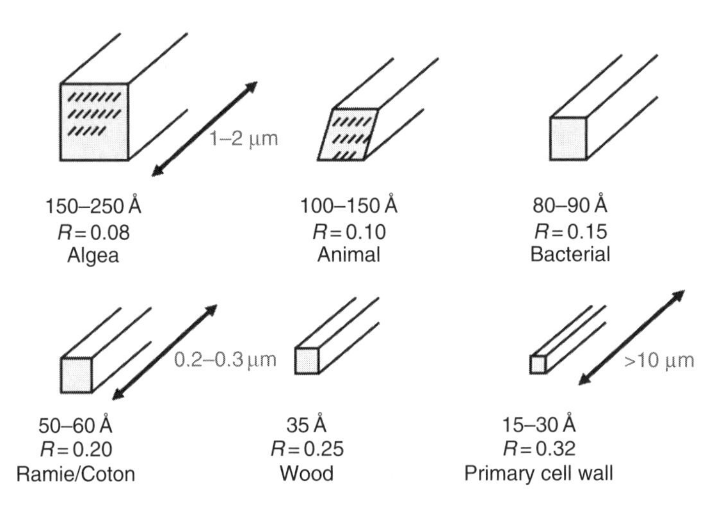

FIG. 20. Range of microfibril sizes for different sources of cellulose; R is the ratio of the number of surface chains over the total number of cellulose chains. The estimated number of cellulose chains in one crystalline microfibril according to the origin of cellulose: *Valonia*, 1200; *Micrasterias*, 900; tunicin, 800; cotton, 80; wood, 35; and parenchyma, 20.

These cellulose whiskers (Fig. 21) possess a mechanical modulus of about 130 GPa, which is close to that calculated for cellulose.[136] These characteristics (microscopic dimensions, form, and exceptional mechanical properties) made "whiskers" a choice ingredient in the manufacture of nanocomposite materials.[137]

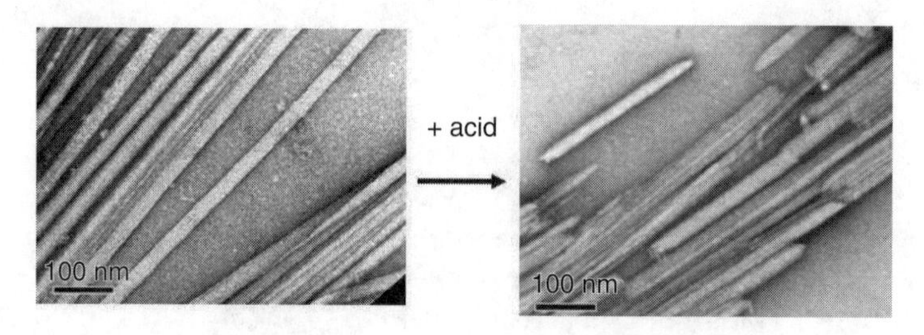

FIG. 21. Whiskers of celluloses (acid: HCl, H_2SO_4, $HClO_4$).

Celluloses of different origin yield whiskers of diverse structural quality and suitable for a range of applications. Hydrolysis of bleached tunicin cellulose (from *Halocynthia roretzi*) with sulfuric acid yields monolithic nanocrystals of smooth appearance and lengths varying from hundreds of nanometers to several micrometers. The lateral dimensions of these monocrystals range from 50 to more than 200 Å. The nanocrystals obtained by hydrolysis of cotton linters are shorter than those from tunicin, and reach lengths of 0.1 μm and widths from 10 to 50 Å, and have an aspect ratio of 20. At the other end of the spectrum there are cellulose microfibrils from parenchyma that are quite different in appearance from those of cotton and tunicin. These microfibrils are produced by a mechanical treatment which, in contrast to hydrolysis, effects disruption of the microfibrils without affecting their original length. As a result, microfibrils several microns long and 20–30 Å wide are obtained.

Analysis of these different specimens by X-ray diffraction allows appreciation of the crystalline variation and the extent to which the amorphous components are present (Fig. 22).

The diffraction patterns of tunicin microcrystals are clearly of the allomorph I; they are detailed with well-defined rings. Those of cotton are less well defined, and the rings are significantly more diffuse. For the parenchyma microfibrils, the resolution of the rings is not as good and they begin to merge; this reflects the decreased lateral order and small diameter of the microfibril. By contrast, longitudinal order is maintained along microfibrils of large dimensions. The amorphous phase increases as a result of the decrease in microfibril diameter and increase in the number of surface chains. The noncrystalline component corresponds essentially to the surface chains, which are comparatively more abundant in microfibrils of smaller lateral dimensions.

An idealized representation of the organization of cellulose chains is depicted in Fig. 23. It displays morphology and dimensions typical of those of microfibrils made up of 36 cellulose chains. In such a case, the total number of cellulose chains would be 36, and among them 20 could be considered as surface chains. This finding has been

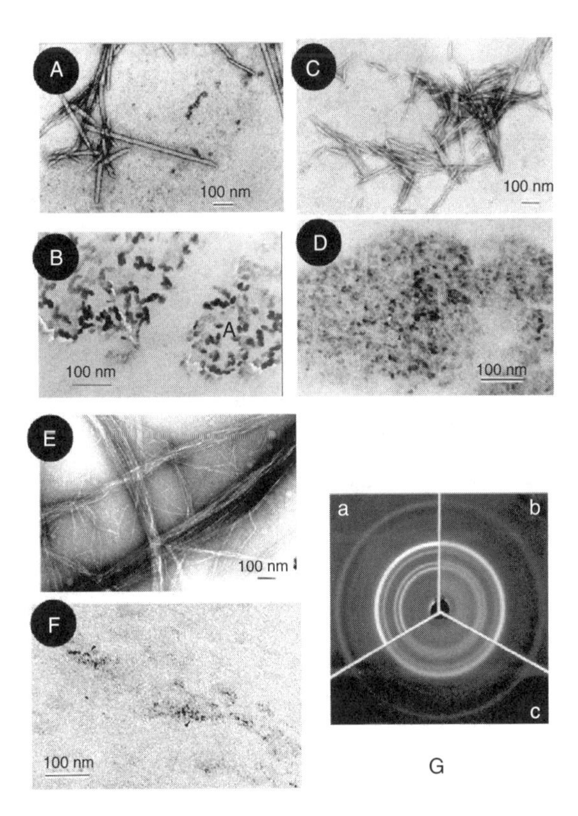

FIG. 22. Transmission electron micrographs of tunicin. (A) Microcrystals negatively stained with uranyl acetate. (B) Ultrathin section of microfibrils in bright field. (C) Transmission electron micrographs of cellulose microcrystals negatively stained with uranyl acetate. (D) Ultrathin section of a cotton fiber in bright field. (E) Transmission electron micrographs of parenchyma cellulose microfibrils negatively stained with uranyl acetate. (F) Ultrathin section of microfibrils in bright field. (G) X-ray diffraction diagrams of cellulose microcrystal: (a) tunicin cellulose; (b) cotton; and (c) parenchyma.[138]

corroborated by CP-MAS NMR studies on ultrathin cellulose microfibrils extracted from sugar-beet pulp.[140]

It may be estimated that the surface chains in tunicin microfibrils constitute no more than 5% of the total number of tunicin chains, whereas surface chains may constitute 70% of the total number in parenchyma microfibrils.[1] The result of increasing number of surface chains and decrease in whisker diameter can also be evidenced by infrared spectroscopy, with a broadening of the absorption bands and loss of resolution. A peak at 1635 cm^{-1} may be attributed to vibrations arising from water molecules absorbed in the noncrystalline regions of cellulose.[141]

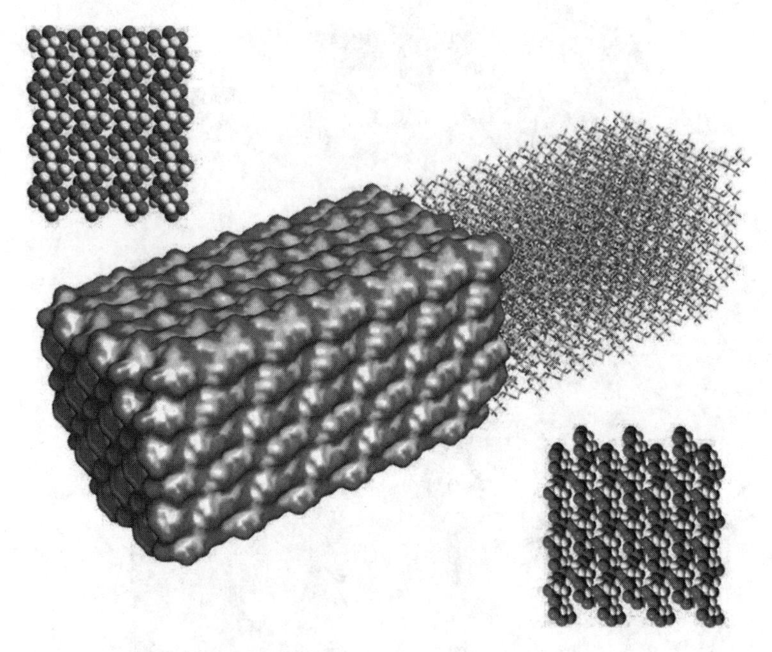

FIG. 23. Computer representation of the crystalline morphology and surfaces of a microfibril of cellulose made up of 36 cellulose chains. (See Color Plate 13.)

4. Surface Features of Cellulose

Many properties of native cellulose depend on the interactions that occur at the surface of the fibrils. As compared to bulk chains, surface chains are accessible and reactive. This is due to the dense packing of the chains within the crystal, wherein all of the hydroxyl groups participate in crystalline cohesion through intra- and intermolecular hydrogen bonds. This structural understanding is supported by the many reported selective modifications of the cellulose fibrils, which occur only at the surface of the cellulose material.

Surface chains play a fundamental role in the interaction processes (adsorption and adhesion) of the cellulose fibrils with other molecules. Such surface interactions play a key role in many areas of science: biology (interaction with the plant cell-wall polymers, adsorption of cellulolytic enzymes), industrial (paper and textile industries), and technology (compatibilization and adhesion: thermoplastic amorphous matrix on cellulose). Unfortunately, very little information has yet been gathered on the organization, conformation, and dynamics of the surface chains. In fact, few experimental techniques can access the morphology of the material and surface interactions; furthermore, they are often delicate to implement and the data obtained are often not readily exploitable.

The organization of the surface chains was first studied by microstructural chemical analysis, testing the reactivity of the exposed hydroxyl groups toward various chemical agents.[142–148] The results were analyzed under the assumption that only the exposed hydroxyl groups are able to react with an external reagent. Furthermore, the hydroxyl group reactivity may be correlated with the degree of organization at the surface. An equivalent reactivity would be expected for the various hydroxyl groups in the case of an amorphous structure, where the surface chains are not organized. This is a direct consequence of similar accessibilities. The specific reactivities of the O-2, O-3, and O 6 hydroxyl groups toward various agents were measured[144] for cellulose samples differing in crystalline content. The studies showed that the 2-hydroxyl group is generally the most accessible, in contrast to the 3-hydroxyl group. The reactivity is strongly dependent on the crystalline index of the cellulose sample studied. The 3-OH group is the least accessible to chemical agents in highly crystalline *Valonia* or bacterial celluloses. This is in contrast to its higher reactivity in cotton cellulose where the degree of organization is far less perfect. The 6-hydroxyl group shows an intermediate reactivity that also depends on the crystallinity index of the cellulose. Such results correspond well with the hydrogen-bonding network revealed from analysis of the neutron-diffraction data of a deuterated tunicin sample.[48] The lack of reactivity of the 3-hydroxyl group suggests that the strong 3-OH...O-5 hydrogen bond observed in the crystalline structure persists at the surface of the cellulose materials, whereas the higher reactivity of the 2-OH and 6-OH groups suggests that the hydrogen bonds involving these groups are partially disrupted at the surface. A greater conformational mobility is therefore expected at the surface than in the interior of the samples.

The organization of the surface chains of cellulose has been studied by atomic force microscopy (AFM).[139,149–152] It could be deduced from the AFM images recorded for *Valonia* samples that the surface chains are organized similarly to those in the interior. A triclinic organization has been observed on *Valonia ventricosa* cellulose whereas *Valonia macrophysa* cellulose displays a monoclinic organization of the surface chains. High-resolution images[139] showed the (1 0 0) face of the Iα triclinic allomorph. Periodicities of 10.4 and 5.2 Å were recorded, corresponding to a cellobiose repeat along the fiber axis and interchain spacing (the distance between interreticular planes was measured at 5.3 Å), respectively. A triclinic organization was confirmed by the typical supermolecular arrangement of the chains, a diagonal shift close to 65° corresponds to this allomorph, whereas discrimination between the (1 0 0) and (0 1 0) surfaces is based on the inter-reticular distance.

Finally, comparison between high-resolution images and a reconstituted AFM image from crystallographic coordinates showed that the surface hydroxymethyl groups adopt a conformation that is different from the *trans–gauche* conformation of the buried groups (Fig. 24). Molecular models of the surface of cellulose having either *gauche–trans* or *gauche–gauche* orientation of the hydroxymethyl groups provide a better agreement with the observed images. Such conformational differences disrupt the intramolecular 2-OH...O-6 hydrogen bond of the chains that are located at the surface of the cellulose. It should be noted that all of these AFM studies require acid pretreatment of the samples to remove the disorganized chains initially present at the surface. Such chemical treatment permits experimental observation of the surface in which the backbone conformation and the supermolecular organization of the chains are close to those of the crystal.

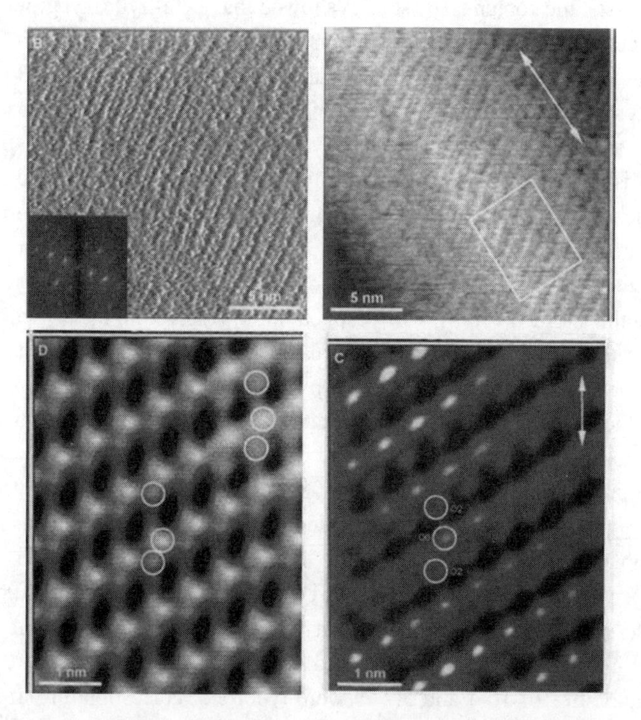

FIG. 24. Real and simulated AFM images of the cellulose Iα surface.[139] (See Color Plate 14.)

Most of the AFM observations are supported by solid-state NMR studies.[140,153] In particular, NMR suggests that the exposed primary (6-OH) groups are in the *trans–gauche* and *gauche–gauche* orientations, in contrast to the *trans–gauche* orientation of the hydroxymethyl groups of the interior. NMR is a powerful tool for determining the conformational disorder of the surface chains; it allows estimation of the relative proportions of the different organizational states of the fibrils of cellulose: crystalline bulk, organized surfaces, less-ordered surfaces, and amorphous domains.[153] It was shown that the purification process, together with acid treatment, affects the ultrastructural organization of the chains.[140]

The various experimental results show that the surface chains are partially disorganized; their conformational freedom is greater than that of the bulk chains. The fewer hydrogen bonds in the surface chains as compared to those in the bulk material are consistent with the higher chemical reactivity of those chains and also affinity for adsorbed species such as water molecules, hemicelluloses, and lignins.

5. Microfibril Organization

Cellulose in nature is most commonly found as part of an architectural complex whose ultrastructural organization depends on the source organism. In a material such as wood, which is rich in cellulose, the cell-walls are composed of cohesive, interlaced crystalline microfibrils that are themselves composed of cellulose. The cellulosic fibers are 1–2 nm long and about 35 Å wide and the microfibrils are composed of 30–40 cellulose chains (Fig. 25).

Applications of new approaches using synchrotron radiation have made a momentous contribution toward characterization of the structural organization of microfibrils.[155]

X-Ray diffraction diagrams have been recorded, using wavelengths of 0.78 Å, on wood sections about 10 μm thick and oriented perpendicular to the incident beam. The specimen under investigation (52 × 42 μm) was scanned in increments of 2 μm, resulting in an accumulation of 26 × 21 diffraction patterns, which provided a distribution map of the orientation of the axes of the cellulose microfibrils. Each diffraction diagram is characterized by strong intensities that were attributed to the (0, 2, 0), (1, 1, 0), and (1, −1, 0) planes, so that the orientation of the microfibril along the direction of propagation could be deduced. The outcome of such an exploration is shown schematically, wherein the dark areas where no diffraction is recorded are considered to arise from the lumen. Analysis of each diagram allows determination of the local orientation of the microfibril axis. Integration of the individual observations gives an image of the degree of disorientation. The large arrows indicate a marked local asymmetry in the microfibrils, and thus that amplitude is of less importance than the local orientation of the microfibril. Translated into three dimensions, these results

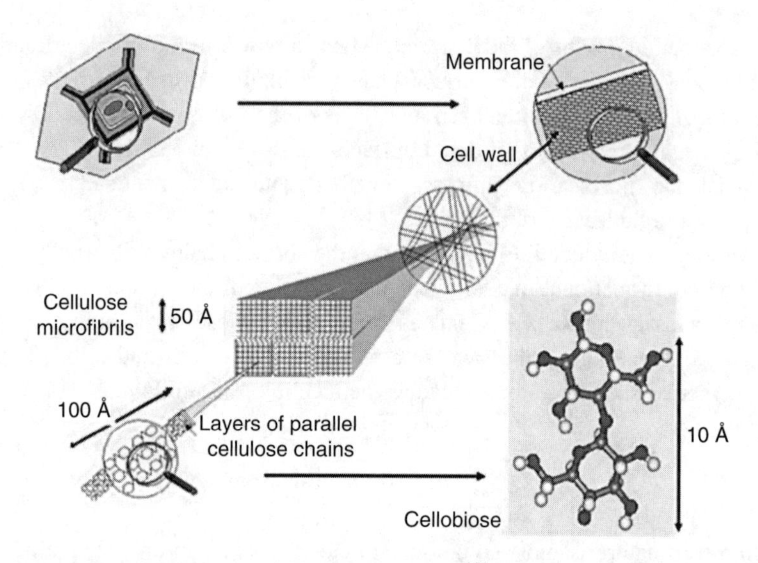

FIG. 25. Structural organization in plants.[154] (See Color Plate 15.)

lead to an ultrastructural model in which the orientation of the cellulose fibrils is aligned with the cell axis in a superhelicoidal fashion (Fig. 26).

V. Chemistry and Topochemistry of Cellulose

The presence of free hydroxyl groups within each glucose residue of the cellulose chain offers the possibility for tailored chemical transformations. These transformations can be performed homogeneously, when cellulose is in solution, or heterogeneously, with cellulose in a fairly swollen state.

1. Conditions for the Reactions of Cellulose

a. Reactions under Homogeneous Conditions.—The advantage of performing reactions homogeneously is the full accessibility of the hydroxyl groups in cellulose. This accessibility is attained by complete dissolution of the cellulose, either in nonderivatizing or derivatizing solvents. The choice between these solvents depends on (i) the type of reaction, (ii) the desired product, (iii) the DP of the starting material,

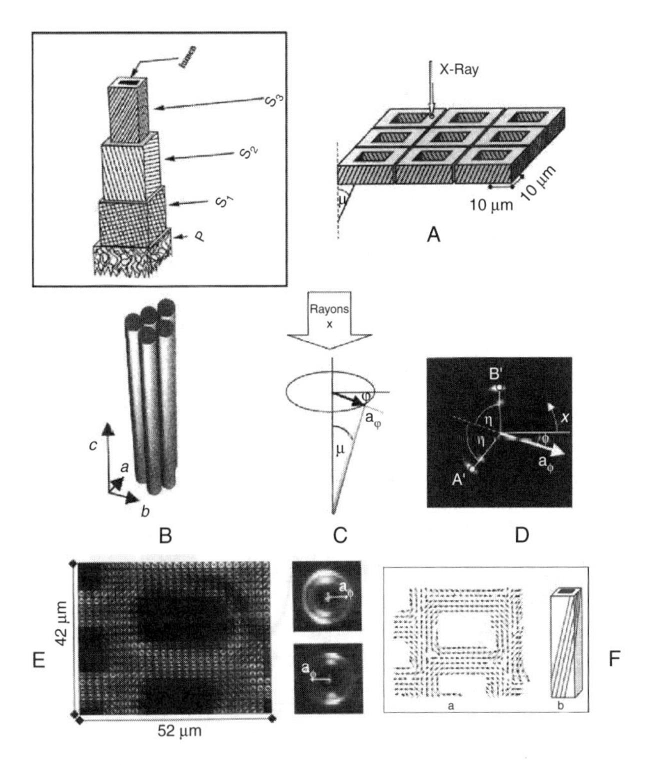

FIG. 26. Schematic representation of wood fiber structure and the helical orientation of cellulose in wood. Inset P: primary cell-wall, S2: middle secondary cell-wall, S1: outer secondary-wall, S3: inner (tertiary) cell-wall. (A) Wood cells in cross section were scanned by a microbeam smaller than the thickness of a single cell-wall. The tilt angle of the cellulose fibrils with respect to the beam corresponds to the microfibril angle μ. (B) A perfect alignment of all crystallographic axes of cellulose would yield sharp diffraction spots corresponding to the (0 2 0), (1 2 0), (1 −1 0) reflections in the plane perpendicular to the fibril axis in the reciprocal space. A random orientation of the fibrils around their longitudinal axis would result in a smearing of the reflections to produce rings. (C) Principle for the measurement of local fibril orientations. The cellulose fibril is tilted by an angle μ with respect to the incoming X-ray beam: αf denotes the orientation of the fibril in the plane perpendicular to the beam. In reciprocal space, the smearing of the reflections (0, 2, 0) large ring; ((1, 1, 0); (1, −1, 0), small ring) is caused by random orientation of the parallel cellulose fibrils around their longitudinal axis. A and B are the points of intersection between the Ewald sphere and the Debye–Scherrer rings. Scattering pattern as it would appear on the area detector in the plane perpendicular to the beam. The scattering pattern is asymmetric. The orientation αf of the cellulose fibrils (indicated by an arrow) can be extracted directly from the peak position. (D) Mesh scan over a complete wood cell in cross section, with part of neighboring cells; pixel size: 2 × 2 mm. The dark region corresponds to lumina, bright region showing a scattering signal corresponds to cell-walls. Two typical diffraction patterns (with greater magnification) show the local orientation of the cellulose fibrils, as denoted by arrows. (E) Map of local cellulose fibril orientations. Following the arrows readily yields the trace of the fibrils around the cell. Longer arrows denote a more pronounced asymmetry of the diffraction patterns corresponding to smaller local fibril angles; shorter arrows denote larger fibril angles. (F) Translation of the arrow map into a three-dimensional model; the cellulose fibrils trace a helix around the cell.

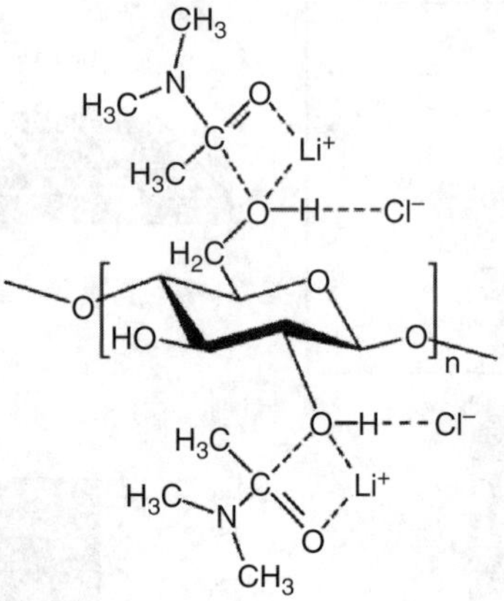

FIG. 27. Scheme of the Li–DMA,Cl^-–cellulose complex.[156]

(iv) the degree of substitution desired for the product, and (v) possible interactions between the reagent and the components of the reaction medium.

Complementary to the traditional xanthate system used to make viscose rayon, films, and sponges, or the use of aqueous NaOH to dissolve cellulose under limited conditions, many nonaqueous solvent systems for cellulose have subsequently been developed. Apart from trifluoroacetic acid, which is the only volatile solvent known for cellulose, the others consist of reagents that react with the hydroxyl groups of cellulose in a polar aprotic solvent, such as dimethyl sulfoxide or *N,N*-dimethylacetamide (DMAC).

Two different procedures may be used to dissolve cellulose in the DMAC–LiCl system. The first proceeds via solvent exchange of cellulose soaked in water, to DMAC through ethanol, followed by stirring in 8% DMAC–LiCl at room temperature. The second one requires heating of a cellulose–DMAC suspension at 165 °C for 30 minutes, whereupon LiCl is added to the cellulose suspension at about 100 °C during the course of cooling, to adjust to 8% DMAC–LiCl (Fig. 27).

N-Methylmorpholine *N*-oxide (NMNO) containing about 20% of water has also been employed as a nonderivatizing organic solvent for cellulose. This is performed at about 90 °C. This solvent is used to make regenerated cellulose fibers (Lyocell) on the industrial scale. According to Maia and Pérez,[157] the interaction between NMNO and cellulose can be interpreted as a hydrogen-bonded complex formation with superimposed ionic interactions (Fig. 28).

Other nonderivatizing solvents, such as LiCl–DMI (1,3-dimethyl-2-imidazolidinone)[158] and tetrabutylammonium fluoride trihydrate–Me_2SO,[159] have also been

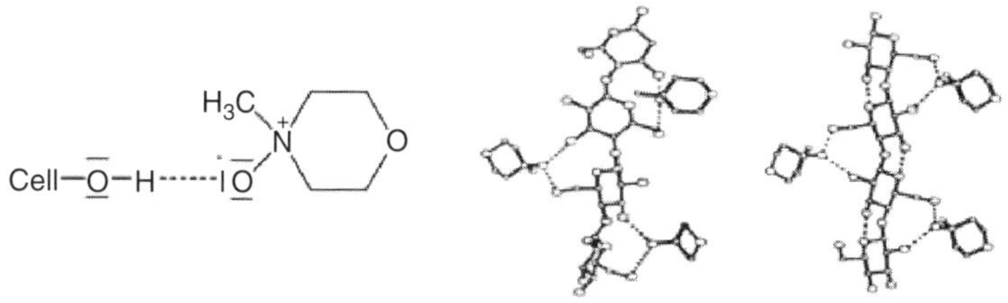

Cell—O—H······O N⁺ H₃C O

FIG. 28. Scheme of the interaction between NMNO and cellulose.[157]

investigated. An example of a derivatizing solvent is the N_2O–DMF (dimethylformamide) system, which forms cellulose nitrite during the dissolution procedure.

Molten hydrates of inorganic salts have attracted attention as new solvents and media for cellulose modification. These are compounds, having the general formula LiX × H_2O (where X = I^-, NO_3^-, $CH_3C_2O^-$, and ClO_4^-) when used in the molten state, are capable of dissolving cellulose having DP values as high as 1500. Interactions of the various cations and anions with the repeating units of the cellulose chains may permit control of the functionalization pattern.[160–164]

b. Reactions under Heterogeneous Conditions.—Under these conditions, the solid-state and microfibrillar nature of the cellulose impedes full access to the hydroxyl functional groups. These conditions can be advantageous if it is necessary to effect reaction of only a limited number of hydroxyl groups located on the cellulose surface in order to limit the extent of modification and preserve the native architecture as well as the morphology and crystallinity of cellulose.

Depending on the type of solvent used, the fibrous morphology of the starting material can either be retained throughout the reaction or be considerably changed through dissolution of the biopolymer, which can occur with swelling systems that promote the disruption of interchain hydrogen-bonds. However, heterogeneity in the reaction can be expected (i) in the amorphous regions as compared to the crystalline zones, (ii) along a single chain, (iii) between chains, and (iv) according to the reactivity of the hydroxyl groups. As compared with reactions under homogeneous conditions, the rate of the reaction, the degree of conversion, and the site of conversion are not altered.

2. Main Reactions Sites in Cellulose

Cellulose has several main sites in the glucose residues that can be considered for its modification (Fig. 29). First, the C-1 and C-4 regions are of interest in degradation

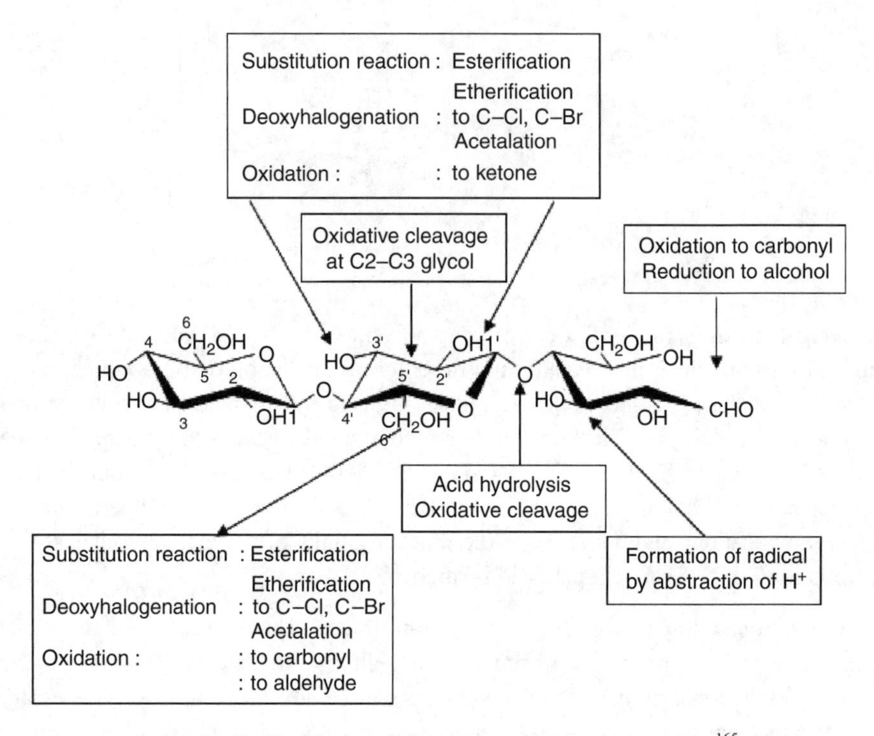

FIG. 29. Positions in the cellulose structure for chemical reactions.[165]

processes and with regard to the reducing end-groups of the chains. Second, the ring oxygen atom and that of the glycosidic linkage play some part in intermolecular interactions, although not in covalent derivatization. Finally, the three hydroxyl groups in each glucose residue unit are particularly relevant. They can enter into all well-known classical reactions, such as oxidation reactions, etherification, acylation, and deoxygenation. The following are the key factors influencing the characteristics of cellulose derivatives: (i) the chemical nature and structure of the substituents introduced; (ii) the degree of substitution, namely, the extent of substitution per glucose residue; (iii) the distribution of substituents; (iv) the DP and its distribution; and (v) the presence and nature of any impurities.

There is no essential difference between the intrinsic reactivity of hydroxyl groups in cellulose and hydroxyl groups in small molecules. The crystallinity and insolubility of cellulose hinder the access of the hydroxyl groups to reagents. The hydroxyl groups in the disordered regions react readily, but the crystalline region, with its close packing

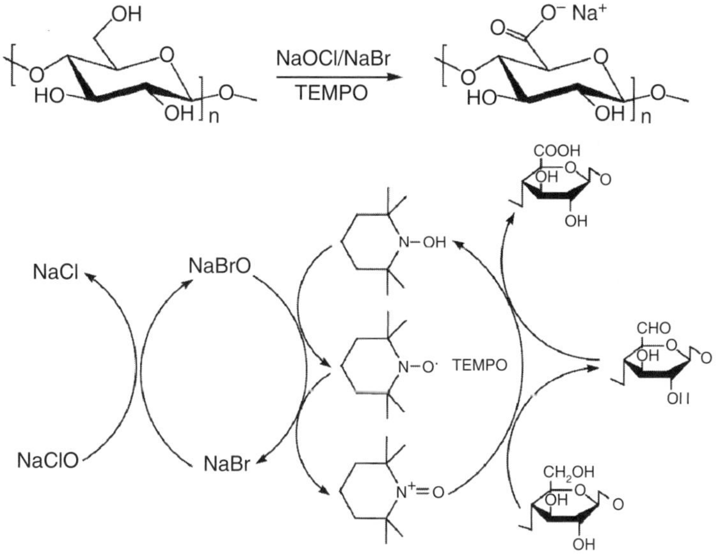

FIG. 30. Scheme of TEMPO-mediated oxidation of cellulose.

and extensive hydrogen bonding, is not readily accessible; the initial reactions occur mainly on the surface of the crystallites.

a. Oxidation Reactions of Cellulose.—The 6-hydroxyl groups of cellulose can be oxidized via an aldehyde group to a carboxyl group. Many authors have used stable water-soluble nitroxyl radicals, such as 2,2,6,6-tetramethylpiperidine-1-oxy radical (TEMPO), to oxidize the primary hydroxyl groups of cellulose to the 6-carboxy derivative[166–169] (Fig. 30). The carboxyl groups obtained by TEMPO oxidation permits subsequent coupling reactions with amine derivatives to generate cellulose substrates bearing new functionalities. Such derivatization performed by peptidic coupling, using a carbodiimide and hydroxysuccimide as amidation agents and catalyst, affords amine-terminated cellulose microfibrils.[170] Native cellulose was first oxidized with catalytic amounts of TEMPO, sodium hypochlorite, and sodium bromide in water. The oxidized cellulose was then coupled with amine derivatives by standard peptide coupling.

Oxidation at the C-2 or/and C-3 positions can generate the respective keto groups. Bond scission between C-2 and C-3 by sodium periodate ($NaIO_4$) leads to the corresponding dialdehyde, oxidizable to the diacid, or reducible to the dialcohol.[171] Varma and Kulkarni[172] examined this reaction in aqueous solution, at various temperatures and with various quantities of reagents. Aldehydes formed via this periodate oxidation can be modified by an amination reaction, leading to an imino-cellulose that can be readily reduced by an agent such as $NaBH_3CN$.[173]

This oxidative cleavage of the C-2–C-3 bond can also be performed by using ceric ammonium nitrate (CAN), which produces an aldehyde group and a radical function on a carbon atom bearing a hydroxyl group. The resulting compound can then be polymerized with acrylonitrile to form cellulose–polyacrylonitrile graft copolymers[174,175] (Fig. 31).

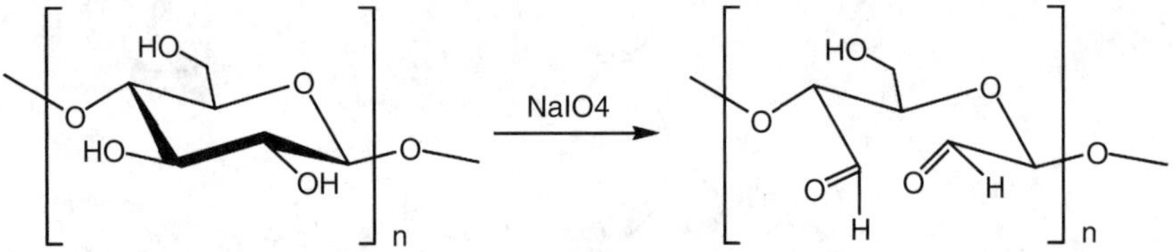

FIG. 31. Oxidation of the cellulose to "2,3-dialdehydecellulose" by the action of sodium periodate.[172]

3. Etherification Reactions of Cellulose

Cellulose ethers can be prepared by various methods, as by using the common Williamson ether synthesis, with alkyl halides in the presence of a strong base (Fig. 32). This procedure is most often used to introduce carboxyl functions [*O*-carboxymethylcellulose (CMC)] or hydroxyl groups [3-hydroxypropylcellulose (HPC) and 2-hydroxyethylcellulose (HEC)].

Addition of acrylic derivative or related unsaturated compounds such as acrylonitrile (Michael addition) is also employed for the synthesis of cellulose ethers (Fig. 33);

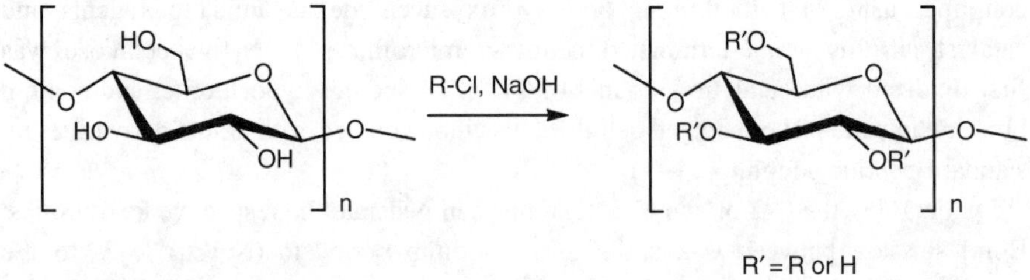

R′ = R or H

FIG. 32. Route for the preparation of cellulose ethers from alkyl halides.

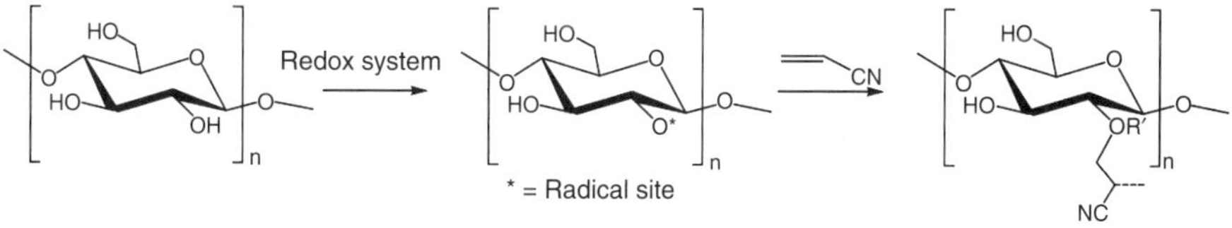

FIG. 33. Scheme of cellulose grafting with acrylonitrile.

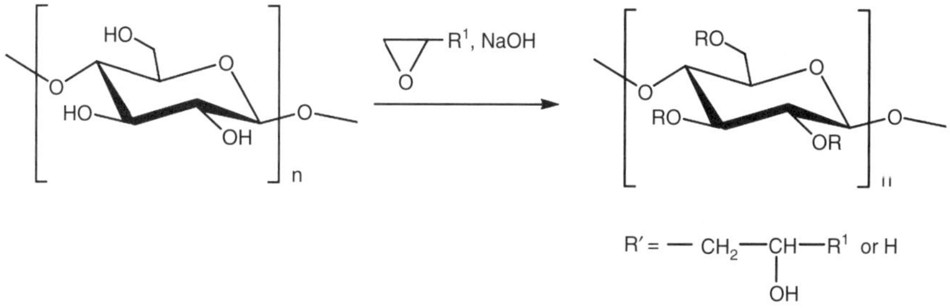

FIG. 34. Route for the preparation of cellulose ethers from alkylene oxides.

moreover, cellulose ethers can be obtained from alkylene oxides reacting in a weakly basic medium (Fig. 34).

Adsorption of alkoxysilanes onto cellulose fibers also holds promise for modification of the cellulose surface. The basic formula of the silane-coupling agents used has an organofunctional group on one side of the chain and an alkoxy group on the other.[176–178] Abdelmouleh[179] studied the adsorption of several prehydrolyzed alkoxysilanes onto the surface of cellulosic fibers in ethanol–water mixtures. This investigation offers the possibility for incorporating cellulosic fibers within a polyalkene matrix for the elaboration of composite materials.

4. Acylation Reactions of Cellulose

Cellulose esters are usually synthesized by acylation with the appropriate acid, its anhydride, or its acid chloride as the reactant. This acylation can be complete or partial, and is the principal route to many commercially useful cellulose derivatives such as cellulose formate, cellulose acetate, or cellulose nitrate.

Acylation can be effected by direct esterification with carboxylic acids. The reaction is performed with activating agents and a base, with *N,N*-dicyclohexylcarbodiimide (DCC) serving as a powerful condensation agent. The homogeneous-phase

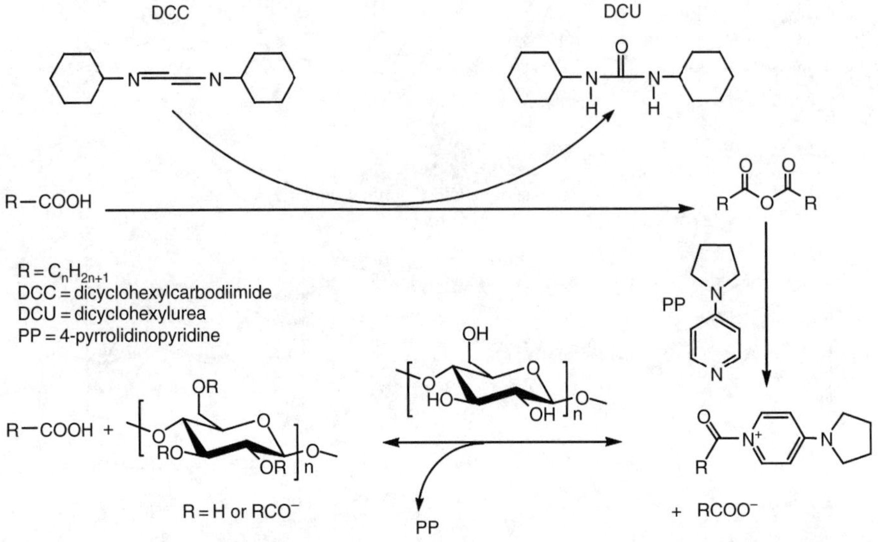

$R = C_nH_{2n+1}$
DCC = dicyclohexylcarbodiimide
DCU = dicyclohexylurea
PP = 4-pyrrolidinopyridine

FIG. 35. Acylation of cellulose according to Samaranayake and Glasser.[174]

acylation of cellulose with acid anhydrides and carboxylic acids performed by Samaranayake and Glasser,[180,181] using DCC and 4-pyrrolidinopyridine (PP), is of interest because of the high reactivity of the mixed anhydride formed with PP (Fig. 35).

Acylation by carboxylic anhydrides is usually performed in the presence of a basic catalyst such as pyridine or triethylamine.

Acylation of cellulose by carboxylic acid chlorides can be performed under homogeneous or heterogeneous conditions. During this reaction, the HCl released is captured by a base, most commonly pyridine or *N,N*-dimethylaminopyridine (DMAP) (Fig. 36).

Acylation of cellulose by transesterification involves reaction of the cellulose alcohol with an ester, leading to an equilibrium that can be displaced toward ester formation by the removal of one of the products obtained. Heinze *et al.*[159] have prepared such cellulose esters by very effective transesterification reactions (Fig. 37).

5. Deoxygenation of Cellulose

When the 2-, 3-, or 6-hydroxyl groups in the glucose-repeating unit of cellulose are replaced by other groups that do not have an oxygen atom directly attached to these

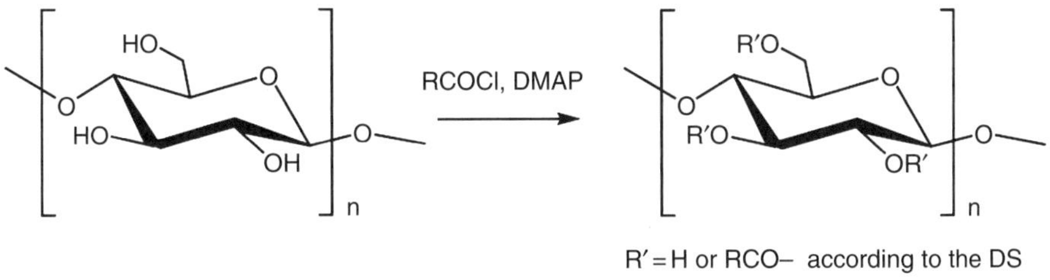

FIG. 36. Acylation of cellulose by carboxylic acid chlorides.

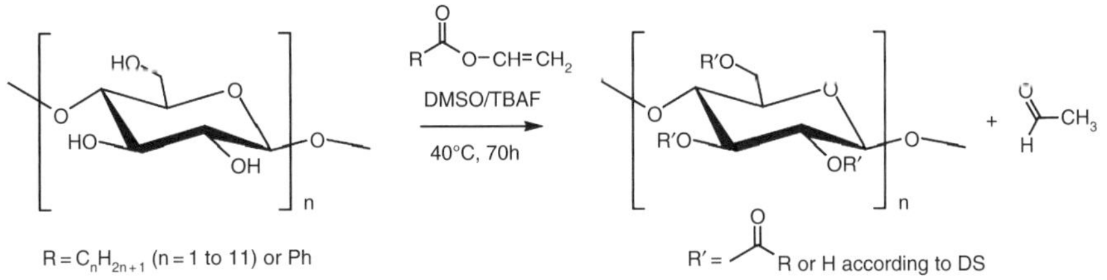

FIG. 37. Scheme of a transesterification reaction.[159]

positions, the products are termed deoxycelluloses. Halodeoxycelluloses prepared by halogenation methods have proved to be useful to further functionalization reactions of cellulose.[171,182,183]

Deoxygenation can also be performed by replacing hydroxyl groups by amino functions. For example, Tiller *et al.*[184] introduced aliphatic diamine residues into cellulose after introducing protecting groups at the C-2 and C-3 positions of the glucose residues and tosylation at C-6. The C-6 position was then substituted by a diamine in basic media to produce aminocellulose derivatives. This method gave aminocellulose derivatives bearing aliphatic diamine residues of different alkyl chain-lengths [$(CH_2)_m$, with $m = 2, 4, 6, 8$, and 12] at C-6 or with an aromatic diamine residue, such as 1,4-phenylenediamine, benzidine, and others, at C-6[185] (Fig. 38).

6. Topochemistry of Cellulose

The availability of well-characterized cellulose samples from different sources, in conjunction with detailed knowledge on the structural features of these crystalline samples, opens the route to the field of cellulose topochemistry.

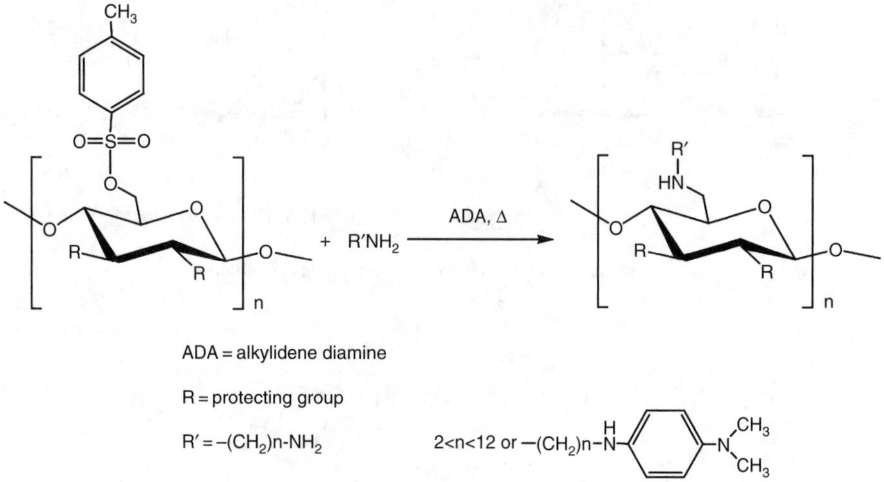

ADA = alkylidene diamine

R = protecting group

R′ = –$(CH_2)n$-NH_2 2<n<12 or –$(CH_2)n$-NH-C_6H_4-$N(CH_3)_2$

Fig. 38. Method for obtaining aminocellulose.[185]

a. Ultrastructural Aspects of the Heterogeneous Acetylation of Native Cellulose.—The acetylation of cellulose is a typical heterogeneous reaction that depends on both the accessibility of the functional groups of cellulose and the susceptibility of the individual crystallites toward acetylation. Detailed ultrastructural characterization of pure crystalline cellulose samples has provided an explanation at the molecular level of the morphological features that accompany the heterogeneous acetylation of cellulose.[138,186]

Under partial acetylation, the size of *Valonia* cellulose crystals diminished in diameter; such decrease is not homogeneous and corresponds to the loss of discrete fragments. At the beginning of the acetylation, the Iα phase is more susceptible to acetylation than the Iβ phase; the latter appears more resistant. The missing fragments correspond to Iα domains, which are solubilized initially. These domains, which are more susceptible to acetylation, are acetylated first leaving behind exposed surfaces somewhat depleted in the Iα phase (Fig. 39). This confirms that in *Valonia* cellulose the Iα and Iβ phases occur as discrete phases within the same microfibril.

b. Solid-State Conversion of Cellulose I into Cellulose III_I.—Treatment with ammonia or amines provides a simple way for increasing the accessibility of crystalline cellulose, thereby enhancing its reactivity when preparing derivatives of cellulose. The solid-state conversion of cellulose I into cellulose III_I is reversible at the crystallographic level. It is not so at the morphological level. The conversion to

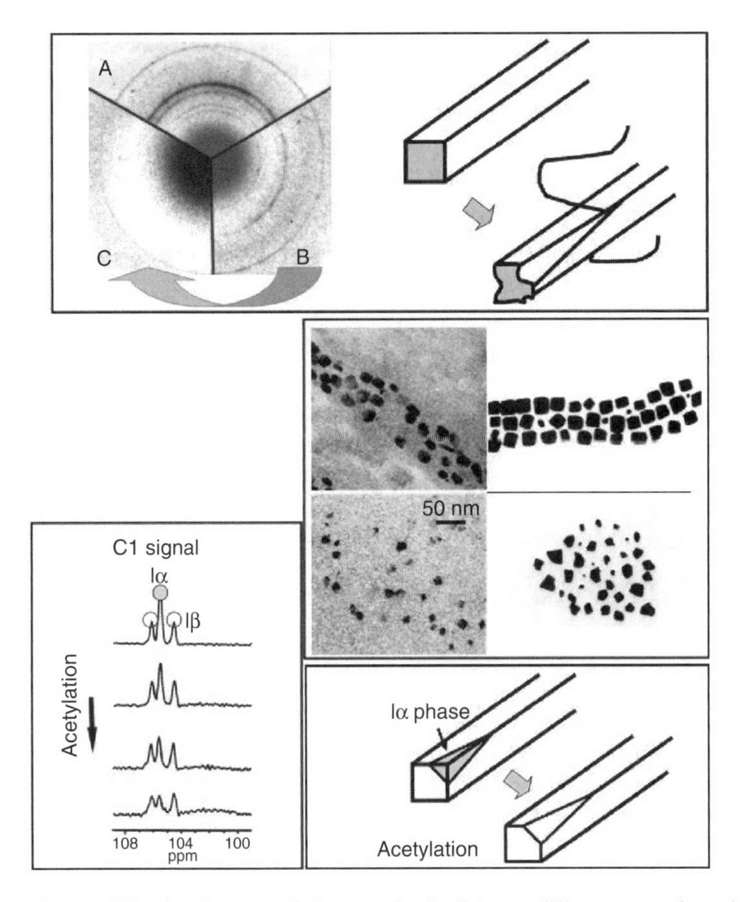

Fig. 39. Electron-diffraction diagrams of microcrystals of cellulose at different states of acetylation and after removal of cellulose acetate by selective hydrolysis: A, initial; B, sample of DS 2.41; C, sample of DS 2.81. Schematic drawing describing the onset of acetylation of a typical crystalline cellulose, showing how chains that are sufficiently acetylated are partially lifted from the crystal. Schematic representation of the change in cross section of the cellulose crystals from *Valonia* during partial acetylation. CP/MAS ^{13}C-NMR spectra of the fraction of cellulose remaining as insoluble at increasing acetylation ratio, showing disappearance of the Iα component. Schematic representation of the localization of one part of the Iα phase in *Valonia* cellulose

cellulose III_1 is accompanied by a dramatic decrease of crystallinity as well as the crystal size. After reconversion into cellulose I, the level of crystallinity is lower than that of the original sample.

A decrease in crystallinity explains why cellulose III_1, and cellulose I reconverted from cellulose III_1, has increased accessibility and reactivity toward external agents.

The ultrastructural characterization of the conversion stages that change cellulose I to cellulose III_1 and back to cellulose I has been thoroughly examined (Fig. 40). In both polymorphs, the cellulose chains are arranged in a parallel fashion, even though the packing of the chains is more dense in cellulose I (1.64 g/cm^3) than in cellulose III_1 (1.55 g/cm^3). One noticeable difference is the orientation of the primary hydroxyl groups, which are in the *gauche–trans* conformation in III_1 as opposed to the *trans–gauche* conformation found in cellulose I (Iα and Iβ). Therefore, in going from this complex solid-state conversion, some molecular changes occur, such as conformational transition of primary hydroxyl groups followed by significant alterations of the native network of intra-chain and inter-chain hydrogen bonds. As a result, the nascent

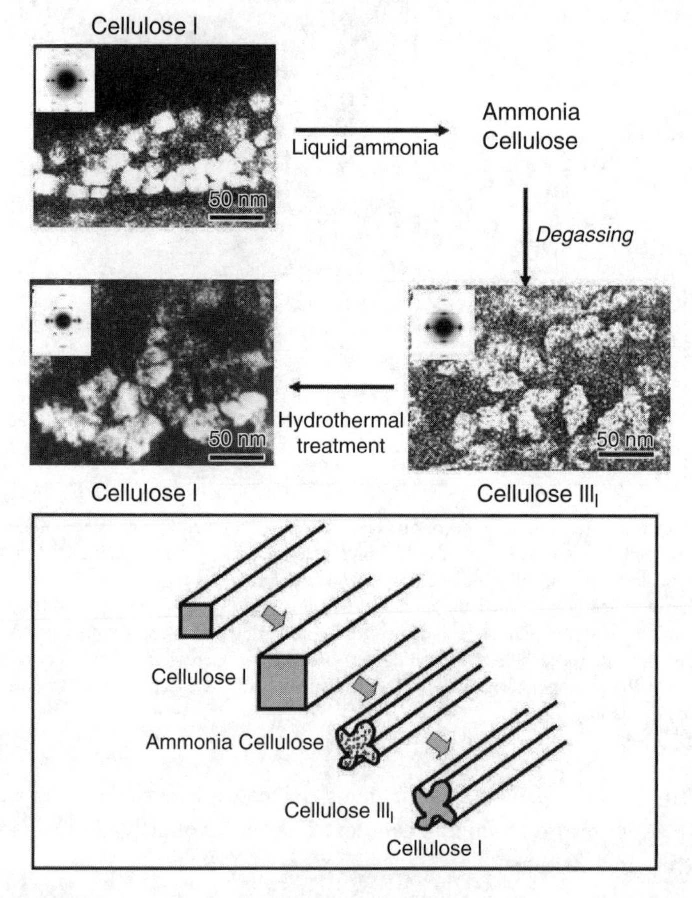

FIG. 40. Morphological changes occurring during the solid-state conversion of cellulose I into cellulose III_1.

architecture of the cellulose microfibril cannot be restored, which induces major alterations of the morphology, although the parallel arrangement of the chains has not been changed.

c. Conversion of Cellulose I into Cellulose II.—Native cellulose (cellulose I) can be converted by treatment in the solid state with NaOH (the so-called mercerization process) into a new allomorph, namely, cellulose II. The unraveling of such a solid-state mechanism has been the base of intense debate. The resolution of the crystalline structures of cellulose I and cellulose II allomorphs establishes unambiguously the parallel arrangement of the cellulose chain in the native state (cellulose I) and the antiparallel arrangement in the regenerated state (cellulose II). An interdiffusion model for explaining such observations was proposed by Okano and Sarko[110–112,127] (Fig. 41). Starting with parallel-packed cellulose I chains of an antiparallel array of nearby microfibrils in a fiber, the uptake of NaOH beginning at the noncrystalline surface of the microfibrils, interdiffusion of chains, and then washing and drying result in an antiparallel cellulose II microfibril.

A prerequisite for this model of structural interconversion is the existence of arrays of parallel-packed chains in a single microfibril, the arrays being oriented in up and down directions. The occurrence of such an arrangement was demonstrated in the highly crystalline and well-organized cell-wall of *Valonia*. Cellulose microfibrils are statistically distributed in opposite polarities within given arrays, where they are packed side by side.[114,187] Further insights into this mechanism were gained through detailed investigations of the effect of mercerization on dispersed cellulose microfibrils, where the interdigitation of the chains cannot occur for isolated individual microfibrils. The mercerization of such samples leads to disappearance of the native microfibrillar morphology.[188]

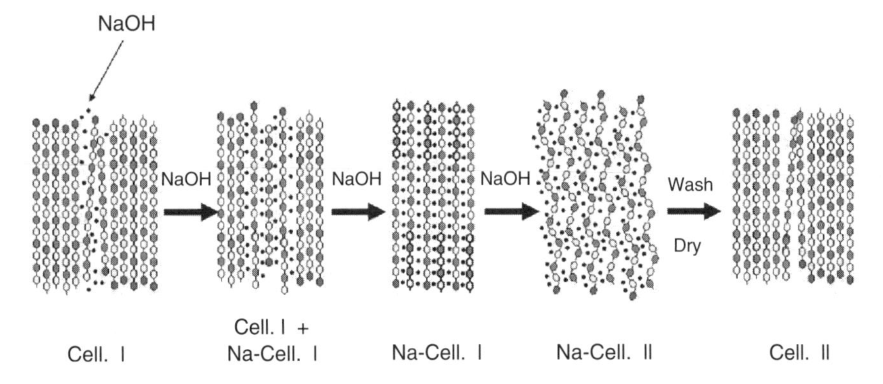

FIG. 41. Model of the conversion of parallel-packed arrays of microfibrils of up and down chains of cellulose I to antiparallel-packed fibrils of cellulose II during mercerization (redrawn from Ref. 108). (See Color Plate 16.)

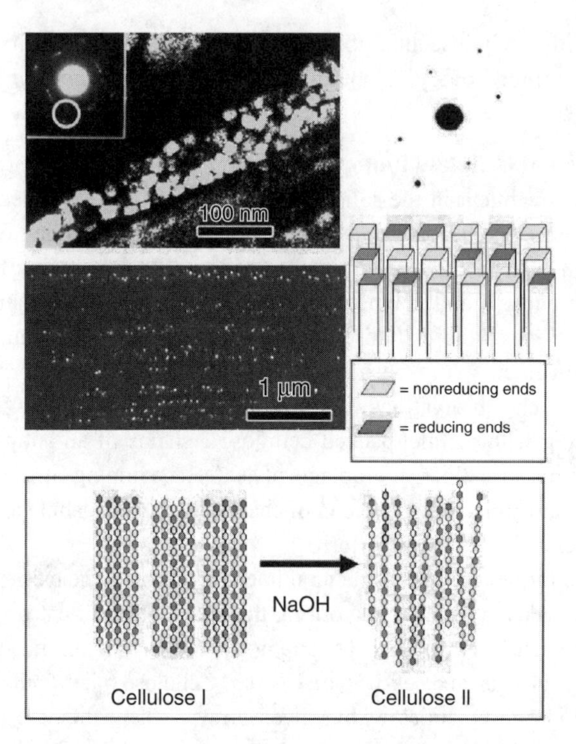

FIG. 42. Diffraction-contrast transmission electron microscopy of a fragment of *Valonia ventricosa* cell-wall cross-sectioned perpendicular to one of the main microfibrillar directions. The picture is printed in reverse contrast, so that the cross-sectioned microfibrils appear as white squares. (See Color Plate 17.)

This change in morphology, which can be monitored by transmission electron microscopy, X-ray diffraction, ^{13}C CP-MAS NMR, and FT-IR spectroscopy, is drastic as soon as the concentrations of NaOH are increased beyond 8%[115] (Fig. 42).

Monitoring of the ultrastructure during the mercerization of isolated microfibrils, along with control of partial mercerization (in sodium ethoxide), illustrates how some of the microfibrils suffer dramatic morphological modifications. In particular, those chains that had become unhinged during the alkali treatment underwent recrystallization on top of the untouched microfibrils. The result of such treatment is the remarkable "shish-kebab" morphologies, where intact microfibril parts of the native cellulose I structure (the "shish") became decorated by recrystallized cellulose II lamellae (the "kebab") (Fig. 43).

d. Topochemistry of Cellulose Nanocrystals.—The hydrolysis of native cellulose with concentrated hydrochloric or sulfuric acids causes a rapid decrease of its DP

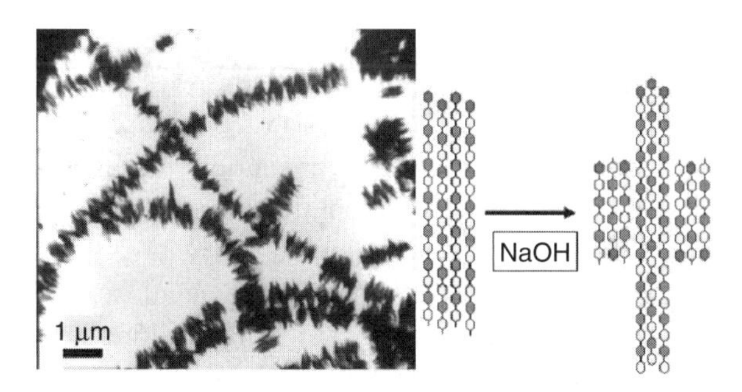

FIG. 43. "Shish-kebab" morphology.[189] (See Color Plate 18.)

toward a lower value that remains constant for a considerable time, even during prolonged hydrolysis. During the hydrolysis, the cellulose microfibrils undergo longitudinal cleavage, yielding shorter crystalline moieties that are referred to as "cellulose whiskers." In view of their very small sizes, these whiskers must be handled as suspensions in aqueous or organic solvents. When prepared by hydrolysis in hydrochloric acid, the dispersion of the whiskers is limited, and consequently the nonaqueous suspensions flocculate. When sulfuric acid is used instead as the hydrolytic reagent, the occurrence of charged sulfate groups interacting with surface hydroxyl groups of cellulose permits a perfect dispersion of the whiskers in water. A disadvantage of this mode of preparation is that the resulting surface moieties at the whisker surfaces are labile, and are rapidly removed under mild alkaline conditions. The TEMPO-mediated oxidation of cellulose whiskers offers another route for converting the surface hydroxyl groups of cellulose into charged carboxyl entities that do not exhibit the labile character of sulfate groups.[190] With a degree of oxidation of up to 0.1, the cellulose samples, while maintaining their native crystallinity and morphological integrity, have the surface hydroxymethyl groups selectively converted into carboxylic groups. In water, the oxidized whiskers become readily dispersible as birefringent nonflocculated suspensions that exhibit a chiral-nematic order.[191] While complementing fundamental structural studies,[90,192] this type of well-controlled modification of the cellulose surfaces opens up routes to further applications, as these whiskers would be amenable to chemical modifications by cross-linking.[193]

7. Enzymatic Alterations and Modifications of Cellulose Fibers

Cellulose undergoes hydrolysis by cellulases under mild conditions, as compared to hydrolysis by inorganic or organic acids. Cellulases consist of core and cellulose-binding modules, and a linker that binds the two enzymic components. The core domain contains an active center that hydrolyzes cellulose in a catalytic manner, together with subsites that interact with the cellulose chain close to the active site. The cellulose-binding modules potentiate the action of cellulolytic enzymes on insoluble substrates and play an important role in the degradation of crystalline cellulose.[194] They consist of amino acids having aromatic residues (tyrosine or tryptophan), which bind strongly to the cellulose chains through van der Waals interactions. Once bound to the crystalline substrate, the active center of the core domains of cellulases can attack the cellulose chains. The cellulases are classified into two types: the exo- and endo-cellulases, depending on whether or not the cellulose can recognize the reducing end of the cellulose chains.

The morphology of the native crystals is of course an essential feature in the enzymatic digestion of crystalline cellulose, and this is a major scientific and industrial question. It is established that the enzymatic breakdown and degradation of cellulose requires a complex of enzymes working together. The general picture that has emerged from earlier investigations in this area indicates that the cooperation of at least three types of enzymes is required for efficient digestion of crystalline cellulose into glucose. These are (i) endoglucanases (EC 3.2.1.4), which cleave the chains randomly, (ii) cellobiohydrolases (EC 3.2.1.91), which recurrently cleave cellobiose from the chain end of cellulose, and (iii) β-glucosidases (EC 3.2.2.21), which hydrolyze cellobiose. Other complementarities are found among these enzymes, as exemplified by the synergistic action of two cellobiohydrolases, namely, Cel6A and Cel7A from organisms such as *Trichoderma reesei* and *Humicola insolens* (Fig. 44). As for the cellulose-binding modules, numerous studies have established that three aromatic residues are needed for binding onto cellulose crystals, and that tryptophan residues contribute to higher binding affinity than tyrosines. However, evidence has accumulated showing that different binding sites for the same cellulose-binding domains could occur.[195]

To elucidate such a detailed synergy, dispersed cellulose ribbons from bacterial cellulose were subjected to enzymatic digestion by cellobiohydrolase, CBH1 (or Cel7A) and CBH II (or Cel6A) from *H. insolens*, either alone or in admixture, and in the presence of an excess of β-glucosidase.[196] Observation by ultrastructural transmission electron microscopy indicated that Cel7A induced a thinning of the cellulose crystalline ribbons, whereas Cel6A cleaved the ribbons into shorter elements, indicating an endomode of action (Fig. 45).

The enzymatic hydrolysis of intractable cellulosic and hemicellulosic substrates of the plant cell-wall is efficiently performed by multienzyme machines designated under the collective name of "cellulosomes."[197] Cellulosomes are large extracellular enzyme

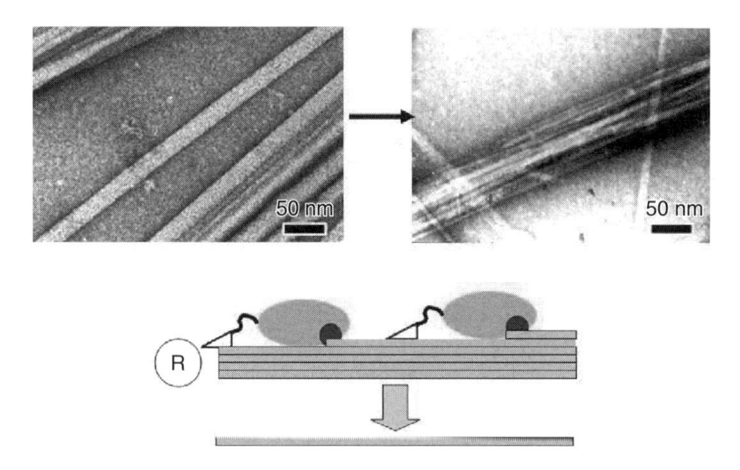

Fig. 44. Microcrystals of *Valonia macrophysa* cellulose subjected to the action of cellulases (Cel7A: from *Humicola insolens*) consisting of a hydrolytic core, a cellulose-binding module, and a linker that binds the two enzymic components. The reducing end of the cellulose chains is indicated. R, transmission electron microscopy of the cellulose microcrystals before and after the enzyme action indicates that Cel7A induced a thinning of the crystals. (See Color Plate 19.)

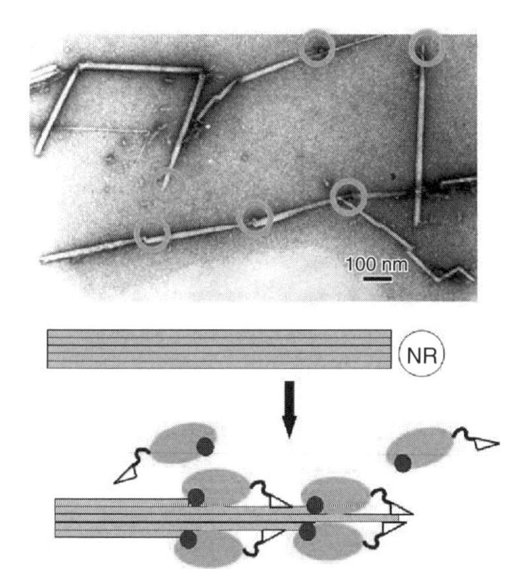

Fig. 45. Microcrystals of *Valonia macrophysa* cellulose subjected to the action of cellulases (Cel6A: from *Humicola insolens*) consisting of a hydrolytic core, a cellulose-binding module, and a linker that binds the two enzymic components. The nonreducing end of the cellulose chain is indicated as NR. Transmission electron microscopy of the cellulose microcrystals before and after enzyme action indicates that the crystals are eroded only on one end of their tips, which corresponds to the nonreducing end of cellulose (marked with circles).[135] (See Color Plate 20.)

complexes that are produced by anaerobic bacteria, and they can efficiently break down plant cell-wall polysaccharides, such as cellulose, hemicellulose, and pectin, into sugars. The cellulosome complex consists of various types of enzymes arranged around a scaffolding protein that does not exhibit catalytic activity, but enables the complex to adhere to cellulose. The organization of the cellulosome is mediated by high-affinity protein–protein interactions between Type-I cohesin domains within the scaffolding proteins and complementary Type-I docking domains carried by cellulosomal enzymes.

Various ways of modifying cellulosic materials by enzymatic treatments have also been investigated. The fiber-modification method exploits the power of enzymes capable of directly introducing new substituents to fibers by covalent bonding. This method of fiber modification could also be a potential tool for coupling bioactive molecules into fiber-based products. A novel chemoenzymatic approach for the efficient incorporation of chemical functionality onto cellulose surfaces has been reported.[198] The modification is brought about by using a transglycosylating enzyme, xyloglucan endotransglycosylase, to join chemically modified xyloglucan oligosaccharides to xyloglucan, which has a naturally high affinity to cellulose. Binding of the chemically modified hemicellulose molecules can thus be used to attach a wide variety of chemical moieties without disruption of the individual fiber or fiber matrix.

VI. Tomorrow's Goal: An Economy Based on Cellulose

Cellulose already occupies a large part of our everyday life through its use in papers, textiles, and, of course, wood. However interesting and important as these uses may be, they constitute only a small part of the potential applications of cellulose in material science. Up to the present, the hydrophilic nature of cellulose and the difficulty of processing the fibers have been stumbling blocks in the development of new applications. The present section shows that it is possible to overcome these difficulties and to conceive new materials with superior properties. To reach this potential, it is essential to take into account what is now known about the fine structure of cellulose. Selected examples will illustrate the type of results that can be obtained when cellulose fibers are no longer viewed as simple inert substrates, but when the polyhydroxyl nature of their surfaces and their entire structural complexity are taken into account. The content of this section presents a prospective view concerning the potential and the innovation that cellulose and ad hoc engineering can offer.

1. Cellulose–Synthetic Polymer Composites

This is an extremely active area of research and most of the work published has been focused on the development of "green" composite materials by replacing mineral-based fibers such as glass by plant fibers such as flax, hemp, or wood while maintaining a polymer matrix of synthetic origin. The philosophy of this research may be arguably challenged on the grounds that, at best, it leads to the "greening" of only one half of the composites. Furthermore, this approach has afforded rather mediocre mechanical properties, which have been traced to the poor quality of the interface between the plant fibers and the matrix polymer.[199–202] Indeed, the evident lack of chemical compatibility between the hydrophilic fibers and hydrophobic matrices gives rise to poor cohesion and rupture at the interface level.

Rather than just using the cellulose fibers as an inert reinforcing moiety within the polymer matrix, and with the problems that this raises, it is thus much more appealing to consider that the hydrophilicity of the fiber surface arises from hydroxyl functional groups which, because of their location, cannot form stable hydrogen bonds. These unattached hydroxyl groups are therefore available for chemical modification, and it has been shown that superficial esterification of cellulose fibers by long-chain fatty acids actually leads to remarkably hydrophobic fibers. The striking efficiency of this process is explained by the disappearance of the hydrophilic hydroxyl functions and their concurrent replacement by hydrophobic long-chain fatty acid ester groups. Taking this reasoning one step further, it would be even more useful to synthesize superficially esterified cellulose fibers in which the simple saturated fatty acids would be replaced by monomeric functions capable of cross-linking with a matrix containing the same type of monomers. The following example illustrates the implementation of this concept for enhancing the anchorage of silicone films at the surface of special paper sheets called "glassine" employed in the production of self-adhesive labels.[203] These labels are multilamellar systems composed of glassine, a film of silicone firmly anchored at the surface of the latter and on which is deposited the bed of adhesive, and then the label itself. The quality of the labels lies in the strength of the adhesion of the silicone film to the glassine. Insufficient adhesion causes dragging of the silicone film by the adhesive, rendering it unusable. Conversely, insufficient covering of the glassine surface by the bed of silicone leads to unwanted adhesion of the adhesive to the glassine.

The film of silicone is deposited in the form of a solution of oligomers associated with a platinum catalyst, which at high temperature allows cross-linking (Fig. 46).

Anchoring of the silicone film is thus achieved through penetration of the oligomer solution into the body of the glassine, followed by cross-linking. A wrapping

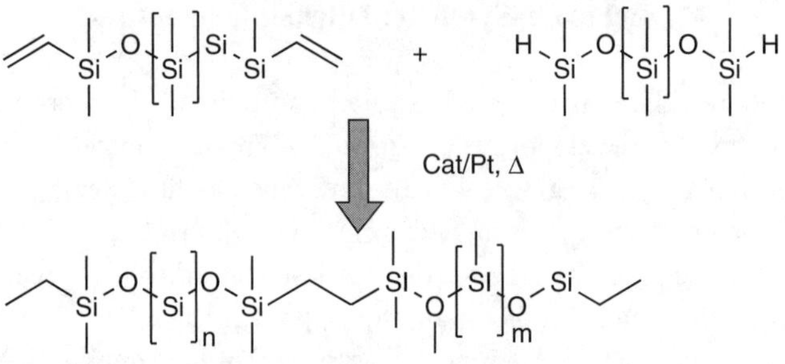

FIG. 46. Polymerizing silicone oligomers throughout hydrosilylation at high temperature, in the presence of a platinum catalyst.

mechanism around the cellulose fibers results, giving rise to physical anchoring of the film. It is easy to understand that deposition of a larger amount of silicone oligomers on the glassine allows a better anchorage and a better barrier effect of the silicone coating. One downside of this approach is that the silicone is the most costly component of the sticky labels, and is also the one that raises the most acute problems for recycling. A major objective of the project was thus to reduce the quantity of silicone oligomers used for the production of the silicone films while maintaining or even improving their adhesion and barrier properties.

This objective was achieved by switching the anchoring procedure from a passive to an active one. The strategy selected involved the introduction, at the surface of the glassine leaflets, of reactive groups capable of bonding covalently with the silicone network. Since the cross-linking proceeded via a hydrosilylation reaction, vinyl substituents were selected as reactive groups.

However, it was still necessary to develop a technology capable of achieving substitution of vinyl residues on the glassine leaflets under industrial conditions. Such substitutions are possible by classical chemistry in organic solvents, or allegedly without solvent, but not on an industrial scale.[202,204–214]

In order to achieve this stage it was then decided to take advantage of a new process coined by its developers "chromatogeny" for the synthesis of barrier films through the application of molecular grafting.[215,216]

In chromatogeny, the reaction occurs in solid–gas conditions akin to the diffusion occurring in gas chromatography. This analogy was responsible in coining the "chromatogeny" terminology. As in gas chromatography, chromatogeny takes advantage of the displacement by a hot gas stream of the gas–liquid equilibrium of a high-boiling-point reagent, which is applied on the surface of a porous medium (such as a paper sheet). This displacement leads to a diffusion mechanism of the reagent across the medium network, which is analogous to the diffusion mechanism of an eluate in gas chromatography. This diffusion mechanism leads to a spreading of the reagent throughout the entire substrate, allowing close contact with the totality of its reactogenic sites, even in the case of high-specific-surface materials. It also gives rise to an increase in activation energy, which promotes an efficient area-coupling reaction between the reagent and the substrate.

This "clean" chemistry procedure (without solvent) is characterized by very rapid reaction kinetics and allows, furthermore, elimination of potential side products such as those arising during the reaction by the gas flow. The speed and simplicity of the reaction procedure have lent itself to a continuous process, as shown in Fig. 47.

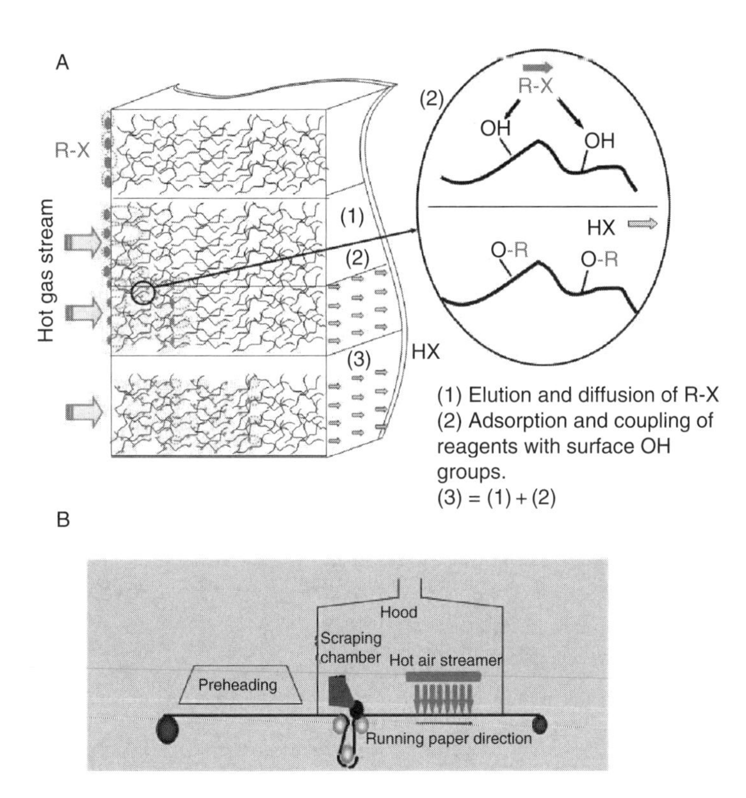

FIG. 47. Chromatogeny principle and engineering. (A) "Chromatogeny" makes use of a novel solvent-free chemical pathway for molecular grafting. It is used to achieve heterogeneous acylation of cellulose fibers through the control of the diffusion properties of acyl chloride reagents and of their by-product, HCl. This reaction occurs in the solid–gas interface, akin to the diffusion occurring in gas chromatography. This analogy is responsible for coining the expression "chromatogeny." (B) Flow sheet of the industrial implementation of "chromatogeny" for cellulose molecular grafting. (See Color Plate 21.)

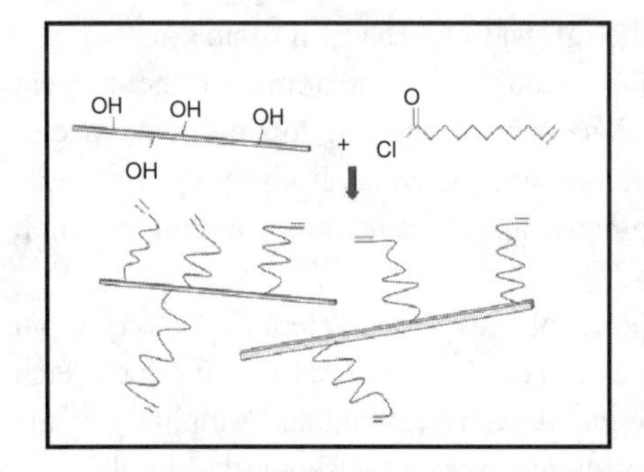

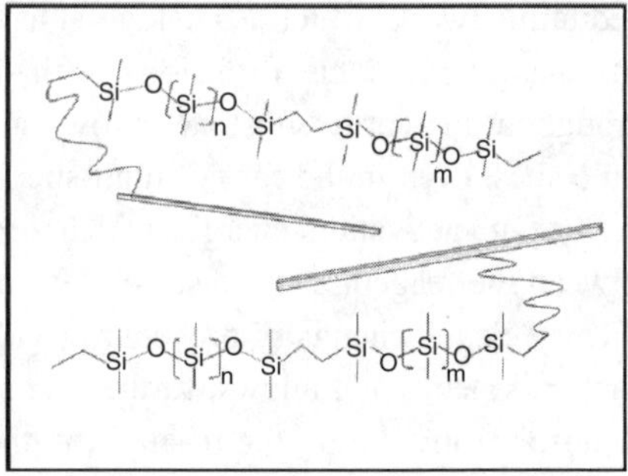

FIG. 48. Grafting cell fibers. (A) Grafting undecylenic acid molecules onto cellulose fibers. (B) Co-reticulation of silicone oligomers and cellulose grafted onto undecylenic acid molecules.

The coupling procedure was developed originally using long-chain saturated fatty acid chlorides. In joint work, the original authors have thus replaced these with 10-undecylenoyl chloride, as shown in Fig. 48A. The latter carries a terminal vinyl function capable of cross-linking the silicone oligomers as shown in Fig. 48B. A special type of glassine, SILICA 2010 Yellow, has been subjected to this procedure and then placed in the presence of silicone oligomers, followed by measurement of the adhesive properties of the silicone bed. Two types of blank experiments were conducted, one in which the virgin glassine was untreated and another in which dodecanoic acid, a saturated fatty acid of analogous chain length, was used. The results are shown in Table II.

The substitution of dodecanoic acid by 10-undecylenic acid led to an impressive six-fold increase in adhesive quality. It was shown that this improvement is fully correlated with the presence of the vinyl function, whereas no improvement in adhesion could be observed with the saturated analogue. The results also indicate that the effect is linked to a surface phenomenon, since no improvement occurs

TABLE II

Efficacy of the Silicone Protective Layer as a Function of the Nature and the Amount of the Fatty Acid Grafted

Reactant	**%w/w**	**% adhesion**
Glassine SILCA 2010 Yellow standard	–	15.3
10-Undecylenic acid	0.1	98.6
10-Undecylenic acid	0.5	99.1
Dodecanoic acid	0.1	14.1
Dodecanoic acid	0.5	12.3

following an increase in the degree of substitution. These results confirm that the introduction of covalent links for anchoring the silicone bed has permitted a considerable decrease in the amount of silicone oligomers required to maintain and even to enhance the properties of the silicone film.

By taking into consideration the presence of hydroxyl groups at the cellulose fiber surface, and their appropriate conversion according to the polymer matrix, it is thus possible to develop a product of improved quality.

2. Protective Films

The preceding section shows how it is possible to decrease the quantity of synthetic polymer while retaining the same functional properties. In this way the environmental footprint is diminished, but the products are still not amenable to recycling processes because of the cross-linking chemistry used in the silicone phase. The fact that these silicone coatings may be used for their barrier properties should motivate a transition toward this important area of packaging.

It is clear that such cellulose products as paper and boards play a major economic role but, because of their poor barrier properties, they have brought little added value to the final products. The latter is actually provided by introduction of hydrophobic synthetic polymers, but these severely compromise the ability for recycling of the products.

It would be interesting, therefore, to devise technologies that would allow protective barriers to be applied to such cellulose materials as paper and cardboard while preserving their original ability to be degraded biologically and to be recycled.

Chromatogeny does offer some possibilities in this context by its ability to confer waterproofing properties through the biocompatible grafting of long-chain fatty acids (octadecanoic acid) at the surface of cellulose fibers. However, the quality of waterproof films is limited by the fact that the chromatogeny relies upon molecular grafting, which retains the porosity of the initial cellulose material. The efficiency of the waterproof barrier is therefore dependent upon Jurin's law, and is affected by the size of the pores as well as the presence of surfactants. As seen earlier in this text, the solution to this problem requires consideration to the presence of hydroxyl groups at the surface of the cellulose fibers. The current objective, however, is to maintain the entire initial biodegradability and recyclability properties of the cellulose materials. This goal is being addressed through modification of the surface hydroxyl functions via the formation of reversible hydrogen-bonds. To this end, the surface of the materials was thus coated with a thin layer of polyvinyl alcohol (PVA), a

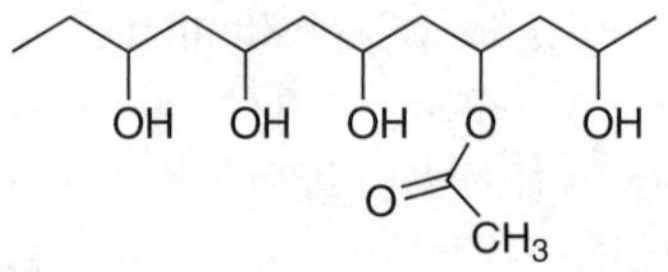

FIG. 49. Molecular structure of polyvinyl alcohol (PVA).

biocompatible polyol film. PVA thus anchored to the fiber surface by hydrogen bonds was then submitted by chromatogeny to the grafting reaction (Fig. 49).

The polyhydroxylated structure of PVA makes it ideal for this purpose because of its solubility in water and its strong adhesion to other polyhydroxylated substances, such as cellulose. In addition to its adhesion properties, PVA is innocuous, biocompatible, and biodegradable.[217–219] Furthermore, it may be applied in extremely thin coatings, which minimizes the amount of material required, thus generating a reduced environmental footprint. PVA forms tight cohesive films that exhibit excellent grease and gas barrier properties, but only in the absence of water.[219,220]

The cellulose substrates, coated with thin layers of PVA and subjected to coupling with long-chain fatty acids according to the chromatogeny process, demonstrated remarkable barrier properties toward grease, gases, and water! An in-depth study of these barrier phenomena suggested that the barrier properties really arose from the PVA coating itself, and that the coupling of the fatty acids actually created a barrier protecting the PVA from water, so that it continued to act as a barrier to gas and grease (Table III).

TABLE III

Grease, Gas, and Water Barrier Properties of PVA-Coated Paper Before and After Grafting with a Long-Chain Fatty Acid

Paper + PVA (9 g/m^2)	**Nongrafted**	**Grafted**
Grease barrier (kit test number)	12	12
(Tappi T559 pm-96)		
Gas barrier (g/m^2/day)	4.1	4.1
(ISO 2582-1995)	*(esd 3.0)*	*(esd 1.9)*
Water barrier (g/m^2)	13	0
(ISO 535: 1991)	*(esd 3.1)*	*(esd 0.2)*

The grease barrier test is a normalized test (Tappi 559 pm-96), which indicates the penetration of a greasy, nonpolar component into paper. The test is graded from 1 to 12 (1 means no barrier against grease and 12 means maximum barrier). The gas barrier test (NF Q 03-075) measures the permeability of paper using the Mariotte flask technique. The water barrier test [also called the Cobb test (NF EN 20535Iso 535)] measures the amount of water absorbed by a given substrate in a given time. Typical printing paper is characterized by a Cobb value of 20, that is, 20 g is absorbed by a square meter of paper in 60 seconds. A Cobb 60 value of 0 indicates a very good barrier property.

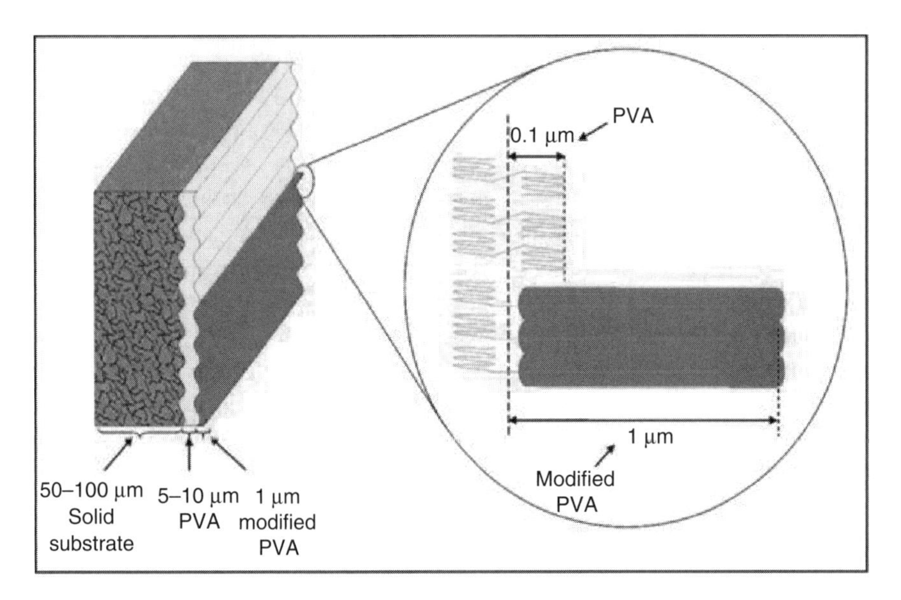

FIG. 50. Schematic representation of the architecture of the cellulosic substrates coated with PVA and grafted with long-chain fatty acid molecules. This cellulosic-based material (BT3 pack TM) exhibits remarkable barrier properties to water, grease, and gases. (See Color Plate 22.)

A detailed structural study showed that interaction between the PVA and the fatty acids was not limited to the surface, but took place through to a depth of about 1 μm. The film impervious to water has a structure in which some chains are chemically modified and some are not. The modified segments have a particular composition in which the molecular weight is increased 10-fold during the coupling reaction. The chains are immobilized by the insertion of the nonmodified segments into the PVA matrix, whereas the modified chain segments are themselves extremely compressed, giving rise to the waterproofing effect (Fig. 50).

3. Cellulose–Cellulose Composites

The substitution of glass fibers by cellulose fibers alone in the design of a composite is not particularly attractive, since the polymer matrix remains conventional and nonrenewable. The ability to produce composite materials that could be entirely renewable and recyclable would be much more appealing. Recyclability could, be further enhanced if both parts of the composite could originate from the same starting material. This concern led to consideration of new approaches to thermoplastic and thermoset cellulose–cellulose composites. The success of this endeavor demands proper understanding of the complex structure of cellulose fibers.

a. Thermoplastic Composites of Cellulose–Cellulose Acetate.—The field of thermoplastics includes several examples wherein mineral fibers are replaced by cellulose fibers.[221] The requirements in terms of overall mechanical properties are lower and the problem of water uptake by the cellulose fibers can be overcome by encapsulation of the fibers within the hydrophobic polymer matrix. This process is rendered economically viable by the current low cost of polymer materials, but this could change with a rise in the cost of fossil resources.

Cellulose acetate stands out among the available polymers that are renewable and suitable for thermoplastic purposes.[222] This substance already has a long history of use as a thermoplastic material and is an important industrial product in the manufacture of materials such as films, wrappings, and cigarette filters.[223] A whole family of related materials such as cellulose acetate butanoate is used for specific applications. Cellulose acetates are not considered to be truly biodegradable, and for a high number of applications, this is viewed as a distinct advantage. They are, however, clearly recyclable through the classical alkaline treatment that converts them back to cellulose. A particular example of such alkaline regeneration occurs in the recycling of printed paper. These cellulose acetates are therefore extremely attractive for applications as matrix thermoplastics, so long as the problems of compatibility of their interface with cellulose fibers can be solved. Indeed, although cellulose acetate and cellulose fibers share the same cellulose backbone, they exhibit quite different physical and chemical properties, and so the problem of their poor compatibility needs to be solved.

An early approach to the resolution of this problem involved the acetylation of "whiskers."[138] The potential of whiskers as nanotech reinforcers for improving the mechanical properties of thermoplastics is well known.[224–227] However, because of their strong hydrophilicity, the dispersion of whiskers in hydrophobic media poses a delicate problem.[227] In order to take full advantage of the specific properties of whiskers, it was therefore necessary to effect their superficial acetylation to render them compatible with the cellulose acetate matrices. The results showed some improvement in dispersion, and the mechanical properties were also enhanced, although only slightly, and disappointing. Another approach more suited to an industrial process involved parenchyma cellulose microfibrils that had been acetylated to a degree of 37%. The improvement of dispersion and mechanical properties of microfibrils were analogous to those obtained on covalent coupling to poly(caprolactone) diol.[228] A degree of acetylation of 37% can only be realized through the presence of pendant cellulose acetate chains of high molecular weight. A substantial improvement in mechanical properties is thus to be expected when the cellulose fibers are surrounded by pendant acetylated cellulose chains having a molecular size sufficiently large to interact effectively with the cellulose acetate chains of the exogenous polymer matrix.[229]

The underlying concept in the production and application of "whiskers" is the elimination by partial acid hydrolysis of the amorphous component of the fiber and the production of perfect nanocrystals that are resistant. However, this process may be redundant if the process of acetylation is considered further. Instead of removing the amorphous phase with mineral acid and acetylating the superficial layers of the nanocrystal, why not acetylate directly the noncrystalline cellulose chains in the amorphous regions to convert them into pendant cellulose acetate chains? The amorphous regions have always been viewed in a negative light since they constitute failure-initiating defects and are the source of the dimensional instability of the fibers. However, they may be considered from a positive point of view as far as acetylation is concerned. A fiber in which the amorphous regions have been specifically acetylated will itself constitute a hybrid cellulosic material composed of crystalline regions dispersed in a matrix of cellulose acetate and bound together by covalent links. The result obtained would have been the same had the amorphous parts of the fibers been removed and the cellulose acetate chains been covalently grafted onto the remaining crystalline skeleton. This hypothesis has been vindicated by performing the partial acetylation of rayon fibers with activation of the amorphous regions by potassium acetate, followed by contact with acetic anhydride.[230,231] The partially acetylated fibers were recovered after washing with chloroform, and the degree of residual acetate was determined by transmission IR (Fig. 51).

The results obtained showed that values of 38% (by weight) can be achieved, corresponding to a mean degree of substitution of about 1.6. It is known that acetylation of the glucose residues proceeds at a low rate toward complete peracetylation. This implies that the partially acetylated fibers are composed of about 50% nonacetylated and 50% fully acetylated cellulose. Since the acetylation procedure favors reaction at the amorphous regions, it may be concluded that the 50% nonacetylation corresponds to the crystalline regions whereas 50% acetylation corresponds to the amorphous regions. This result is in good agreement with literature data that indicate a value of 50% crystallinity in rayon.

The proportion of cellulose extracted by the chloroform treatment has also been determined, and it turns out to be very small (less than 4%). The cellulose molecules must thus be arranged alternately in crystalline and amorphous zones. The proportion of cellulose that is wholly amorphous is very small, since only 4% of free acetylated

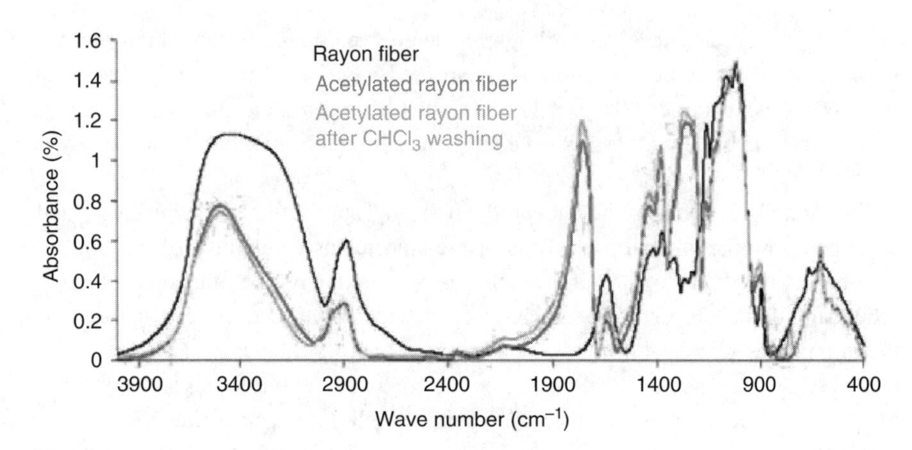

FIG. 51. Transmission infrared spectra of partially acetylated rayon fibers (A) before and (B) after washing with $CHCl_3$.

cellulose can be extracted. At the molecular level, this alternation of crystalline and amorphous segments leads, after acetylation, to a sequence of alternate crystalline cellulose and acetylated cellulose zones as in a block copolymer. In the final analysis, the acetylation of amorphous segments is equivalent to the grafting of oligomers of cellulose acetate onto a framework of crystalline cellulose. Moreover, the acetylation of the amorphous zones allows the production of a material with a very high density of pendant chains, which would be impossible to obtain with conventional grafting procedures. The compatibility of the fibers obtained by acetylation of the amorphous segments with exogenous cellulose acetate has been determined by testing dispersions of partially acetylated fibers in matrices of cellulose acetate. The results are shown in Fig. 52.

These results show that the fibers of native rayon are not at all dispersed in a cellulose acetate matrix, but appear there as tangled bundles. On the other hand, acetylated fibers are able to disperse fully into the cellulose acetate matrix, into

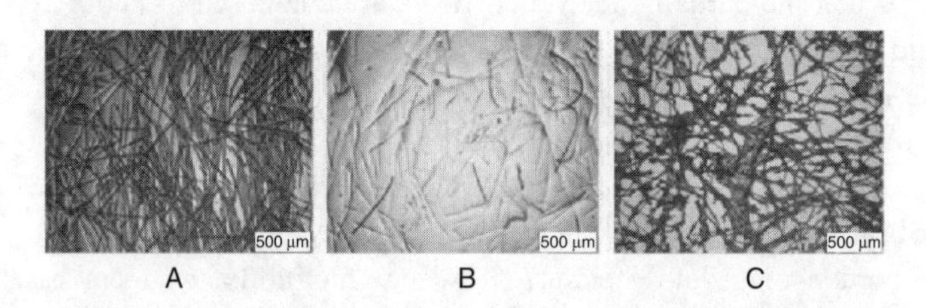

FIG. 52. Optical microscopy of (A) cellulose acetate matrix and rayon fibers; (B) cellulose acetate and partially acetylated (38%) rayon fibers; and (C) partially acetylated (38%) rayon fibers after partial removal of the cellulose acetate matrix. (See Color Plate 23.)

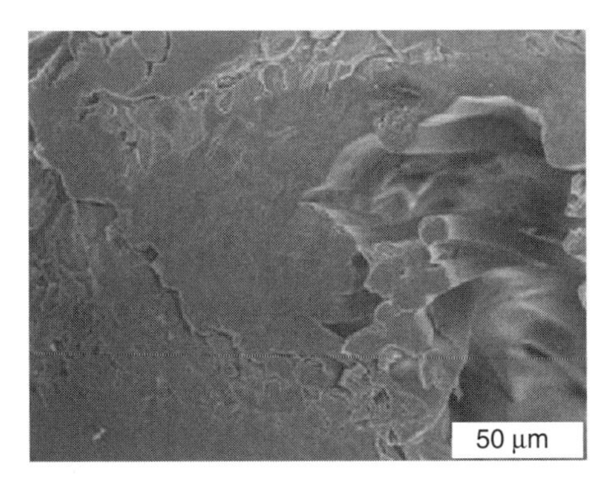

FIG. 53. Scanning electron micrograph of the composite acetylated rayon fiber–cellulose acetate matrix.

which they even appear to melt. However, this is only an optical effect, since extraction of the acetate matrix with chloroform yields back the starting fibers. Analysis of the features of thin sections of the acetylated composite by electron microscopy confirms the good quality of the interface of the fiber matrix (Fig. 53).

The "grafting" of cellulose acetate chains should now allow the production of thermoplastic composites made from one material (cellulose–cellulose acetate) with superior bonding properties between the matrix and the reinforcing fibers. Characterization of the mechanical properties (resistance to shock and surface damage) is being conducted. It should also be noted that, since the acetylated zones correspond to the previously amorphous regions, they contribute only marginally to the mechanical properties of the fibers (especially in the presence of water). Their conversion into acetate should therefore not cause a weakening of the mechanical properties. Actually, strengthening may even occur, because some of the defects that initiate rupture would be eliminated. The precise extent of this strengthening is under current study. It must be emphasized that the partial acetylation concerns the amorphous regions, which are the origin of the sensitivity to water and the structural instability. The partially acetylated fibers are therefore much less sensitive to water and show much less structural instability than the starting fibers.[231] They are "molecularly" encapsulated.

These partially acetylated fibers are already themselves a thermoplastic material; they can be further assembled by means of an exogenous link. They are also able to form links directly by compression at a temperature close to the melting point of cellulose acetate (Fig. 54).

b. Cellulose–Cellulose Composites.—This section illustrates the potential of cellulose for the production of thermostable cross-linked composites. The term

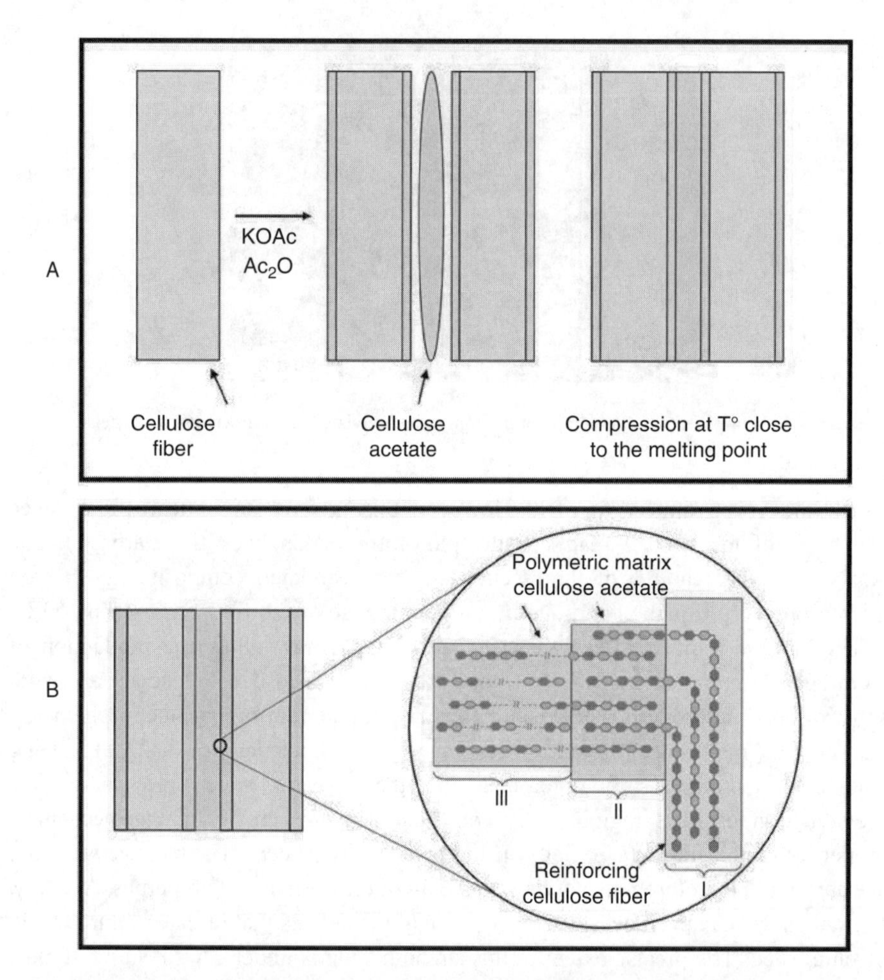

FIG. 54. Principle of superficial acetylation of cellulose fibers, followed by compression at a temperature close to the melting point of cellulose acetate. Schematic representation of the intimate macromolecular interaction involved in the thermoplastic composites of cellulose–cellulose acetate. (See Color Plate 24.)

"thermostable" refers to the mechanical performance of these materials at relatively high temperature.

Cellulose–cellulose composites have already been described in the literature. For example, they can be prepared by using cellulose solvents such as DMAC/LiCl. From a mixture of partially dissolved and undissolved cellulose, it is possible to regenerate the dissolved cellulose at any stage in the dissolution process. This leads to the formation of composites in which the undissolved cellulose constitutes the reinforcing component while the regenerated material forms the matrix. Characteristically, this type of composite is a network not held together by irreversible covalent links, but rather reversible hydrogen bonds. Since the cellulose is not denatured, the properties of biodegradability and recyclability are preserved. It is, however, unpractical, as an industrial process, (i) to control the time required for the preparation of the product and (ii) to keep the sample under compression during the period for removal of the solvent.[232]

However, another approach has been proposed that should lend itself to industrial production. Such cellulose solvents as DMAC/LiCl allow regeneration of cellulose in the form of a homogeneous product. Other processes, such as alkaline saponification of cellulose acetate, have also been used in the manufacture of rayon.[233,234]

A blank experiment has used this approach to produce a cellulose–cellulose composite, using cellulose acetate to effect aggregation of rayon fibers. Analysis of the product obtained upon alkaline hydrolysis showed that the cellulose acetate had been fully regenerated as cellulose, but with no cohesion with preexisting cellulose at the interface level. The poor quality of the interface is clearly observable at the surface of the thin sections (Fig. 55). The same experiment was performed using the partially acetylated cellulose fibers that had been previously synthesized for the production of

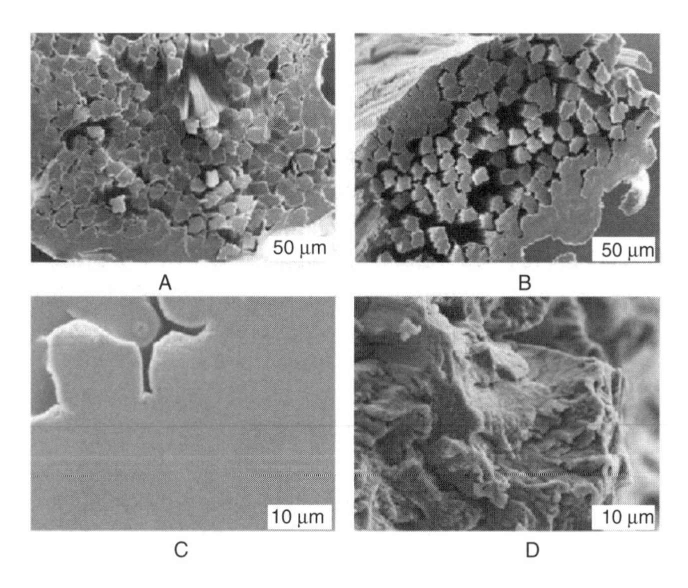

FIG. 55. Scanning electron micrograph of a cellulose–cellulose composite. (A) and (B) Rayon fiber–cellulose acetate composites, after regeneration of cellulose acetate in alkaline conditions, showing the poor quality of the interface. (C) and (D) Acetylated rayon fiber–cellulose acetate composites after co-regeneration in alkaline conditions, showing disappearance of the interface.

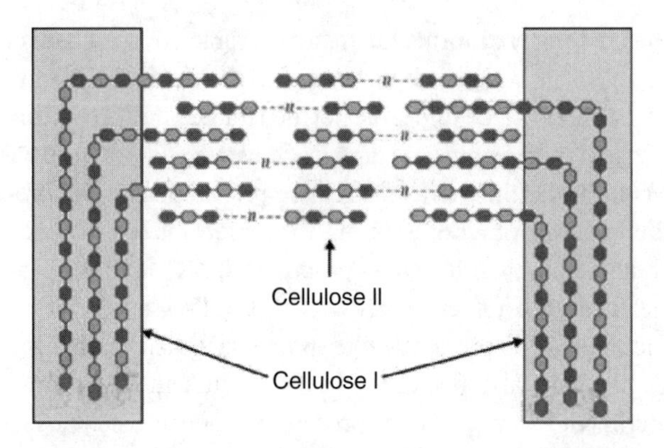

FIG. 56. Schematic representation of the macromolecular interaction involved in the cellulose–cellulose composite. (See Color Plate 25.)

thermoplastic composites. In this case, the co-regeneration of the two acetates of cellulose indeed led to the disappearance of the interface on the face of the thin section (Fig. 55).[235]

A composite was then made in which cellulose provides both the reinforcement and the matrix, and in which the network is wholly maintained by co-operative hydrogen bonds. Besides its intrinsic properties of being a single material, biodegradable and recyclable, this type of composite has the advantage of being able to be formed as a thermoplastic, and so has industrial potential. It has the appeal of requiring only mild chemical procedures involving acetic acid.

Cellulose–cellulose composites can be expected to show exceptional mechanical properties, depending on the quality of the interface between the fibers and the matrix, and on the reaction conditions for their preparation. This approach lends itself particularly well to assemblies with high hierarchical orders in which various types of fibers could be brought into play on different industrial scales (Fig. 56).

VII. CONCLUSIONS

Upon looking at the history of cellulose science since the beginning of the 19th century, the past 10 years have witnessed significant advances in our understanding of cellulose structure, biosynthesis, chemistry, and engineering. These advances have been made possible by methodologies and instrumentations that have been developed over the past few years. They have provided new insights into the complexity of the biogenesis pathway whereby polymerization and crystallization, albeit being distinct, are strongly linked in such a way as to influence each other. As for the structural and ultrastructural features, we now have responses to many of the long-outstanding questions that are now well understood.

The polarity of the cellulose chains and their parallel orientations within the microfibrils are now well characterized. This is true for the two allomorphs of native cellulose, Iα and Iβ, for which the differences and similarities have been clearly established. Unexpected features are emerging that outline the metastable character of the native cellulose assembly. Understanding of this is based on the complex biosynthetic pathway that polymerizes glucose at a rate ranging from 300 to 1000 units per minute, with a concomitant crystallization process. Among the two components of metastability that can be underlined, it is noteworthy that the orientation of the primary hydroxyl groups in crystalline native cellulose does not correspond to any of the thermodynamically stable orientations. Presumably these hydroxyl groups are folded onto the cellulosic chain as a "nascent" structure (which is likely to offer the minimum interchain hydrogen-bonding; they have no time to relax and orient themselves in a more-favored conformation). This trapped conformation may explain the disordered network of interchain hydrogen-bonds observed unambiguously in the crystal of native celluloses. Another feature of such a "nascent morphology" and its structural metastability is the parallel arrangement of polar chains within the microfibrils, since native cellulose undergoes an irreversible transition to cellulose II, the stable crystalline form (mercerized and regenerated cellulose II exhibit the same crystal structure). Despite the occurrence of a number of interchain hydrogen-bonds (all hydroxyl groups are involved in hydrogen bonds), the significance of van der Waals interactions, expressing complementarities of the shapes of neighboring cellulose chains, should not be underestimated; it may indeed be a determining factor in inducing and very strongly maintaining the chain association, so that cellulose is largely crystalline. This crystallinity is expressed at the microfibril level, which appears to be the basic structural component of all cellulose samples, irrespective of their origin. Each of them consists of a monocrystalline continuous cellulose "whisker," having a cross section and a lateral width that depends on the species origin, ranging from 20 nm for *Valonia* to 2 nm for cellulose from parenchyma. In certain cases, the same microfibril may incorporate the Iα and Iβ allomorphs; the transition between these crystalline forms implying some localized amorphous regions. Controlled degradation to remove such amorphous regions yields highly crystalline fractions, the components referred to as "cellulose whiskers." These components, albeit exhibiting the same structural makeup, display a range of shapes that translate into morphological differences, such as "shape factor" or the ratio of surface chains to core chains. As native components they offer some striking properties (liquid crystal-like) that can be modulated and amplified throughout fine chemical modifications (grafting), taking into account the difference of reactivity of the various crystalline surfaces. The "ideal" crystalline surfaces have concurrently been characterized at as conformational dynamics that the surface chains may undergo. This picture provides a conceptual framework that opens the areas of topochemistry and topoenzymology. The latter is likely to prove a factor in the rational optimization of the enzymatic digestion of cellulose I toward the production of "second-generation bioethanol" as a fuel.

The puzzling question about the irreversible solid-state interconversion of the parallel chain-based crystalline arrangement in cellulose I into an antiparallel arrangement of chains in cellulose II has now received an answer. This is more than an academic question. An understanding of the entire structural complexity of cellulose is providing rational ways for design of new chemical modifications of the cellulosic surfaces, such as that exemplified by "chromatogenic chemistry." More generally, they indicate the wealth of innovation that the ad hoc engineering of cellulosic material can provide, while preserving the unique biodegradable character of the most abundant biomacromolecule on earth.

Acknowledgments

The authors are indebted to the many colleagues who have, over the years, contributed to the advancement of the science of cellulose. They express their most sincere thanks to colleagues at CERMAV who have shared their results. Dr. Henri Chanzy deserves special mention for his enlightening dedication to unraveling the structural features of cellulose, and his willingness to provide illustrations and materials. The authors dedicate this chapter to the memory of Dr. Jean-François Revol for his significant contributions to the field.

References

1. E. Ott, H. M. Spurlin, and M. Grafflin, *Cellulose and Cellulose Derivatives, 2nd edn., Parts I, II, and III*, Interscience, New York, 1954.
2. H. Chanzy, Aspects of cellulose structure, in J. F. Kennedy, G. O. Phillips, and P. A. Williams (Eds.), *Cellulose Sources and Exploitation*, Ellis Horwood Ltd., New York, 1990, pp. 3–12.
3. A. D. French, Structure and biosynthesis of cellulose. Part I: Structure, in S.-D. Kung and S.-F. Yang (Eds.), *Discovery in Plant Biology*, World Scientific, Singapore, 2000, pp. 163–197.

4. R. H. Marchessault and P. R. Sundararajan, Cellulose, in G. O. Aspinall (Ed.), *The Polysaccharides*, Academic, New York, 1983, pp. 11–95.
5. Y. Nishiyama, Structure and properties of the cellulose microfibril, *J. Wood Sci.*, 55 (2009) 241–249.
6. K. Okamura, Structure of cellulose, in D. N. S. Hon and N. Shiraishi (Eds.), *Wood and Cellulose Chemistry*, Marcel Dekker, New York, 1991, pp. 80–111.
7. A. O'Sullivan, Cellulose: The structure slowly unravels, *Cellulose*, 4 (1997) 173–207.
8. R. D. Preston, E. Nicolai, R. Reed, and A. Mallard, Electron-microscopic study of cellulose in the wall of *Valonia ventricosa, Nature*, 162 (1948) 665–667.
9. A. Sarko, Cellulose—how much do we know about its structure? in J. F. Kennedy (Ed.), *Wood and Cellulosic Industrial Utilization Biotechnology, Structures and Properties*, Elis Horwood, Chichester, UK, 1987, pp. 55–70.
10. P. Zugenmaier, *Crystalline Cellulose and Cellulose Derivatives: Characterization and Structures*, Springer, Berlin Heidelberg, 2008.
11. A. Payen, Mémoire sur la composition du tissu propre des plantes et du ligneux, *Compt. Rend.*, 7 (1838) 1052–1056.
12. P. Ross, R. Mayer, and M. Benziman, Cellulose biosynthesis and function in bacteria, *Microbiol. Rev.*, 55 (1991) 35–58.
13. X. Zogaj, M. Nimtz, M. Rohde, W. Bokranz, and U. Römling, The multicellular morphotypes of *Salmonella typhimurium* and *Escherichia coli* produce cellulose as the second component of the extracellular matrix, *Mol. Microbiol.*, 39 (2001) 1452–1463.
14. M. R. J. Brown, Cellulose microfibril assembly and orientation: Recent developments, *J. Cell Sci. Suppl.*, 2 (1985) 13–32.
15. R. L. Blanton, D. Fuller, N. Iranfar, M. J. Grimson, and W. F. Loomis, The cellulose synthase gene of *Dictyostelium*, *Proc. Natl. Acad. Sci. U.S.A.*, 97 (2000) 2391–2396.
16. D. R. Nobles, D. K. Romanovicz, and R. M. J. Brown, Cellulose in cyanobacteria. Origin of vascular plant cellulose synthase? *Plant Physiol. Biochem.*, 127 (2001) 529–542.
17. M. McCann, B. Weels, and K. Roberts, Direct visualization of cross links in the primary cell-walls, *J. Cell. Sci.*, 96 (1990) 322–334.
18. M. C. Jarvis, Structure and properties of pectin gels in plant cell walls, *Plant Cell Environ.*, 7 (1984) 153–164.
19. S. J. MacQueen-Mason, D. M. Durachko, and D. J. Cosgrove, Two endogenous proteins that induce cell wall extension in plants, *Plant Cell*, 4 (1992) 1425–1433.

20. M. R. Brown and I. M. J. Saxena, Cellulose biosynthesis: A model for understanding the assembly of biopolymers, *Plant Physiol. Biochem.*, 38 (2000) 57–67.
21. S. K. Cousin and R. Malcom Bown, Cellulose I microfibril assembly: Computational mechanics energy analysis favour bonding by van der Waals forces as the initial step in crystallization, *Polymer*, 36 (1995) 3885–3888.
22. A. M. C. Emmons, H. Hofte, and B. M. Mulder, Microtubules and cellulose microfibrils: How intimate is their relationship? *Trends Plant Sci.*, 12 (2007) 279–281.
23. S. DeBolt, R. Gutierrez, D. W. Ehrhardt, and C. Somerville, Nonmotile cellulose synthase subunits repeatedly accumulate within localized regions at the plasma membrane in *Arabidopsis* hypocotyl cells following 2,6-dichlorobenzonitrile treatment, *Plant Physiol.*, 145 (2007) 334–338.
24. T. Desprez, M. Juraniec, E. F. Crowell, H. Jouy, Z. Pochylova, F. Parcy, H. Hofte, M. Gonneau, and S. Vernhettes, Organization of cellulose synthase complexes involved in the primary cell wall synthesis in *Arabidopsis thaliana*, *Proc. Natl. Acad. Sci. U.S.A.*, 104 (2007) 15572–15577.
25. F. Diotallevi and B. M. Mulder, The cellulose synthase complex: A polymerization driven supramolecular motor, *Biophys. J.*, 92 (2007) 2666–2673.
26. A. R. Paredez, C. R. Sommerville, and D. W. Erhradt, Visualization of cellulose synthase demonstrates functional association with microtubules, *Science*, 312 (2006) 1491–1495.
27. H. Braconnot, Sur la conversion du corps ligneux en gomme, en sucre, et en un acide d'une nature particulière, par le moyen de l'acide sulfurique conversion de la même substance ligneuse en ulmine par la potasse, *Ann. Chim.* (1819) 172–195.
28. R. Willstätter, Zur Kenntnis der Hydrolyse von cellulose. I, *Ber. Deut. Chem. Ges.*, 46 (1913) 2401–2412.
29. H. Staudinger, Die Chemie der hochmolekularen organischen Stoffe im Sinne der Kekuléschen Strukturlehre, *Ber. Deut. Chem. Ges.*, 59 (1926) 3019–3043.
30. J. C. Irvine and E. L. Hirst, The constitution of polysaccharides. Part VI. The molecular structure of cotton cellulose, *J. Chem. Soc.*, 123 (1923) 518–532.
31. K. Freudenberg and E. Braun, Methyl cellulose, *Justus Liebigs Ann. Chem.*, 460 (1928) 288–304.
32. W. Charlton, W. N. Haworth, and S. Peat, A revision of the structural formula of glucose, *J. Chem. Soc.* Part 1. (1926) 89–101.
33. W. N. Haworth, Revision of the structural formula of dextrose, *Nature*, 116 (1925) 430.
34. W. N. Haworth, The structure of carbohydrates, *Helv. Chim. Acta*, 11 (1928) 534–548.

35. S. S. C. Chu and G. A. Jeffrey, The refinement of the crystal structures of β-D-glucose and cellobiose, *Acta Crystallogr.*, 24 (1968) 830–838.
36. R. E. Mark, Molecular and cell wall structure of wood, *J. Educ. Modules Mater. Sci. Eng.*, 2 (1980) 251–308.
37. B. Philipp and K.-J. Linow, Untersuchungen zur Kettenlangendifferenz zwischen Nitrat- und Cuoxam-DP der Cellulose and ihrer Änderung im Viskoseprozelβ, *Papier*, 20 (1960) 649–657.
38. M. Marx-Figini, Über die Kinetik der Biosynthese der Cellulose in der Baumwolle, *Papier*, 18 (1964) 546–549.
39. S. Nishikawa and S. Ono, Transmission of X-rays through fibrous, lamellar and granular substances, *Proc. Tokyo Math. Phys. Soc.*, 7 (1913) 131–138.
40. R. H. Marchessault and S. Pérez, Conformations of the hydroxymethyl group in crystalline aldopyranoses, *Biopolymers*, 18 (1979) 2379–2374.
41. Z. Peralta-Inga, G. P. Johnson, M. K. Dowd, J. A. Rendelman, E. D. Stevens, and A. D. French, The crystal structure of the alpha-cellobiose: 2 NaI:2 H_2O complex in the context of related structures and conformational analysis, *Carbohydr. Res.*, 337 (2002) 851–861.
42. S. Raymond, B. Henrissat, D. T. Qui, A. Kvick, and H. Chanzy, The crystal structure of methyl beta-cellotrioside monohydrate -0.25 ethanolate and its relationship to cellulose II, *Carbohydr. Res.*, 277 (1995) 208–229.
43. K. Gessler, N. Krauss, T. Steiner, C. Betzel, A. Sarko, and W. Saenger, β-D-Celloteraose hemihydrate as a structural model for cellulose II. An X-ray diffraction study, *J. Am. Chem. Soc.*, 17 (1995) 11397–11406.
44. A. Rencurosi, J. Röhrling, J. Pauli, A. Potthast, C. Jäger, S. Pérez, P. Kosma, and A. Imberty, Polymorphism in the crystal structure of the cellulose fragment analogue methyl-4-*O*-methyl-β-D-glucopyranosyl-(1→4)-β-D-glucopyranoside, *Angew. Chem. Int. Ed.*, 22 (2002) 4277–4281.
45. J. T. Ham and D. G. Williams, The crystal and molecular structure of methyl-β-cellobioside methanol, *Acta Crystallogr. B*, 26 (1970) 1373–1383.
46. P. Langan, Y. Nishiyama, and H. Chanzy, The X-ray structure of mercerized cellulose II at 1 Ang. resolution, *Biomacromolecules*, 2 (2001) 410–416.
47. P. Langan, N. Sukumar, Y. Nishiyama, and H. Chanzy, Synchrotron X-ray structures of cellulose Iβ and regenerated cellulose II at ambient temperature and 100 K, *Cellulose*, 12 (2005) 551–562.
48. Y. Nishiyama, P. Langan, and H. Chanzy, Crystal structure and hydrogen bonding system in cellulose Iβ from synchrotron X-ray and neutron fiber diffraction, *J. Am. Chem. Soc.*, 124 (2002) 9074–9082.

49. Y. Nishiyama, J. Sugiyama, H. Chanzy, and P. Langan, Crystal structure and hydrogen bonding system in cellulose Iα from synchrotron X-ray and neutron fiber diffraction, *J. Am. Chem. Soc.*, 125 (2003) 14300–14306.
50. S. Pérez, C. Gautier, and A. Imberty, Oligosaccharide conformation by diffraction method, in B. Ernst, G. Hart, and P. Sinaÿ (Eds.), *Oligosaccharides in Chemistry and Biology*, Wiley VCH, Weinheim, 2000, pp. 969–1001.
51. P. E. Hermans, *Physics and Chemistry of Cellulose Fibers with Particular Reference to Rayon*, Elsevier, New York, 1949.
52. D. Fengel, Characteristics of cellulose by deconvoluting the OH valency range in FTIR spectra, *Holzforschung*, 46 (1992) 283–288.
53. J. A. Howsmon and W. A. Sisson, Structure and properties of cellulose fibers. B-Submicroscopic structure, in cellulose and cellulose derivatives, Part I, *High Polym.*, V (1963) 231–346.
54. R. H. Atalla and D. L. VanderHart, Native cellulose: A composite of two distinct crystalline forms, *Science*, 223 (1984) 283–285.
55. H. Chanzy, K. Imada, and R. Vuong, Electron diffraction from the primary wall of cotton fibers, *Protoplasma*, 94 (1978) 299–306.
56. H. Chanzy, K. Imada, A. Mollard, R. Vuong, and F. Barnoud, Crystallographic aspects of sub-elementary cellulose fibrils occurring in the wall of rose cells cultured in vitro, *Protoplasma*, 100 (1979) 303–316.
57. H. Ambronn, Über das Zusammenwirken von Stäbchendoppelbrechung und Eigendoppelbrechung. II, *Kolloid Z.*, 18 (1916) 273–281.
58. H. Ambronn, Über das Zusammenwirken von Stäbchendoppelbrechung und Eigendoppelbrechung. III, *Kolloid Z.*, 20 (1917) 173–185.
59. R. H. Hermans and A. Weidinger, X-ray studies on the crystallinity of cellulose, *J. Polym. Sci.*, 4 (1949) 135–144.
60. T. Kubo, Untersuchungen über die Umwandlung von Hydratcellulose in natürliche Cellulose. VII. Die Kristallstruktur des Umwandlungs-produktes sowie eines höchst orientierten natürlichen Cellulosepräparates, *Z. Phys. Chem.*, 187 (1940) 297–312.
61. H. J. Wellard, Variation in the lattice spacing of cellulose, *J. Polym. Sci.*, 13 (1954) 471–476.
62. K. H. Meyer and L. Misch, Position des atomes dans le nouveau modéle spatial de la cellulose, *Helv. Chim. Acta*, 20 (1937) 232–244.
63. K. H. Gardner and J. Blackwell, The structure of native cellulose, *Biopolymers*, 13 (1974) 1975–2001.
64. A. Sarko and R. Muggli, Packing analysis of carbohydrates and polysaccharides. III. *Valonia* cellulose and cellulose II, *Macromolecules*, 7 (1974) 486–494.

65. W. Claffey and J. Blackwell, Electron diffraction of *Valonia* cellulose. A quantitative interpretation, *Biopolymers*, 15 (1976) 1903–1915.
66. D. G. Fischer and J. Mann, Crystalline modifications of cellulose. Part VI. Unit cell and molecular symmetry of cellulose I, *J. Polym. Sci.*, 62 (1960) 189–194.
67. I. A. Nieduszynski and R. D. Preston, Crystallite size in natural cellulose, *Nature*, 225 (1970) 273–274.
68. D. L. VanderHart and R. H. Atalla, Studies of microstructure in native celluloses using solid state ^{13}C NMR, *Macromolecules*, 17 (1984) 1465–1472.
69. D. L. A. VanderHart and R. H. Atalla, Further ^{13}C NMR evidence for the co-existence of two crystallines forms in native celluloses, *The Structures of Cellulose, Characterization of the Solid States*, ACS Symposium Ser., 1987, pp. 88–118.
70. P. S. Belton, S. F. Tanner, N. Cartier, and H. Chanzy, High-resolution solid-state ^{13}C nuclear magnetic resonance spectroscopy of tunicin, an animal cellulose, *Macromolecules*, 22 (1989) 1615–1617.
71. R. H. Atalla, R. E. Whitmore, and D. L. VanderHart, A highly crystalline cellulose from *Rhizoclonium hieroglyphicum*, *Biopolymers*, 24 (1985) 421–423.
72. H. Chanzy, B. Henrissat, M. Vincendon, S. Tanner, and P. B. Belton, Solid-state ^{13}C-NMR and electron microscopy study on the reversible cellulose I → cellulose III_I transformation in *Valonia*, *Carbohydr. Res.*, 160 (1987) 1–11.
73. A. Hirai, H. Yamamoto, M. Tsuji, and F. Horii, Electron microscopic observation of the formation processes of the microfibrils and the composite crystals for bacterial cellulose, *Proc. '94 Cellulose R&D, 1st Ann. Meet. Cellulose Soc.* (1994) 41–42.
74. M. Tanahashi, T. Goto, F. Horii, A. Hirai, and T. Higuchi, Characterization of steam-exploded wood. III. Transformation of cellulose crystals and changes of crystallinity, *Mokuzai Gakkaishi*, 35 (1989) 654–662.
75. G. Honjo and M. Watanabe, Examination of cellulose fibre by the low-temperature specimen method of electron diffraction and electron microscopy, *Nature*, 181 (1958) 326–328.
76. H. J. Marrinan and J. Mann, Infrared spectra of the crystalline modifications of cellulose, *J. Polym. Sci.*, 21 (1956) 301–311.
77. J. Mann and H. J. Marrinan, Crystalline modifications of cellulose. Part II. A study with plane-polarized infrared radiation, *J. Polym. Sci.*, 32 (1958) 357–370.
78. J. Wiley and R. H. Atalla, Band assignments in the Raman spectra of celluloses, *Carbohydr. Res.*, 160 (1987) 113–129.
79. T. T. Imai, J. Sugiyama, T. Itoh, and F. Horii, Almost pure Iα cellulose in the cell wall of *Glaucocystis*, *J. Struct. Biol.*, 127 (1999) 248–257.

80. J. Sugiyama, R. Vuong, and H. Chanzy, Electron diffraction study of the two crystalline phases occurring in native cellulose from an algal cell wall, *Macromolecules*, 24 (1991) 4168–4175.
81. M. Koyama, W. Helbert, T. Imai, J. Sugiyama, and B. Henrissat, Parallel-up structure evidences the molecular directionality during biosynthesis of bacterial cellulose, *Proc. Natl. Acad. Sci. U.S.A.*, 94 (1997) 9091–9095.
82. A. P. Heiner, J. Sugiyama, and O. Teleman, Crystalline cellulose Iα and Iβ studied by molecular dynamics simulation, *Carbohydr. Res.*, 273 (1995) 207–223.
83. L. M. Kroon-Batenburg, B. Bouma, and J. Kroon, Stability of cellulose structures studied by MD simulations. Could mercerized cellulose II be parallel? *Macromolecules*, 29 (1996) 5695–5699.
84. I. Simon, H. A. Scheraga, and R. S. J. Manley, Structure of cellulose. 2: Low energy crystalline arrangements, *Macromolecules*, 21 (1998) 990–998.
85. R. J. Vietor, K. Mazeau, M. Lakin, and S. Pérez, A priori crystal structure prediction of native celluloses, *Biopolymers*, 54 (2000) 342–354.
86. V. L. Finkendstadt and R. P. Milane, Crystal structure of *Valonia* cellulose Iβ, *Macromolecules*, 31 (1998) 7776–7783.
87. Y. Nishiyama, A. Isogai, O. Isogai, T. M. Müller, and H. Chanzy, Intracrystalline deuteration of native cellulose, *Macromolecules*, 32 (1999) 2078–2081.
88. Y. Nishiyama, T. Okano, P. Langan, and H. Chanzy, High resolution neutron fiber diffraction data on hydrogenated and deuterated cellulose, *Int. J. Biol. Macromol.*, 26 (1999) 279–283.
89. R. H. Attala, in B. M. Pinto (Ed.), *Comprehensive Natural Products Chemistry, Carbohydrates and Their Natural Derivatives Including Tanins, Cellulose and Related Lignins*, Elsevier, Cambridge, 1999, pp. 529–598.
90. J. Sugiyama, J. Persson, and H. Chanzy, Combined infrared and electron diffraction study of the polymorphism of native cellulose, *Macromolecules*, 24 (1991) 2461–2466.
91. M. Wada, J. Sugiyama, and T. Okano, Native celluloses on the basis of the two crystalline phase (Iα/Iβ) system, *J. Appl. Polym. Sci*, 49 (1993) 1491–1496.
92. J. Hayashi, T. Yamada, and Y.-L. Shimizu, Memory phenomenon of the original crystal structure in allomorphs of Na-cellulose, in C. Schuerch and A. Sarko, (Eds.), *Cellulose and Wood: Chemistry and Technology*, Wiley, New York, 1989, pp. 77–102.
93. K. R. Andress, The X-ray diagram of mercerized cellulose, *Z. Physik. Chem. Abt. B Chem. Elem. Prozess. Aufbau Mater.*, 4 (1929) 190–206.
94. A. U. Ahmed, Neutron diffraction studies of the unit cell of cellulose I, *J Polym. Sci. Polym. Lett. Ed.*, 14 (1976) 561–564.

95. F. J. Kolpak and J. Blackwell, Determination of the structure of cellulose II, *Macromolecules*, 9 (1976) 273–278.
96. S. C. Nyburg, Fibrous macromolecular substances, in L.F. Fieser and M. Fieser (Eds.), X-Ray Analysis of Organic Structures, Harvard University, Academic Press, Inc. New York, 1961, pp. 302–314.
97. S. Kuga, S. Takagi, and R. M. J. Brown, Native folded-chain cellulose II, *Polymer*, 9 (1993) 3293–3297.
98. A. J. Stipanovich and A. Sarko, Packing analysis of carbohydrates and polysaccharides. 6. Molecular and crystal structure of regenerated cellulose II, *Macromolecules*, 9 (1976) 851–857.
99. B. J. Poppleton and A. Mathieson, Crystal structure of β-D-cellotetraose and its relationship to cellulose, *Nature*, 219 (1968) 1046–1049.
100. S. Raymond, A. Heyraud, D. Tran Qui, Å. Kvick, and H. Chanzy, Crystal and molecular structure of β-D-cellotetraose hemihydrate as a model of cellulose II, *Macromolecules*, 28 (1995) 2096–2100.
101. P. Langan, Y. Nishiyama, and H. Chanzy, A revised structure and hydrogen bonding system in cellulose II from a neutron fiber diffraction analysis, *J. Am. Chem. Soc.*, 121 (1999) 9940–9946.
102. A. Sarko, What is the crystalline structure of cellulose? *Tappi*, 61 (1978) 59–61.
103. J. Sugiyama and T. Okano, Electron microscopic and X-ray diffraction study of cellulose III_I and cellulose I, in C. Schuerch and A. Sarko, (Eds.), *Cellulose and Wood: Chemistry and Technology, Proceedings of the Tenth Cellulose Conference*, Wiley, New York, 1989, pp. 119–127.
104. E. Roche and H. Chanzy, Electron microscopy study of the transformation of cellulose I into cellulose III_I in *Valonia*, *Int. J. Biol. Macromol.*, 3 (1981) 201–206.
105. A. Hirai, F. Horii, and R. Kitamaru, Transformation of native cellulose crystals from cellulose Iβ to cellulose Iα through solid-state chemical reactions, *Macromolecules*, 20 (1987) 1440–1442.
106. A. Sarko, J. Southwick, and J. Hayashi, Packing analysis of carbohydrates and polysaccharides 7. Crystal structure of cellulose III, and its relationship to other cellulose polymorphs, *Macromolecules*, 9 (1976) 857–863.
107. M. Wada, H. Chanzy, Y. Nishiyama, and P. Langan, Cellulose III_1 crystal structure and hydrogen bonding by synchrotron X-ray and neutron diffraction, *Macromolecules*, 37 (2004) 8548–8555.
108. J. Hayashi, A. Sufoka, J. Ohkita, and S. Watanabe, Confirmation of existence of cellulose III1, III2, IV1, and IV2 by X-ray method, *J. Polym. Sci.*, 13 (1975) 23–27.

109. E. S. Gardiner and A. Sarko, Packing analysis of carbohydrates and polysaccharides. 16. The crystal structures of cellulose IV_1 and IV_n, *Can. J. Chem.*, 63 (1985) 173–180.
110. T. Okano and A. Sarko, Mercerization of cellulose. II. Alkali-cellulose intermediates and a possible mercerization mechanism, *J. Appl. Polym. Sci*, 30 (1985) 325–332.
111. T. Okano and A. Sarko, Mercerization of cellulose. I. X-ray diffraction evidence for intermediate structures, *J. Appl. Polym. Sci.*, 29 (1984) 4175–4182.
112. H. Nishimura, T. Okano, and A. Sarko, Mercerization of cellulose. 5. Crystal and molecular structure of Na-cellulose 1, *Macromolecules*, 24 (1991) 759–770.
113. H. Nishimura, T. Okano, and A. Sarko, Mercerization of cellulose. 6. Crystal and molecular structure of Na-cellulose 1, *Macromolecules*, 24 (1991) 771–778.
114. J. F. Revol and D. A. I. Goring, Directionality of the fiber c-axis of cellulose crystallites in microfibrils of *Valonia ventricosa*, *Polymer*, 24 (1983) 1547–1550.
115. E. Dinand, M. Vignon, and H. Chanzy, Mercerization of the primary wall cellulose and its implication from the conversion of cellulose I–cellulose II, *Cellulose*, 9 (2002) 3311–3314.
116. D. M. Lee and J. Blackwell, Cellulose–hydrazine complexes, *J. Polym. Sci. B*, 19 (1981) 459–465.
117. D. M. Lee and J. Blackwell, Structure of cellulose II hydrate, *Biopolymers*, 20 (1981) 2165–2179.
118. D. M. Lee, K. E. Burnfield, and J. Blackwell, Structure of a cellulose I–ethylenediamine complex, *Biopolymers*, 23 (1984) 111–126.
119. K. Hess and C. Trogus, Zur Kenntnis der Reaktionsweise der Cellulose, *Z. Physik. Chem.*, B15 (1931) 157–222.
120. B. S. Sprague, J. L. Riley, and H. D. Noether, Factors influencing the crystal structure of cellulose triacetate, *Text. Res. J.*, 28 (1958) 275–287.
121. A. J. Stipanovich and A. Sarko, Molecular and crystal structure of cellulose triacetate I: A parallel chain structure, *Polymer*, 19 (1978) 3–8.
122. W. J. Dulmage, The molecular and crystal structure of cellulose triacetate, *J. Polym. Sci. B*, 26 (1957) 277–288.
123. E. Roche, H. Chanzy, M. Boudeulle, and R. H. Marchessault, Three-dimensional crystalline structure of cellulose triacetate II, *Macromolecules*, 11 (1978) 86–94.
124. O. Kratky and H. Mark, Zur Frage der individuellen Cellulosemicellen, *Z. Physik. Chem.*, B36 (1937) 129–139.
125. A. Frey-Wyssling, K. Miihlethaler, and R. W. G. Wyckoff, Zur Frage der individuellen Cellulosemicellen, *Z. Physik. Chem.*, B36 (1948) 129–139.

126. R. D. Preston, X-ray analysis and the structure of the components of plant cell walls, *Phys. Rep.*, 21 (1975) 183–226.
127. J. F. Revol, Change of the d spacing in cellulose crystals during lattice imaging, *J. Mater. Sci. Lett.*, 4 (1985) 1347–1349.
128. J. Sugiyama, H. Harada, Y. Fujiyoshi, and N. Uyeda, High resolution observations of cellulose microfibrils, *Mokuzai Gakkaishi*, 30 (1985) 98–99.
129. R. H. Marchessault, Cellulosics as advanced materials, in C. Schuerch (Eds.), *Cellulose and Wood-Chemistry and Technology*, Wiley, New York, 1989, pp. 1–20.
130. J. R. Colvin, Oxidation of cellulose microfibril segments by alkaline silver nitrate and its relation to the fine structure of cellulose, *J. Appl. Polym. Sci.*, 8 (1964) 2763–2774.
131. J. R. Colvin, Tip-growth of bacterial cellulose microfibrils and its relation to the crystallographic fine structure of cellulose, *J. Polym. Sci. B*, 4 (1966) 747–754.
132. K. Hieta, S.Kuga, and M. Usuda, Electron staining of reducing ends evidences a parallel-chain structure in *Valonia* cellulose, *Biopolymers*, 10 (1984) 1807–1810.
133. S. Kuga and R. M. J. Brown, Silver labeling of the reducing ends of bacterial cellulose, *Carbohydr. Res.*, 180 (1988) 345–350.
134. A. Maurer and D. Fengel, Parallel orientation of the molecular chains in cellulose I and cellulose II deriving from higher plants, *Holz. Roh. Werkstoff*, 50 (1992) 493.
135. H. Chanzy and B. Henrissat, Unidirectional degradation of *Valonia* cellulose microcrystals subjected to cellulase action, *FEBS Lett.*, 184 (1985) 285–288.
136. L. Sakurada, Y. Nukushina, and T. Ito, Experimental determination of the elastic modulus of crystalline regions in oriented polymers, *J. Polym. Sci.*, 57 (1962) 651–660.
137. A. Dufresne, M. B. Kellerhals, and B. Witholt, Transcrystallization in Mcl-PHAs/cellulose whiskers composites, *Macromolecules*, 32 (1999) 7396–7401.
138. J. F. Sassi and H. Chanzy, Ultrastructural aspects of the acetylation of cellulose, *Cellulose*, 2 (1995) 111–127.
139. A. A. Baker, W. Helbert, J. Sugiyama, and M. J. Miles, New insight into cellulose structure by atomic force microscopy shows the Iα crystal phase at near-atomic resolution, *Biophys. J.*, 79 (2000) 1139–1145.
140. L. Heux, E. Dinand, and M. Vignon, Structural aspects in ultrathin cellulose microfibrils followed by ^{13}C CP-MAS NMR, *Carbohydr. Polym.*, 40 (1999) 115–124.
141. C. Y. Liang and R. H. Marchessault, Infrared spectra of crystalline polysaccharides. I. Hydrogen bonds in native celluloses, *J. Polym. Sci.*, 37 (1959) 385–395.

142. S. O. Rowland and P. S. Howley, Hydrogen bonding on accessible surface from various sources and relationship to other crystalline regions, *J. Polym. Sci. A Polym. Chem.*, 26 (1988) 1769–1778.
143. S. O. Rowland and P. S. Howley, Structure in amorphous regions, accessible segments of fibrils, of the cotton fiber, *Text. Res. J.*, 58 (1988) 96–101.
144. S. O. Rowland and E. J. Roberts, Selective accessibilities of hydroxyl groups in the microstructure of cotton cellulose, *Text. Res. J.*, 39 (1969) 530–542.
145. S. O. Rowland and E. J. Roberts, The nature of accessible surfaces in the microstructure of native cellulose, *J. Polym Sci. A*, 10 (1972) 2447–2461.
146. S. O. Rowland and E. J. Roberts, Disposition of D-glucopyranosyl units on the surface of crystalline elementary fibrils of cotton cellulose, *J. Polym. Sci. Polym. Chem. Ed.*, 10 (1972) 867–879.
147. S. Tasker, J. P. S. Badyal, S. C. E. Backson, and R. W. Richards, Hydroxyl accessibility in celluloses, *Polymer*, 35 (1994) 4717–4721.
148. C. Verlhac, J. Dedier, and H. Chanzy, Availability of surface hydroxyl groups in *Valonia* and bacterial cellulose, *J. Polym. Sci. A Polym. Chem.*, 28 (1990) 1171–1177.
149. A. A. Baker, W. Helbert, J. Sugiyama, and M. J. Miles, High resolution atomic force microscopy of native *Valonia* cellulose I microcrystals, *J. Struct. Biol.*, 119 (1997) 129–138.
150. A. A. Baker, W. Helbert, J. Sugiyama, and M. J. Miles, Surface structure of native cellulose microcrystals by AFM. Part 1. Scanning tunneling microscopy/spectroscopy and related techniques, *Appl. Phys. Mater. Sci. Process A*, 66 (1998) 559–563.
151. S. J. Hanley, J. Giasson, J. F. Revol, and D. G. Gray, Atomic force microscopy of cellulose microfibrils: Comparison with transmission electron microscopy, *Polymer*, 33 (1992) 4639–4642.
152. L. Kuutti, J. Peltonen, J. Pere, and O. Teleman, Identification and surface structure of crystalline cellulose studied by atomic force microscopy, *J. Microsc.*, 178 (1995) 1–6.
153. R. H. Newman and J. A. Hemmingson, Carbon-13 NMR distinction between categories of molecular order and disorder in cellulose, *Cellulose*, 2 (1995) 95–110.
154. S. Pérez and K. Mazeau, Conformations, structures, and morphologies of celluloses, in S. Dimitriu (Ed.), *Polysaccharides: Structural Diversity and Functional Versatility*, 2nd edn., Dekker, New York, 2004, pp. 41–68.
155. H. Lichtenegger, M. Müller, O. Paris, C. Riekel, and P. Fratzl, Imaging of the helical arrangement of cellulose fibrils in wood by synchrotron X-ray microdiffraction, *J. Appl. Cryst.*, 32 (1999) 1127–1133.

156. B. Morgenstern and H. W. Kammer, Solvation in cellulose-LiCl-DMAc solutions, *Trends Polym. Sci.*, 4 (1996) 87–92.
157. E. Maia and S. Pérez, Solvants organiques de la cellulose: IV Modélisation des interactions des molécules de N-oxyde de méthylmorpholine (MMNO) et d'une chaîne de cellulose, *Nouv. J. Chim.*, 7 (1983) 89–100.
158. A. Takaragi, M. Minoda, T. Miyamoto, H. Q. Liu, and L. N. Zhang, Reaction characteristics of cellulose in the LiCl/1,3-dimethyl-2-imidazolidinone solvent system, *Cellulose*, 6 (1999) 93–102.
159. T. Heinze, R. Dicke, A. Koschella, A. H. Kull, E. A. Klohr, and W. Koch, Effective preparation of cellulose derivatives in a new simple cellulose solvent, *Macromol. Chem. Phys.*, 201 (2000) 627–631.
160. S. Fischer, H. Leipner, E. Brendler, W. Voigt, and K. Fischer, Molten inorganic salt hydrates as cellulose solvents, in M. A. El-Nokaly and H. A. Soini (Eds.), *Polysaccharide Applications, Cosmetics and Pharmaceuticals*, American Chemical Society, Washington DC, 1999, pp. 143–150.
161. S. Fischer, W. Voigt, and K. Fischer, The behaviour of cellulose in hydrate melts of composition $LiX \times nH_2O$, *Cellulose*, 6 (1999) 213.
162. H. Leipner, S. Fischer, E. Brendler, and W. Voigt, Structural changes of cellulose dissolved in molten salt hydrates, *Macromol. Chem. Phys.*, 201 (2000) 2041.
163. R. P. Swatloski, S. K. Spear, J. D. Holbrey, and R. D. Rogers, Dissolution of cellulose with ionic liquids, *J. Am. Chem. Soc.*, 124 (2002) 4974.
164. L. Taiz and E. Zeiger, *Plant Physiology*, The Benjamin/Cumming Publishing Company, San Francisco, 1991.
165. R. H. Attala and A. Isogai, Recent developments in spectroscopic and chemical characterization of cellulose, in S. Dumitriu (Eds.), *Polysaccharides, Structural Diversity & Functional Diversity*, Marcel Dekker, New York, 2005, pp. 123–157.
166. P. L. Bragd, H. van Bekkum, and A. C. Besemer, TEMPO-mediated oxidation of polysaccharides: Survey of methods and applications, *Top. Catal.*, 27 (2004) 49–66.
167. P. S. Chang and J. F. Robyt, Oxidation of primary alcohol groups of naturally occurring polysaccharides with 2,2,6,6-tetramethyl-1-piperidine oxoammonium ion, *J. Carbohydr. Chem.*, 15 (1996) 819–830.
168. A. E. J. Denooy, A. C. Besemer, and H. Vanbekkum, Highly selective nitroxyl radical-mediated oxidation of primary alcohol groups in water-soluble glucans, *Carbohydr. Res.*, 269 (1995) 89–98.
169. Y. Kato, J. Kaminaga, R. Matsuo, and A. Isogai, TEMPO-mediated oxidation of chitin, regenerated chitin and N-acetylated chitosan, *Carbohydr. Polym.*, 58 (2004) 421–426.

170. E. Lasseuguette, D. Roux, and Y. Nishiyama, Rheological properties of microfibrillar suspension of TEMPO-oxidized pulp, *Cellulose*, 15 (2008) 425–433.
171. A. S. Perlin, Glycol cleavage oxidation, *Adv. Carbohydr. Chem. Biochem.*, 60 (2006) 183–250.
172. A. J. Varma and M. P. Kulkarni, Oxidation of cellulose under controlled conditions, *Polym. Degrad. Stab.*, 77 (2002) 25–27.
173. U. J. Kim and S. Kuga, Reactive interaction of aromatic amines with dialdehyde cellulose gel, *Cellulose*, 7 (2000) 287–297.
174. S. Farag and E. Al-Afaleq, Preparation and characterization of saponified delignified cellulose polyacrylonitrile-graft copolymer, *Carbohydr. Polym.*, 48 (2002) 1–5.
175. K. C. Gupta, S. Sahoo, and K. Khandekar, Graft copolymerization of ethyl acrylate onto cellulose using ceric ammonium nitrate as initiator in aqueous medium, *Biomacromolecules*, 3 (2002) 1087–1094.
176. N. M. Bikales and L. Segal, in N. M. Bikalesand L. Segal (Eds.), *Cellulose and Cellulose Derivatives*, Wiley-Interscience, New York, 1971, pp. 877–905.
177. A. K. Bledzki, S. Reihmane, and J. Gassan, Properties and modification methods for vegetable fibers for natural fiber composites, *J. Appl. Polym. Sci.*, 59 (1998) 1329–1336.
178. R. Karnani, M. Krishnan, and R. Narayan, Biofiber-reinforced polypropylene composites, *Polym. Eng. Sci.*, 37 (1997) 476–483.
179. M. Abdelmouleh, S. Boufi, A. ben Salah, M. N. Belgacem, and A. Gandini, Interaction of silane coupling agents with cellulose, *Langmuir*, 18 (2002) 3203–3208.
180. G. Samaranayake and W. G. Glasser, Cellulose derivatives with low DS. I. A novel acylation system, *Carbohydr. Polym.*, 22 (1993) 1–7.
181. G. Samaranayake and W. G. Glasser, Cellulose derivatives with low DS. II. Analysis of alkanoates, *Carbohydr. Polym.*, 22 (1993) 79–86.
182. K. Furuhata, N. Aoki, S. Suzuki, M. Sakamoto, Y. Saegusa, and S. Nakamura, Bromination of cellulose with tribromoimidazole, triphenylphosphine and imidazole under homogeneous conditions in LiBr–dimethylacetamide, *Carbohydr. Polym.*, 26 (1995) 25–29.
183. O. Varela, Oxidative reactions and degradations of sugars and polysaccharides, *Adv. Carbohydr. Chem. Biochem.*, 58 (2003) 307–369.
184. J. Tiller, D. Klemm, and P. Berlin, Designed aliphatic aminocellulose derivatives as transparent and functionalized coatings for enzyme immobilization, *Design. Monom. Polym.*, 4 (2001) 315–328.

185. P. Berlin, D. Tiller, and R. Rieseler, A novel soluble aminocellulose derivative type: Its transparent film-forming properties and its efficient coupling with enzyme proteins for biosensors, *Macromol. Chem. Phys.*, 201 (2000) 2070–2082.
186. J. F. Sassi, P. Tekely, and H. Chanzy, Relative susceptibility of the Iα and Iβ phases of cellulose towards acetylation, *Cellulose*, 7 (2000) 119–132.
187. H. Chanzy, B. Henrissat, R. Vuong, and J.-F. Revol, Structural changes of cellulose crystals during the reversible transformation cellulose I-V_{II} in *Valonia*, *Holzforshung*, 40 (1986) 25–30.
188. E. Dinand, H. Chanzy, and M. Vignon, Suspension of cellulose microfibrils from sugar beet pulp, *Food Hydrocolloids*, 13 (1999) 275–283.
189. A. Buleon, H. Chanzy, and E. Roche, Shish-kebab-like structure of cellulose, *Polym. Lett. Ed.*, 15 (1977) 265–270.
190. S. Montanari, M. Roumani, L. Heux, and M. Vignon, Topochemistry of carboxylated cellulose nanocrystals resulting from TEMPO-mediated oxidation, *Macromolecules*, 38 (2005) 1165–1671.
191. Y. Habibi, H. Chanzy, and M. Vignon, TEMPO-mediated surface oxidation of cellulose whiskers, *Cellulose*, 13 (2006) 679–687.
192. W. Helbert, J. Y. Cavaillé, and A. Dufresne, Thermoplastic nanocomposites filled with wheat straw cellulose whiskers. Part I: Processing and mechanical behavior, *Polym. Compos.*, 17 (1996) 604–611.
193. M. A. S. Azizi Samir, F. Alloin, and A. Dufresne, Review of recent research into cellulose whiskers, their properties and their applications in nanocomposite field, *Biomacromolecules*, 6 (2005) 612–626.
194. M. Linder and T. T. Teeri, The roles and function of cellulose-binding domains, *J. Biotechnol.*, 57 (1997) 15–28.
195. J. Lehtlö, J. Sugiyama, M. Gustavsson, L. Fransson, M. Linder, and T. T. Teeri, The binding specificity and affinity determinants of family 1 and family 3 cellulose binding modules, *Proc. Natl. Acad. Sci.*, 100 (2003) 484–489.
196. C. Boisset, C. Fraschini, M. Schülein, B. Henrissat, and H. Chanzy, Imaging the enzymatic digestion of bacterial cellulosc ribbons reveals the endo character of the cellobiohydrolase Cel6A from *Humicola insolens* and its mode of synergy with cellobiohydrolase Cel7A, *Appl. Environ. Microbiol.*, 66 (2000) 1444–1452.
197. E. A. Bayer, J.-P. Belaich, Y. Shoham, and R. Lamed, The cellulosomes: Multienzyme machines for degradation of plant cell wall polysaccharides, *Annu. Rev. Microbiol.*, 58 (2004) 521–524.

198. H. Brumer, Q. Zhou, M. Baumann, K. Carlsson, and T. T. Teeri, Activation of crystalline cellulose surfaces through the chemoenzymatic modification of xyloglucan, *J. Am. Chem. Soc.*, 126 (2004) 5715–5721.
199. A. D. Beshay, B. V. Kokta, and C. Daneault, Use of wood fibers in thermoplastic composite II: Polyethylene, *Polym. Compos.*, 6 (1985) 261–271.
200. B. V. Kokta, R. Chen, C. Daneault, and J. L. Valade, Use of wood fibers in thermoplastic composites, *Polym. Compos.*, 4 (1983) 229–232.
201. B. V. Kokta, C. Daneault, and A. D. Beshay, Use of grafted aspen fibers in thermoplastic composites: IV. Effect of extreme conditions on mechanical properties of polyethylene composites, *Polym. Compos.*, 7 (1986) 337–348.
202. D. Pasquini, M. N. Belgacem, A. Gandini, and A. A. D. Curvelo, Surface esterification of cellulose fibers: Characterization by DRIFT and contact angle measurements, *J. Colloid Interface Sci.*, 295 (2006) 79–83.
203. M. Dufour and G. Gauthier, Vol. FR2865482-B1, France, 2004.
204. G. Chauvelon, N. Gergaud, L. Saulnier, D. Lourdin, A. Buléon, J.-F. Thibault, and P. Krausz, Esterification of cellulose-enriched agricultural by-products and characterization of mechanical properties of cellulosic films, *Carbohydr. Polym.*, 42 (2000) 385–392.
205. A. G. Cunha, C. Freire, A. Silvestre, C. P. Neto, and A. Gandini, Reversible hydrophobization and lipophobization of cellulose fibers via trifluoroacetylation, *J. Colloid Interface Sci.*, 301 (2006) 333–336.
206. C. S. R. Freire, A. J. D. Silvestre, C. Pascoal Neto, M. N. Belgacem, and A. Gandini, Controlled heterogeneous modification of cellulose fibers with fatty acids: Effect of reaction conditions on the extent of esterification and fiber properties, *J. Appl. Polym. Sci.*, 100 (2006) 1093–1102.
207. C. S. R. Freire, A. J. D. Silvestre, C. Pascoal Neto, A. Gandini, P. Fardim, and B. Holmbom, Surface characterization by XPS, contact angle measurements and ToF-SIMS of cellulose fibers partially esterified with fatty acids, *J. Colloid Interface Sci.*, 301 (2006) 205–209.
208. H. Gauthier, A.-C. Coupas, P. Villemagne, and R. Gauthier, Physicochemical modifications of partially esterified cellulose evidenced by inverse gas chromatography, *J. Appl. Polym. Sci.*, 69 (1998) 2195–2203.
209. P. Jandura, B. V. Kokta, and B. Riedl, Fibrous long-chain organic acid cellulose esters and their characterization by diffuse reflectance FTIR spectroscopy, solid-state CP/MAS ^{13}C-NMR, and X-ray diffraction, *J. Appl. Polym. Sci.*, 78 (2000) 1354–1365.
210. H. S. Kwatra, J. M. Caruthers, and B. Y. Tao, Synthesis of long chain fatty acids esterified onto cellulose via the vacuum-acid chloride process, *Ind. Eng. Chem. Res.*, 31 (1992) 2647–2651.

211. H. Matsumura, J. Sugiyama, and W. G. Glasser, Cellulosic nanocomposites. I. Thermally deformable cellulose hexanoates from heterogeneous reaction, *J. Appl. Polym. Sci.*, 78 (2000) 2242–2253.
212. J. Peydecastaing, S. Girardeau, C. Vaca-Garcia, and M. E. Borredon, Long chain cellulose esters with very low DS obtained with non-acidic catalysts, *Cellulose*, 13 (2005) 95–103.
213. C. Vaca-Garcia and M. E. Borredon, Solvent-free fatty acylation of cellulose and lignocellulosic wastes. Part 2: Reactions with fatty acids, *Bioresour. Technol.*, 70 (1999) 135–142.
214. H. Yuan, Y. Nishiyama, and S. Kuga, Surface esterification of cellulose by vapor-phase treatment with trifluoroacetic anhydride, *Cellulose*, 12 (2005) 543–549.
215. D. Samain, Procédés de traitement d'un matériau solide pour le rendre hydrophobe, matériau obtenu et applications, WO9908784, 1998.
216. S. Berlioz, C. Stinga, J.-S. Cordoret, and D. Samain, Investigation of a novel principle of chemical grafting for modification of cellulose fibers, *Int. J. Chem. Reactor Eng.*, 6 (2008) A2.
217. W. Amass, A. Amass, and B. Tighe, A review of biodegradable polymers: Uses, current developments in the synthesis and characterization of biodegradable polyesters, blends of biodegradable polymers and recent advances in biodegradation studies, *Polym. Int.*, 47 (1998) 89–144.
218. R. Chandra and R. Rustgi, Biodegradable polymers, *Prog. Polym. Sci.*, 23 (1998) 1273–1335.
219. S. Matsumura, Biodegradation of poly(vinyl alcohol) and its copolymers, *Biopolymer*, 9 (2003) 329–361.
220. G. D. Miller, J. R. Boylan, and R. B. Jones, Poly(vinyl alcohol)—a versatile polymer for paper and paper board applications, in A. Macnair (Ed.), *Synthetic Coating Adhesives—A Project of the Coating Binders Committee of Tappi's Coating and Graphic Arts Division*, Tappi Press, Atlanta, 1998, pp. 33–47.
221 F. Michaud, Vol. PhD, p. 256. Université Laval et Université Bordeaux I, 2003.
222. P. Zugenmaier, 4. Characteristics of cellulose acetates—4.1. Characterization and physical properties of cellulose acetates, *Macromol. Symp.*, 208 (2004) 81–166.
223. D. A. Cerqueira, G. Rodrigues Filho, R. M. N. Assunção, Cd. S. Meireles, L. C. Toledo, M. Zeni, K. Mello, and J. Duarte, Characterization of cellulose triacetate membranes, produced from sugarcane bagasse, using PEG 600 as additive, *Polym. Bull.*, 60 (2008) 397–404.
224. A. Dufresne, Polysaccharide nanocrystal reinforced nanocomposites, *Can. J. Chem.*, 86 (2008) 484–494.

225. V. Favier, H. Chanzy, and J. Y. Cavaillé, Polymer nanocomposites reinforced by cellulose whiskers, *Macromolecules*, 28 (1995) 6365–6367.
226. P. Hajji, J. Y. Cavaillé, V. Favier, C. Gauthier, and G. Vigier, Tensile behavior of nanocomposites from latex and cellulose whiskers, *Polym. Compos.*, 17 (1996) 612–619.
227. L. Petersson, I. Kvien, and K. Oksman, Structure and thermal properties of poly (lactic acid)/cellulose whiskers nanocomposites materials, *Compos. Sci. Technol.*, 67 (2007) 2535–2544.
228. H. Lönnberg, L. Fogelström, M. A. S. Azizi Samir, L. Berglund, E. Malmström, and A. Hult, Surface grafting of microfibrillated cellulose with poly(ε-caprolactone). Synthesis and characterization, *Eur. Polym. J.*, 44 (2008) 2991–2997.
229. B. Ly, W. Thielemans, D. A. D. Chaussy,, and M. N. Belgacem, Surface functionalization of cellulose fibers and their incorporation in renewable polymeric matrices, *Compos. Sci. Technol.*, 68 (2008) 3193–3201.
230. K. Nagai and M. Saito, Rapid acetylation of cellulose fibers, JP35013248, 1960.
231. J. Fraizy, Acétylation des fibres de cellulose régénérée et notamment des polynosiques, *Teintex*, 11 (1966) 781–790.
232. T. Nishino and N. Arimoto, All-cellulose composite prepared by selective dissolving of fiber surface, *Biomacromolecules*, 8 (2007) 2712–2714.
233. I. S. Kim, J. S. An, and B. H. Kim, Cellulosic materials having composite crystalline structures, US6361862B1, USA, 2002.
234. J. Koh, I. S. Kim, S. S. Kim, W. S. Shim, J. P. Kim, S. Y. Kwak, S. W. Chun, and Y. K. Kwon, Dyeing properties of novel regenerated cellulosic fibers, *J. Appl. Polym. Sci.*, 91 (2004) 3481–3488.
235. C. Stinga and D. Samain, Patent pending (2010).

CHEMICAL STRUCTURE ANALYSIS OF STARCH AND CELLULOSE DERIVATIVES

BY PETRA MISCHNICK[a] and DANE MOMCILOVIC[b]

[a] Technische Universität Braunschweig, Institute of Food Chemistry, Schleinitzstr. 20, D-38106 Braunschweig, Germany

[b] Department of Fibre and Polymertechnology, Royal Institute of Technology, Teknikringen 56-68, SE-100 44 Stockholm, Sweden

I. Introduction 118
1. History 118
2. Polysaccharide Derivatives as Functional Polymers of Renewable Resources 119
II. Chemical Modification of (1→4)-Glucans 121
1. Cellulose 121
2. Starch 125
3. Kinetically Controlled Reactions 127
4. Thermodynamically Controlled Reactions 131
5. Structural Complexity of Glucan Derivatives 132
III. Structure Analysis of (1→4)-Glucan Derivatives 144
1. Average Degree of Substitution 144
2. Monomer Composition: Substituent Distribution in the Glucose Residue 150
3. Distribution of Monomer Residues along the Polymer Chain 159
4. Heterogeneities in the Bulk Material 180
IV. Summary and Perspectives for the Future 183
Acknowledgment 184
References 184

ABBREVIATIONS

ABN, aminobenzonitrile; AFM, atomic force microscopy; AGU, D-glucose residue, "anhydroglucose unit"; CE, capillary electrophoresis; CEC, *O*-(2-cyanoethyl)cellulose; CID, collision-induced dissociation; CMC, *O*-(carboxymethyl)cellulose; CMS, *O*-(carboxymethyl)starch; DE, dextrose equivalents; DEAEC, *O*-(2-diethylaminoethyl) cellulose; DMAc, *N,N*-dimethylacetamide; DMAP, 4-dimethylaminopyridine; DMF, *N,N*-dimethylformamide, DP, degree of polymerization; DS, degree of substitution; EC, *O*-ethylcellulose; ECR, effective carbon response; ESI, electrospray ionization;

ISBN: 978-0-12-380854-7
DOI: 10.1016/S0065-2318(10)64004-8

FAB, fast-atom bombardment; FID, flame ionization detection; GLC, gas-liquid chromatography; HEC, *O*-(2-hydroxyethyl)cellulose; HEMC, *O*-(2-hydroxyethyl)methylcellulose; HP, 2-hydroxypropyl; HPAEC, high-performance anion exchange chromatography; HPC, *O*-(2-hydroxypropyl)cellulose; HPLC, high-performance liquid chromatography; HPMC, *O*-(2-hydroxypropyl)methylcellulose; IT, ion trap; MALDI, matrix-assisted laser desorption/ionization; MALS, multiangle light scattering; MC, *O*-methylcellulose; Me_2SO, dimethyl sulfoxide; MDS, molar degree of substitution; MS, mass spectrometry; MS^n, tandem mass spectrometry; NMR, nuclear magnetic resonance; PAD, pulsed-amperometric detection; qTOF, quadrupole time-of-flight; RI, refractive index; SEC, size-exclusion chromatography; TOF, time of flight; UV, ultraviolet.

I. Introduction

1. History

The polysaccharides starch and cellulose have been put to many uses, either as such or fermented for nutrition, as an energy source, for building material, paper, and textiles, long before their structures had been elucidated,[1,2] their polymeric character recognized[3,4] and confirmed,[5,6] and chemical modification performed in a purposive way.[7–9] In 1846 Schönbein[10,11] reported the preparation of cellulose nitrate, which became the "smokeless powder" to replace gunpowder as a propellant explosive, and this was followed by the technical production of the first thermoplastic semisynthetic polymer, "celluloid." The limitations of available analytical techniques in the late 19th and the early 20th century, along with the wide structural diversity of polysaccharide derivatives, led to the development of such empirical methods as the Zeisel method[12a,b–15] for the determination of the average degree of substitution (DS) of cellulose ethers. Periodate oxidation[16] and the Smith degradation[17] were used to quantify the amount of unsubstituted and 6-*O*-substituted glucose residues, which are the only ones having a free vicinal diol structure in (1→4)-glucans. Selective modification of the primary 6-OH group is another chemical transformation that was used for structure elucidation in earlier days.[16] Elemental analysis, since Liebig's (1837) development of the "Fünfkugelapparat" method,[18] and such wet-chemical methods as titration or measurement of the change in iodine-binding capacity in modified starches[19,20] were also earlier procedures. Topochemical influences were already addressed around 1960 by electron micrographs of stained starch-granule sections.[21] The series of monographs on *Cellulose and Cellulose Derivatives: Their Preparation, Properties, and Analysis*, first edited by Spurlin, Ott, and Grafflin in 1943, remains a significant milestone in the field of chemically modified cellulose.[20]

Following the suppression, or at least neglect, of polysaccharide research during the decades of petroleum chemistry, with its major emphasis on synthetic polymers, the oil crisis of the 1970s stimulated a resurgence of interest in polysaccharides as renewable resources. This increased awareness for the potential of polysaccharides as raw material, together with the newer analytical techniques established in the 1950s and 1960s, notably gas chromatography,[22a,b] mass spectrometry (MS),[23a,b] and nuclear magnetic resonance (NMR) spectroscopy,[24a,b] also stimulated new analytical work on starch and cellulose derivatives. Furthermore, a major progress in the permethylation procedure,[25] an important step in structure elucidation of polysaccharides,[26a,b] also accelerated the analysis of polysaccharide derivatives, since branching of a linear glycan may be considered as a special case of substitution. Since the early work on enzymatic digestion of cellulose ethers,[27–31] the enhanced knowledge of and the availability of enzymes for degrading cellulose and starch has provided an additional tool for comparison of samples with respect to their degradability and range of products. Research groups during recent decades have elucidated new procedures and products, found new solvents and reagents, and have enhanced our understanding of the relationship between derivatization and properties. This has been summarized and discussed in a number of recent reviews.[32–37] At the same time our knowledge has been extended on the link between the structure and composition of the modified glucans.[38–41] Analysis of substitution patterns at various hierarchical levels of structure has greatly benefitted from developments in instrumental methodology and data processing, as in NMR spectroscopy, in infrared (IR) and Raman spectroscopy, and especially from major progress in soft-ionization MS (electrospray ionization MS (ESI-MS); matrix-assisted laser desorption/ionization time-of-flight MS (MALDI-TOF-MS)).[42] With respect to topochemical aspects and structure formation, high-resolution microscopic techniques and such surface analytical methods as transmission electron microscopy (TEM), scanning electron microscopy (SEM), atomic force microscopy (AFM), or—in combination with labeling strategies—laser fluorescence microscopic techniques, have become of increasing interest. This article surveys the state of the art in the wide field of substitution-pattern analysis, with a focus on the concepts and interpretability of experimental results. For cellulose and starch *per se*, modification procedures and properties of products are treated only insofar as required as background for the analytical questions.

2. Polysaccharide Derivatives as Functional Polymers of Renewable Resources

Development of new polysaccharide derivatives and improvements of those with a long history, such as acetates and methyl ethers, are still ongoing. Along with the long-established commercial products, newer special derivatives, such as amphiphilic or bioactive ones, have been developed, incorporating such functionalities as unsaturated reactive intermediates, bulky groups, photosensitive residues,[42,43] and others. Protecting-group strategies and *de novo* synthesis from suitable building blocks have been aimed at developing uniformly regioselective,[44,45] block-like,[46–52] or other unusual patterns[53] of substitution. Along with starch and cellulose, other glucans such as dextrans have been used as substrates for chemical modification targeting new applications.[54] This field requires separate treatment and can be mentioned here only to emphasize the importance of further progress in the field of substitution-pattern analysis, since many of these different glucan derivatives have not been comprehensively characterized.

Cyclodextrins, a special class of starch-derived cyclic oligomeric (1→4)-glucans, have also been modified in various ways.[55] Their cyclic and conformationally more-rigid character allows regioselective substitution with respect to the individual glucose residues and the entire cyclodextrin molecule (topochemical control).[56,57] This higher uniformity, together with a defined degree of polymerization (DP), extends the analytical methods that are applicable, but in principle, they can be analyzed by methods similar to those used for other polymeric glucan derivatives.

The properties of modified glucans are effectively the manifestation of their molecular chemical structure and composition on the macroscopic scale. One important goal of cellulose etherification is to achieve water solubility. Viscosity, adhesiveness, water binding, film forming, thickening, swelling, and emulsifying ability are properties used in the building and construction industries, as well as excipients for pharmaceuticals, and also in food applications. Oil drilling and manufacturing of paper, textiles, membranes, and chromatographic supports, as well as bioanalytical applications, constitute further fields of uses. Reproducibility and fine-tuning of properties are required for many of these applications. However, because of the natural origin of the biopolymer and small differences in reaction processes, this goal is still difficult to achieve. For better understanding and control, the factors influencing the bulk properties, namely, the large assembly of molecules with different size and substitution patterns, need to be identified and their interactions studied. One prominent example of such properties is the thermoreversible gelation of cellulose ethers. The hydrophobic effect enables the less-polar cellulose ethers to dissolve in water at low temperature. With increasing temperature, the highly structured water cages around the hydrophobic domains break down at a certain temperature (gel point or flocculation point), and the viscosity increases tremendously. The shape of the gel curves and the gel point are very important parameters that are influenced by the DP and the DS of ether residues, but cannot be adjusted in a simple manner. The decoding of the chemical structure with respect to physical and biological functions, and tracing the molecular composition back to the modification process, is the goal of and motivation for research in this field. Finally, degradability by enzymes has often been emphasized as an environmentally friendly argument for the use of renewable resources. However, this aspect is not demanded for all fields of application, and by contrast sometimes must be excluded because of stability requirements. Enzymatic degradability depends on the degree and distribution of substituents and is therefore also a property of analytically diagnostic value.

II. Chemical Modification of (1→4)-Glucans

The remarks on chemical modification of (1→4)-glucans are restricted to aspects that are of interest for the main topic of this article, namely, the analysis of the substitution patterns of this class of compounds. For more details, the reader is referred to the respective literature.

1. Cellulose

The most abundant polysaccharide produced in nature is cellulose. About 2×10^{11} tonnes are synthesized annually by green plants using solar energy. Soft and hard wood, and cotton, are sources for celluloses of different quality as regards accompanying hemicelluloses, and cellulose of high molecular weight is also produced by bacteria.[36,37] Cellulose has a linear β-(1→4)-linked glucan structure, enabling strong cooperative intra- and intermolecular hydrogen-bond patterns, stiffening the chain, and forming mechanically stable, insoluble fibrils and fibers. Fig. 1 illustrates the hierarchical order of parallel-oriented cellulose I according to Kroon-Batenburg *et al.*[58]

Although the crystal structure of cellulose has been investigated for almost a century, certain details remain to be deciphered.[59,60] For an overview of the complexity concerning different crystal structures and conformations of cellulose and their transitions, see Ref. 37 and the article by Pérez and Samain in this volume.

As a consequence of this molecular and supramolecular structure, cellulose is insoluble in water, despite its many hydroxyl groups and in contrast to other more-flexible glucans, such as dextran, a mainly α-(1→6)-linked glucan. Therefore, modification of cellulose is commonly performed in heterogeneous processes, in which an efficient activation is a very important step. In the production of cellulose ethers, this is achieved by treatment with strong alkali. Crystallinity is thus destroyed, and in the

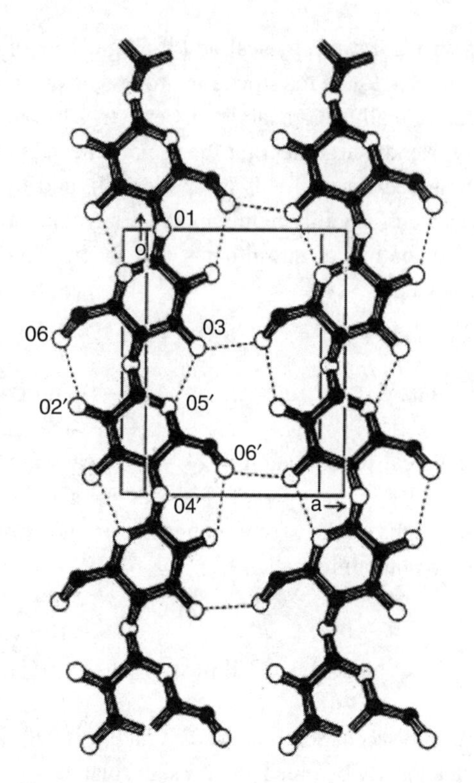

FIG. 1. Hydrogen-bond patterns in cellulose I according to Kroon-Batenburg *et al.* Reprinted from Ref. 58 with permission from Elsevier, Copyright 1986.

ideal case all glucose residues should become equally accessible.[61,62] Such activation, which causes swelling and can be modified by the addition of various organic solvents (diluents),[63] is followed by the addition of an alkyl halide or by an oxirane. Fig. 2 shows a selection of some very common and more specialized examples of chemical modification of (1→4)-glucans: namely, nonionic, cationic, anionic ether formation, organic and inorganic esterifications, substitutions, and additions.

The addition of oxiranes requires only catalytic amounts of base, and therefore the excess of base required for the activation of cellulose may be quenched after a certain period.[61c] By these kinetically controlled reactions, which are treated in more detail in Section III, carboxymethyl- (CMC), methyl- (MC), ethyl- (EC), hydroxyethyl- (HEC), and hydroxypropyl cellulose (HPC), and such mixed derivatives as HEMC or HPMC are produced. The nucleophlic nitrogen in 2-(diethylaminoethyl)cellulose (DEAEC), prepared by substitution of 1-chloro-2-*N,N*-diethylaminoethane (via *N,N*-diethylaziridinium

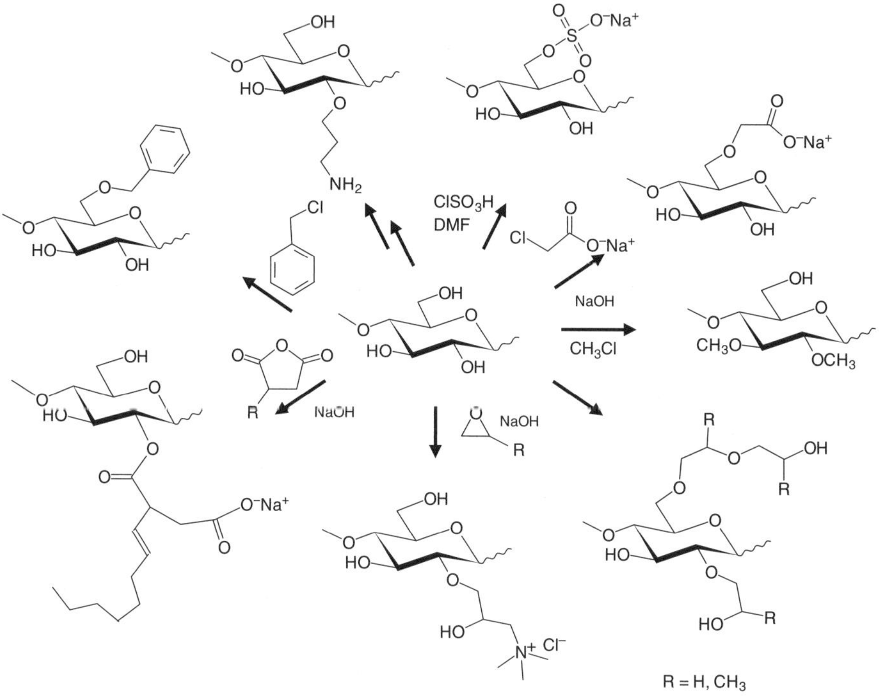

FIG. 2. Common and specialized examples of chemical modification of (1→4)-glucans.

chloride), as well as the hydroxyl functions formed by oxirane addition reactions can compete with the carbohydrate OH groups and undergo tandem reactions, thus adding a further dimension of complexity to structure analysis. Furthermore, the viscosity, a property important for many applications of these ethers, is dependent on the molecular weight. Progress in chemical modification of cellulose, including existing industrial aspects as well as new synthetic concepts, has been the subject of various reviews.[34,37,53,64,65] These have described protecting groups, reactive intermediates, the influence of solvent, and other parameters of these polymer-based reactions on chemical structure. It is possible to tune the regioselectivity and random or uncommon patterns of substitution to afford new products with novel properties. Characterization of the products and thus control of the reaction has been achieved mainly by means of chromatographic [high-performance liquid chromatography (HPLC)] and NMR-spectroscopic techniques.[32–34,53,66] Such unsaturated ethers as allyl cellulose have been prepared for further cross-linking studies.[67] Cross-linked celluloses are a special group of derivatives, with their own challenges in analysis, and are not treated in this article.

To describe the extent of chemical modification, the DS is defined as the average number of substituted hydroxyl groups of glucose residue and is thus in the range of 0–3 as long as end groups can be neglected (see Section III.1.a). For modifications, including tandem reactions, a molar degree of substitution (MS or MDS) is additionally defined as the average number of substituent groups linked to glucose residues, that is, in the DS range of up to a theoretically unlimited level.

The most prominent cellulose ester produced on the industrial scale is cellulose acetate.[68] The reaction is usually performed with acetic anhydride and with sulfuric acid as a catalyst. To minimize heterogeneities, acetylation is allowed to run nearly to completion, and subsequently partial ester hydrolysis is initiated by the addition of water until a desirable solubility is achieved that corresponds to a DS of about 2.5. Such higher acyl homologues as propanoyl or butanoyl exhibit more thermoplastic properties. Many specialized esters such as chiral (–)-menthyloxyacetates, furan-2-carboxylates, or crown-ether-containing acylates have been prepared on the laboratory scale and characterized by NMR spectroscopy.[69] Various procedures have been applied, using anhydrides and acyl chlorides as acylating agents in combination with such bases as pyridine, 4-dimethylaminopyridine (DMAP), or *N,N'*-carbonyldiimidazole.[35] The substitution pattern of cellulose acetates has also been modified by postchemical enzymatic deacetylation.[70] Cellulose 6-tosylates have been used as activated intermediates for nucleophilic substitution to afford 6-amino-6-deoxy, 6-deoxy, or 6-deoxy-6-halo-celluloses.[71]

Along with organic esters, inorganic cellulose esters are also of interest. Cellulose nitrate is used in lacquers because of its film-forming properties, but as already mentioned, it is useful as an explosive at a high degree of esterification. This was a focus of interest in Alfred Nobel's research and thus the basis for the Nobel foundation and award. Various procedures for sulfation of cellulose have been studied because of similarities to such biologically significant polysaccharides as heparin or chondroitin sulfate.[72–77]

Finally, oxidation with TEMPO or periodate, respectively yielding 6-carboxy derivatives (glucuronans) and "dialdehyde cellulose," should be mentioned.[78,79] 6-Aldehydo cellulose has been prepared via 6-azido-6-deoxy-cellulose and analyzed by gas–liquid chromatography (GLC)-MS of its deuterium-labeled reduced constituents.[80]

The long-known complexing solvent, dimethylacetamide (DMAc)/LiCl,[81,82] along with dimethyl sulfoxide (Me_2SO)–tetrabutylammonium fluoride (TBAF),[83,84] ionic liquids,[85–87] and others have broadened the spectrum of solvents for cellulose.[88] For a more comprehensive overview on cellulose chemistry, the reader is referred to the chapter by Heinze and Petzold, and for structural aspects to the chapter by Pérez and Samain in this volume.[34]

2. Starch

After cellulose and chitin, starch is the third most abundant biopolymer and is one of the most important renewable resources. The main sources in the United States and Europe are corn, potatoes, and wheat. Other sources include rice, tapioca (cassava), sorghum, arrowroot, and sago palm.[89] In contrast to cellulose, the glucose residues in starch are mainly α-(1→4)-linked, and this apparently small configurational difference causes a major change in macromolecular structure, leading to a helical instead of a linear chain conformation, with all of its resultant consequences for supramolecular organization and properties. The two constituents of starch are amylose, the linear component, and amylopectin, the α-(1→6)-branched component. They are organized in starch granules of highly sophisticated architecture in a layered structure (Fig. 3). Amylopectin, nonrandomly branched at an extent of about 5% and of higher molecular weight than amylose ($M_w = 10^7$–10^8 for amylopectin and 10^4–10^6 g mol^{-1} for amylose), forms the crystalline layers, while amylose is located in the amorphous domains. Furthermore, the growth rings of amylopectin show a substructure: the

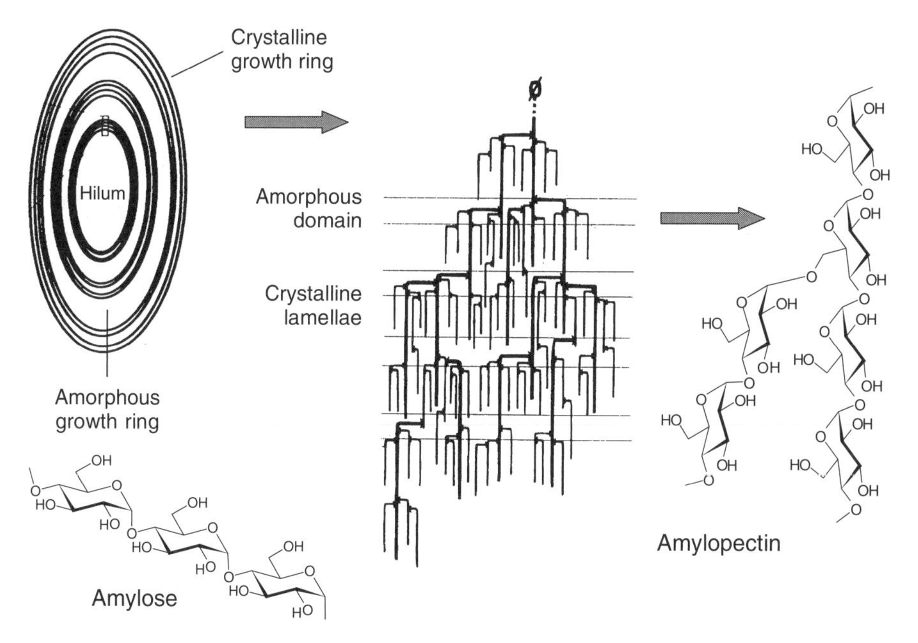

FIG. 3. Starch granule structure according to Jenkins and Donald[91] and Baldwin *et al.*[92] Reprinted from Refs. 91 and 92 with permission from Elsevier, Copyright 1995 and 1997.

levels of branching points are less dense and ordered (amorphous), while crystallinity results from clustered pendant side chains with DP ~15 from the populations of so-called A and B chains with different lengths.[90–92]

Starch granules show different shapes, sizes, and morphologies depending on the plant source, but still exhibit a certain heterogeneity for granules from a single source.[93] The granule surfaces of some starch types (such as corn) also exhibit channels to the interior, which are important for the diffusion of solvents and chemical reagents (see Section II.5).

Chemical modification of starch has been comprehensively reviewed in this series, most recently by Tomasik and Schilling.[94] It is usually performed in heterogeneous processes and therefore, the topochemical structure of the granule cannot be neglected. The question thus arises as to whether there are differences between the amylose and amylopectin components, between branching points and linear domains, and whether there is a gradient of substitution from the outer to the inner part of the granule, the surface, and the interior.[95–100]

The range of ethers, inorganic and organic esters, and oxidized starches is closely similar to that described for cellulose (Section II.1), but DS values obtained are usually much lower (often in the range of 0.01–0.2) to avoid dissolution or "pasting" of the products, since purification is generally performed by washing the modified granules.

One of the most important starch derivatives is "cationic starch," namely, *O*-(3-trimethylammonium-2-hydroxy)propyl starch, which is mainly used in paper manufacturing. It is produced by addition of starch to the corresponding quaternary aminopropyloxirane (QUAB). The hydroxyalkyl groups can also undergo tandem reaction, forming oligoethers. As in the simpler hydroxypropyl ethers, each substituent adds a new stereogenic carbon to the glucose and thus forms diastereomeric isomers, which again complicates analysis.

Methyl-, hydroxyethyl-, hydroxypropyl-, and carboxymethyl starches, starch acetates, succinates, alkenyl succinates (Fig. 2), adipates, and phosphates,[9,89] are all well-known products. Furthermore, special derivatives have also been prepared, such as vinyl-,[101] silyl-,[102] or propargyl starches,[103] as reactive intermediates for further functionalization. Unusual substitution patterns can also be established by highly selective deacetylation with alkyldiamines and subsequent introduction of such functional groups as sulfates.[104] From the analytical point of view, the most important aspects are: stability under alkaline (methylation) and acidic or Lewis-acidic (depolymerization) conditions, reactivity (such as migration, rearrangement, further substitution or addition reactions, or any intramolecular reaction), and polarity (lipophilic/hydrophilic, ionic/nonionic, acidic/basic). These properties mainly determine the analytical strategy to be applied, appropriate steps, for sample preparation and sources of discrimination.

3. Kinetically Controlled Reactions

a. Alkylation.—Alkylation, including carboxymethylation of starch and cellulose, is usually performed by Williamson-type etherification, namely, a nucleophilic substitution of the corresponding alkyl halide (bromides and iodides on the laboratory scale, and chlorides in industrial production). To form the alcoholate as an active nucleophile, the polysaccharide is treated with a base (B^-), required in stoichiometric amount. In the heterogeneous process, starch granules are suspended in an aqueous slurry, and low molecular-weight alcohols and salts as anti-swelling agents may be added. Cellulose is activated with concentrated alkali to break hydrogen bonds and promote swelling. For homogeneous alkylations on the laboratory scale, the anion of Me_2SO, prepared with NaH ("Na-dimsyl") or MeLi ("Li-dimsyl"), is often used. Furthermore, depending on the solvent, NaH and NaOH are introduced, and the latter is the base of choice for industrial processes. Since primary alkyl (alkenyl, alkynyl) halides (RX) are usually used, the reaction follows the S_N2 mechanism. Formation of primary ethers is irreversible and is thus a kinetically controlled reaction. This means that the regioselectivity with respect to the three available OH groups in (1→4)-glucans depends on the distinct activation energies (ΔE_a) of the reaction and on the relative concentrations of the alcoholate of the various OH-positions (PS-O$^-$), and thus upon their pK_a values. In addition, the reaction time (t), temperature (T), solvent, and any additives influence the outcome of the reaction.[105–108]

Dependence of the rate constants k_i (of OH in position i) on the parameters just mentioned are given by the Arrhenius equation (3) (polysaccharide (PS) backbone). The kinetics are of second order, but due to the excess of reagent RX are usually pseudo-first order, that is dependent on the concentration of the available hydroxyl functions in the glucose residue.

$$k = A \cdot \mathrm{e}^{-\Delta E_a/RT} \tag{3}$$

and

$$\frac{\mathrm{d[PS\text{-}OR]}}{\mathrm{dt}} = k_i[\mathrm{PS\text{-}O^-}]\,[\mathrm{RX}] = k_i'[\mathrm{PS\text{-}O^-}] \tag{4}$$

and

$$[\mathrm{PS\text{-}O^-}] = K_a \frac{[\mathrm{PS\text{-}OH}][\mathrm{B^-}]}{[\mathrm{BH}]} \tag{5}$$

As long as the base is not used in high excess, [PS-O$^-$], and hence the rate of the reaction at position *i*, also depends on the pK_a value of the various OH groups; there is a preference for the most acidic OH group. Hydroxyl groups of sugars exhibit much higher acidity than aliphatic alcohols. While pK_a values of aliphatic alcohols are in the range of 16–18, they are between 12 and 14 for OH groups in glucans.[109] Aside from the hemiacetal OH group, the most acidic hydroxyl group is located in position 2, due to the electron withdrawing and thus anion-stabilizing effect of the anomeric center. However, this does not explain the large differences in O-2 selectivity of kinetically controlled reactions of amylose and cellulose. In α-, but not in β-glucans, deprotonation of 2-OH can be supported by the glucosidic oxygen, coordinating with the counter ion (Fig. 4), and O-3 and O-3′ might also be involved.[110]

Acidity of the remaining positions follows the rule primary OH > secondary OH. Sterically demanding bulky reagents favor the more accessible and conformationally more flexible primary OH group in position 6. In practice, however, the situation is not so simple, since, for example, in starch the reactivity at O-6 is often unexpectedly low, even lower than that of the secondary 3-OH. This reveals the significant influence of the macromolecular structure and architecture, since relative reactivities of the corresponding monosaccharides cannot simply be extended to the α- and β-glucans. Already in the 1960s, Roberts and Rowlands summarized kinetic data and investigated the influence of the reaction conditions (solvent, base concentration, steric demands) on the substituent distribution, but due to the limited technical opportunities available at that time for a differentiated substitution-pattern analysis, some of these interesting early results must nowadays be critically reconsidered.[111–113] Timell and Spurlin, for instance, concluded from results obtained for CMC that O-3 does not react if position 2 has been carboxymethylated, since they could not identify 2,3-di-*O*-carboxymethyl-D-glucose as a constituent at that time. Because of the anionic character of this group, it was plausibly interpreted as the result of electrostatic repulsion.[114] Besides the kinetics of modification of the glucose residue, there might be

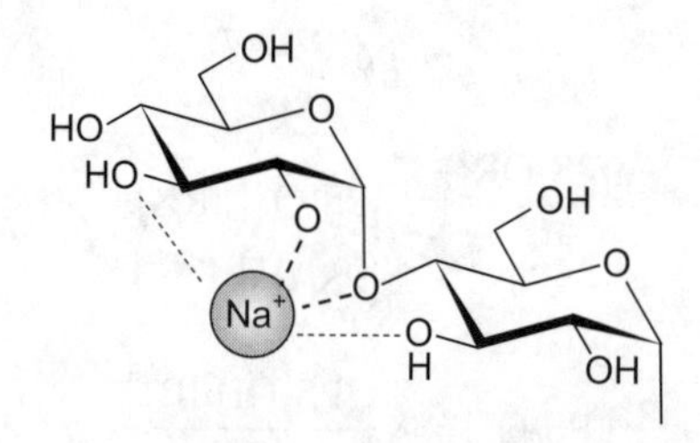

Fig. 4. Probable support for 2-OH deprotonation by multiple Na-coordination in an α-glucan.

different accessibilities caused by partly crystalline, partly amorphous domains in cellulose fibrils and starch granules. Thus diffusion control of the reagents becomes significant as a superimposing rate-determining process. If the material is not uniformly activated, different local concentrations of reactive alcoholate will lead to different rates of modification. There are also several studies on the kinetics of cellulose ether formation, for instance, with CMC, showing that the overall reaction can be the result of more than one reaction having different rate constants: higher in the more-accessible areas and lower in the more-ordered regions.[115–118] It has to be kept in mind that a certain derivatization does not always occur in the same manner and with the same regioselectivity, but is the result of several interrelated parameters that can vary over a very wide range. Therefore, it is in most cases misleading to state any general preference for a certain position without considering the conditions applied. As an example, the relative rate constants determined for CMC, and reported in the literature, are summarized in Table I.

TABLE I
Rate Constants Reported for Carboxymethylation of Cellulose

Relative rate constants				
k_2	k_3	k_6	**Analytical method applied**	**Reference**
3.0	1.0	2.1	^{13}C NMR after ultrasonic degradation	Baar *et al.*[119]
1.8	1.0	1.3	HPAEC after acidic ($HClO_4$) degradation	Kragten *et al.*[120]
2.35	1.00	1.82	GLC after methylation and reductive cleavage	Zeller *et al.*[121]
2.1	1.0	1.5	^{1}H and ^{13}C NMR of undegraded CMC	Abdel-Malik and Yalpani[122]
3.1	1.0	2.6	GLC after acidic (HCl) degradation, trimethyl-silyl derivatization	Niemelä and Sjöström[123]
2.14	1.00	1.58	^{13}C NMR after acidic ($HClO_4$) degradation	Reuben and Connor[124]
2.0	1.0	1.5	^{1}H NMR after acidic (H_2SO_4) degradation	Ho and Klosiewicz[125]
2.5^{a}	1.0	1.8	Gas–liquid chromatography	Buytenhuys and Bonn[126]
4.6^{b}	1.0	3.6		
2.0	1.0	2.5	Paper chromatography and electrophoresis after hydrolysis, methanolysis, and reduction ($LiAlH_4$)	Croon and Purves[127]
1	1	2	Tosylation–iodination and tritylation–periodate cleavage	Timell and Spurlin[114]

Molar ratio of water to cellulose during synthesis: ^{a}: 7:1; ^{b}: 14:1 with equal base concentrations.

b. Hydroxyalkylation.—Various hydroxyalkyl ethers—non-ionic and cationic—are produced by the addition of starch or cellulose to oxiranes. This is also an irreversible, kinetically controlled reaction, because of rapid protonation and thus deactivation of the comparably less-acidic alcoholate as the addition product primarily formed.[9,64,89,128] Again the alcoholate is the reactive nucleophile, which attacks asymmetric oxiranes almost exclusively at the unsubstituted position. For instance, addition to methyloxirane (propylene oxide) gives mainly the (2′-hydroxy)-propyl ethers, and only traces of the isomeric (1′-methyl-2-hydroxy)ethyl derivatives have been detected.[129] In contrast to Williamson etherification, oxirane addition does not consume base, but requires only catalytic amounts. Therefore, the reaction is performed at substantially lower concentrations of sodium hydroxide and, as a consequence, differences in the acidities of hydroxyl groups are reflected much more distinctly in the regioselectivity of etherifcation. For instance, in cationic starches, usually produced by addition to (*N*-trimethylammoniummethyl)oxirane in an aqueous slurry,[9,89] 80–90% of the substituents are located at the most acidic position (O-2) in spite of the steric demand of this group, followed by nearly equal substitution at 3-OH and 6-OH.[130–132] As already mentioned, the alcoholate formed by addition to an oxirane can again react directly with further oxirane to give oligoether chains (the so-called tandem-reaction) or is protonated. Addition to unsymmetrical oxiranes, which are chiral, also enhances the number of stereoisomers, since racemates are generated in the substituents. Thus addition reactions of oxiranes lead to much more structural diversity than simple alkylations, as outlined in more detail in Section II.5.

The solvent has an interesting effect on the substituent distribution in the glucose residue. In aprotic solvents, which neither offer protons nor shield the alcoholate by solvation, the alcoholate formed by oxirane opening can detach a proton from the adjacent sugar OH and thus enhance its reactivity. For instance, for hydroxyalkylations in ionic liquids and without base, a very pronounced increase of 2,3-di-*O*-substitution was observed.[87]

c. Miscellaneous Reactions.—Besides the two most important kinetically controlled reaction types in polysaccharide etherifications just outlined, there are some special reactions that merit brief mention. Sulfopropylation and sulfobutylation are performed either by addition of sulfite to allyl ethers or by ring-opening addition of the polysaccharide alcoholate to the corresponding 5- and 6-membered ring sultones.[133–136] They are therefore comparable to oxirane additions. The difference is that sulfonic acids are such strong acids that the base is not recycled by protonation of the primary products and is thus consumed as in Williamson-type etherifications.

Another group of reactions is trialkylsilylation, reviewed by Klemm *et al.*[44,137–139] Their classification as kinetically controlled reactions must be guarded, since silyl groups can migrate through the possibility of octet extension at Si, favored by appropriate stereochemistry and alkaline conditions.[140]

Further parameters influencing the regioselectivity of analogous polymer modifications of polysaccharides, within the mechanistic regime already described, are solvent, temperature, and reagent concentrations, as mentioned at the outset of this section. Since the polarity of the product usually changes more or less drastically during the course of the reaction, choice of an appropriate solvent can arrest the reaction at a certain point through precipitation of the product, which thus escapes further modification. Yokota has discussed the role of such diluents as low molecular-weight alcohols, for example, isopropyl alcohol, which are poorer solvents for NaOH than water. Due to the higher affinity of such solvents to the modified and thus more hydrophobic areas of cellulose, a redistribution of base to the less-substituted areas is proposed.[60,61] Additional influence can be exerted by coordinating or chelating cations. Thus barium hydroxide is reported to enhance the regioselectivity of position 2 in α-glucans.[141] Copper(II) can favor more-selective alkylation at O-3, which is less deactivated than O-2 by chelation of the 2,3-diol structure.[142,143] Special cases are observed for cyclodextrins, where prepositioning of the reagent in the cavity can also turn the common reactivity order in favor of O-3.[136] Esterification has also been performed under kinetic control by using vinyl acetate.[144] The vinyl alcohol released isomerizes to the more-stable acetaldehyde, thus preventing the reverse reaction.

4. Thermodynamically Controlled Reactions

a. Acylations.—Reversible reactions, or conditions under which the introduced groups can migrate, essentially allow equilibration to the energetically most-stable distribution of substituents in the glucose residues as well as in the polysaccharide chains. Acylation performed with acyl anhydrides or chlorides, and a base which acts as acyl-transfer vehicle and proton scavenger (e.g., DMAP, *N*-methylimidazole), and sometimes also as solvent, is one example. Since primary esters are the most stable, acyl groups preferably locate at O-6, and this tendency becomes even more pronounced for sterically demanding esters. However, the acid-catalyzed acetylation process earlier mentioned, going through the triacetate followed by partial hydrolysis, does not yield the thermodynamically favored pattern, but gives predominantly the 2,3-diacetates because of favored hydrolysis of the primary 6-acetate.[145a,b] In contrast to cellulose, acetylation of aqueous starch systems with acetic anhydride at moderate pH still favors the order at O-2 ≈ O-3 > O-6.[146a,b] The preference for O-3 is probably due to acyl migration. For more detailed information on the developing field of esterification reactions of polysaccharides, the reader is referred to the comprehensive review of Heinze *et al.*[35,147]

b. Sulfation.—Sulfation of cellulose and starches has been thoroughly investigated by Wagenknecht *et al.*[74,76] Soluble intermediates activated (trimethysilyl ethers) as well as protected (acetates, nitrites), in combination with different sulfating agents, have been applied to tune the pattern of sulfation. Recently, cellulose sulfates, water-soluble at a DS as low as 0.2, have been obtained in ionic liquid–*N*,*N*-dimethylformamide (DMF) solution.[77] If reactions are performed under thermodynamically controlled conditions without any protecting group involved, 6-sulfation clearly dominates the substitution pattern. Equilibration to the thermodynamically most stable distribution has been demonstrated by the fact that, independent of the specific procedure used, and even with heterogeneously initiated sulfation of cellulose, the final result is 6-sulfation followed by the 2- and 3-sulfates and a random pattern along the chain, based on these relative positional stabilities.[148]

c. Cyanoethylation.—Not only can esters be produced under thermodynamic control, but also certain ethers through nucleophilic addition to activated alkenes. Cyanoethylation is the most prominent example of this Michael-addition reaction type.[149,150] Cyanoethyl groups have been further transformed into aminopropyl residues,[151] since the direct introduction of aminoalkyl residues by Williamson-type etherification suffers from tandem reaction of the nucleophilic amino group. In cyanoethyl starch of DS < 1, approximately 50% of substituents are located at O-6, 37% at O-2, and 13% at O-3.[152] Cyanoethylation of cellulose and its relation to the reaction conditions, DS, and solubility of *O*-(2-cyanoethyl)celluloses (CECs) have been comprehensively reported.[153]

5. Structural Complexity of Glucan Derivatives

In the analysis of starch and cellulose derivatives, we are confronted with very complex mixtures. But how complex are they? As already outlined, it is necessary to consider different hierarchical levels, from supramolecular structures to molecular substructures, as well as molecular-weight distributions. Furthermore, defects of the ideal glucan chain caused by "peeling" processes, and oxidation of hydroxyl groups under alkaline conditions, must be considered. As a consequence, deoxyhexonic acid residues are formed at the former reducing end of polymer chains, and carbonyl groups are created within the chain as "hot spots," from which further degradation can be initiated. These minor structural deviations (carboxyl groups in a typical range of 10–60 μmol/g; in extreme cases up to 300 μmol/g)[154] might also influence the bulk properties; they have been thoroughly studied by Potthast *et al.*[155–159] Sensitive methods have been developed that allow quantitative carbonyl and carboxyl analysis and their dependence on molecular weight, which gives further valuable information. Depending on the source and quality of cellulose, various proportions of hemicelluloses, xylans, and galactoglucomannans may be present and might act as additives. The following sections focus on the hierarchical levels of glucan structures.

a. Distribution in the Glucose Residues: The Models of Spurlin and Reuben.— At the level of the smallest entity, the (1→4)-linked glucopyranosyl group [glucose residue, anhydroglucose unit (AGU)], the analogous modification in the polymer may be considered as a number of consecutive reactions with three options. The number of resulting patterns in a monomer unit is a^n with a = number of different functional groups (such as OH and OMe in MC) and n = number of OH groups available/AGU, which is 3 in cellulose and starch. In the case of *one* type of substituent, $2^3 = 8$ different patterns are possible in the glucose residue (Fig. 5).

For such mixed ethers as the amphiphilic carboxymethyl benzyl starch ($a = 3$: OH, OCM, OBn), there are $3^3 = 27$ monomer patterns. The widely used 2-hydroxypropylmethyl and 2-hydroxyethylmethylcelluloses (HPMC and HEMC) already contain four different functionalities/AGU, since methylation and hydroxyalkylation are performed simultaneously ($a = 4$: O-$[CH_2CH_2O]_mH$ and O-$[CH_2CH_2O]_mMe$ with $m = 0$ and 1 for HEMC), and thus $4^3 = 64$ different patterns result, which in practice are enhanced by tandem reaction ($m > 1$) and formation of diastereoisomers in the case of 2-hydroxypropyl (HP). In starch, the situation is complicated by branching of the amylopectin component. Thus a certain number of 6-OH groups are blocked and compensated by the additional 4-OH of terminal residues of the chains, so the average number of OH groups available is also 3, but of course, the pattern complexity is enhanced. Terminal residues yield $2^4 = 16$ different patterns, while the number of patterns in the branched units is decreased to $2^2 = 4$ for a simple derivative.

Already in 1938 Spurlin had framed the basic principles of the substituent distribution on cellulose derivatives in an clear and comprehensive way.[161] In 1939 he introduced a model (Spurlin I) to calculate the resulting pattern of the consecutive reactions of glucose residues in cellulose within a DS range of 0–3.[162] Spurlin defined the following boundary conditions to decrease complexity: (i) All AGUs are equally accessible; (ii) chains are considered to be of unlimited length, and thus terminal groups can be neglected; (iii) all OH groups of an AGU react independently of each other, that is, primary substitution does not influence the reactivity of other OH groups and the relative rate constants are maintained over the whole course of the reaction.

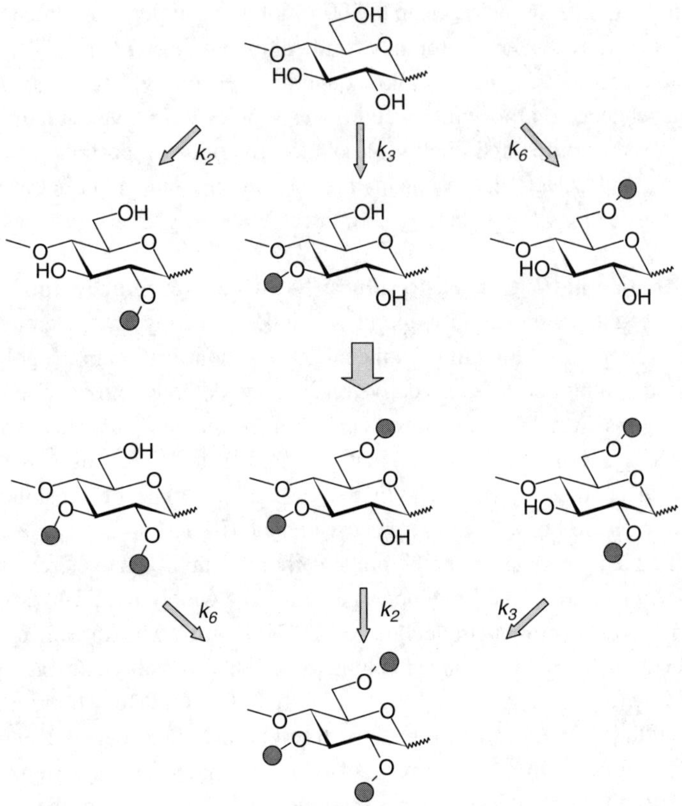

Fig. 5. Monomer patterns of (1→4)-glucans formed by etherification or esterification of OH groups. Reprinted from Ref. 160 with permission from Wiley-VCH Verlag GmbH & Co, Copyright 2008.

He defined a first-order kinetic model and a thermodynamic model, which allow calculation of the monomer composition for every stage of the reaction for a given ratio of relative reactivities k_2:k_3:k_6 or equilibrium constants K_2, K_3, and K_6. However, when applied in analysis, considering a specific sample, the calculation of the expected monomer composition according to this model and its assumptions for a certain set of determined partial DS values x_i is of more interest. In the following treatment, the nomenclature of Spurlin is used. For equal reactivities of the three OH groups (k_2:k_3:k_6 = 1:1:1), the diagram showing the mol fractions c_i ($\sum c_i = 1$) of un- (c_0), mono- (c_1), di- (c_2), and tri-substituted AGU (c_3) in dependence on total DS is symmetrical (Fig. 6). Fig. 7 shows the course of regioisomeric mono- and disubstituted AGU (s_i, i = substituted position) during carboxymethylation of cellulose as

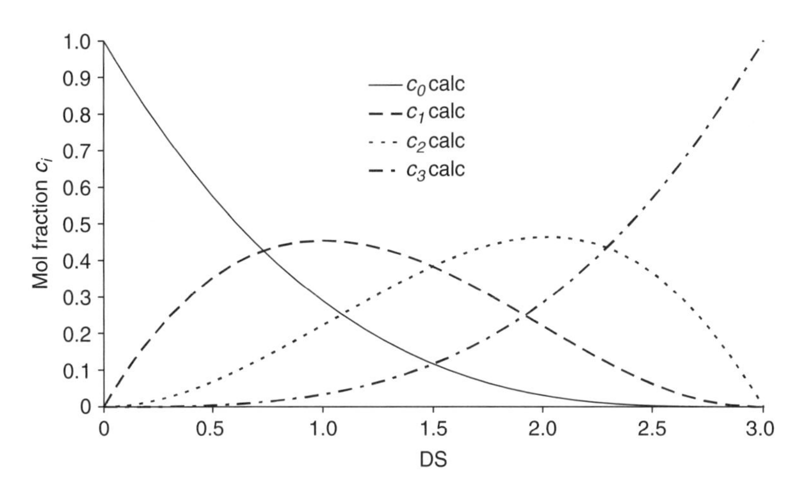

FIG. 6. Diagram (k_2:k_3:k_6=1:1:1) of mol fractions c_i of un-, mono-, di-, and trisubstituted AGU depending on DS.

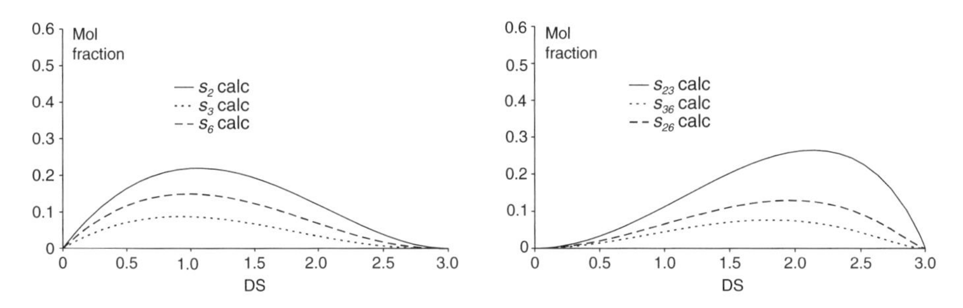

FIG. 7. Diagram (k_2:k_3:k_6=2.14:1.00:1.58) of mol fractions of regioisomeric mono- and disubstituted AGU (s_i) depending on DS, as determined for CMC by Reuben and Conner.[124]

determined by Reuben and Conner.[124] They found the relative reactivities to be k_2:k_3:k_6=2.14:1.00:1.58. Therefore, regioisomers are formed at various extents. Although the difference is not so pronounced for these sets of relative rate constants, it should be noted that the summarized profiles (c_1 and c_2) are no longer mirror-images, as depicted in Fig.6. The more a certain position is favored, the more the curves deviate from symmetry, going through maxima at significantly higher mol fraction c_1, and also c_2. Consequently, a steeper decrease of c_0 and a delayed increase of c_3 are obtained.[162]

It is very important that relative reactivities are weighted (i.e.,k_2:k_3:k_6 $\neq$ 1:1:1), otherwise deviations from the model are preassigned and lead to misinterpretations, especially for reactions of pronounced regioselectivity. When the model is applied to calculate the corresponding random monomer composition for a sample having certain partial DS values x_i ($i=2, 3, 6$; $\sum x_i = \text{DS}$), it is not necessary to differentiate between thermodynamic and kinetic control. Agreement of experimentally determined monomer composition ($s_0, s_2, s_3, s_6, s_{23}, s_{26}, s_{36}, s_{236}$; $\sum s_i = 1$ or 100%) and the calculated result of a randomly acting reaction indicates that Spurlin's model is fulfilled.

Deviations, which can be expressed by an average heterogeneity parameter $H_1 = (\sum \Delta s_i^2)^{1/2}$, only indicate that the model is not fulfilled, but do not answer why (Δs_i is the difference between the experimental and calculated s_i values; it can also be defined as standard deviation, thus dividing summarized Δs_i by n, the number of data.). Various reasons can be differentiated: (i) The assumption of the Spurlin model that primary substitution within an AGU does not influence the reactivities of the other units does not apply. This caused Spurlin to reconsider his model and to extend it,[163] and this was later applied by Reuben to HEC ("Reuben model").[164] (ii) The reaction has followed the Spurlin I model in microdomains, but the reactivities of the macromolecules and certain domains within them have been different, and therefore, several random patterns are superimposed and the resulting average pattern deviates from the random model (broadened/distorted pattern).

This is illustrated in Fig. 8. The reasons for differences in local reactivities can be different accessibilities because of the supramolecular architecture of starch granules or retained crystalline domains in cellulose fibers, and also different local

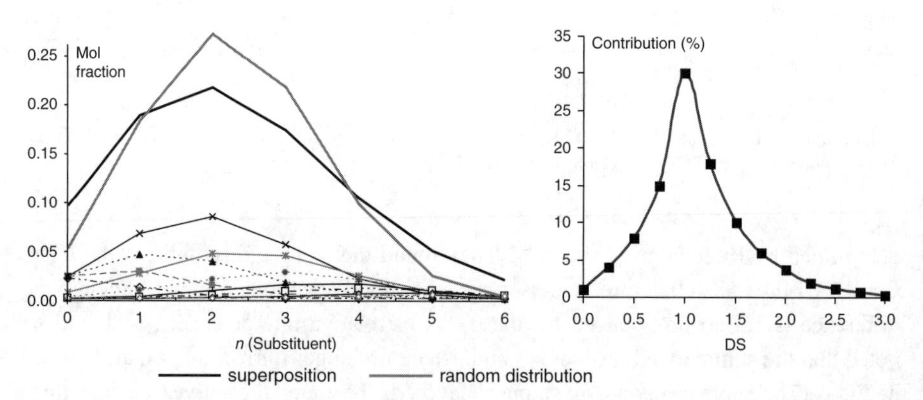

FIG. 8. Left: Broadened substituent distribution (bold black line) in dimers as the result of cumulative superposition of random distributions of various DS ($k_2 : k_3 : k_6 = 1:1:1$, solid and dashed lines with markers); average DS = 1.12; the bold gray line (left figure) presents the random pattern for this DS. Right: Contribution in mol% of glucan derivatives with various DS, from distributions which are superimposed in the left graphic.

concentrations of reagents in heterogeneous reactions or solvent mixtures.[61,62,115,165,166] (iii) Besides these differences in *space*, reactivities can change with *time* through the influence of primary substitution on solubility (thus nucleophilic reactivity) and hydrogen-bonding pattern (rigidity versus conformational flexibility). This time-dependent change is most pronounced for reactions starting heterogeneously and ending up homogeneously. The situation is further complicated by a combination of all these aspects.

If the partial DS values x_i are determined at different times during the course of the reactions, the relative rate constants can be determined from the decrease of concentration of AGU with free OH in position i with time (d[OH-i]/dt).[124] In Fig. 9 $-\ln(1-x_i)$ is plotted against $-\ln(c_0)$, with c_0 being the mol fraction of unsubstituted AGU. The slopes of the graphs determined for a set of carboxymethyl starches (CMSs) represent the relative rate constants of the reactions.[167]

In the case of a defined reactivity-enhancing effect on a hydroxyl group in an AGU by substitution of the vicinal OH (see Section II.3.b), this *conditional probability* can be included in the Spurlin I model. A positive deviation of 2,3-disubstitution in certain cellulose ethers led to further development of the Spurlin model by Reuben.[20] In methyl,[163] ethyl,[168] and hydroxyethyl cellulose,[169] the reactivity of O-3 was found to be enhanced by a factor of 3–5 by primary substitution at O-2. Therefore, an additional rate constant k_3' was introduced (Model II). In contrast to the Spurlin model, a

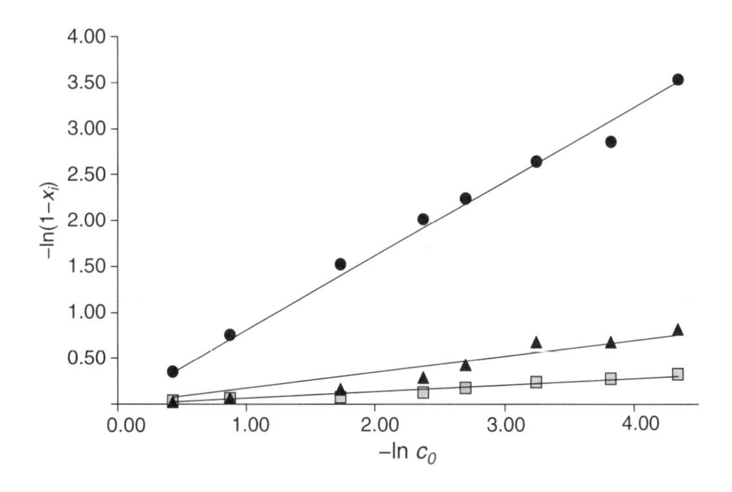

FIG. 9. Plot of $-\ln(1-x_i)$ against $-\ln c_o$ for carboxymethylation of starch for determination of relative rate constants k_i (slope); c_0 is the molar fraction of unsubstituted AGU and x_i the partial DS values in position i. ● O-2, ■ O-3, ▲ O-6. Reprinted from Ref. 167 with permission from John Wiley & Sons, Ltd, Copyright 2002.

complete monomer composition (s_i) is needed to calculate the monomer composition according to the Reuben model. For equations see the original literature.[164] Deviation from the Spurlin model, but agreement with Reuben model, indicates that the reaction can be described as statistic, for which enhancement of k_3 ($\rightarrow k_3'$) has to be regarded as a conditional probability. An inverse effect, namely enhancement of k_2 by substitution at k_3 (Model III $\rightarrow k_2'$), was not found.[164] According to Reuben's NMR studies, this model applies to hydroxyalkylations and alkylations, while carboxymethylations follow the Spurlin I model,[124] as long as requirements for equal accessibility are fulfilled. The enhancement of the reactivity of the adjacent OH in one AGU is a type of neighboring-group effect, but in this context it can be named an *intramonomeric effect*, in contrast to *intermonomeric effects* that act on vicinal AGUs, as illustrated in Fig. 10. Whether specific AGUs intermonomeric effects exist, or whether these are more-local effects, is not known at present. There could also be conceived intermonomeric effects acting through space, for example, on glucose residues of a proximate chain (*interchain effect*) or on an AGU in the next coil of a helix.

For tandem reactions of substituents with such functional groups as hydroxyalkyl, Reuben introduced a further rate constant k'.[170] Oligoether formation does not play any significant role at low MS and is less pronounced for hydroxypropyl as compared to hydroxyethyl. In a comprehensive study Arisz *et al.* have evaluated a set of 16 HECs in the MS range of 0.2–2.4.[171,172] From analysis of these data they concluded that the Reuben model for the hydroxyethylation process is false.[173] However, the models of Reuben or Spurlin are not mere hypotheses, but are calculations for the outcome of a reaction under certain idealizing or simplifying assumptions, as just outlined. They serve as a reference model, and the analytical data may fit in or not. This depends on whether, or to what extent, all the ideal assumptions could be fulfilled. Arisz *et al.* concluded from their own data that oligoether formation (tandem reaction) does not occur with the same relative rate constant as the reactions on the glucose core, and claimed from that to falsify the Reuben model.[174] A key point of this

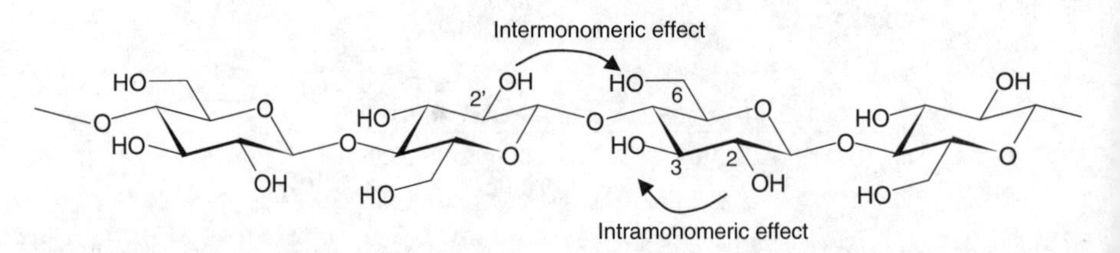

FIG. 10. Possibilities of inter- and intra-monomeric effects refining statistical models by causing conditional probabilities.

data analysis was that the MS/DS ratio decreases with increasing MS of the sample. It was admitted that there is a certain systematical bias, since higher-substituted glucose residues having larger portions of tandem products partially escape determination by gas chromatography. However, it was not considered that, within the detected fractions, those isomers with tandem patterns are discriminated, since they are distributed on many more isomers. More generally, it should be kept in mind that reaction rates depend not only on rate constants but also on concentrations. In hydroxyethylation, there are two different situations: (i) The concentration of glucose hydroxyl groups decreases during the course of the reaction until the maximum DS 3 is reached. (ii) Oligoether formation is a consecutive reaction, the initiator for which is only formed *in situ* in the glucose-etherification step. Therefore, the concentration of starting substrate for this process is 0 and increases up to 3 (at the same time it is 0 for the glucose OH) and remains constant. Consequently, the probability of this reaction (not the rate constant) increases and after a short incubation period reaches a limiting value. Analytical studies of hydroxyalkyl cellulose also show that tandem reaction does not occur to the same extent for all three positions, and that reactivity is not directly decoupled from the sugar core positions in the first tandem step.[175,176] However, it is difficult to interpret this final static result with respect to the dynamic of the reaction, since the substrates for chain elongation (2-, 3-, or 6-*O*-hydroxyethyl-glucose residues) are also formed at various rates for the individual positions. In the comprehensive data-analysis of Arisz,[173] the experimental results imply a decrease of rate constant k_x of oligoether formation with increasing MS, which is at least in contradiction to the increasing statistic probabilities. For explanation, the authors referred to the NaOH redistribution hypothesis of Yokota in water–alcohol mixtures, as already mentioned.[61]

b. Substituent Distribution in the Glucan Chains: Models and Terminology.— The various substituted glucose residues present in certain molar ratios can theoretically be arranged in a nearly unlimited number of sequences. For a sequence of 20 glucose residues, the order of magnitude is already 10^{18}. While taking into account that, in practice, many of these sequences do not exist in significant amount, it can easily be imagined that there are no two identical molecules in such a mixture. No sequence analysis is feasible, but the composition can best be described by statistical models, or by patterns and profiles.

If the polymer is degraded to its monomeric constituents to determine their molar ratios and the average DS, the results cannot be directly related to the structural features of the bulk material, since the information about the monomer arrangement in and over the polymer chains is lost. However, knowing the partial DS values x_i,

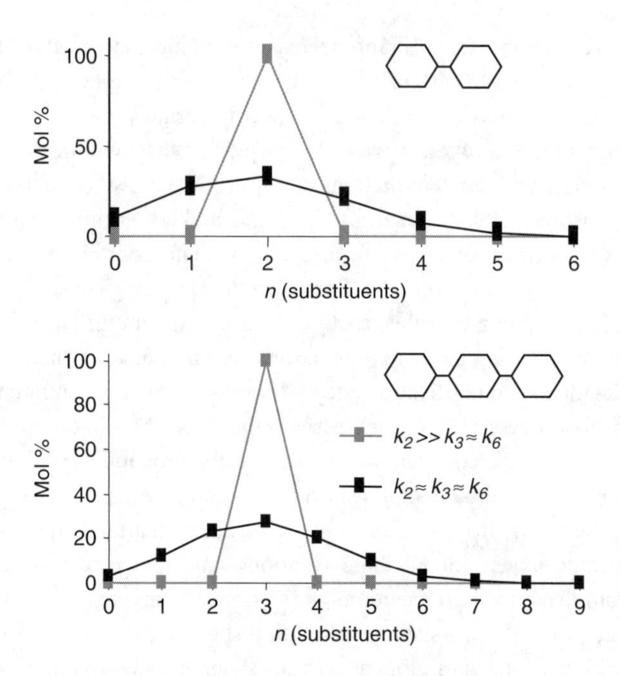

FIG. 11. Substituent distribution in dimers and trimers, calculated for very high (gray lines) and no (black line) regioselectivities at DS 1. Reprinted from Ref. 160 with permission from Wiley-VCH Verlag GmbH & Co, Copyright 2008.

presenting the reactivities of positions 2, 3, and 6, it is possible to calculate the composition of the polymeric material, including the chemical dispersity over the polymer chains. For a material that fulfills the Spurlin assumptions, the latter is determined by the regioselectivity of the derivatization reaction as the result of kinetic or thermodynamic control (Fig. 11).[160]

There is some confusion in the literature concerning the terminology used to classify the substitution pattern of polysaccharide derivatives. *Homogeneous, uniform*, and even *regular* have been used to describe patterns resulting from reactions performed homogeneously. But it is evident, as already outlined, that such reactions will produce a *random* or nearly random ("by chance") pattern, the individual substituent-distribution of which depends strongly, and (if the Spurlin model is fulfilled) exclusively, on the regioselectivity of the reaction. Since glucan derivatives are always mixtures of molecules having a DP and DS distribution, that is, with a certain dispersity with respect to size and chemistry, they are never uniform or homogeneous in the strict sense of having one structurally defined compound at hand. However, if a glucan is separated from a xylan, this is usually termed "purification to homogeneity." Spurlin also addressed this problem and defined uniformity as "equality of ease of access of the reagents to the individual anhydroglucose units," while from a technological point of view uniformity means uniform behavior with respect to a certain property, such as complete solubility in a certain solvent.[163] Following these terminologies, a random mixture in accordance with the Spurlin model on an extended structural level could be assigned as *homogeneous* or *uniform*, but to avoid misinterpretations, *random* is preferred. Since these random distributions can be calculated, they can be used as a reference model in analysis (see next). Deviations from this random patterns were defined as *more heterogeneous, more regular* (than random), or *bimodal*, to differentiate them from the inherent or natural heterogeneity of polysaccharide derivatives.[166] These types are illustrated in Fig. 12. As already mentioned, (Section II.5.a, Fig. 8), heterogeneity may be the result of superimposing random patterns from different local reactivities, which give a distorted graph.

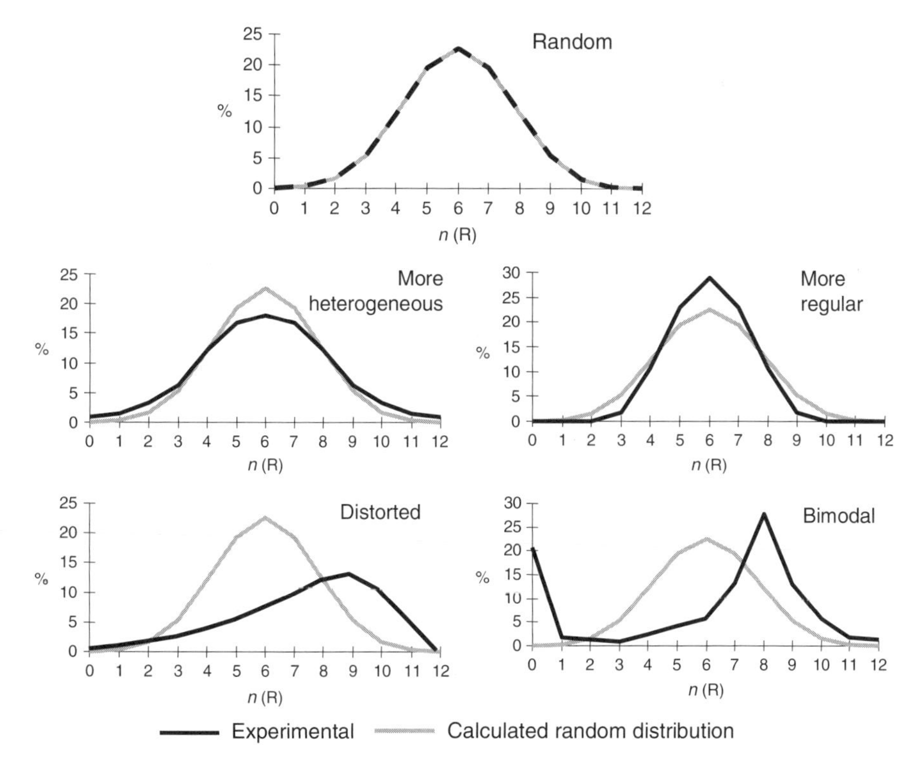

FIG. 12. Patterns (random, more heterogeneous, more regular, distorted, and bimodal). DP4 for a sample with DS 1.5 is shown; n (R) is the number of substituents R. Amounts are given in mol% AGU.

Bimodal patterns are obtained if certain domains within the material, such as crystalline areas, show very low reactivity, and thus remain nearly unaffected,[175–178] or two competing reaction mechanisms, such as heterogeneous and homogeneous, are followed in parallel. As long as no permethylation is performed, bimodal patterns with a very high substituted fraction, besides a more-randomly substituted one, are formed in alkylations with solid base in polar aprotic solvents.[165,166] The most prominent example is pulverized NaOH as used in the Ciucanu permethylation procedure.[179] This is a very significant observation, which was explained as follows: The NaOH particles are dispersed but not dissolved in Me_2SO. Dissolved polysaccharide chains are adsorbed on these particles according to the model of Fleer and Scheutjens[180] forming trains, loops, and tails hanging in the solution. These molecules are thus exposed to a high local base concentration and are presumably deprotonated in the adsorbed domains (trains) and subsequently alkylated, while pendant molecules or those staying in solution do not come in contact with base and thus remain inactive.

In addition to the distribution along a polymer chain, there is always a distribution of DS over the macromolecules in a mixture. These structural levels are denoted as *heterogeneities of the first order* (over the various polymer molecules) and *second order* (along a single polymer molecule). These heterogeneities are closely interrelated, and their theoretical profile can be calculated, as long as there is no distorting effect from the hierarchical order of structure. To differentiate experimentally between these two types of heterogeneities, the material must be fractionated with respect to DS prior to further analysis, since investigation of the whole mixture always yields only average data. As already mentioned, the random distribution naturally implies a certain dispersity in chemical structure, which depends on regioselectivity of the reactions and is interrelated on all structural levels.

This differentiation is very important in the case of the bimodal patterns just mentioned. These types of products from Me_2SO–NaOH–RX reaction systems show an unusual substituent distribution and have been described as block-like structures.[53,165] As long as bimodality is located in the same polymer molecule, and thus presents a heterogeneity of second order, this description fits. However, if the bimodality is of first order, we have a blend. Fractionation studies showed that bimodality is observed in both respects.[39]

c. Topochemical Effects and Heterogeneities in the Bulk Material.—As mentioned in Sections II.1 and II.2, both cellulose and starch feature supramolecular structures, fibers and granules, respectively, that may cause different susceptibilities. In the industrial modification of cellulose, the usual aim is to swell and destroy crystallinity and thus activate the fibers and fibrils down to the molecular level by treatment with strong alkali (aqueous NaOH) to achieve equalized starting conditions. However, minor proportions of insoluble particles are often retained in water-soluble cellulose ethers. Germgård *et al.* identified such undissolved residues in CMC mainly as swollen cell-wall parts from thickwalled compression wood, which are activated and penetrated much more slowly because of diffusion resistance.[181] On the laboratory scale, derivatizations are also performed in solution (with DMAc–LiCl, Me_2SO–TBAF, or ionic liquids)[32-34,36,37,81-84] to overcome topochemical effects. However, by use of structure-selective derivatization of cellulose, it was claimed to achieve a block-like substitution mode.[178] By using the less efficient alkali activation, only the amorphous domains should be affected and consequently react with chloroacetic acid, while the higher-ordered and denser crystalline domains should be retained.

Besides the differentiation between crystalline and amorphous regions, the question arises as to whether there is a gradient from the outer to the inner part of the fibers or granules. As already mentioned, starch granules show various kinds of heterogeneity, depending on the botanical source and specific cultivar, and also exhibit variations in morphology for a single source.[93,95] Whether or not there is finally a DS-gradient over the granule depends on the relative rates of diffusion and penetration of the reagents into the granule and the chemical reaction.[182,183] Pores and channels connecting the granule surface with an internal cavity are present in corn and sorghum starch granules, but not in potato-starch granules. They facilitate the transport of reagents into the granule interior. Different rates of dye diffusion into swollen granules have been demonstrated. The highly reactive phosphoryl chloride reacted to a large extent on the peripheral and cavital granule surfaces.[93,184] In contrast, sulfomethyloxirane, a less reactive agent, traveled through the channels and diffused from the inside throughout the whole granule, thus reacting more homogeneously and uniformly with respect to the granule size.

Starch granules on the whole are about 70% amorphous. In some cases there is good agreement between double-helical content (amylopectin side-chains) and crystallinity, while for some, such as rice and potato starch, the degree of crystallinity is lower, indicating that not all double-helical side-chains are involved in extended crystalline areas.[185] Results obtained by various groups indicate that the amorphous amylose and the less-dense area of branching points in amylopectin layers, as well as the terminal residues of the pendant chains, are more accessible. They are more highly substituted than the crystalline double helices in such wet processes as the slurry or paste modifications.[96a,b,186–190] Different kinetics due to a change from diffusion to kinetically controlled reaction and the possible influence of diluents on local concentrations of base have already been mentioned (see Section II.3.c).[61,62,115–118]

III. Structure Analysis of (1→4)-Glucan Derivatives

The complexity of glucan derivatives and the structural parameters of influence on the reactivity and properties of products have now been outlined. Next discussed is how this challenging complexity is faced and handled in analysis. Concepts and experimental approaches to achieve quantitative information on various structural levels and interpretability of experimental data are also discussed.

1. Average Degree of Substitution

a. Definition of DS and MS.—The degree of substitution, DS, of polysaccharide derivatives is defined as follows:

$$\mathrm{DS} = \frac{n(\text{substituted OH})}{\mathrm{AGU}} \tag{1}$$

$$\mathrm{DS} = (3{\cdot}c_3 + 2{\cdot}c_2 + c_1) \tag{2}$$

$$c_0 = s_0 \tag{3}$$

$$c_1 = s_2 + s_3 + s_6 \tag{4}$$

$$c_2 = s_{26} + s_{36} + s_{23} \tag{5}$$

$$c_3 = s_{236} \tag{6}$$

$$\mathrm{DS} = 3{\cdot}s_{236} + 2 \cdot (s_{26} + s_{36} + s_{23}) + (s_2 + s_3 + s_6) \tag{7}$$

$$\mathrm{DS} = \Sigma x_i = x_2 + x_3 + x_6 \tag{8}$$

$$x_2 = (s_2 + s_{23} + s_{26} + s_{236}) \tag{9}$$

$$x_3 = (s_3 + s_{23} + s_{36} + s_{236}) \tag{10}$$

$$x_6 = (s_6 + s_{26} + s_{36} + s_{236}) \tag{11}$$

where s_i are the mol fractions of the AGU substituted in positions *i,* and $\sum_{s_i} = 1$. Usually, the monomer composition of a glucan derivative is given in mol%, namely, $\sum_{s_i} = 100$, which can be simply converted. As already introduced, x_i are the partial DS values in position *i*. For derivatives undergoing tandem reaction, a molar degree of substitution (MDS), is defined as:

$$\mathrm{MDS} = \frac{i(\text{substituents})}{\mathrm{AGU}} \tag{12}$$

$$\mathrm{MDS} = \sum_{i=0}^{i=\infty} i \cdot ci \tag{13}$$

The number of substituents (*i*) means, for example, the number of ethylene oxide molecules belonging to one AGU. For the following component of HEC, the DS is 2 and the MDS is 3:

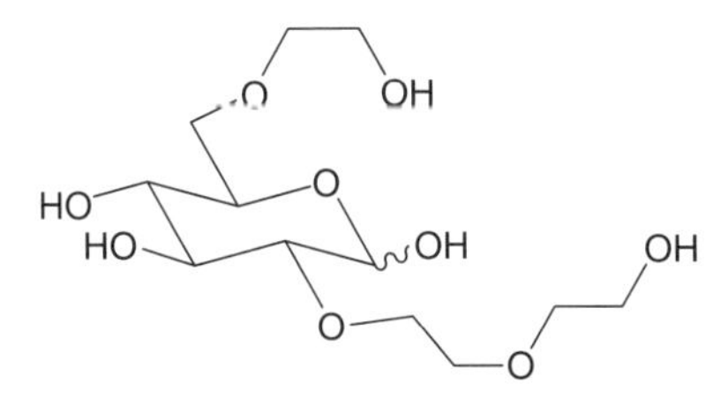

The MDS/DS ratio is an important structural parameter in the more-highly substituted HECs and HPCs.

b. Zeisel Method.—The average DS of a cellulose or starch derivative can be determined by various methods. The so called Zeisel method is very commonly used in industry to determine the DS or MDS of alkyl and hydroxyalkyl starches, and cellulose ethers. It is based on ether cleavage with hydriodic acid. Alkenes and alkyl iodides are produced and are quantified by GLC, from which the amount of methoxy or hydroxyalkoxy groups is calculated. This method is widely used and is described in the United States' and the Japanese Pharmacopoeias.[15] This technique has been found for HPC and HPMC to give results in good accord with determinations by ^{13}C-NMR and Raman spectroscopy.[14]

c. Elemental Analysis.—Elemental analysis is especially used for derivatives containing heteroatoms (N, S, Si, or halogen). However, the literature often only gives the content of this heteroatom, and this is taken as an absolute measure for calculating the DS according to the following equation, with $M_{\text{Subst.}}$ = mass of the substituent:

$$\mathrm{DS} = \frac{162 \cdot \%\ \text{Subst}}{M_{\text{Subst.}} - [(M_{\text{Subst.}} - 1) \cdot \%\ \text{Subst.}]} \tag{14}$$

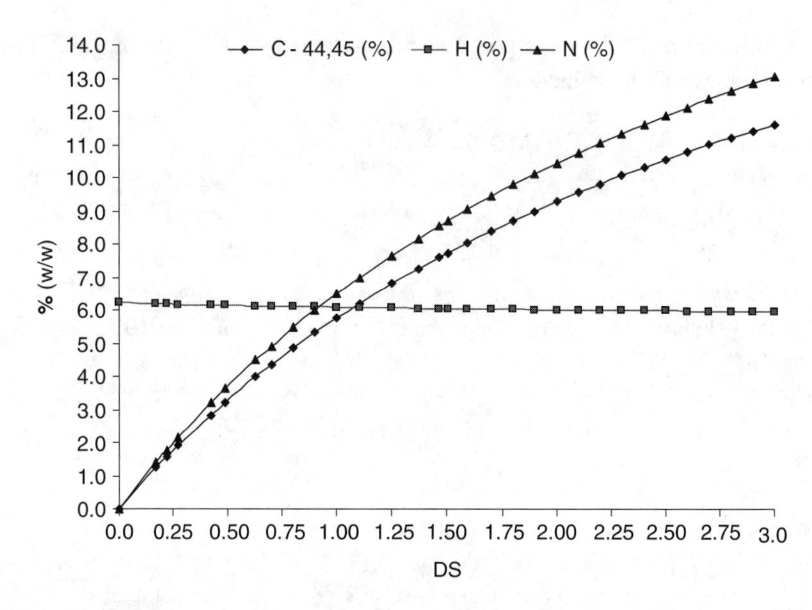

FIG. 13. C, H, and N (%) of *O*-cyanoethyl glucans, depending on DS. For better comparison the graph for C (44.45% at DS 0) is shifted to 0%. (See Color Plate 26.)

A more sophisticated computer-based approach for estimation of DS from elemental analysis, especially for derivatives with more than one type of substituent, was reported in 1971.[191]

However, the elemental composition can be influenced by humidity, by-products of the reaction (such as homo-oligomers or hydrolysis products of the reagent) or residual reagents adsorbed or embedded in the polymer, by salts, or by different forms of ionic derivatives (free amine, acid, or salt, various counter ions). As long as only inorganic salts, which do not contain the atoms of relevance, or humidity, are present, the ratio of carbon to the heteroatom, such as C/N, can be used for calculation of DS. To confirm the validity of elemental analysis data, all elemental measures and their ratios should be considered, becoming aware of inconsistencies. Fig. 13 shows the development of absolute C, H, and N content in cyanoethyl glucans as function of DS. As seen in Fig. 14, there is a steep decline of the C/N ratio in the low DS range, while at DS >1 the sensitivity becomes very low.

d. NMR Spectroscopy.—^{1}H- and ^{13}C-NMR spectroscopy are widely used tools for determinations of DS, but mainly in the field of the non-branched and higher-substituted celluloses, and especially for esters. The great advantage of NMR spectroscopy is that the intact polysaccharide can be analyzed, avoiding any bias of

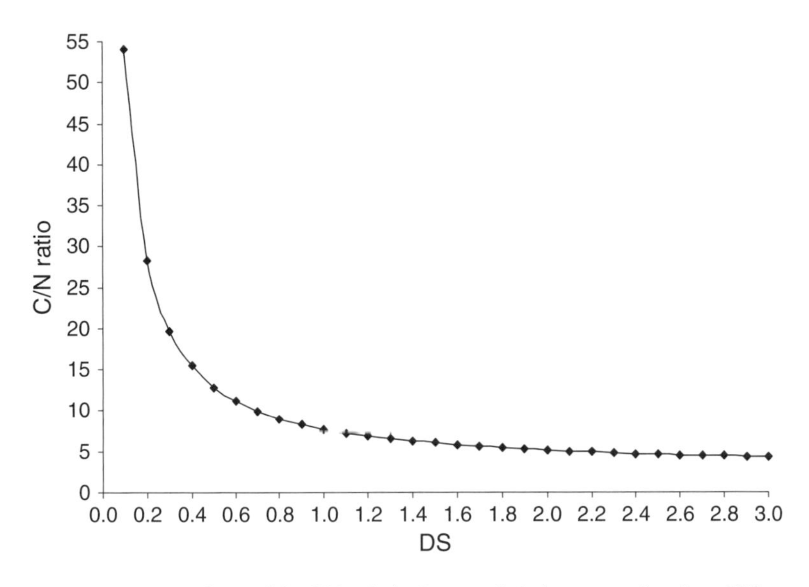

FIG. 14. Dependence of the C/N ratio in *O*-cyanoethyl glucans as a function of DS.

depolymerization and further sample-preparation steps. However, at the same time, the polymeric nature also is a drawback. Because of higher viscosity and the character of a mixture, signals are broad, often poorly resolved, and unsymmetrical. Viscosity and peak broadening can be partly compensated by employing low concentrations, but at the expense of a lower signal-to-noise-ratio. Therefore, polysaccharides have been partially or even completely degraded by enzymes, ultrasonic treatment,[192] or acid hydrolysis,[124,163,164,168,169,193] resulting in better resolved, but more-complex NMR spectra. As in chemical analysis, miscalculations due to impurities may occur if signals of interest are superimposed upon signals of side products, as, for example, glycolate (2-hydroxyacetate), the hydrolysis-product of sodium chloroacetate in carboxymethylation (Fig. 15).[167]

Nehls *et al.*[194] investigated solvents suitable for NMR and differentiated between nonderivatizing and derivatizing solvents, such as trifluoroacetic acid (TFA), which in the absence of water dissolves cellulose and converts it into its trifluoroacetate. While ^{1}H NMR profits from higher sensitivity and quantifiability under standard measurement conditions, signal assignment is much easier in ^{13}C NMR. ^{1}H-NMR spectroscopy is applied when the substituent offers a well resolved signal that can be related to the (usually separated) signal of the anomeric protons. Since the latter is shifted downfield by 2-*O*-substitution, but is not sensitive to changes at the 3- and 6-positions, the partial DS at O-2 (x_2) can additionally be obtained. Substituents suitable for DS

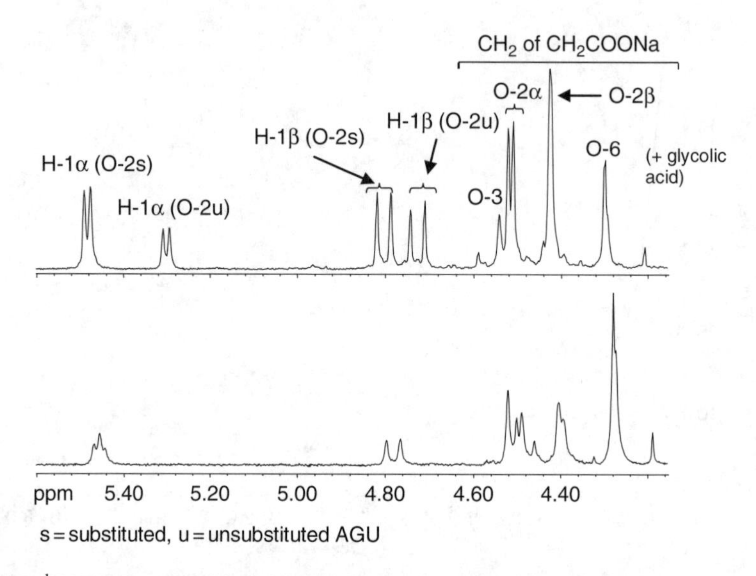

FIG. 15. ^{1}H-NMR spectra of hydrolyzed sodium carboxymethyl starches with low (top) and high (bottom) DS. Reprinted from Ref. 167 with permission from John Wiley & Sons, Ltd, Copyright 2002.

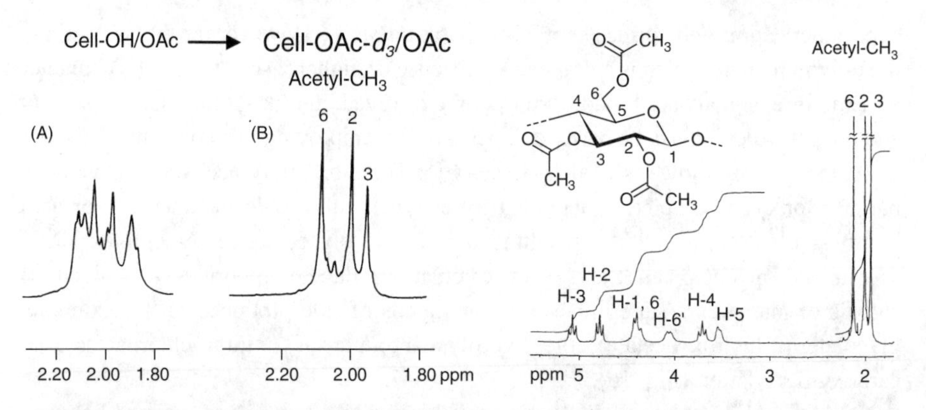

FIG. 16. ^{1}H NMR of cellulose triacetates. Left: Effect of deuteroacetylation on chemical shifts of CH_3 (Ac) signals (A→B). Right: Unlabelled triacetate. Reprinted from Ref. 145 with permission from Wiley-VCH Verlag GmbH & Co, Copyright 1991.

estimations are quaternary ammonium in cationic starches or hydroxypropyl residues. Methylene groups of cyanoethyl-, aminopropyl-,[151] or carboxymethyl ethers can also be used.[124] Beside the CH_3 group of acetates, the glucosyl ring protons are shifted downfield by acylation and are therefore much better resolved (Fig. 16). Deus *et al.*[145a,b] improved the resolution of the methyl signals by deuteroperacylation, thus compensating the slight differences of resonance due to different substitution patterns of the neighboring glucose residue. An analogous approach had been used by Miyamoto *et al.* in ^{13}C NMR spectroscopy of cellulose acetates.[195] Chemical compensation of differences has successfully been achieved by Liebert *et al.* by carbamoylation, especially with ethoxycarbonylmethyl isocyanates prior to ^{1}H-NMR spectroscopic analysis of cellulose esters.[196] Carbamates are stable toward acid hydrolysis and can additionally be applied for HPLC analysis of monomers, presenting the complementary substitution pattern.

^{13}C NMR spectra must be recorded under conditions that allow quantitative evaluation of the signals. Partial DS values x_i are obtained from the relative ratios of a certain carbon signal and the corresponding signal shifted by substitution. The C-6 signal is shifted downfield and that of C-1 is shifted to higher field by substitution of the adjacent O-2. The value of x_3 is often difficult to determine because of the low reactivity at this position and poor resolution of the C-3′ signal. These problems have been overcome by full acylation of the free OH groups of cellulose derivatives.[145a,b,194,197–199] Solubility is improved; viscosity is decreased; and sensitivity, uniformity of chemical shifts, and resolution are all enhanced. In addition, the carbonyl C=O signals are separated for all three positions and their ratios can also be used to calculate the distribution on the three positions available. This approach is most successful in the case of cellulose esters, using a different acylating reagent, such

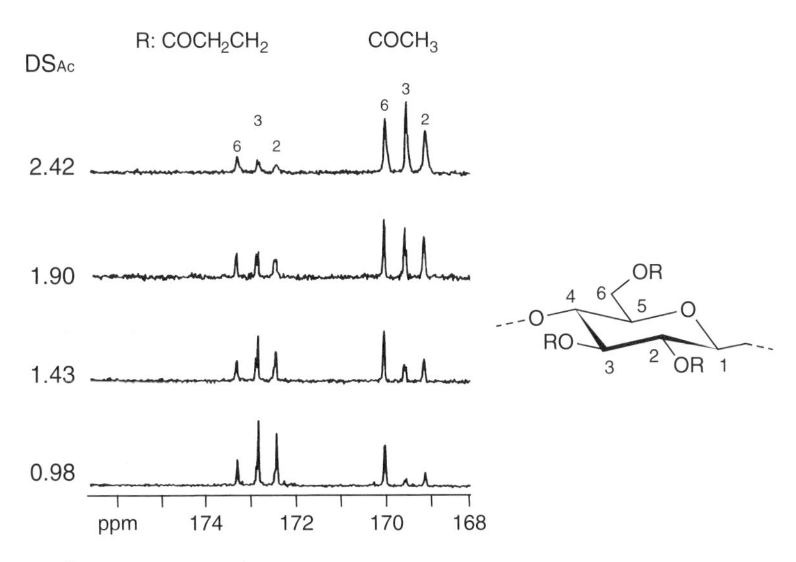

Fig. 17. ^{13}C spectra of cellulose acetates after perpropanoylation. Reprinted from Ref. 197 with permission from Elsevier, Copyright 1995.

as propanonyl chloride for acetates and vice versa (Fig. 17).[35,145a,b,197–200] For more details and applications, the reader is referred to the cited literature.

Regioselectively functionalized 2,3-*O*-hydroxyalkyl ethers are mentioned as an example of a detailed one- and two-dimensional ^{13}C-NMR spectroscopic study of cellulose ethers.[67,200,201] Tri-MC has also been thoroughly studied.[202,203] While NMR spectroscopy in solution clearly dominates this field, solid-state NMR has also been applied, but mainly for the analysis of crystallinity and interaction with plasticizers or water.[178,204,205] A fundamental study of solid-state NMR of selectively *O*-methylated celluloses obtained from *de novo*-synthesis[206,207] has been published, which establishes the basis for rapid analysis of methyl patterns in methyl celluloses.[208a,b]

e. Miscellaneous.—Depending on the nature of the substituent, additional methods can be applied. Acidic derivatives, such as CMC, can be titrated. The pK_a value of CMC is between 4 and 5, depending on the DS.[209,210] The average DS of acetates can also be determined, after saponification, by titration.[195,211] DS of derivatives bearing a chromophore (benzyl ethers, benzoates) can, in principle, be analyzed photometrically, but the method requires calibration. The content of amino groups can be measured after reaction with amine sensitive reagents that give colored products.[212] The average DS may also be obtained from monomer analysis, as is outlined in the next section.

2. Monomer Composition: Substituent Distribution in the Glucose Residue

The basis for investigating the substituent distribution at all structural levels is the analysis of the monomer composition, and thus the molar proportions of the eight, or more, different glucosyl monomers. While NMR spectroscopy offers in many cases the possibility for deducing partial DS values x_i (see III.1.d), starch or cellulose derivatives have usually to be depolymerized to their monomeric constituents for subsequent separation by chromatographic or electrophoretic techniques. There have been numerous publications employing cleavage of all glucosidic linkages and subsequent separation and quantification of appropriate derivatives. Aqueous acid hydrolysis, methanolysis, and reductive depolymerization with Lewis acids and triethylsilane according to Gray[213] are commonly applied. The methods used must always be adapted to the stability and nature of the functional group.[39] HPLC coupled with various detectors, high-pH anion-exchange chromatography with pulsed amperometric detection (HPAEC/PAD), and capillary electrophoresis (CE) with detection by UV (ultraviolet) or induced fluorescence have been applied. Together with these techniques, gas–liquid chromatography with flame-ionization detection (GLC/FID) is still the method of choice, as the well-established coupling with MS allows the identification of the constituents. In addition, the effective-carbon-response (ECR) concept[214–217] allows for quantification without the need for calibration with standards that may not be readily available.

a. Separation after Complete Depolymerization.—The separation efficiency of HPLC is usually not sufficient for separation of all mono- and di-substituted regioisomers. In addition, the diastereomeric α and β anomers may interfere. Therefore, it is often better to restrict selectivity to a DS-dependent separation. Thus, for estimation of the DS from the mol fractions c_i, HPLC was successfully applied with methyl-cellulose and carboxymethyl-cellulose.[66,218] As an alternative, sugars can be labeled with a chromophore by reductive amination, thereby changing both the retention behavior on a reversed-phase column and the detection abilities.[219] In particular, HPAEC/PAD has been applied to such anionic ethers as CMC[120,220] and sulfoethyl-cellulose.[221] Furthermore, since sugar OH groups are deprotonated under the strongly alkaline conditions, separation of such nonionic water-soluble ethers as *O*-methyl glucoses can also be achieved.[222] A drawback of pulsed amperometric detection is the strong decrease in sensitivity caused by the blocking of OH groups by substitution. Thus, the relative response of the trisubstituted glucose is only 10–15% of that of free glucose and, depending on the potentials used in the PAD, it can be even more suppressed. The aldoses are degraded by stepwise oxidation from C-1 to formic acid,[223] and this oxidation is obviously blocked by substituents. Thus, standards are required for calibration, and the linear range of the PAD is narrow. However, for routine analysis of same type of samples, the method is successful.

CE requires a charged analyte, and usually a chromophore or fluorophore for detection. As with HPAEC, polar and ionic derivatives are applicable with this method. However, with carbohydrates, ionic character is not required, since the formation of borate esters of diol groups (stability: *cis* > *trans,* and 1,2 >> 1,3) in an alkaline buffer provides the ionic character.[224,225] The differences in average charge of regioisomeric analytes depend on the relative dissociation constants of the available diol-borates. Thus, separation follows a different and more regiospecific interaction than in GLC, which offers an attractive alternative for regioisomers that are sometimes difficult to separate (Fig. 18). The carbonyl function of the free glucose can be used for labeling with a chromophore by reductive amination (Fig. 19). Fortunately, there is no significant influence of the substitution pattern on the molar absorption of the labeled analytes.[167,226] Therefore, for quantitative evaluation, peak areas need only to be corrected for their migration (and hence residence time) in the detector window. Assignment of analytes requires standard compounds or reference methods. Although CE–MS coupling is available, borate buffer interferes with electrospraying.

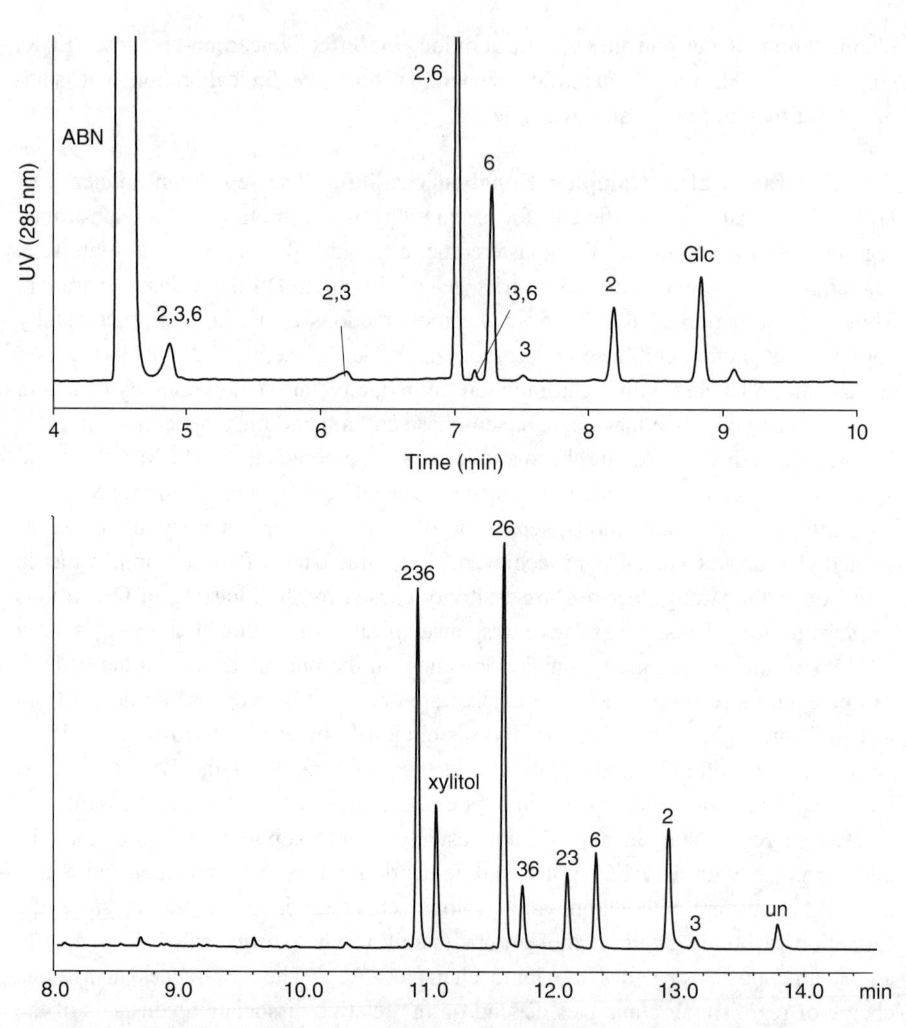

FIG. 18. Separation of glucitol derivatives labeled with aminobenzonitrile (ABN), obtained from methyl starch by capillary electrophoresis (CE/UV, top) and from glucitol acetates obtained from methyl cellulose by gas chromatography (GLC/FID, bottom; xylitol = internal standard).

When hydroxyalkyl ethers, or their combinations with alkyl groups, are the object of investigation, the complexity of monomer composition requires capillary GLC coupled with MS. This procedure usually entails a higher demand on sample preparation, since volatile, thermally stable analytes, and ideally only *one* peak for *one* pattern, are required. Sequential hydrolysis, reduction, and acetylation, either direct[218]

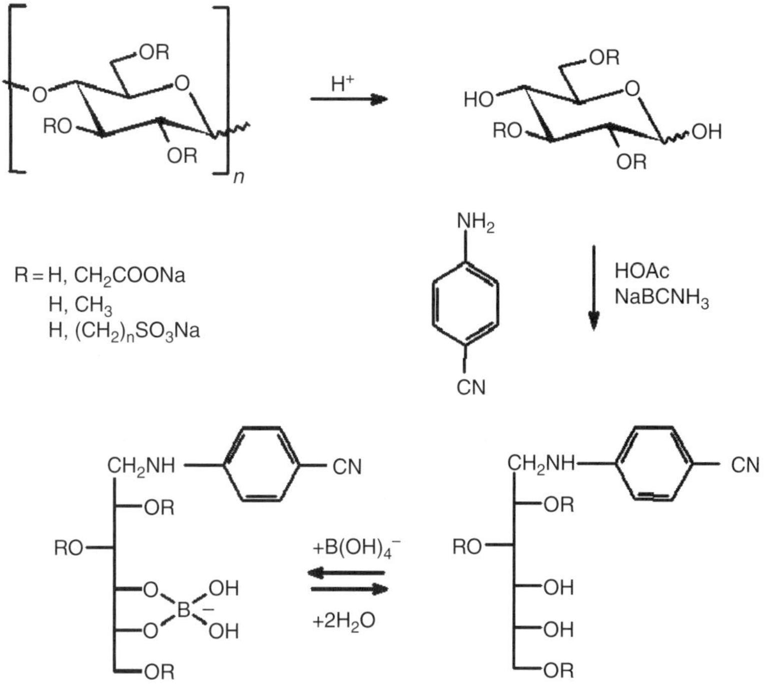

FIG. 19. Reductive amination for separation of carbohydrate derivatives in an alkaline borate buffer by capillary electrophoresis.

or after permethylation, have been widely applied, for example, also for derivatives with acid-labile substituents as esters (acetates, sulfates)[227,228] or silyl ethers,[140] thus enabling indirect determination of acid-labile groups from the complementary methyl pattern.

Reductive cleavage, introduced by Gray *et al.* in the early eighties as a more convenient one-potalternative to conventional methylation analysis, has also been successfully applied to a number of permethylated cellulose ethers and esters[121,229–235] (see also Fig. 22).

Special methods have been developed for such ionic compounds as sulfobutyl[236] ethers and cationic starches.[131,132] To identify the substituted position from typical shifts in mass spectra, it is important to avoid coincidence of the mass contribution of functional groups as, for example, with propyl and acetyl. For instance, glucan allyl ethers cannot be hydrolyzed directly, since the 2-*O*-allyl group forms an intramolecular 6-membered cyclic adduct with the carbenium ion formed during hydrolysis.

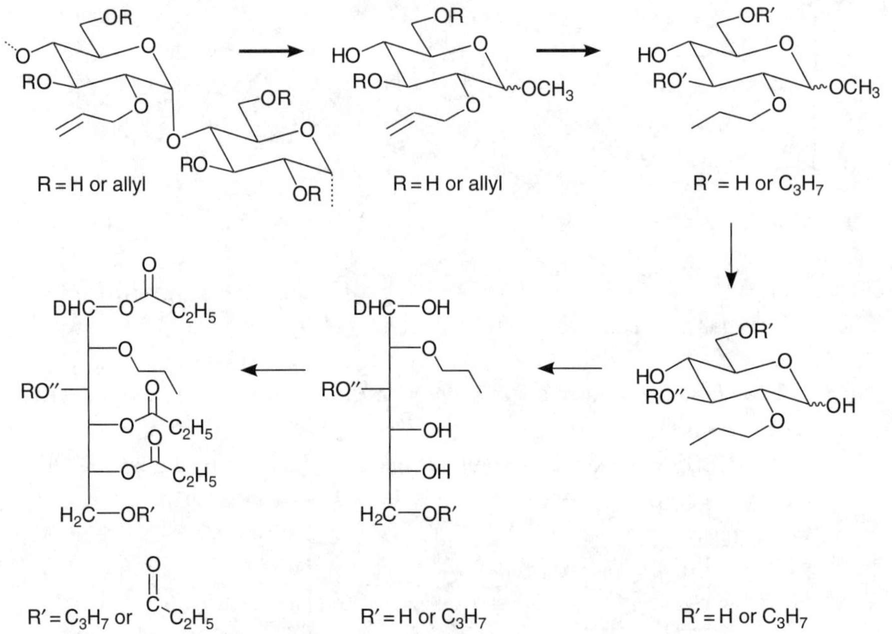

FIG. 20. Modified sample preparation of alditol acylate for GLC monomer analysis of *O*-allyl glucans.[237]

Therefore, such derivatives are methanolyzed, and then hydrogenated (→*O*-propyl), hydrolyzed, and reduced to glucitols, which finally are propanoated (Fig. 20)

b. Sample Preparation.—The details for certain derivatives can be found in the references cited and have also been comprehensively reviewed,[39,238] but some general requirements and critical points should be addressed here. Each step of sample preparation comprises a potential bias of the analyte mixture. It is not essential to achieve a yield of 100% per step, although this would be the ideal case, but it is more important to avoid specific losses with respect to polarity (extraction steps), volatility (evaporation steps), or side reactions (such as incomplete methylation, formation of anhydro sugars, or formation of reversion products). Concerning the latter, hydrolysis and subsequent work-up is a critical step. Dehydrations (eliminations and condensations) might occur, forming 2-(hydroxymethyl)furfural, which can polymerize to colored products. During evaporation of acid, the TFA often used concentrates in the residual water and finally can cause formation of reversion products, a thermodynamically controlled reaction, favoring (1→6)-glycosidic linkages. Formation of 1,6-anhydro-β-D-glucose can also cause a bias with constituents having free primary OH groups.[239,240] Besides TFA, the most widely used hydrolytic agent, perchloric acid has also been employed, especially for hydrolysis of CMC [241] and has been compared with the traditional hydrolysis by sulfuric acid.[242] Further information can be found in the cited literature.[35,39,243,244] Hydrolysis yields α,β-glucose derivatives, which are most commonly reduced to the glucitol derivatives and acetylated for GLC or reductively aminated for CE/UV. Utilization of aqueous hydrolysis is limited by the hydrophobicity of substituents, causing a severe underrating of more highly substituted glucose residues, and by intramolecular reactions of reactive groups in appropriate positions, for example, allowing the formation of a 6-membered ring.

Examples for the latter are 2-(2-hydroxy)alkyl (Fig. 21) and 2-allyl ethers (see also Figs. 20 and 21), forming intramolecuar acetals and addition products, or derivatives with strongly nucleophilic functions at C-6, such as 6-amino-6-deoxy derivatives, which are even more prone to 1,6-anhydroglucose formation than glucose itself.

In such cases, methanolysis with dry methanol/HCl is preferred, as this inhibits most of the intramolecular side-reactions mentioned[245] and enables better dissolution and accessibility of less-polar samples. The methyl glucosides obtained are less sensitive to side reactions, but cannot be further reduced to alditols or labeled by reaction of their carbonyl function.

While hydrolysis and methanolysis do not essentially require full protection of the OH groups, permethylation is a prerequisite for reductive depolymerization

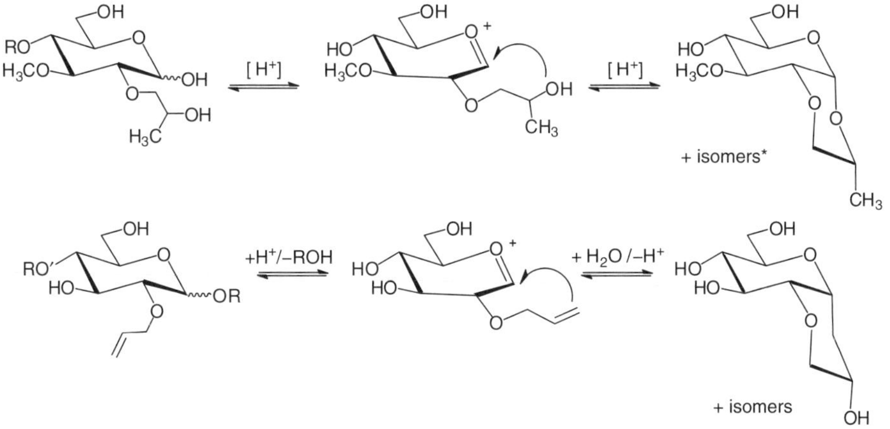

FIG. 21. Intramolecular side reactions during hydrolysis of 2-*O*-hydroxalkyl and 2-*O*-allyl ethers of glucans. Intramolecular acetal or addition product formation. *, favored isomer and conformation according to Lee and Perlin.[245] 1,2-*O*-[1,2-methyl-(*S,R*)-1,2-ethanediyl]-α-D-glucofuranoses and 1,2-*O*-[1,2-methyl-(*S*)-1,2-ethanediyl]-α-D-glucopyranose are also formed in equilibria.

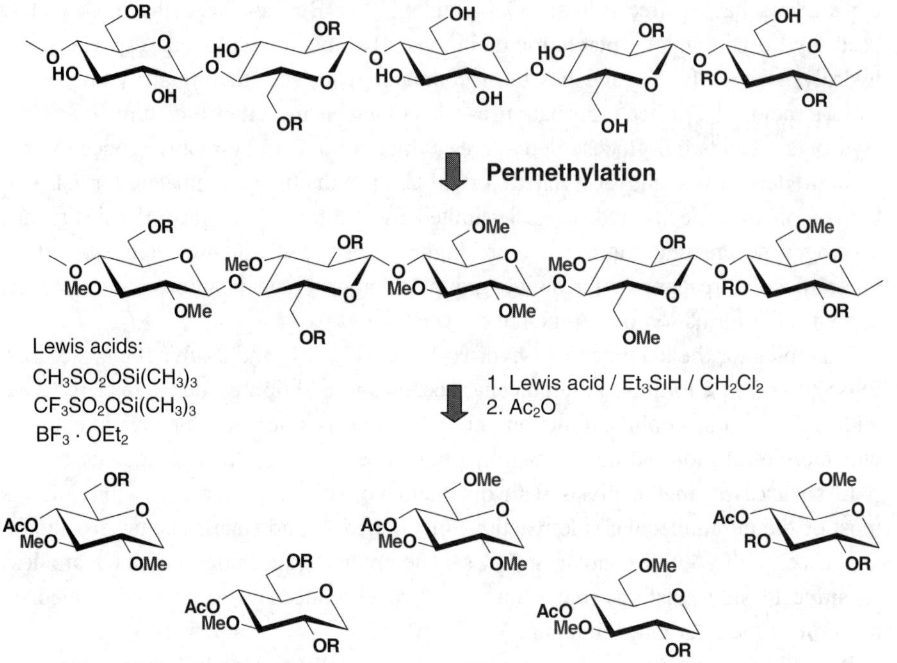

Permethylation

Lewis acids:
$CH_3SO_2OSi(CH_3)_3$
$CF_3SO_2OSi(CH_3)_3$
$BF_3 \cdot OEt_2$

1. Lewis acid / Et_3SiH / CH_2Cl_2
2. Ac_2O

FIG. 22. Reductive depolymerization of cellulose derivatives after permethylation.

(Fig. 22),[121,229,230,232–235] where the Lewis acids employed will otherwise lead to serious side reactions. Advantages of the reductive cleavage method in the analysis of starch and cellulose derivatives are (i) speed of this two-step one pot-procedure performed at room temperature, (ii) preservation of ring size and acyl residues under optimized conditions, and (iii) optional introduction of trialkylsilyl groups (TMS, TES) instead of acyl residues at the released OH positions. Drawbacks are (i) the essential requirement of permethylation, (ii) moisture sensitivity of the reagents, (iii) potential promotion/catalysis of rearrangement and isomerization reactions, (iv) potential discrimination of intermediate higher *O*-silylated products (from branched sugars, silyl ethers), and (v) less straightforward interpretability of mass spectra. The otherwise very attractive and fast reductive cleavage method has therefore not become a widely established alternative to the alditol acetate method for routine analysis of polysaccharide derivatives.

Differences in the sensitivity toward side-reactions are probably more pronounced between partially substituted glucoses having different numbers of free OH groups than between fully protected sugars with different patterns. Therefore, the additional permethylation step, although not essential prior to hydrolysis of hydrolytically stable polysaccharide ethers, is eventually worthwhile. While the Hakomori method[25] using Na-dimsyl was the original breakthrough for routine application of methylation analysis,[26a,b] nowadays more often Li-dimsyl[246,247] or pulverized NaOH are used with methyl iodide in Me_2SO.[179] In the latter case, it should be noted that oxidation of terminal alditols may occur.[248] In particular, oligosaccharides of low DP are reduced prior to methylation to avoid chain degradation by alkali-promoted peeling reactions from the reducing end. Avoidance of alditol oxidation during this methylation procedure has also been addressed.[249,250] For such alkali-sensitive groups as sulfates, which can undergo intramolecular nucleophilic displacement, special methylation conditions had to be established, to give only one example.[148] Acetates require nonalkaline methylation conditions. Instead of preparing an alkoxide as a strong nucleophile, use of a strong electrophile, offering "CH_3^+" in combination with a proton scavenger, may be applied. Reagents that fulfill these requirements are methyl triflate[251] or Meerwein salts.[252] For problems inherent to these procedures, such as acyl migration, the reader is referred to the literature.[253] Methylation methods have been reviewed recently.[254a,b]

If the substituents are not stable during acid hydrolysis or methanolysis, the complementary substitution pattern can be determined after permethylation, as already noted. Examples are organic (acetates)[227] and inorganic esters (sulfates),[228] and benzyl[255] and silyl ethers.[140] Such an approach was applied as early as 1964 to vinyl starches.[101] In contrast to the direct determination of the substituent pattern, incomplete methylation and consequently erroneous evaluations cannot be recognized from the product pattern obtained in the indirect methods. If the average DS is known from an independent determination, it allows at least a certain control. Therefore, the absence of OH absorption should be carefully monitored by IR spectroscopy.

Since standard compounds may not be readily available, control of these parameters and validation of the analytical procedure can be very difficult. Usually, agreement with the independently determined average DS, the absence of side products, and high recovery of material (determined by addition of an internal standard) are used for control. GLC/FID allows calculation of the relative molar composition of monomers by correcting their peak areas according to the ECR concept, as already mentioned.[214–217]

c. ESI-MS/CID.—In the special case of methyl ethers, an independent method for monomer analysis was developed, not requiring total depolymerization and the attendant risks of bias already mentioned.[256]

The fragmentation pattern of 2,3,6-*O*-methylated maltooligosaccharides in ESI-MS/collision-induced dissociation (CID) has been thoroughly studied with positionally deuteriomethyl-labeled compounds. It was found to give reproducible fragment-ion patterns, following a defined mechanism of cross-ring cleavages, that were assigned according to Domon and Costello.[257] From the partial shift of a certain fragment, the relative contribution of a methyl group in a specific position could be calculated.[258] This approach is illustrated in Figs. 23 and 24. Methyl cellulose or starch is perdeuteriomethylated for chemical uniformity. After partial hydrolysis, the oligomeric mixture is examined by ESI-MS. The pattern for the disaccharide fraction is shown in Figs. 23 und 24. Depending on the number of methyl residues originally present and the number of deuteriomethyl groups introduced for isotopic labeling of free OH groups, the *m/z* of the analytes, $[M+Na]^+$, lies between 449 and 467.

The border peaks represent a single isomer, the hexamethylated and the originally unsubstituted dimer. The next two signals, at *m/z* 452 and 464, represent 6 isomers of monomethylated and monodeuteriomethylated disaccharides, respectively. Fig. 24 illustrates schematically how the intensity of this signal is distributed between the

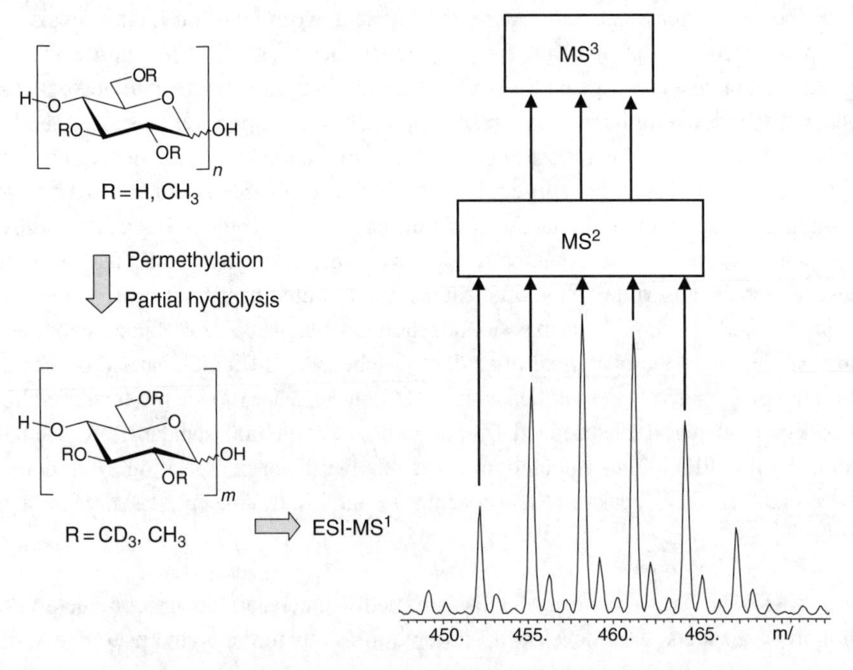

FIG. 23. Monomer analysis by ESI-MS/CID of methylated glucans after perdeuteriomethylation and partial hydrolysis (according to Ref. 256).

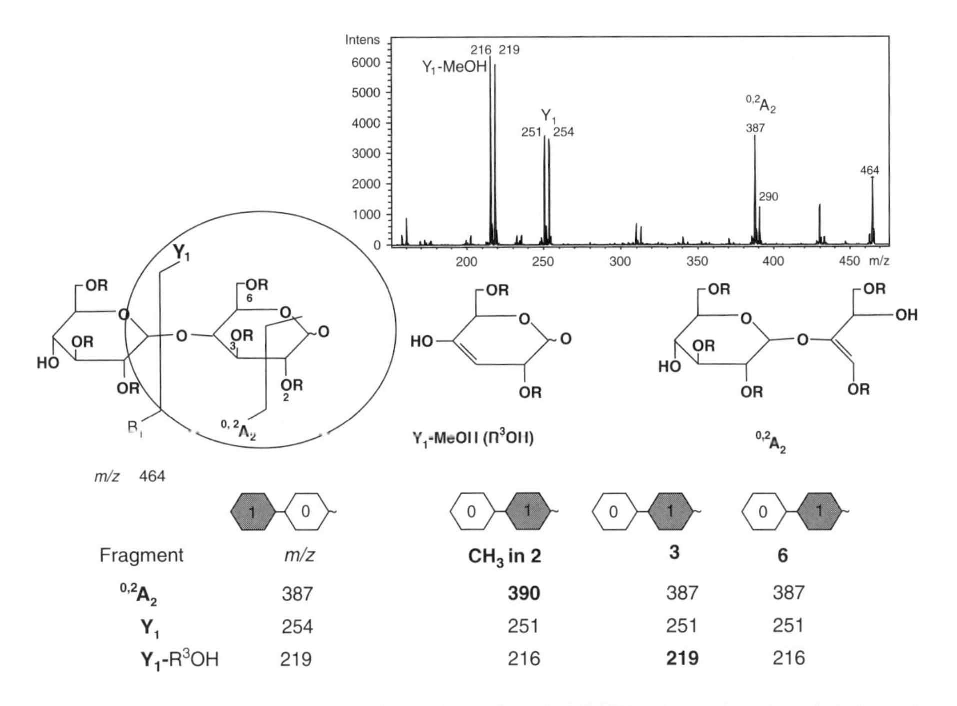

FIG. 24. Monomer analysis of methylated glucans by ESI-MS/CID after perdeuteriomethylation and partial hydrolysis. As an example, evaluation of daughter mass spectrum of monomethylated disaccharides, *m/z* 464, is demonstrated; for details see Ref. 256.

positional glucosyl isomers. Starting from the first quantitative information, apportionment of the other signals is evolved step by step. Since the three inner signals comprise 15 or even 20 isomers (originally tri-*O*-methylated disaccharides), ESI-MS^2 no longer gives a sufficient number of independent data, but ESI-MS^3 is required to furnish the additional information.

This procedure is also not free from shortcomings in comparison with the procedure based on chromatographic separation after total depolymerization and is therefore of value as a reference method. In many cases results are in good agreement with monomer data after hydrolysis.

3. Distribution of Monomer Residues along the Polymer Chain

On the next hierarchical level, it is necessary to know how the monomers, the unsubstituted to fully substituted glucose residues present in certain molar ratios, are

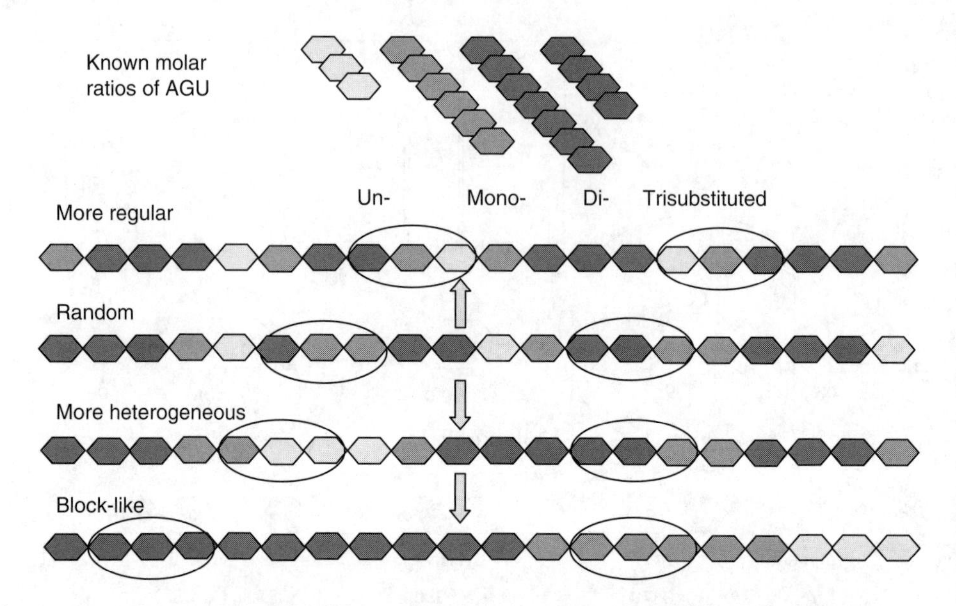

FIG. 25. Schematic view of types of distribution of monomer units (AGU) in the polysaccharide chains (see also Fig. 12); "random" is used as defined reference distribution; composition of trimers from partial depolymerization are determined by MS and compared. (See Color Plate 27.)

organized in the polysaccharide chains (Fig. 25). Again, the reference is the random-distribution model, based on constant relative reactivities of the three hydroxyl groups (see Section II.5.b). Since sequence analysis is neither practicable nor reasonable for such a complex mixture of macromolecules, the substituent pattern is analyzed on the oligomeric level, which comprises the relation between adjacent glucosyl units.

a. Oligomer Analysis After Random Degradation.—This approach, which was first reported by Arisz *et al.*[174] and Kühn and Mischnick,[166] includes the following steps: (i) random depolymerization of glucan derivatives into oligosaccharides, (ii) quantitative analysis of oligosaccharide fractions of a certain DP, and (iii) comparison of the experimental results with the corresponding random distribution for this DP as calculated from the monomer composition (compare Fig. 12). To achieve representative quantitative data of monomer distributions in oligomeric sequences, which can be compared with the random model, some requirements must be fulfilled: (i) a dissolved or dispersed structure, which allows equal accessibility to all glycosidic linkages, (ii) no influence of the substitution pattern on the rate of glycosidic-linkage cleavage, and (iii) no bias of relative signal intensities in MS for the constituents of a certain DP. These are exacting demands, which at present can only be fulfilled with acceptable accuracy for a limited number of types of derivatives. Thus, appropriate sample preparation is a crucial step, and a better understanding of the processes remains an object of further research.

(i) Sample Preparation. Direct partial hydrolysis or methanolysis of a starch or cellulose ether cannot implicitly be expected to proceed statistically. Fundamental studies on the kinetics of glycoside hydrolysis were performed in the sixties of the last century and have been comprehensively reviewed by BeMiller[259] and Capon.[260] Depending on polarity and size, substituents can change the steric and electronic situation of the glucosides and the oxacarbenium ion intermediate, and therefore the rate of hydrolysis. There are few data available for hydrolytic rate-constants of substituted glycosides. De and Timell compared the hydrolysis of regioisomeric methyl *O*-methyl-β-D-glucopyranosides and 6-*O*-(mono-*O*-methyl-β-D-glucopyranosyl) D-glucose with 0.5 M H_2SO_4 at 60, 70, and 80 °C.[261] For both aglycones, methyl and glucose, first-order constants of hydrolysis decreased in the order Glc > 3 > 2 ≈ 4 > 6-*O*-methyl-glucosyl, while the disaccharides were cleaved faster than the methyl glucosides. Höök and Lindberg reported the hydrolytic rate-constants of methyl *O*-isopropyl α- and β-D-glucopyranosides with sulfuric acid at various temperatures and found a stability order of 3 > 2 > 4 ≈ Glc > 6-*O*-iPr-Glc in the α series and 2,3 > 2 ≈ 3 > 4 ≈ Glc > 6-*O*-iPr-Glc in the β series.[262] There are many studies of various glycosides indicating that the aglycone also has a significant, albeit minor and different, influence on the rates of hydrolysis.[259,260,263] With respect to cellulose derivatives, the order of release for DEAEC was reported to be Glc > 3 > 6 ≈ 2-*O*-diethylaminoethyl-Glc.[264] Recently, we found for CMC that 2-*O*-substitution caused the most pronounced delay in acid hydrolysis.[265] In 2003 Jensen and Bols[266] reported that, in contrast to the old and plausible explanation of Edwards from 1955,[267] the relief of steric constraints by conformational change is not responsible for rate differences in hydrolysis of various methyl glycosides; instead field effects are the controlling factors. Thus electron-withdrawing groups destabilize the intermediate oxacarbenium ion. Besides electronic and steric effects, it should be noted that neighboring-group participation can be of major influence. Thus sulfate groups at O-2 strongly accelerate the hydrolysis of the β-glycosidic linkage in cellulose, on account of its ability to form a cyclic sulfate.[148]

Differences in hydrolysis rate can, in principle, be diminished at higher temperatures or by using higher acid concentrations,[268] working as nonselectively as possible. If feasible, however, the preparation of chemically more-uniform substrates is the method of choice. In the case of methyl ethers, this can be achieved simply by perdeuteriomethylation, as already mentioned (Section II.2.b), following the principle of isotopic labeling. While CD_3 is slightly larger and more hydrophobic than CH_3, no significant differences in hydrolysis have been found.[166]

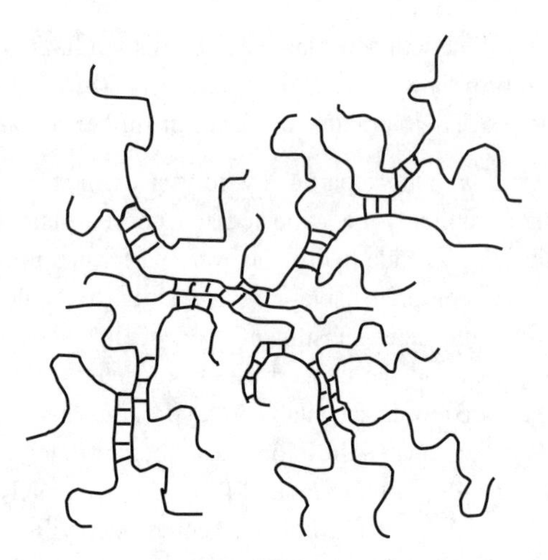

FIG. 26. Fringed micelle model of cellulose derivatives in solution, according to Schulz and Burchard.[82] Reprinted from Ref. 82 with permission from Wiley-VCH Verlag GmbH & Co, Copyright 2000.

A number of derivatives having labile substituents have been transformed into methyl/deuteriomethyl ethers by the sequence: permethylation, cleavage of labile substituents/perdeuteromethylation, and partial hydrolysis (or methanolysis).[145a,b,148] The isotopic-labeling approach fails in the case of hydroxyalkyl ethers, where the OH groups of the substituents take part in the same reactions as the remaining hydroxyl groups of the carbohydrate backbone. However, to achieve random hydrolysis, per(deuterio)-methylation for minimizing the chemical differences is sufficient. Peralkylation prior to partial hydrolysis or methanolysis influences not only the reactivity of the glucosidic linkage, but also the accessibility of the polysaccharide chains. As is known from the thorough studies by Burchard,[82] the so-called "dissolved" cellulose derivatives are not molecularly dispersed as long as they are not completely substituted, which is illustrated by the fringed micelle model in Fig. 26. This is a further potential source of bias during partial hydrolysis, which is overcome by permethylation.

The situation is, therefore, more critical for derivatives that cannot simply be peralkylated, as, for example, CMSs and CMCs. To establish whether a partial degradation proceeds in a random manner, the composition of new reducing-end units formed by cleavage of glucosidic linkages can be analyzed. After labeling them in an appropriate way (as by reduction)[174] and subsequent complete hydrolysis, they can be analyzed as a separate set of constituents. If there is not any bias, their molar ratios should be equal to the average composition of the entire sample. If degradation cannot be performed in homogeneous solution, or if the sample is not molecularly dissolved (probably favoring the hydrolysis of the loose ends of polymer molecules), the oligosaccharides released should be separated from the bulk material, and their composition can be analyzed. Again, the average DS should be in accordance, to confirm that the oligosaccharides released represent the starting material. As outlined in the next paragraph, mass-spectral analysis might require further derivatization steps.

(ii) Mass-Spectrometric Analysis. Mass Spectrometry is the method of choice for analyzing the composition of oligosaccharide mixtures rapidly and with high sensitivity. Analysis of oligosaccharides has benefited significantly from developments in mass-spectrometric instrumentation during the past two or three decades.[40] The first studies on methylated cellulose and amylose following the strategy just outlined were performed with fast-atom bombardment (FAB)-MS.[166,174] These analyses required background correction and were limited to derivatives of low DPs, but were enhanced by further developments of ESI- and MALDI-MS techniques. A special challenge for substitution-pattern analysis is that accurate quantitative data are required. In contrast, most of the work reported on carbohydrate analysis by mass-spectrometric methods is related to structure elucidation and sequencing, where a rough estimation of molar ratios is sufficient, and quantitative data are available from methylation analysis and NMR. Haebel *et al.* have reported a very interesting approach for quantitative analysis of partially deacetylated chitin oligosaccharides, employing multistage MS techniques in combination with appropriate sample preparation.[269] MS, however, is not inherently a quantitative method. The relation of signal intensities to concentration of compounds is determined by ion yields in the ESI and MALDI processes, and by ion stability and transfer into the mass analyzer; both steps can cause biases. Detail of current knowledge on the ionization mechanism is beyond the scope of this article, and the reader is referred to the related literature.[40,270–280a,b] Differences in polarity and solvation energies of the constituents influence their distributions between the outer and inner parts of the droplets formed in electrospray and in their desolvation, and favor those compounds that are relatively enriched at the droplet surface. The basicity, number, and orientation of coordinating sites influence the competition for cations to form quasimolecular ion-adducts ($[M+Na]^+$, $[M+H]^+$). The molar signal-intensity usually decreases in ESI-MS but increases in MALDI-MS over a certain range of molecular weight (namely DP) of homooligosaccharides. For instance, disaccharides are detected at a much higher sensitivity than monosaccharides, since the additional coordination sites and the conformational flexibility of the glycosidic linkage favor the formation of quasimolecular ions (Fig. 27).[281] Hydroxyalkylation also strongly enhances the ion yield, especially for monomers (Fig. 28).[282]

Naven and Harvey studied the effect of structure on the signal strength of oligosaccharides in MALDI-TOF-MS and found no influence on the signal intensity at $m/z > 1000$,

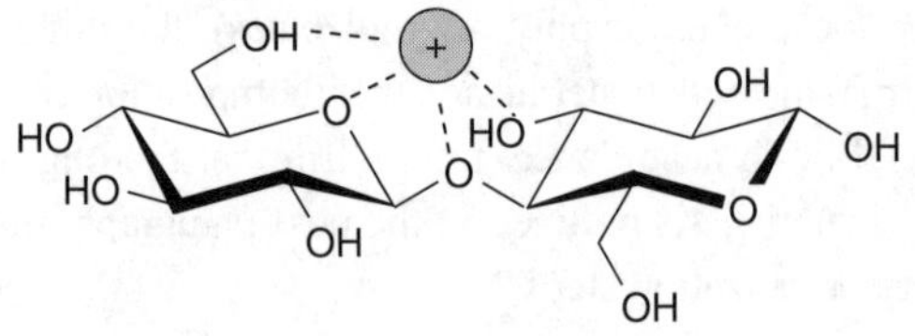

FIG. 27. Complexation of the cation by basic atoms of a disaccharide, according to Hofmeister.[281]

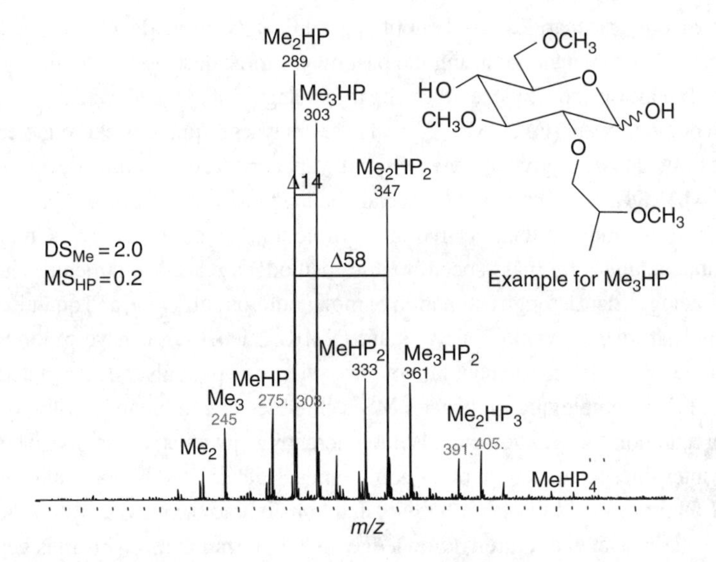

FIG. 28. ESI-MS of completely hydrolyzed HPMC; signals are assigned according to the number of Me and HP residues. Reprinted from Ref. 282 with permission from Wiley-VCH Verlag GmbH & Co, Copyright 2005.

but a progressive diminution with decreasing molecular weight. They attributed this to temporary saturation of the detector by matrix ions.[283] This observation from natural oligosaccharides corresponds with a leveling effect of the average DS with increasing DP,[282,284,285] where the main constituents become more similar. Dilution of samples was also observed to diminish the bias in ESI-MS, presumably by alleviating competition for droplet surface and counter ions. Furthermore, such instrument parameters as the skimmer voltages and the target mass of ion traps (ITs) strongly influence the ion yield.

From this short overview of crucial points in quantitative MS, it is obvious that isotopic labeling, if applicable, is the best strategy [see III.3.a (i)]. Quantitative evaluation of *O*-methyl/*O*-methyl-d_3 derivatives gives correct average DS values for the particular oligosaccharide mixture by use of FAB-, ESI-IT-, and MALDI-TOF-MS, which indicates that there is no bias for constituents of higher or lower substitution. Permethylation is no longer sufficient for hydroxyalkyl ethers, because of the additional coordinating sites added by the hydroxyalkyl residues.

Differences in ion yields and surface activities can be overcome by introducing a permanent charge. After partial hydrolysis, reductive amination has been performed with propylamine and subsequent quaternization with methyl iodide (Fig. 29). This labeling, in combination with MALDI-TOF-MS, was successful for HPMC, HEMC, and HEC.[175–177] Fig. 30 shows the ESI and MALDI mass spectra for an HEMC prepared as described and the average DS values/DP calculated from these. While for ESI there is still a decrease of the MS with DP, MS becomes constant by MALDI at DP ≥ 3, while at the same time showing a higher sensitivity for the larger oligosaccharides.

Momcilovic *et al.* were able to significantly enhance sensitivity by labeling partially depolymerized nonionic cellulose ethers with dimethylamine.[286,287] It is also possible to introduce a permanent charge directly, as reported for oligosaccharides, for instance, with Girard's T.[288,289] This acid hydrazide of betaine, an "*N*-acyl hydrazine," adds to the sugar carbonyl group and forms an acyl hydrazone that does not require additional reduction. Gil *et al.* demonstrated that combination of labeling and MALDI-TOF-MS can successfully be applied for quantitative analysis of

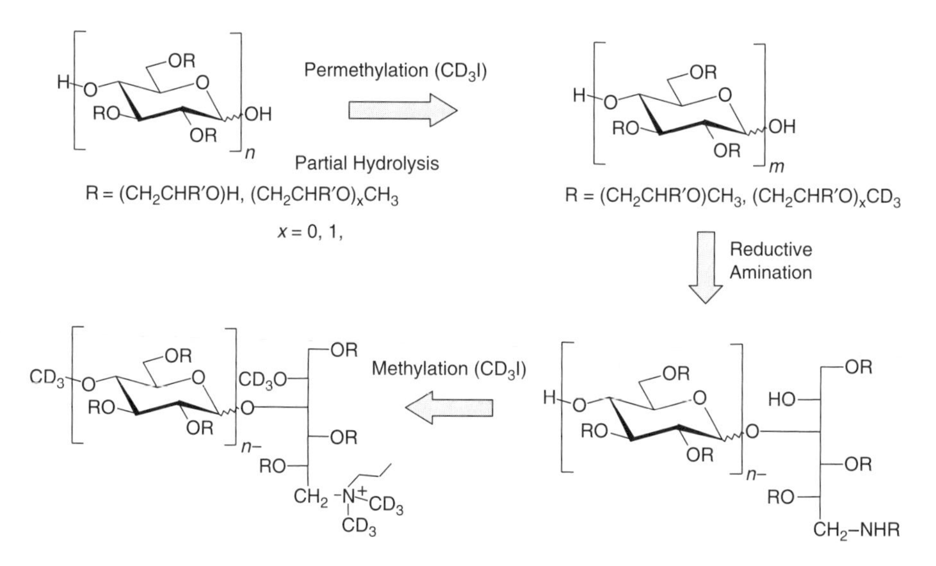

FIG. 29. Sample preparation for quantitative mass spectrometry of hydroxyalkyl ethers. R′=H or CH_3, $m = 1, 2, 3, \ldots, <n$.

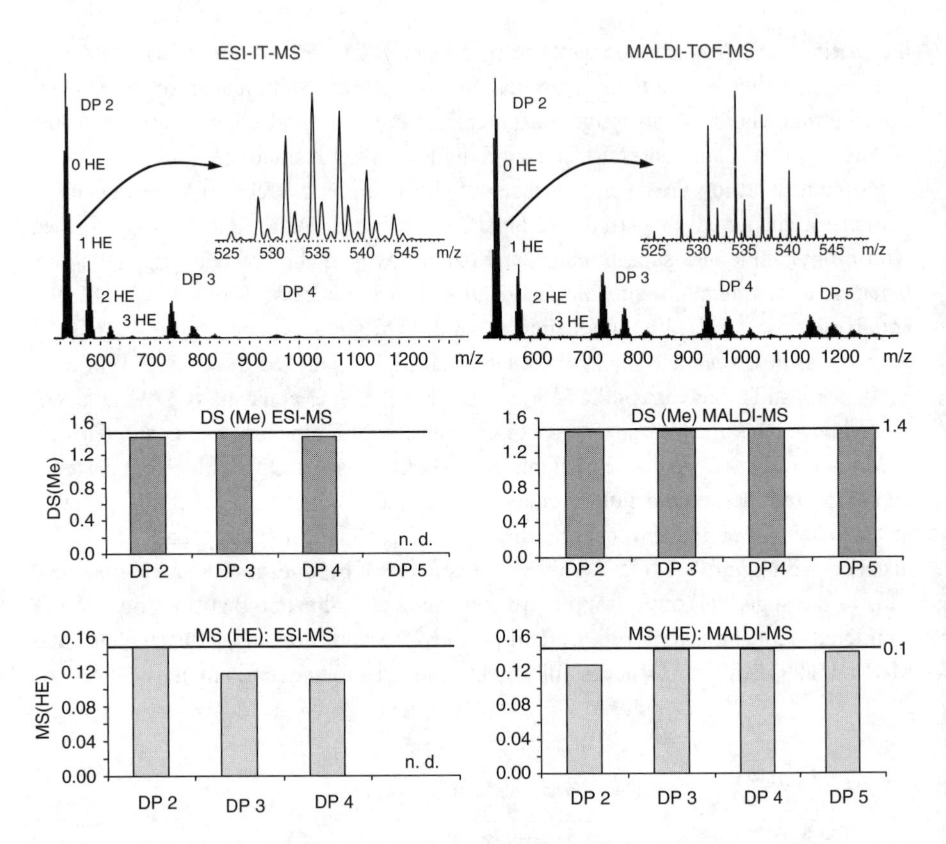

FIG. 30. ESI-MS and MALDI-MS of an HEMC (DS 1.45, MS 0.15) prepared according to Fig. 29. Reprinted from Ref. 177 with permission from the American Chemical Society, Copyright 2006. (See Color Plate 28.)

oligosaccharides.[290,291] As a reference method they applied HPLC-UV of the same oligosaccharde mixture, which was reductively aminated with aminobenzoic acid. Negatively charged tags can also be introduced.[292]

The instrumental setups, ESI-IT-MS, ESI-triple quadrupole-MS, and MALDI-TOF-MS, applied in comparison for measuring a mixture of methylated cellooligosaccharides, with and without perdeuteriomethylation of free OH groups (isotopic labeling), showed that the MALDI-TOF was less susceptible to differences in polarity and mass, thus giving nearly the same correct average DS for the permethylated and partially methylated sample.[293]

In contrast, the methylated higher oligosaccharides were overestimated by the other instruments. In the study already mentioned, Naven and Harvey found that there was no discrimination of lower molecular weight analytes when a MALDI ion source and a magnetic-sector field mass analyzer were combined.[283]

(iii) Quantitative Evaluation. From mass spectrometric analysis, the DS/DP distribution is deduced, which then is compared with the reference model, namely the random pattern calculated from the monomer composition. The average DS calculated from the signal intensities in MS must be equal for all DPs and agree within experimental error with the average DS of the material. Fig.12 already showed various types of deviations for the DP4 fraction (0–12 substituents) of a sample of DS 1.5.

In the example shown in Fig. 31, bimodality is observed for HEMC, probably as a consequence of areas of very low substitution resulting from insufficient activation.

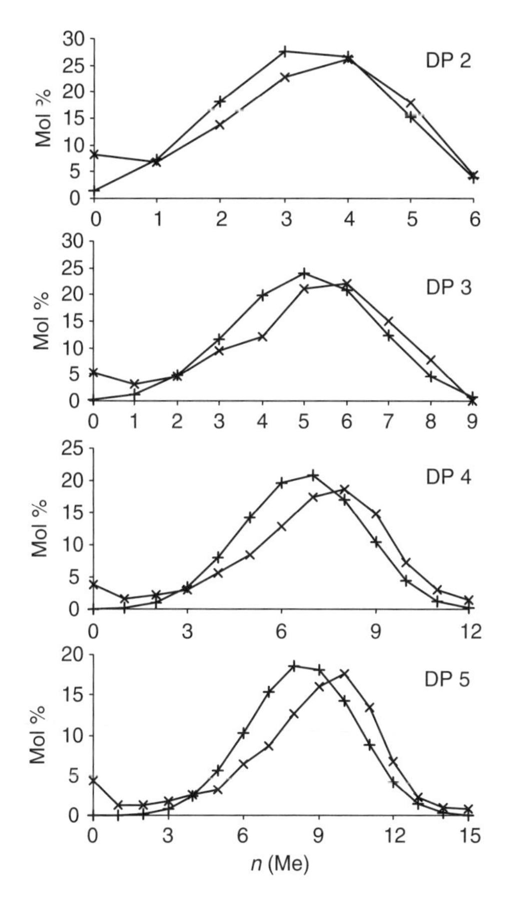

FIG. 31. Bimodal methyl-group distribution in HEMC, DS 1.66. x, experimental data; +, calculated random distribution. The nearly constant portion of AGU organized in non- to low-substituted sequences indicates the presence of non- to low-substituted domains in this sample. Reprinted from Ref. 177 with permission from the American Chemical Society, Copyright 2006.

This is also confirmed by the fact that the amount of unsubstituted sequences (DS = 0) remains constant at about 5 mol% with increasing DP. If the pattern is strictly bimodal, the monomers can be distributed on two groups presenting subpatterns, which can be reassembled to the observed substituent distribution. As at the monomer level, a heterogeneity parameter H_i can be calculated as a measure for the average deviation of the experimental from the random reference pattern. However, for interpretation of the substitution pattern, not only the average deviation but also the characteristics of deviation from the random model need to be considered. In the case of deviation caused by enhanced reactivity at the proximate glucose residue, namely by interruption of hydrogen bonds or change of solvation stage, it should be possible to integrate this *intermonomeric effect*, similar to the *intramonomeric* one regarded in Reuben model, where enhanced reactivity of O-3 by primary substitution of O-2 in the same glucose residue is considered (see Fig. 10, Section II.5.a).

The probabilities for a dimeric sequence P_{ij} is $p_i p_j$, with p_i and p_j being the probabilities of the glucosyl moieties involved. The extended model introduces a conditional probability which considers that p_j depends on whether its neighbor is substituted ($p_{j|+}$) or not ($p_{j|0}$), and thus P_{ij} is $p_i p_{j|+}$ or $p_i p_{j|0}$. The conditional probabilities are estimated by fitting the substituent-distribution curves to the experimental data (→ “dependence model”). Fig. 32 shows an example for an MC with bimodal substitution pattern. Under the assumption of the dependence model, the experimental data fitted much better than by the random model.[294]

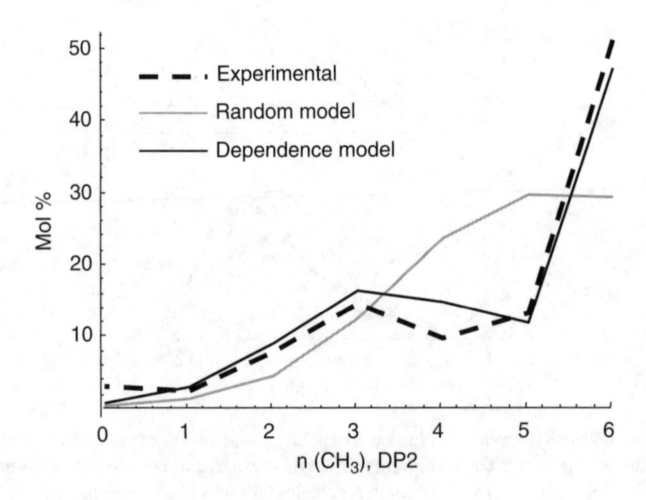

FIG. 32. Application of the “dependence model” to a heterogeneous bimodal MC, DS 2.32.

b. Enzymesas Selective Tools.—*(i) General.* As mentioned in the introductory part of this article, along with chemical methods, enzymes are attractive tools in the analysis of chemical structure. The application of enzymes for unmodified polysaccharides was surveyed in this series by McCleary and Matheson in 1986.[295] Chemical modification of polysaccharides directly affects their enzymatic digestibility with respect to type, degree, and patterns of substitution. To understand these relationships it is necessary to provide some general information on enzymes selective for starch and cellulose.

Enzymes that catalyze hydrolysis of such *O*-glycosyl compounds as cellulose and starch are termed glycosidases. Some enzymes act more efficiently on polymeric substrates, whereas others are more efficient on oligomeric substrates and hydrolyze these into mono- or di-saccharides (which may then be further metabolized). The enzymes that are described here are named according to the "Enzyme List" of the Joint Commission on Biochemical Nomenclature (JCBN) of the International Union of Pure and Applied Chemistry and the Nomenclature Committee of the International Union of Biochemistry and Molecular Biology.

Glycosidases selective for starch and cellulose are commonly referred to as glucanases, since they catalyze depolymerization of glucose-based polysaccharides. Glucanases are subdivided as either *endo*- or *exo*-enzymes. *endo*-Enzymes randomly cleave the glucosidic linkages of the polysaccharide, whereas *exo*-enzymes act from the nonreducing and, in some cases, from the reducing end of the substrate. The *endo*-enzymes produce polymeric and/or oligomeric compounds (depending on the DP of the substrate and cleavage point), which may be further hydrolyzed to smaller compounds. *exo*-Enzymes, on the other hand, produce monomeric or oligomeric compounds by successive cleavage, usually from the nonreducing end. It should be pointed out that some enzymes are not distinct with respect to *endo*- or *exo*-mode of action, but rather have a preference for one of them.[296,297]

Common for all cellulose- and starch-selective glucanases is their substrate-binding active site, which normally contains several subsites (Fig. 33). A subsite is comprised

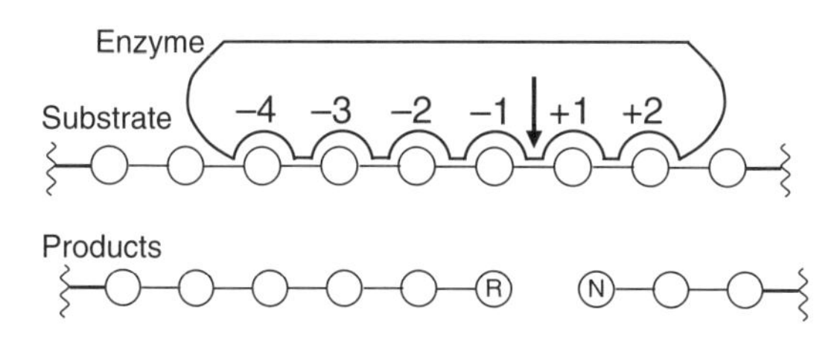

FIG. 33. Schematic representation of the active site of a glucanase comprised of six subsites.

of one to several amino acid residues, which interact with one of the glucose residues in the substrate. During hydrolysis the substrate is bound to the subsites, as by hydrogen bonds between the substrate OH groups and the amino acid residues on the enzyme. The number of subsites varies between different enzymes. In addition, it has been shown that, for some enzymes it is not necessary for *all* subsites to be occupied in order for hydrolysis to occur.[298,299]

Subsites are numbered according to Davies *et al.*[300] The arrow (Fig. 33) indicates the glucosidic linkage that is hydrolyzed and "N" and "R" refer to the nonreducing and the reducing end, respectively.

In the classification that has been proposed by Henrissat, glycosidases are grouped into families on the basis of similarities in their amino acid sequence.[301,302] This classification was later applied in a scheme for enzymes that hydrolyze β-(1→4)-linked plant cell-wall polysaccharides.[303] According to this scheme, an enzyme is named *Xx YyyNZ*, where *Xx* refers to the organism of origin, *Yyy* is the favored substrate, *N* is the enzyme family, and *Z* indicates which one in the order of family N enzymes that is being referred to (if more than one are known). For example, the fungus *Trichoderma reesei* produces two family 7 enzymes, cellobiohydrolase I and endoglucanase I. According to the scheme they are named Tr Cel7A and Tr Cel7B, respectively.

The mechanism of action of glucanases is strongly influenced by the shape and structure of the active site. *exo*-Enzymes normally have a pocket or a tunnel-shaped active site, which "forces" the enzyme to attack a chain end of the polysaccharide substrate.[304] The active site of *endo*-enzymes has the shape of a cleft. *endo*-Enzymes can therefore bind to the substrate randomly along the polymeric backbone (provided that the active site can accommodate the sequence of the polysaccharide). Furthermore, some glucanases also contain a binding domain, which enables reversible sorption onto (and, thus, hydrolysis of) nondissolved polysaccharides, such as granular starch and fibrous cellulose.[305,306]

Fig. 34 shows schematic structures of amylopectin, amylose, and cellulose and also indicates where different glucanases attack these polysaccharides. Alpha-amylases (EC 3.2.1.1) are *endo*-enzymes, which hydrolyze α-(1→4) glucosidic linkages in amylopectin and amylose (Fig. 34A). In amylopectin, the A, B, and C chains are hydrolyzed by these enzymes, whereas the α-(1→6) glucosidic linkages in the branching points of amylopectin are not attacked. These linkages can be hydrolyzed by pullulanases (EC 3.2.1.41) and isoamylases (EC 3.2.1.68).

Beta-amylases (EC 3.2.1.2) (Fig. 34A,B) are *exo*-enzymes that attack the substrate from the nonreducing end and hydrolyze α-(1→4) glucosidic linkages. Maltose (disaccharide) molecules are successively detached from the non-reducing end during hydrolysis by beta-amylase. In addition, glucan (1→4)-α-glucosidases (EC 3.2.1.3,

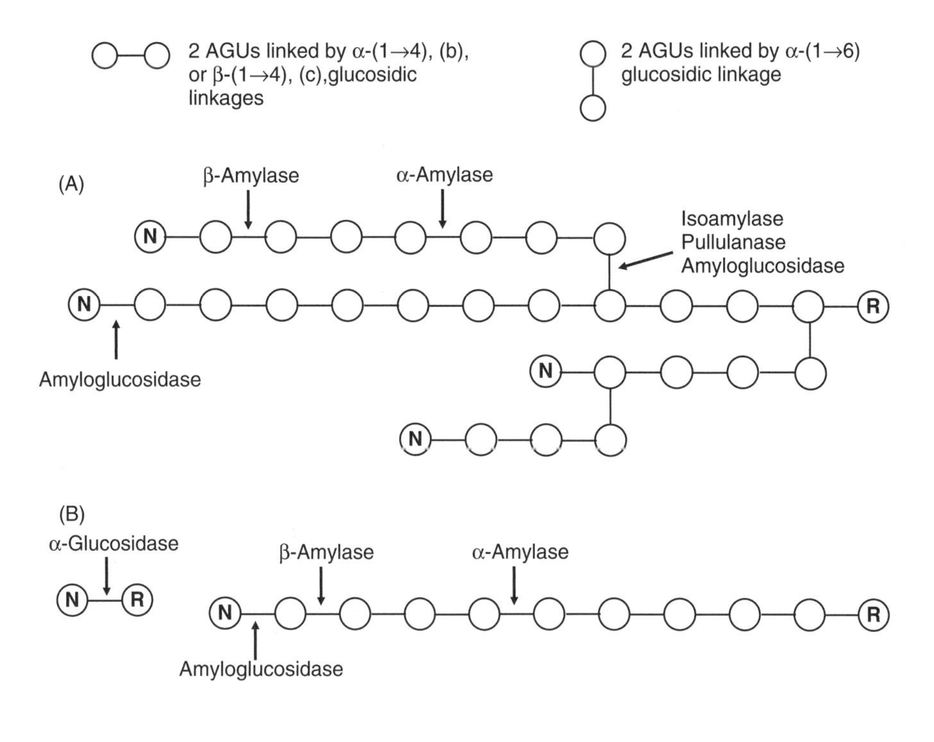

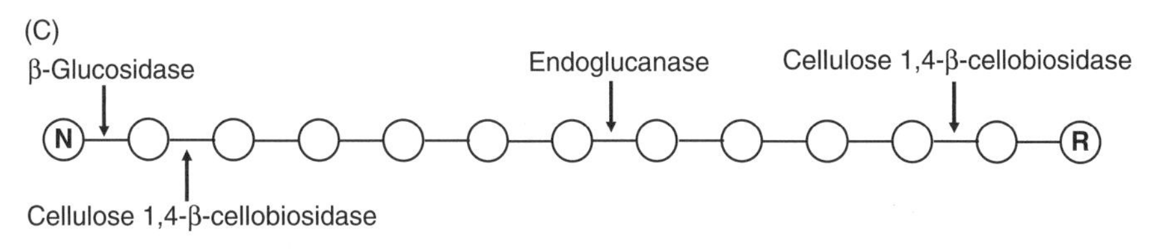

FIG. 34. Schematics of amylopectin (A), amylose (B), and cellulose (C), and the selectivity of various glucanases on these polysaccharides.

also known as amyloglucosidases or glucoamylases) hydrolyze α-(1→4)- and α-(1→6) glucosidic linkages from the nonreducing end, yielding D-glucose.[38] (Note that the term "amyloglucosidase" is used throughout this article, since it is more frequently encountered in the literature.) Whereas beta amylases and amyloglucosidases have higher activity on polysaccharides as compared to oligosaccharides, the reverse is the case for α-glucosidases (EC 3.2.1.20). These *exo*-enzymes efficiently degrade maltose and other α-(1→4)-linked [and in some cases α-(1→6)-linked] oligosaccharides to glucose.

Cellulose-selective *endo*-enzymes are sometimes referred to as cellulases (EC 3.2.1.4, Fig 34c), which is also the definition according to JCBN. However, recognizing that "cellulase" is not always equivalent to "cellulose selective *endo*-enzyme" in the literature, we use the name "endoglucanase" throughout this article. Examples of *exo*-enzymes are β-(1→4)-cellobiosidases (cellobiohydrolases) (EC 3.2.1.91) and β-glucosidases (cellobiases) (EC 3.2.1.21). β-(1→4)-Cellobiosidases may attack the substrate on the nonreducing or reducing terminus, whereas β-glucosidases only attack on the nonreducing end, with the production of glucose. β-Glucosidases also efficiently hydrolyze cellobiose to glucose.

(ii) General Aspects of Enzyme-Aided Structure Analysis of Starch/Cellulose Derivatives. The main objective of employing glucanases in the chemical structure analysis of starch and cellulose derivatives is to investigate the substituent distribution heterogeneity. Since the active site of these enzymes is designed to accommodate the native (unsubstituted) polysaccharide, introduction of substituents will affect the activity of the enzymes. Depending on the size, charge, and position of the substituents, formation of the enzyme–substrate complex is weakened by, for instance, hindrance of hydrogen-bond formation or by steric and/or electrostatic repulsion. Heterogeneity of both the *first* and the *second order* of the substituent distribution may influence the degradability. The chemical composition of the products from a certain enzyme will therefore be related to the substituent distribution. Intrepretation of this is, however, not always straightforward, since enzymatic hydrolysis of the glucosidic linkages cannot be regarded simply as "hydrolysis" or "no hydrolysis." Hydrolysis occurs on a rate scale, where the rate of hydrolysis of a glucosidic linkage in a specific situation is influenced by interference by the substituent. Enzymes act by lowering the activation energy, thereby increasing the reaction rate, and any interference will decrease the rate. The glucosidic linkages in nonsubstituted regions are therefore hydrolyzed at higher rate than those in substituted regions. In addition, the position(s) of substitution on the glucose residues involved in the enzyme–substrate complex also affect the hydrolysis.[307–310] Furthermore, other more-complex factors, such as interdependence between the enzyme subsites, may also have an influence. Therefore, up to the present, it has not been possible to predict the outcome of the hydrolysis in relation to the distribution of substituents, because relatively little is known about the enzyme selectivity. Although individual subsites, especially –1 and +1, have been studied for some enzymes, there is a lack of knowledge about subsites that are more distant from the cleavage site. Furthermore, a better overview of the whole active site and, especially, improved understanding of the interdependence between subsites must be taken into perspective for further developments in this field.

An additional problem arises from variations of activity in commercially available enzyme preparations, which in the case of glucanases are often mixtures (cocktails of *endo*-enzymes and/or *exo*-enzymes), while pure "monocomponent" enzymes require isolation and purification in the laboratory. Common sources for both commercial and noncommerical enzymes are bacteria, fungi, and various plant crops. Enzymes from *T. reesei* (also known as *Trichoderma longibrachiatum*), *Humicola insolens*, and *Trichoderma viride* have frequently been applied for structure analysis of cellulose derivatives, whereas starch derivatives have mainly been analyzed employing amyloglucosidases from *Aspergillus niger* and alpha-amylases from *Bacillus licheniformis*, but enzymes from many other sources have also been utilized.[38]

Along with such side activities and cocktail composition, commercial enzymes are sometimes stabilized with mono- and/or oligosaccharides, which must be removed prior to hydrolysis in order to minimize systematic errors. This can be achieved by, for instance, ultrafiltration or dialysis. Impurities that cause side-activity, such as trace amounts of *exo*-enzymes in an *endo*-enzyme preparation, or vice versa, are not always readily detected and are more tedious to remove.[293] Other practical parameters that should be controlled in enzyme hydrolysis are, for example, pH, temperature, ionic strength, and salt type, since all these may have substantial effect on the rate of hydrolysis.[299,311]

Fig. 35 provides an overview of applications of different techniques in the chemical analysis of enzymatic hydrolyzates of starch and cellulose derivatives. The degree of degradation of glucan derivatives can be monitored by size-exclusion chromatography (SEC) and by analysis of dextrose equivalent (DE, which is essentially the quantification of reducing-end groups).[38,312] The latter methods are based on the reducing power of the chain end and are traditionally employed for reduction of a Cu(II) or Fe(III) salt, which can then further react or form complexes, affording spectrophotometrically quantifiable products. Thus, the number of cleavages of glucosidic linkages can be quantified and, for example, the hydrolysis endpoint determined. A sophisticated approach for reducing-end analysis was presented by Melander *et al.*, who developed a flow-injection analytical method for real-time quantification of reducing sugars formed during enzymatic hydrolysis of MC.[313] The $Fe(CN)_6^{4-}$ ion, obtained from the reaction between $Fe(CN)_6^{3-}$ and reducing sugar, was electrochemically oxidized and the resulting current measured by PAD.

A classical approach for analysis of the heterogeneity of substituent distribution is to quantify the glucose that is liberated from enzymatic hydrolysis.[29,30,38,314] Increased heterogeneity, as in derivatives of the same type and DS, is displayed as increased amounts of liberated glucose, since the enzymes hydrolyze unsubstituted regions more efficiently. For this type of analysis, often referred to as "exhaustive digestion," both *endo*- and *exo*-enzymes are applied.

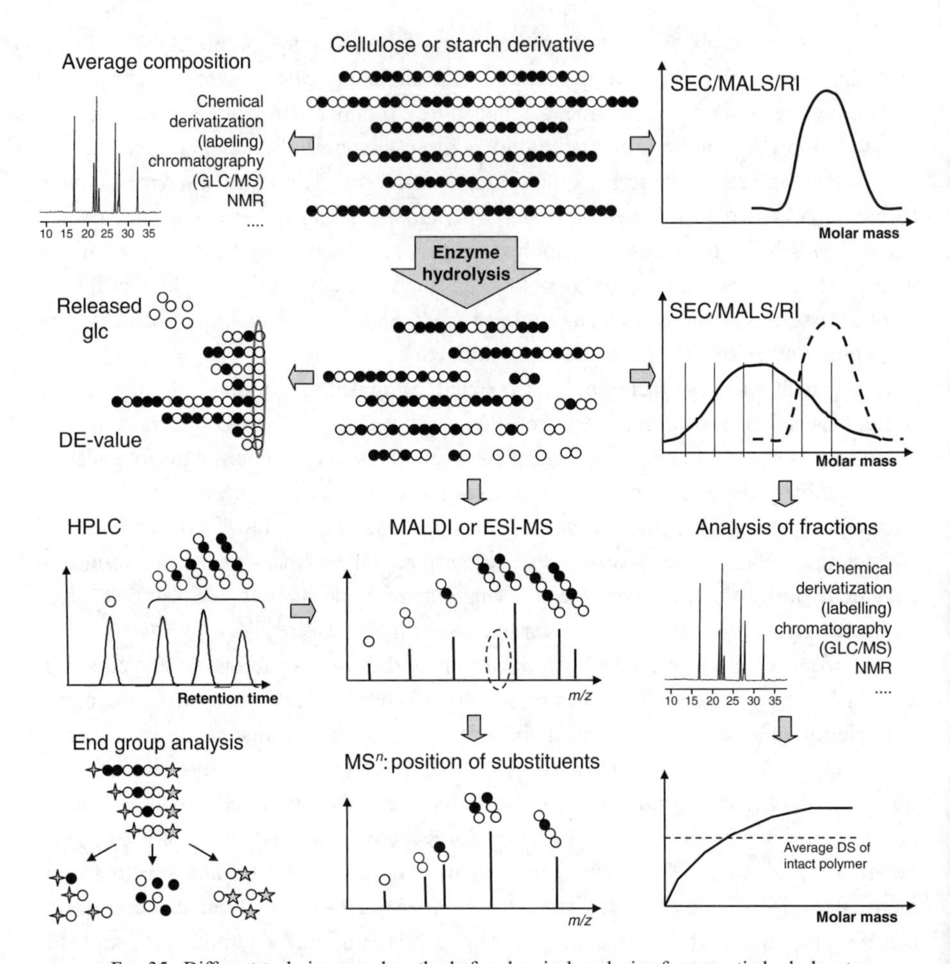

FIG. 35. Different techniques and methods for chemical analysis of enzymatic hydrolyzates.

As shown in Fig. 35, along with these average parameters, analysis providing more differentiation can be performed, including chromatographic fractionation and mass-spectrometric structure analysis. The oligomeric fraction up to a certain DP, which depends on the MS instrument and on the type of substitution, can be analyzed directly by MS and tandem mass spectrometry (MS^n) techniques. As already described for randomly degraded glucan derivatives, the DS/DP pattern can be obtained. If quantification is required, postenzymatic chemical modification may be necessary. To consider which part of the sample is represented by the fraction amenable to MS techniques, quantification by addition of an internal standard has been performed.[293,313] MS also provides a detailed insight into the chemical compositions of the oligosaccharides released. Use of MS^n, with or without further sample preparation (as by labeling or total hydrolysis) and with and without chromatographic separation, provides information on the selectivity of the enzyme(s) with respect to the location of substituents.

Current knowledge indicates that glucanases are most sensitive to substitution on the glucose residues involved with the −1 subsite and, especially, O-2-substitution.[99,189,287,309,313,315–318] In general, the +1 subsite is less sensitive (Fig. 33). Relatively little is known, however, on the interaction with subsites ±2, 3, and others. Together with the parameters already mentioned, such as the chemical structure of the substrate, the microbial source of the enzyme and the enzyme family are also of influence. Husemann, as early as in the 1950s, found that the hydrolysis rate of MC and CMC decreased with increasing DS, and that this effect was stronger with carboxymethyl than with methyl substituents.[28] More recently, Schagerlöf *et al.* revealed a higher degradability of MC by Cel5A from *T. reesei* (Tr) than by Cel5A from *H. insolens.*[319] Furthermore, for Tr Cel5A a higher number of substituents were always detected in the oligosaccharides obtained from MC than in those from CMC.[308] A prerequisite for the plausible conclusion that the enzyme has a higher tolerance for methyl than for carboxymethyl is that such other parameters as DS and regioselectivity are comparable.

Nowadays, more-detailed analysis of the substituent distribution can be achieved by employing pure *endo*-enzyme preparations and quantifying the oligosaccharides obtained from the hydrolysis, by the standard addition method.[313,320] *endo*-Enzymes that are unable to hydrolyze small oligomers (DP < ~6) are beneficial in this approach, since an indication on the DP of the unsubstituted sequences in the polymer can be obtained.

To examine the entire material, enzymatic digests of glucan derivatives have been fractionated by SEC and the fractions obtained analyzed further. The whole assembly of analytical methods already described can be applied to these fractions to provide insight into DS development and substituent distribution as a function of molecular weight (see Fig. 35).[187,188,220,321–326]

The following paragraphs (*iii* and *iv*) give a selection of pioneering work and more recent applications of enzyme hydrolysis for structure analysis of starch and cellulose derivatives.

(iii) Application to Starch Derivatives. As described in Section II.2, starch is often chemically modified while retaining its granular form. The substituent distribution will, therefore, among other factors, be influenced by the hierarchical structure of the granules. Sequential hydrolysis by enzymes having different modes of action (*endo* or *exo*) and/or selectivity [α-(1→4) or α-(1→6)] is an attractive approach for characterizing the substituent distribution of starch derivatives. By employing a

debranching enzyme, such as pullulanase or isoamylase, with subsequent hydrolysis by beta-amylase, it is possible to analyze the substituent density near the nonreducing ends of amylopectin.[187,327] Since beta-amylases are *exo*-enzymes, this will be related to the amount of maltose released. Enzymes that hydrolyze α-(1→6) glucosidic bonds can be employed to investigate the substitution pattern near the branching points in amylopectin. A high substituent density in these areas will hinder the enzyme and thereby decrease the concentration of products from debranching. Interestingly, it has been observed that low substitution near the branching points may actually promote enzyme hydrolysis, as cationized amylopectin was more efficiently hydrolyzed by pullulanases and isoamylases than native amylopectin (at $DS < 0.05$ and $DS < 0.005$, respectively).[188]

The importance of the enzyme selectivity in substituent distribution analysis of starch derivatives has been addressed by several groups.[96a,b–100,189,307,309] Chan *et al.* investigated the selectivity of porcine pancreatic alpha-amylase on hydroxyethyl amylose, employing paper chromatography and GLC of size-fractionated enzyme products.[309] It was found that the active site, which is comprised of 5 subsites, tolerated only unsubstituted glucose residues at the –1 subsite. Furthermore, O-3 substitution was tolerated to a higher extent on the other subsites than was O-2 and O-6 substitution. Selectivity of subsequent digestion of methyl and cationic starches by alpha-amylase (*endo*) (*B. licheniformis*) and amyloglucosidase (*exo*) (*A. niger*) was studied by GLC.[307] By appropriate chemical-sample preparation, the substituent patterns could be differentiated for three types of glucose groups: from the nonreducing end, the internal glucose residues, and from the reducing end. Neither of the substituent types was tolerated on the residue interacting at the –1 subsite. Furthermore, both substituent types were accepted by the +1 subsite if localized on positions O-2 or O-6. For methyl amylose, O-2/O-6 di-substitution was also accepted at +1. Substituents were detected at all positions on the internal residues. The oligosaccharides obtained from cationic starches under such conditions were further studied by ESI-MS^n analysis[189] to afford additional sequence information. This enabled the differentiation of various wet and dry preparation methods.

Use of MALDI-TOF-MS allows relatively straightforward separation and identification of oligosaccharides with DPs up to ~30. This feature makes it possible to identify products from enzymatic debranching of amylopectin. Richardsson *et al.* employed MALDI-TOF-MS and ESI-MS for enzyme-aided structure analysis of cationic potato amylopectin.[314] By MALDI-TOF-MS, identification of the oligosaccharides with DP 6–20 obtained from hydrolysis with pullulanase could be performed. Use of ESI-MS^n, on the other hand, provided detailed information on the chemical structure of oligosaccharides with DP <5. However, the lack of suitable standards prevented quantification by MALDI-TOF-MS and ESI-MS.

In a later study, the substitution patterns in methylated potato starches obtained by methylation in granular suspension and in solution were compared.[99] The starch derivatives were successively hydrolyzed by alpha-amylase and amyloglucosidase, and the products were size-fractionated and further studied by MALDI-TOF-MS and GLC-MS. It was found by MALDI-TOF-MS that the oligosaccharides obtained from enzymatic hydrolysis of the starch modified in granular suspension had a higher DS and broader range of methyl substitution than those modified in solution. These findings, accompanied with quantification of glucose and the oligosaccharide and polysaccharide fractions, were indicative of a more-heterogeneous substituent distribution in methyl starch prepared in granular suspension.

(iv) Application to Cellulose Derivatives. Cellulose derivatives, in general, have higher DS values than their starch counterparts. Therefore, they are more resistant to enzymatic hydrolysis, and products range from glucose up to DPs comparable to that of the original polymer. Consequently, and in contrast to most starch derivatives, analysis of the chemical composition by, for instance, MS or HPAEC-PAD, is only possible on the small (low molecular weight) fraction (< ~5%) of the products from enzymatic hydrolysis of cellulose derivatives. To study the fraction of high molecular weight, additional chemical degradation is necessary.

Pioneer efforts in the application of enzymes for chemical structure analysis of cellulose derivatives include the work of Husemann,[28] Reese,[27] Bhattacharjee *et al.*,[31] Erikson *et al.*,[328,329] and Wirick[29,30] and Gelman.[330] Techniques available at those times employed, reducing sugar analysis, intrinsic viscosity, paper chromatography, and ultracentrifugation.

Enzyme-aided structure analysis of cellulose derivatives can be combined with size-based fractionation, as by SEC, and subsequent chemical structure analysis of the fractions with respect to molar mass (SEC), refractive index (RI), multi-angle light-scattering (MALS), and viscometric detection, or monomer composition (acid hydrolysis followed by HPAEC-PAD).[220] This approach permits detailed analysis of all products (both low and high mass) in the complex mixtures obtained from enzymatic hydrolysis. Horner *et al.* found that, for both CMC with DS 1.2 and DS 0.6, polymeric material of low DS remained after treatment with endoglucanase (*H. insolens*).[220] X-Ray analysis revealed that this material was non-crystalline. This finding might be explained by the formation of aggregates resistant to enzymatic attack, but at least indicates larger unsubstituted sequences. In addition, data showed that the particular enzyme used tolerated substitution at O-6 on at least one of the residues next to the hydrolyzed linkage, that is, at the −1 and/or +1 subsite. This conclusion is also supported by the studies of Kondo *et al.* on enzymatic degradability of regioselectively substituted MCs.[310]

CMCs prepared by the "reaction in reactive microstructures" concept, with solid NaOH in aprotic polar solvents, were also studied alongside conventional CMCs by this analytical approach, and differences with respect to heterogeneity ("blockiness") were detected in the material.[321] For cellulose sulfates, the efficiency of hydrolysis by endoglucanase (*H. insolens* Cel7B) has been found to be comparable to CMC at DS <1.[322] However, at higher DS, the enzyme is less efficient than for CMCs. Interestingly, O-6 sulfation hindered the enzyme most efficiently. Similar approaches for enzyme-aided substituent distribution analysis of MC and cellulose acetates can be found in the literature.[323–325] Fitzpatrick el al. presented an alternative approach, where enzymatic hydrolyzates of MC were fractionated by SEC, followed by DS analysis of the fractions by NMR spectroscopy.[326] The DS increased with molecular size up to a DS higher than the average of the original sample.

The group of Tjerneld has investigated the hydrolytic efficiency of endoglucanases on CMC, MC, HPC, and HPMC.[317,331–333] Significant differences were demonstrated, by MALDI-TOF-MS, reducing-sugar analysis, and SEC, not only between endoglucanases from the same organism, but also between endoglucanases of the same family but from different organisms. For some endoglucanases certain characteristics are known, such as active-site structure and number of sugar-binding subsites, but more detailed information on endoglucanase requirements, namely substituent position and type, remains to be explored. Momcilovic *et al.* analyzed the chemical structure of methylated oligosaccharides obtained by hydrolysis of MC with a highly efficient endoglucanase (*Bacillus agaradhaerens* Cel5A) by MALDI-qTOF-MS[2] and ESI-triple-quadrupole-MS.[2,287] Substitution at O-2 on the glucose residues in the −1 subsite hindered enzyme hydrolysis. In addition, the results suggested that the residue at the +1 subsite could even be trisubstituted (Fig.36). The latter finding was supported by Melander *et al.*, who employed MALDI-TOF-MS and ESI-IT-MS for the per-*O*-deuteriomethylated hydrolysis products.[313] In addition, it was shown that the reducing end of the enzyme products, namely that involved with the −1 subsite during hydrolysis, was preferentially un- or monosubstituted (Fig. 36).

Enebro *et al.* measured the concentration of unsubstituted oligosaccharides from hydrolysis by Tr Cel45A of two model CMCs.[320] This enzyme is unable to hydrolyze cellooligosaccharides of DP ≤5 and can, therefore, be used to analyze unsubstituted sequences of up to 5 glucose residues in cellulose derivatives. The CMCs, which were of similar molar mass and had almost identical DS, displayed large differences in rheological properties. For the CMC having the highest viscosity in solution, the concentration of unsubstituted DP4 and DP5 was double that of the CMC displaying lower viscosity, as determined by the standard addition method and MALDI-TOF-MS. In addition, a higher polydispersity was detected by SEC-MALS-RI for the high-

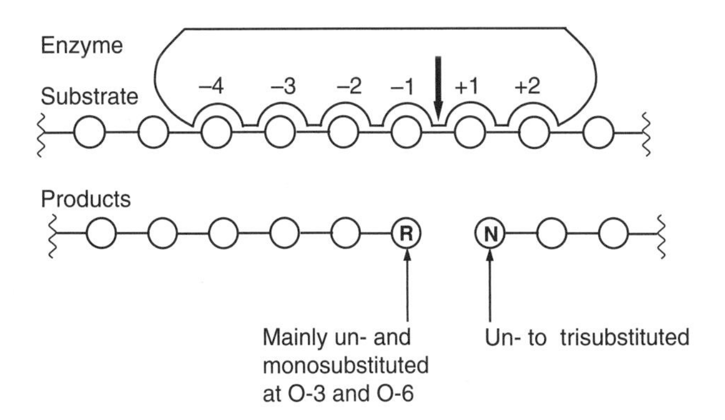

FIG. 36. Selectivity of the −1 and +1 subsites in the active site of *Bacillus agaradhaerens* Cel5A, as determined by Momcilovic *et al.*[287] and Melander *et al.*[313]

viscosity CMC after enzymatic hydrolysis. These findings indicate a more heterogeneous substituent distribution in the high-viscosity CMC.

Further investigations on Tr Cel45A selectivity revealed that the enzyme could tolerate one carboxymethyl substituent on the glucose residue interacting at the −1 subsite if it was located on the O-3 position, that is to say, carboxymethylation on O-6 and O-2 hindered hydrolysis.[308,334] In contrast, Tr Cel74A was found to tolerate monosubstitution on the O-6 position only. This is very important to take into consideration, as it implies that not only substituent distribution along polymer molecules, but also the position in the glucose residue affects the release of oligosaccharides during enzyme hydrolysis.

(v) Concluding Remarks. Although our knowledge on the tolerance of enzymes with respect to the substituent location has significantly increased during the past decade, it has still to be enhanced, especially with respect to patterns of sequences tolerated instead of just single subunits. In addition, the two points just mentioned, heterogeneity and regioselectivity of substitution, affect the enzyme degradation in independent and possibly contradictory ways. Two cellulose derivatives, having the same DS and a random substitution pattern along the chain, but one with preferred substitution at O-2, the other at O-6, may show completely different degrees of degradation. From Fig. 11 it is evident that the larger the differences of the rate constants, k_2, k_3, and k_6 are, the narrower is the substitution profile, and, thus the lower the probability of an unsubstituted dimeric sequence.

In contrast to the random chemical degradation approach just outlined, SEC of enzymatic digests reveals heterogeneities in the bulk material, as this should cause a much higher dispersity of the enzymatic hydrolyzate.

As a consequence of the strengths and weaknesses, enzyme-aided analysis is especially suited for starch derivatives. Their complex structure, comprising the high molecular weight and branched amylopectin and the lower molecular weight amylose, complicates the chemical approach. For this reason, the chemical approach has been applied preferably to amylose ethers as model compounds, while all studies of whole starch employ enzymatic or a mix of chemical and enzymatic methods in combination with fractionation, MS, and other subsequent analyses. A second reason for the application of enzymes with starch is the typically low DS of starch derivatives. On one hand, this enables complete degradation to low molecular weight compounds (in the mono- to oligosaccharide range), and at the same time it allows enrichment of the substituted sequences. In contrast, a nonselective chemical depolymerization is not sensitive enough to recognize deviations from the calculated random pattern (compare Fig. 12).

Tandem mass spectrometry and, especially, HPLC-MS^n have been shown to be valuable tools in these investigations, and future applications will probably include refined HPLC-MS^n methods. One interesting approach would be to employ our knowledge on enzyme selectivity in modeling the hydrolysis of glucan derivatives. This could allow for development of reference models that might improve interpretation of the substituent distribution.

Furthermore, sequential hydrolyses using various enzymes having different selectivities would be an interesting approach for further resolving structural differences by a kind of "glucan derivative mapping." Combining enzymes with known prerequisites, for instance, of the –1 subsite, could provide more-reliable data with respect to substituent distribution, since effects of regiosubstitution could be minimized. Although time-consuming and challenging, comprehensive analysis of such digests employing all the methods mentioned is currently the most promising way to trace back the pieces of the puzzle to the original picture.

4. Heterogeneities in the Bulk Material

a. Fractionation of Cellulose Derivatives.—It has already been emphasized at the beginning of this article that structural data obtained for the whole material can be the result of various superimposed sub-patterns, which cannot simply be recognized and differentiated by their averaged data. To obtain better insight into bulk heterogeneities, it is necessary to fractionate the material with respect to DS and/or DP, or at least record profiles, as can be determined for the molecular-weight distribution by SEC.[335] As already outlined, there is always a certain normal DS distribution over the polysaccharide chains, which in the ideal case is symmetrical around the average DS and becomes narrower with increasing DP, as can already be recognized for the di- to penta-saccharide fractions in Fig. 31. At the same average DS, normal distribution becomes narrower with increasing regioselectivity (see Fig. 11). In reality, we are faced with deviations from this model behavior. The goal of fractionation of the bulk material and subsequent analysis is to reveal such deviations, to quantify and interpret them, and thus differentiate between heterogeneities of the first (over the chains) and second order (on the chains).

While SEC aims to separate a polymeric mixture only with respect to size (hydrodynamic volume), investigations of substituent distribution requires a separation with respect to the chemistry of the constituents. Spurlin, very early on, recognized this problem and fractionated cellulose nitrate by repeated dissolution and precipitation.[20,336] The fractions obtained showed increasing viscosities and, related to this, increasing flexibility of solution-cast thin films. Saake *et al.* also made use of this common principle in polysaccharide isolation, in particular the (stepwise in this instance) precipitation of CMC by ethanol from aqueous solution.[321]

Chromatographic methods using solid phases might cause loss of insoluble or strongly adsorbed material and are difficult to perform on a preparative scale. On the analytical scale, Oudhoff *et al.* have determined DS-dependent profiles of CMC by CE-UV,[337] which allows comparison of samples, but does not give absolute information. Fitzpatrick *et al.* fractionated MC prior to and after enzymatic digestion by SEC and were able to show significant differences in gelation temperature and behavior for the fractions from the enzyme digest, the DS values of which increased as expected with DP. No significant dependence of DS on DP was observed in this study for the undegraded material.[326] Röhrling *et al.* established a method that enables detection of the amount of carbonyl and carboxyl groups (and thus the degree of oxidation) in cellulose derivatives and its in dependence on DP.[156,157]

A stepwise extraction procedure was performed with a set of MCs of DS 1.8, starting with the solid material.[160,338] With chloroform, small fractions of more highly substituted material was obtained (DS 2.1), and the main portion was extracted with a 1:10 mixture of MeOH and $CHCl_3$, and finally a residue was obtained that was no longer soluble in water, whereas the original sample could be completely dissolved in water at low temperature. This result emphasizes the pronounced interaction and interdependency of the wide range of polymer molecules present. Certain molecules, presumably those highly substituted, act as solubility enhancers (emulsifier) for the less soluble ones. This effect has also been observed in model studies of mixtures of "diblock"-oligosaccharides that consist of a permethylated cellotetraose and 1 to 4 unsubstituted glucose residues.[50,51]

b. Analysis of Fractions.—To finally recompose the puzzle, all pieces, namely the fractions, have to be further analyzed following the schedules here outlined. This is tedious but necessary, at least in basic research, to learn more about the rules and interactions responsible for the final outcome of a polymer analogous reaction (post-polymerization chemical modification). Thus the fractions obtained by stepwise extraction of MCs with solvents of increasing polarity showed increasing heterogeneity in the less-substituted fractions, a result that would have been overlooked in the overall averaged data.[160,338] Thus the fractionation procedure, with subsequent analysis of monomer compositions and oligomeric patterns, has the potential to shed light on certain border fractions in a material, at a much higher level of sensitivity than direct approaches.

c. Topochemical Differentiation of Starch Derivatives.—Starch is usually modified in a slurry of intact granules, since recovery and purification are facilitated by its insolubility in cold water, but semidry methods are also in use. Studies dealing with topochemical effects often compare the influence of heterogeneous and homogeneous, wet and "dry" methods on the substituent distribution.[96a,b–100,189] Heterogeneities with respect to granule size, channeling, and pore patterns, with regard to crystalline and amorphous domains, and between branched and linear structures are of interest. The latter two aspects are both partially related to the differentiation between amylose and amylopectin. A direct approach to observe the sites of reaction uses microscopic techniques in combination with staining procedures. SEM and reflectance confocal laser-scanning microscopy (R-CLSM), respectively, have been applied after cation exchange with thallium or silver, followed by reduction. It turned out that the region most highly reacted was that surrounding the central cavity, which is less organized, and concentric staining of the more amorphous growth rings has also been nicely detected for a good penetrating reagent.[183] Unfortunately, these methods are restricted to anionic derivatives.

In their investigations on starch methylation as a slow, kinetically controlled model reaction, Steeneken *et al.* found that the patterns of substitution are markedly influenced by the integrity of the granular structure.[96a,b] While there was no difference between the inner and outer part of the granules, reaction of the amorphous amylose componemt was favored and was most pronounced for the suspension process. Interestingly, the semi-dry process caused uneven distribution of base, which resulted in pasted lumps with higher DS, indicating a bimodal distribution in the material. Amylose and amylopectin can be differentiated by leaching the granules by mild acid treatment, hydrolyzing and dissolving mainly the amorphous amylose layers.[339] Alternatively, selective precipitation of amylose as a butanol complex (applicable only at very low DS), or of amylopectin with concanavalin A after complete dissolution,[340] or fractionation of both components with aqueous $MgSO_4$ has been applied.[96a,b] Branched and unbranched structural features can be separated following enzymatic digestion with alpha-amylase.[190] Separation by SEC as a consequence of the higher molecular weight of amylopectin has also been applied. The problem with all these procedures is that they are based on the behavior of *unmodified* amylose and amylopectin. Substitution changes such properties as solubility, enzymatic digestibility, or the hydrodynamic radius, in an unknown manner and extent. From all of these investigations, it can be concluded that the amorphous areas, the amylose component, and the branching-point areas of amylopectin are more accessible without gradient over the granule in slow aqueous reactions.

For cationic potato starches from various preparation technologies (slurry, paste, semi-dry, extruder), similar effects have been found: Wet procedures enable an even distribution over the whole granule or dissolved material, whereas dry procedures are more harsh and cause a less-selective reaction in the outer parts of the granules. In the extruder, where plasticization and reaction of the starch occur more or less simultaneously, regioselectivity of this kinetically controlled oxirane addition is drastically decreased, especially in the very early stage of the reaction.[130,189]

In contrast to methylation and cationization, acetylation of starch under aqueous alkaline conditions yields granules having high surface functionalization. Obviously, the acetylation is significantly faster than diffusion-controlled granule penetration. This topochemical effect has been analyzed[100] by a surface peeling method in concentrated $CaCl_2$ solution.[341] Binding of Ca^{2+} causes peripheral heat development and thus gelatinization and subsequent dissolution—as with a piece of candy in the mouth. The DS decreases strongly with distance from the granule surface. Another analytical approach is based on fractionation with respect to granule size. If the surface is preferably substituted, the DS will decrease with the size (diameter) of the granule, while at full penetration there is no size dependence.

Kok *et al.* have reported on the separation and quantification of amylose and amylopectin by CE, making use on the different iodine-binding behavior of these two components. It is to be expected that this procedure might also be applicable to starch derivatives.[342]

IV. Summary and Perspectives for the Future

It is to be expected that chemical modification of polysaccharides will become of even higher importance in the near future. In addition to the limitations of petroleum-based alternatives and the requirement of sustainability, one reason is the wide spectrum of size, structures, and functionalities available with polysaccharides and their derivatives. In addition to starch and cellulose, chitin, xylan, guaran, or other glucans such as dextran, pullulan, or schizophyllan can be modified to seek out new applications or to establish useful properties that combine their various native structural features with new functionalities. This implies also the increasing importance of structure-elucidation methodologies for better control of analogous polymer reactions and the understanding of structure–properties relationships. The past decades have shown enormous progress, made possible by improved instrumental techniques in the fields of separation, NMR, and especially soft-ionization MS, as well as by enhanced knowledge on the chemistry of carbohydrates and the action of enzymes on these biomaterials. Greater awareness of the limited meaning of such average structural parameters as the DS has led to focus on structural hierarchies of fibers and granules, and sophisticated analytical strategies have been developed. The frequently complex character of glucan derivatives has required mathematical models for interpretation of the results. At present this field has yet undeveloped potential for more advanced data elucidation in the future. Molecular modeling and fundamental studies regarding the influence of substituents on chemical or enzymatic degradability would also be valuable. Since one of the great shortcomings in development of methodology and its validation is the lack of standard compounds, the analysis of polysaccharide derivatives would significantly profit from further progress in the solid-phase synthesis of complex but defined oligosaccharides. And finally, young dedicated scientists should be attracted into this interdisciplinary field, to bring it forward in a creative and innovative manner, based on knowledge, controlled by intelligence, and driven by affection.

Acknowledgment

We thank Peter A. M. Steeneken, Bodo Saake, and P.M.'s research group for carefully reading the manuscript.

References

1. M. Payen, Mémoire sur la composition du tissue proper des plantes et du ligneux, *C. R. Hebd. Seances Acad. Sci.*, 7 (1838) 1052–1056.
2. A. Brogniart, T. J. Pelouze, and R. Dumas, Rapport sur un memoire de M. Payen relative à la composition de la matière ligneuse, *C. R. Hebd. Seances Acad. Sci.*, 8 (1839) 51–53.

3. H. Staudinger, Die Chemie der hochmolekularen organischen Stoffe im Sinne der Kekuléschen Strukturlehre/The structure of organic materials of high molecular weight interpreted according to Kekule's theory, *Helv. Chim. Acta*, 12 (1929) 1183–1197.
4. K. H. Meyer and H. Mark, Über den Bau des krystallisierten Anteils der Cellulose, *Ber. Dtsch. Chem. Ges*, 61 (1928) 593–614.
5. K. Freudenberg, Nachtrag zu der Mitteilung über Methylcellulose, *Ann. Chem.*, 461 (1928) 130–131.
6. W. N. Haworth, H. Machemer, and P. X. Polysaccharides., molecular structure of cellulose, *J. Chem. Soc.* (1932) 2270–2277.
7. K. Balser, L. Hoppe, T. Eichler, M. Wendel, and A. -J. Astheimer, W. Gerhartz, Y. S. Yamamoto, F. T. Campbell, R. Pfefferkorn, J. F. Rounsaville (Eds.), *Ullmanns Encyclopedia of Industrial Chemistry*, Vol. A5, VCH, Weinheim, 1986, pp. 419–459.
8. H. Krässig, J. Schurz, R. G. Steadman, K. Schliefer, and W. Albrecht, Cellulose, *Ind. Polym. Handb.*, 3 (2001) 1423–1500.
9. O. B. Wurzburg (Ed.), *Modified Starches: Properties and Uses*, CRC Press, Boca Raton FL, 1986.
10. C. F. Schönbein, Improvement in preparation of cotton-wool and other substances as substitutes for gunpowder, US-patent 1846, US 00004874 A 18461205.
11. C. F. Schönbein, *Ber. Naturforsch. Ges. Basel*, 7 (1847) 27.
12. (a) S. Zeisel, Über ein Verfahren zum quantitativen Nachweise von Methoxyl, *Monatshefte Chem.* 6 (1885) 989-997; (b) S. Zeisel, Zum quantitativen Nachweise von Methoxyl, *Monatshefte Chem.* 7 (1886) 406–409.
13. D. L. Miller, E. P. Samsel, and J. G. Cobler, Determination of acrylate and maleate esters in polymers by combined Zeisel and gas-chromatographic analysis, *Anal. Chem.*, 33 (1961) 677–680.
14. C. Alvarez-Lorenzo, R. A. Lorenzo-Ferreira, J. L. Gomez-Amoza, R. Martinez-Pacheco, C. Souto, and A. Concheiro, A comparison of gas-liquid chromatography, NMR spectroscopy and Raman spectroscopy for determination of the substituent content of general nonionic cellulose ethers, *J. Pharmaceut. Biomed. Anal.*, 20 (1999) 373–383.
15. *The United States Pharmacopeia*, 2006, Rockville, MD.
16. E. Husemann and M. Kafka, Über natürliche und synthetische Amylose. XII. Über die Verteilung der Substituenten in wasserlöslichen Amyloseäthern, *Macromol. Chem.*, 41 (1960) 208–217.
17. H. C. Svrivastava and K. V. Ramalingham, Distribution of hydroxyethyl groups in commercial hydroxyethyl starch, *Stärke/Starch*, 19 (1967) 295–300.

18. Liebig's Kali-Apparat zu organischen Analysen. In J. Liebig: *Anleitung zur Analyse organischer Körper.* Verlag Vieweg, Braunschweig, 1837 (1. Aufl.).
19. R. L. Whistler and W. W. Spencer, Distribution of substituents in corn starch granules with low degrees of substitution, *Arch. Biochem. Biophys.*, 87 (1960) 137–139.
20. High Polymers, Vol 5, Cellulose, Part I—III, E. Ott, H. M. Spurlin, and M. W. Grafflin (Eds.), Interscience Publishers New York, 1st ed. 1943, 2nd ed. 1954, 3rd ed. 1963.
21. R. L. Whistler, J. L. Goatley, and W. W. Spencer, Effect of drying on the physical properties and chemical reactivity of corn-starch granules, *Cereal Chem.*, 36 (1959) 84–90.
22. (a) G. C. S. Dutton, Applications of gas-liquid chromatography of carbohydrates: Part I, *Adv. Carbohydr. Chem.*, 28 (1973) 11-160; (b) G. C. S. Dutton, Applications of gas-liquid chromatography of carbohydrates: Part II, *Adv. Carbohydr. Chem.*, 30 (1974) 9–110.
23. (a) N. K. Kochetkov and O. S. Chizov, Mass spectrometry of carbohydrate derivatives, *Adv. Carbohydr. Chem.*, 21 (1966) 39-93; (b) J. Lönngren, and S. Svensson, Mass Spectrometry and structural analysis of natural carbohydrates, *Adv. Carbohydr. Chem.*, 29 (1974) 41–106.
24. (a) L. D. Hall, Nuclear magnetic resonance, *Adv. Carbohydr. Chem.*, 19 (1964) 51–93; (b) P. A. J. Gorin, Carbon-13 Nuclear magnetic resonance spectroscopy of polysaccharides, *Adv. Carbohydr. Chem. Biochem.*, 38 (1981) 13–104.
25. S. -I. Hakomori, A rapid permethylation of glycolipids and polysaccharides catalyzed by methylsulfinyl carbanion in dimethyl sulfoxide, *J. Biochem. (Tokyo)*, 55 (1964) 205–208.
26. (a) H. Björndal, C. G. Hellerqvist, B. Lindberg, and S. Svensson, Gas–liquid chromatography and mass spectrometry in methylation analysis of polysaccharides, *Angew. Chem. Int. Ed.*, 9 (1970) 610-619; (b) Gas-Flüssigkeits-Chromatographie und Massenspektrometrie bei der Methylierungsanalyse von Polysacchariden, *Angew. Chem.*, 82 (1970) 643–674.
27. E. T. Reese, Biological degradation of cellulose derivatives, *Ind. Eng. Chem.*, 49 (1957) 89–93.
28. E. Husemann, Über den sauren, fermentativen und oxydativen Abbau von Cellulose und Cellulosederivaten, *Das Pap.*, 8 (1954) 157–162.
29. M. G. Wirick, Study of the substitution pattern of hydroxyethylcellulose and its relationshipf to enzymic degradation, *J. Polym. Sci. A1*, 6 (1968) 1705–1718.
30. M. G. Wirick, A study of the enzymic degradation of cmc and other cellulose ethers, *J. Polym. Sci. A1*, 6 (1968) 1965–1974.

31. S. S. Bhattacharjee and A. S. Perlin, Enzymatic degradation of carboxymethylcellulose and other cellulose derivatives, *J. Polym. Sci. C*, 36 (1971) 509–521.
32. A. Koschella, D. Fenn, N. Illy, and Th. Heinze, Regioselectively functionalized cellulose derivatives: A mini review, *Macromol. Symp.*, 244 (2006) 59–73.
33. B. Volkert and W. Wagenknecht, Substitution patterns of cellulose ethers—influence of the synthetic pathway, *Macromol. Symp.*, 262 (2006) 97–118.
34. Th. Heinze and K. Petzold, Cellulose chemistry: Novel products and synthetic paths, in M. N. Belgacem, A. Gandini (Eds.), Monomer, Polymers and Composites from Renewable Resources, Elsevier Ltd, Oxford, 2008, pp. 434–368.
35. Th. Heinze, T. Liebert, and A. Koschella, *Esterification of Polysaccharides*, Springer-Verlag, Heidelberg, Germany, 2006.
36. D. Klemm, B. Heublein, H. -P. Fink, and A. Bohn, Cellulose: Fascinating biopolymer and sustainable raw material, *Angew. Chem. Int. Ed.*, 44 (2005) 3358–3393.
37. D. Klemm, B. Philipp, Th. Heinze, U. Heinze, and W. Wagenknecht, Comprehensive cellulose chemistry, Wiley VCH, Weinheim, 1998, Vol 1: Fundamentals and analytical methods, Vol 2: Functionalization of cellulose.
38. S. Richardson and L. Gorton, Characterisation of the substituent distribution in starch and cellulose derivatives, *Anal. Chim. Acta*, 497 (2003) 27–65.
39. P. Mischnick, J. Heinrich, M. Gohdes, O. Wilke, and N. Rogmann, Structure Analysis of 1,4-Glucan Derivatives, *Macromol. Chem. Phys.*, 201 (2000) 1985–1995.
40. J. A. Rodrigues, A. M. Taylor, D. P. Sumpton, J. C. Reynolds, R. Pickford, and J. Thomas-Oates, Mass spectrometry of carbohydrates: Newer aspects, *Adv. Carbohydr. Chem. Biochem.*, 61 (2008) 58–141.
41. F. Hillenkamp and J. Peter-Katalinic (Eds.), *MALDI MS. A practical guide to instrumentation, methods and applications*, Wiley-VCH Verlag GmbH, Weinheim, Germany, 2007.
42. F. X. Redl, O. Kothe, K. Rockl, W. Bauer, and J. Daub, Azulene-appended cellulose: Synthesis, optical and chiroptical properties, film formation by electrochemical oxidation, *Macromol. Chem. Phys.*, 201 (2000) 2091–2100.
43. W. Holzer, A. Penzkofer, F. X. Redl, M. Lutz, and J. Daub, Excitation energy density dependent fluorescence behavior of a regioselectively functionalized tetraphenylporphyrin–cellulose conjugate, *Chem. Phys.*, 282 (2002) 89–99.
44. K. Petzold, A. Koschella, D. Klemm, and B. Heublein, Silylation of cellulose and starch—selectivity, structure analysis, and subsequent reactions, *Cellulose*, 10 (2003) 251–269.
45. H. Kern, S. W. Choi, G. Wenz, J. Heinrich, L. Ehrhardt, P. Mischnick, P. Garidel, and A. Blume, Synthesis, control of substitution pattern and phase transitions of 2,3-di-*O*-methyl-cellulose, *Carbohydr. Res.*, 326 (2000) 67–79.

46. R. Adden, A. Bösch, and P. Mischnick, Novel possibilities by cationic ring-opening polymerisation of cyclodextrin derivatives: Preparation of a copolymer bearing block-like sequences of tri-*O*-methylglucosyl units, *Macromol. Chem. Phys.*, 205 (2004) 2072–2079.
47. A. Bösch, M. Nimtz, and P. Mischnick, Mechanistic studies on cationic ring-opening polymerization of cyclodextrin derivatives using various Lewis acids, *Cellulose*, 13 (2006) 493–507.
48. A. Bösch and P. Mischnick, Bifunctional building blocks for glyco-architectures by $TiCl_4$-promoted ring opening of cyclodextrin derivatives, *Biomacromolecules*, 8 (2007) 2311–2320.
49. H. Kamitakahara, A. Yoshinaga, H. Aono, F. Nakatsubo, D. Klemm, and W. Burchard, New approach to unravel the structure-property relationship of methylcellulose, *Cellulose*, 15 (2008) 797–801.
50. H. Kamitakahara, F. Nakatsubo, and D. Klemm, Block- copolymers of tri-*O*-methylated and unmodified cello-oligosaccharides as model compounds for methylcellulose and its dissolution /gelation behaviour, *Cellulose*, 13 (2006) 375–392.
51. H. Kamitakahara, F. Nakatsubo, and D. Klemm, New class of carbohydrate-based non-ionic surfactants: Diblock co-oligosaccharides of tri-*O*-methylated and unmodified cellooligosaccharides, *Cellulose*, 14 (2007) 513–528.
52. H. Kamitakahara, A. Koschella, Y. Mikawa, F. Nakatsubo, Th. Heinze, and D. Klemm, Syntheses and comparison of 2,6-di-*O*-methyl celluloses from natural and synthetic celluloses, *Macromol. Biosci.*, 8 (2006) 690–700.
53. Th. Heinze and T. Liebert, Unconventional methods in cellulose functionalization, *Prog. Polym. Sci.*, 26 (2001) 1689-1762.
54. Th. Heinze, T. Liebert, B. Heublein, and S. Hornig, Functional polymers based on dextrans, *Adv. Polym. Sci.*, 205 (2006) 199–291.
55. J. M. Garcia Fernandez, C. Ortiz Mellet, and J. Defaye, Glyconanocavities: Cyclodextrins and beyond, *J. Incl. Phenom. Macrocycl. Chem.*, 56 (2006) 149–159.
56. T. Lecourt, A. Herault, A. J. Pearce, M. Sollogoub, and P. Sinaÿ, Triisobutylaluminium und diisobutylaluminium hydride as molecular scalpels: The regioselective stripping of perbenzylated sugars and cyclodextrins, *Chem. Eur. J.*, 10 (2004) 2960–2971.
57. T. Lecourt, J. -M. Mallet, and P. Sinaÿ, An efficient preparation of $6^{I,IV}$ dihydroxy permethylated β-cyclodextrin, *Carbohydr. Res.*, 338 (2003) 2417–2419.
58. L. M. J. Kroon-Batenburg, J. Kroon, and M. G. Nordholt, Chain modulus and intramolecular hydrogen bonding in native and regenerated cellulose fibers, *Polym. Commun.*, 27 (1986) 290–292.

59. L. M. J. Kroon-Batenburg and J. Kroon, The crystal and molecular structures of cellulose I and II, *Glycoconj. J.*, 14 (1997) 677–690.
60. Y. Nishiyama, J. Sugiyama, H. Chanzy, and P. Langan, Crystal structure and hydrogen bonding system in cellulose I from synchrotron X-ray and neutron fiber diffraction, *J. Am. Chem. Soc.*, 125 (2003) 14300–14306.
61. H. Yokota, The mechanism of cellulose alkalization in the isopropyl alcohol-water sodium hydroxide-cellulose system, *J. Appl. Polym. Sci.*, 30 (1985) 263–277.
62. H. Yokota, Alkalization mechanism of cellulose in hydroxypropyl alcohol-water-sodium hydroxide-cellulose system, ***J. Appl. Polym. Sci.***, 32 (1986) 3423–3433.
63. S. Kamel and K. Jahngier, Optimization of carboxymethylation of starch in organic solvents, ***Int. J. Polym. Mat.***, 56 (2007) 511-519.
64. R. Dönges, Non-ionic cellulose ethers, *Br. Polym. J.*, 23 (1990) 315–326.
65. J. F. Kennedy, Chemically reactive derivatives of polysaccharides, *Adv. Carbohydr. Chem. Biochem.*, 29 (1974) 305–405.
66. Th. Heinze and A. Koschella, Carboxymethyl ethers of cellulose and starch—A review, *Macromol. Symp.*, 223 (2005) 13–39.
67. N. D. Sachinvala, D. L. Winsor, O. A. Hamed, K. Maskos, W. Niemczura, G. J. Tregre, W. Glasser, and N. R. Bertoniere, The physical and NMR characterization of allyl- and crotylcelluloses, *J. Polym. Sci. A Polym. Chem.*, 38 (2000) 1889–1902.
68. W. G. Glasser, Prospects for future applications of cellulose acetate, ***Macromol. Symp.***, 208 (2004) 371-394.
69. T. F. Liebert and Th. Heinze, Tailored cellulose esters: Synthesis and structure determination, *Biomacromolecules*, 6 (2005) 333–340.
70. C. Altaner, B. Saake, M. Tankanen, J. Eyzaguirre, C. B. Faulds, P. Biely, L. Viikari, M. Siika-aho, and J. Puls, Regioselective deacetylation of cellulose acetates by acetyl xylan esterases of different families, *J. Biotech.*, 105 (2003) 95–104.
71. A. Koschella and Th. Heinze, Unconventional cellulose products by fluorination of tosyl cellulose, *Macromol. Symp.*, 197 (2003) 243–254.
72. H. Baumann, A. Richter, D. Klemm, and V. Faust, Concepts for preparation of novel regioselective modified cellulose derivatives sulfated, aminated, carboxylated and acetylated for hemobompatible ultrathin coatings on biomaterials, *Macromol. Chem. Phys.*, 201 (2000) 1950-1962.
73. K. T. Andrews, N. Klatt, Y. Adams, P. Mischnick, and R. Schwartz-Albiez, Inhibition of chondroitin-4-sulfate-specific adhesion of *Plasmodium falciparum*

infected erythrocytes by sulfated polysaccharides, *Infect. Immun.*, 73 (2005) 4288–4294.

74. K. Hettrich, W. Wagenknecht, B. Volkert, and S. Fischer, New possibilities of the acetosulfation of cellulose, *Macromol. Symp.*, 262 (2008) 162–169.
75 A. Richter and D. Klemm, Regioselective suflation of trimethylsilyl cellulose using different SO_3-complexes, *Cellulose*, 10 (2003) 133–138.
76. W. Wagenknecht, I. Nehls, A. Stein, D. Klemm, and B. Philipp, Synthesis and substituent distribution of sodium cellulose sulfates via trimethylsilyl cellulose as intermediate, *Acta Polym.*, 43 (1992) 266–269.
77. M. Gericke, T. Liebert, and Th. Heinze, Interaction of ionic liquids with polysaccharides, 8—synthesis of cellulsoe sulfates suitable for polyelectrolyte complex formation, *Macromol. Biosci.*, 9 (2009) 343–353.
78. T. Isogai, M. Yanagisawa, and A. Isogai, Degrees of polymerization (DP) and DP distribution of cellouronic acids prepared from alakli-treated celluloses and ball-milled native celluloses by TEMPO-mediated oxidation, *Cellulose*, 16 (2009) 117–127.
79. B. Ding, Y. Ye. J. Cheng, K. Wang, J. Luo, and B. Jiang, TEMPO-mediated selective oxidation of substituted polysaccharides-an efficent approach for the determination of the degree of substitution, *Carbohydr. Res.*, 343 (2009) 3112–3116.
80. D. Horton and D. M. Clode, Synthesis of the 6-aldehydo derivative of cellulose, and a mass spectrometric method for determining position and degree of substitution by carbonyl groups in oxidzed polysaccharides, *Carbohydr. Res.*, 19 (1971) 329–337.
81. W. Burchard, Solubility and solution structure of cellulose derivatives, *Cellulose*, 10 (2003) 213–225.
82. L. Schulz, B. Seger, and W. Burchard, Structures of cellulose in solution, *Macromol. Chem. Phys.*, 201 (2000) 2008–2022.
83. S. Köhler and T. Heinze, New solvents for cellulose: Dimethyl sulfoxide/ammonium fluorides, *Macromol. Biosci.*, 7 (2007) 307–314.
84. Th. Heinze, T. Lincke, D. Fenn, and A. Koschella, Efficient allylation of cellulose in dimethyl sulfoxide/tetrabutylammonium fluoride trihydrate, *Polym. Bull.*, 61 (2008) 1–9.
85. O. A. El Seoud, A. Koschella, L. C. Fidale, S. Dorn, and T. Heinze, Applications of ionic liquids in carbohydrate chemistry: A window of opportunities, *Biomacromolecules*, 8 (2007) 2629–2647.
86. S. Köhler, T. Liebert, and T. Heinze, Interaction of ionic liquids with polysaccharides. VI. pure cellulose nanoparticles from trimethylsilyl cellulose synthesized in ionic liquids, *J. Polym. Sci. A Polym. Chem.*, 46 (2008) 4070–4080.

87. S. Köhler, T. Liebert, Th. Heinze, A. Vollmer, P. Mischnick, E. Möllmann, and W. Becker, Interaction of ionic liquids with polysaccharides 9. Hydroxyalkylation of cellulose without additional inorganic bases, *Cellulose*, 17 (2010) 437–448.
88. L. C. Fidale, N. Ruiz, Th. Heinze, and O. A. El Seoud, Cellulose swelling by aprotic and protic sovlents: What are the similarities and differences?, *Macromol. Chem. Phys.*, 209 (2008) 1240–1254.
89. J. R. Daniel, R. L. Whistler, and H. Röper, Starch, in E. S. Wilks (Ed.), *Industrial Polymers Handbook*, Wiley-VCH Verlag GmbH, Weinheim, Germany, 4, 2001, pp. 2227-2264.
90. D. French, Organization of starch granules, in R. L. Whistler, J. N. BeMiller-*Starch: Chemistry and Technology*, Academic Press, Orlanda, FL, 1984, pp. 184–247.
91. P. J. Jenkins and A. M. Donald, The influence of amylose on starch granule structure, *Int. J. Biol. Macromol.*, 17 (1995) 315–321.
92. D. J. Gallant, B. Bouchet, and P. M. Baldwin, Microscopy of starch: Evidence of a new level of granule organization, *Carbohydr. Polym.*, 32 (1997) 177–191.
93. J. E. Fannon, J. A. Gray, N. Gunawan, K. C. Huber, and J. N. BeMiller, Heterogeneity of starch granules and the effect of granule channelization on starch modification, *Cellulose*, 11 (2004) 247–254.
94. P. Tomasik and C. H. Schilling, Chemical modification of starch, *Adv. Carbohydr. Chem. Biochem.*, 59 (2004) 175–403.
95. R. L. Whistler, M. A. Madson, J. Zhao, and J. R. Daniel, Surface derivatization of corn starch granules, *Cereal Chem.*, 75 (1998) 72–74.
96. (a) P. A. M. Steeneken and E. Smith, Topochemical effects in the methylation of starch, *Carbohydr. Res.*, 209 (1991) 239-249; (b) J. Muetgeert, Fractionation of starch, *Adv. Carbohydr. Chem.*, 16 (1961) 299–333.
97. P. A. M. Steeneken, Reactivity of amylose and amylopectin in potato starch, *Stärke/Starch*, 36 (1984) 13–18.
98. P. A. M. Steeneken and A.J.J. Woortman, Substitution pattern in methylated starch as studied by enzymic degradation, *Carbohydr. Res.*, 258 (1994) 207–221.
99. P. A. M. Steeneken, A. C. Tas, A. J. J. Woortman, P. Sanders, P. J. H. C. Mijland, and L. G. R. de Weijs, Substitution patterns in methylated starch as revealed from the structure and composition of fragments in enzymatic digests, *Carbohydr. Res.*, 343 (2008) 2411–2416.
100. P. A. M. Steeneken and A. J. J. Woortman, Surface effects in the acetylation of granular potato starch, *Carbohydr. Res.*, 343 (2008) 2278–2284.

101. J. W. Berry, A. J. Deutschman Jr., and J. P. Evans, The distribution of substituents in vinyl starch, *J. Org. Chem.*, 29 (1964) 2619–2620.
102. K. Petzold, L. Einfeldt, W. Günther, A. Stein, and D. Klemm, Regioselective functionalization of starch: Synthesis and ^{1}H NMR characterization of 6-*O* silylethers, *Biomacromolecules*, 2 (2001) 965–969.
103. P. F. Tankam, R. Müller, P. Mischnick, and H. Hopf, Alkynyl polysaccharides: Synthesis of propargyl potato starch followed by subsequent derivatizations, *Carbohydr. Res.*, 14 (2007) 2049–2060.
104. A. Richter and W. Wagenknecht, Synthesis of amylose acetates and amylose sulfates with high structural uniformity, *Carbohydr. Res.*, 338 (2003) 1397–1401.
105. V. Stigsson, G. Kloow, and U. Germgard, The influence of the solvent system used during manufacturing of CMC, *Cellulose*, 13 (2006) 705–712.
106. F. Cheng, L. Guifeng, F. Jianxin, and Z. Jingwu, Characterisation of carboxymethyl cellulose synthesized in two phase medium C_6H_6 -C_2H_5OH. I. Distribution of substituent groups in the anhydroglucose unit, *J. Appl. Polym. Sci.*, 61 (1996) 1831–1838.
107. N. Olaru, O. Olaru, A. Stoleriu, and D. Timpu, Carboxymethylcellulose synthesis in organic media containing ethanol and/or acetone, *J. Appl. Polym. Sci.*, 67 (1998) 481–486.
108. Th. Heinze, T. Liebert, P. Klüfers, and F. Meister, Carboxymethylation of cellulose in unconventional media, *Cellulose*, 6 (1999) 153–165.
109. P. Adams, A. Zegeer, H. J. Bohnert, and R.G.- Jensen, Anion exchange separation and pulsed amperometric detection of inositols from flower petals, *Anal. Biochem.*, 214 (1993) 321–324.
110. G. Von Helden, T. Wyttenbach, and M. T. Bowers, Inclusion of a MALDI ion source in the ion chromatography technique: Conformational information on polymer and biomolecular ions, *Int. J. Mass Spectrom. Ion Process.*, 146/147 (1995) 349–364.
111. E. J. Roberts and S. P. Rowland, Effects of selected reaction conditions and structural variations on the distribution on 2-(diethylamino)ethyl groups on D-glucopyranosyl residues of 2-(diethylamino)ethylated cellulose, starch, and the anomers of methyl 4,6-*O*-benzylidene-D-glucopyranoside, *Carbohydr. Res.*, 5 (1967) 1–12.
112. E. J. Roberts, C. P. Wade, and S. P. Rowland, Effect of base concentration upon the reactivities of the hydroxyl groups in methyl D-glucopyranosides, *Carbohydr. Res.*, 17 (1971) 393–399.

113. E. J. Roberts, C. P. Wade, and S. P. Rowland, Neighbouring group effects in the reactivities of hydroxyl groups in D-glucopyranosides, *Carbohydr. Res.*, 21 (1972) 357–367.
114. T. E. Timell, H. M. Spurlin, and I. I. Carboxymethylcellulose, The distribution of the substituents in partially substituted carboxymethylcelluloses, *Sven. Papper-stidning*, 55 (1952) 700–708.
115. X. Lin, T. Qu, and S. Qi, Kinetics of carboxymethylation of cellulose in the isopropyl alcohol system, *Acta Polym.*, 41 (1990) 220–222.
116. T. Salmi, D. Valtakari, E. Paatero., B. Holmbom, and R. Sjöholm, Kinetic study of the carboxymethylation of cellulose, *Ind. Eng. Chem. Res.*, 33 (1994) 1454–1459.
117. N. Olaru and O. Olaru, Mathematical models for the synthesis of carboxymethylation in the isopropyl alcohol system, *Cellulose Chem. Technol.*, 26 (1992) 685–690.
118. A. Hedlund and U. Germgård, Some aspects on the kinetics of etherification in the peparation of CMC, *Cellulose*, 14 (2007) 161–169.
119. A. Baar, W. -M. Kulicke, K. Szablikowski, and R. Kiesewetter, Nuclear magnetic resonance spectroscopic characterization of carboxymethylcellulose, *Macromol. Chem. Phys.*, 195 (1994) 1483–1492.
120. E. A. Kragten, J. P. Kamerling, and J. F. G. Vliegenthart, Composition analysis of carboxymethylcellulose by high-pH anion-exchange chromatography with pulsed amperometric detection, *J. Chromatogr.*, 623 (1992) 49–53.
121. S. G. Zeller, G. W. Griesgraber, and G. R. Gray, Analysis of positions of substitution of *O*-carboxymethyl groups in partially *O*-carboxymethylated cellulose by the reductive-cleavage method, *Carbohydr. Res.*, 211 (1991) 41–45.
122. M. M. Abdel-Malik and M. Yalpani, Determination of substitiuent distribution in cellulose ethers by means of carbon-13 NMR spectroscopy, in J. F. Kennedy, G. O. Phillips, and P. A. Williams (Eds.), *Cellulose*, Horwood, Chichester, UK, 1990, pp. 263–268.
123. K. Niemelä and E. Sjöström, Characterization of hardwood-derived carboxymethyl cellulose by gas-liquid chromatography and mass spectrometry, *Polym. Comm.*, 30 (1989) 254–256.
124. J. Reuben and H. T. Conner, Analysis of the carbon-13 N.M.R. spectrum of hydrolyzed *O*-(carboxymethyl)cellulose: Monomer compositions and substitution pattern, *Carbohydr. Res.*, 115 (1983) 1–13.
125. F. F. L. Ho and D. W. Klosiewicz, Proton NMR spectrometry for determination of substituents and their distribution in carboxymethylcellulose, *Anal. Chem.*, 52 (1980) 913–916.
126. F. A. Buijtenhuijs and R. Bonn, Distribution of substituents in CMC, *Papier*, 31 (1977) 525–527.

127. I. Croon and C. B. Purves, The distribution of the substituents in partially substituted carboxymethyl cellulose, *Sven. Papperstidning*, 62 (1959) 876–882.
128. A. N. Jyothi, S. N. Moorthy, and K. N. Rajasekharan, Studies on the synthesis and properties of hydroxypropyl derivatives of cassava (*Manihot esculenta* Crantz) starch, *J. Sci. Food Agric.*, 87 (2007) 1964–1972.
129. P. Mischnick, Determination of the pattern of substitution of hydroxyethyl and hydroxypropyl cyclomaltoheptaose derivatives, *Carbohydr. Res.*, 192 (1989) 233–241.
130. S. Radosta, W. Vorwerg, A. Ebert, A. Haji Begli, D. Grülc, and M. Wastyn, Properties of low-substituted cationic starch derivatives prepared by different derivatisation processes, *Starch/Stärke*, 56 (2004) 277–287.
131. O. Wilke and P. Mischnick, Analysis of cationic starches: Determination of the substitution pattern of *O*-(2-hydroxy-3-trimethylammonium)propyl ethers, *Carbohydr. Res.*, 275 (1995) 309–318.
132. V. Goclik and P. Mischnick, Determination of the DS and the substituent distribution of cationic alkyl polyglycosides and cationic starch ethers by GLC after dealkylation with morpholine, *Carbohydr. Res.*, 338 (2003) 733–741.
133. P. M. van der Velden, B. Rijpkema, C. A. Smolders, and A. Bantjes, Reactions with 1,3-propane sultone for the synthesis of cation-exchange membranes, *Eur. Polym. J.*, 13 (1977) 37–39.
134. S. Knop, H. Thielking, and W. -M. Kulicke, Simplex membranes of sulphoethylcellulose and poly(diallyl-dimethylammonium chloride) for the pervaporation of water-alcohol mixtures, *J. Appl. Polym. Sci.*, 77 (2000) 3169–3177.
135. M. C. Vieira, D. Klemm, L. Einfeldt, and G. Albrecht, Dispersing agents for cement based on modified polysaccharides, *Cement Concr. Res.*, 35 (2005) 883–890.
136. N. Rogmann, J. Seidel, and P. Mischnick, Formation of unexpected substitution patterns in sulfonylbutylation of cyclomaltoheptaose promoted by host-guest interaction, *Carbohydr. Res.*, 327 (2000) 269–274.
137. A. Stein and D. Klemm, Syntheses of cellulose derivatives via *O*-triorganosilyl cellulose, 1: Effective synthesis of organic cellulose esters by acylation of trimethylsilyl celluloses, *Macromol. Chem. Rapid*, 9 (1988) 569–573.
138. W. Mormann, J. Demeter, and T. Wagner, Partial silylation of cellulose with predictable degree of silylation—Stoichiometric silylation with hexamethyldisilazane, *Macromol. Chem. Phys.*, 200 (1999) 693–697.
139. W. Mormann and J. Demeter, Controlled desilylation of cellulose with stoichiometric amounts of water in the presence of ammonia, *Macromol. Chem. Phys.*, 201 (2000) 1963–1968.

140. P. Mischnick, M. Lange, M. Gohdes, A. Stein, and K. Petzold, Trialkylsilyl derivatives of cyclomaltoheptaose, cellulose, and amylose: Rearrangement during methylation analysis, *Carbohydr. Res.*, 277 (1995) 179–187.
141. Y. S. Ovodoc and E. V. Evthushenko, Partial methylation of carbohydrates, *Carbohydr. Res.*, 27 (1973) 169–174.
142. S. J. Angyal, Complexes of metal cations with carbohydrates in solution, *Adv. Carbohydr. Chem. Biochem.*, 47 (1989) 1–43.
143. H. M. I. Osborn, V. A. Brome, L. M. Harwood, and W. G. Suthers, Regioselective C- 3-O-acylation and *O*-methylation of 4,6-*O*-benzylidene-β-D-gluco- and galactopyranosides displaying a range of anomeric substituents, *Carbohydr. Res.*, 332 (2001) 157–166.
144. R. Dicke, A straight way to regioselectively functionalized polysaccharide esters, *Cellulose*, 11 (2004) 255–263.
145. (a) C. Deus, H. Friebolin, and E. Siefert, Partially acetylated cellulose. Synthesis and determination of the substituent distribution via proton NMR spectroscopy, *Makromol. Chem.*, 192 (1991) 75-83; (b) J. Heinrich and P. Mischnick, Determination of the substitution pattern in the polymer chain of cellulose acetates, *J. Polym. Sci. A: Polym. Chem.*, 37 (1999) 3011–3016.
146. (a) P. A. M. Steeneken, personal communicaiton; (b) D. Heins, W. M. Kulicke, P. Käuper, and H. Thielking, Characterization of acetyl starch by means of NMR spectroscopy and SEC/MALLS in comparison with hydroxyethyl starch, *Starch/Stärke*, 50 (1998) 431–437.
147. Th. Heinze and T. Liebert, Chemcial characteristics of cellulose acetate, *Macromol. Symp.*, 208 (2004) 167–237.
148. M. Gohdes and P. Mischnick, Determination of the substitution pattern in the polymer chain of cellulose sulfates, *Carbohydr. Res.*, 309 (1998) 109–115.
149. T. Morooka, M. Norimoto, and T. Yamada, Cyanoethylated cellulose prepared by homogeneous reaction in paraformaldehyde-DMSO system, *J. Appl. Polym. Sci.*, 32 (1986) 3575–3587.
150. A. Hebeish, T. Abd El-Thalouth, and M. A. El-Kashouti, Chemical modification of starch II. Cyanoethylation, *J. Appl. Polym. Sci.*, 26 (1981) 171–176.
151. D. L. Verraest, E. Zitha-Bovens, J. A. Peters, and H. van Bekkum, Preparation of *O*-(aminopropyl)inulin, *Carbohydr. Res.*, 310 (1998) 109–115.
152. A. Gonera, V. Goclik, M. Baum, and P. Mischnick, Preparation and structural characterisation of *O*-aminopropyl starch and amylose, *Carbohydr. Res.*, 337 (2002) 2263–2272.
153. B. Volkert, W. Wagenknecht, and M. Mai, Structure-property relationship of cellulose ethers—Influence of the synthetic pathway on cyanoethylation, in:

Cellulose Solvents, T. Liebert, Th. Heinze, and K. Edgar (Eds.), ACS publisher, Washington, ACS Symp. Ser. 1033 (2010), in press (doi: 10.1021/bk-2010-1033.ch018).

154. H. Schleicher and H. Lang, Carbonyl and carboxyl groups in pulps and cellulose products, *Das Pap.*, 12 (1994) 765–768.

155. A. Potthast, T. Rosenau, J. Sartori, H. Sixta, and P. Kosma, Hydrolytic processes and condensation reactions in the cellulose solvent system *N,N*-diemthylacetamide/lithium chloride. Part 2: Degradation of cellulose, *Polymer*, 44 (2003) 7–17.

156. J. Röhrling, A. Potthast, T. Rosenau, T. Lange, G. Ebner, H. Sixta, and P. Kosma, A novel method for the determination of carbonyl groups in cellulosics by flourescence labeling. 1. Method development, *Biomacromolecules*, 3 (2002) 959–968.

157. J. Röhrling, A. Potthast, T. Rosenau, T. Lange, A. Borgards, H. Sixta, and P. Kosma, A novel method for the determination of carbonyl groups in cellulosics by flourescence labeling. 2. Validation and application, *Biomacromolecules*, 3 (2002) 969–975.

158. A. Potthast, J. Röhrling, T. Rosenau, H. Sixta, and P. Kosma, A novel method for the determination of carbonyl groups in cellulosics by flourescence labeling. 3. Monitoring oxidative processes, *Biomacromolecules*, 4 (2003) 743–749.

159. A. Bohrn, A. Potthast, S. Schiehser, T. Rosenau, H. Sixta, and P. Kosma, The FDAM method: Determination of carboxyl profiles in cellulosic materials by combining groups selective fluorescence labeling with GPC, *Biomacromolecules*, 7 (2006) 1743–1750.

160. P. Mischnick and R. Adden, Fractionation of polysaccharide derivatives and subsequent analysis to differentiate heterogeneities on various hierarchical levels, *Macromol. Symp.*, 262 (2008) 1–7.

161. H. M. Spurlin, Mechanisms of cellulose reactions, *Trans. Electrochem. Soc.*, 73 (1938) 95–109.

162. H. M. Spurlin, Arrangement of substituents in cellulose derivatives, *J. Am. Chem. Soc.*, 61 (1939) 2222–2227.

163. H. M. Spurlin, Reactivity and reactions of cellulose, in E. Ott, H. M. Spurlin, and M. W. Grafflin, Eds., *Cellulose and Cellulose Derivatives, Part II*, 2nd edn Interscience, New York, 1954, pp. 673–712.

164. J. Reuben, Analysis of the ^{13}C-N.M.R. Spectra of hydrolyzed and methanolyzed *O*-methylcelluloses; monomer compositions and models for their description, *Carbohydr. Res.*, 157 (1986) 201–213.

165. T. Liebert and Th. Heinze, Induced phase separation: a new synthetic concept in cellulose chemistry, in: Cellulose Derivatives, Modification, Characterization

and Nanostructures, Th. J. Heinze, and W. G. Glasser (Eds.), ACS publisher, Washington, ACS Symp. Ser., 688, (1998), 61-72.

166. P. Mischnick and G. Kühn, Correlation between reaction conditions and primary structure: Model studies on methyl amylose, *Carbohydr. Res.*, 290 (1996) 199–207.
167. W. Lazik, Th. Heinze, K. Pfeiffer, G. Albrecht, and P. Mischnick, Starch derivatives of high degree of functionalization—6. Multi step carboxymethylation, *J. Appl. Polym. Sci.*, 86 (2002) 743–752.
168. J. Reuben, Analysis of the carbon-13 N.M.R. spectrum of methanolyzed *O*-ethylcellulose; monomer compositions and models for its description, *Carbohydr. Res.*, 161 (1987) 23–30.
169. J. Reuben and T. E. Casti, Distribution of substituents in *O*-(2-hydroxyethyl) cellulose: A ^{13}C-N.M.R. approach, *Carbohydr. Res.*, 163 (1987) 91–98.
170. J. Reuben, Description and analysis of hydroxyethyl cellulose, *Macromolecules*, 17 (1984) 156–161.
171. P. W. Arisz, J. A. Lomax, and J. J. Boon, Structure of *O*-(2-hydroxyethyl) celluloses with high degrees of molar substitution: A critical evaluation, *Carbohydr. Res.*, 243 (1993) 99–114.
172. B. Lindbberg, U. Lindquist, and O. Stenberg, Distribution of substituents in *O*-(2-hydroxyethyl)cellulose, *Carbohydr. Res.*, 5 (1987) 207–214.
173. P. W. Arisz, H. T. T. Thai, and J. J. Boon, Changes in substituent distribution patterns during the conversion of cellulose to *O*-(2-hydroxyethyl)celluloses, *Cellulose*, 3 (1996) 45–61.
174. P. W. Arisz, H. J. J. Kauw, and J. J. Boon, Substituent distribution along the cellulose backbone in *O*-methylcelluloses using GC and FAB MS for monomer and oligomer analysis, *Carbohydr. Res.*, 271 (1995) 1–14.
175. R. Adden, R. Müller, and P. Mischnick, Analysis of the substituent distribution in the glucosyl units and along the polymer chain of hydroxypropylmethylcelluloses and statistical evaluation, *Cellulose*, 13 (2006) 459–476.
176. R. Adden, R. Müller, G. Brinkmalm, R. Ehrler, and P. Mischnick, Comprehensive analysis of the substituent distribution in hydroxyethyl celluloses by quantitative MALDI-TOF-MS, *Macromol. Biosci.*, 6 (2006) 435–444.
177. R. Adden, W. Niedner, R. Müller, and P. Mischnick, Comprehensive analysis of the substituent distribution in the glucosyl units and along the polymer chain of hydroxyethylmethylcelluloses and statistical evaluation, *Anal. Chem.*, 78 (2006) 1146–1157.
178. G. Mann, J. Kunze, F. Loth, and H. -P. Fink, Cellulose ethers with a block-like distribution of the substituents by structure-selective derivatization of cellulose, *Polymer*, 39 (1998) 3155–3165.

179. I. Ciucanu and F. Kerek, A simple and rapid method for the permethylation of carbohydrates, *Carbohydr. Res.*, 131 (1984) 209–217.
180. J. M. H. M. Scheutjens and G. J. Fleer, Statistical theory of the adsorption of interacting chain molecules. 2. Train, loop, and tail size distribution, *J. Phys. Chem.*, 84 (1980) 178–190.
181. K. Jardeby, H. Lennholm, and U. Germgård, Characterization of undissolved residuals in CMC-solutions, *Cellulose*, 11 (2004) 195–202.
182. J. -A. Han and J. N. BeMiller, Influence of reaction conditions on MS values and physical properties of waxy maize starch derivatized by reaction with propylene oxide, *Carbohydr. Polym.*, 64 (2006) 158–162.
183. J. A. Gray and J. N. BeMiller, Development and utilization of reflectance confocal laser scanning microscopy to locate reaction sites in modfied starch granules, *Cereal Chem.*, 81 (2004) 278–286.
184. K. C. Huber and J. N. BeMiller, Location of sites of reaction within starch granules, *Cereal Chem.*, 78 (2001) 173–180.
185. M. J. Gidley and S. M. Bociek, Molecular organization in starches: A ^{13}C CP/MAS NMR study, *J. Am. Chem. Soc.*, 107 (1985) 7040–7044.
186. S. Richardson, G. Nilsson, K.-E. Bergquist, L. Gorton, and P. Mischnick, Characterization of the substituent distribution in hydroxypropylated potato amylopectin starch, *Carbohydr. Res.*, 328 (2000) 365–373.
187. R. Manelius, A. Buleon, K. Nurmi, and E. Bertoft, The substitution pattern in cationised and oxidised potato starch granules, *Carbohydr. Res.*, 329 (2000) 621–633.
188. R. Manelius, K. Nurmi, and E. Bertoft, Characterization of dextrins obtained by enzymatic treatment of cationic potato starch, *Starch*, 57 (2005) 291–300.
189. W. Tüting, K. Wegemann, and P. Mischnick, Enzymatic degradation and electrospray tandem mass spectrometry as tools for determining the structure of cationic starches prepared by wet and dry methods, *Carbohydr. Res.*, 339 (2004) 637–648.
190. Y. E. M. van der Burgt, J. Bergsma, I. P. Bleeker, P. J. H. C. Mijland, A. van der Kerk-van Hoof, J. P. Kamerling, and J. F. G. Vliegenthart, Distribution of methyl substituents over branched and linear regions in methylated starches, *Carbohydr. Res.*, 312 (1998) 201–208.
191. D. Horton and W. D. Pardoe, Calculation of degree of substitution of polysaccharides: A FORTRAN computer program, *Carbohydr. Res.*, 12 (1970) 269–272.
192. P. Käuper, W. -M. Kulicke, S. Horner, B. Saake, J. Puls, J. Kunze, H. -P. Fink, U. Heinze, Th. Heinze, E.-A. Klohr, H. Thielking, and W. Koch, Development and evaluation of methods for determine the pattern of functionalization in sodium carboxymethylcelluloses, *Angew. Macromol. Chem.*, 260 (1998) 53–63.

193. J. Reuben, Carbon-13 n.m.r. spectroscopy of *O*-(2-hydroxyethyl) derivatives of D-glucose, *Carbohydr. Res.*, 197 (1990) 257–261.
194. I. Nehls, W. Wagenknecht, B. Philipp, and D. Stscherbina, Characterization of cellulose and cellulose derivatives in solution by high resolution ^{13}C-NMR spectroscopy, *Prog. Polym. Sci.*, 19 (1994) 29–78.
195. T. Miyamoto, Y. Sato, T. Shibata, and H. Inagaki, Carbon-13 nuclear magnetic resonance studies of cellulose acetate, *J. Polym. Sci.*, 22 (1984) 2363–2370.
196. T. Liebert, K. Pfeiffer, and Th. Heinze, Carbamoylation applied for structure determination of cellulose derivatives, *Macromol. Symp.*, 223 (2005) 93–108.
197. Y. Tezuka and Y. Tsuchiya, Determination of substituent distribution in cellulose acetate by means of a ^{13}C NMR study on its propanoated derivative, *Carbohydr. Res.*, 273 (1995) 83–91.
198. Y. Tezuka, K. Imai, M. Oshima, and T. Chiba, Determination of substituent distribution in cellulose ethers by means of a carbon-13 NMR study on their acetylated derivatives. 1. Methylcellulose, *Macromolecules*, 20 (1987) 2413–2418.
199. Y. Tezuka, K. Imai, M. Oshima, and K. -I. Ito, Carbon-13 NMR structural studies on cellulose ethers by means of their acylated derivatives. 6. Carbon-13 NMR structural study on an enteric pharmaceutical coating cellulose derivative having ether and ester substituents, *Carbohydr. Res.*, 222 (1991) 255–259.
200. T. Liebert, M. A. Hussain, and Th. Heinze, Structure determination of cellulose esters via subsequent functionalization and NMR spectroscopy, *Macromol. Symp.*, 223 (2005) 79–91.
201. J. Schaller and T. Heinze, Studies on the synthesis of 2,3-*O*-hydroxyalkyl ethers of cellulose, *Macromol. Biosci.*, 5 (2005) 58–63.
202. N. D. Sachinvala, K. Maskos, D. V. Parikh, W. Glasser, U. Becker, E. J. Blanchard, and N. R. Bertoniere, characterization of tri-*O*-methylcellulose by one- and two-dimensional NMR methods, *J. Polym. Sci. A Polym. Chem.*, 37 (1999) 4019–4032.
203. N. D. Sachinvala, K. Maskos, W. P. Niemczura, T. L. Vigo, and N. R. Bertoniere, Review of 2-dimensional NMR techniques and results with cellulose ethers. Abstracts of Papers, 221st ACS National Meeting, San Diego, CA, US, (2001) CELL–041.
204. C. M. Keely, X. Zhang, and V. J. McBrierty, Hydration and plasticization effects in cellulose acetate: A solid-state NMR study, *J. Mol. Struct.*, 355 (1995) 33–46.
205. S. Hesse-Ertelt, R. Witter, A. S. Ulrich, T. Kondo, and Th. Heinze, Spectral assignments and anisotropy data of cellulose Iα: ^{13}C-NMR chemical shift data of cellulose Iα determined by INADEQUATE and RAI techniques applied to

uniformly ^{13}C-labeled bacterial celluloses of different *Gluconobacter xylinum* strains, *Magn. Reson. Chem.*, 46 (2008) 1030–1036.
206. T. Kondo, A. Koschella, B. Heublein, D. Klemm, and Th. Heinze, Hydrogen bond formation in regioselectively functionalized 3-mono-*O*-methyl cellulose, *Carbohydr. Res.*, 343 (2008) 2600–2604.
207. K. Petzold, D. Klemm, B. Heublein, W. Burchard, and G. Savin, Investigation on structure of regioselectively functionalized celluloses in solution exemplified by using 3-*O*-alkyl ethers and light scattering, *Cellulose*, 11 (2004) 177–193.
208. (a) A. Karrasch, C. Jäger, M. Karakawa, F. Nakatsubo, A. Potthast, and T. Rosenau, Solid-state NMR studies of methyl celluloses. Part 1: Regioselectively substituted celluloses as standards for establishing an NMR data basis, *Cellulose*, 16 (2009) 129-137; (b) A. Karrasch, C. Jäger, B. Saake, A. Potthast, and T. Rosenau, Solid-state NMR studies of methyl celluloses. Part 2: Determination of degree of substitution and O-6 vs. O-2/O-3 substituent distribution in commercial methyl cellulose samples, *Cellulose*, 16 (2009) 1159–1166.
209. H. Thielking and M. Schmidt, Cellulose Ethers. *Ullmann's Encyclopedia of Industrial Chemistry*, Wiley-VCH Verlag GmbH, Weinheim, Germany, 2006.
210. H. C. Trivedi, C. K. Patel, and R. D. Patel, Studies on carboxymethylcellulose: Potentiometric titrations, 3;, *Macromol. Chem.*, 182 (1981) 243–245.
211. T. P. Nevell and S. H. Zeronian, Action of ethylamine on cellulose. I. Acetylation of ethylamine-treated cotton, *Polymer*, 3 (1962) 187–194.
212. J. L. Groff and R. Cherniak, The incorporation of amino groups into cross-linked Sepharose by use of (3-aminopropyl)triethoxysilane, *Carbohydr. Res.*, 87 (1980) 302–305.
213. D. Rolf and G. R. Gray, Reductive cleavage of glycosides, *J. Am. Chem. Soc.*, 104 (1982) 3539–3541.
214. R. G. Ackman, Fundamental groups in the response of flame ionization detectors to oxygenated aliphatic hydrocarbons, *J. Gas Chromatogr.*, 2 (1964) 173–179.
215. R. F. Addison and R. G. Ackman, Flame ionization detector molar responses for methyl esters of some polyfunctional metabolic acids, *J. Gas Chromatogr.*, 6 (1968) 135–138.
216. P. Mares, J. Skorepa, and M. Friedrich, Computerized quantitative analysis of methyl and ethyl esters of long chain fatty acids by gas-liquid chromatography using relative molar response, *J. Chromatogr.*, 42 (1969) 435–441.
217. D. P. Sweet, R. H. Shapiro, and P. Albersheim, Quantitative analysis by various GLC [gas-liquid chromatography] response-factor theories for partially methylated and partially ethylated alditol acetates, *Carbohydr. Res.*, 40 (1975) 217–225.

218. U. Erler, P. Mischnick, A. Stein, and D. Klemm, Determination of the substitution pattern of cellulose methyl ethers by HPLC and GLC- comparison of methods, *Polym. Bull.*, 29 (1992) 349–356.
219. O. Mahot and D. Momcilovic, *unpublished results* (Master Thesis, KTH Stockholm, 2009).
220. S. Horner, J. Puls, B. Saake, E. -A. Klohr, and H. Thielking, Enzyme-aided characterization of carboxymethyl cellulose, *Carbohydr. Polym.*, 40 (1999) 1–7.
221. E. A. Kragten, J. P. Kamerling, J. F. G. Vliegenthart, H. Botter, and J. G. Batelaan, Composition analysis of sulfoethylcelluloses by high-pH anion-exchange chromatography with pulsed amperometric detection, *Carbohydr. Res.*, 233 (1992) 81–86.
222. J. Heinrich and P. Mischnick, A rapid method for the determination of the substitution pattern of *O*-methylated 1,4-glucans by high-pH anion-exchange chromatography with pulsed amperometric detection, *J. Chromatogr. A*, 749 (1996) 41–45.
223. D. C. Johnson, D. Dobberpuhl, R. Roberts, and P. Vandeberg, Pulsed amperometric detection of carbohydrates, amines and sulfur species in ion chromatography—the current state of research, *J. Chromatogr.*, 640 (1993) 79–96.
224. Z. El Rassi, Capillary electrophoresis of carbohydrates, *Adv. Chromatogr.*, 34 (1994) 177–250.
225. M. van Duin, J. A. Peters, A. P. G. Kieboom, and H. van Bekkum, Studies on borates esters II, *Tetrahedron*, 41 (1985) 3411–3421.
226. W. Tüting, G. Albrecht, B. Volkert, and P. Mischnick, Structure analysis of carboxymethyl starch by enzymic degradation and capillary electrophoresis, *Starch/Stärke*, 56 (2004) 315–321.
227. P. Mischnick, Determination of the substitution pattern of cellulose acetates, *J. Carbohydr. Chem.*, 10 (1991) 711–722.
228. M. Gohdes, P. Mischnick, and W. Wagenknecht, Methylation analysis of cellulose sulphates, *Carbohydr. Polym.*, 33 (1997) 163–168.
229. N. Yu and G. R. Gray, Analysis of the positions of substitution of acetate and propionate groups in cellulose acetate-propionate by the reductive-cleavage method, *Carbohydr. Res.*, 313 (1998) 29–36.
230. N. Yu and G. R. Gray, Analysis of the positions of substitution of acetate and butyrate groups in cellulose acetate butyrate by the reductive-cleavage method, *Carbohydr. Res.*, 312 (1998) 225–231.
231. J. Zheng, J. L. Gore, and G. R. Gray, Electrochemically-promoted reductive cleavage of glycosides, *J. Am. Chem. Soc.*, 120 (1998) 2684–2685.

232. C. K. Lee and G. R. Gray, Analysis of positions of substitution of *O*-acetyl groups in partially *O*-acetylated cellulose by the reductive-cleavage method, *Carbohydr. Res.*, 269 (1995) 167–174.
233. J. S. Sherman and G. R. Gray, Studies of model compounds for the analysis of ester-containing polysaccharides by the reductive-cleavage method, *Carbohydr. Res.*, 231 (1992) 221–235.
234. S. G. Zeller, A. J. D'Ambra, M. J. Rice, and G. R. Gray, Synthesis and mass spectra of 4-*O*-acetyl-1,5-anhydro-2,3,6-tri-*O*-ethyl-D-glucitol and the positional isomers of 4-*O*-acetyl-1,5-anhydro-di-*O*-ethyl-*O*-methyl-D-glucitol and 4-*O*-acetyl-1,5-anhydro-*O*-ethyl-di-*O*-methyl-D-glucitol, *Carbohydr. Res.*, 182 (1988) 53–62.
235. A. J. D'Ambra, M. J. Rice, S. G. Zeller, P. R. Gruber, and G. R. Gray, Analysis of positions of substitution of *O*-methyl or *O*-ethyl groups in partially methylated or ethylated cellulose by the reductive-cleavage method, *Carbohydr. Res.*, 177 (1988) 111–116.
236. N. Rogmann, P. Jones, and P. Mischnick, Determination of the substituent distribution in *O*-sulfonylbutyl-1→4-glucans, *Carbohydr. Res.*, 327 (2000) 275–285.
237. A. Vollmer, Alkenylether von Glucose und Dextran als reaktive Intermediate, Thesis, Braunschweig 2010.
238. P. Mischnick, Challenges in structure analysis of polysaccharide derivatives, *Cellulose*, 8 (2001) 245–257.
239. M. Černý and J. Staňek Jr., 1,6-Anhydro derivatives of aldohexoses, *Adv. Carbohydr. Chem. Biochem.*, 34 (1977) 23–177.
240. S. J. Angyal and K. Dawes, Conformational analysis in carbohydrate chemistry. II. Equilibria between reducing sugars and their glycosidic anhydrides, *Aust. J. Chem.*, 21 (1968) 2747–2760.
241. B. Saake, S. Horner, J. Puls, T. Heinze, and W. Koch, A new approach in the analysis of the substituent distribution of carboxymethyl celluloses, *Cellulose*, 8 (2001) 59–67.
242. R. R. Selvendran, J. F. March, and S. -G. Ring, Determination of aldoses and uronic acid content of vegetable fiber, *Anal. Biochem.*, 96 (1979) 282–292.
243. N. C. Carpita and E. M. Shea, in C. J. Biermann and G. D. McGinnis (Eds.), *Analysis of Carbohydrates by GLC and MS*, CRC Press, Boca Raton FL, USA, (1989), 157–216.
244. Th. Heinze, T. Liebert, and A. Koschella, *Esterification of Polysaccharides*, Springer-Verlag, Heidelberg, Germany, 2006, Tab. 3.17.

245. D. -S. Lee and A. S. Perlin, Formation, and Stereochemistry, of 1,2-*O*-(1-methyl-1,2-ethanediyl)-D-glucose acetals formed in the acid-catalyzed hydrolysis of *O*-(2-hydroxypropyl)cellulose, *Carbohydr. Res.*, 126 (1984) 101–114.
246. P. A. Sandford and H. E. Conrad, The structure of the Aerobacter aerogenes A3 (SI) polysaccharide. I. A reexamination using improved procedures for methylation analysis, *Biochemical*, 5 (1966) 1508–1517.
247. B. Blakeney and B. A. Stone, Methylation of carbohydrates with lithium methylsulphinyl carbanion, *Carbohydr. Res.*, 140 (1985) 319–324.
248. W. S. York, L. L. Kiefer, P. Albersheim, and A. G. Darvill, Oxidation of oligoglycosyl alditols during methylation catalyzed by sodium hydroxide and iodomethane in methyl sulfoxide, *Carbohydr. Res.*, 208 (1990) 175–182.
249. P. W. Needs and R. R. Selvendran, Avoiding oxidative degradation during sodium hydroxide/methyl iodide-mediated carbohydrate methylation in dimethyl sulfoxide, *Carbohydr. Res.*, 245 (1993) 1–10.
250. I. Ciucanu and E. C. Costello, Elimination of oxidative degradation during the per-*O*-methylation of carbohydrates, *J. Am. Chem. Soc.*, 125 (2003) 16213–16219.
251. P. Prehm, Methylation of carbohydrates by methyl trifluoromethanesulfonate in trimethyl *phosphate, Carbohydr. Res.*, 78 (1980) 372–374.
252. H. Meerwein, Triethyloxonium fluoroborate, *Org. Syn.* (1966) 46.
253. K. Agoston, A. Dobo, J. Rako, J. Kerekgyarto, and Z. Szurmai Anomalous, Zemplén deacylation reactions of alpha- and beta-D-mannopyranoside derivatives, *Carbohydr. Res.*, 330 (2001) 183–190.
254. (a) I. Ciucanu, Per-O-methylation reaction for structural analysis of carbohydrates by mass spectrometry. *Anal. Chim. Acta*, 576 (2006) 147-155; (b) A. Jay, The methylation reaction in carbohydrate analysis, *J. Carbohydr. Chem.*, 15 (1996) 897–923.
255. P. Mischnick-L. and W. A. König, Determination of the substitution pattern of modified polysaccharides. Part I. Benzyl starches, *Carbohydr. Res.*, 185 (1989) 113–118.
256. R. Adden and P. Mischnick, A novel method for the analysis of the substitution pattern of *O*-methyl-α- and β-1,4-glucans by means of electrospray ionisation-mass spectrometry/collision induced dissociation, *Int. J. Mass Spectrom.*, 242 (2005) 63–73.
257. B. Domon and C. E. Costello, A systematic nomenclature for carbohydrate fragmentations in FAB-MS/MS spectra of glycoconjugates, *Glycoconj. J.*, 5 (1988) 397–409.

258. W. Tüting, R. Adden, and P. Mischnick, Fragmentation pattern of regioselectively *O*-methylated maltooligosaccharides in electrospray ionisation-mass spectrometry/collision induced dissociation, *Int. J. Mass Spectrom.*, 232 (2004) 107–115.
259. J. N. BeMiller, Acid-catalyzed hydrolysis of glycosides, *Adv. Carbohydr. Chem.*, 22 (1967) 25–108.
260. B. Capon, Mechanism in carbohydrate chemistry, *Chem. Rev.*, 69 (1969) 407–498.
261. K. K. De and T. E. Timell, Acid hydrolysis of glycosides III. Hydrolysis of *O*-methylated glucosides and disaccharides, *Carbohydr. Res.*, 4 (1967) 72–77.
262. J. E. Höök and B. Lindberg, Acid hydrolysis of the monoisopropyl ethers of methyl alpha- and beta-D-glucopyranoside, *Acta Chem. Scand.*, 20 (1966) 2363–2369.
263. N. Roy and T. E. Timell, Acid hydrolysis of glycosides VIII. Sythesis and hydrolysis of three aldotriuronic acids, *Carbohydr. Res.*, 6 (1968) 475–481.
264. S. R. Rowland and P. S. Howley, Influence of the location of the substituents in substituted celluloses on solution hydrolysis of the D-glucosidic linkages, *Carbohydr. Res.*, 165 (1987) 69–76.
265. A. Adden, Substitution Patterns in and over Polymer Chains – New Approaches for Carboxymethyl Cellulose, Thesis, Braunschweig 2009.
266. H. H. Jensen and M. Bols, Steric effects are not the cause of the rate difference in hydrolysis of stereoisomeric glycosides, *Org. Lett.*, 5 (2003) 3419–3421.
267. J. T. Edward, Stability of glycosides to acid hydrolysis, *Chem. Ind. (London)* (1955) 1102–1104.
268. T. Narui, S. Iwata, K. Takahashi, and S. Shibata, Partial hydrolysis of α-D-glucans with acid in the presence of 1,1,3,3-tetra-methylurea, *Carbohydr. Res.*, 170 (1987) 269–273.
269. S. Haebel, S. Bahrke, and M. G. Peter, Quantitative sequencing of complex mixtures of heterochitooligosaccharides by MALDI-linear ion trap mass spectrometry, *Anal. Chem.*, 79 (2007) 5557–5566.
270. A. Gomez and K. Tang, Charge and fission of droplets in electrostatic sprays, *Phys. Fluids*, 6 (1994) 404–414.
271. D. Duft, T. Achtzehn, R. Müller, B. A. Huber, and T. Leisner, Coulomb fission. Rayleigh jets from levitated microdroplets, *Nature*, 421 (2003) 128.
272. M. Dole, L. L. Mack, R. L. Hines, R. C. Mobley, L. D. Ferguson, and M. B. Alice, Molecular beams of macroions, *J. Chem. Phys.*, 49 (1968) 2240–2249.
273. J. V. Iribarne and B. A. Thomson, On the evaporation of small ions from charged droplets, *J. Chem. Phys.*, 64 (1976) 2287–2294.

274. J. B. Fenn, Ion formation from charged droplets: Roles of geometry, energy, and time, *J. Am. Soc. Mass Spectrom.*, 4 (1993) 524–535.
275. R. Zenobi and R. Knochenmuss, Ion formation in MALDI mass spectrometry, *Mass Spectrom. Rev.*, 17 (1998) 337–366.
276. M. Karas and R. Krüger, Ion formation in MALDI: The cluster ionization mechanism, *Chem. Rev.*, 103 (2003) 427–439.
277. D. J. Harvey, B. Küster, and T.J.P. Naven, Perspectives in the glycosciences—matrix-assisted laser desorption/ionization (MALDI) mass spectrometry of carbohydrates, *Glycoconj. J.*, 15 (1998) 333–338.
278. D. J. Harvey, Matrix-assisted laser desorption/ionization mass spectrometry of carbohydrates, *Mass Spectrom. Rev.*, 18 (1999) 349–450.
279. D. J. Harvey, Matrix-assisted laser desorption/ionization mass spectrometry of carbohydrates and glycoconjugates, *Int. J. Mass Spectrom.*, 226 (2003) 1–35.
280. (a) D. J. Harvey, Analysis of carbohydrates and glycoconjugates by matrix-assisted laser desorption/ ionization mass spectrometry: An update covering the period 1999-2000, *Mass Spectrom. Rev.*, 25 (2006) 595–661; (b) D. J. Harvey, Analysis of carbohydrates and glycoconjugates by matrix-assisted laser desorption/ ionization mass spectrometry: An update for 2003-2004, *Mass Spectrom. Rev.*, 28 (2009) 273–361.
281. G. Hofmeister, Z. Zhou, and J. A. Leary, Linkage position determination in lithium-cationized disaccharides: Tandem mass spectrometry and semiempirical calculations, *J. Am. Chem. Soc.*, 113 (1991) 5964–5970.
282. P. Mischnick, W. Niedner, and R. Adden, Possibilities of mass spectrometry and tandem-mass spectrometry in the analysis of cellulose ethers, *Macromol. Symp.*, 223 (2005) 67–77.
283. T. J. P. Naven and D. J. Harvey, Effect of structure on the signal strength of oligosaccharides in matrix-assisted laser desorption/ionization mass spectrometry on time-of-flight and magnetic sector instruments, *Rapid. Commun. Mass Spectrom.*, 10 (1996) 1361–1366.
284. D. Momcilovic, B. Wittgren, K. -G. Wahlund, J. Karlsson, and G. Brinkmalm, Sample preparation effects in matrix-assisted laser desorption/ionisation time-of-flight mass spectrometry of partially depolymerized carboxymethyl cellulose, *Rapid. Commun. Mass. Spectrom.*, 17 (2003) 1107–1115.
285. J. Enebro and S. Karlsson, Improved matrix-assisted laser desorption/ ionisation time-of-flight mass spectrometry of carboxymethylcellulose, *Rapid Commun. Mass Spectrom.*, 20 (2006) 3693–3698.

286. D. Momcilovic, H. Schagerlöf, B. Wittgren, K.-G. Wahlund, and G. Brinkmalm, Improved chemical analysis of cellulsoe ethers using dialkylamine derivatization and mass spectrometry, *Anal. Chem.*, 77 (2005) 2948–2959.
287. D. Momcilovic, H. Schagerlöf, D. Röme, M. Jörntén-Karlsson, K.-E. Karlsson, B. Wittgren, F. Tjerneld, K.-G. Wahlund, and G. Brinkmalm, Novel derivatization using dimethylamine for tandem mass spectrometric structure analysis of enzymatically and acidically depolymerized methyl cellulose, *Biomacromolecules*, 6 (2005) 2793–2799.
288. T. J. P. M. Naven and D. J. Harvey, Cationic derivatization of oligosaccharides with Girard's T reagent for improved performance in matrix-assisted laser desorption/ionization and electrospray mass spectrometry, *Rap. Commun. Mass. Spectrom.*, 10 (1996) 829–834.
289. D. W. Johnson, A modified Girard derivatizing reagent for universal profiling and trace analysis of aldehydes and ketones by electrospray ionization tandem mass spectrometry, *Rapid. Commun. Mass. Spectrom.*, 21 (2007) 2926–2932.
290. G. -C. Gil, Y. -G. Kim, and B. -G. Kim, A relative and absolute quantification of neutral N-linked oligosaccharides using modification with carboxymethyl trimethylammonium hydrazide and matrix-assisted laser desorption/ionization time-of-flight mass spectrometry, *Anal. Biochem.*, 379 (2008) 45–59.
291. K. -S. Jang, Y. -G. Kim, G. -C. Gil, S. -H. Park, and B. -G. Kim, Mass spectrometric quantification of neutral and sialylated N-glycans from a recombinant therapeutic glycoprotein produced in the two Chinese hamster ovary cell lines, *Anal. Biochem.*, 386 (2009) 228–236.
292. D. J. Harvey, L. Royle, C. Radcliffe, P. M. Rudd, and R. A. Dwek, Structural and quantitative analysis of N-linked glycans by matrix-assisted laser desorption ionization and negative ion nanospray mass spectrometry, *Anal. Biochem.*, 376 (2008) 44–60.
293. R. Adden, C. Melander, G. Brinkmalm, L. Gorton, and P. Mischnick, new approaches to the analysis of enzymatically hydrolyzed methyl cellulose. Part 1. investigation of structural parameters on the extent of degradation, *Biomacromolecular*, 7 (2006) 1399–1409.
294. P. Mischnick and C. Hennig, A new model for the substitution patterns in the polymer chain of polysaccharide derivatives, *Biomacromolecules*, 2 (2001) 180–184.
295. B. V. McCleary and N. K. Matheson, Enzymic analysis of polysaccharide structure, *Adv. Carbohydr. Chem. Biochem.*, 44 (1986) 147–276.
296. M. Srisodsuk, J. Lehtio, M. Linder, E. Margolles-Clark, T. Reinikainen, and T. T. Teeri, *Trichoderma reesei* cellobiohydrolase I with an endoglucanase

cellulose-binding domain: Action on bacterial microcrystalline cellulose, *J. Biotechnol.*, 57 (1997) 49–57.
297. J. Stahlberg, G. Johansson, and G. Pettersson, *Trichoderma reesei* has no true exo-cellulase: All intact and truncated cellulases produce new reducing end groups on cellulose, *Biochim. Biophys. Acta Gen. Subj.*, 1157 (1993) 107–113.
298. T. Desmet, T. Cantaert, P. Gualfetti, W. Nerinckx, L. Gross, C. Mitchinson, and K. Piens, An investigation of the substrate specificity of the xyloglucanase Cel74A from *Hypocrea jecorina, FEBS J.*, 274 (2007) 356–363.
299. J. Karlsson, M. Siika-aho, M. Tenkanen, and F. Tjerneld, Enzymatic properties of the low molecular mass endoglucanases Cel12A (EG III) and Cel45A (EG V) of Trichoderma reesei, *J. Biotechnol.*, 99 (2002) 63–78.
300. G. J. Davies, K. S. Wilson, and B. Henrissat, Nomenclature for sugar-binding subsites in glycosyl hydrolases, *Biochem. J.*, 321 (1997) 557–559.
301. B. Henrissat, A classification of glycosyl hydrolases based on amino acid sequence similarities, *Biochem. J.*, 280 (1991) 309–316.
302. B. Henrissat, M. Claeyssens, P. Tomme, L. Lemesle, and J. P. Mornon, Cellulase families revealed by hydrophobic cluster analysis, *Gene*, 81 (1989) 83–95.
303. B. Henrissat, T. T. Teeri, and R.A.J. Warren, A scheme for designating enzymes that hydrolyze the polysaccharides in the cell walls of plants, *FEBS Lett.*, 425 (1998) 352–354.
304. G. Davies and B. Henrissat, Structures and mechanisms of glycosyl hydrolases, *Structure*, 3 (1995) 853–859.
305. M. Linder and T. T. Teeri, The roles and function of cellulose-binding domains, *J. Biotechnol.*, 57 (1997) 15–28.
306. H. M. Jespersen, E. A. MacGregor, M. R. Sierks, and B. Svensson, Comparison of the domain-level organization of starch hydrolases and related enzymes, *Biochem. J.*, 280 (1991) 51–55.
307. P. Mischnick, Specificity of microbial α-amylase and amyloglucosidase hydrolysis of methyl amyloses, *Starch*, 53 (2001) 110–120.
308. J. Enebro, D. Momcilovic, M. Siika-aho, and S. Karlsson, Liquid chromatography combined with mass spectrometry for investigation of endoglucanase selectivity on carboxymethyl cellulose, *Carbohydr. Res.*, 344 (2009) 2173–2181.
309. Y. C. Chan, P. J. Braun, D. French, and J. F. Robyt, Porcine pancreatic alpha.-amylase hydrolysis of hydroxyethylated amylose and specificity of subsite binding, *Biochemistry*, 23 (1984) 5795–5800.
310. M. Nojiri and T. Kondo, Application of regioselectively substituted methylcelluloses to characterize the reaction mechanism of cellulase, *Macromolecules*, 29 (1996) 2392–2395.

311. M. Schulein, Enzymic properties of cellulases from *Humicola insolens*, *J. Biotechnol.*, 57 (1997) 71–81.
312. B. Saake, S. Horner, and J. Puls, Progress in the enzymic hydrolysis of cellulose derivatives, *ACS Symp. Ser.*, 688 (1998) 201–216.
313. C. Melander, R. Adden, G. Brinkmalm, L. Gorton, and P. Mischnick, New approaches to the analysis of enzymatically hydrolyzed methyl cellulose. Part 2. comparison of various enzyme preparations, *Biomacromolecules*, 7 (2006) 1410–1421.
314. S. Richardson, G. Nilsson, A. Cohen, D. Momcilovic, G. Brinkmalm, and L. Gorton, Enzyme-aided investigation of the substituent distribution in cationic potato amylopectin starch, *Anal. Chem.*, 75 (2003) 6499–6508.
315. A. Parfondry and A. S. Perlin, Carbon-13 NMR spectroscopy of cellulose ethers, *Carbohydr. Res.*, 57 (1977) 39–49.
316. A. Varrot and G. J. Davies, Direct experimental observation of the hydrogen-bonding network of a glycosidase along its reaction coordinate revealed by atomic resolution analyses of endoglucanase Cel5A, *Acta Crystallogr. D Biol. Crystallogr.*, D59 (2003) 447–452.
317. V. Notenboom, C. Birsan, M. Nitz, D. R. Rose, R. A. J. Warren, and S. G. Withers, Insights into transition state stabilization of the β-1,4-glycosidase Cex by covalent intermediate accumulation in active site mutants, *Nat. Struct. Biol.*, 5 (1998) 812–818.
318. M. N. Namchuk and S. G. Withers, Mechanism of *Agrobacterium* β-glucosidase: Kinetic analysis of the role of noncovalent enzyme/substrate interactions, *Biochemistry*, 34 (1995) 16194–16202.
319. U. Schagerlöf, H. Schagerlöf, D. Momcilovic, G. Brinkmalm, and F. Tjerneld, Endoglucanase sensitivity for substituents in methyl cellulose hydrolysis studied using MALDI-TOFMS for oligosaccharide analysis and structural analysis of enzyme active sites, *Biomacromolecules*, 8 (2007) 2358–2365.
320. J. Enebro, D. Momcilovic, M. Siika-aho, and S. Karlsson, A new approach for studying correlations between the chemical structure and the rheological properties in carboxymethyl cellulose, *Biomacromolecules*, 8 (2007) 3253–3257.
321. B. Saake, S. Horner, T. Kruse, J. Puls, T. Liebert, and T. Heinze, Detailed investigation on the molecular structure of carboxymethyl cellulose with unusual substitution pattern by means of an enzyme-supported analysis, *Macromol. Chem. Phys.*, 201 (2000) 1996–2002.
322. B. Saake, J. Puls, and W. Wagenknecht, Endoglucanase fragmentation of cellulose sulfates derived from different synthesis concepts, *Carbohydr. Polym.*, 48 (2002) 7–14.

323. S.-J. Lee, C. Altaner, J. Puls, and B. Saake, Determination of the substituent distribution along cellulose acetate chains as revealed by enzymatic and chemical methods, *Carbohydr. Polym.*, 54 (2003) 353–362.
324. B. Saake, C. Altaner, S. -J. Lee, and J. Puls, Endoglucanase degradation and enzyme-aided characterization of cellulose acetates, *Macromol. Symp.*, 223 (2005) 137–150.
325. B. Saake, S. Lebioda, and J. Puls, Analysis of the substituent distribution along the chain of water-soluble methyl cellulose by combination of enzymatic and chemical methods, *Holzforschung*, 58 (2004) 97–104.
326. F. Fitzpatrick, H. Schagerlöf, T. Anderson, S. Richardson, F. Tjernheld, K.-G. Wahlund, and B. Wittgren, NMR, Cloud-point measurements and enzymatic depolymerization: Complementary tools to investigate substituent patterns in modified celluloses, *Biomacromolecules*, 7 (2006) 2909–2917.
327. L. F. Hood and C. Mercier, Molecular structure of unmodified and chemically modified manioc starches, *Carbohydr. Res.*, 61 (1978) 53–66.
328. K. E. Almin and K. E. Eriksson, Influence of carboxymethyl cellulose properties on the determination of cellulase activity in absolute terms, *Arch. Biochem. Biophys.*, 124 (1968) 129–134.
329. K. E. Eriksson and B. H. Hollmark, Kinetic studies of the action of cellulase upon sodium carboxymethyl cellulose, *Arch. Biochem. Biophys.*, 133 (1969) 233–237.
330. R. A. Gelman, Characterization of carboxymethyl cellulose: Distribution of substituent groups along the chain, *J. Appl. Polym. Sci.*, 27 (1982) 2957–2964.
331. J. Karlsson, D. Momcilovic, B. Wittgren, M. Schulein, F. Tjerneld, and G. Brinkmalm, Enzymatic degradation of carboxymethyl cellulose hydrolyzed by the endoglucanases Cel5A, Cel7B, and Cel45A from *Humicola insolens* and Cel7B, Cel12A and Cel45Acore from *Trichoderma reesei*, *Biopolymers*, 63 (2002) 32–40.
332. H. Schagerlöf, M. Johansson, S. Richardson, G. Brinkmalm, B. Wittgren, and F. Tjerneld, Substituent distribution and clouding behavior of hydroxypropyl methyl cellulose analyzed using enzymatic degradation, *Biomacromolecules*, 7 (2006) 3474–3481.
333. H. Schagerlöf, S. Richardson, D. Momcilovic, G. Brinkmalm, B. Wittgren, and F. Tjerneld, Characterization of chemical substitution of hydroxypropyl cellulose using enzymatic degradation, *Biomacromolecules*, 7 (2006) 80–85.
334. J. Enebro, D. Momcilovic, M. Siika-aho, and S. Karlsson, Investigation of endoglucanase selectivity on carboxymethyl cellulose by mass spectrometric techniques, *Cellulose*, 16 (2009) 271–280.

335. H. Pasch, Analytical techniques for polymers with complex architectures, *Macromol. Symp.*, 178 (2002) 25–37.
336. H. M. Spurlin, Homogeneity and properties of nitrocellulose, *J. Ind. Eng. Chem.*, 30 (1938) 538–542.
337. K. A. Oudhoff, F. A. Buijtenhuijs, P. H. Wijnen, P. J. Schoenmakers, and W. T. Kok, Determination of the degree of substitution and its distribution of carboxymethylcelluloses by capillary zone electrophoresis, *Carbohydr. Res.*, 330 (2004) 1917–1924.
338. R. Adden, R. Müller, and P. Mischnick, Fractionation of methyl cellulose according to polarity—a tool to differentiate first and second order heterogeneity of the substituent distribution, *Macromol. Chem. Phys.*, 207 (2006) 954–965.
339. Y. E. M. van der Burgt, J. Bergsma, I. P. Bleeker, P. J. H. C. Mijland, A. van der Kerk-van Hoof, J. P. Kamerling, and J.F.G. Vliegenthart, Distribution of methyl substituents over crystalline and amorphous domains in methylated starches, *Carbohydr. Res.*, 320 (1999) 100–107.
340. Y. E. M. van der Burgt, J. Bergsma, I. P. Bleeker, P. J. H. C. Mijland, A. van der Kerk-van Hoof, J. P. Kamerling, and J. F. G. Vliegenthart, Distribution of methyl substituents in amylose and amylopectin from methylated potato starches, *Carbohydr. Res.*, 325 (2000) 183–191.
341. J. L. Jane and J. J. Shen, Internal structure of the potato starch granule revealed by chemical gelatinization, *Carbohydr. Res.*, 247 (1993) 279–290.
342. J. M. Herrero-Martinez, P. J. Schoenmakers, and W. Th. Kok, Determination of the amylose-amylopectin ratio of starches by iodine-affinity capillary electrophoresis, *J. Chromatogr. A*, 1053 (2004) 227–234.

GLYCONANOPARTICLES: POLYVALENT TOOLS TO STUDY CARBOHYDRATE-BASED INTERACTIONS

BY MARCO MARRADI, MANUEL MARTÍN-LOMAS, and SOLEDAD PENADÉS

Laboratory of GlycoNanotechnology, Biofunctional Nanomaterials Unit, CIC biomaGUNE/CIBER-BBN, San Sebastian, 20009, Spain

I. Introduction 212
1. Background 212
2. Glyconanoparticles and Glyconanotechnology 217
II. Glyconanoparticles: Synthesis, Characterization, and Types 218
1. Noble-Metal Glyconanoparticles 219
2. Magnetic GNPs 239
3. Glyco Quantum Dots 245
III. Applications of GNPs 250
1. GNPs in Carbohydrate–Carbohydrate Interaction Studies 251
2. GNPs in Carbohydrate–Protein Interactions 254
3. Other Interaction Studies 258
4. Glyconanoparticles as Antiadhesion Agents 260
5. GNPs in Cellular and Molecular Imaging 264
6. Other Applications of GNPs 269
IV. Future and Perspectives 269
References 270

ABBREVIATIONS

3D, three-dimensional; AFM, atomic force microscopy; ATP, adenosine triphosphate; BclA, *Burkholderia cenocepacia* lectin A; ConA, concanavalin A; DC, dendritic cell; DC-SIGN, dendritic cell-specific ICAM 3-grabbing nonintegrin; DLS, dynamic light scattering; DNA, deoxyribonucleic acid; DO3A, tetraazacyclododecane triacetic acid; EDTA, ethylendiaminetetraacetic acid; ELISA, enzyme-linked immunosorbent assay; ELLA, enzyme-linked lectin assay; FITC, fluorescein isothiocyanate; GNP,

ISBN: 978-0-12-380854-7
DOI: 10.1016/S0065-2318(10)64005-X

glyconanoparticle; GSL, glycosphingolipid; HA, hyaluronic acid; HER2, human epidermal growth factor receptor 2; HIV, human immunodeficiency virus; ICP-AES, inductively coupled plasma atomic emission spectroscopy; IR, infrared spectroscopy; ITC, isothermal titration calorimetry; MALDI-TOF, matrix assisted laser desorption/ionization time-of-flight; MEND, multifunctional envelope-type nano-devices; MGNP, magnetic glyconanoparticle; MNP, magnetic nanoparticle; MPA, 3-mercaptopropanoic acid; MRI, magnetic resonance imaging; MS, mass spectrometry; NIRF, near-infrared fluorescence; NMMO, *N*-methylmorpholine-*N*-oxide; NMR, nuclear magnetic resonance; NP, nanoparticle; PEG, poly(ethylene glycol); PLGA, poly(lactic-*co*-glycolic acid); *p*-MBA, *p*-mercaptobenzoic acid; QCM, quartz crystal microbalance; QD, quantum dot; RCA, *Ricinus communis* agglutinin; RNA, ribonucleic acid; SAM, self-assembled monolayer; SDS-PAGE, sodium dodecyl sulfate polyacrylamide gel electrophoresis; siRNA, small interfering RNA; SPR, surface plasmon resonance; TEM, transmission electron microscopy; TOP, tri-*n*-octylphosphine; TOPO, tri-*n*-octylphosphine oxide; UV-Vis, ultraviolet-visible spectroscopy; XPS, X-ray photoelectron spectroscopy.

I. Introduction

1. Background

This article surveys the state of the art and current advances in the synthesis and applications of nanoparticles functionalized with carbohydrates and discusses perspectives for their use as tools for addressing current challenges at the glycobiology, biotechnology, and the materials science interface. The fabrication of hybrid materials from inorganic nanostructures and biomolecules constitutes a significant development in the fields of nanoscience and nanotechnology.[1–5] These biofunctionalized nanostructures, where materials science, nanotechnology, and chemical biology meet,[3,6] have attracted much attention for applications in bioassays and in interaction studies. New hybrid systems based on nanoclusters and biomolecules are being designed, synthesized, and used to approach various research problems.[3–5] The high interest of scientists from different disciplines in these biofunctional nanomaterials is reflected in the increasing number of new journals and book series that have been launched during the past few years, dealing with various aspects of nanobiomaterials and nanobiotechnology.[1–6] The wide variety of materials that can be chosen for the inorganic core of these nanobiomaterials permits access to nanostructures possessing a range of optical, electronic, mechanical, and magnetic properties.[7] Their size can be modulated and their surface engineered to insert multivalency and multifunctionality.[8] The nanometric size of these

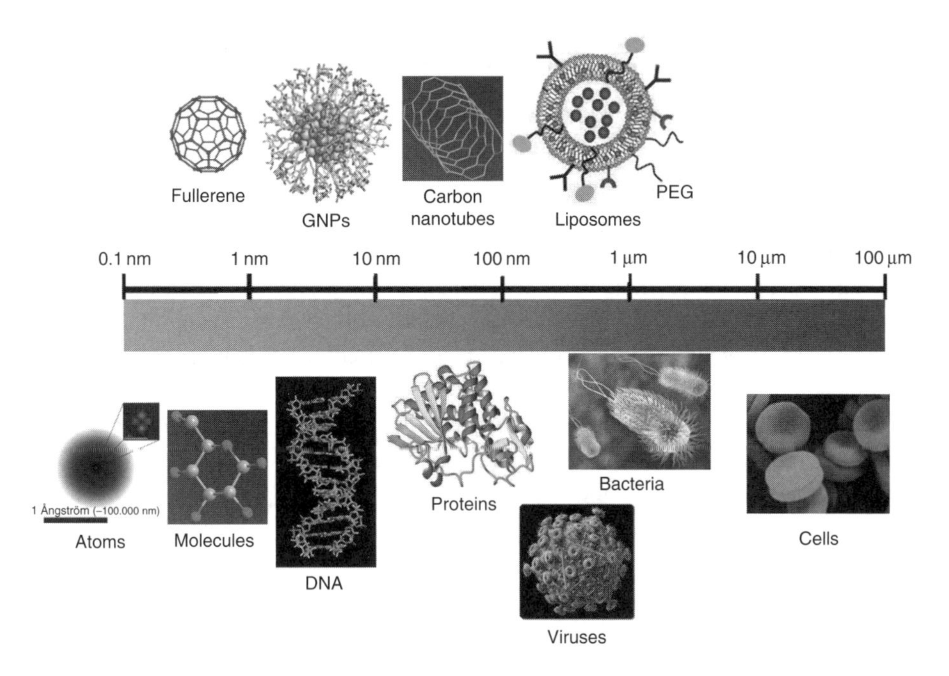

FIG. 1. Sizes of glyconanoparticles (GNPs) in relation to other chemical and biological objects. (See Color Plate 29.)

biofunctional materials, in the same order of magnitude as such biological molecules as proteins and nucleic acids (Fig. 1), permits their use for biomimetic purposes and to intervene in biorecognition processes at the molecular level.

Gold–biomolecule nanoparticles are perhaps the best known among the hybrid nanobiomaterials.[9] The first scientific report on stable colloidal gold was published by Faraday in 1857.[10] In the nineteen fifties and seventies, Turkevich and Frens, independently, reported on the nucleation and formation of colloidal gold,[11–13] and an interesting historical overview was published by Turkevich in the nineteen eighties.[14] Numerous procedures based on solution-phase synthesis and size-selective separation methods for the production of monodisperse metal nanoparticles and the different techniques used for their characterization, have been reported.[15–18] Gold nanoparticles are nowadays an integral part of research in nanoscience, as a result of their unique chemical and physical properties[19] and because of the wide variety of molecules that can be used for their functionalization, especially through well-known chemistry involving thiol-mediated attachment to the gold surface.[20]

Although the use of gold colloids in biology was published in 1970 when immunogold-staining procedures were first reported,[15,21] it was in the late nineteen nineties when the first reports on gold nanoparticles functionalized with proteins, antibodies, peptides, and DNA appeared.[4,8,22,23] The first reports reviewing the work on gold nanoparticles bearing carbohydrates were published in 2004 and 2006.[24,25] Since then, the number of reports on the synthesis, characterization, and application of carbohydrate–nanoparticle hybrid materials has grown exponentially.

This survey is focused on glyconanoparticles in which the carbohydrate ligands are "covalently" linked to the metallic nucleus, and especially highlights examples where the sugar ligands confer biological functionality to the inorganic core of the nanomaterials. However, before discussing the preparation and characterization of the glyconanoparticles, a short overview on the methods used for obtaining protected and biofunctional noble-metal nanoparticles is presented.

During more than fifty years, the use of sodium citrate for the reduction/stabilization of a gold salt ($HAuCl_4$) has permitted the preparation of stable and uniform gold nanoparticles.[12] This "citrate method" for producing citrate-capped gold nanoparticles of uniform size (in the diameter range of 20–150 nm) was proposed by Turkevich *et al.* in the nineteen fifties, when interest in the formation (nucleation and growth processes) of gold colloids was developing.[12,13] The reduction of chloroauric acid by sodium citrate at 100 °C produces a colloidal solution having excellent stability and uniform particle-size of about 20 nm.[14] Frens also reported a method whereby the reducing–stabilizing agent (citrate/gold salt ratio) was varied[11] to afford particles with a broad range of sizes (diameters between 10 and 150 nm), although the observed monodispersity was very low for nanoparticles whose diameter is more than 30 nm.

An interesting application of the citrate method is the preparation of monolayer-protected nanoclusters by *citrate–ligand exchange* with a thiolated molecule under mild conditions (the exchange does not need any additional reagent) and with the possibility for choosing the nanocluster size in advance. This is generally achieved by incubation at room temperature of the prepared solution of metal nanoparticles with the desired ligands (usually for no less than 24 hours in the presence of a great excess of the ligand). The mildness of the reaction conditions renders this method especially useful when manipulating biomolecules that are sensitive to reducing agents. DNA-oligonucleotides functionalized with alkanethiols at one of their termini were attached to nanometric gold colloids through this method.[26] The group of Mirkin especially has demonstrated the great utility of the DNA-functionalized gold nanoparticles (DNA-AuNPs) thus prepared for selective detection of polynucleotides based on colorimetric changes, conductivity variation, and other detection methods.[27,28] Other groups also used DNA-AuNPs for the detection of proteins,[29] oligonucleotides,[30] metal ions,[31–33] and cancer cells.[34]

The citrate method permits synthesis of biofunctional gold nanoparticles larger than those prepared by a direct method (see later).[35] However, it requires longer reaction times and does not allow the controlled construction of multifunctional nanoparticles. It was used to prepare the first alkanethiol-stabilized gold colloids.[36] In addition to citrate, other capping agents have also been used for the preparation of stable gold colloids.[37]

Most of the thiol monolayer-protected nanoparticles reported have been prepared by different modifications of the "direct *in situ* formation" method proposed by Brust and Schiffrin.[38,39] This method consists of the one-pot two-phase (water–toluene) reduction of a Au(III) salt by sodium borohydride in the presence of an alkanethiol, using tetraoctylammonium bromide as the phase-transfer catalyst. The thiol ligand binds the metal strongly in the organic phase, because of the "soft" character of both gold and sulfur, and thus stable solutions of 1–3 nm thiol-coated gold particles can be obtained, which can be isolated and re-dissolved without aggregation or decomposition. The method allows the facile synthesis of thermally and air-stable gold nanoparticles of reduced dispersion and controlled size. The nanoparticles thus prepared, which conserve to a great extent the chemical properties of the ligands, can be fully characterized by a variety of techniques, including ultraviolet-visible spectroscopy (UV-Vis), infrared spectroscopy (IR), elemental analysis, nuclear magnetic resonance (NMR), transmission electron microscopy (TEM), and X-ray photoelectron spectroscopy (XPS).

The solubility of the protected nanoparticles depends on the peripheral functionalities of the ligands. Water-soluble metallic nanoparticles have been prepared from water-soluble thiolate ligand components such as thiolated poly(ethylene glycol) polymer electrolytes,[40] tiopronin,[41,42] glutathione,[43] or mercaptosuccinic acid[44] using different modifications of the Brust and Shiffrin method. Mercaptoammonium derivatives have also been used to obtain water-soluble gold, silver, and palladium nanoclusters.[45] Leff *et al.*[46] have shown that the diameter of the gold nanocrystals can be controlled by the initial $AuCl_4^-$/thiol ratio, producing a wide range (from 1.5 to 20 nm) of particle sizes. Following this method, a *p*-mercaptobenzoic acid-protected gold nanoparticle has been prepared and crystallized. Its structure was determined by X-ray crystallography at 1.15 Å resolution.[47] This result is very important because it has provided, for the first time, experimental insight in the structure of a monolayer-protected gold nanoparticle. An electron-density map (red mesh in Fig. 2A) reveals the presence of 102 gold atoms and 44 ligands. The particles proved to be chiral, with half of one enantiomer in the asymmetric unit of the crystal (Fig. 2B).

Murray and co-workers have demonstrated that further functionalization of these nanoparticles can be achieved by "place exchange reactions," which occur when an excess of functionalized alkanethiol is added to an alkanethiolate–cluster solution.[48] In this way, a new chemical functionality can be introduced without changes in the

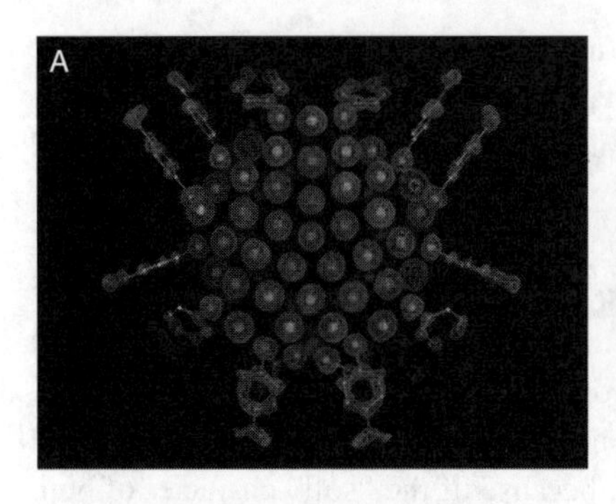

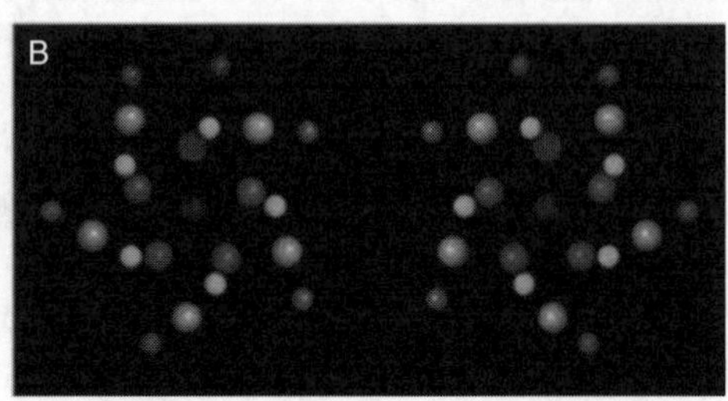

FIG. 2. X-Ray crystal-structure determination of the $Au_{102}(p\text{-MBA})_{44}$ nanoparticle. (A) Electron density map (red mesh) and atomic structure (gold atoms depicted as yellow spheres, and *p*-MBA shown as framework and with small spheres [sulfur in cyan, carbon in gray, and oxygen in red]). (B) View down the cluster axis of the two enantiomeric particles. Colour scheme as in (A), except only sulfur atoms of *p*-MBA are shown. Adapted from Ref. 47. (See Color Plate 30.)

gold-core dimension. By application of the "place exchange reactions" concept, Rotello and co-workers demonstrated the feasibility of glutathione-mediated release of fluorescent payloads from nanoparticle carriers in living cells.[49] However, "place exchange reactions" are subjected to dynamics that depend essentially on the chain length and the steric volume of the ligands,[50] and it is therefore difficult to establish general procedures for the preparation of nanoparticles having differing density of several ligands by this procedure.

The possibility of manipulating thiol-derivatized gold nanoparticles as simple chemical entities was demonstrated by (partial) esterification of the surface-bound hydroxyl groups of *p*-mercaptophenol-functionalized nanoparticles.[39] Further transformations of the functionalities inserted on the surface of nanoparticles can be performed under reaction conditions similar to those employed in conventional organic synthesis.

The feasibility of passivation of other metals by alkanethiolate monolayers was later demonstrated with silver,[51,52] alloys of gold with other transition metals,[53] palladium and iridium,[54] and platinum.[55]

The synthesis of gold nanoparticles is not limited to the methods just mentioned but comprises also new thermal and photochemical techniques. Some examples are the preparation of stable carbon-protected gold and platinum nanoparticles stabilized by metal–carbon bonds through decomposition of aryldiazonium salts,[56] and the one-pot photochemical synthesis of naked gold nanoparticles whose size can be controlled by altering the illumination intensity.[57] An exhaustive account on the variety of conditions (including reducing agents and stabilizers) reported in the literature for the preparation of metal nanoparticles is beyond the scope of this article.

2. Glyconanoparticles and Glyconanotechnology

Glycoconjugates are bioactive molecules that play a key role in many normal and pathological processes.[58–60] The mechanisms that govern their biological behavior at the molecular level involve, to a great extent, carbohydrate–protein interactions[59,61,62] and carbohydrate–carbohydrate interactions.[63,64] The key characteristic feature of carbohydrate-based interactions is an extremely low affinity. Therefore, in biological systems, carbohydrates usually appear arranged in multivalent clusters with proper orientation and spacing to achieve high affinity for the corresponding ligands. For this reason, an important challenge in current glycoscience research is the design and synthesis of multivalent model systems[65] that may be able to mimic the natural presentation of biomolecules. Since the pioneer work of Lee on the cluster effect,[59] a plethora of multivalent model systems based on peptides, proteins, liposomes, dendrimers, or polymers as scaffolds[66–69] have been prepared to study and evaluate carbohydrate–protein associations.

Gold nanoparticles constructed by "covalent" linkage of thiol-ended carbohydrate derivatives to the gold surface were first developed in the laboratory of Soledad Penadés.[70] These gold glyconanoparticles were used as a new multivalent chemical tool to investigate the existence and properties of carbohydrate–carbohydrate interactions in water between oligosaccharide epitopes responsible for cell association *in vivo*. The search for a suitable multivalent model system resulted in a new integrated approach based on the use of carbohydrate self-assembled monolayers (SAMs) on two- and three-dimensional surfaces that was termed the *glyconanotechnology strategy*.[24] Glyconanoparticles were first named and defined in a 2001 publication presenting "a new water-soluble three-dimensional (3D) polyvalent model system based on sugar-modified gold nanoclusters. These so-called glyconanoparticles (GNPs) provide a glycocalyx-like surface with a globular carbohydrate display, and chemically well-defined compositions for study of carbohydrate interactions and to interfere with cell–cell adhesion processes."[70]

The suitably oriented carbohydrate coating provides stability and solubility in biological media and confers biocompatibility and non-toxicity to these glyconanomaterials. In comparison with other multivalent carbohydrate-functionalized systems, such as dendrimers,[68,71] polymers,[72] and liposomes,[73] the glyconanoparticle platform offers some additional advantages. Thus, the glyconanoparticle technology permits the preparation of a great variety of water-soluble glycoclusters having different carbohydrate ligand densities (high and low loading) and different linkers attaching the carbohydrates to the gold surface, the structure of which can modulate the rigidity, flexibility, and accessibility of the ligands. The nature (hydrophilic or hydrophobic), the length, and the flexibility of these spacers are key factors in controlling the presentation of the carbohydrates and their behavior during molecular-recognition events.

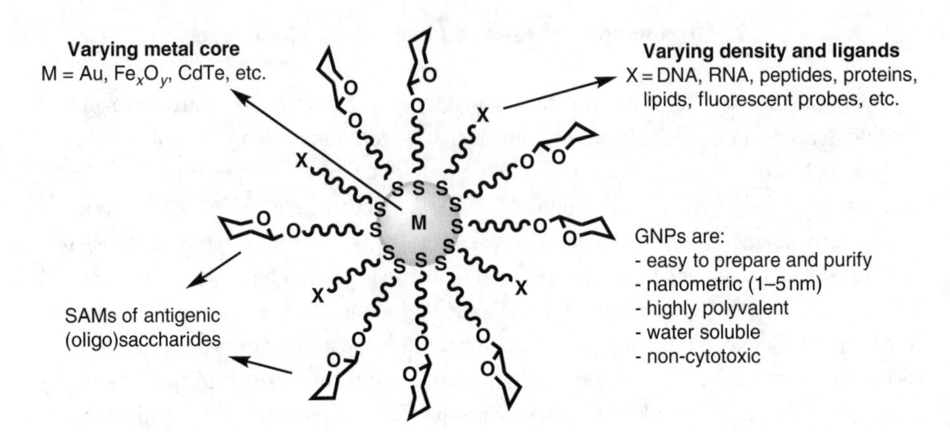

FIG. 3. The glyconanoparticle (GNP) concept: from simple GNPs to complex "nanocells."

This technology permits the facile preparation of a great variety of water-soluble biofunctional nanoclusters having globular shapes and chemically well-defined composition to intervene in cell–cell adhesion and recognition processes. Multifunctional GNPs incorporating simultaneously, on a single gold cluster, carbohydrates, peptides, lipids, DNA, RNA, or fluorescent probes permit effective creation of what may be called "artificial nanocells." Furthermore, the nature of the metallic core and its dimensions can be varied, thereby modulating both intrinsic and size-related electronic, magnetic, and optical properties (quantum size effect) (Fig. 3).

II. Glyconanoparticles: Synthesis, Characterization, and Types

There are essentially two methods for preparing sugar-functionalized nanoclusters: (i) a direct method that consists in controlling the growth of nascent metal by *in situ* reduction of a metallic salt in the presence of carbohydrate-based ligands bearing a thiol end-group, and (ii) a stepwise method that utilizes either a ligand exchange of previously prepared nanoclusters with other thiol-derivatized carbohydrates, or a chemical reaction between functional groups on the nanoparticle surface and suitably derivatized carbohydrates (Fig. 4). This section surveys the preparation and characterization of glyconanoparticles with noble (especially gold and silver), magnetic, and semiconductor (cadmium, indium, etc.) metallic cores.

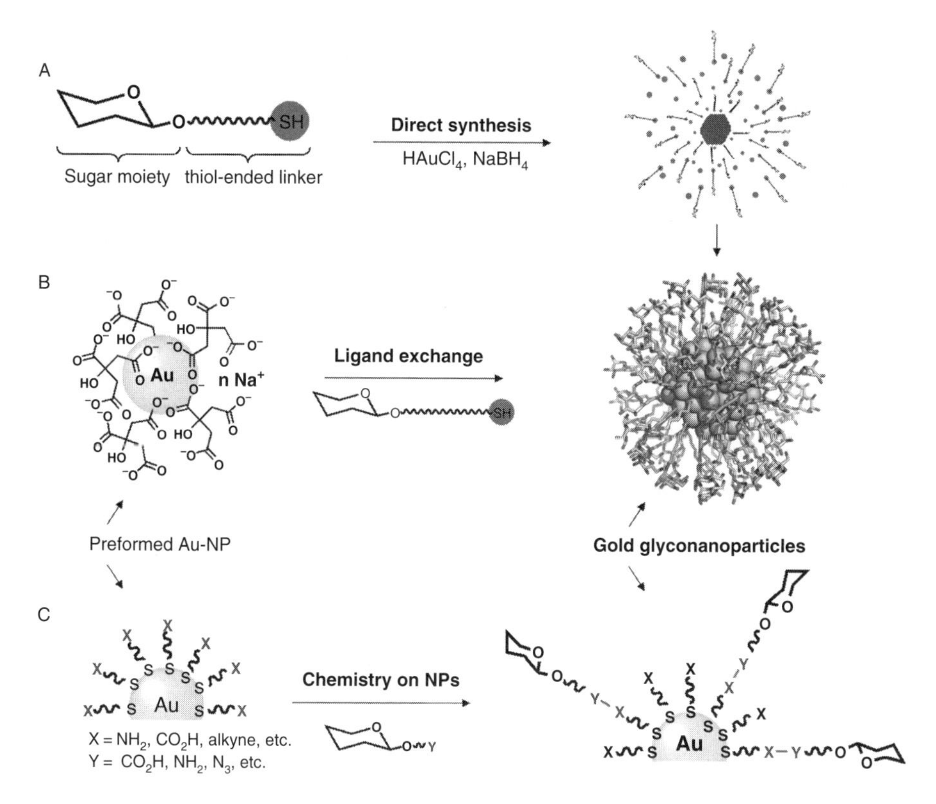

FIG. 4. Main methods for the preparation of gold GNPs. (A) Direct synthesis based on the reduction of a Au(III) salt and *in situ* protection of the nascent gold nanocluster with thiol-armed glycoconjugates. (B) Ligand place exchange reactions based on the treatment of preformed gold nanoparticles with thiol-derivatized glycoconjugates. (C) Functionalization of gold nanoparticles by reaction between functional groups on gold surface and suitably derivatized carbohydrates. (See Color Plate 31.)

1. Noble-Metal Glyconanoparticles

a. *Direct* in situ *Formation of Glyconanoparticles*.—Based on Brust's method, Penadés and co-workers reported in 2001 the first synthesis of gold GNPs.[70] These carbohydrate-passivated nanoclusters were prepared in water in the presence of neoglycoconjugates bearing thiol end-groups to afford 2-nm-sized water-soluble and stable gold GNPs. This direct *in situ* synthesis permits control of the growth of the metal cluster by varying the organic materials/Au(III) molar ratio, to give diameters ranging approximately from 1 to 10 nm. Gold GNPs of 2-nm average diameter were obtained by adding a methanolic or aqueous solution of thiol-derivatized conjugates of the disaccharide lactose [β-Gal-(1→4)-Glc] and the Lewis X trisaccharide [β-Gal-(1→4)-[α-Fuc-(1→3)]-β-GlcNAc, Le^X] to an aqueous solution of $HAuCl_4$ and subsequent reduction with $NaBH_4$. These GNPs were prepared to investigate the selective self-recognition of the Le^X antigen via carbohydrate–carbohydrate interaction.[70] Following this method, the synthesis of different GNPs functionalized with thiol-functionalized α-D-mannopyranoside,[74] β-D-glucopyranoside,[75,76] β-D-galactopyranoside,[76] β-maltoside [α-Glc-(1→4)-β-Glc],[75] the tetrasaccharide Le^Y [[α-Fuc-(1→2)]-β-Gal-(1→4)-[α-Fuc-(1→3)]-β-GlcNAc],[77] the Thomsen–Friedenreich disaccharide [β-Gal-(1→3)-GalNAc],[78,79] globotriose derivatives,[80] the sulfated disaccharide β-GlcNAc3S-(1→3)-Fuc,[81] and the pyruvated trisaccharide β-Gal4,6(*R*) Pyr-(1→4)-β-GlcNAc-(1→3)-Fuc[82] involved in the self-recognition process of the marine sponge *Microciona prolifera* were reported.

A series of GNPs have also been prepared from thiol-ending derivatives of the products obtained by reductive amination of glucose- and mannose-oligosaccharides.[83] In this case, the structural integrity of the pyranoid ring at the reducing end is not maintained. GNPs coated with C-glycosyl compounds have also been synthesized.[84]

Various thiolated linkers of hydrophobic alkanes and hydrophilic polyethylene glycol derivatives have been used to bind the carbohydrate moiety to the gold core.[75] The GNPs prepared in this way are water-soluble, stable in solution, non-cytotoxic, and have a metal core diameter of few nanometers (Table I).[25] The preparation of the corresponding neoglycoconjugates was effected by various conjugation strategies (direct glycosylation, peptide coupling, thiourea linkage, and others).

Biologically active multivalent GNPs incorporating galactosyl-grafted glycopolymers, obtained by reversible addition–fragmentation chain-transfer (RAFT) polymerization, were also prepared by this methodology.[85] GNPs of larger dimensions (40–80 nm) were synthesized by using glycopolymers,[86] and a novel variant of this method based on an *in situ* photochemical process has also been reported.[87]

GNPs in Table I contain only carbohydrate ligands; however, *hybrid GNPs* incorporating other type of molecule (fluorescent probes, peptides, DNA, RNA, and others) can also be prepared in the same one-pot fashion (Table II). Manipulation of the ligand ratios allows preparation of GNPs having varied carbohydrate density at the surface, providing controllable models for investigating the influence of density and presentation on molecular-recognition features. It was demonstrated that the lactoside density on a GNP gold surface can be controlled by using mixtures of lactose conjugates and thiol-derivatized linkers in different molar ratios.[75] The density of the carbohydrate modulates degradation by β-glycosidases and recognition by lectins of the gold GNPs.[88] Fluorescence-labeled GNPs have also

Table I

Glycoconjugates Used in Direct Syntheses of Gold GNPs

Core size (diameter)	Glycosyl	Thiol-derivatized conjugate	Reference
Au 1.8 nm	Lewis X (Le^X)	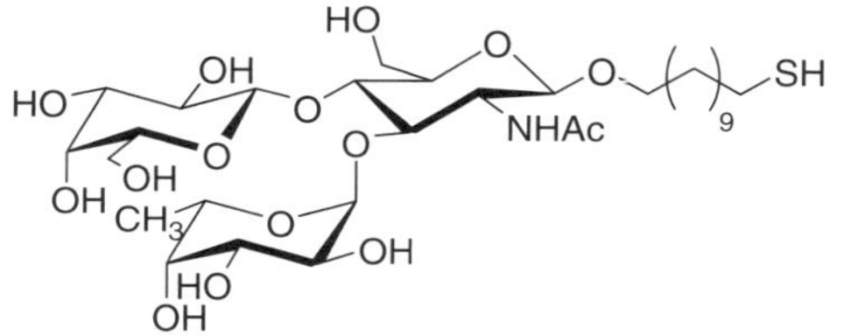 HO, O, OH, HO, O, O, NHAc, O, $(\ )_9$, SH, O, OH, OH, CH_3, O, OH, HO, OH	70,75
Au 1–2 nm	Lactose, maltose, glucose	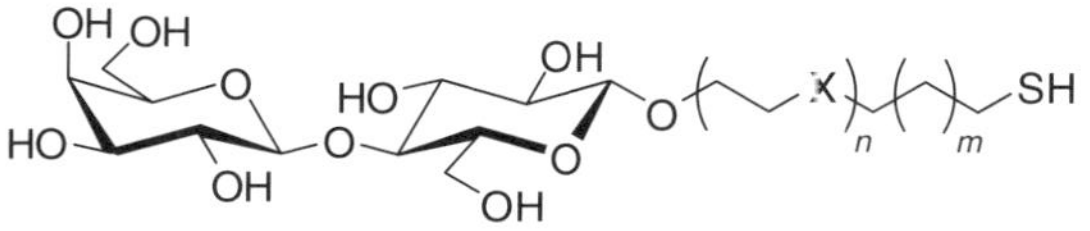 OH, OH, O, HO, O, OH, OH, HO, O, O, $(\ )_n$, X, $(\ )_m$, SH, OH $X = CH_2$, $n = 3$, $m = 0$ or $X = O$, $n = 3$, $m = 0$ or $X = O$, $n = 6$, $m = 9$ 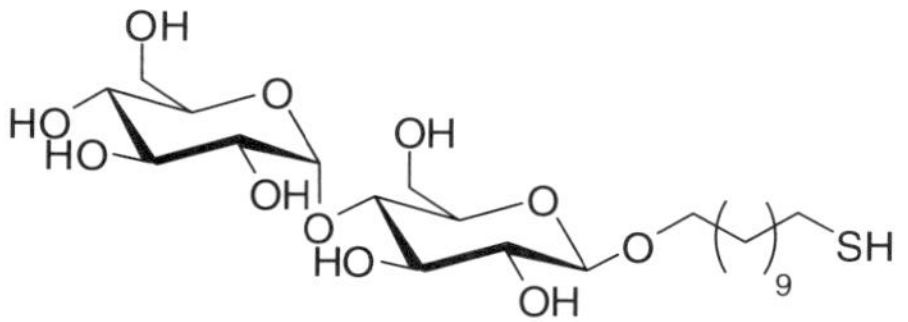 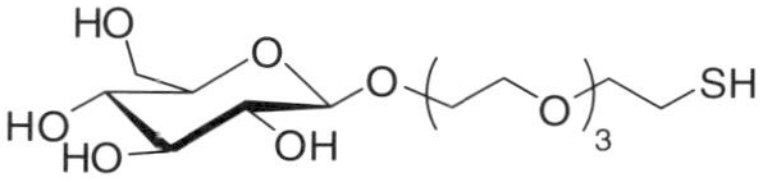 OH, HO, HO, O, OH, O, OH, O, HO, OH, O, $(\ )_9$, SH; HO, O, HO, HO, OH, O, $(\ )_3$, O, SH	75
Au 6 and/ or 20 nm	Mannose, glucose, galactose	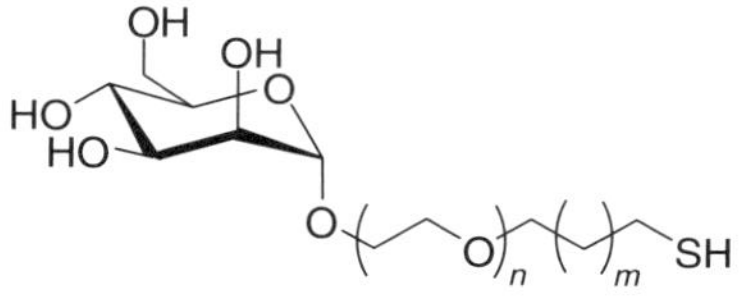 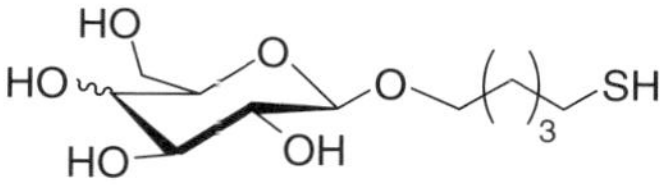 OH, OH, HO, HO, O, O, $(\ )_n$, O, $(\ )_m$, SH; HO, HO, O, HO, OH, O, $(\ )_3$, SH $n = 0$, $m = 3$ or $n = 4$, $m = 1$	74,76

(*Continued*)

TABLE I
(*Continued*)

Core size (diameter)	Glycosyl	Thiol-derivatized conjugate	Reference
Au 1.5 nm	Lewis Y (Le^Y)	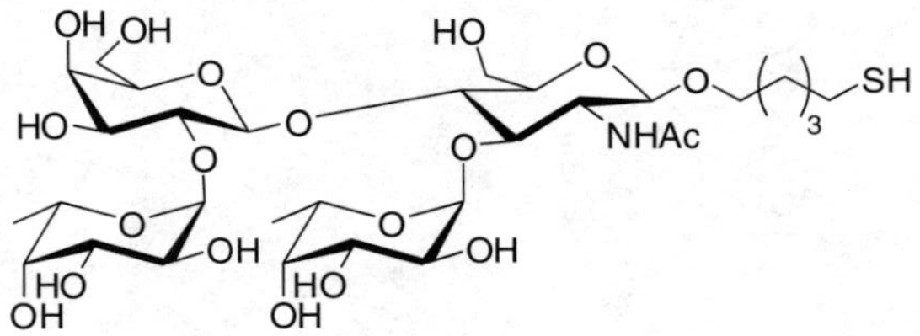	77
Au 1–2 nm	Thomsen–Friedenreich antigen	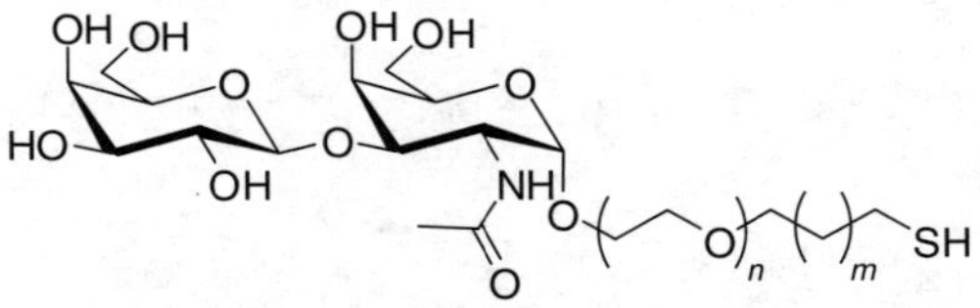 $n=0$, $m=3$ or $n=6$, $m=3$	78,79
Au 1.5–2 nm	Disaccharide from marine sponge and analogues	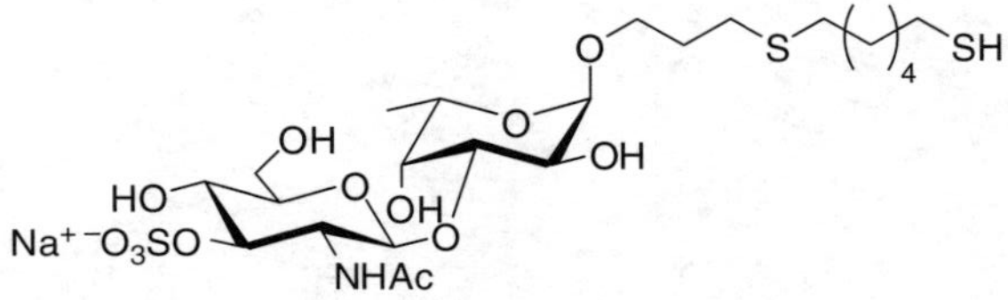	81
Au 1.6 nm	Pyruvated trisaccharide from marine sponge	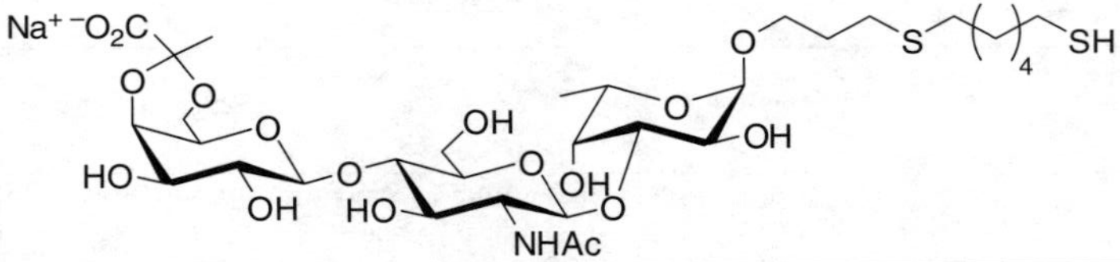	82

(*Continued*)

Table I
(*Continued*)

Core size (diameter)	Glycosyl	Thiol-derivatized conjugate	Reference
Au 2–4 nm	(Oligo)manno- and glucoses	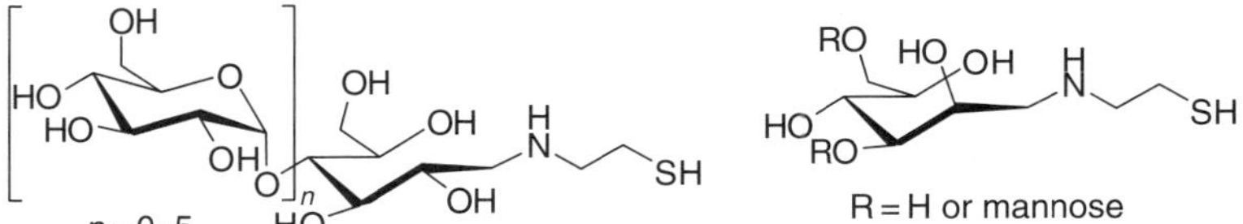 $n = 0–5$; R = H or mannose	83
Au 2 nm	Galactose, glucose	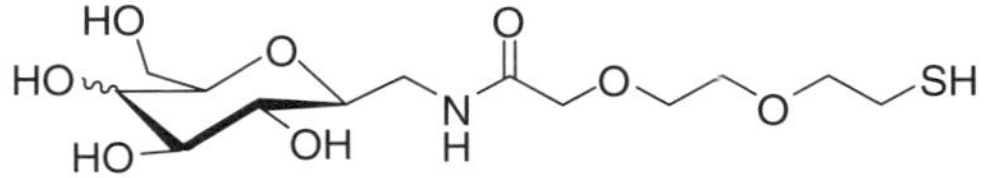	84
Au 4 nm	Globotriose	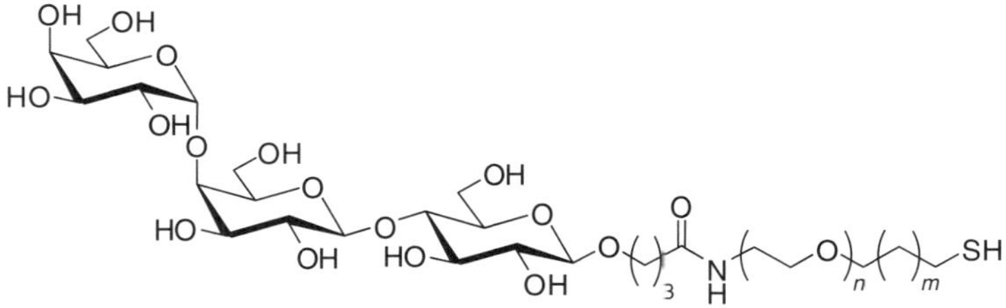 $n = 0$, $m = 0$ or $n = 3$, $m = 1$	80
Au 5–30 nm	Cellulose, lactose, maltose, chitosan	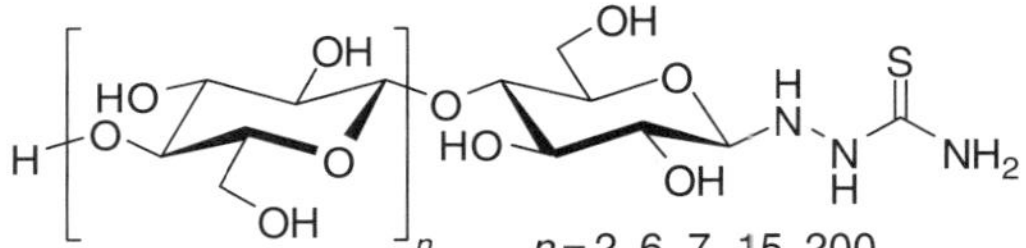 $n = 2, 6, 7, 15, 200$; 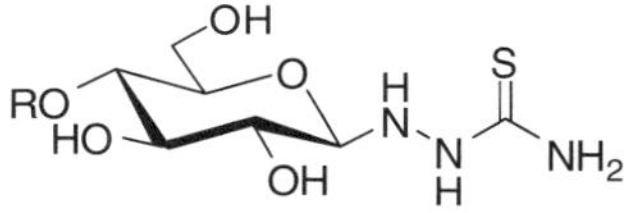 R = β-galactose or α-glucose	93

(*Continued*)

Table II
Hybrid Gold GNPs by Direct *In Situ* Formation

Core size (diameter)	N[a]	Ligands	Reference
Au 1–2 nm	2	HO, OH, HO, O, O, O, O, SH, 9, NHAc, OH, OH, CH_3, O, O, OH, HO, OH or OH OH, O, HO, OH, HO, O, O, O, SH, 9, OH, OH and O, CO_2H, HN, O, 6, 9, SH, N, H, S	75
Au 1–2 nm	2	OH OH, O, OH, HO, O, HO, O, O, X, n, m, SH, OH, OH and HO, X, n, m, SH; $X = CH_2$, $n = 3$, $m = 0$ or $X = O$, $n = 6$, $m = 9$	75
Au 1–2.5 nm	4	OH, OH, HO, CO_2H, AcHN, O, HO, OH, O, O, HO, O, NH, O, 3, SH, O and OH OH, O, HO, O, HO, O, O, O, SH, 3, NHAc, O, O, OH, HO, OH, O, OH, HO, OH and HO, O, HO, HO, OH, O, SH and HS, m, O, O, n, O, O, FKLQTMVKLFNRIKNNVA; $n = 0$, $m = 8$ or $n = 5$, $m = 10$	92

(*Continued*)

TABLE II
(Continued)

Core size (diameter)	N[a]	Ligands	Reference
Au 2–3 nm	2	HO HO HO O OH O SH n $n = 0$ or 3 or OH OH HO O OH O HO OH O OH O 3 SH or HO OH HO O OH O 5 SH and HO_2C N N m SH HO_2C N N CO_2H $m = 3$ or 9	91
Au2–3 nm	2	O O H N O O O 5 O 9 SH Manα1-2Manα Manα1-3Manα Manα1-2Manα1-2Manα or Manα1-3(Manα1-6)Manα and HOOC O O 5 O 4 SH O O N H S N H O 4 9 SH Manα Manα1-2Manα Manα1-2Manα1-2Manα Manα1-2Manα1-2Manα1-3Manα (Manα1-2Manα1-3)(Manα1-2Manα1-6)Manα or (Manα1-2Manα1-2Manα1-3)(Manα1-2Manα1-2Manα1-6)Manα and HO HO HO O OH O 5 SH	89

[a] Number of components (ligands) on the same nanoplatform.

been prepared by mixing a thiol-derivative of fluorescein isothiocyanate (FITC) (5%) and lactose or Lewis X conjugates (95%) prior to formation of the nanoparticles.[75] A small library of multivalent gold GNPs, displaying various densities (10, 50, and 100%) of truncated (oligo)mannoside structures of the high-mannose undecasaccharide $Man_9GlcNAc_2$ [present in the human immunodeficiency virus (HIV) envelope glycoprotein gpl20], has been prepared and characterized (Fig. 5).[89] Three families of "*manno*-GNPs" were generated from thiol-ended (oligo)mannose conjugates, obtained either by direct glycosylation (Fig. 5A), or by peptidic coupling (Fig. 5B) or thiourea coupling (Fig. 5C). The mannose (oligo) saccharides were functionalized with long amphiphilic linkers to ensure a suitable presentation of the antigens. The density of the (oligo)saccharides on the gold surface was controlled by preparing the GNPs in the presence of different ratios of (oligo)mannosides and thiol-ended alkyl β-glucoside $GlcO(CH_2)_5SH$ or alkyl-hexaethylene glycol carboxylic acid $HO_2CCH_2O(CH_2CH_2O)_5C_{11}SH$. Also prepared were "*manno*-GNPs" fluorescently labeled to study their interaction with cellular systems (Fig. 5D). GNPs incorporating glucose and small interfering RNA (siRNA) have also been obtained for transfection and silencing.[90] The synthesis of gold GNPs coated with glycosides of glucose, galactose, or lactose bearing short aliphatic linker-arms, and gadolinium complexes of tetraazacyclododecane triacetic acid (DO3A) derivatives has also been reported.[91]

To our knowledge, the most complex hybrid GNPs prepared thus far contain four different ligands on the same gold nanocluster (Fig. 6).[92] Gold GNPs incorporating the sialyl Tn epitope [α-Neu*p*5Ac-(2→6)-α-GalNAc], Lewis Y antigen, a peptide from tetanus toxoid (T-cell helper peptide), and glucose in well-defined average proportions and with differing densities have been synthesized in one step and characterized by ^1H NMR and TEM. The proportions of the different ligands on the nanocluster surface could be assessed by comparison of the ^1H NMR spectra of the initial mixtures, the GNPs formed, and the mother liquors recovered after the self-assembly process.

Modifications of the reaction conditions (reducing agent, temperature, equivalents) in the preparation of GNPs have been also reported. A variation of the synthesis of GNPs in water which makes use of *N*-methylmorpholine-*N*-oxide (NMMO) as solvent and reducing agent has been proposed.[93] Thiosemicarbazones of maltose, lactose, chitosan, cellobiose, cello-oligomers, and cellulose were prepared in a hot aqueous solution of NMMO (Fig. 7A) and then an Au(III) salt was added to furnish the corresponding gold GNPs (Fig. 7B). NMMO oxidizes the chloride ions of $HAuCl_4$ to hypochlorite ions (with simultaneous disproportionation into chloride and chlorate), which then trigger the reductive formation of Au(0). The solution immediately

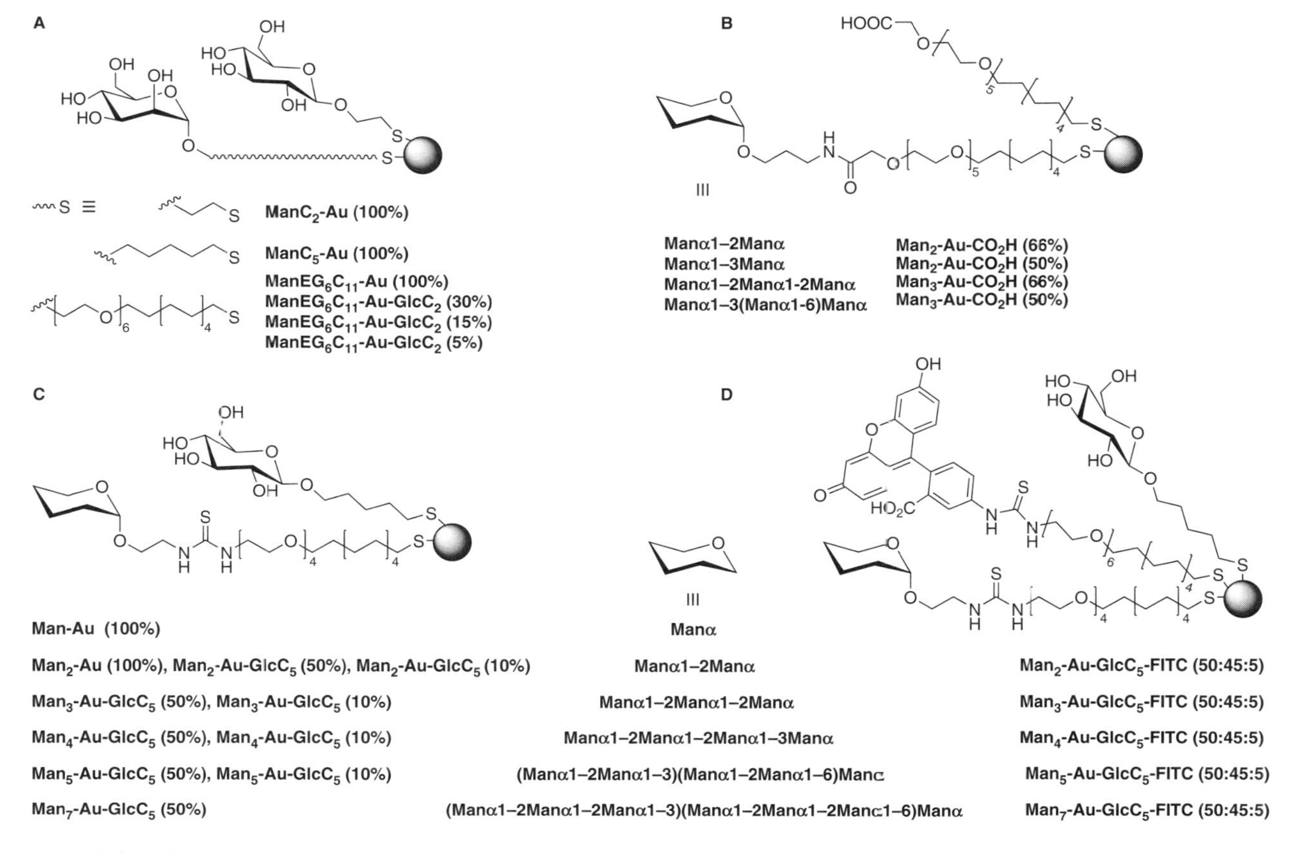

FIG. 5. Schematic representation of hybrid gold "*manno*-GNPs". "*Manno*-GNPs" prepared from (oligo)mannose conjugates obtained by direct glycosylation (A), peptidic coupling (B), and thiourea coupling (C). The percentages in brackets refer to the (oligo)mannoside density. (D) Three-components "*manno*-GNPs" containing (oligo)mannose, glucose, and fluorescein conjugates in 50:45:5 ratio, respectively. Adapted from Ref. 89.

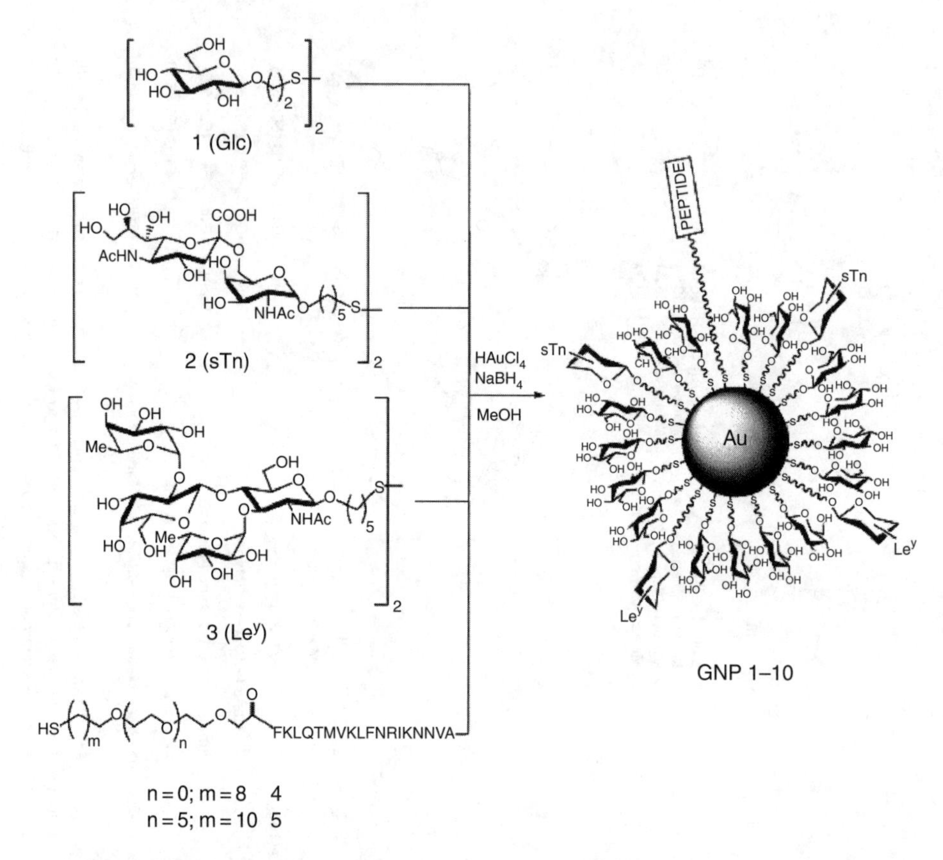

FIG. 6. Hybrid gold GNPs incorporating up to four different ligands in variable ratio. Reprinted from Ref. 92 with permission.

changes from a light-yellow (blank NMMO solution) to a reddish color (Fig. 7C). The UV-visible spectrum confirmed a surface plasmon resonance band around 530 nm. This unusual mechanism of GNPs formation avoids the use of other reducing agents, and the use of NMMO as solvent allows the formation of nanoparticles coated with polysaccharides that are insoluble in water.

b. Two-Step Formation of GNPs.—*(i) Chemical modification on the gold surface*. Multistep syntheses of GNPs that involve chemical reactions over the nanoparticles are also described in the literature (Table III). Gold nanoparticles protected with acetal-terminated mercaptopoly(ethylene glycol) were converted, after unmasking the aldehyde functionality under acidic conditions, into GNPs by

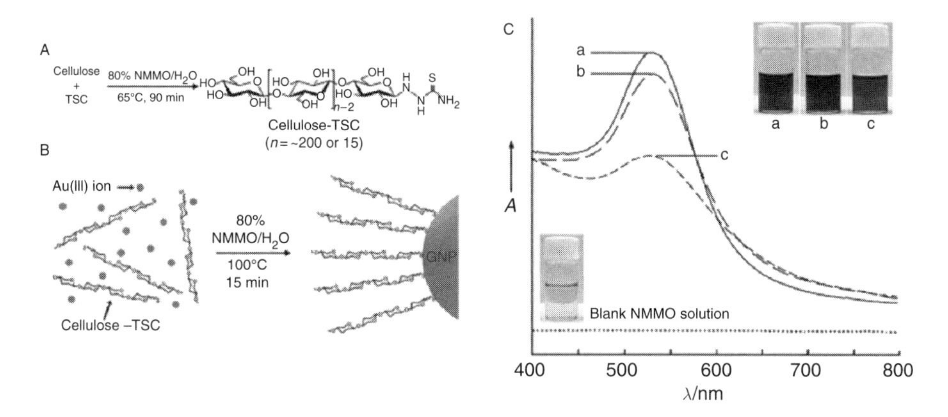

FIG. 7. Gold GNPs by *in situ* protection with thiosemicarbazones carbohydrate derivatives and their UV-Vis spectra. (A) Synthesis of cellulose thiosemicarbazone (Cellulose-TSC). (B) Preparation and possible structure of a cellulose-conjugated gold nanoparticle. (C) UV/Vis spectra and true-colour images of a GNP-free NMMO blank solution and GNPs/NMMO solutions: a) Cellulose-free GNPs, b) Cel_{200}-GNPs, and c) Cel_{15}-GNPs. Adapted from Ref. 93. (See Color Plate 32.)

reductive amination with *p*-aminophenyl glycosides (lactoside and mannoside) in the presence of $(CH_3)_2NHBH_3$.[94] These nanoparticles were the first monolayer-protected GNPs prepared by a three-step synthesis. A three-step procedure was also used by the group of Lakowicz[95] to prepare silver GNPs. Nanoclusters capped with boronic acids were prepared from silver nanoparticles protected with [(2-mercaptopropanoyl)-amino]-acetic acid (tiopronin) by ligand exchange with 3-[2-(2-mercaptopropanamido)-acetamido]phenylboronic acid and subsequently coupled to a polysaccharide (dextran 3000) or a monosaccharide (glucose). "Click chemistry" was also used to convert gold nanoclusters into GNPs: gold nanoparticles electrodeposited on a carbon electrode were functionalized with SAMs of an alkynyl-terminated symmetric disulfide (4,7,10,13,38,41,44,47-octaoxa-25,26-dithiapentaconta-1,49-diyne) by thiol chemistry so that an azide-terminated sialic acid derivative could be immobilized by 1,3-dipolar cycloaddition.[96] Another example of chemical modifications over nanoclusters makes use of silica-coated silver nanoparticles functionalized with primary amine groups. Dextran randomly activated with *N,N'*-disuccinimidyl carbonate was then coupled to the amino groups through peptidic coupling.[97]

In subsequent work, aminooxy-functionalized gold nanoparticles were brought into reaction with glycosphingolipid (GSL) aldehydes to afford GNPs that reproduce the self-assembled microdomain model of multivalent GSLs present at the cell surface.[98] The synthesis involves the selective transformation of the double bond in the ceramide

TABLE III
Functionalization of Metal Nanoparticles Using Different Linking Methodologies

Core size (diameter)	Glycosyl	Linking methodology	Ligand glycoconjugate[a]	Reference
Au 9 nm	Lactose, mannose	Reductive amination	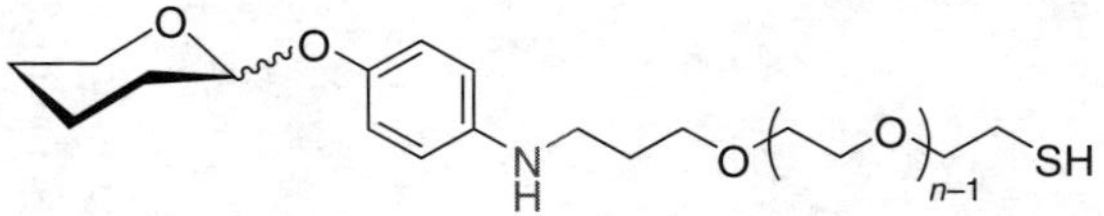	94
Ag 5 nm	Glucose, dextran	Boronate complex	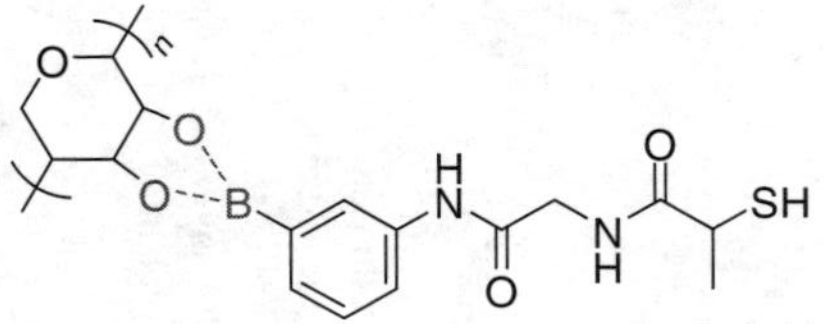	95
Silica-coated Ag 2–10 nm	Dextran	Peptidic coupling	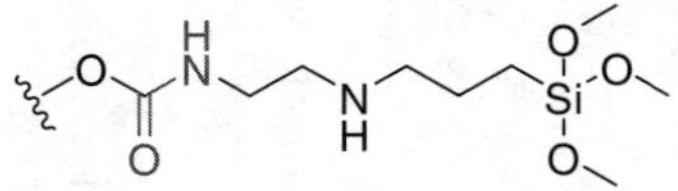	97
Au 12 nm	GalCer, Gg3Cer, GM1, GM3, GD1a, GD1b, GT1b	Oxime formation	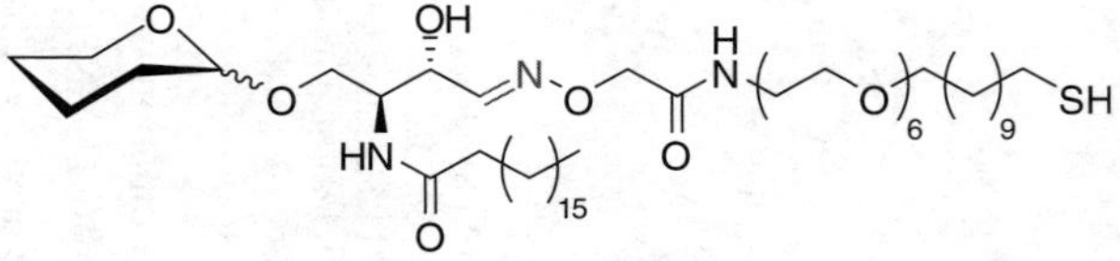	98

[a] The new bond formation in red.

moiety of various GSLs into an aldehyde group by ozonolysis to provide a reactant with the aminooxy groups of the previously prepared nanoparticles. These nanoparticles were prepared in two steps. Gold nanoparticles incorporating an *N-tert*-butoxycarbonyl-protected aminooxy functionality were prepared by direct synthesis, and then the *N-tert*-butoxycarbonyl group was removed by treatment with an aqueous HCl at 40 °C. Direct coupling of the GSL aldehydes with the aminooxy- NPs proceeded directly under mild conditions.

(ii) Thiolated ligand exchange on citrate-capped gold nanoparticles. Water-soluble gold nanoparticles, prepared according to the classical "citrate method" of Turkevich, have been used to prepare GNPs by displacement of the citrate with thiol-derivatized carbohydrate conjugates (Table IV). Nanoparticles bearing mannose ligands were prepared by this method after incubating citrate-capped gold nanoparticles with 2-mercaptoethyl α-D-mannopyranoside for 48 hours.[99] After replacing the salts, the excess of unbound mannose derivative was removed by centrifugal filtering. The mercaptoethyl tether of the mannose conjugate employed for the construction of these Man-GNPs was later replaced by thioctic acid-derived linkers in order to avoid disordered SAMs on the gold surface.[100] Gold nanoparticles coated with lactose derivatives bearing thiolated ethylene glycol or ethylene anchor chains were also obtained by this methodology.[101]

The functionalization of the carboxylic groups of hyaluronic acid (HA) with cystamine by peptidic coupling allowed assembly of this polysaccharide on gold nanoparticles by displacing the citrate capping.[102] HA labeled with a near-infrared fluorescence (NIRF) dye was also immobilized on gold nanoparticles by using the same principle.[103] In this instance, the fluorescent dye was incorporated by peptidic coupling via the acid functionalities of the polysaccharide, while the thiol-ended linker was inserted at the reducing end.

To synthesize globotriose-capped gold glyconanoparticles of larger size (13 and 20 nm) than those obtained by the direct synthesis with $NaBH_4$ (4 nm), sodium citrate was chosen as the reducing agent and thiol-ended globotriose conjugates were used as the exchanging ligands.[80]

Subsequently, an *O*-tetraethylene glycol hydroxylamine-type reagent bearing a terminal trityl-protected thiol group was chemoselectively coupled to carbohydrates (glucose, maltose, and maltotriose) through the aminooxy functionality (oxime formation). The resulting oximes were also reduced with sodium cyanoborohydride to the corresponding oxyamines. After detritylation, both *N*-glycosyl oximes and *N*-glycosyl oxyamines were attached to previously generated 12 nm citrate-stabilized gold nanoclusters.[104] These oxime- and oxyamine-GNPs were tested with proteins to compare their molecular recognition behavior. The use of *N*-glycosyl oximes instead

TABLE IV

Glycoconjugates Used in Thiolated Ligand Exchange on Citrate-Capped (or Analogues) Gold Nanoparticles

Core size (diameter)	Glycosyl	Glycoconjugate ligand	Reference
Au 16–17 nm	(Oligo)mannoside	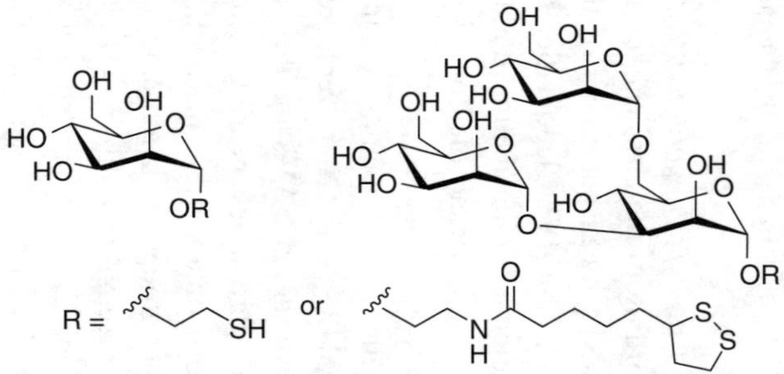 OH OH HO O HO OR; OH OH HO O HO OH OH HO O HO O OH HO O O OR; R = ~SH or ~N(H)C(=O)~S-S	99,100
Au 16 nm	Glucose, lactose	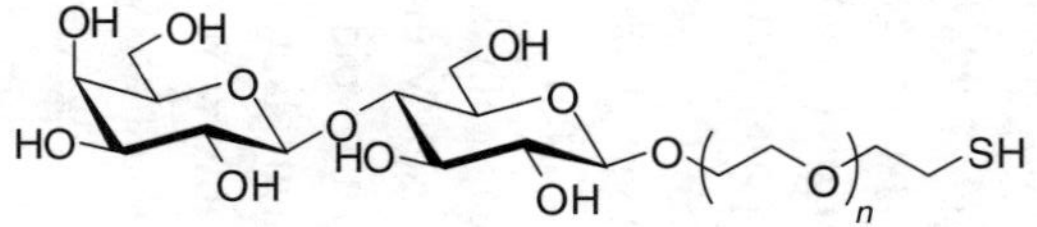 OH OH O HO OH O OH HO O OH O(~O)$_n$~SH; n = 0, 1, 2, 3; 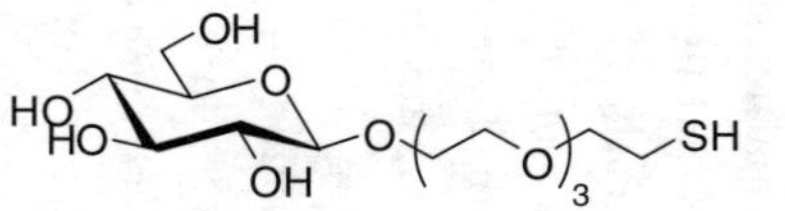 OH HO HO O OH O(~O)$_3$~SH	101
Au 20 nm	Hyaluronic acid	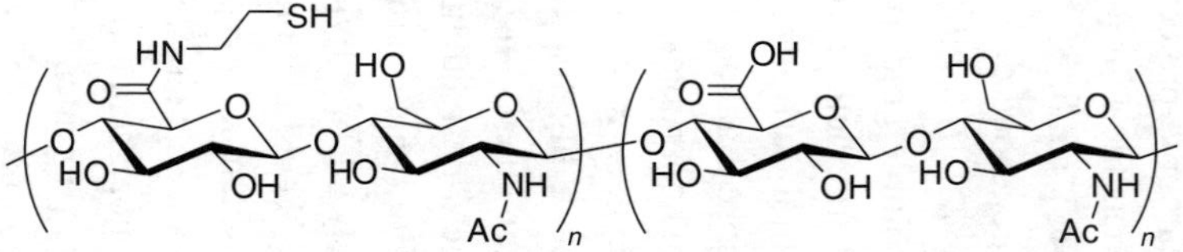 (O HO O HN ~SH O OH O HO HO O NH Ac)$_n$ (O HO OH O O OH O HO HO O NH Ac)$_n$	102

(*Continued*)

Table IV
(*Continued*)

Core size (diameter)	Glycosyl	Glycoconjugate ligand	Reference
Au 16 nm	Hilyte-labeled hyaluronic acid	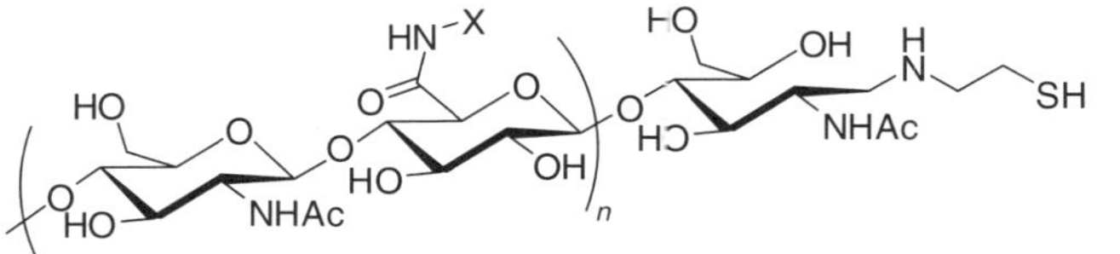 HN–X, HO, OH, H, N, SH, HO, O, O, O, O, O, HO, OH, NHAc, HO, NHAc, n X = Hilyte Fluor™ 647	103
Au 13/20 nm	Globotriose	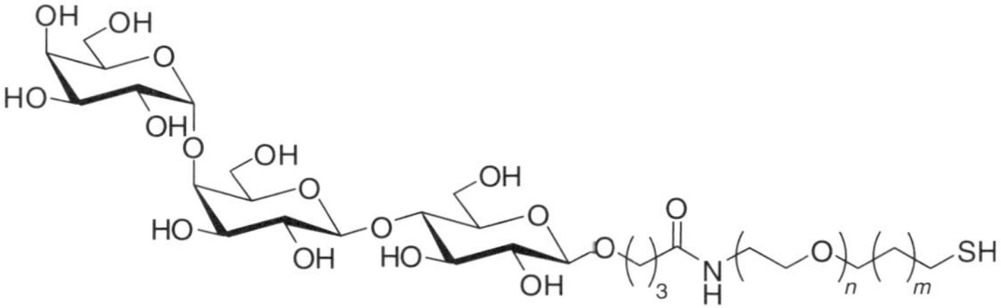 OH, OH, O, HO, OH, O, OH, O, HO, OH, O, OH, O, HO, OH, O, O, N, H, O, SH, 3, n, m $n = 0$, m = 0 or $n = 3$, $m = 1$	80
Au 12 nm	Glucose, maltose,maltotriose	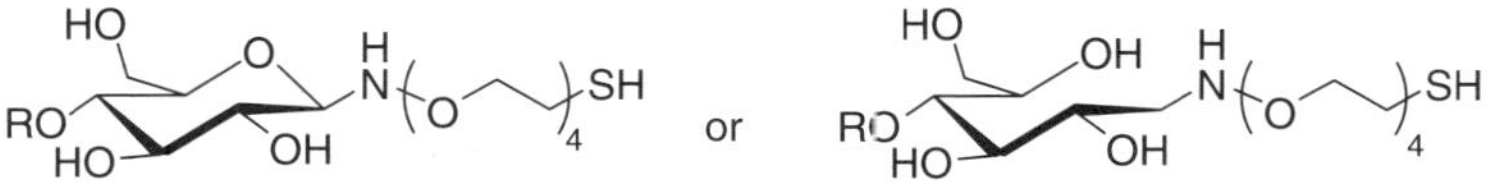 HO, O, H, N, O, SH, RO, HO, OH, 4 or HO, OH, H, N, O, SH, RO, HO, OH, 4 R = H, α-glucose, α-maltose	104

(*Continued*)

Table IV
(*Continued*)

Core size (diameter)	Glycosyl	Glycoconjugate ligand	Reference
Au 2–5 nm	Mono-, di, tri-*N*-acetyl mannosaminides	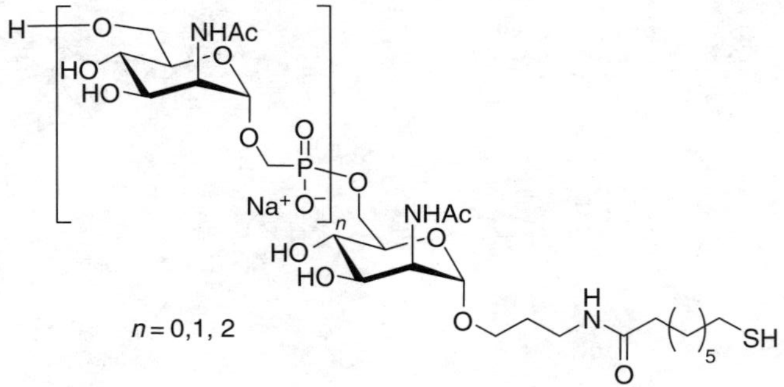 $n = 0,1, 2$	105,106
Au 1.5–3 nm	Mannose	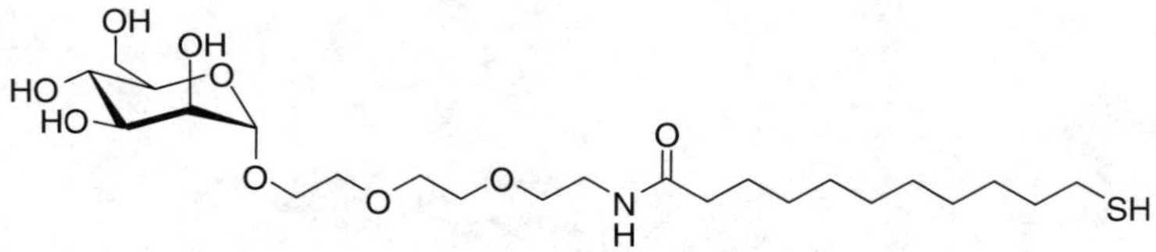	107
Au 4–5 nm	Vancomycin	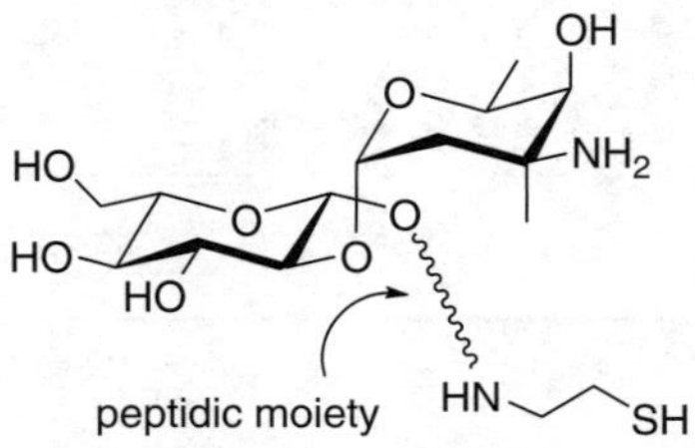	109

of conjugates derived by reductive amination (usually employed to functionalize sugars at their free reducing-end) can be a valid alternative when the integrity of the terminal sugar ring needs to be preserved, although oximes undergo hydrolysis at pH < 2, as in the stomach, or at pH > 11.

Alternative protocols that make use of the thiolated ligand exchange of preformed gold nanoparticles and based on stabilizing/reducing salts other than citrate have also been proposed for the preparation of GNPs. For example, water-soluble, thiol-protected gold GNPs equipped with (oligo)saccharide synthetic analogues of type A *Neisseria meningitides* antigens were prepared by using dioctylamine[105] as a stabilizing agent of the gold cluster instead of citrate, before the final passivation with thiols.[106] Dioctylamine reduces Au(III) to Au(I) and stabilizes the growing cluster when $NaBH_4$ is added to complete the reduction process.

Tetrakis(hydroxymethyl)phosphonium chloride (THPC) is another reducing/capping agent which has been employed for the synthesis of gold nanoparticles prior to biofunctionalization with thiol-ended mannose conjugates.[107]

Polyvalent presentation of the antibiotic vancomycin on gold nanoparticles was achieved starting from a dispersion of gold nanoparticles (5 nm) in toluene obtained by the methodology of Schiffrin *et al.*[108] (the "naked" colloidal gold solutions were prepared in toluene using a quaternary ammonium bromide salt as phase-transfer reagent), and treating them with a water solution of bis(vancomycin) cystamide under vigorous stirring for 12 hours.[109] The formation of Au–S bonds allowed the glycopeptide-capped nanoparticles to dissolve in the aqueous phase, which was readily separated from the organic phase. In this case, the sugar moieties are not involved in the recognition properties of the nanoparticles, as the antibacterial activity is due to interactions between the amino acid residues and the bacteria.

The methodology of thiolated ligand-exchange on salt-capped gold nanoparticles was also used to obtain *hybrid glycoclusters* (Table V).

Gold GNPs of various saccharide percentages on the covering surface (100, 75, 50%) were obtained from di-*n*-octylamine-stabilized gold nanoparticles by thiol passivation with different mole ratios of thiol-armed *N*-acetyl-mannosaminide and 8-mercapto-*N*-{2-[2-(2-methoxyethoxy)ethoxy]ethyl}octanamide.[105]

Gold GNPs bearing different percentages of lactose derivatives ranging from 0 to 65% were prepared to explore the carbohydrate density effect on biorecognition processes. Commercially available gold nanoparticles (diameter 20 nm) were treated with mixtures of different molar ratios of a poly(ethylenglycol) derivative containing both diethylacetal and mercapto terminal groups $(\text{acetal-PEG-S-})_2$ and *p*-aminophenyl-β-D-lactopyranoside conjugated to the deprotected acetal $(\text{HC(O)-PEG-S-})_2$.[110]

TABLE V
Hybrid Gold GNPs by Ligand Place Exchange

Core size (diameter)	N[a]	Ligands	Reference
Au 20 nm	2	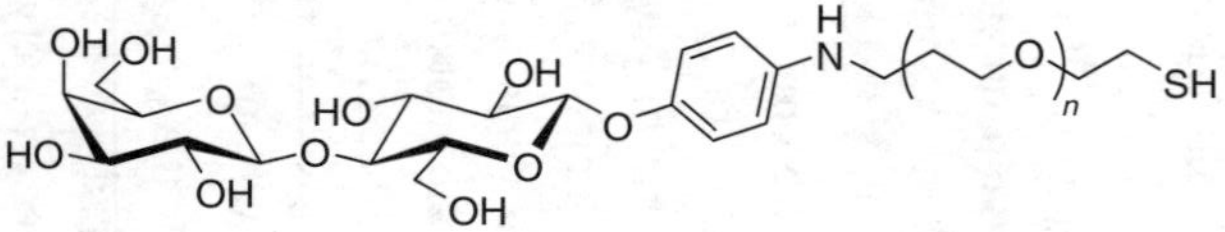 and 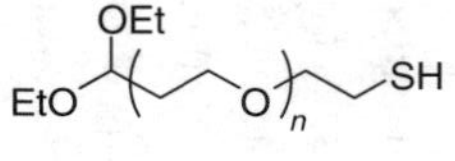	110
Au 5 nm	2	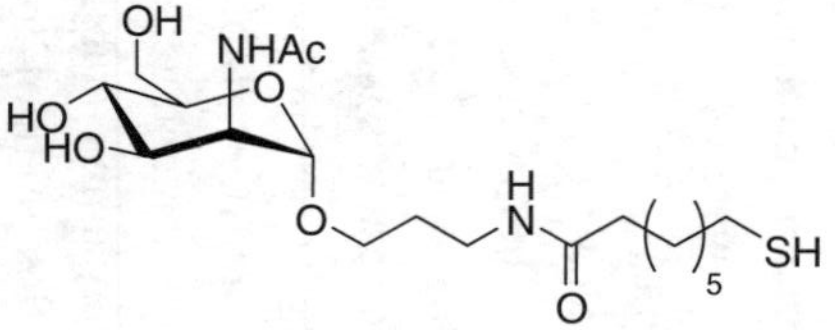 and 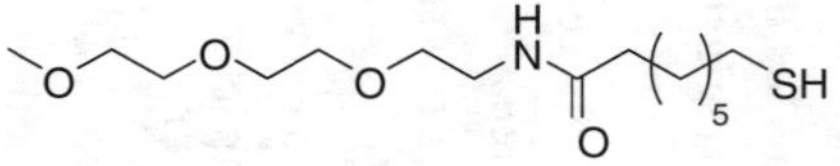	105
Au 16 nm	2	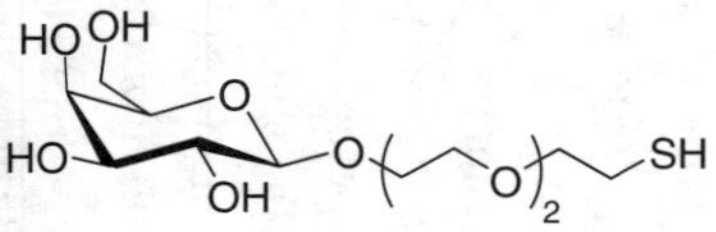 or 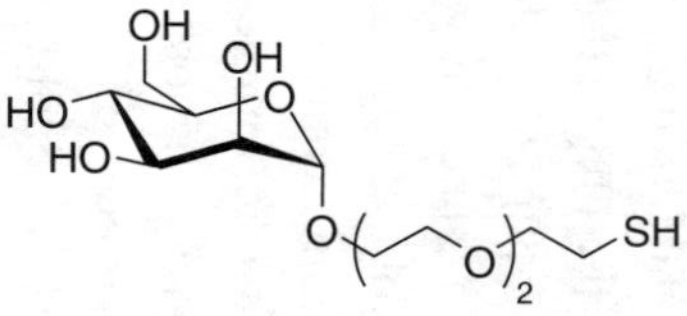 / and 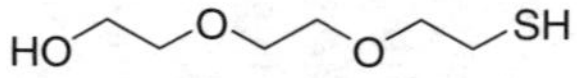	111

[a] Number of components (ligands) on the same nanoplatform.

Hybrid gold GNPs coated with either a thiolated β-galactose derivative or a thiolated α-mannose derivative and varying ratios of a thiolated triethylene glycol derivative were also reported.[111]

The use of suitable glycopolymers as protecting agents for metallic nanoclusters has also been examined. Glucose-carrying polymer chains bearing a disulfide group were used to construct SAMs onto colloidal silver with 50 nm hydrodynamic size.[112]

(iii) Non-covalent approaches to gold glyconanoparticles. Other protocols have been reported for the synthesis of GNPs in which the carbohydrates are non-covalently attached to the metal cluster. Saccharide-modified hyperbranched poly(ethylenimines) were used to elaborate copper, silver, gold, and platinum nanoparticles.[113,114] 12-α-*C*-Ribofuranosyl and ribopyranosyl dodecanoic acids were heated with silver nitrate in dilute alkaline solution to afford water-soluble 15 nm silver GNPs.[115]

Gold nanoparticles coated with natural polysaccharides have also been reported. Usually, the polysaccharide works as stabilizing agent and it is not covalently linked to the gold cluster. The group of Yang[116–118] prepared gold and silver nanoparticles incorporating chitosan and heparin, using the polysaccharides as reducing–stabilizing agents. One of the main advantages of coating of metal nanoparticles with polysaccharides is to enhance their biocompatibility.[119] This allows their use for diagnostic purposes and drug delivery. For example, oral and nasal administration of chitosan-reduced gold nanoparticles for transmucosal delivery of insulin to diabetic rats improved pharmacodynamic activity.[120]

It has also been reported that gellan gum, a linear, anionic heteropolysaccharide, can be employed both as reducing and stabilizing agent for the synthesis of gold GNPs.[121] These nanoparticles display stability to addition of electrolytes and pH changes. The resultant anionic gold nanoparticles were subsequently loaded with doxorubicin, the anticancer agent, by electrostatic interactions.

Gum arabic is another biocompatible polymer that has been used to encapsulate gold nanoparticles. It is a complex mixture of polysaccharides (mainly natural arabinogalactan) and glycoproteins. It is used in the food industry as a stabilizer, and is thus a good candidate for *in vivo* use. Gold nanoparticles coated with gum arabic were prepared by mixing an aqueous solution of sodium tetrachloroaurate with the conjugate of alanine and tris(hydroxymethyl)phosphine [$P(CH_2NHCH(CH_3)COOH)_3$] as a reducing agent and *in situ* protectant of the nanoparticles.[122]

These are some of the many reported examples of drug-delivery systems where polysaccharides are used to encapsulate metal nanoparticles.

c. Characterization of Gold GNPs.—GNPs can be isolated and purified by membrane filtration and by dialysis. The analytical techniques for characterization

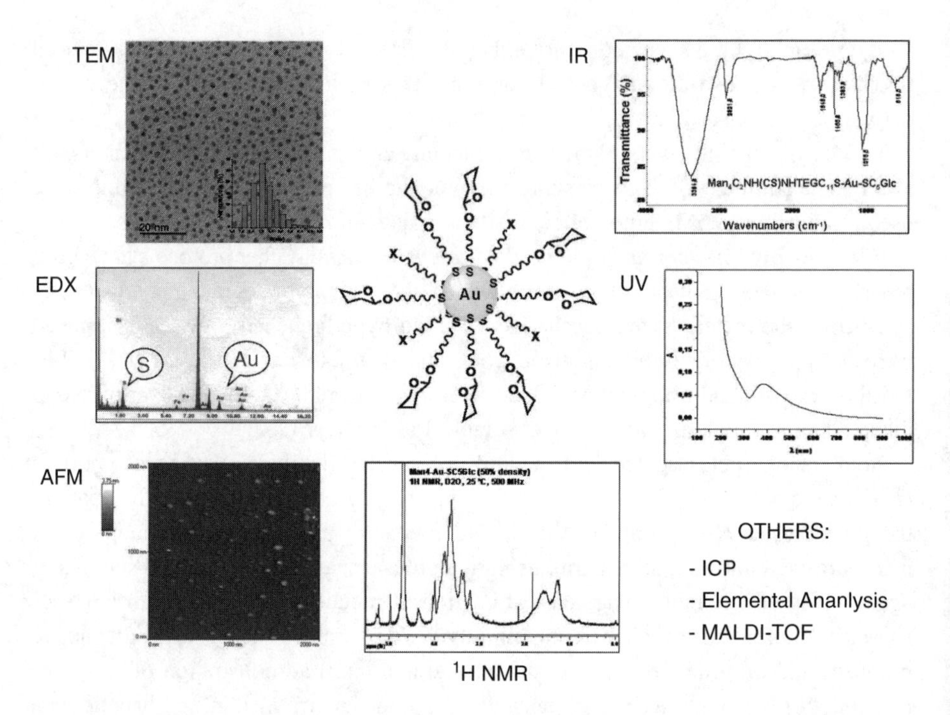

FIG. 8. Different techniques employed for the characterization of GNPs. (See Color Plate 33.)

of GNPs address both the metallic core and the protecting organic layers (Fig. 8).[18] Transmission electron microscopy (TEM) and energy-dispersion X-ray (EDX) give information about both the size distribution and the nature of the metallic core. The surface plasmon resonance band in UV-visible spectrum is also useful for establishing the size of the gold cluster.[123] ^{1}H NMR, IR, UV-Vis, and XPS spectroscopies, and matrix assisted laser desorption/ionization time-of-flight mass spectrometry (MALDI-TOF MS) can be used to determine the nature of the organic ligands on the surface. By combining data obtained by elemental analysis with data from inductively coupled plasma atomic-emission spectroscopy (ICP-AES), an average molecular formula for the gold GNPs can be established. Dynamic light scattering (DLS) gives the hydrodynamic diameter in solution.

As previously mentioned, crystallography is essential for providing detailed information of the whole structure of monolayer-protected gold nanoparticles.[47] Although scarcely a routine technique, because of the difficulty in preparing nanoparticles sufficiently uniform in size, crystallography may develop as the most important tool for disclosing unknown structural features of GNPs.

2. Magnetic GNPs

Nanoparticles that can be manipulated by magnetic fields are generally termed magnetic nanoparticles (MNPs). A typical MNP contains a magnetic metallic core (iron, nickel, manganese, cobalt, and related iron- or cobalt-containing alloys). MNPs are valuable probes for magnetic resonance imaging (MRI), hyperthermic treatment of malignant cells, site-specific drug delivery, and the manipulation of cell membranes.[124–129] In particular, the combination of the magnetic characteristics with properties that can be incorporated in the MNPs by means of nanotechnology opens up the availability of synergetic multifunctional nanoplatforms. Among the different types of MNPs, iron oxide nanoparticles are probably the most studied and used. Parallel to these superparamagnetic systems, there is also growing attention toward paramagnetic nanoparticles based on Gd(III) ions.[130] The main application of these inorganic magnetic nanoparticles is as contrast agents in MRI (iron oxide nanoparticles usually being T_2-agents, and Gd-based probes being T_1-agents).[131]

During recent years, great progress has been made in the synthesis and characterization of MNPs in terms of stability, dispersability, size, and shape control.[132,133] Co-precipitation, thermal decomposition and/or reduction, laser-pyrolysis techniques, and others, in tandem with core-protection processes (surfactant/polymer-, silica-, and carbon-coating) are well-established methods for the preparation of high-quality MNPs.[134] Many strategies have been used to conjugate ligands to these nanoparticles.[135] The surface of MNPs can be coated with functional moieties (radionuclides, biomolecules, fluorescence tags, and the like) for targeted imaging, gene delivery, and cellular trafficking.[136] MNPs coated with targeting biomolecules (for instance, antibodies, aptamers) are effective probes for specific imaging of different diseases.[135] Essential requisites for the biological applications of these nanomaterials are high magnetization values, small sizes (less than 20 nm in diameter), narrow particle-size distribution, and a special surface coating for both avoiding toxicity and allowing the coupling of biomolecules. The most widely used MNPs for applications *in vitro* are the iron oxide-based superparamagnetic nanoparticles. These are usually prepared by alkaline co-precipitation of ferrous and ferric salt solutions in a single-step process. The composition of the iron oxide cores varies from magnetite to maghemite, and the size of the particles has a great influence on such properties as their biodistribution. They are thus usually classified into superparamagnetic iron oxide (SPIO; hundreds of nanometers size), and ultrasmall superparamagnetic iron oxide[137] (USPIO; <50 nm). Synthetic polymers based on poly(lactic-*co*-glycolic acid) (PLGA) and/or natural polysaccharides have been used to stabilize iron oxide nanoparticles.[127,138] Several commercially available MRI contrast-agents consist of a dextran shell, which stabilizes iron oxide cores, ensures water-solubility, and enhances biocompatibility and biodegradability. Magnetic iron oxide nanoparticles have been obtained by alkaline co-precipitation of iron(II) and iron(III) salts in aqueous solutions of sugar-based macromolecules, such as dextran, *O*-carboxymethyldextran, chitosan, starch, and heparin.[127]

Dextran- and *O*-carboxymethyldextran-coated SPIO nanoparticles have been marketed since the early 1990s and their physical and chemical properties have been reviewed.[139,140] Although the encapsulation of the magnetic core within a polysaccharide matrix is based on adsorption processes, further chemical reactions can be effected once the polysaccharide has been suitably functionalized. In particular, dextran-stabilized iron oxide nanoparticles have attracted much interest for their possibility of cross-linking with epichlorhydrin (cross-linked iron oxide, CLIO). The functionalization of dextran-stabilized iron oxide nanoparticles by amine groups has enabled their conjugation with various types of molecules. The group of Weissleder has been very active in the covalent binding of biomolecules to the aminodextran matrix. Conjugation with a peptide from HIV-Tat protein,[141] selected oligonucleotide sequences,[142] an anti-human E-selectin antibody fragment,[143] the indocyanine optical dye Cy5.5,[144] and various synthetic small molecules[145] are some of the examples reported.

In this way, it has been possible to conduct specific targeting and/or multimodal imaging otherwise not feasible with non-functionalized MNPs. While dextran-coated nanoparticles cannot properly be considered GNPs, as here the carbohydrates play a stabilizing role more than an antigenic role, functionalization of the dextran-MNPs by a specific probe for targeting carbohydrate receptors is one way to prepare magnetic glyconanoparticles (MGNPs) as will be discussed in detail next.

a. Superparamagnetic and Paramagnetic GNPs.—MGNPs protected with SAMs of simple neoglycoconjugates have been obtained by following the direct *in situ* procedure used for the synthesis of the gold GNPs. Two approaches have been used to convert noble-metal GNPs directly into magnetic or paramagnetic GNPs. One of them modifies the gold core by introducing a magnetic metal (iron, manganese, cobalt) to furnish superparamagnetic GNPs (T_2-contrast agents). In addition, modification of the organic shell by introduction of a gadolinium complex as an additional ligand allows the preparation of paramagnetic GNPs (T_1-contrast agents) (Fig. 9).

The first water-soluble gold–iron glyconanoparticles covalently functionalized with glucose, maltose, and lactose derivatives and having different gold/iron ratio were prepared by reduction in water of a mixture of gold and iron salts in the presence of thiol-ending neoglycoconjugates.[146] The new magnetic GNPs were fully characterized by TEM and ICP-AES and their magnetic properties assessed by a superconducting quantum-interference device (SQUID). A remarkable observation was made when comparing the magnetic behavior of Au/Fe-containing GNPs with that of the

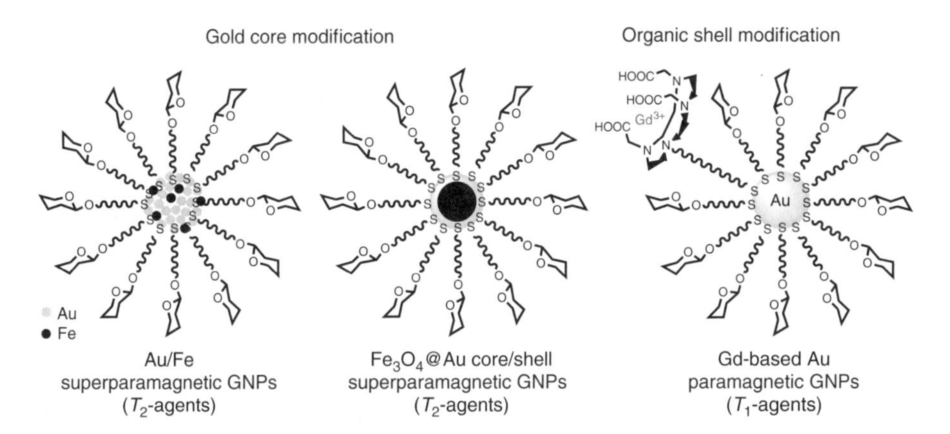

FIG. 9. Strategies for converting gold GNPs into superpara- and paramagnetic GNPs. (See Color Plate 34.)

corresponding pure gold GNPs. Unexpectedly, a permanent magnetism at room temperature was observed in the iron-free gold GNPs, that was not present in the corresponding gold–iron oxide GNPs (TEM micrographs of globular *maltose*-Au and *maltose*-Au/Fe GNPs and the comparison of their magnetic properties are shown in Fig. 10A).[148] The thiol-capped gold nanoparticles display magnetic hysteresis while the iron–gold GNPs showed a superparamagnetic behavior. The small particle size (below 2 nm) and the change in the electronic structure of the nanoparticles due to the covalent gold–sulfur bond seem to be responsible for their ferromagnetism. This behavior was confirmed by the controlled introduction of Fe impurities into gold nanoparticles: magnetic impurities decrease the high local anisotropic field responsible for the ferromagnetic properties of thiol-capped gold GNPs.[149]

Furthermore, when an amphiphilic maltose neoglycoconjugate (equipped with an eleven-carbon lipophilic spacer) was used in preparation of the GNPs,[146] supramolecular aggregates soluble in methanol were also formed (Fig. 10B). The proposed model of aggregation shown in Fig. 10B (middle), and based on TEM micrographs, was confirmed by atomic force microscopy (AFM) imaging (right).

Full characterization of this novel material by TEM and AFM has been reported.[147] Non-contact dynamic AFM showed polymeric aggregates packed in units of about 65 nm in length and 40 nm in width on gold surfaces, indicating encapsulation of the GNPs by the organic material.

The fluorescence spectrum of the supramolecular aggregates (*maltose*-Au/Fe polymeric GNPs) showed an emission band at 600 nm when the sample was irradiated at 254 nm. This fluorescence emission was not observed in the case of the non-ordered

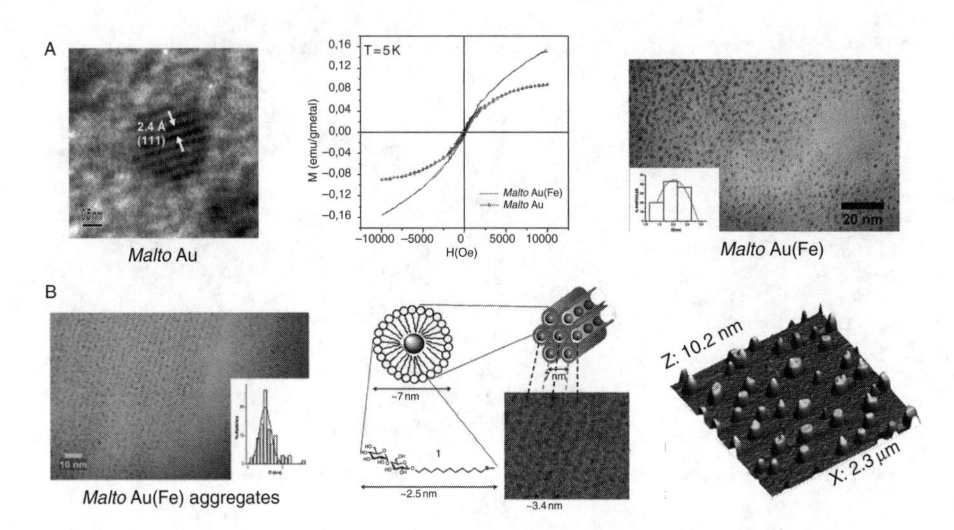

FIG. 10. Different types of maltose-coated Au and Au/Fe GNPs and their properties. (A) TEM micrographs of globular *maltose*-GNPs (*Malto* Au, left) and *maltose*-Au/Fe GNPs (*Malto* Au(Fe), right), and corresponding magnetization curves measured at 5 K (middle). (B) TEM (left) and AFM (right) images of non-globular *maltose*-Au/Fe GNPs (*Malto* Au(Fe) aggregates), and cartoon showing the micelle model for *Malto* Au(Fe) aggregates with the corresponding dimensions superposed on a TEM micrograph (middle). Adapted from Refs. 146 and 147.

maltose-Au/Fe GNPs. This phenomenon is probably related to the size of these glyconanoparticles, since the polymeric GNPs present a smaller diameter (~1.6 nm; Fig. 10B) than the non-ordered ones (2.2 nm; Fig. 10A, right). Chang and co-workers later reported that Man-GNPs having a core diameter of 1.8 nm fluoresced at 545 nm, when excited at 375 nm.[107]

The biocompatibility of *maltose*-, *glucose*-, and *lactose*-MGNPs was tested on human dermal fibroblasts *in vitro* by fluorescence and scanning electron microscopy.[150] Depending on the nature of the carbohydrate, these MGNPs elicited different cell responses. While *lactose*- and *maltose*-MGNPs were endocytosed (the *maltose*-MGNPs also inducing cell death), the *glucose*-MGNPs were not.

In a second approach, hybrid GNPs bearing on the same gold nanoplatform, sugar conjugates and Gd(III) chelates have been prepared by the one-step procedure.[91] Thiol-terminated neoglycoconjugates of glucose, galactose, or lactose were used in tandem with an *N*-alkyl (pentyl or undecyl)-DO3A derivative to coat the gold nanoclusters (Fig. 11). The longitudinal (T_1) and transverse (T_2) relaxation times were determined, and the relaxivity ($r_{1(2)}$) values calculated from T_1 (the slope of $1/T$ as a function of millimolar Gd(III) concentration), and compared with Dotarem®, a

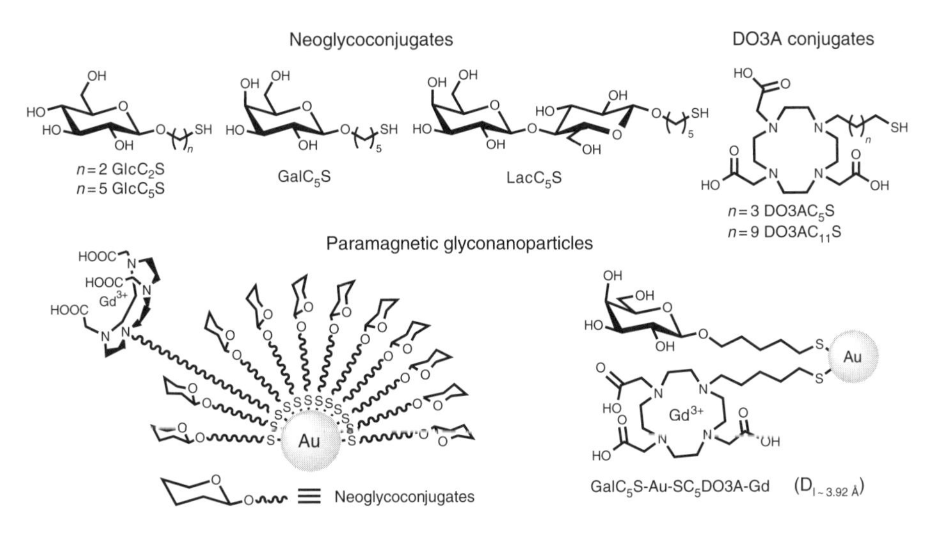

FIG. 11. Paramagnetic Gd-based gold glyconanoparticles. Adapted from Ref. 91.

product in clinical use. Both the nature of the carbohydrate and the relative position (Δl) of the sugar with respect to the Gd(III) ion seem to control the relaxation values of these GNPs. The best GNPs yielded relaxivities above 20 mM^{-1} s^{-1}, which are over six times the values for Dotarem®. These GNPs were used for *in vivo* imaging of gliomas in mice.

There appear to be no other examples of GNPs that incorporate Gd(III) in the organic shell, although dextran-coated $GdPO_4$[151] or Gd_2O_3[152] nanoparticles have been reported.

b. Iron Oxide-Based GNPs.—The protection of a magnetic core with materials that can be further functionalized via chemical reactions opens up the possibility of inserting antigenic glycoconjugates (Fig. 12). Iron oxide-based magnetic GNPs have also been prepared by modifying the organic outer shell with suitable functionalities capable of reacting with the desired glycoconjugates.

The insertion of antigenic carbohydrates in this type of construction was elegantly demonstrated by the group of Davis.[153] The synthesis of amine-reactive complex carbohydrates employed a versatile masked *S*-cyanomethyl group in a combined chemoenzymatic synthetic strategy. Dextran-coated iron oxide nanoparticles were equipped with high levels of amine groups per particle and further functionalized with the glycan ligands GlcNAc, β-D-galactopyranosyl-(1→4)-*N*-acetyl-D-glucosamine (LacNAc), sialyl LacNAc, and sialyl Lewis X (sLe^X). The sLe^X-MGNP targeted to selectins allowed direct detection by MRI of endothelial markers in acute

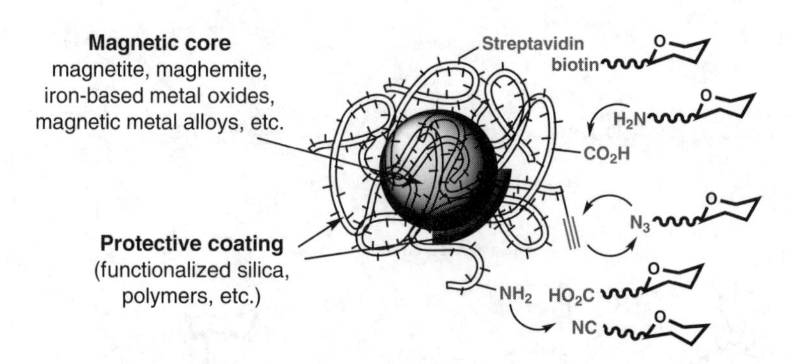

FIG. 12. Strategies to prepare iron oxide-based magnetic glyconanoparticles. (See Color Plate 35.)

inflammation. This is an excellent example wherein a combination of a magnetic nanomaterial and a bioactive carbohydrate has allowed the direct observation of a biological process *in vivo*.

MGNP-based systems have also been constructed through covalent functionalization of silica-coated magnetite nanoparticles with α-mannose or α-galactose derivatives.[154] Tetraethoxysilane was used to build up the first silica shell. Triethoxysilane, derivatized either with a terminal alkyne or with an amine group, was further used to functionalize the shell. In this way, it was possible to immobilize on the nanoparticle surface suitably derivatized carbohydrates via "click chemistry" or by peptidic coupling, that is, through a covalent approach. A similar approach was used to obtain silica-coated amine-armed Fe_3O_4 nanoparticles, which were allowed to react with dextran activated with *N,N′*-disuccinimidyl carbonate.[97] Chitosan has also been used for the functionalization of previously prepared MNPs.[155] Treating amine-functionalized magnetite nanoparticles with glutaraldehyde allowed the introduction of a carbonyl group in the outer shell of the nanoparticle. The carbonyl-magnetite nanoparticles thus obtained reacted with chitosan. These magnetic nanocomposites were employed to remove heavy-metal ions from water, utilizing the reversible coordination properties of chitosan. Biotinylated bi-antennary glycoconjugates of a novel sialic acid-containing trisaccharide, mannoside, lactoside, or thiosialioside, and a biotinylated tetra-antennary conjugate of α-mannoside were also used to functionalize avidin-coated magnetic beads.[156]

Ligand exchange-like protocols have also been used for attachment of carbohydrates onto protected-MNPs. Magnetic Fe–Pt nanoparticles functionalized with the glycopeptide vancomycin[157–159] were prepared analogously to the corresponding gold nanoparticles.[109] Chemically stable, water-soluble, and highly magnetic anisotropic FePt magnetic nanoparticles (4 nm) were obtained.

Other approaches based on non-covalent functionalization of the magnetic cores have been reported. D-Mannose-modified iron oxide nanoparticles were prepared by random absorption, and used for labeling of stem cells.[160] The synthesis was reported of stable and water-redispersible 50 nm cobalt superparamagnetic nanoparticles. It employed $NaBH_4$ reduction of cobalt chloride using oleic acid-derived sophorolipids (a class of glycolipids obtainable by incubation of glucose and fatty acids with suitable yeasts) as a new water-soluble capping agent.[161] Electrostatic interactions between the carboxylic functionalities of the sophorolipids and the metallic core are presumably responsible for the robust glycolipid capping of the nanoparticle surface. The saturation magnetization value of sophorolipid-capped cobalt GNPs (23 emu/g) was higher than that of simple oleic acid-capped cobalt nanoparticles (3 emu/g) prepared under similar conditions, although differences in size, chemical composition, and capping nature may play a key role in tuning the magnetic properties.

3. Glyco Quantum Dots

Fluorescent semiconductor nanocrystals (CdSe, CdTe, PbSe, and others), otherwise included in the term "quantum dots" (QDs),[162] have attracted much attention in various research fields for more than 20 years[163,164] owing to their chemical and physical properties, which differ markedly from those of the bulk solid (quantum size effect).[165,166] Quantum dots have size-tuneable light emission (usually with a narrow emission band), bright luminescence (high quantum yield), long stability (photobleaching resistance), and broad absorption spectra for simultaneous excitation of multiple fluorescence colors compared with classical organic fluorescent dyes.

As an example, Fig. 13A shows the size- and material-dependent emission spectra of surfactant-coated QDs.[167] The blue series shows different sizes of CdSe nanocrystals, with diameters of 2.1, 2.4, 3.1, 3.6, and 4.6 nm (from right to left); the green series consists of InP nanocrystals with diameters of 3.0, 3.5, and 4.6 nm; and the red series is of InAs nanocrystals with diameters of 2.8, 3.6, 4.6, and 6.0 nm. A true-color image of a series of silica-coated core (CdSe)-shell (ZnS or CdS) nanocrystal probes in aqueous buffer, all illuminated simultaneously with a handheld ultraviolet lamp, is depicted in Fig. 13B. ZnS-capped CdSe-QDs, covalently coupled to a protein (for example, transferrin and immunoglobulin G) by mercaptoacetic acid (Fig. 13C), were designed and prepared, and various techniques were used for characterization.[168] For example, QD-transferrin conjugates showed a CdSe core size of 4.2 nm as measured by TEM (Fig. 13D). These seminal works demonstrated the feasibility of using quantum dot bioconjugates as fluorescent probes in biological staining, non-isotopic detection, and diagnostics.

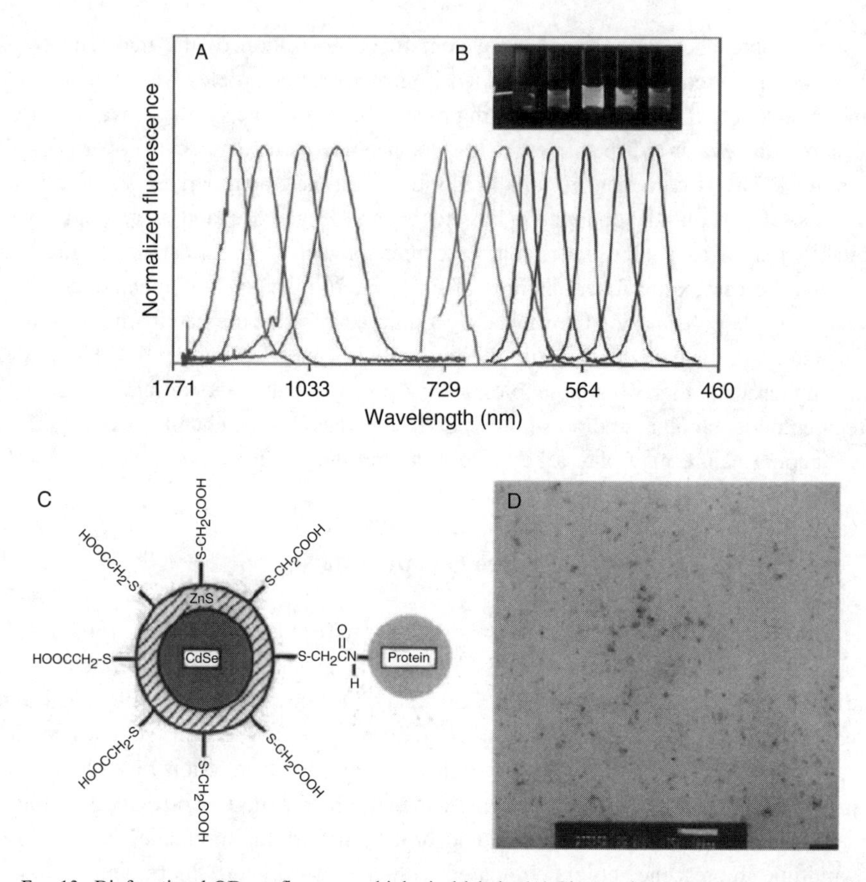

FIG. 13. Biofunctional QDs as fluorescent biological labels. (A) Size- and material-dependent emission spectra of different surfactant-coated semiconductor nanocrystals. The blue series represents CdSe nanocrystals with diameters of 2.1, 2.4, 3.1, 3.6, and 4.6 nm (from right to left). The green series represents InP nanocrystals with diameters of 3.0, 3.5, and 4.6 nm. The red series represents InAs nanocrystals with diameters of 2.8, 3.6, 4.6, and 6.0 nm. (B) True-colour images of silica coated core (CdSe)-shell (ZnS or CdS) nanocrystals in aqueous buffer, illuminated with an ultraviolet lamp. (C) Schematic representation of a core (CdSe)-shell (ZnS)quantum dot that is covalently coupled to a protein by mercaptoacetic acid. (D) TEM of QD-transferrin conjugates. Scale bar, 100 nm. Adapted from Refs. 167 and 168. (See Color Plate 36.)

Research on QDs has evolved from electronic materials science to biological applications.[167–171] A broad variety of synthetic methods has been developed for the preparation of QDs, conjugation with biomolecules, and application as bioluminescent probes for live-cell and/or *in vivo* animal imaging.[172–180] In general, capping the outer surface of core semiconductors prevents aggregation and oxidation, stabilizes nanoparticles in solution, and electronically isolates the particles by surface passivation. Tri-*n*-octylphosphine oxide (TOPO), in tandem with tri-*n*-octylphosphine (TOP)[181] or tri-*n*-butylphosphine (TBP),[182] is commonly used as capping agent for the crystallites, but many other compounds have been employed.[172,183] To use QDs in bio-applications, it is recommended to protect the semiconductor cluster with polymers of high molecular weight (such as polyethylene glycols), or to use "core-shell" systems in which ZnS is usually used as the shell component (because is less toxic and also because sulfur chemistry can be performed). Typically, QDs are prepared in organic solvents; their hydrophobic surface makes them insoluble in water. The hydrophobic ligands can be replaced by suitable molecules through ligand-exchange processes, or can be chemically modified to permit water solubility.[175,184] Many types of QDs have become commercially available, and are routinely used either as starting materials or for various applications.

Glyco-quantum dots have been made by various strategies. Carbohydrates have been capped onto QD surfaces by means of electrostatic interactions, template systems, and thiol chemistry. The first synthesis of QD–polysaccharide nanocomposites involved mixing the (negatively charged) *O*-carboxymethyldextran and mercaptosuccinate-capped CdSe–ZnS QDs with the (positively charged) polylysine.[185] Dextran-grafted luminescent nanospheres, which maintained the emission properties of individual QDs, were thereby obtained. CdSe–ZnS QD–chitosan nanoparticles, used in bioapplications, were prepared by variations of this procedure, starting from mercaptosuccinate-capped CdSe–ZnS QDs.[119,186–189] CdSe–ZnS-labeled *O*-carboxymethylchitosan was also prepared by carboxyl-group chelation of the metal ions on the metal surface.[190] Monodisperse 100-nm chitosan nanoparticles[191] encapsulating QDs had a fluorescence quantum yield almost 12% higher than that of the free QDs. These nanoparticles are susceptible to further (bio)functionalization through chemical bonding, as their chitosan layer contains amine and hydroxyl groups.[186]

Non-covalent coating of calix[4]arene-based glycocluster amphiphiles was used to biofunctionalize CdSe QDs capped by TOPO.[192] The alkyl chains of the glycocluster amphiphiles were inserted into the TOPO-coated QDs, and the terminal saccharide moieties (α- and β-glucoside, β-galactoside, and α-maltohexaoside) were exposed to bulk water to ensure solubility in aqueous buffers. Subsequently, gum arabic and TOPO-QDs (CdSe/CdS) were used to produce non-toxic and biocompatible QD nanocolloids by non specific physical interactions.[193] These glyco-QDs showed good photochemical stabilities and quantum yields.

Coupling of commercially available streptavidin-coated QDs with a biotin-ended lactose glycopolymer took place in one hour at room temperature to produce fluorescent bioprobes.[194] Likewise, biotinylated Le^X, or HIV-1 envelope glycoprotein gp120, were

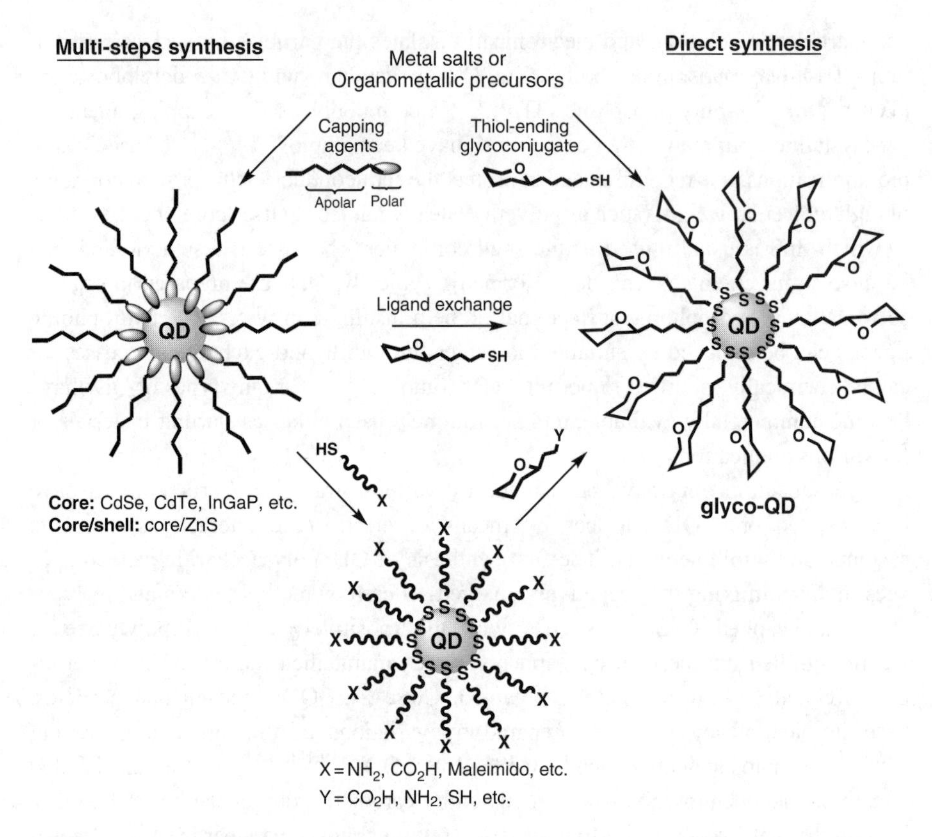

FIG. 14. Main strategies for the synthesis of glyco-QDs by means of thiol chemistry. (See Color Plate 37.)

bound to commercially available PEG-coated fluorescent QDs (Qdot® 605 ITK™) conjugated with streptavidin to generate systems of ~40 nm overall diameter.[195]

Direct methods and multistep indirect synthesis have been used to prepare glyco-quantum dots by means of thiol chemistry (Fig. 14). Covalent coupling to thiol-substituted sugar derivatives has been used to functionalize QDs, both in a single-step solution procedure (direct method) and in two- or three-step protocols (indirect method). The former is based on direct functionalization of nano-crystallites formed *in situ* with thiol-ended glycoconjugates. The indirect methods consist in mixing preformed quantum dots (usually pyridine-[196] or 3-mercaptopropanoic acid (MPA)-capped QDs) with free-thiol glycoconjugates in the presence of reducing agents (usually $NaBH_4$). A third strategy starts from protected QDs, which by thiol chemistry incorporate a linker containing functional groups for subsequent reaction with a suitable carbohydrate molecule. Biofunctionalization, solubilization, and stabilization of quantum dots in aqueous solution are thus readily achieved.

QDs conjugated to carbohydrate antigens for specific cell targeting have been prepared by the direct method, using aqueous solutions of thiol-functionalized neoglycoconjugates and metal salts. The preparation of cadmium sulfide (CdS) nanocrystals covalently bound to Le^X and maltose derivatives was accomplished by adding sodium sulfide to an aqueous solution of the thiol-ending neoglycoconjugates and cadmium nitrate in a single-step procedure at room temperature and pH 10.[197] The QDs obtained were water-soluble, stable (in the absence of light at 4 °C), and emitted light at 550 nm when excited at 360 nm. This type of aqueous self-assembly procedure has also been used for the generation of mannose-conjugated CdS-QDs.[198] QDs coated with the tumor-associated Thomsen–Friedenreich antigen conjugated to various linkers were similarly prepared by the use of small percentages of different surface-passivating agents and a suitable glycoconjugate.[199] Binding and agglutination assays confirmed that the functional characteristics of the sugar were intact on the particles.

Two-step protocols have also been used to prepare glyco-QDs. They usually consist in preparing QDs bearing classic capping agents, which are then subjected to ligand exchange with suitably functionalized thiol-ended glycoconjugates. Pyridine-encapsulated QDs were treated with 11-mercaptoundecyl *N*-acetyl-β-glucosaminide (in its disulfide form) and $NaBH_4$ in aqueous solution to give the water-soluble GlcNAc-QDs.[200] The sulfur atom mediated the linkage to the colloidal CdSe–ZnS core-shell QDs by displacement of pyridine. In later work, a thiol-ended triantennary dendritic β-galactoside ligand was anchored to pyridine-capped ZnS–CdSe nanoparticles by a similar procedure.[201] Likewise, CdSe–ZnS QDs having lactose, melibiose, and maltotriose derivatives on their surface have been obtained by using tetramethylammonium hydroxide as the initiator.[202]

The preparation of 6-nm GlcNAc-, Glc-, Gal-, and Man-displaying CdTe QDs was also achieved by surface exchange in water from MPA to the thiol-ended neoglycoconjugates.[203] Using the same protocol, oligosaccharides (maltose, maltotriose, cellotriose, and panose) were also incorporated into CdTe QDs.[204]

Glyco-quantum dots have also been prepared by multistep protocols, starting from commercially available (or readily prepared) QDs that are not soluble in water. Most of these strategies are based on replacement of the hydrophobic ligands by water-soluble bifunctional molecules in which one terminus is a thiol anchored to the surface of a QD and the other end is a reactive group (such as amine, carboxylic acid) that can be used to couple suitably derivatized glycoconjugates. TOPO-capped InGaP–ZnS QDs were exchanged with 16-mercaptohexadecanoic acid and then activated with 1-ethyl-3-(3-dimethylaminopropyl)carbodiimide hydrochloride (EDC) for covalent coupling with *N*-deacetylated (85%) chitosan.[205] ZnS–CdSe nanocrystals coated by amino-functionalized silica were coupled to dextran randomly activated with *N,N'*-disuccinimidyl carbonate to prepare water-soluble and stable QDs.[97]

Highly biocompatible mannosylated quantum dots were prepared starting from commercially available amino-polyethyleneglycol-coated QDs (Qdot® 655 ITK™ incorporating polyethyleneglycol PEG_{2000}).[206] The amino functionality was used for coupling with 2-imino-2-methoxyethyl 1-thiomannoside, which allowed mannosylation of the QDs. The "PEGylation" of the toxic nucleus decreases the cytotoxicity of these systems. CdSe–ZnS QDs (15–20 nm) capped with galactose, mannose, and galactosamine derivatives were prepared in three steps.[207] Ligand exchange of TOPO-capped QDs with thioctic (6,8-dithiooctanoic) acid, coupled with diamino-PEG_{2000} under reducing conditions afforded the intermediate QD-PEG_{2000}-NH_2. The terminal amine was then allowed to react with a carboxylic group of a spacer bearing a maleimide group. The thiol-ended glycoconjugates were inserted by maleimide conjugation.

Commercially available QDs containing carboxylic groups at their surface were treated with an amino-functionalized biotinylated glycopolymer, or with the mixture of amine-functionalized biotin and *N*-(2-aminoethyl)gluconamide.[208] In this work, however, the pyranose ring of glucose was disrupted by chemical conjugation with the linkers.

III. Applications of GNPs

The previous section discussed the design, synthesis, and characterization of different types of GNPs. These systems constitute a good biomimetic model for carbohydrate presentation at the cell surface, and have therefore been used as biofunctional nanomaterials in several fields of glycobiology and biotechnology.[25] GNPs have been widely applied not only for study of carbohydrate-mediated interactions, but also as biolabels or biosensors, and *in vivo* imaging experiments.

The controllable geometry and size, which can range from few nanometers (the order of magnitude of glycoproteins) to hundreds of nanometers (sizes of viruses and small bacteria), the multivalent display of carbohydrates (providing glycocalix-like presentation), the multifunctionality that can be readily inserted on the same metal platform by well-developed surface chemistry (allowing the construction of synergic and/or multimodal systems), the chemical stability, and the physical properties because of the quantum-size effect make GNPs very useful systems for studying and intervening in interactions where sugars are involved.

This section presents the main applications of GNPs in carbohydrate-based interaction studies: carbohydrate–carbohydrate interactions, carbohydrate–protein interactions, adhesion processes, and cellular and molecular imaging.

1. GNPs in Carbohydrate–Carbohydrate Interaction Studies

Characteristic features of carbohydrate–carbohydrate interactions are their specificity, their strong dependency on divalent cations, and their extremely low affinities, which have to be compensated by multivalent presentation of the ligands. Because of their low affinity, carbohydrate–carbohydrate interactions can serve as an initial step in cell recognition, modulating a type of preliminary matching state that is required before carbohydrate–protein interactions (between carbohydrate conjugates and their binding proteins), and/or protein–protein interactions (between adhesion receptors and their ligands) can effectively give rise to the necessary cascade of events that makes the biological interaction occur. In agreement with the well-established "glycoside-cluster effect",[59,62] multivalent presentation is an essential requirement for study of carbohydrate–carbohydrate interactions.[63] Although the concept of carbohydrate–carbohydrate interactions in cell adhesion was introduced independently by Hakomori (in mammalian cells)[209] and Burger (in sponge cells)[210] at the end of the 1980s, systematic studies for their quantification with model systems appeared in the literature only ten years later.[24,64,211] The existence of this interaction is nowadays generally accepted, but clarification of the mechanisms, and further investigation for exploring its role in a variety of biological processes, is still needed. The challenging task is the design and synthesis of adequate chemically defined systems that can allow the analysis of these weak-affinity interactions. Several studies of carbohydrate–carbohydrate interactions using such model systems as giant vesicles,[212] apposed bilayers,[213] micelles,[214] and artificial glycoviruses[215] have been reported.[64] The first application of GNPs as chemical multivalent models to demonstrate and quantify carbohydrate–carbohydrate interactions at the molecular level was reported by Penadés' group: gold GNPs coated with the trisaccharide Lewis X (Le^X) were used to study the calcium-dependent Le^X self-aggregation.[24] The antigen determinant Le^X seems to mediate morula compaction in the mouse via a carbohydrate self-interaction.[209] Thiol-ended conjugates of lactose (used as a control) and Le^X having a linear aliphatic (eleven carbon atoms) linker were employed for the construction of 2-nm diameter GNPs. The aggregation of *lactose*- and *Le^X*-GNPs was studied in the presence and absence of calcium cations (10 mM $CaCl_2$ solutions) by TEM.[70] *Le^X*-GNPs in calcium solution showed reversible aggregation (the aggregates being dispersible by addition of ethylendiaminetetracetic acid (EDTA)) at all concentrations tested (0.1–0.9 mg/mL). In these cases, the aggregation was so strong that TEM micrographs showed three-dimensional morula-type aggregates (Fig. 15A). In contrast, *lactose*-GNPs showed little aggregation in the presence of calcium.

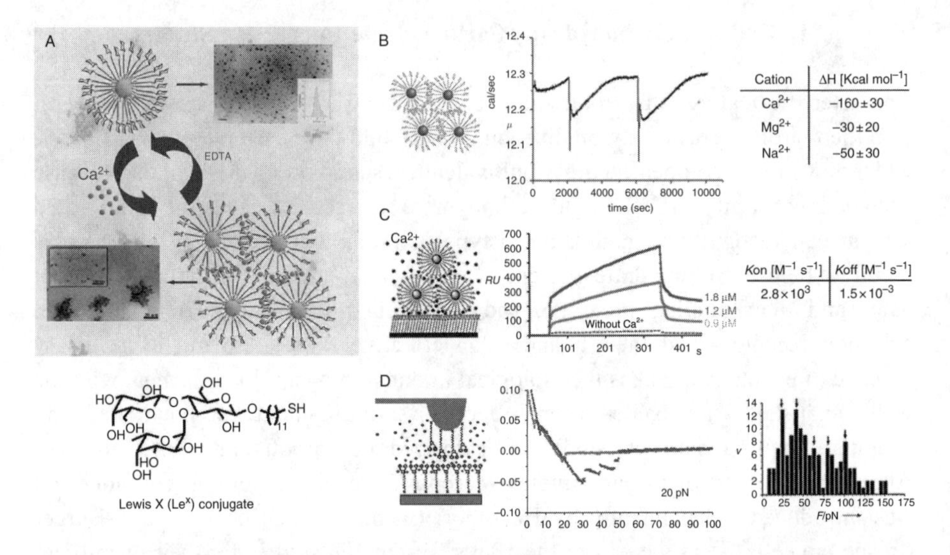

FIG. 15. Glyconanotechnology in carbohydrate–carbohydrate interactions. (A) TEM images showing that the self-aggregation of *Le*^{X}-GNPs is calcium-dependent (0.1 mg/mL of *Le*^{X}-GNPs in 10 mM $CaCl_2$ solution causes 3D *Le*^{X}-GNPs aggregates, bottom-left) and reversible (treating the previous solution with EDTA causes disgregation of the aggregates, up-right). (B) ITC curves showing thermodynamic evidence for self-aggregation of *Le*^{X}-GNPs added to 10 mM CaCl2. (C) SPR sensorgrams for the interaction of *Le*^{X}-GNPs to SAMs of *Le*^{X} conjugates on gold surfaces without (- - -) and with Ca^{2+} ions (—) (middle) and rate constants (k_{on} and k_{off}) obtained from the analysis of the sensorgrams (right). (D) AFM force-distance curves obtained between a *Le*^{X}–functionalized tip and SAMs of *Le*^{X} conjugates in 10 mM $CaCl_2$ solution (middle) and histogram of the corresponding unbinding force measured through 300 force-distance curves (right). Adapted from Refs. 70, 216, 217, and 218. (See Color Plate 38.)

Thermodynamic evidence for calcium-mediated self-aggregation of *Le*^{X}-GNPs was obtained by isothermal titration calorimetry (ITC).[216] A slow aggregation process, with a favorable enthalpy term of about 160 kcal per mole of GNP injected, was observed when *Le*^{X}-GNPs were added to a 10 mM $CaCl_2$ solution. The use of magnesium instead of calcium led to a five-times lower heat emission, indicating that the Le^{X} aggregation is selectively mediated by Ca^{2+} (Fig. 15B). The heat evolved upon addition of *lactose*-GNPs to a calcium solution was rather low, the thermal equilibrium being quickly achieved, and the thermal signal confirmed a weaker inter-particle interaction. This behavior was not observed with maltose-functionalized GNPs. The selectivity of the aggregation process of *Le*^{X}-GNPs was further confirmed by surface plasmon resonance (SPR).[217] Kinetic data of Le^{X}–Le^{X} interaction were obtained by using a combination of SAMs of a Le^{X} neoglycoconjugate on a biosensor gold surface as substrate and *Le*^{X}-GNPs as analyte. The sensorgrams indicated a slow association phase (k_{on}~10^3 M^{-1} s^{-1}) and a gradual dissociation phase (k_{off} ~10^{-3} s^{-1}) in the presence of calcium cations (Fig. 15C). The binding was of high affinity, with a dissociation constant in the micromolar range (K_d ~10^{-7} M). The Le^X–Le^X interactions were also confirmed by AFM.[218] The adhesion forces measured in Ca^{2+} solution between individual molecules of Le^X conjugates amounted to 20 pN (Fig. 15D), whereas in experiments between SAMs of lactose conjugates, no detectable interactions were observed. These results demonstrated the self-recognition capability of this antigen, and confirmed that carbohydrate–carbohydrate interactions in water are specific, stabilizing, and may be considered as a mechanism for cell adhesion and recognition.

A similar approach was followed by Kamerling *et al.*[219] to explore the carbohydrate-mediated self-recognition of marine sponges. GNPs were used as multivalent probes for investigating the role of two repetitive carbohydrate epitopes (a sulfated disaccharide and a pyruvated trisaccharide), which are present in the extracellular proteoglycans of marine sponge cells and are involved in species-specific self-recognition and adhesion. Gold GNPs coated with the thiol-derivatized sulfated disaccharides were employed for studying the Ca^{2+}-dependent self-recognition of this epitope by TEM[219,220] and AFM.[221] Similar AFM and TEM experiments were also conducted with the pyruvated trisaccharide. No binding was detected by AFM, and no visible aggregation was identified by TEM in the presence of calcium cations, suggesting that the self-interaction between the sulfated disaccharide fragments is stronger than that between the pyruvate trisaccharide.[221] The same group used transferred nuclear Overhauser enhancement NMR spectroscopy (TR-NOESY) and diffusion-ordered spectroscopy (DOSY) to detect the carbohydrate–carbohydrate self-recognition in solution.[222] The pyruvated trisaccharide-decorated gold nanoparticles were used as the "macromolecule", and the same carbohydrate as the ligand. Changes in the diffusion coefficient in Ca^{2+} solution of the free carbohydrate in the presence of the GNP, as well as changes in the sign of the sugar nuclear Overhauser enhancement (NOE) peaks suggested weak Ca^{2+}-mediated carbohydrate–carbohydrate interactions in solution.

A series of lactosyl derivatives bearing thiol-ended oligoethylene glycol chains of different lengths were used by Russell and co-workers[101] to prepare larger gold nanoparticles (~16 nm diameter) and to study their behavior upon addition of calcium ions by TEM, and also by UV-Vis, based on the red shift of the surface plasmon absorption band. The length and nature of the linker chain had a remarkable effect on the calcium-induced aggregation. UV-Vis experiments indicated that the aggregation process was very rapid, as the change in absorbance intensity was measured 30 seconds after addition of the cation. In previous works with different carbohydrates and smaller gold nanoclusters, the aggregation process required hours.[70,219] Curiously, lactosides bearing a short two carbon atoms aliphatic chain favoured a three-dimensional aggregation of the corresponding *lactose*-GNPs after addition of 1.4 mM of calcium ions, while lactosides with triethylenglycol-based chains caused only bi-dimensional agglomerates even at 10 mM of calcium ions, as showed by TEM images.

2. GNPs in Carbohydrate–Protein Interactions

Carbohydrate–protein interactions are usually highly specific.[59] As with carbohydrate–carbohydrate interactions, multivalence is essential, because multiple interactions between carbohydrates and proteins are necessary to achieve strong binding or enhanced inhibition.[65] In pursuit of synthetic targets where carbohydrates are multiply displayed over suitable scaffolds, a number of model systems have been used to study multivalent interactions between carbohydrates and proteins.[67,223] GNPs have also been used in this field, and the results have been reviewed.[25]

In general, colorimetric biosensing assays based on aggregation of gold nanoparticles and dispersion stages that are guided by interparticle interactions have attracted considerable attention because of their simplicity and versatility.[224] The group of Kataoka was the first to use gold GNPs to study carbohydrate–protein interactions,[94] taking advantage of the special characteristics of the UV-visible absorption spectra of colloidal metallic elements.[225] The specific aggregation of GNPs in the presence of a suitable analyte produces a color change associated with a shift in the surface plasmon absorption-band of the metal cluster, and this can be monitored using UV-visible spectrophotometry. The authors employed lactoside-coated gold GNPs in selective bioassays to target the bivalent *Ricinus communis* agglutinin (RCA_{120}). This lectin specifically recognizes β-D-galactopyranose residues, thus inducing selective and reversible (upon an excess addition of galactose) aggregation of *lactose*-GNPs, which can be monitored by color changes (from pinkish red to purple) in the UV-visible adsorption spectrum.[94] The degree of aggregation is proportional to lectin concentration. Furthermore, systematic variation of lactoside coating (0–65%) on the gold surface demonstrated that the aggregation induced by RCA_{120} depends also on the ligand density: a critical lactoside density (>20%) was necessary to trigger aggregation.[110] The same group later developed a novel method for the quantitative SPR-assay of low molecular-weight analytes (in this case, galactose) using these GNPs.[226] Briefly, preadsorbed *lactose*-GNPs on an RCA_{120}-functionalized sensor chip were eluted by injection of galactose in a concentration-dependent way. The method is applicable to a wide range of galactose concentrations (0.1–50 ppm) and the operating time is very short (five minutes).

Since Kataoka *et al.* first disclosed the possibility of studying carbohydrate–protein interactions by means of *lactose*-GNPs-based bioassays, other groups have used carbohydrate-based nanoparticles, as multivalent and chemically well-defined systems, in selective aggregation studies. Russell and co-workers have used mannose-modified GNPs for selectively sensing concanavalin A (ConA) lectin.[99] A rapid (within 30 seconds) colorimetric method for determination of sub-micromolar concentrations of the protein was achieved. The mercaptoethyl tether in these Man-GNPs was later replaced by thioctic (6,8-dithiooctanoic) acid-derived linkers to minimize nonspecific protein binding.[100] They also employed gold and silver GNPs (having a 16-nm diameter of the metal core) in colorimetric tests for the analytical detection of RCA_{120}[111,227] and cholera toxin.[228] All of these studies show that nanoparticle size, linker length, and carbohydrate density on the metal surface are factors to be considered when designing GNPs to obtain stable systems and to achieve high analytical sensitivity.

The group of Chen conducted a competitive colorimetric assay for the indirect detection of protein–protein interactions, using ~30 nm diameter Man-GNPs–ConA agglomerates in solution (blue-colored solution).[229] The addition of suitable proteins (such as thyroglobulin) capable of interacting with ConA breaks the agglomeration, which is reflected in a color change of the solution to reddish-purple. In this way, protein–protein interactions can be evaluated indirectly in solution in a very sensitive (only nanomolar concentrations of proteins are required) and straightforward way (no protein labeling, no use of special analytical instruments).

The first quantitative study of multivalent binding between carbohydrate-based nanoparticles and lectins was published by Lin *et al.*[76] They prepared a series of mannoside-encapsulated gold nanoparticles and investigated by SPR their interaction with ConA. These GNPs had previously proved to bind mannose-specific FimH proteins on the type 1 pili in *Escherichia coli.*[74] Later, globotriose-functionalized GNPs were employed by the same group in SPR competition-binding assays with the pentameric B-subunit of Shiga-like toxins that specifically recognize globotriaosylceramide (the globotriose blood group antigen).[80] *Globotriose*-GNPs showed size- and linker length-dependent affinity for the protein, and were used as multivalent probes for the purification of the B-subunit from cell lysates. The group of Kamerling has also evaluated the interaction between ConA and different (oligo)manno- and glucoside GNPs by SPR, TEM, and UV,[230] confirming the multivalent binding character of this interaction.

A ConA biosensor based on a microgravimetric quartz crystal microbalance (QCM) has been developed in a sandwich-type experiment, using mannoside-stabilized GNPs as a signal amplifier.[231] The addition of ConA over gold QCM electrode functionalized with thiol-modified mannosides resulted in a frequency decrease because of the specific association of the protein. The Man-GNPs were then added and their specific binding with the remaining free sites of ConA produced a quantifiable frequency change. The sensitivity was 13 times higher than that obtained with the simple detection method (without the signal amplification upon using GNPs). The advantage of using QCM-based biosensors is that labeling of the analytes is not required. This technique complements other analytical methods, such as enzyme-linked immunosorbent assay (ELISA).

The oxime- and oxyamine-linked (oligo)glucoside-GNPs prepared by Jensen and co-workers[104] were tested with ConA to compare their recognition properties. In aggregation assays, *N*-glucosyl oximes on gold GNPs permit recognition of ConA (in contrast to the corresponding reduced oxyamines) as a consequence of the different glucose structures on the GNPs. The *N*-glucosyl oximes are in fact in equilibrium with their ring-closed tautomers, and can be recognized by ConA, whereas the *N*-glucosyl oxyamines (with their acyclic, open-chain form) cannot. The recognition of oxime- and oxyamine-*maltose*-GNPs by glucoamylase was also studied to assess the accessibility of the glucan ligands. While the latter GNPs displayed complete stability toward enzyme degradation, the former ones were partially hydrolyzed. This result was ascribed to the three times lower ligand density of oxime GNPs with respect to the oxyamine GNPs, which could allow easier accessibility of the sugars to the enzyme.

These results are in agreement with the previous observation by Penadés and collaborators that β-galactosidase was unable to cleave the galactosyl moiety of *lactose*-GNPs.[70] The influence of carbohydrate density and presentation on the recognition by protein receptors was further evaluated by examining the interaction of lactoside-functionalized GNPs with two different galactose-specific, carbohydrate-binding proteins: *E. coli* β-galactosidase and *Viscum album* agglutinin.[88] Modulating the density of lactose conjugates on the gold surfaces results in different rates of enzymatic hydrolysis. These observations suggest that the correct selection of ligand densities and spacers in GNP functionalization is an important prerequisite for matching the topological requirements of the target receptor while escaping degradation by glycosidases.

Penadés and co-workers used gold "*manno*-GNPs"[89] to target the mannose-binding lectin DC-SIGN (dendritic cell-specific ICAM 3-grabbing nonintegrin). The DC-SIGN receptor is present on dendritic cells (DCs), and mediates the interaction between the glycans of the HIV envelope gp120 and DCs.[232] The HIV-DC interaction is one of the early steps involved in viral infection. SPR experiments with GNPs containing α-Man-(1→2)-α-Man, α-Man-(1→3)-α-Man, or α-Man-(1→2)-α-Man-(1→2)-α-Man were used to test their inhibition potency toward binding of DC-SIGN to immobilized gp120. These "*manno*-GNPs" completely inhibited the binding at the micro- to nano-molar range, while the corresponding monovalent mannosides required millimolar concentrations. GNPs containing the disaccharide α-Man-(1→2)-α-Man were the best inhibitors tested showing a more than 20,000-fold increased activity (100% inhibition at 115 nM) as compared to the corresponding monomeric disaccharide (100% inhibition at 2.2 mM).[89] These SPR experiments demonstrate that multivalent "*manno*-GNPs" are able to intervene efficiently in DC-SIGN–gp120 interactions.

Gold nanoparticles deposited onto a carbon electrode and modified with sialic acid derivatives by "click" chemistry were used for electrochemical sensing of interactions between sialic acid and Alzheimer's amyloid-beta (Aβ) peptides.[96] The attachment of Aβ peptides to the sialic acid layer was confirmed by electrochemistry and AFM imaging. The peak oxidation-current response of Tyr residue of Aβ was utilized as the analytical signal.

GNPs with metal cores different from gold have also been employed in recognition experiments with proteins. Silica-coated Ag, CdSe–ZnS, and Fe_3O_4 nanoparticles functionalized with dextran were tested using ConA by the group of Jana and Ying.[97] Upon addition of the lectin, the nanoparticles formed aggregates that precipitated over time from the solution. With the silver GNPs, the precipitation process was associated with red-shifting of the plasmon absorption. QDs showed a decrease in absorbance and fluorescence over time. In the case of magnetic GNPs, removal of the aggregates by using a magnetic field was possible only after aggregation through particle–particle cross-linking induced by ConA. Particle aggregation was not observed when a sufficient amount of free glucose was present.

QDs have been extensively used to evaluate carbohydrate–protein interactions. Rosenzweig's group has reported that the dextran residues on the nanosphere surface of CdSe–ZnS QDs protected with carboxymethyldextran and polylysine show high affinity toward the glucose-binding protein ConA.[185] Fang and co-workers demonstrated the specific affinity of *N*-acetyl-β-glucosamine-encapsulated QDs for wheat-germ agglutinin (WGA), using fluorescence titration, TEM, DLS, and flow cytometry.[200] Sun and Chaikof used streptavidin-bound QDs coupled with a biotin-functionalized lactose glycopolymer for the fluorescent staining of RCA_{120}-immobilized agarose beads.[194] Surolia and co-workers[202] employed glyco-QDs in agglutination assays with several lectins. The specific and multivalent carbohydrate–protein interactions were studied by monitoring the light scattered at 600 nm. Selective binding of soybean agglutinin (SBA) with *melibiose*-QD, as compared to *lactose*-QD, and specific deagglutination caused by α-galactose (α-mannose being ineffective) were observed. *Maltotriose*-QDs were tested with ConA. All of these studies demonstrate that the recognition properties of the sugars are retained, and usually enhanced because of multivalence, when incorporated in a QD platform. In addition, the sensitivity in the detection of proteins can be enhanced by the fluorescence properties of QDs.

Non-metallic GNPs based on diblock copolymers functionalized with mannose residues have been used by Rieger *et al.*[233] for recognition of surface-exposed mannose by recombinant *Burkholderia cenocepacia* lectin A (BclA), which is dimeric and specifically binds terminal α-D-mannopyranosyl residues with high affinity. BclA was labeled with biotin. For the recognition assays (referred to as modified enzyme-linked lectin assay, ELLA), NPs were incubated in the presence of Ca^{2+} with the biotinylated lectin, then with streptavidin phosphatase, and finally with the substrate of the enzyme (*p*-nitrophenyl phosphate). In this way, the interaction of the nanoparticles with biotin-labeled BclA was spectrometrically revealed by measuring the absorbance of the colored end-complex. ITC was then used as a complementary technique to measure directly the enthalpy of interaction between the lectin and its carbohydrate ligand, avoiding the use of intermediate complexation agents as in the modified ELLA. The addition of BclA to the nanoparticle dispersion in the microcalorimeter cell generated multiple responses, depending both on the concentration of mannose and on the mannose coating of the nanoparticles.

Pun and co-workers used β-galactoside-coated gold nanoparticles of various sizes (from 50 to 150 nm) for liver targeting in mice.[234] Only 50 nm Gal-GNPs accumulated in the liver, mainly by uptake into hepatocytes via the asialoglycoprotein receptor. *In vivo* blood and liver distribution of the nanoparticles were assessed by determination of the gold content.

3. Other Interaction Studies

Gene therapy, as a therapeutic treatment for genetic or acquired diseases, is attracting much interest in terms of design and construction of viral vectors. Non-viral gene carriers have also received increasing attention in an effort to avoid the safety problems related to their viral counterparts.[235] AFM, TEM, and gel electrophoresis were used by Penadés' group to study the binding of a model DNA plasmid to gold GNPs that present carbohydrate and amino group motifs at their surface.[236] The GNPs tested showed different affinities for the DNA depending on their sugar functionalization, with implications for their potential use as non-viral gene-delivery agents. The interaction of hybrid amino–α-Gal glyconanoparticles and DNA deposited from solution has been imaged by AFM (Fig. 16A) and by TEM (Fig. 16B). The aggregation of these GNPs around DNA fixed on mica is shown in Fig. 16C. Only a few individual, unattached GNPs are visible. The very thin white lines are "uncoated" DNA

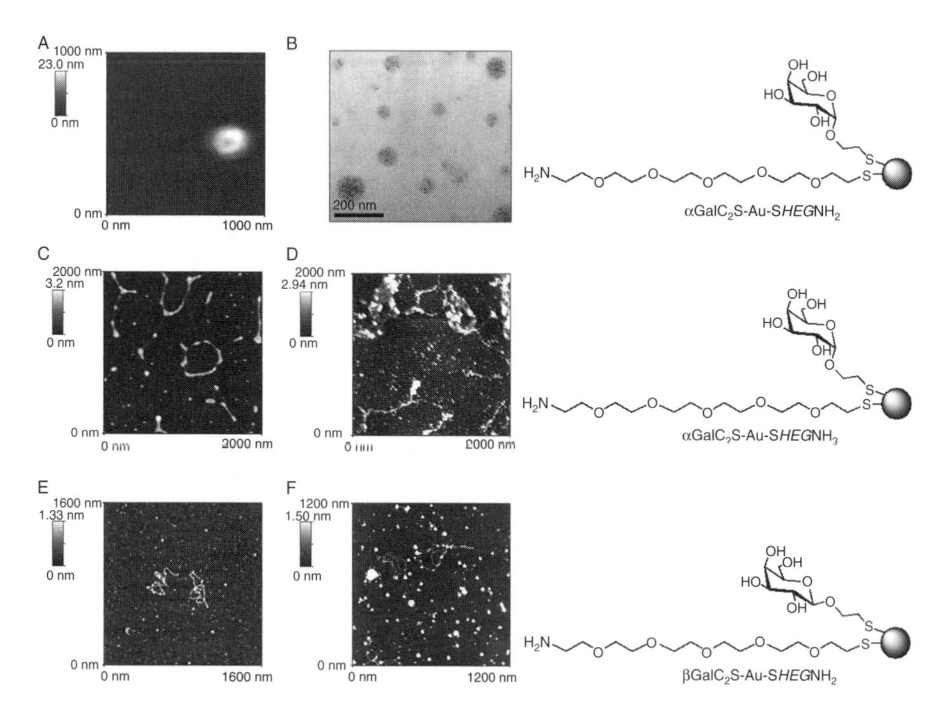

FIG. 16. AFM and TEM images of the interactions of gold GNPs with DNA. AFM (A) and TEM (B) images of hybrid amino–α-Gal gold glyconanoparticles and DNA deposited from solution. AFM images of hybrid amino–α-Gal gold glyconanoparticles deposited onto DNA fixed on mica before (C) and after (D) washing with water. AFM images of hybrid amino–β-Gal gold glyconanoparticles deposited onto DNA fixed on mica before (E) and after (F) washing with water. Adapted from Ref. 236.

molecules. Upon washing with water, the previous aggregates break and large compact globular aggregates are formed. More unattached GNPs are also visible (Fig. 16D). In comparison with α-Gal-containing GNPs, β-Gal-containing GNPs exhibit much less DNA binding (Fig. 16E). After washing with water, the aggregates also break in this case, forming several large aggregates (Fig. 16F).

The influence of sugars in cell transfections has also been demonstrated by means of non-metallic nanoparticles. Sugar–cholesterol conjugates were anchored to lipid-encapsulated plasmid DNA.[237] These novel non-metallic glyconanoparticles (with sizes ranging from 180 to 220 nm and with different contents of cholesterol) can be considered as a subclass of nano-systems termed multifunctional envelope-type nanodevices (MEND).[238] The functionalization of MEND with GlcNAc–, Gal–, or Man–cholesterol conjugates allowed the use of sugars as nuclear targeting ligands in non-dividing and dividing cells.[237] It was demonstrated that the presence of the sugars affected the nuclear transfer, but not the cellular uptake. In particular, the partial disruption of the lipid envelope caused by sugar-improved nuclear import enhanced the transfection process. By means of confocal laser-scanning microscopy, the signals of fluorescently labeled glyco-MEND containing pDNA were detected in the nuclei, and were much more intense than those of unmodified MEND, demonstrating that sugar-modified MEND show higher transfection efficiency.

4. Glyconanoparticles as Antiadhesion Agents

GNPs constitute a good biomimetic model to intervene in carbohydrate-mediated biological processes. Carbohydrates are responsible for many normal and pathogenic cell-adhesion processes. The adhesion of microbes to host cells can occur via carbohydrate–protein and/or carbohydrate–carbohydrate interactions. The discovery of the important role of carbohydrates in recognition processes has suggested a new antimicrobial therapy based on their anti-adhesion potential. The use of antimicrobial adhesion agents, based on carbohydrate analogues of host glycoconjugates, has been envisaged as a possible alternative to current antibiotic-based treatments.[239] The construction of GNPs for this purpose is still in its infancy, but some examples of different types of carbohydrate-functionalized nanomaterials have been reviewed.[240]

The first demonstration that gold GNPs behave as anti-adhesion agents against progression of lung metastasis in mice was reported in 2004.[241] A set of experiments was designed to evaluate the anti-metastatic potential of the GNPs (Fig. 17A). Mice were injected with melanoma cells (B16F10) pre-incubated with gold glyconanoparticles, and after three weeks the animals were sacrificed. The lungs were observed under the microscope for analysis of tumor foci. *Ex vivo* pre-incubation of tumor cells with *lactose*-GNPs at 90 μM concentration inhibited cancer-cell metastasis up to 70%. A comparison with tumor inhibition in animal groups inoculated only with melanoma cells or with melanoma cells, incubated with Glc-GNPs, was also reported. The image analysis at two different magnifications (×8 and ×80) showed that at the same GNPs concentration (90 μM), both small tumoral foci (<1 mm; black arrows in Fig. 17B) and large foci (>1 mm; blue arrows) were scarcely present when *lactose*-GNPs were used, in contrast with the result obtained using Glc-GNPs. The specific anti-metastatic effect of *lactose*-GNPs on the B16F10-dependent development of lung tumoral foci is macroscopically shown in Fig. 17C, where pictures of the lungs corresponding to treated animals with Glc-GNPs and *lactose*-GNPs, in comparison with the lungs obtained from a control animal not injected with B16F10 cells, are reported.

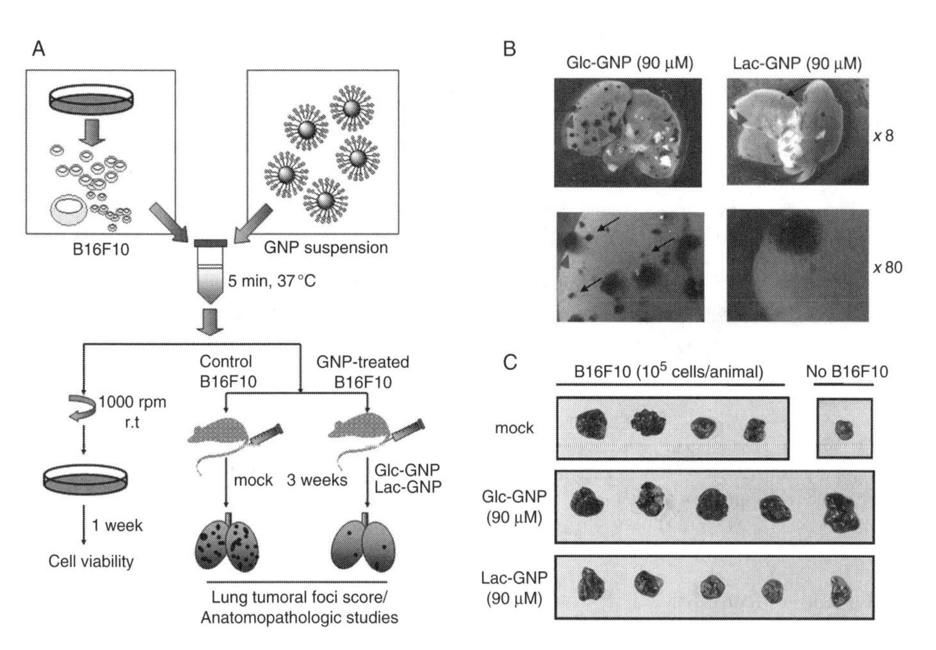

FIG. 17. Anti-metastatic effect of gold *lactose*-GNPs on lung tumor development in mice. (A) Schematic representation of *ex vivo* experiments for evaluating cell viability and anti-metastatic potential of glyconanoparticles coated with lactose (Lac-GNP) or glucose (Glc-GNP). (B) Pictures of the lungs at two different magnifications (×8, ×80) corresponding to mice treated with B16F10 cells and Glc-GNP (left) or Lac-GNP (right). Black arrows indicate the small foci (<1 mm) and blue triangles denote the presence of large foci (>1 mm). (C). Lungs corresponding to mice treated with B16F10 cells (mock, up-left), Glc-GNP (middle), or Lac-GNP (bottom) in comparison with the lungs obtained from a control animal (not injected with B16F10 cells; mock, up-right). The specific anti-metastatic effect of Lac-GNPs is evident from (B) and (C). Adapted from Ref. 241. (See Color Plate 39.)

Silica-coated magnetite nanoparticles functionalized with mannose or galactose derivatives (Fig. 18A) were employed by the group of Huang[154] for rapid detection of *E. coli* and for removing up to 88% of the target bacteria from the medium. After incubation of Man-MGNPs with solutions of an *E. coli* strain, a magnetic field was applied for separating MGNP–*E. coli* aggregates (Fig. 18B), which were then analyzed. Fluorescence microscopic images of *E. coli* captured with Man-MGNPs showed that the detection limit is 10^4 cells/mL (Fig. 18C). MGNP aggregates were observed by TEM on the surface, at the lateral ends, and along the pili of *E. coli* cells (Fig. 18D). These MGNPs showed higher capture efficiency than antibody- or lectin-functionalized magnetic particles. Furthermore, the identities of three different *E. coli* strains were determined on the basis of the response patterns to Gal- and Man-MGNPs.

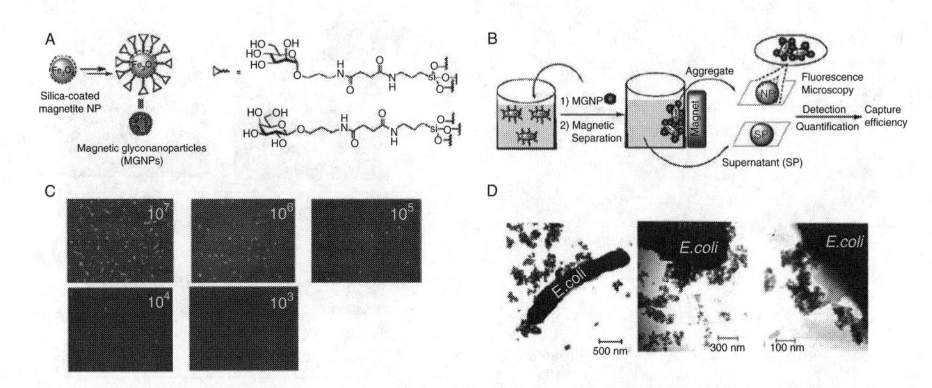

FIG. 18. *Escherichia coli* detection and decontamination with Man-MGNPs. (A) Silica-coated magnetite GNPs and neoglycoconjugates used to protect the magnetic core. (B) Schematic representation of pathogen detection by MGNPs. (C) Fluorescence microscopic images of captured *E. coli*. The concentration (cells/mL) of bacteria incubated with Man-MGNPs is indicated on each image. (D) TEM images of Man-MGNPs/*E. coli* complexes. Adapted from Ref. 154. (See Color Plate 40.)

Scrimin's group used GNPs functionalized with synthetic analogues of the repeating unit of the capsular polysaccharide of type A *Neisseria meningitidis* to inhibit binding of the natural serotype A of this bacterium to its specific mouse polyclonal antibodies, as determined by ELISA assays.[105]

In an attempt to intervene in the entry processes of HIV, GNPs have been used to investigate the role of multivalent interactions of the HIV envelope glycoprotein gp120 with host cellular receptors. Galactosylceramide (GalCer) expressed on mucosal membrane cells has been reported to be implicated in viral entry.[242] The group of Gervay-Hague investigated the polyvalent interactions of gold Gal- and Glc-GNPs with recombinant gp120 (rgp120).[84] A biotin–neutravidin adhesion assay was employed to evaluate the ability of the GNPs to displace gp120 from plate-bound GalCer. Divalent disulfides of the tested glycoconjugates were much less active than biotinylated GalCer, but when these conjugates were multivalently presented by GNPs they were two orders of magnitude more active than the disulfides and at least one order of magnitude more active than biotinylated GalCer.

To construct potential microbicides against HIV-1 infection, Penadés' group has employed GNPs coated with oligomannosides that are present on gp120 ("*manno*-GNPs"), with the aim of mimicking the virus itself.[89] Dendritic cells (DCs) can transfer the virus to T lymphocytes where viral replication occurs through a mechanism named trans-infection.[243] The HIV–DC interaction is mediated by the high-mannose type glycans of gp120 and the C-type lectin DC-SIGN expressed on DCs.[232] "*Manno*-GNPs" coated with different percentages of high-mannose

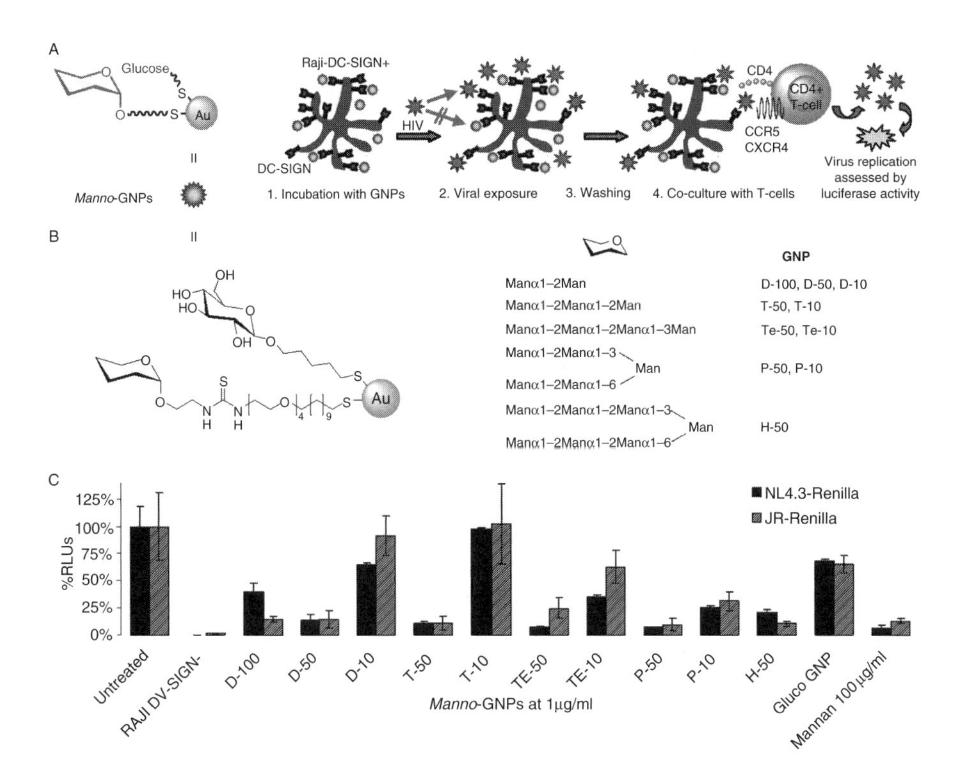

FIG. 19. "*Manno*-GNPs" as inhibitors of HIV trans-infection of human lymphocites. (A) Schematic representation of cellular experiments for evaluating the potential of gold glyconanoparticles coated with oligomannosides ("*Manno*-GNPs") as inhibitors of DC-SIGN-mediated HIV-1 trans-infection of human T cells. (B) Schematic representation of "*manno*-GNPs". D, T, Te, P, and H stand for di- tri-, tetra-, penta-, and heptamannose conjugates, respectively; the numbers indicate the percentages of mannose oligosaccharides on GNP, the rest being the stealthy glucoside component. (C) Anti-HIV evaluation of "*manno*-GNPs" at 1 μg/mL in DC-SIGN-mediated trans-infection of human T cells. HIV-1 recombinant viruses NL4.3-Renilla X4 (black) or JR-Renilla R5 (striped) were used. Raji cells not expressing DC-SIGN (Raji DC-SIGN-) were used as control to allow for DC-SIGN-independent viral transfer. Mannan (100 μg/mL) was used as a positive control. Results are expressed as percentages of infection related to untreated control. Adapted from Ref. 244. (See Color Plate 41.)

oligosaccharides, ranging from a di- to a hepta-saccharide, were tested in cell-based experiments for evaluating their anti-HIV potential as inhibitors of DC-SIGN-mediated HIV-1 trans-infection of human T cells (Fig. 19A and B).[244] Raji cells expressing the receptor DC-SIGN were incubated with these GNPs and then pulsed with HIV recombinant viruses (JR-Renilla R5 or NL4.3-Renilla X4). After washing, the cell cultures were co-cultured with human activated peripheral blood mononuclear cells (PBMCs) that would be infected through transfer of the virus bound to DC-SIGN in Raji cells. This experimental setting tries to mimic the natural route of virus transmission from DCs to T lymphocytes (Fig. 19A). Viral replication was assessed by luciferase activity in cell lysates. At 1 μg/mL (nano-molar range), "*manno*-GNPs" were able to inhibit the DC-SIGN-mediated trans-infection (Fig. 19C; results are expressed as percentages of infection related to untreated control). Raji cells not expressing DC-SIGN (Raji DC-SIGN-) were used as a control to allow for DC-SIGN-independent viral transfer. (1→4)-β-Mannan (100 μg/mL) was used as a positive control. In this way, it was shown that synthetic, carbohydrate-based multivalent systems can prevent viral attachment to DC-SIGN-expressing cells, and thus may function as an anti-adhesive barrier at an early stage of HIV infection.

5. GNPs in Cellular and Molecular Imaging

Current advances in the engineering of optical imaging techniques have revolutionized both basic research and clinical applications. There is an increasing demand for new probes that can guarantee not only high stability, specificity, and low toxicity but which can also be employed simultaneously with different imaging techniques of high sensitivity. Nanotechnology offers an outstanding contribution to our understanding of biological processes at a molecular level, and in the development of diagnostic tools and innovative therapies. Dye-doped silica nanoparticles, QDs, and gold nanoparticles have been prepared and applied for non-invasive bio-imaging.[245] Bioconjugated QDs have been used both for live-cell labeling and *in vivo* animal imaging.[176] Absorbance and emission in the near-infrared window are expected to allow real time and deep tissue optical imaging. In the area of MRI, magnetic nanoparticles have been demonstrated to be highly sensitive and target-specific for observing biological events, both at cellular and molecular level.[246] The development of multimodal nanoparticles that are both optically- and MRI-active is probably one of the most important advances in applied nanotechnology.[130]

Gold, magnetic, and semiconductor nanoparticles can also be useful probes in multimodal imaging techniques. Both glycobiology and glycobiophysics can profit from these biofunctional nano-tools.

Carbohydrate-functionalized QDs have been applied in cell imaging. Aoyama and co-workers have used CdSe QDs coated with glycocalix[4]resorcarenes (obtained by condensation of resorcinol with dodecanal, functionalization of the 3,5-dihydroxyphenyl groups of the resulting macrocycle with 2-aminoethyl groups and final coupling with "cellobiolactone") as endosome markers.[192] These 15-nm systems mark endosomes less efficiently than previously prepared 50-nm sized artificial "glycoviruses", suggesting that the endocytosis is size-dependent.

Fang's group used *N*-acetylglucosamine-conjugated CdSe–ZnS core-shell QDs to stain lectin and sperm by fluorometry.[200] Confocal microscopy showed that these GlcNAc-encapsulated QDs bind to the head of the sperm, whereas mannose-encapsulated QDs spread over the whole sperm body, revealing site-specific interactions in dependence of different distribution of sugar receptors on the sperm surface.

Niikura and Ijiro's group demonstrated that GlcNAc-displaying CdTe QDs specifically accumulate in the endoplasmic reticulum of digitonin-permeabilized HeLa cells, but only in the presence of adenosine triphosphate (ATP).[203] The same group has reported that QDs coated with oligoglucosides (Fig. 20A andB) can be transported into the nucleus of digitonin-permeabilized HeLa cells without the use of cationic nuclear-localization signals.[204] Monoglucopyranoside-displaying QDs (irrespective of the anomeric configuration of the neoglycoconjugate) were to be retained in the cytosol and not translocated into the nucleus, as shown by confocal microscopy. Although only a few *maltose*-QDs were observed in the nucleus, the images of

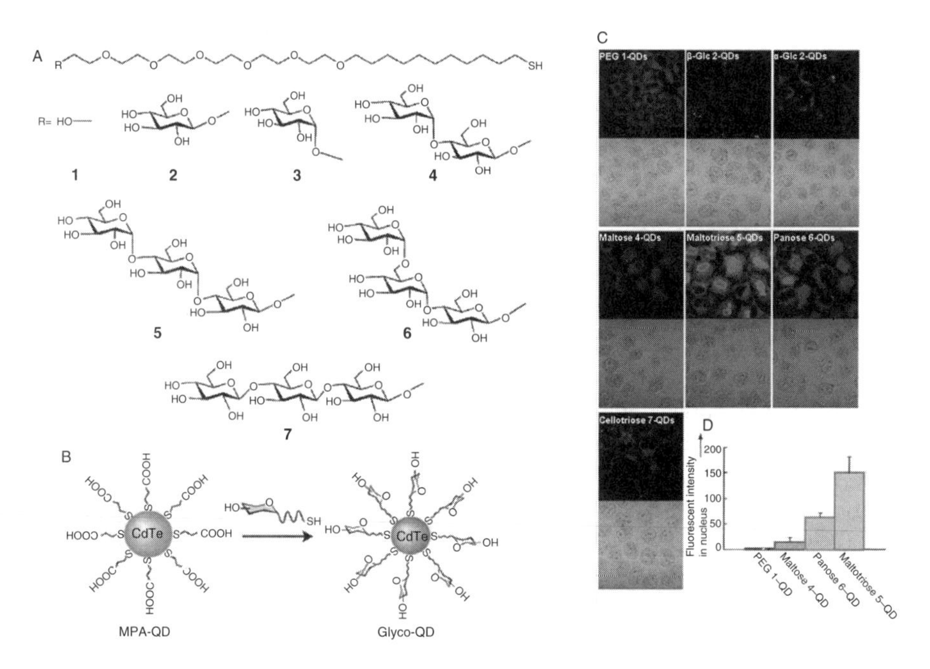

FIG. 20. Oligosaccharide-mediated nuclear transport of glyco-QDs. (A) Chemical structures of neoglycoconjugates. (B) Synthesis of glyco-QDs by thiol exchange reaction. (C) Confocal fluorescence images (top panels) and differential interference contact (lower panels) of digitonin-permeabilized HeLa cells incubated with glyco-QDs. (D) Digitalized fluorescence intensity of different QDs in the nucleus. Adapted from Ref. 204. (See Color Plate 42.)

digitonin-permeabilized HeLa cells incubated with sugar-displaying QDs showed a marked accumulation of *maltotriose*- and *panose*-QDs in the nucleus (Fig. 20C). The fluorescent intensities in the nuclei from the sugar-displaying QDs followed the order: trisaccharide>disaccharide>>PEG (Fig. 20D).

Chen *et al.* delivered fluorescent CdSe–ZnS QDs coated with galactoside dendrons inside HeLa, kidney, and metastatic lung cancer cells, studying the trans-membrane photodynamic profiles. The more asialoprotein receptors that are present at cell surface, the more endocytosis was observed.[201]

Hashida and co-workers used mannosylated QDs for labeling macrophages.[206] A mannose receptor-mediated uptake of these QDs in primary cultured peritoneal macrophages was confirmed by inhibition of the uptaking process when an excess of mannose was used.

Chitosan nanoparticles encapsulating CdSe–ZnS QDs were used by Y. Zhang and co-workers as a self-tracking and non-viral vehicle to deliver human epidermal growth factor receptor 2 (HER2) siRNA into cells *in vitro*.[188] The delivery, internalization, and transfection of the siRNA were monitored by confocal microscopy. Targeted delivery of HER2 siRNA nanoparticles was specific toward over-expressing HER2 breast cancer cells, as indicated by labeling with HER2 antibody. Addition to the encapsulated QDs of paramagnetic diethylenetriaminepentacetate (DTPA)–Gd complex converts them in dual optical–magnetic probes.[189,247]

CdSe–ZnS-labeled carboxymethylchitosan was used by Xie *et al.* as a biocompatible and non-cytotoxic probe for live-cell imaging.[190] Following incubation with live yeast cells, these QDs were found inside the cells, whereas CdSe–ZnS-labeled mercaptoacetic acid (used as a control) not only made the cells die, but also was not taken up.

Covalently chitosan-capped InGaP–ZnS QDs of 30–50 nm size were first used for *in vivo* bioimaging by Tabrizian's group.[205] The chitosan coating not only enhanced biocompatibility, but also improved cellular uptake, and allowed deeper tissue penetration with respect to PEGylated QDs.

QDs conjugated with fluorescently labeled PLGA nanoparticles were applied to image cells at two different emission wavelengths by using the same excitation energy.[248] Iron oxide-conjugated PLGA nanoparticles were also prepared and showed high r_2 relaxivity values, suggesting potential use in MRI.

A very interesting work with glyco-QDs for targeting the lectin DC-SIGN in cellular models has been reported by Cambi *et al.*[195] Quantum dots conjugated to polymeric Lewis X (pLe^{x}-QDs) or HIV envelope gp120 (gp120-QDs) were used to monitor by fluorescence antigen-binding and uptake by cells that express DC-SIGN, and to investigate the intracellular destiny of the antigens. Confocal imaging demonstrated that DC-SIGN-enriched vesicles were involved in QDs internalization by DC-

SIGN-expressing cells. The internalized QDs were not retained in DC-SIGN-enriched vesicles, but were able to reach the lysosomes. The fluorescent signal of endocyted gp120-QDs in DCs was still detectable after 2 days and co-localized in major histocompatibility complex class II (MHCII)-enriched compartments. In this way, QDs could be used as sensitive systems to track antigen-presenting cells *in vivo* and thus elucidate the pathways of DCs-mediated immune response.

In an attempt to obtain versatile tools to address site-specific drug delivery to the liver, Seeberger's group tested PEGylated QDs capped with mannose, galactose, and galactosamine derivatives for specific liver sequestration.[207]

A significant application of fluorescent gold GNPs has been reported by Park.[103] *In vivo* imaging of arthritic inflammation and tumors upon systemic injection into mice was performed with multifunctional gold GNPs capped with HA conjugates labeled with the NIRF dye Hilyte-647 (Fig. 21). The fluorescence quenching by nanoparticle surface-energy transfer (NSET) between Hilyte-647-labeled HA oligomers and gold (Fig. 21A, left) is deactivated when the GNP reach the target zones (arthritic joints and tumors). These zones over-express reactive oxygen species (ROS) and hyaluronidase (HAdase), which synergistically cleaves and releases dye-labeled oligo-HA fragments (Fig. 21A, right). The dye release allows recovery of fluorescence and thus provides ultra-sensitive detection of these disease states by *in vivo* fluorescence imaging. For example, fluorescence images of collagen-induced rheumatoid arthritis in mice are shown in Fig. 21B. The GNPs were injected via the tail vein in normal (upper images) and arthritic (bottom images) mice, and were able to detect the inflammation (becoming a red color indicates

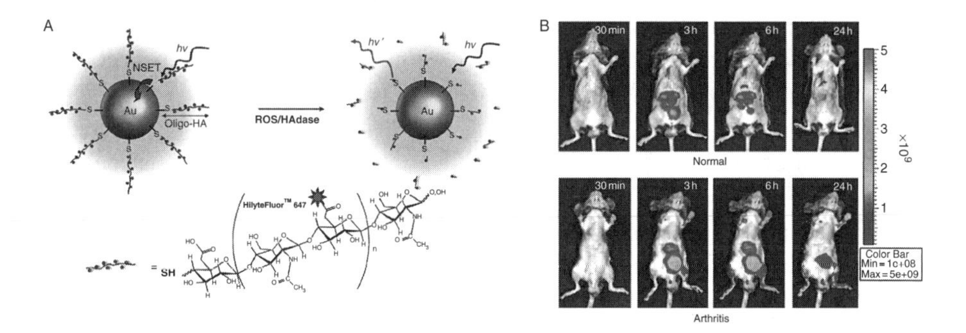

FIG. 21. *In vivo* diagnostic application of hyaluronic acid immobilized gold nanoprobes. (A) The fluorescence quenching by nanoparticle surface-energy transfer between Hilyte-647 dye labelled oligo-HA and gold nanocluster (left) is followed by fluorescence recovery after addition of reactive oxygen species/HAdase which release the dye labeled oligo-HA fragments. (B) Tail vein injection of GNPs capped with HA conjugates labelled with Hilyte-647 in normal (up) and arthritis (bottom) mice. Adapted from Ref. 103. (See Color Plate 43.)

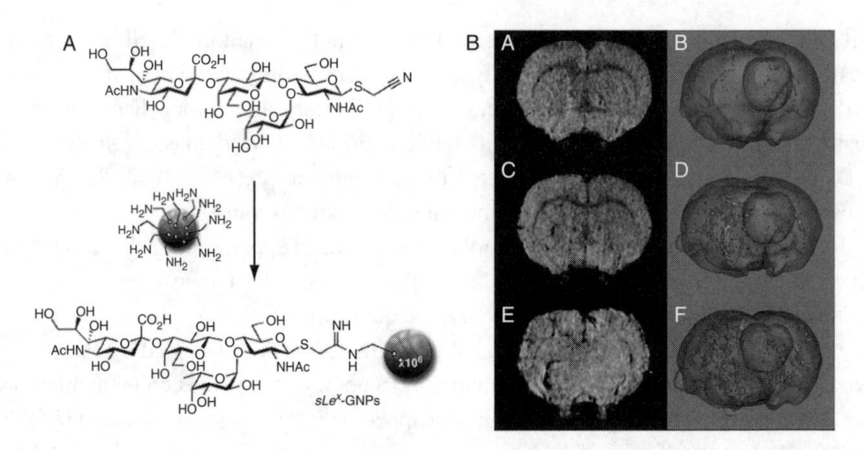

FIG. 22. *In vivo* imaging of brain inflammation by means of *sLe*^{X}-GNPs. (A) Example of construction of *sLe*^{X}-GNPs. (B) T_2^*-weighted images (*A, C*, and *E*) and 3D reconstructions of the GNPs accumulation (*B, D*, and *F*) reveal that *sLe*^{X}-GNPs enables detection of lesions in models of multiple sclerosis (*C* and *D*) and stroke (*E* and *F*) in contrast to unfunctionalized control-NP (*A* and *B*). Adapted from Ref. 153. (See Color Plate 44.)

higher NIRF intensity at 671 nm). These results suggest that gold nanoprobes can be exploited not only as *in vitro* molecular and cellular imaging sensors, but also as *in vivo* optical imaging agents for detection of local HA-degrading diseases.

Magnetic GNPs have been used as contrast agents in MRI. Noteworthy is the application of superparamagnetic GNPs for presymptomatic *in vivo* imaging of brain diseases, as reported by Anthony, Sibson, and Davis.[153] MRI-active dextran-coated iron oxide nanoparticles functionalized with sialyl Lewis X (sLe^X) (Fig. 22A) were used for *in vivo* detection of inflammation. The *sLe*^{X}-GNP constructs were highly sensitive and selective T_2 contrast agents for detecting the endothelial markers E-/P-selectin (CD62E/CD62P) in acute inflammation. Selected images taken from the T_2*-weighted 3D datasets (Fig. 22B; *A, C*, and *E*) and 3D reconstructions of the accumulation of contrast agent (*B, D*, and *F*) showed that *sLe*^{X}-GNP enable clear detection of lesions in clinically relevant models of multiple sclerosis (*C* and *D*) and stroke (*E* and *F*), in contrast to unfunctionalized control-NP (*A* and *B*). T_1-weighted images with Omniscan, a gadolinium-based contrast agent, verified a lack of blood–brain barrier breakdown at the end of the GNP protocol. The good correlation observed between the *in vitro* data for those particles that bind well to E-selectin (*sLe*^{X}-GNP and *sLe*^{X}-GNP-FITC) and those that do not (*LacNAc*-GNP) with the binding observed *in vivo* suggests that this may be mediated by expressed selectin biomarkers.

Paramagnetic Gd-based GNPs prepared by Penadés' group have been also tested as T_1-contrast agents in MRI of murine gliomas generated with GL261 tumoral cells.[91]

Glc-GNPs enhance the contrast in the tumoral zones better than the contrast agents in clinical use, while *lactose*-GNPs do not reach the tumor although they highly enhance the contrast outside the brain. It seems that the nature of the sugar also influences the behavior of GNPs as *in vivo* contrast agents.

6. Other Applications of GNPs

Gold GNPs have also been used as probes for analytical techniques. They constitute a model system for mimicking GSL clusters to quantify carbohydrate self-interactions. The first application of GNPs as analytical probes was reported in the study of Le^X–Le^X interactions by means of SPR.[217] Attempts to evaluate carbohydrate–carbohydrate interactions of immobilized Gg3Cer and GSL-GNPs by SPR were later reported.[98] SPR studies with lectins and a series of carbohydrate-containing GNPs were also conducted.[76,80,226,229,230]

Based on carbohydrate–protein interactions, gold GNPs have been used for rapid identification and isolation of target proteins.[249] A one-step purification of lectins using sugar-immobilized gold nanoparticles has also been proposed.[250] For example, a banana lectin was isolated from the crude extracts of plant materials by mixing with Glc-GNPs. The GNP–lectin aggregate (detected visually) was separated by centrifugation. The aggregate was dissolved by adding glucose and the protein analyzed by sodium dodecyl sulfate polyacrylamide gel electrophoresis (SDS-PAGE) and MS. The isolated lectin was of high purity.

Gold nanoparticles have also been used, by Nishimura[251] and Tseng,[252] as an assisted matrix for determining neutral small carbohydrates, employing MALDI-TOF MS. GNP-assisted MALDI-TOF MS enables an enrichment analysis of living-cell-surface glycosphingolipids.[98]

The conjugation of mass-limited small molecules onto high molecular-weight gold clusters, the efficient ionization of thiol compounds chemisorbed onto gold, and the great ionization efficiency of gold nanoparticles[253] are major advantages for their use in MALDI-MS-based structural characterization.

IV. Future and Perspectives

The application of nanotechnology in the fields of biology and medicine is starting to become a reality. The conjugation of biomolecules with nanomaterials has been explored for specific recognition of target cells and tissues. The unique chemical and physical properties of nanomaterials, including magnetism and/or optical effects, are particularly attractive for their application as sensitive probes in basic research and biomedicine. Multifunctional nanoparticles provide powerful tools for imaging, diagnostic use, and therapy, even if most of the clinical trials are based on non-metallic nanoparticles.[254–256] Nanoparticles can function as drug carriers, and the use of gold and magnetic nanoparticles in drug delivery and therapy have been the subject of reviews.[257–259] These advances in the development of bio-nanomaterials have also stimulated great interest in their applications as biosensors. They permit sensitive real-time detection of pathogens and simultaneous identification of several microorganisms.[260]

Multifunctional GNPs are attractive tools for glycobiology studies. They are coming into use as anti-HIV agents, as anti-cancer vaccines, and as magnetic probes for application in cellular labeling and MRI.[261]

GNPs offer some unique advantages as solid-supported chemical tools for glycobiology together with other multivalent carbohydrate-functionalized systems.[262] The three-dimensional presentation of sugars on metal or semiconductor platforms can be matched with ready insertion of multifunctionality on the metal surface, and the combined properties of the metal cluster and the organic shell can be exploited simultaneously. Their nanometric size, high water solubility, stability in biological and buffered aqueous media, stability to enzymes, and tracing via electron microscopy or fluorescence emission offer outstanding potential for future applications.[263]

References

1. C. S. S. R. Kumar (Ed.), *Nanotechnologies for Life Sciences*, Wiley–VCH Verlag GmbH & Co. KGaA, Weinheim, 2005–2006, Vol. 1–10.
2. V. Vogel (Ed.), *Nanotechnology*, Wiley–VCH Verlag GmbH & Co. KGaA, Weinheim, 2009, Vol. 5.
3. U. Drechsler, B. Erdogan, and V. M. Rotello, Nanoparticles: Scaffolds for molecular recognition, *Chem. Eur. J.*, 10 (2004) 5570–5579.
4. E. Katz and I. Willner, Integrated nanoparticle-biomolecule hybrid systems: Synthesis, properties, and applications, *Angew. Chem. Int. Ed. Engl.*, 43 (2004) 6042–6108.
5. J. Wang, Nanomaterial-based amplified transduction of biomolecular interactions, *Small*, 1 (2005) 1036–1043.
6. C. M. Niemeyer and C. A. Mirkin (Eds.), *Nanobiotechnology: Concepts, Applications and Perspectives*, Wiley–VCH Verlag GmbH & Co. KGaA, Weinheim, 2004.

7. A. S. Edelstein and R. C. Cammarata (Eds.), *Nanomaterials: Synthesis, Properties and Applications*, Institute of Physics Publishing, Bristol, UK, 1996.
8. C.-C. You, A. Chompoosor, and V. M. Rotello, The biomacromolecule-nanoparticle interface, *Nano Today*, 2 (2007) 34–43.
9. M.-C. Daniel and D. Astruc, Gold nanoparticles: Assembly, supramolecular chemistry, quantum-size-related properties, and applications toward biology, catalysis, and nanotechnology, *Chem. Rev.*, 104 (2004) 293–346.
10. M. Faraday, *Philos. Trans. R. Soc. Lond.*, 147 (1857) 145–181.
11. G. Frens, Controlled nucleation for regulation of particle-size in monodisperse gold suspensions, *Nat. Phys. Sci.*, 241 (1973) 20–22.
12. J. Turkevich, P. C. Stevenson, and J. Hillier, The formation of colloidal gold, *J. Phys. Chem.*, 57 (1953) 670–673.
13. J. Turkevich, P. C. Stevenson, and J. Hillier, A study of the nucleation and growth processes in the synthesis of colloidal gold, *Discuss. Faraday Soc.*, 11 (1951) 55–75.
14. J. Turkevich, Colloidal Gold. Part I, *Gold Bull.*, 18 (1985) 86–91.
15. M. A. Hayat (Ed.), *Colloidal Gold: Principles, Methods and Applications*, Academic Press, Inc., San Diego, CA, 1989, Vol. 1, pp. 13–32.
16. C. B. Murray, C. R. Kagan, and M. G. Bawendi, Synthesis and characterization of monodisperse nanocrystals and close-packed nanocrystal assemblies, *Annu. Rev. Mater. Sci.*, 30 (2000) 545–610.
17. G. Schmid, Large clusters and colloids—metals in the embryonic state, *Chem. Rev.*, 92 (1992) 1709–1727.
18. A. C. Templeton, M. P. Wuelfing, and R. W. Murray, Monolayer protected cluster molecules, *Acc. Chem. Res.*, 33 (2000) 27–36.
19. C. Burda, X. Chen, R. Narayanan, and M. A. El-Sayed, Chemistry and properties of nanocrystals of different shapes, *Chem. Rev.*, 105 (2005) 1025–1102.
20. J. C. Love, L. A. Estroff, J. K. Kriebel, R. G. Nuzzo, and G. M. Whitesides, Self-assembled monolayers of thiolates on metals as a form of nanotechnology, *Chem. Rev.*, 105 (2005) 1103–1169.
21. J. M. Polak, and I. M. Varndell (Eds.), *Immunolabelling for Electron Microscopy*, Elsevier Science Publishers, Amsterdam-New York, 1984.
22. C. M. Niemeyer, Nanoparticles, proteins, and nucleic acids: Biotechnology meets materials science, *Angew. Chem. Int. Ed. Engl.*, 40 (2001) 4128–4158.
23. J. J. Storhoff and C. A. Mirkin, Programmed materials synthesis with DNA, *Chem. Rev.*, 99 (1999) 1849–1862.
24. J. M. de la Fuente and S. Penadés, Understanding carbohydrate-carbohydrate interactions by means of glyconanotechnology, *Glycoconj. J.*, 21 (2004) 149–163.

25. J. M. de la Fuente and S. Penadés, Glyconanoparticles: Types, synthesis and applications in glycoscience, biomedicine and material science, *Biochim. Biophys. Acta Gen. Subj.*, 1760 (2006) 636–651.
26. C. A. Mirkin, R. L. Letsinger, R. C. Mucic, and J. J. Storhoff, A DNA-based method for rationally assembling nanoparticles into macroscopic materials, *Nature*, 382 (1996) 607–609.
27. N. L. Rosi and C. A. Mirkin, Nanostructures in biodiagnostics, *Chem. Rev.*, 105 (2005) 1547–1562.
28. C. S. Thaxton and C. A. Mirkin, DNA-gold-nanoparticle conjugates, in C. M. Niemeyer, and C. A. Mirkin, (Eds.), *Nanobiotechnology: Concepts, Applications and Perspectives*, Wiley–VCH Verlag GmbH & Co. KGaA, Weinheim, 2004, pp. 2288–2307.
29. V. Pavlov, Y. Xiao, B. Shlyahovsky, and I. Willner, Aptamer-functionalized Au nanoparticles for the amplified optical detection of thrombin, *J. Am. Chem. Soc.*, 126 (2004) 11768–11769.
30. L. He, M. D. Musick, S. R. Nicewarner, F. G. Salinas, S. J. Benkovic, M. J. Natan, and C. D. Keating, Colloidal Au-enhanced surface plasmon resonance for ultrasensitive detection of DNA hybridization, *J. Am. Chem. Soc.*, 122 (2000) 9071–9077.
31. S.-Y. Lin, S.-H. Wu, and C.-H. Chen, A simple strategy for prompt visual sensing by gold nanoparticles: General applications of interparticle hydrogen bonds, *Angew. Chem. Int. Ed. Engl.*, 45 (2006) 4948–4951.
32. J. Liu and Y. Lu, Stimuli-responsive disassembly of nanoparticle aggregates for light-up colorimetric sensing, *J. Am. Chem. Soc.*, 127 (2005) 12677–12683.
33. J. Liu and Y. Lu, Accelerated color change of gold nanoparticles assembled by DNAzymes for simple and fast colorimetric Pb^{2+} detection, *J. Am. Chem. Soc.*, 126 (2004) 12298–12305.
34. C. D. Medley, J. E. Smith, Z. Tang, Y. Wu, S. Bamrungsap, and W. Tan, Gold nanoparticle-based colorimetric assay for the direct detection of cancerous cells, *Anal. Chem.*, 80 (2008) 1067–1072.
35. R. Levy, N. T. K. Thanh, R. C. Doty, I. Hussain, R. J. Nichols, D. J. Schiffrin, M. Brust, and D. G. Fernig, Rational and combinatorial design of peptide capping ligands for gold nanoparticles, *J. Am. Chem. Soc.*, 126 (2004) 10076–10084.
36. M. Giersig and P. Mulvaney, Preparation of ordered colloid monolayers by electrophoretic deposition, *Langmuir*, 9 (1993) 3408–3413.
37. D. A. Handley, Methods for synthesis of colloidal gold, in M. A. Hayat, (Ed.), *Colloidal Gold: Principles, Methods and Applications*, Academic Press Inc., San Diego, CA, 1989, Vol. 1, pp. 13–32.

38. M. Brust, M. Walker, D. Bethell, D. J. Schiffrin, and R. Whyman, Synthesis of thiol-derivatized gold nanoparticles in a 2-phase liquid–liquid system, *J. Chem. Soc. Chem. Commun.*, (1994) 801–802.
39. M. Brust, J. Fink, D. Bethell, D. J. Schiffrin, and C. Kiely, Synthesis and reactions of functionalized gold nanoparticles, *J. Chem. Soc. Chem. Commun.*, (1995) 1655–1656.
40. W. P. Wuelfing, S. M. Gross, D. T. Miles, and R. W. Murray, Nanometer gold clusters protected by surface-bound monolayers of thiolated poly(ethylene glycol) polymer electrolyte, *J. Am. Chem. Soc.*, 120 (1998) 12696–12697.
41. A. C. Templeton, D. E. Cliffel, and R. W. Murray, Redox and fluorophore functionalization of water-soluble, tiopronin-protected gold clusters, *J. Am. Chem. Soc.*, 121 (1999) 7081–7089.
42. A. C. Templeton, S. W. Chen, S. M. Gross, and R. W. Murray, Water-soluble, isolable gold clusters protected by tiopronin and coenzyme A monolayers, *Langmuir*, 15 (1999) 66–76.
43. T. G. Schaaff, G. Knight, M. N. Shafigullin, R. F. Borkman, and R. L. Whetten, Isolation and selected properties of a 10.4 kDa gold: Glutathione cluster compound, *J. Phys. Chem. B*, 102 (1998) 10643–10646.
44. S. H. Chen and K. Kimura, Synthesis and characterization of carboxylate-modified gold nanoparticle powders dispersible in water, *Langmuir*, 15 (1999) 1075–1082.
45. D. E. Cliffel, F. P. Zamborini, S. M. Gross, and R. W. Murray, Mercaptoammonium-monolayer-protected, water-soluble gold, silver, and palladium clusters, *Langmuir*, 16 (2000) 9699–9702.
46. D. V. Leff, P. C. Ohara, J. R. Heath, and W. M. Gelbart, Thermodynamic control of gold nanocrystal size—experiment and theory, *J. Phys. Chem.*, 99 (1995) 7036–7041.
47. P. D. Jadzinsky, G. Calero, C. J. Ackerson, D. A. Bushnell, and R. D. Kornberg, Structure of a thiol monolayer-protected gold nanoparticle at 1.1 angstrom resolution, *Science*, 318 (2007) 430–433.
48. M. J. Hostetler, S. J. Green, J. J. Stokes, and R. W. Murray, Monolayers in three dimensions: Synthesis and electrochemistry of omega-functionalized alkanethiolate-stabilized gold cluster compounds, *J. Am. Chem. Soc.*, 118 (1996) 4212–4213.
49. R. Hong, G. Han, J. M. Fernandez, B. J. Kim, N. S. Forbes, and V. M. Rotello, Glutathione-mediated delivery and release using monolayer protected nanoparticle carriers, *J. Am. Chem. Soc.*, 128 (2006) 1078–1079.

50. M. J. Hostetler, A. C. Templeton, and R. W. Murray, Dynamics of place-exchange reactions on monolayer-protected gold cluster molecules, *Langmuir*, 15 (1999) 3782–3789.
51. C. P. Collier, R. J. Saykally, J. J. Shiang, S. E. Henrichs, and J. R. Heath, Reversible tuning of silver quantum dot monolayers through the metal-insulator transition, *Science*, 277 (1997) 1978–1981.
52. J. R. Heath, C. M. Knobler, and D. V. Leff, Pressure/temperature phase diagrams and superlattices of organically functionalized metal nanocrystal monolayers: The influence of particle size, size distribution, and surface passivant, *J. Phys. Chem. B*, 101 (1997) 189–197.
53. M. J. Hostetler, C. J. Zhong, B. K. H. Yen, J. Anderegg, S. M. Gross, N. D. Evans, M. Porter, and R. W. Murray, Stable, monolayer-protected metal alloy clusters, *J. Am. Chem. Soc.*, 120 (1998) 9396–9397.
54. C. K. Yee, R. Jordan, A. Ulman, H. White, A. King, M. Rafailovich, and J. Sokolov, Novel one-phase synthesis of thiol-functionalized gold, palladium, and iridium nanoparticles using superhydride, *Langmuir*, 15 (1999) 3486–3491.
55. C. Yee, M. Scotti, A. Ulman, H. White, M. Rafailovich, and J. Sokolov, One-phase synthesis of thiol-functionalized platinum nanoparticles, *Langmuir*, 15 (1999) 4314–4316.
56. F. Mirkhalaf, J. Paprotny, and D. J. Schiffrin, Synthesis of metal nanoparticles stabilized by metal-carbon bonds, *J. Am. Chem. Soc.*, 128 (2006) 7400–7401.
57. K. L. McGilvray, M. R. Decan, D. S. Wang, and J. C. Scaiano, Facile photochemical synthesis of unprotected aqueous gold nanoparticles, *J. Am. Chem. Soc.*, 128 (2006) 15980–15981.
58. R. A. Dwek, Glycobiology: Toward understanding the function of sugars, *Chem. Rev.*, 96 (1996) 683–720.
59. Y. C. Lee and R. T. Lee, Carbohydrate–protein interactions—basis of glycobiology, *Acc. Chem. Res.*, 28 (1995) 321–327.
60. A. Varki, Biological roles of oligosaccharides—all of the theories are correct, *Glycobiology*, 3 (1993) 97–130.
61. C. R. Bertozzi and L. L. Kiessling, Chemical glycobiology, *Science*, 291 (2001) 2357–2364.
62. J. J. Lundquist and E. J. Toone, The cluster glycoside effect, *Chem. Rev.*, 102 (2002) 555–578.
63. S. Hakomori, Carbohydrate–carbohydrate interaction as an initial step in cell recognition, *Pure Appl. Chem.*, 63 (1991) 473–482.
64. J. Rojo, J. C. Morales, and S. Penadés, Carbohydrate–carbohydrate interactions in biological and model systems, *Top Curr. Chem.*, 218 (2002) 45–92.

65. M. Mammen, S.-K. Choi, and G. M. Whitesides, Polyvalent interactions in biological systems: Implications for design and use of multivalent ligands and inhibitors, *Angew. Chem. Int. Ed. Engl.*, 37 (1998) 2755–2794.
66. H. C. Hang and C. R. Bertozzi, Chemoselective approaches to glycoprotein assembly, *Acc. Chem. Res.*, 34 (2001) 727–736.
67. B. T. Houseman and M. Mrksich, Model systems for studying polyvalent carbohydrate binding interactions, *Top Curr. Chem.*, 218 (2002) 1–44.
68. Y. M. Chabre and R. Roy, Design and creativity in synthesis of multivalent neoglycoconjugates, *Adv. Carbohydr. Chem. Biochem.*, 63 (2010) 165–393.
69. T. K. Lindhorst, Artificial multivalent sugar ligands to understand and manipulate carbohydrate-protein interactions, *Top Curr. Chem.*, 218 (2002) 201–235.
70. J. M. de la Fuente, A. G. Barrientos, T. C. Rojas, J. Rojo, J. Cañada, A. Fernández, and S. Penadés, Gold glyconanoparticles as water-soluble polyvalent models to study carbohydrate interactions, *Angew. Chem. Int. Ed. Engl.*, 40 (2001) 2258–2261.
71. C. C. Lee, J. A. MacKay, J. M. J. Fréchet, and F. C. Szoka, Designing dendrimers for biological applications, *Nat. Biotechnol.*, 23 (2005) 1517–1526.
72. L. L. Kiessling, J. E. Gestwicki, and L. E. Strong, Synthetic multivalent ligands as probes of signal transduction, *Angew. Chem. Int. Ed. Engl.*, 45 (2006) 2348–2368.
73. V. P. Torchilin, Recent advances with liposomes as pharmaceutical carriers, *Nat. Rev. Drug Discov.*, 4 (2005) 145–160.
74. C.-C. Lin, Y.-C. Yeh, C.-Y. Yang, C.-L. Chen, G.-F. Chen, A. C. Chen, and Y.-C. Wu, Selective binding of mannose-encapsulated gold nanoparticles to type 1 pili in *Escherichia coli*, J. Am. Chem. Soc., 124 (2002) 3508–3509.
75. A. G. Barrientos, J. M. de la Fuente, T. C. Rojas, A. Fernández, and S. Penadés, Gold glyconanoparticles: Synthetic polyvalent ligands mimicking glycocalyx-like surfaces as tools for glycobiological studies, *Chem. Eur. J.*, 9 (2003) 1909–1921.
76. C.-C. Lin, Y.-C. Yeh, C.-Y. Yang, G.-F. Chen, Y.-C. Chen, Y.-C. Wu, and C.-C. Chen, Quantitative analysis of multivalent interactions of carbohydrate-encapsulated gold nanoparticles with concanavalin A, *Chem. Commun.*, (2003) 2920–2921.
77. J.-L. de Paz, R. Ojeda, A. G. Barrientos, S. Penadés, and M. Martín-Lomas, Synthesis of a Le(y) neoglycoconjugate and Le(y)-functionalized gold glyconanoparticles, *Tetrahedron Asymmetry*, 16 (2005) 149–158.
78. S. A. Svarovsky, Z. Szekely, and J. J. Barchi, Synthesis of gold nanoparticles bearing the Thomsen-Friedenreich disaccharide: A new multivalent presentation of an important tumor antigen, *Tetrahedron Asymmetry*, 16 (2005) 587–598.

79. A. Sundgren and J. J. Barchi, Varied presentation of the Thomsen-Friedenreich disaccharide tumor-associated carbohydrate antigen on gold nanoparticles, *Carbohydr. Res.*, 343 (2008) 1594–1604.
80. Y.-Y. Chien, M.-D. Jan, A. K. Adak, H.-C. Tzeng, Y.-P. Lin, Y.-J. Chen, K.-T. Wang, C.-T. Chen, C.-C. Chen, and C.-C. Lin, Globotriose-functionalized gold nanoparticles as multivalent probes for Shiga-like toxin, *ChemBioChem*, 9 (2008) 1100–1109.
81. A. Carvalho de Souza, K. M. Halkes, J. D. Meeldijk, A. J. Verkleij, J. F. G. Vliegenthart, and J. P. Kamerling, Synthesis of gold glyconanoparticles: Possible probes for the exploration of carbohydrate-mediated self-recognition of marine sponge cells, *Eur. J. Org. Chem.*, (2004) 4323–4339.
82. A. Carvalho de Souza, J. F. G. Vliegenthart, and J. P. Kamerling, Gold nanoparticles coated with a pyruvated trisaccharide epitope of the extracellular proteoglycan of *Microciona prolifera* as potential tools to explore carbohydrate-mediated cell recognition, *Org. Biomol. Chem.*, 6 (2008) 2095–2102.
83. K. M. Halkes, A. Carvalho de Souza, C. E. P. Maljaars, G. J. Gerwig, and J. P. Kamerling, A facile method for the preparation of gold glyconanoparticles from free oligosaccharides and their applicability in carbohydrate–protein interaction studies, *Eur. J. Org. Chem.*, (2005) 3650–3659.
84. B. Nolting, J.-J. Yu, G.-Y. Liu, S.-J. Cho, S. Kauzlarich, and J. Gervay-Hague, Synthesis of gold glyconanoparticles and biological evaluation of recombinant Gp120 interactions, *Langmuir*, 19 (2003) 6465–6473.
85. S. G. Spain, L. Albertin, and N. R. Cameron, Facile *in situ* preparation of biologically active multivalent glyconanoparticles, *Chem. Commun.*, (2006) 4198–4200.
86. A. Housni, H. J. Cai, S. Y. Liu, S. H. Pun, and R. Narain, Facile preparation of glyconanoparticles and their bioconjugation to streptavidin, *Langmuir*, 23 (2007) 5056–5061.
87. R. Narain, A. Housni, G. Gody, P. Boullanger, M.-T. Charreyre, and T. Delair, Preparation of biotinylated glyconanoparticles via a photochemical process and study of their bioconjugation to streptavidin, *Langmuir*, 23 (2007) 12835–12841.
88. A. G. Barrientos, J. M. de la Fuente, M. Jiménez, D. Solís, J. Cañada, M. Martín-Lomas, and S. Penadés, Modulating glycosidase degradation and lectin recognition of gold glyconanoparticles, *Carbohydr. Res.*, 344 (2009) 1474–1478.
89. O. Martínez-Ávila, K. Hijazi, M. Marradi, C. Clavel, C. Campion, C. Kelly, and S. Penadés, Gold manno-glyconanoparticles: Multivalent systems to block HIV-1 gp120 binding to the lectin DC-SIGN, *Chem. Eur. J.*, 15 (2009) 9874–9888.

90. T. W. Rademacher, K. Gumaa, M. Martín-Lomas, S. Penadés, R. Ojeda, and A. G. Barrientos, Nanoparticles comprising RNA ligands, Pub. No.: WO/2005/116226 International Application No.: PCT/GB2005/002058.
91. M. Marradi, D. Alcántara, J. M. de la Fuente, M. L. García-Martín, S. Cerdán, and S. Penadés, Paramagnetic Gd-based gold glyconanoparticles as probes for MRI: Tuning relaxivities with sugars, *Chem. Commun.*, (2009) 3922–3924.
92. R. Ojeda, J. L. de Paz, A. G. Barrientos, M. Martín-Lomas, and S. Penadés, Preparation of multifunctional glyconanoparticles as a platform for potential carbohydrate-based anticancer vaccines, *Carbohydr. Res.*, 342 (2007) 448–459.
93. S. Yokota, T. Kitaoka, M. Opietnik, T. Rosenau, and H. Wariishi, Synthesis of gold nanoparticles for *in situ* conjugation with structural carbohydrates, *Angew. Chem. Int. Ed. Engl.*, 47 (2008) 9866–9869.
94. H. Otsuka, Y. Akiyama, Y. Nagasaki, and K. Kataoka, Quantitative and reversible lectin-induced association of gold nanoparticles modified with α-lactosyl-ω-mercapto-poly(ethylene glycol), *J. Am. Chem. Soc.*, 123 (2001) 8226–8230.
95. J. Zhang, C. D. Geddes, and J. R. Lakowicz, Complexation of polysaccharide and monosaccharide with thiolate boronic acid capped on silver nanoparticle, *Anal. Biochem.*, 332 (2004) 253–260.
96. M. Chikae, T. Fukuda, K. Kerman, K. Idegami, Y. Miura, and E. Tamiya, Amyloid-β detection with saccharide immobilized gold nanoparticle on carbon electrode, *Bioelectrochemistry*, 74 (2008) 118–123.
97. C. Earhart, N. R. Jana, N. Erathodiyil, and J. Y. Ying, Synthesis of carbohydrate-conjugated nanoparticles and quantum dots, *Langmuir*, 24 (2008) 6215–6219.
98. N. Nagahori, M. Abe, and S.-I. Nishimura, Structural and functional glycosphingolipidomics by glycoblotting with an aminooxy-functionalized gold nanoparticle, *Biochemistry*, 48 (2009) 583–594.
99. D. C. Hone, A. H. Haines, and D. A. Russell, Rapid, quantitative colorimetric detection of a lectin using mannose-stabilized gold nanoparticles, *Langmuir*, 19 (2003) 7141–7144.
100. R. Karamanska, B. Mukhopadhyay, D. A. Russell, and R. A. Field, Thioctic acid amides: Convenient tethers for achieving low nonspecific protein binding to carbohydrates presented on gold surfaces, *Chem. Commun.*, (2005) 3334–3336.
101. A. J. Reynolds, A. H. Haines, and D. A. Russell, Gold glyconanoparticles for mimics and measurement of metal ion-mediated carbohydrate-carbohydrate interactions, *Langmuir*, 22 (2006) 1156–1163.
102. H. Lee, S. H. Choi, and T. G. Park, Direct visualization of hyaluronic acid polymer chain by self-assembled one-dimensional array of gold nanoparticles, *Macromolecules*, 39 (2006) 23–25.

103. H. Lee, K. Lee, I. K. Kim, and T. G. Park, Synthesis, characterization, and *in vivo* diagnostic applications of hyaluronic acid immobilized gold nanoprobes, *Biomaterials*, 29 (2008) 4709–4718.
104. M. B. Thygesen, J. Sauer, and K. J. Jensen, Chemoselective capture of glycans for analysis on gold nanoparticles: Carbohydrate oxime tautomers provide functional recognition by proteins, *Chem. Eur. J.*, 15 (2009) 1649–1660.
105. F. Manea, C. Bindoli, S. Fallarini, G. Lombardi, L. Polito, L. Lay, R. Bonomi, F. Mancin, and P. Scrimin, Multivalent, saccharide-functionalized gold nanoparticles as fully synthetic analogs of Type A *Neisseria meningitidis* antigens, *Adv. Mater.*, 20 (2008) 4348–4352.
106. F. Manea, C. Bindoli, S. Polizzi, L. Lay, and P. Scrimin, Expeditious synthesis of water-soluble, monolayer-protected gold nanoparticles of controlled size and monolayer composition, *Langmuir*, 24 (2008) 4120–4124.
107. C.-C. Huang, C.-T. Chen, Y.-C. Shiang, Z.-H. Lin, and H.-T. Chang, Synthesis of fluorescent carbohydrate-protected Au nanodots for detection of concanavalin A and *Escherichia coli, Anal. Chem.*, 81 (2009) 875–882.
108. J. Fink, C. J. Kiely, D. Bethell, and D. J. Schiffrin, Self-organization of nano-sized gold particles, *Chem. Mater.*, 10 (1998) 922–926.
109. H. W. Gu, P. L. Ho, E. Tong, L. Wang, and B. Xu, Presenting vancomycin on nanoparticles to enhance antimicrobial activities, *Nano Lett.*, 3 (2003) 1261–1263.
110. S. Takae, Y. Akiyama, H. Otsuka, T. Nakamura, Y. Nagasaki, and K. Kataoka, Ligand density effect on biorecognition by PEGylated gold nanoparticles: Regulated interaction of RCA_{120} lectin with lactose installed to the distal end of tethered PEG strands on gold surface, *Biomacromolecules*, 6 (2005) 818–824.
111. C. L. Schofield, B. Mukhopadhyay, S. M. Hardy, M. B. McDonnell, R. A. Field, and D. A. Russell, Colorimetric detection of *Ricinus communis* agglutinin 120 using optimally presented carbohydrate-stabilised gold nanoparticles, *Analyst*, 133 (2008) 626–634.
112. A. Yoshizumi, N. Kanayama, Y. Maehara, M. Ide, and H. Kitano, Self-assembled monolayer of sugar-carrying polymer chain: Sugar balls from 2-methacryloyloxyethyl D-glucopyranoside, *Langmuir*, 15 (1999) 482–488.
113. M. Krämer, N. Pérignon, R. Haag, J. D. Marty, R. Thomann, N. Lauth-de Viguerie, and C. Mingotaud, Water-soluble dendritic architectures with carbohydrate shells for the templation and stabilization of catalytically active metal nanoparticles, *Macromolecules*, 38 (2005) 8308–8315.

114. A. Köth, J. Koetz, D. Appelhans, and B. Voit, "Sweet" gold nanoparticles with oligosaccharide-modified poly(ethyleneimine), *Colloid Polym. Sci.*, 286 (2008) 1317–1327.
115. C. V. Ramana, K. A. Durugkar, V. G. Puranik, S. B. Narute, and B. L. V. Prasad, C-glycosides of dodecanoic acid: New capping/reducing agents for glyconanoparticle synthesis, *Tetrahedron Lett.*, 49 (2008) 6227–6230.
116. H. Huang and X. Yang, Synthesis of chitosan-stabilized gold nanoparticles in the absence/presence of tripolyphosphate, *Biomacromolecules*, 5 (2004) 2340–2346.
117. H. Huang and X. Yang, Synthesis of polysaccharide-stabilized gold and silver nanoparticles: A green method, *Carbohydr. Res.*, 339 (2004) 2627–2631.
118. H. Huang, Q. Yuan, and X. Yang, Preparation and characterization of metal-chitosan nanocomposites, *Colloids Surf. B*, 39 (2004) 31–37.
119. W. B. Tan and Y. Zhang, Surface modification of gold and quantum dot nanoparticles with chitosan for bioapplications, *J. Biomed. Mater. Res. A*, 75A (2005) 56–62.
120. D. R. Bhumkar, H. M. Joshi, M. Sastry, and V. B. Pokharkar, Chitosan reduced gold nanoparticles as novel carriers for transmucosal delivery of insulin, *Pharm. Res.*, 24 (2007) 1415–1426.
121. S. Dhar, E. M. Reddy, A. Shiras, V. Pokharkar, and B. L. V. Prasad, Natural gum reduced/stabilized gold nanoparticles for drug delivery formulations, *Chem. Eur. J.*, 14 (2008) 10244–10250.
122. V. Kattumuri, K. Katti, S. Bhaskaran, E. J. Boote, S. W. Casteel, G. M. Fent, D. J. Robertson, M. Chandrasekhar, R. Kannan, and K. V. Katti, Gum arabic as a phytochemical construct for the stabilization of gold nanoparticles: *In vivo* pharmacokinetics and X-ray-contrast-imaging studies, *Small*, 3 (2007) 333–341.
123. W. Haiss, N. T. K. Thanh, J. Aveyard, and D. G. Fernig, Determination of size and concentration of gold nanoparticles from UV-Vis spectra, *Anal. Chem.*, 79 (2007) 4215–4221.
124. See the special issue: *Int. J. Hyperthermia* 2008; volume 24(6); special issue: Hyperthermia and Nanotechnology; editor: M. W. Dewhirst.
125. C. C. Berry and A. S. G. Curtis, Functionalisation of magnetic nanoparticles for applications in biomedicine, *J. Phys. D Appl. Phys.*, 36 (2003) 38650–38654.
126. D. L. Hüber, Synthesis, properties, and applications of iron nanoparticles, *Small*, 1 (2005) 482–501.
127. S. Mornet, S. Vasseur, F. Grasset, and E. Duguet, Magnetic nanoparticle design for medical diagnosis and therapy, *J. Mater. Chem.*, 14 (2004) 2161–2175.

128. Q. A. Pankhurst, J. Connolly, S. K. Jones, and J. Dobson, Applications of magnetic nanoparticles in biomedicine, *J. Phys. D Appl. Phys.*, 36 (2003) 40035–40041.
129. P. Tartaj, M. D. Morales, S. Veintemillas-Verdaguer, T. Gonzalez-Carreno, and C. J. Serna, The preparation of magnetic nanoparticles for applications in biomedicine, *J. Phys. D Appl. Phys.*, 36 (2003) 37541–37542.
130. J. Kim, Y. Piao, and T. Hyeon, Multifunctional nanostructured materials for multimodal imaging, and simultaneous imaging and therapy, *Chem. Soc. Rev.*, 38 (2009) 372–390.
131. H. B. Na, I. C. Song, and T. Hyeon, Inorganic nanoparticles for MRI contrast agents, *Adv. Mater.*, 21 (2009) 1–16.
132. C. B. Murray, S. H. Sun, W. Gaschler, H. Doyle, T. A. Betley, and C. R. Kagan, Colloidal synthesis of nanocrystals and nanocrystal superlattices, *IBM J. Res. Dev.*, 45 (2001) 47–56.
133. T. Hyeon, Chemical synthesis of magnetic nanoparticles, *Chem. Commun.*, (2003) 927–934.
134. A.-H. Lu, E. L. Salabas, and F. Schüth, Magnetic nanoparticles: Synthesis, protection, functionalization, and application, *Angew. Chem. Int. Ed. Engl.*, 46 (2007) 1222–1244.
135. J. R. McCarthy and R. Weissleder, Multifunctional magnetic nanoparticles for targeted imaging and therapy, *Adv. Drug Deliv. Rev.*, 60 (2008) 1241–1251.
136. J. Cheon and J.-H. Lee, Synergistically integrated nanoparticles as multimodal probes for nanobiotechnology, *Acc. Chem. Res.*, 41 (2008) 1630–1640.
137. R. Weissleder, G. Elizondo, J. Wittenberg, C. A. Rabito, H. H. Bengele, and L. Josephson, Ultrasmall superparamagnetic iron-oxide—characterization of a new class of contrast agents for MR imaging, *Radiology*, 175 (1990) 489–493.
138. A. K. Gupta and M. Gupta, Synthesis and surface engineering of iron oxide nanoparticles for biomedical applications, *Biomaterials*, 26 (2005) 3995–4021.
139. C. W. Jung and P. Jacobs, Physical and chemical-properties of superparamagnetic iron-oxide MR contrast agents—ferumoxides, ferumoxtran, ferumoxsil, *Magn. Reson. Imaging*, 13 (1995) 661–674.
140. Y.-X. J. Wang, S. M. Hussain, and G. P. Krestin, Superparamagnetic iron oxide contrast agents: Physicochemical characteristics and applications in MR imaging, *Eur. Radiol.*, 11 (2001) 2319–2331.
141. L. Josephson, C.-H. Tung, A. Moore, and R. Weissleder, High-efficiency intracellular magnetic labeling with novel superparamagnetic-tat peptide conjugates, *Bioconjug. Chem.*, 10 (1999) 186–191.

142. L. Josephson, J. M. Perez, and R. Weissleder, Magnetic nanosensors for the detection of oligonucleotide sequences, *Angew. Chem. Int. Ed. Engl.*, 40 (2001) 3204–3206.
143. H. W. Kang, L. Josephson, A. Petrovsky, R. Weissleder, and A. Bogdanov, Magnetic resonance imaging of inducible E-selectin expression in human endothelial cell culture, *Bioconjug. Chem.*, 13 (2002) 122–127.
144. L. Josephson, M. F. Kircher, U. Mahmood, Y. Tang, and R. Weissleder, Near-infrared fluorescent nanoparticles as combined MR/optical imaging probes, *Bioconjug. Chem.*, 13 (2002) 554–560.
145. R. Weissleder, K. Kelly, E. Y. Sun, T. Shtatland, and L. Josephson, Cell-specific targeting of nanoparticles by multivalent attachment of small molecules, *Nat. Biotechnol.*, 23 (2005) 1418–1423.
146. J. M. de la Fuente, D. Alcántara, P. Eaton, P. Crespo, T. C. Rojas, A. Fernández, A. Hernando, and S. Penadés, Gold and gold-iron oxide magnetic glyconanoparticles: Synthesis, characterization and magnetic properties, *J. Phys. Chem. B*, 110 (2006) 13021–13028.
147. M. Fuss, M. Luna, D. Alcántara, J. M. de la Fuente, S. Penadés, and F. Briones, Supramolecular self-assembled arrangements of maltose glyconanoparticles, *Langmuir*, 24 (2008) 5124–5128.
148. P. Crespo, R. Litrán, T. C. Rojas, M. Multigner, J. M. de la Fuente, J. C. Sánchez-López, M. A. García, A. Hernando, S. Penadés, and A. Fernández, Permanent magnetism, magnetic anisotropy, and hysteresis of thiol-capped gold nanoparticles, *Phys. Rev. Lett.*, 93 (2004) 087204.
149. P. Crespo, M. A. García, E. F. Pinel, M. Multigner, D. Alcántara, J. M. de la Fuente, S. Penadés, and A. Hernando, Fe impurities weaken the ferromagnetic behavior in Au nanoparticles, *Phys. Rev. Lett.*, 97 (2006) 177203.
150. J. M. de la Fuente, D. Alcántara, and S. Penadés, Cell response to magnetic glyconanoparticles: Does the carbohydrate matter? *IEEE Trans. Nanobiosci.*, 6 (2007) 275–281.
151. H. Hifumi, S. Yamaoka, A. Tanimoto, D. Citterio, and K. Suzuki, Gadolinium-based hybrid nanoparticles as a positive MR contrast agent, *J. Am. Chem. Soc.*, 128 (2006) 15090–15091.
152. M. A. McDonald and K. L. Watkin, Investigations into the physicochemical properties of dextran small particulate gadolinium oxide nanoparticles, *Acad. Radiol.*, 13 (2006) 421–427.
153. S. I. van Kasteren, S. J. Campbell, S. Serres, D. C. Anthony, N. R. Sibson, and B. G. Davis, Glyconanoparticles allow presymptomatic *in vivo* imaging of brain disease, *Proc. Nat. Acad. Sci. U.S.A.*, 106 (2009) 18–23.

154. K. El-Boubbou, C. Gruden, and X. Huang, Magnetic glyco-nanoparticles: A unique tool for rapid pathogen detection, decontamination, and strain differentiation, *J. Am. Chem. Soc.*, 129 (2007) 13392–13393.
155. X. Liu, Q. Hu, Z. Fang, X. Zhang, and B. Zhang, Magnetic chitosan nanocomposites: A useful recyclable tool for heavy metal ion removal, *Langmuir*, 25 (2009) 3–8.
156. D. M. Hatch, A. A. Weiss, R. R. Kale, and S. S. Iyer, Biotinylated bi- and tetra-antennary glycoconjugates for *Escherichia coli* detection, *ChemBioChem*, 9 (2008) 2433–2442.
157. H. Gu, P.-L. Ho, K. W. T. Tsang, L. Wang, and B. Xu, Using biofunctional magnetic nanoparticles to capture vancomycin-resistant enterococci and other Gram-positive bacteria at ultralow concentration, *J. Am. Chem. Soc.*, 125 (2003) 15702–15703.
158. H. Gu, P.-L. Ho, K. W. T. Tsang, C.-W. Yu, and B. Xu, Using biofunctional magnetic nanoparticles to capture Gram-negative bacteria at an ultra-low concentration, *Chem. Commun.*, (2003) 1966–1967.
159. A. J. Kell, G. Stewart, S. Ryan, R. Peytavi, M. Boissinot, A. Huletsky, M. G. Bergeron, and B. Simard, Vancomycin-modified nanoparticles for efficient targeting and preconcentration of Gram-positive and Gram-negative bacteria, *ACS Nano*, 2 (2008) 1777–1788.
160. D. Horak, M. Babic, P. Jendelova, V. Herynek, M. Trchova, Z. Pientka, E. Pollert, M. Hajek, and E. Sykova, D-Mannose-modified iron oxide nanoparticles for stem cell labeling, *Bioconjug. Chem.*, 18 (2007) 635–644.
161. M. Kasture, S. Singh, P. Patel, P. A. Joy, A. A. Prabhune, C. V. Ramana, and B. L. V. Prasad, Multiutility sophorolipids as nanoparticle capping agents: Synthesis of stable and water dispersible Co nanoparticles, *Langmuir*, 23 (2007) 11409–11412.
162. M. A. Reed, J. N. Randall, R. J. Aggarwal, R. J. Matyi, T. M. Moore, and A. E. Wetsel, Observation of discrete electronic states in a zero-dimensional semiconductor nanostructure, *Phys. Rev. Lett.*, 60 (1988) 535–537.
163. R. Rossetti, S. Nakahara, and L. E. Brus, Quantum size effects in the redox potentials, resonance Raman-spectra, and electronic-spectra of CdS crystallites in aqueous-solution, *J. Chem. Phys.*, 79 (1983) 1086–1088.
164. A. P. Alivisatos, Birth of a nanoscience building block, *ACS Nano*, 2 (2008) 1514–1516.
165. H. Gleiter, Nanostructured materials, *Adv. Mater.*, 4 (1992) 474–481.
166. A. P. Alivisatos, Semiconductor clusters, nanocrystals, and quantum dots, *Science*, 271 (1996) 933–937.

167. M. Bruchez, M. Moronne, P. Gin, S. Weiss, and A. P. Alivisatos, Semiconductor nanocrystals as fluorescent biological labels, *Science*, 281 (1998) 2013–2016.
168. W. C. W. Chan and S. M. Nie, Quantum dot bioconjugates for ultrasensitive nonisotopic detection, *Science*, 281 (1998) 2016–2018.
169. J. K. Jaiswal, H. Mattoussi, J. M. Mauro, and M. Simon, Long-term multiple color imaging of live cells using quantum dot bioconjugates, *Nat. Biotechnol.*, 21 (2003) 47–51.
170. X. Michalet, F. F. Pinaud, L. A. Bentolila, J. M. Tsay, S. Doose, J. J. Li, G. Sundaresan, A. M. Wu, S. S. Gambhir, and S. Weiss, Quantum dots for live cells, *in vivo* imaging, and diagnostics, *Science*, 307 (2005) 538–544.
171. X. Wu, H. Liu, J. Liu, K. N. Haley, J. A. Treadway, P. Larson, N. Ge, F. Peale, and M. P. Bruchez, Immunofluorescent labeling of cancer marker Her2 and other cellular targets with semiconductor quantum dots, *Nat. Biotechnol.*, 21 (2003) 41–46.
172. V. Biju, T. Itoh, A. Anas, A. Sujith, and M. Ishikawa, Semiconductor quantum dots and metal nanoparticles: Syntheses, optical properties, and biological applications, *Anal. Bioanal. Chem.*, 391 (2008) 2469–2495.
173. X. Gao and S. Nie, Luminescent quantum dots for biological labeling, in C. M. Niemeyer, and C. A. Mirkin, (Eds.), *Nanobiotechnology: Concepts, Applications and Perspectives*, Wiley–VCH Verlag GmbH & Co. KGaA, Weinheim, 2004, Part IV, Chapter 22 pp. 343–352.
174. M. Howarth, W. Liu, S. Puthenveetil, Y. Zheng, L. F. Marshall, M. M. Schmidt, K. D. Wittrup, M. G. Bawendi, and A. Y. Ting, Monovalent, reduced-size quantum dots for imaging receptors on living cells, *Nat. Methods*, 5 (2008) 397–399.
175. I. L. Medintz, H. T. Uyeda, E. R. Goldman, and H. Mattoussi, Quantum dot bioconjugates for imaging, labelling and sensing, *Nat. Mater.*, 4 (2005) 435–446.
176. A. M. Smith, H. W. Duan, A. M. Mohs, and S. M. Nie, Bioconjugated quantum dots for *in vivo* molecular and cellular imaging, *Adv. Drug Deliv. Rev.*, 60 (2008) 1226–1240.
177. M. Stroh, J. P. Zimmer, D. G. Duda, T. S. Levchenko, K. S. Cohen, E. B. Brown, D. T. Scadden, V. P. Torchilin, M. G. Bawendi, D. Fukumura, and R. K. Jain, Quantum dots spectrally distinguish multiple species within the tumor milieu *in vivo*, *Nat. Med.*, 11 (2005) 678–682.
178. A. J. Sutherland, Quantum dots as luminescent probes in biological systems, *Curr. Opin. Solid State Mater. Sci.*, 6 (2002) 365–370.

179. J. van Embden, J. Jasieniak, D. E. Gómez, P. Mulvaney, and M. Giersig, Review of the synthetic chemistry involved in the production of core/shell semiconductor nanocrystals, *Aust. J. Chem.*, 60 (2007) 457–471.
180. M. Zhou and I. Ghosh, Quantum dots and peptides: A bright future together, *Biopolymers*, 88 (2007) 325–339.
181. C. B. Murray, D. J. Norris, and M. G. Bawendi, Synthesis and characterization of nearly monodisperse Cde (E = S, Se, Te) semiconductor nanocrystallites, *J. Am. Chem. Soc.*, 115 (1993) 8706–8715.
182. J. E. Bowen Katari, V. L. Colvin, and A. P. Alivisatos, X-ray photoelectron-spectroscopy of CdSe nanocrystals with applications to studies of the nanocrystal surface, *J. Phys. Chem.*, 98 (1994) 4109–4117.
183. W. W. Yu, Y. A. Wang, and X. Peng, Formation and stability of size-, shape-, and structure-controlled CdTe nanocrystals: Ligand effects on monomers and nanocrystals, *Chem. Mater.*, 15 (2003) 4300–4308.
184. W. W. Yu, E. Chang, R. Drezek, and V. L. Colvin, Water-soluble quantum dots for biomedical applications, *Biochem. Biophys. Res. Commun.*, 348 (2006) 781–786.
185. Y. Chen, T. Ji, and Z. Rosenzweig, Synthesis of glyconanospheres containing luminescent CdSe-ZnS quantum dots, *Nano Lett.*, 3 (2003) 581–584.
186. D. K. Chatterjee and Y. Zhang, Multi-functional nanoparticles for cancer therapy, *Sci. Technol. Adv. Mater.*, 8 (2007) 131–133.
187. W. B. Tan, N. Huang, and Y. Zhang, Ultrafine biocompatible chitosan nanoparticles encapsulating multi-coloured quantum dots for bioapplications, *J. Colloid Interface Sci.*, 310 (2007) 464–470.
188. W. B. Tan, S. Jiang, and Y. Zhang, Quantum-dot based nanoparticles for targeted silencing of HER2/neu gene via RNA interference, *Biomaterials*, 28 (2007) 1565–1571.
189. W. B. Tan and Y. Zhang, Multifunctional quantum-dot-based magnetic chitosan nanobeads, *Adv. Mater.*, 17 (2005) 2375–2380.
190. M. Xie, H.-H. Liu, P. Chen, Z.-L. Zhang, X.-H. Wang, Z.-X. Xie, Y.-M. Du, B.-Q. Pan, and D.-W. Pang, CdSe/ZnS-labeled carboxymethyl chitosan as a bioprobe for live cell imaging, *Chem. Commun.*, (2005) 5518–5520.
191. Q. L. Nie, W. B. Tan, and Y. Zhang, Synthesis and characterization of monodisperse chitosan nanoparticles with embedded quantum dots, *Nanotechnology*, 17 (2006) 140–144.
192. F. Osaki, T. Kanamori, S. Sando, T. Sera, and Y. Aoyama, A quantum dot conjugated sugar ball and its cellular uptake on the size effects of endocytosis in the subviral region, *J. Am. Chem. Soc.*, 126 (2004) 6520–6521.

193. C. Park, K. H. Lim, D. Kwon, and T. H. Yoon, Biocompatible quantum dot nanocolloids stabilized by gum arabic, *Bull. Korean Chem. Soc.*, 29 (2008) 1277–1279.
194. X.-L. Sun, W. Cui, C. Haller, and E. L. Chaikof, Site-specific multivalent carbohydrate labeling of quantum dots and magnetic beads, *ChemBioChem*, 5 (2004) 1593–1596.
195. A. Cambi, D. S. Lidke, D. J. Arndt-Jovin, C. G. Figdor, and T. M. Jovin, Ligand-conjugated quantum dots monitor antigen uptake and processing by dendritic cells, *Nano Lett.*, 7 (2007) 970–977.
196. X. Peng, M. C. Schlamp, A. V. Kadavanich, and A. P. Alivisatos, Epitaxial growth of highly luminescent CdSe/CdS core/shell nanocrystals with photostability and electronic accessibility, *J. Am. Chem. Soc.*, 119 (1997) 7019–7029.
197. J. D. M. de la Fuente and S. Penadés, Glyco-quantum dots: A new luminescent system with multivalent carbohydrate display, *Tetrahedron Asymmetry*, 16 (2005) 387–391.
198. B. Mukhopadhyay, M. B. Martins, R. Karamanska, D. A. Russell, and R. A. Field, Bacterial detection using carbohydrate-functionalised CdS quantum dots: A model study exploiting *E. coli* recognition of mannosides, *Tetrahedron Lett.*, 50 (2009) 886–889.
199. S. A. Svarovsky and J. J. Barchi, *De novo* synthesis of biofunctional carbohydrate-encapsulated quantum dots, in A. V. Demchenko, (Ed.), *Frontiers in Modern Carbohydrate Chemistry*, Oxford University Press, New York, NY, 2007, pp. 375–394.
200. A. Robinson, J.-M. Fang, P.-T. Chou, K.-W. Liao, R.-M. Chu, and S.-J. Lee, Probing lectin and sperm with carbohydrate-modified quantum dots, *ChemBioChem*, 6 (2005) 1899–1905.
201. C.-T. Chen, Y. S. Munot, S. B. Salunke, Y.-C. Wang, R.-K. Lin, C.-C. Lin, C.-C. Chen, and Y.-H. Liu, A triantennary dendritic galactoside-capped nanohybrid with a ZnS/CdSe nanoparticle core as a hydrophilic, fluorescent, multivalent probe for metastatic lung cancer cells, *Adv. Funct. Mater.*, 18 (2008) 527–540.
202. P. Babu, S. Sinha, and A. Surolia, Sugar-quantum dot conjugates for a selective and sensitive detection of lectins, *Bioconjug. Chem.*, 18 (2007) 146–151.
203. K. Niikura, T. Nishio, H. Akita, Y. Matsuo, R. Kamitani, K. Kogure, H. Harashima, and K. Ijiro, Accumulation of *O*-GlcNAc-displaying CdTe quantum dots in cells in the presence of ATP, *ChemBioChem*, 8 (2007) 379–384.
204. K. Niikura, S. Sekiguchi, T. Nishio, T. Masuda, H. Akita, Y. Matsuo, K. Kogure, H. Harashima, and K. Ijiro, Oligosaccharide-mediated nuclear transport of nanoparticles, *ChemBioChem*, 9 (2008) 2623–2627.

205. M. G. Sandros, M. Behrendt, D. Maysinger, and M. Tabrizian, InGaP-ZnS-Enriched chitosan nanoparticles: A versatile fluorescent probe for deep-tissue imaging, *Adv. Funct. Mater.*, 17 (2007) 3724–3730.
206. Y. Higuchi, M. Oka, S. Kawakami, and M. Hashida, Mannosylated semiconductor quantum dots for the labeling of macrophages, *J. Controll. Release*, 125 (2008) 131–136.
207. R. Kikkeri, B. Lepenies, A. Adibekian, P. Laurino, and P. H. Seeberger, *In vitro* imaging and *in vivo* liver targeting with carbohydrate capped quantum dots, *J. Am. Chem. Soc.*, 131 (2009) 2110–2112.
208. X. Jiang, M. Ahmed, Z. Deng, and R. Narain, Biotinylated glyco-functionalized quantum dots: Synthesis, characterization, and cytotoxicity studies, *Bioconjug. Chem.*, 20 (2009) 994–1001.
209. I. Eggens, B. Fenderson, T. Toyokuni, B. Dean, M. Stroud, and S.-I. Hakomori, Specific interaction between Le^x and Le^x determinants—a possible basis for cell recognition in preimplantation embryos and in embryonal carcinoma-cells, *J. Biol. Chem.*, 264 (1989) 9476–9484.
210. G. N. Misevic, J. Finne, and M. M. Burger, Involvement of carbohydrates as multiple low affinity interaction sites in the self-association of the aggregation factor from the marine sponge *Microciona prolifera*, *J. Biol. Chem.*, 262 (1987) 5870–5877.
211. S. Hakomori, Carbohydrate-to-carbohydrate interaction in basic cell biology: A brief overview, *Arch. Biochem. Biophys.*, 426 (2004) 173–181.
212. F. Pincet, T. Le Bouar, Y. Zhang, J. Esnault, J.-M. Mallet, E. Perez, and P. Sinaÿ, Ultraweak sugar-sugar interactions for transient cell adhesion, *Biophys. J.*, 80 (2001) 1354–1358.
213. J. M. Boggs, A. Menikh, and G. Rangaraj, *Trans* interactions between galactosylceramide and cerebroside sulfate across apposed bilayers, *Biophys. J.*, 78 (2000) 874–885.
214. P. V. Santacroce and A. Basu, Probing specificity in carbohydrate-carbohydrate interactions with micelles and Langmuir monolayers, *Angew. Chem. Int. Ed. Engl.*, 42 (2003) 95–98.
215. T. Nakai, T. Kanamori, S. Sando, and Y. Aoyama, Remarkably size-regulated cell invasion by artificial viruses. Saccharide-dependent self-aggregation of glycoviruses and its consequences in glycoviral gene delivery, *J. Am. Chem. Soc.*, 125 (2003) 8465–8475.
216. J. M. de la Fuente, P. Eaton, A. G. Barrientos, M. Menéndez, and S. Penadés, Thermodynamic evidence for Ca^{2+}-mediated self-aggregation of Lewis X gold

glyconanoparticles. A model for cell adhesion via carbohydrate-carbohydrate interaction, *J. Am. Chem. Soc.*, 127 (2005) 6192–6197.

217. M. J. Hernáiz, J. M. de la Fuente, A. G. Barrientos, and S. Penadés, A model system mimicking glycosphingolipid clusters to quantify carbohydrate self-interactions by surface plasmon resonance, *Angew. Chem. Int. Ed. Engl.*, 41 (2002) 1554–1557.
218. C. Tromas, J. Rojo, J. M. de la Fuente, A. G. Barrientos, R. García, and S. Penadés, Adhesion forces between Lewis^X determinant antigens as measured by atomic force microscopy, *Angew. Chem. Int. Ed. Engl.*, 40 (2001) 3052–3055.
219. A. Carvalho de Souza, K. M. Halkes, J. D. Meeldijk, A. J. Verkleij, J. F. G. Vliegenthart, and J. P. Kamerling, Gold glyconanoparticles as probes to explore the carbohydrate-mediated self-recognition of marine sponge cells, *ChemBioChem*, 6 (2005) 828–831.
220. A. Carvalho de Souza and J. P. Kamerling, Analysis of carbohydrate-carbohydrate interactions using gold glyconanoparticles and oligosaccharide self-assembling monolayers, *Meth. Enzymol.*, 417 (2006) 221–243.
221. A. Carvalho de Souza, D. N. Ganchev, M. M. E. Snel, J. van der Eerden, J. F. G. Vliegenthart, and J. P. Kamerling, Adhesion forces in the self-recognition of oligosaccharide epitopes of the proteoglycan aggregation factor of the marine sponge *Microciona prolifera, Glycoconj. J.*, 26 (2009) 457–465.
222. J. I. Santos, A. Carvalho de Souza, F. J. Cañada, S. Martín-Santamaría, J. P. Kamerling, and J. Jiménez-Barbero, Assessing carbohydrate-carbohydrate interactions by NMR spectroscopy: The trisaccharide epitope from the marine sponge *Microciona prolifera, ChemBioChem*, 10 (2009) 511–519.
223. R. J. Pieters, Maximising multivalency effects in protein-carbohydrate interactions, *Org. Biomol. Chem.*, 7 (2009) 2013–2025.
224. W. Zhao, M. A. Brook, and Y. Li, Design of gold nanoparticle-based colorimetric biosensing assays, *ChemBioChem*, 9 (2008) 2363–2371.
225. J. A. Creighton and D. G. Eadon, Ultraviolet visible absorption-spectra of the colloidal metallic elements, *J. Chem. Soc., Faraday Trans.*, 87 (1991) 3881–3891.
226. S. Takae, Y. Akiyama, Y. Yamasaki, Y. Nagasaki, and K. Kataoka, Colloidal Au replacement assay for highly sensitive quantification of low molecular weight analytes by surface plasmon resonance, *Bioconjug. Chem.*, 18 (2007) 1241–1245.
227. C. L. Schofield, A. H. Haines, R. A. Field, and D. A. Russell, Silver and gold glyconanoparticles for colorimetric bioassays, *Langmuir*, 22 (2006) 6707–6711.
228. C. L. Schofield, R. A. Field, and D. A. Russell, Glyconanoparticles for the colorimetric detection of cholera toxin, *Anal. Chem.*, 79 (2007) 1356–1361.

229. C.-S. Tsai, T.-B. Yu, and C.-T. Chen, Gold nanoparticle-based competitive colorimetric assay for detection of protein-protein interactions, *Chem. Commun.*, (2005) 4273–4275.
230. K. M. Halkes, A. Carvalho de Souza, C. E. P. Maljaars, G. J. Gerwig, and J. P. Kamerling, A facile method for the preparation of gold glyconanoparticles from free oligosaccharides and their applicability in carbohydrate-protein interaction studies, *Eur. J. Org. Chem.*, (2005) 3650–3659.
231. Y.-K. Lyu, K.-R. Lim, B.-Y. Lee, K. S. Kim, and W.-Y. Lee, Microgravimetric lectin biosensor based on signal amplification using carbohydrate-stabilized gold nanoparticles, *Chem. Commun.*, (2008) 4771–4773.
232. Y. van Kooyk and T. B. H. Geijtenbeek, DC-SIGN: Escape mechanism for pathogens, *Nat. Rev. Immunol.*, 3 (2003) 697–709.
233. J. Rieger, H. Freichels, A. Imberty, J.-L. Putaux, T. Delair, C. Jérôme, and R. Auzély-Velty, Polyester nanoparticles presenting mannose residues: Toward the development of new vaccine delivery systems combining biodegradability and targeting properties, *Biomacromolecules*, 10 (2009) 651–657.
234. J. M. Bergen, H. A. von Recum, T. T. Goodman, A. P. Massey, and S. H. Pun, Gold nanoparticles as a versatile platform for optimizing physicochemical parameters for targeted drug delivery, *Macromol. Biosci.*, 6 (2006) 506–516.
235. A. Ragusa, I. García, and S. Penadés, Nanoparticles as nonviral gene delivery vectors, *IEEE Trans. Nanobiosci.*, 6 (2007) 319–330.
236. P. Eaton, A. Ragusa, C. Clavel, C. T. Rojas, P. Graham, R. V. Durán, and S. Penadés, Glyconanoparticle-DNA interactions: An atomic force microscopy study, *IEEE Trans. Nanobiosci.*, 6 (2007) 309–318.
237. T. Masuda, H. Akita, T. Nishio, K. Niikura, K. Kogure, K. Ijiro, and H. Harashima, Development of lipid particles targeted via sugar-lipid conjugates as novel nuclear gene delivery system, *Biomaterials*, 29 (2008) 709–723.
238. H. Akita and H. Harashima, Advances in non-viral gene delivery: Using multifunctional envelope-type nano-device, *Expert Opin. Drug Deliv.*, 5 (2008) 847–859.
239. N. Sharon, Carbohydrates as future anti-adhesion drugs for infectious diseases, *Biochim. Biophys. Acta Gen. Subj.*, 1760 (2006) 527–537.
240. P. G. Luo and F. J. Stutzenberger, Nanotechnology in the detection and control of microorganisms, *Adv. Appl. Microbiol.*, 63 (2008) 145–181.
241. J. Rojo, V. Díaz, J. M. de la Fuente, I. Segura, A. G. Barrientos, H. H. Riese, A. Bernade, and S. Penadés, Gold glyconanoparticles as new tools in antiadhesive therapy, *ChemBioChem*, 5 (2004) 291–297.

242. K. D. McReynolds and J. Gervay-Hague, Chemotherapeutic interventions targeting HIV interactions with host-associated carbohydrates, *Chem. Rev.*, 107 (2007) 1533–1552.
243. T. B. H. Geijtenbeek, D. S. Kwon, R. Torensma, S. J. van Vliet, G. C. F. van Duijnhoven, J. Middel, I. Cornelissen, H. Nottet, V. N. KewalRamani, D. R. Littman, C. G. Figdor, and Y. van Kooyk, DC-SIGN, a dendritic cell-specific HIV-1-binding protein that enhances trans-infection of T cells, *Cell*, 100 (2000) 587–597.
244. O. Martínez-Ávila, L. M. Bedoya, M. Marradi, C. Clavel, J. Alcamí, and S. Penadés, Multivalent manno-glyconanoparticles inhibit DC-SIGN-mediated HIV-1 trans-infection of human T cells, *ChemBioChem*, 10 (2009) 1806–1809.
245. P. Sharma, S. Brown, G. Walter, S. Santra, and B. Moudgil, Nanoparticles for bioimaging, *Adv. Colloid Interface Sci.*, 123 (2006) 471–485.
246. Y.-W. Jun, J.-H. Lee,, and J. Cheon, Chemical design of nanoparticle probes for high-performance magnetic resonance imaging, *Angew. Chem. Int. Ed. Engl.*, 47 (2008) 5122–5135.
247. W. B. Tan and Y. Zhang, Multi-functional chitosan nanoparticles encapsulating quantum dots and Gd-DTPA as imaging probes for bio-applications, *J. Nanosci. Nanotechnol.*, 7 (2007) 2389–2393.
248. F.-Y. Cheng, S. P.-H. Wang, C.-H. Su, T.-L. Tsai, P.-C. Wu, D.-B. Shieh, J.-H. Chen, P. C.-H. Hsieh, and C.-S. Yeh, Stabilizer-free poly(lactide-*co*-glycolide) nanoparticles for multimodal biomedical probes, *Biomaterials*, 29 (2008) 2104–2112.
249. Y.-J. Chen, S.-H. Chen, Y.-Y. Chien, Y.-W. Chang, H.-K. Liao, C.-Y. Chang, M.-D. Jan, K.-T. Wang, and C.-C. Lin, Carbohydrate-encapsulated gold nanoparticles for rapid target-protein identification and binding-epitope mapping, *ChemBioChem*, 6 (2005) 1169–1173.
250. S. Nakamura-Tsuruta, Y. Kishimoto, T. Nishimura, and Y. Suda, One-step purification of lectins from banana pulp using sugar-immobilized gold nanoparticles, *J. Biochem. (Tokyo)*, 143 (2008) 833–839.
251. N. Nagahori and S.-I. Nishimura, Direct and efficient monitoring of glycosyltransferase reactions on gold colloidal nanoparticles by using mass spectrometry, *Chem. Eur. J.*, 12 (2006) 6478–6485.
252. C.-L. Su and W.-L. Tseng, Gold nanoparticles as assisted matrix for determining neutral small carbohydrates through laser desorption/ionization time-of-flight mass spectrometry, *Anal. Chem.*, 79 (2007) 1626–1633.
253. J. A. McLean, K. A. Stumpo, and D. H. Russell, Size-selected (2-10 nm) gold nanoparticles for matrix assisted laser desorption ionization of peptides, *J. Am. Chem. Soc.*, 127 (2005) 5304–5305.

254. K. B. Hartman, L. J. Wilson, and M. G. Rosenblum, Detecting and treating cancer with nanotechnology, *Mol. Diagn. Ther.*, 12 (2008) 1–14.
255. D. Peer, J. M. Karp, S. Hong, O. C. FaroKhzad, R. Margalit, and R. Langer, Nanocarriers as an emerging platform for cancer therapy, *Nat. Nanotechnol.*, 2 (2007) 751–760.
256. N. Sanvicens and M. P. Marco, Multifunctional nanoparticles—properties and prospects for their use in human medicine, *Trends Biotechnol.*, 26 (2008) 425–433.
257. E. Boisselier and D. Astruc, Gold nanoparticles in nanomedicine: Preparations, imaging, diagnostics, therapies and toxicity, *Chem. Soc. Rev.*, 38 (2009) 1759–1782.
258. P. Ghosh, G. Han, M. De, C. K. Kim, and V. M. Rotello, Gold nanoparticles in delivery applications, *Adv. Drug Deliv. Rev.*, 60 (2008) 1307–1315.
259. S. C. McBain, H. H. P. Yiu, and J. Dobson, Magnetic nanoparticles for gene and drug delivery, *Int. J. Nanomedicine*, 3 (2008) 169–180.
260. H. Yang, H. Li, and X. Jiang, Detection of foodborne pathogens using bioconjugated nanomaterials, *Microfluid. Nanofluidics*, 5 (2008) 571–583.
261. S. Penadés, J. M. de la Fuente, A. G. Barrientos, C. Clavel, O. Martínez-Ávila, and D. Alcántara, Multifunctional glyconanoparticles: Applications in biology and biomedicine, in M. Giersig and G. B. Khomutov (Eds.), *Nanomaterials for Applications in Medicine and Biology*, Springer Science + Business Media, Dordrecht, Netherlands, 2008, pp. 93–101.
262. K. Larsen, M. B. Thygesen, F. Guillaumie, W. G. T. Willats, and K. J. Jensen, Solid-phase chemical tools for glycobiology, *Carbohydr. Res.*, 341 (2006) 1209–1234.
263. I. García, M. Marradi, and S. Penadés, Glyconanoparticles: multifunctional nanomaterials for biomedical applications, *Nanomedicine*, 5 (2010) 777–792.

1-AMINO-1-DEOXY-D-FRUCTOSE ("FRUCTOSAMINE") AND ITS DERIVATIVES

By Valeri V. Mossine and Thomas P. Mawhinney

Department of Biochemistry, University of Missouri, Columbia, Missouri, 65211, USA

Abbreviations

AGE, advanced glycation end product; CML, N^{ε}-carboxymethyl-L-lysine; DMF, *N,N*-dimethylformamide; DOTA, 1,4,7,10-tetraazacyclododecane-1,4,7,10-tetraacetate; DTPA, diethylenetriamine pentaacetate; FAD, flavin adenine dinucleotide; FN3K, fructosamine-3-kinase; GLUT5, glucose transporter-5; GTP, guanidine triphosphate;

ISBN: 978-0-12-380854-7
DOI: 10.1016/S0065-2318(10)64006-1

HbA_{1c}, hemoglobin A_{1c}; HMF, 5-(hydroxymethyl)furfural; HPLC, high-performance liquid chromatography; IR, infrared; LC, liquid chromatography; LDL, low-density lipoprotein; MDP-1, magnesium-dependent phosphatase-1; MOP, mannopine; MS, mass-spectrometry; NBT, Nitroblue Tetrazolium dye; NF-κB, nuclear factor κ-light-chain-enhancer of activated B cells; NMR, nuclear magnetic resonance; PE, phosphatidylethanolamine; RNase, ribonuclease; ROS, reactive oxygen species; SDS-PAGE, sodium dodecyl sulfate polyacrylamide gel electrophoresis; SOP, santhopine; T-DNA, transfer DNA; TGF-β1, Transforming Growth Factor-β1; THF, tetrahydrofuran; THT1, trypanosome hexose transporter-1; TLC, thin-layer chromatography.

I. Introduction

A number of reviews have been dedicated, partly or wholly, to the chemistry of fructosamines, or Amadori rearrangement products, including a seminal review by John Hodge early in this series.[1–4] Since the last comprehensive review[5] was released in 1994, the total number of publications covering fructosamines has more than doubled, reaching about 4800 by early 2010, with the emphasis shifting toward biochemical studies. The main purpose of this account is to reflect the scope of fructosamine significance and applications in chemical, food, and health sciences, by focusing on the major findings in the field accomplished largely within the past ten years.

1. Historical Perspective

The timeline of studies on fructosamine can be divided into three periods. The compound was initially described in 1886 by Emil Fischer, who treated D-glucose phenylosazone with zinc dust and isolated "isoglucosamine" in form of its crystalline acetate salt.[6] Fischer correctly suggested that isoglucosamine must be structurally related to fructose, in a similar way as glucosamine is related to glucose. In 1925, Mario Amadori demonstrated that interaction between glucose and an aromatic amine at elevated temperatures typically produced two distinct products, an acid-labile glucosylamine and a relatively stable isomer.[7] Amadori believed that the structure of the latter was related to a Schiff base. However, Richard Kuhn reinvestigated this reaction in 1936 and concluded that the *N*-aryl-glucosylamine must have undergone a rearrangement into the *N*-aryl-fructosamine.[8] Kuhn named the rearrangement reaction after Amadori and offered a mechanism, which enjoys general acceptance to date. The significance of the Amadori rearrangement as the initial phase of the Maillard reaction in food was realized at the beginning of 1950s and was summarized by John Hodge in his classic review.[9] It appeared that aroma, taste, and color formation in baked, fried, or dried foods could be traced, to a large extent, to fructosamines which, in turn, originated through condensation reactions between glucose and amino groups in free amino acids or polypeptides and followed by the Amadori rearrangement during thermal food processing, dehydration, or storage. During this period, much of research on fructosamine derivatives was inspired by their importance for food chemistry and focused on analysis, reactivity, and nutritional significance of Amadori rearrangement products.[10] By the end of the 1970s, a new era in fructosamine research began with the establishment[11] of the structure of a long-term blood glucose marker in diabetes, hemoglobin A_{1c}. In HbA_{1c}, the N-terminal valine is decorated with 1-deoxy-D-fructose, as a result of non-enzymatic modification with D-glucose through the Amadori rearrangement. This discovery caused an explosion of research on "non-enzymatic glycosylation" (glycation) of proteins and other biomolecules *in vivo*, as soon as it became clear that such modification is a common event in many diabetes-related pathologies and in the process of aging.[12,13] As of today, D-fructosamine, along with its derivatives, remains perhaps the only non-enzymatically catalyzed carbohydrate modification of a major significance to both food and health sciences.

2. Nomenclature

In the course of more than one hundred years, a number of systematic and trivial names were used for D-fructosamine and its derivatives. We have compiled these in Table I, in order to aid those readers who are concerned with completeness of the literature searches, since databases may omit a significant portion of relevant references due to the name variability used for D-fructosamine core structures, such as **1** or **2**.

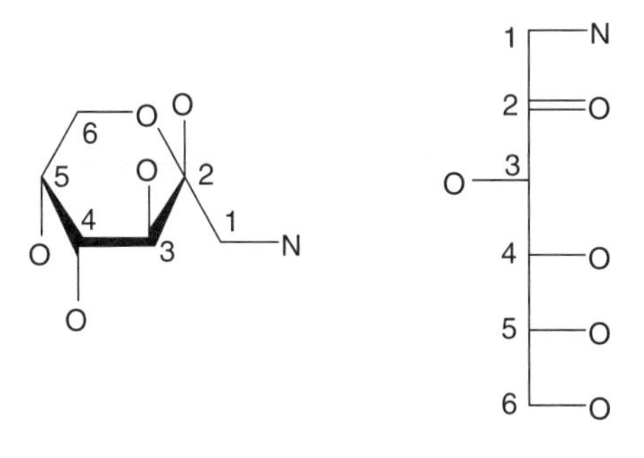

TABLE I
Names for Fructosamine (1-Amino-1-deoxyfructose) Structures Most Often Encountered in the Literature

Name	Comment	Representative examples	Reference
[D-]Fructosamine	General name, most common in clinical chemistry	Fructosamine assay	14
		Fructosamine oxidase	15
Amadori rearrangement product, Amadori compound	General name, incorporates other 1-amino-1-deoxyketose derivatives		5
[D-]Isoglucosamine	The first trivial name, outdated		6
1-[Amino]-1-deoxy-D-fructose, *N*-(1-Deoxy-D-fructos-1-yl)-[amine], *N*-(1-Deoxy-D-*arabino*-hexulos-1-yl)-[amine][a]	Recommended systematic names	1-Deoxy-1-propylamino-D-fructose, *N*-(1-deoxy-β-D-fructopyranos-1-yl)-L-alanine	16
[D-]Fructose–[amine]	Most common in food chemistry	Fructose–amino acids	17
Fructosyl[amine]	Ambiguous, incorrect name	Fructosylamine oxidase	18

[a] Note that in this article, this fully systematic name is abbreviated, most frequently, to the "fructose–[amine]" term.

Some of trivial names embodying the fructosamine structure that refer to the carbohydrate modifications and are mentioned in this chapter, include:

- L-Rhamnulosamine = 6-deoxy-L-fructosamine
- Lactulosamine = β-D-galactopyranosyl-(1→4)-D-fructosamine
- Cellobiulosamine = α-D-glucopyranosyl-(1→4)-D-fructosamine
- Maltulosamine = β-D-glucopyranosyl-(1→4)-D-fructosamine
- Maltotriulosamine = β-D-glucopyranosyl-(1→4)-β-D-glucopyranosyl-(1→4)-D-fructosamine

II. METHODS OF FORMATION

The commonly occurring formation of fructosamine derivatives in foods and *in vivo*, as well as their value as synthetic intermediates, scaffolds, and bioactive end products, has prompted numerous mechanistic studies of the Amadori rearrangement, while other methods of their preparation have also been developed. Although fructosamine formation is generally seen as a typical non-enzymatic carbohydrate modification *in vivo*, a number of reports have also identified enzymes involved in synthetic pathways of physiologically significant Amadori compounds.

1. Non-Enzymatic

a. Reduction of D-Glucose Phenylosazone.—Historically the first developed method for preparation of fructosamine, Fischer's approach, retains its practical value to date. According to the original protocol,[6] as shown in Scheme 1, crystalline **3** is reduced by zinc dust in acetic acid, affording the readily isolable acetate salt of 1-deoxy-β-D-fructopyranos-1-ylamine (**4**).

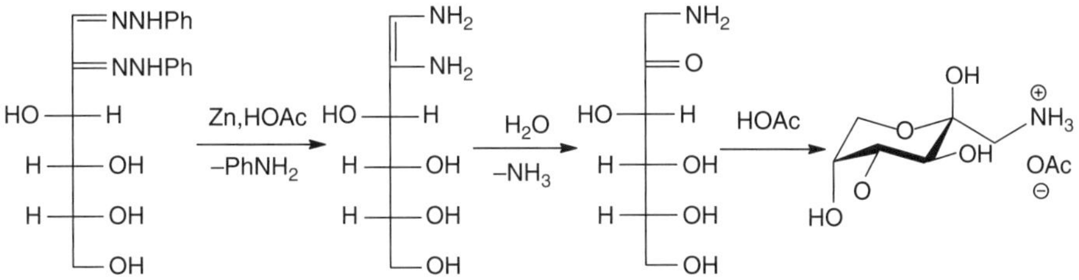

SCHEME 1

The original synthetic protocol was further optimized in terms of yield and workup procedure by Maurer and Schiedt, who applied Pd-catalyzed hydrogenation to break the hydrazine N–N bonds.[19] The acetate salt of D-fructosamine can be further used to prepare numerous *N*-substituted D-fructosamines. For example, **4** could be acylated with fatty acid anhydrides[20] or vinylated with 1-ethoxy-2-nitroethene[21] or methyl 3-methoxy-2-methoxycarbonylacrylate,[22] in order to obtain respective derivatives **5, 6**, and **7**. The 2,2-dimethoxycarbonylvinyl group in **7** proved to be a useful amino protective group, and allowed Gómez-Sánchez and colleagues[23] an access to alkyl glycosides of D-fructosamine, such as methyl 1-amino-1-deoxy-α-D-fructofuranoside (**8**). A convenient route to D-fructose-substituted hydroxyurea employed the reaction of unprotected **4** with 1-(4-nitrophenol)-*N*-(*O*-benzylhydroxy)carbamate, followed by cleavage of the benzyl group by catalytic hydrogenolysis.[24] Levi *et al.*[25] started with the oxalate salt of D-fructosamine when they prepared fluorescent conjugates of fructosamine and the fluorophores 7-nitro-1,2,3-benzadiazole or Cy5.5 for imaging breast cancer cells. Both conjugations were performed through one-step arylation or acylation of the amino group, respectively.

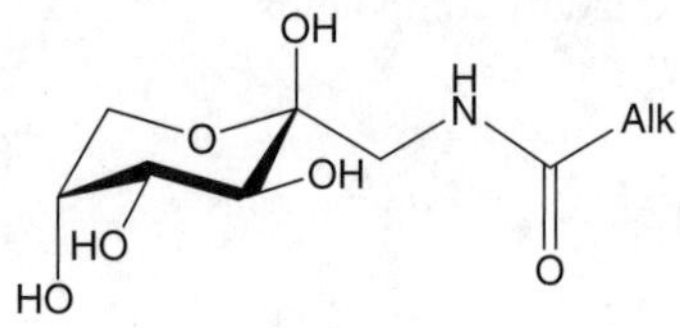

Another important value of the osazone-based approach is its utility in preparation of other ketosamines.[26,27] Phenylosazones of many mono- and di-saccharides are readily prepared in high yields as pure crystalline materials from both aldoses and ketoses, and such ketosamines as 1-amino-1-deoxy-L-sorbose, for example, can be synthesized most rationally from commercially available L-sorbose.[28]

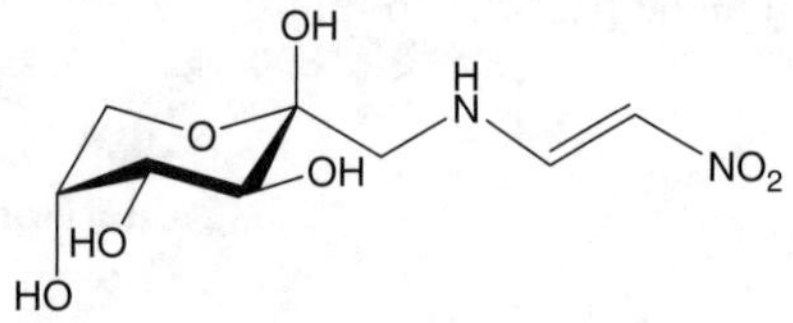

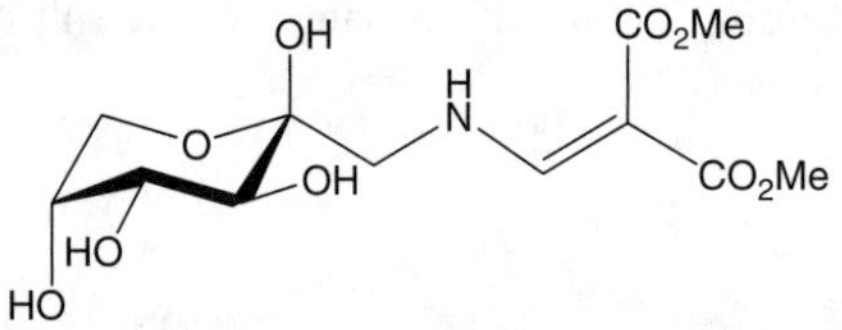

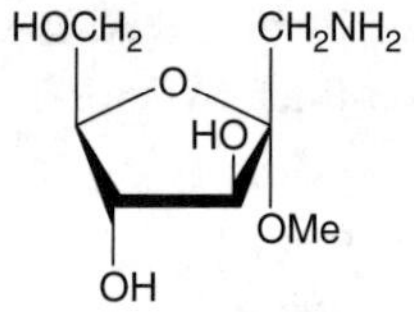

b. Amadori Rearrangement of Glucosylamines.—Overwhelming evidence accumulated during decades of research established the condensation reaction between free D-glucose and amino groups of amino acids or proteins as the main source of D-fructosamine derivatives in foods and *in vivo*. It is generally agreed that the initial condensation results in the formation of a labile *N*-substituted D-glucosylamine (**9**), which may undergo the Amadori rearrangement to form the respective *N*-substituted D-fructosamine **10** (Scheme 2). The condensation reaction between D-mannose and amines proceeds along the same scenario; however, its

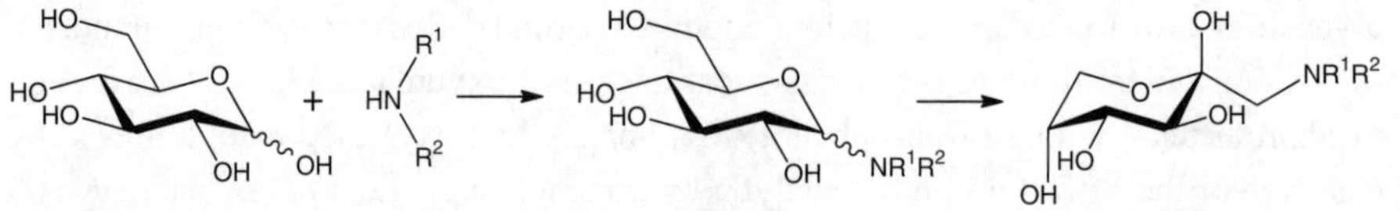

SCHEME 2

practical importance is relatively marginal. Because of the significance of the reaction to food and health sciences, a considerable number of studies have been undertaken in order to understand its mechanism in detail, and several different mechanisms have been offered, which are reviewed in this section. A situation where there is a disagreement on the exact mechanism of the glycosylamine–ketosamine transformation indicates that *there is more than one single Amadori rearrangement pathway* and that such pathways must be sensitive to the nature of the participating carbohydrate, amine, catalyst, solvent, and temperature.

(i) Glucosylamine Formation. The condensation reaction between the carbohydrate carbonyl group and an amino group was established early in the history of carbohydrate chemistry.[29] An acyclic Schiff base, cyclic glycopyranosylamine, and other structural forms were initially proposed for the isolated crystalline hexose–amine adducts. Kuhn and other investigators convincingly demonstrated on the basis of IR and other spectral analyses that *N*-aryl and *N*-alkyl D-glucosylamines in solution and in the crystalline state exist exclusively in pyranose forms, with predominance of the β anomer.[30–32] No signals of acyclic glucosylamine tautomers were detectable in solution. Although formation of the Schiff base from the acyclic *aldehydo*-glucose and the amine was suggested as the initial step of D-glucosylamine formation, experimental data from later studies prompted the conclusion that such an intermediate was of minor significance, and favored pyranoside-derived carbocations **11** and **12** as the main targets for nucleophilic attack by the amine (Scheme 3).[33]

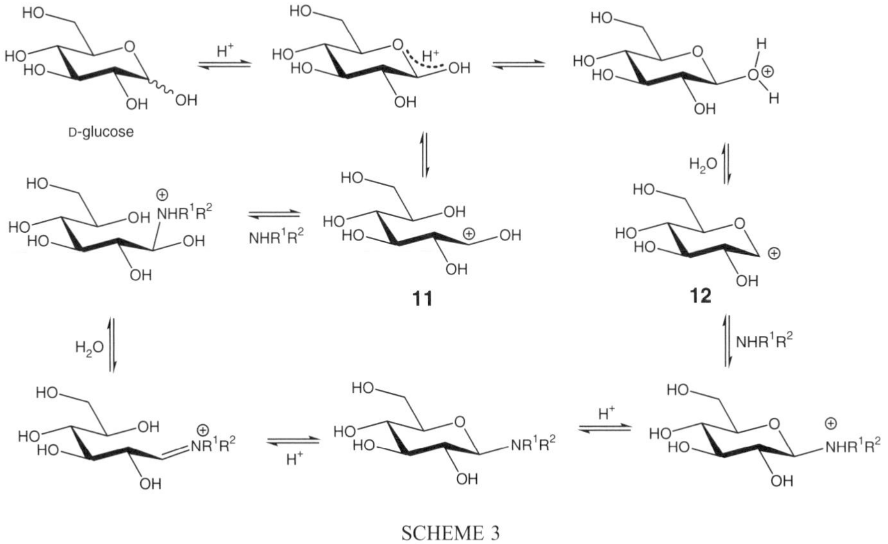

SCHEME 3

This mechanism implies that formation of glucosylamines is acid-catalyzed and that the stereo configuration at the anomeric carbon of glucose contributes to its reactivity in this scheme. Indeed, according to a standard procedure for preparation of crystalline *N*-aryl D-glucosylamines, a concentrated aqueous solution of D-glucose is mixed with an amine in ethanol, and the reaction may be promoted by heating or by a catalytic amount of strong acid, such as hydrochloric or sulfuric.[29]

Strong bases probably can also catalyze rapid formation of D-glucosylamines at ambient temperatures. A classic method for preparation of *N*-(D-glucosyl)-amino acids employs conversion of an amino acid into its sodium salt in water or methanol, followed by introduction of D-glucose into the alkaline solution at room temperature.[34] The reaction is complete as soon as the solid carbohydrate dissolves. Basic conditions, however, may favor a rather direct attack of an amine at the anomeric carbon, with water or another weak acid synchronously providing donation of a proton to O-5.[33] Such a mechanism may explain the inefficiency of *N*-aryl D-glucosylamine formation that is observed under alkaline conditions.

(ii) Diaminals and N,N-*Aminodiglucosides.* The aldose carbonyl group readily reacts with two aromatic amines to form diaminals. As shown here, diaminals may be important intermediates of the acid-catalyzed Amadori rearrangement. Several stable diaminals of D-glucose have been isolated and structurally characterized.[35,36]

Similarly, D-glucosylamines can further react with D-glucose to form the respective bis(D-glucosyl)amines. These compounds are not stable, and only bis(D-glucosyl)-amine itself has been isolated and well characterized.[37] However, the formation of bis-D-glucosylamine *N*-derivatives as intermediates in the preparation of *N,N*-di-(1-deoxy-D-fructos-1-yl)-amino acids[38] in alkaline media may also be suggested.

(iii) Acid Catalysis. The Amadori rearrangement of glycosylamines, as a particular case of sequential Schiff base–enamine and keto–enol tautomerizations, would be expected to follow general acid–base catalysis. The most widely accepted mechanism[39] was formulated in the early 1960s (Scheme 4) and includes initial proton-catalyzed opening of the glucopyranosyl ring, followed by proton abstraction and concomitant enolization at carbon-2 of glucose. The enaminol **15** eventually tautomerizes into a more stable ketosamine (**16**).

This mechanism was developed mainly on the basis of experiments with *N*-aryl D-glycosylamines and, served a purpose to explain: (a) the catalytic effect of carboxylic acids; (b) the absence of the Amadori rearrangement in neutral medium[40] or for *N*-aryl D-glucosylamines derived from weakly basic amines such as dinitroaniline or indole;[41] (c) enolization, rather than a 2,1-hydride shift, as the rate-determining step.[42]

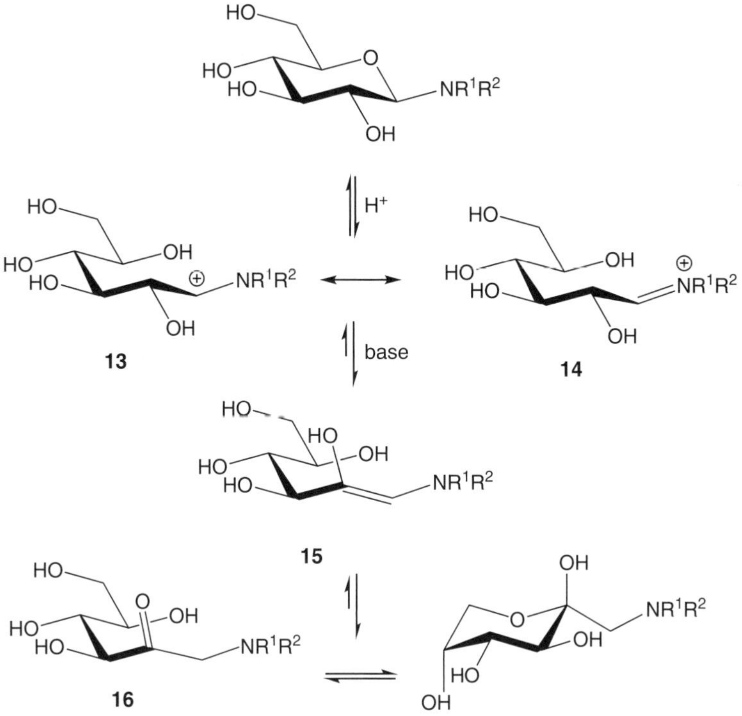

SCHEME 4

The rearrangement of *N*-aryl D-glucosylamines under conditions of anhydrous 1,4-dioxane solvent and elevated temperatures is accelerated 2–4 times in the presence of acid catalysts, such as acetic, oxalic acids, but an even stronger effect was achieved with such aprotic Lewis acids as $ZnCl_2$ and $AlCl_3$, and other anhydrous metal chlorides[43,44] It should be stressed that these experiments have been performed with model *N*-aryl D-glucosylamines and conducted in anhydrous aprotic medium, where the concurrent acid-catalyzed hydrolysis of *N*-D-glucosylamines was not significant. In foods and *in vivo*, however, water is an active ingredient in the medium, and the significance of acid catalysis as the rate-limiting step in such systems may be drastically diminished by other, more relevant mechanisms.

Indeed, hydrochloric acid failed to catalyze formation of 1-deoxy-1-*p*-tolylamino-D-fructose from *N*-*p*-tolyl-D-glucosylamine in methanol at 100 °C, while all carboxylic acids tested gave 8–40% yields of the Amadori rearrangement product, and even better yields were provided by the respective carboxylate potassium salts.[45]

The Amadori rearrangement of D-glucosylamines derived from 4,6-*O*-benzylidene-D-glucose and alkyl glycinates in dry 1,4-dioxane was much less promoted by a strong mineral acid, such as sulfuric, as compared to the weaker carboxylic acid catalysts.[46] Complete protonation of the amino group in *N*-alkyl-D-glucosylamines does not change their tautomeric patterns in water or organic solvents, with the dominating β-pyranose and minor α-pyranose being the only detectable forms in the equilibria,[32] thus making the significance of the proton-induced ring opening (Scheme 4) for this group open to discussion. An argument against the importance of Lewis acids as catalysts of the Amadori rearrangement of *N*-alkyl and *N*-carboxyalkyl D-glucosylamines may be drawn from the well documented stability of their complexes with copper(II), nickel(II), and other metal ions in the β-pyranose rather than acyclic aldimine form.[47,48] As a matter of fact, metal complexation may thus far be the only known method for stabilizing D-glucosylamines derived from amino acids.[34]

Micheel and Dijong proposed an alternative acid-catalyzed mechanism of the Amadori rearrangement, with an aldose diaminal as an intermediate.[35] They investigated reaction products between 3,4,5,6-tetra-*O*-benzyl-D-glucose and *p*-toluidine, and showed that the acyclic Schiff base **17** was mostly hydrolyzed when treated with oxalic acid in anhydrous conditions, whereas the acyclic diaminal **18** under similar conditions quickly rearranged into 3,4,5,6-tetra-*O*-benzyl-*N*-*p*-tolyl-D-fructosamine (Scheme 5). According to the mechanism suggested, the acid catalyzes elimination of one amine and formation of the carbonium ion **19**, which may undergo a 2,1-hydride shift to form a more stable carbocation **20**, while the latter deprotonates into the ketosamine **21**.

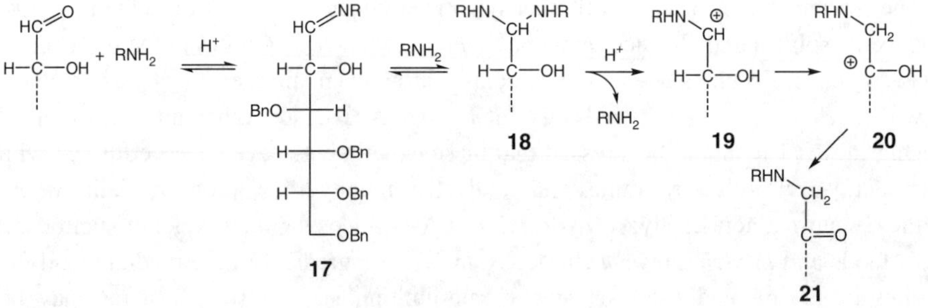

SCHEME 5

An important conclusion concerning the secondary role of acid-catalysis kinetics in the Amadori rearrangement of protein-derived D-glucosylamines was put forward by Gil *et al.*[49] The rate of hemoglobin glycation by D-glucose at 37 °C was found to be independent of proton transfer, since the reaction exhibited identical rates in phosphate-buffered H_2O or D_2O.

(iv) Base Catalysis: The Prevalent Rate-Determining Step of the Amadori Rearrangement of D*-Glucosylamines in Foods and in vivo.* The lack of practical efficiency of strong acids, as compared to weak carboxylic acids or carboxylate and phosphate ions, as catalysts of the Amadori rearrangement, as well as the similarity of the reaction to the well established Lobry de Bruyn–Alberda van Ekenstein transformation,[50] implies the relevance of the base catalysis for production of D-fructosamine derivatives found in foods and *in vivo*. Such biologically significant amines as α-amino acids or the ε-amino groups of lysine side chains in proteins are basic enough ($pK_a > 8$) to abstract a proton from surrounding water or other weakly acidic sources, thus making its availability for the amino–carbonyl reaction prearranged. Enolization of the intermediate Schiff base **14** or its resonance carbocation **13** then becomes the rate-limiting step of the reaction, which requires a nucleophilic catalyst or elevated temperature in order to proceed.

According to a classic view, the role of nucleophilic enolization catalyst is to accelerate the proton abstraction at carbon-2 in the acyclic form of a D-glucosylamine. The mechanism is conventionally established by observing a large kinetic isotope effect upon comparing the rate parameters of D-fructosamine formation for 2-(^{1}H)- and 2-(^{2}H)-D-glucose. Thus, formation of D-fructose–albumin at 37 °C and pH 7.3 is autocatalytic and its $^{H}k_0/^{D}k_0 = 4.43$.[51] A similar parameter determined for glycation of hemoglobin by D-glucose is $^{H}k_0/^{D}k_0 = 2.2–1.96$.[49,52] The proton-abstracting base can be an inorganic anion (hydroxide, phosphate, arsenate, carbonate), a weak organic acid or its anion (carboxylate), organic phosphate (glucose phosphate, phospholipid), organic base (imidazole, pyridine, amine), or even water.[52–54] The mechanism may also rely on formation of a hydrogen-bonding system facilitated by solvent molecules, and offers an alternative between the acyclic azomethine and β-D-glucopyranosylamine forms (Scheme 6) as the proton donors.

In proteins, the proton-abstracting base is probably represented by side-chain residues located in spatial proximity to the D-glucose–amine initial adducts, since external nucleophilic buffer ions (phosphate, arsenate, carbonate) did not show the isotope effect.[49,52] In hemoglobin, for example, the imidazole of His-2 provides the base-catalysis assistance for the Amadori rearrangement at Val-1, while most, if not all, other proteins are glycated site-specifically (see Section II.1.b.iv), as well.

Phosphate, as a physiological pH-buffering species, has been widely used in model Maillard reaction systems containing proteins (or other biological amines) and glucose.[53] Although phosphate was largely credited for catalyzing the Amadori

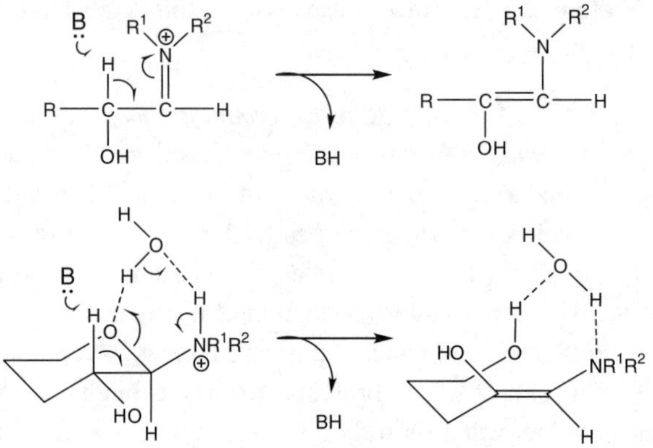

SCHEME 6

rearrangement and consequent fructosamine degradation reactions, the Amadori modification of serum albumin and ovalbumin by D-glucose was not significantly affected by phosphate under physiological conditions,[51] while phosphate-catalyzed modification of lysine residues in RNase was site-specific.[53] Phosphate, as well as other anions, can accelerate glycation of proteins nevertheless, but these effects may be related to anion-induced conformational changes at the glycation sites.[49,52]

2-Amino-2-deoxy-D-glucosyl- and -mannosylamines may also afford D-fructosamines through the base-catalyzed Amadori rearrangement. Thus, interaction of D-fructose, the second most abundant monosaccharide in foods, with amines brings about formation of 2-amino-2-deoxy-D-glucose (**22**) and 2-amino-2-deoxy-D-mannose (**23**) derivatives, through the so-called Heyns rearrangement reaction, which is essentially an analogue of the Amadori rearrangement.[4,55–57] The reaction may proceed further through the condensation of a free amine with the carbonyl group of Heyns' compounds, followed by rearrangement and hydrolysis of the adducts (Scheme 7).[55,57]

The nature of the solvent has an expectably significant influence on rates and yields of the Amadori rearrangement. Solubility is perhaps the most obvious factor, as the number of solvents suitable for the highly hydrophilic carbohydrates is limited. Common alcohols are the most traditional medium for reactions of D-glucose with lower aliphatic and aromatic amines. Reaction of D-glucose with amino acids of lower solubility proceeds more slowly and may be aided by the addition of Me_2SO or glycerol.[58] Me_2SO or DMF are solvents appropriate for lactose or oligopeptides,[59] although DMF may cause accumulation of dimethylamine glycoconjugates over extended reaction times.

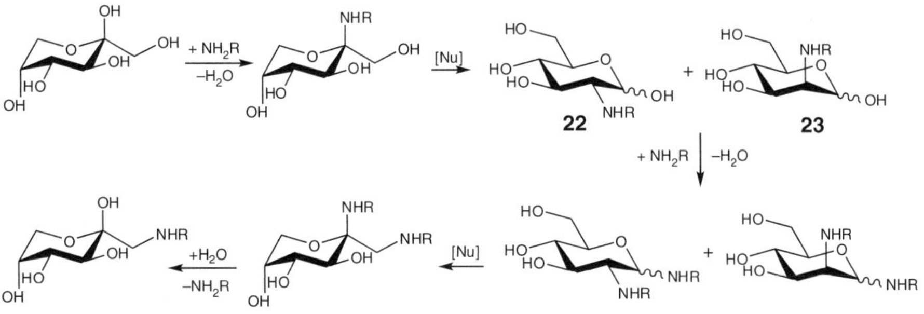

SCHEME 7

Water is necessarily a medium of choice for glycation of proteins, although aqueous syrups were successfully employed for syntheses of small D-fructosamines as well.[60] Food proteins, most often casein or lactoglobulin, have been purportedly modified by reducing saccharides, in order to improve their solubility, thermal stability, emulsifying properties, or to reduce the allergenicity. The reaction is carried out in the solid phase and with controlled humidity.[61,62] Optimal conditions for formation of the protein Amadori product from lyophilized protein–carbohydrate mixtures are 55–60 °C, water activity $a_w = 0.44$–0.65, carbohydrate–protein molar ratio >2.

(v) Oxazolidin-5-one Intermediates. Several workers have demonstrated the significance of an intramolecular cyclization (Scheme 8) in the Schiff base form of aldosyl-α-amino acids, as an intermediate step of the Amadori rearrangement, which is accompanied by decarboxylation of the amino acid. One important example of the reaction is a pathway for acrylamide formation from D-glucose and L-asparagine in fried foods, such as potatoes.[63,64]

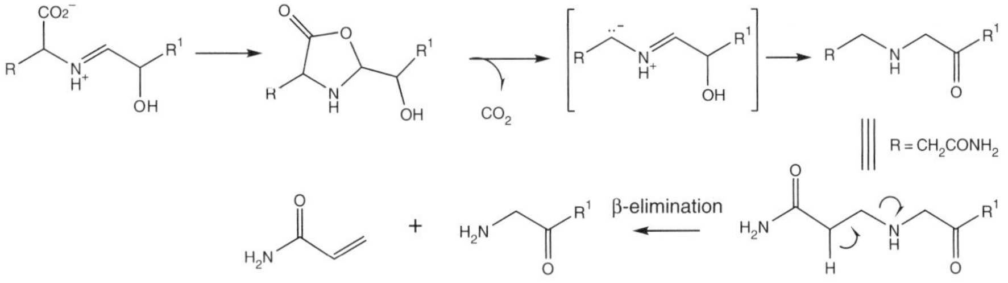

SCHEME 8

(vi) Site Specificity of Protein Glycation by Glucose. Post-translational *in vivo* modification of proteins by D-glucose proceeds through the site-specific Amadori rearrangement, as has been realized since the discovery of the hemoglobin A_{1c} structure.[11] Hemoglobin αβ dimer contains twenty lysine side-chain and two *N*-terminal amino groups that are theoretically available for the amino-carbonyl reaction with D-glucose. However, there is an order of prevalence for formation of D-fructosamine at specific amino acids, and this was estimated as the order β-Val-1 > α-Lys-16 > β-Lys-66 > β-Lys-17 > α-Val-1 > α-Lys-7 > β-Lys-120 *in vitro* and β-Val-1 > β-Lys-66 > α-Lys-61 > β-Lys-17 > α-Val-1 *in vivo*.[65] On average, the content of εFruLys in hemoglobin of both healthy and diabetic humans is about twice as high as the FruVal content.[66] The site specificity of non-enzymatic protein glycation has been confirmed later for a number of other proteins (RNase,[53,67] other enzymes,[68] serum albumin,[69–71] immunoglobulins,[72] casein,[73,74] lactoglobulin[75,76]) and other reducing sugars as well, and is now confidently established as a rule. Two apparent parameters that might influence the kinetics of Amadori product formation were considered: the availability of an amino group for the initial condensation reaction with D-glucose and the availability of the Amadori rearrangement catalyst in the microenvironment.[76,77] A comparison of rates of D-glucosylamine and D-fructosamine formation at particular lysine or *N*-terminal amino acid residues was made by trapping the glycoamino-conjugates with borohydride and subsequent analysis of tryptic peptides.[77] No correlation was found between the respective glycation rates, indicating that the reversible formation of D-glucosylamine (the Schiff base) is a much faster reaction than its Amadori rearrangement, and that the catalytic activity of the microenvironment is likely the single most influential spatial factor that determines relative distribution of D-fructosamine residues in a glycated protein. In turn, the catalytic power of the protein microenvironment is considered to be determined by relative abundance of such nucleophilic species as carboxylate side-chains of Glu and Asp, imidazole of His, or even clusters of positively charged Lys and Arg residues, which lower the pK_a's of the amino groups and may attract such nucleophilic anions as phosphate or citrate.[77,78] As expected, the tertiary/quaternary protein structure has a considerable influence on such catalytical power through stabilization of three-dimensional structural arrangements.[76]

Glycation of milk proteins by lactose also proceeds in the site-specific manner. Native β-casein, at a low glycation level, contains lactulose residues attached exclusively to side chains of lysine-107.[74] Two sites that are primarily modified by lactose in β-lactoglobulin upon heating of the milk are Lys-47 and Lys-138/141.[75]

Short peptides also react with D-glucose and related sugars unevenly. In a series of LysPxx dipeptides, the rates of FruLysPxx formation in methanol at 64 °C were in the following order for Pxx: Leu>Arg>Ile>Phe>Ser>Val>Tyr>Lys>Ala>Gly, and the reactivity of sugars followed the order: Glc=Gal>Mal>Lac.[79] In the model peptide AcLysLysβAlaLysβAlaLysGly, only two lysine residues neighboring each other reacted with D-glucose or lactose to form the respective Amadori compounds, of which the εFruLys-1 moiety was relatively more stable toward degradation.[80]

(vii) Intramolecular Amadori Rearrangements. Horvat and colleagues[81,82] explored peculiarities of the intramolecular cyclization of amino acid or peptide conjugates with D-glucose linked through its C-6 hydroxyl group (**24**). The intramolecular Amadori compounds of amino acid esters do not seem to be stable, possibly because of steric constraints, and degrade further to form pyrrol-lactones (**25**) as a major detectable product structure (Scheme 9).

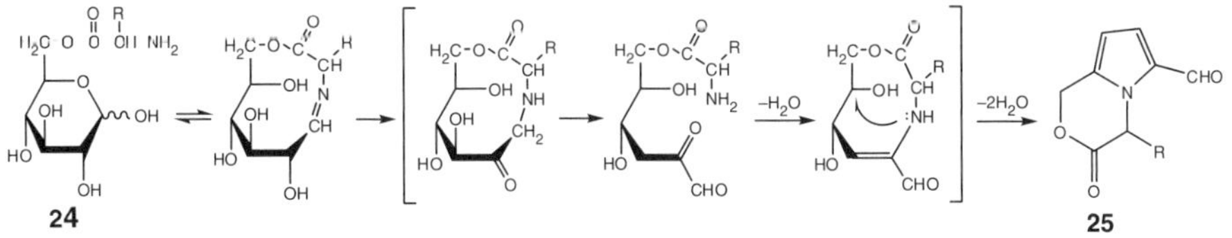

SCHEME 9

On the other hand, 6-esters of D-glucose and Tyr-terminal peptides rearrange in pyridine–acetic acid (but not in dry methanol, see Section II.1.b.viii later) into stable bicyclic D-fructosamine structures (**26**) (Scheme 10). The length and rigidity of the bridging peptide, however, defines its ability to form the Amadori rearrangement products and their tautomeric equilibria in solutions.[83,84]

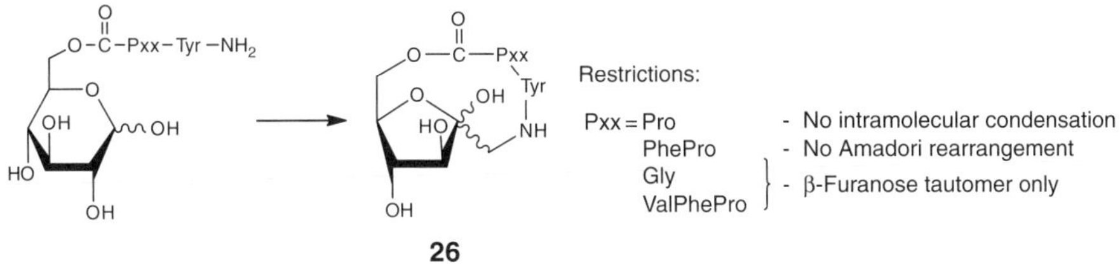

SCHEME 10

(viii) Concurrent Reactions. In some cases, the expected Amadori rearrangement of glucosylamines is accompanied by, or completely substituted with, alternative reactions. To an extent, decarboxylation of the oxazolidin-5-one intermediates

(Section II.1.b.v) falls into this category. It has long been acknowledged that the Amadori rearrangement of unsubstituted D-glucosylamine does not readily afford D-fructosamine, because of extensive degradation and side reactions. Instead, practical routes to unsubstituted D-fructosamine through the Amadori rearrangement included preparation of the readily isolable 1-arylamino- or 1-dibenzylamino-1-deoxy-D-fructose and their subsequent hydrogenation in the presence of a palladium catalyst.[27,85]

Formation of thiazolidines (**27**) in reactions of D-glucose with cysteine and other β-aminothiols (Scheme 11)[86] (which inhibits formation of the respective fructosamines) is thought to proceed mainly through the hemithioacetal intermediate, but slow accumulation of tetrahydro-1,4-thiazines (**28**) suggests formation of the Schiff bases and Amadori products in the equilibria, as well.[87] The thiol group does not per se impede formation of the Amadori rearrangement products in reactions of D-glucose with reduced or oxidized forms of L-glutathione.[88]

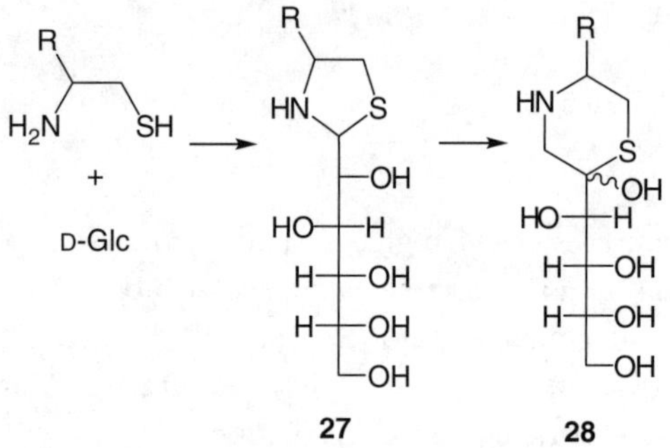

SCHEME 11

Tryptophan reacts with D-glucose in presence of nucleophilic catalysts, with FruTrp as the main reaction product, which is accompanied by formation (Scheme 12) of a tetrahydro-β-carboline derivative **29**.[89] Its yields may be dramatically increased in acidic methanol; NMR data suggest that one of two possible enantiomers is predominantly formed. Reactions of *m*-tyrosine, DOPA, or catecholamines with D-glucose start with formation of the respective Schiff bases, but the subsequent Amadori rearrangement is accompanied, to varied degrees, by the Pictet–Spengler phenolic condensation (Scheme 12).[90] The amino acid may[91] or may not undergo decarboxylation, prior to cyclization of the Schiff base to the tetrahydroquinoline derivatives **30**.

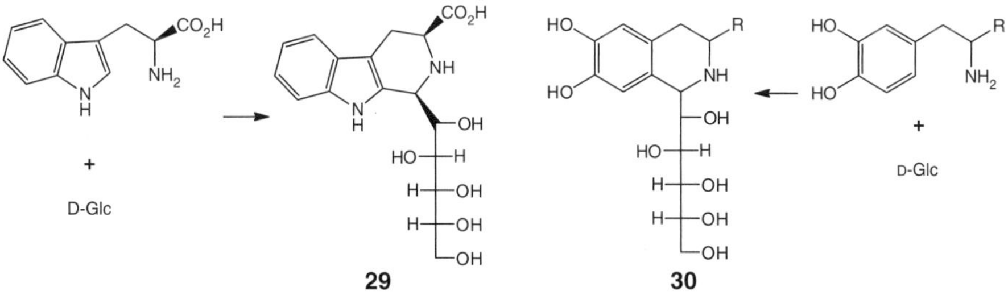

SCHEME 12

α-Amino groups of peptides usually react smoothly with D-glucose, with relatively small amounts of side products at the Amadori rearrangement stage. However, glycosylamines of Glc, Man, or Gal and enkephalin-related peptides with N-terminal tyrosine may produce, depending on reaction conditions, imidazolidinones (**31**) in amounts comparable to those of the D-fructose-peptides (Scheme 13).[92] Specifically, in presence of phosphate, imidazolidinones (**31**) were favored over the Amadori products, while autocatalysis in methanol led to almost exclusive formation of D-fructose-Leu-enkephalin. The structure of the peptides was also a critical determinant for the reaction's direction, causing in some cases exclusive formation of the imidazolidinones.[84,92,93]

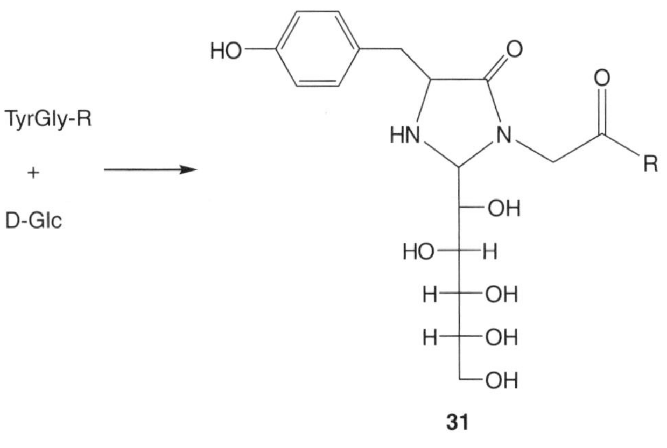

SCHEME 13

c. Reductive Amination of "D-Glucosone".—Walton *et al.*[94] evaluated the utility of this approach for the preparation of several D-fructose–amino acids (Scheme 14). "D-Glucosone" (D-*arabino*-hexos-2-ulose, **32**) or its 2,3:4,5-di-*O*-isopropylidene derivative **35** are readily available from **3** or 2,3:4,5-di-*O*-isopropylidene-β-D-fructopyranose, respectively. When D-glucosone was treated with a free amino acid in the presence of $NaCNBH_3$ in water, the main product was the *N*-(1-deoxy-D-fructos-1-yl)-amino acid. The reaction was not optimally controlled, because a significant proportion of the product, about 30%, underwent further reduction to give a mixture of the respective 1-amino-1-deoxy-D-glucitol (**33**) and -mannitol (**34**), which complicated chromatographic separation of pure Amadori compounds. This problem was solved when **35** was chosen as the starting material. Its reductive amination with amino acids was effected with practically no side reactions. The intermediate *N*-(1-deoxy-2,3:4,5-di-*O*-isopropylidene-β-D-fructopyranos-1-yl)-amino acids (**36**) were isolated on a strongly acidic cation-exchange resin and characterized, but this step could be skipped when the same resin was used for both acid hydrolysis of the intermediate and chromatographic isolation of the product D-fructose–amino acids on it.

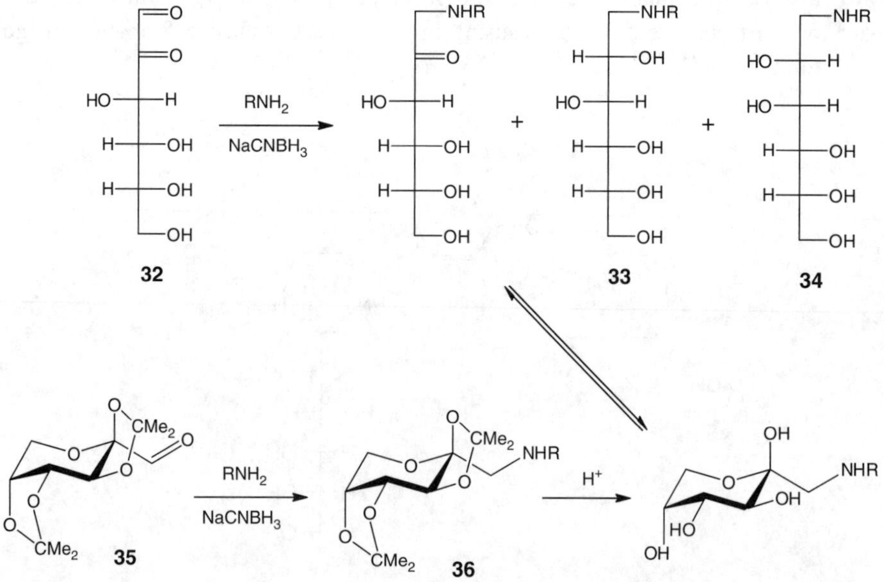

SCHEME 14

The latter protocol has some advantages over the methods employing direct D-glucose–amino acid coupling through the Amadori rearrangement, such as more universal reaction conditions, stability of the protected diacetonated intermediates, lower amount of side products and higher yields. A number of research groups employed the method to obtain D-fructose–peptides in solution,[56] on a solid support,[95,96] or through protected D-fructose–L-amino acids for automated D-fructose–peptide synthesis;[97] to synthesize D-fructose conjugates with thioflavin and other bioactive molecules.[98]

It is not clear whether secondary amines (excluding such compact pyrrolidine derivatives as proline) could be successfully coupled to **35** using this reaction because of steric constraints inflicted by the neighboring bulky 2,3-*O*-isopropylidene substituent; it was possible, however, to alkylate the secondary amine nitrogen in *N*-(1-deoxy-2,3:4,5-di-*O* isopropylidene-β-D-fructopyranos-1-yl)-peptides by reductive amination of small aldehydes with **36**.[99] Of note is the fact that, one of the first conversions of **35** into a ketosamine derivative was carried by Tronchet *et al.*,[100] who employed hydrogenation over Pd catalyst to perform reductive amination and deprotection simultaneously, resulting in a 92% yield of bis-(1-deoxy-2,3:4,5-di-*O*-isopropylidene-β-D-fructopyranos-1-yl)amine (**37**) (Scheme 15).

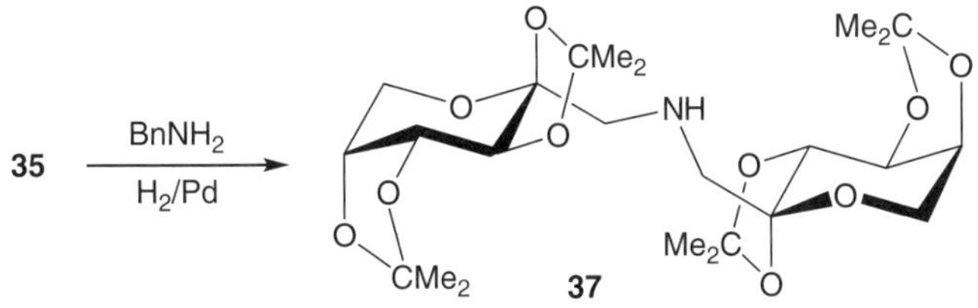

SCHEME 15

Since "D-glucosone" is an α-dicarbonyl derivative, it reacts with α-amino acids along the Strecker degradation pathway[101] and may be converted into D-fructosamine (Scheme 16), although this reaction is unlikely to have any synthetic value, on account of a number of concomitant transformations.

d. Nucleophilic Substitution Reactions.—In D-fructose derivatives with appropriately protected hydroxyl groups in positions 2, 3, 4, and 5 or 6, the hydroxyl group in position 1 can be replaced by a halogen or, more conveniently, activated with a sulfonic acid anhydride and then allowed to react with primary or secondary amines. This pathway to *N*-substituted 1-amino-1-deoxy-D-fructose has

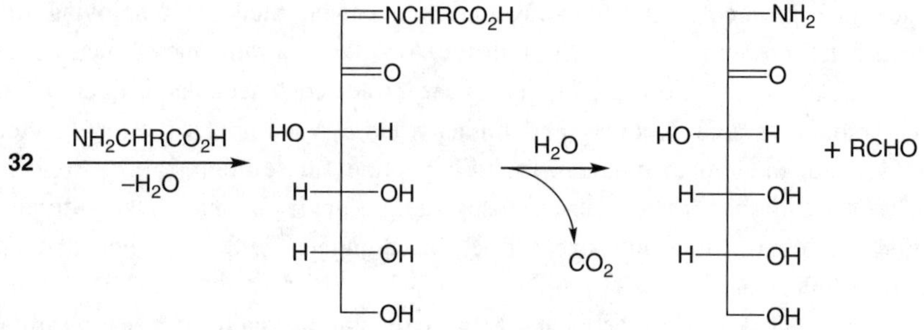

SCHEME 16

been successfully employed for over 50 years. Vargha and co-workers,[102] in a search for antitumor agents, activated 2,3:4,5-di-*O*-isopropylidene-β-D-fructopyranose by converting it into tosylate **38** and then brought this into the substitution reaction with a secondary amine (Scheme 17), which afforded an over 75% yield of the product **39**.

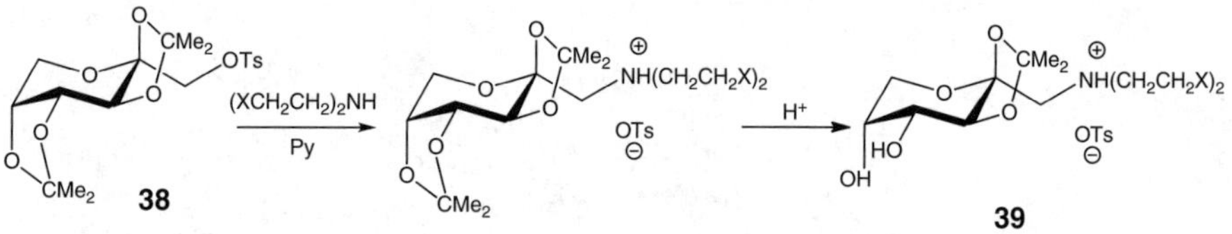

SCHEME 17

Interestingly, tosylate **38**, when treated with hydrazine hydrate, afforded mainly the 1-deoxy-1-hydrazino-di-*O*-isopropylidene-D-fructose, but also a small proportion of the 1-deoxy-di-*O*-isopropylidene-D-fructose.[103]

In more-recent publications, activation of 2,3:4,5-di-*O*-isopropylidene-β-D-fructopyranose was performed preferentially by trifluoromethylsulfonylation. Triflate proved to be an excellent leaving group in the S_N2 reactions, and a significant number of D-fructosamine derivatives were prepared by using **40** as a 1-deoxy-D-fructosylation reagent (Scheme 18). Its activity, however, imposes certain demands on the conditions of coupling, such as protection of potentially reactive groups, use of anhydrous solvent, and so on. For example, to prepare D-fructose–amino acids through this approach, **40** was typically brought into the reaction with amino acid methyl or benzyl esters in dry DMF at elevated temperatures.[104,105] After separation, the Me/Bn esters

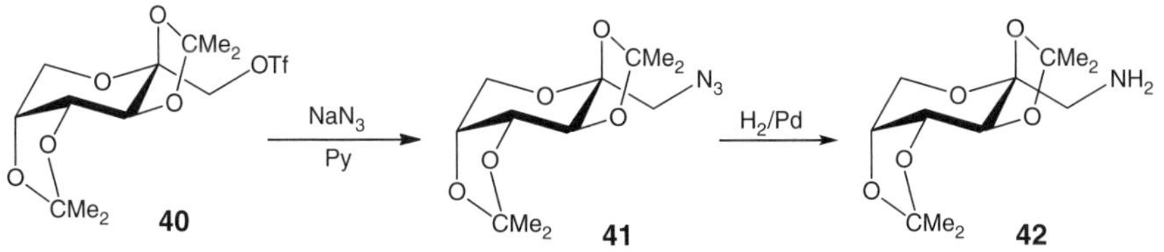

SCHEME 18

of *N*-(1-deoxy-2,3:4,5-di-*O*-isopropylidene-β-D-fructopyranos-1-yl)-amino acids could be deprotected by hydrogenolysis and trifluoroacetic acid-catalyzed hydrolysis. The method may be a suitable alternative to the Amadori rearrangement protocol in cases when amino acids, such as Tyr, Gln, or Trp, are poorly soluble or sensitive to conditions typical for a direct D-glucose–amino acid coupling. To prepare conjugates of D-fructose with such biological lipophilic amines, as aminophospholipids, it is also a preferred synthetic option.[106]

Nucleophilic substitution of triflate by azide is an important access route to fructosamine intermediates in syntheses of glucosidase and carbonic anhydrase[107] inhibitors.

The azide **41** has also been obtained by substitution of the respective iodide in Me_2SO, and its reduction to **42** is readily achieved by treatment with triphenylphosphine in THF.[108]

e. Other Methods.—Catalytic hydrogenation of “D-fructosazine” (**43**), which can be prepared from D-fructose or D-glucosamine and ammonia,[109] affords D-fructosamine in good yield[110] (Scheme 19); this method could be a useful approach to labeled fructosamines on a small scale. Other “D-fructosazine” derivatives, such as 3′-deoxy-D-fructosazine (**44**) or 3′,4′-dideoxy-4′-sulfonyl-D-fructosazine[111] in similar conditions afford both D-fructosamine and 3-deoxy-D-fructosamine (**45**) or 3,4-dideoxy-4-sulfonyl-D-fructosamine, respectively.

Regioselective epoxide opening in 1,2-anhydro-3,4:5,6-di-*O*-isopropylidene-D-mannitol (**46**) to amine or azide constitutes another interesting synthetic pathway to D-fructosamine derivatives.[4] As shown in Scheme 20, this reaction furnishes protected 1-amino-1-deoxy-D-mannitol derivatives **47**, which can be subsequently oxidized and deprotected into respective *N*-substituted 1-amino-1-deoxy-D-fructose.

The Mitsunobu condensation of 2,3,4,6-tetra-*O*-acetyl-D-glucose with amino acid-derived 2-nitrobenzenesulfonamides led to formation of the protected

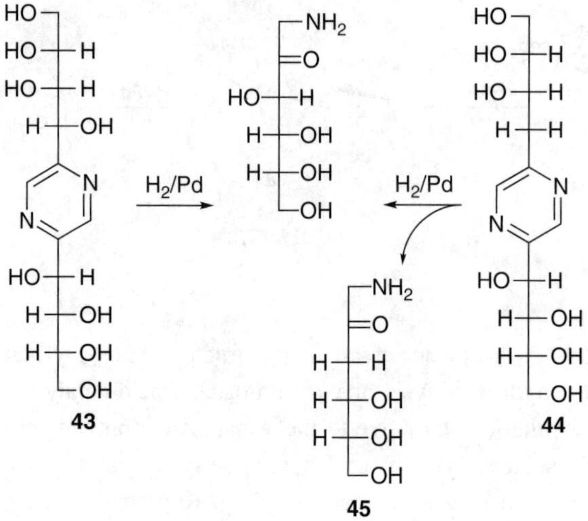

SCHEME 19

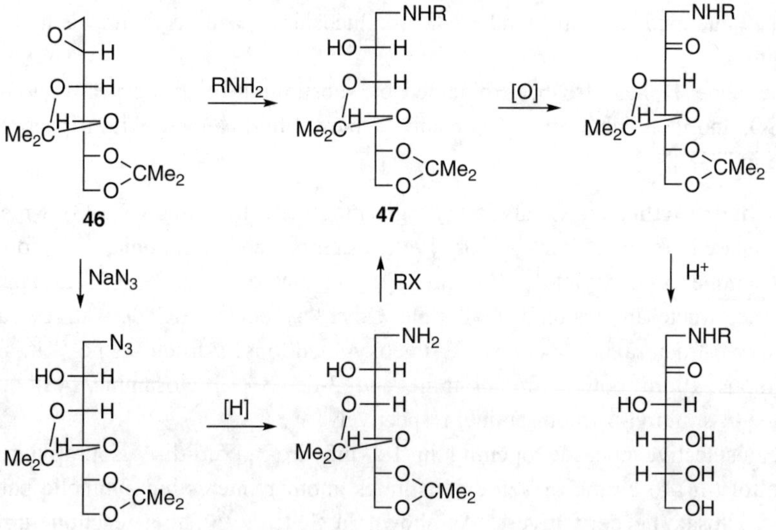

SCHEME 20

glucosylamines in high yields.[112] Following smooth deacetylation, the cleavage conditions for the *o*-nitrophenylsulfenyl protective group afforded quantitative conversion of the glucosylamines into the respective Amadori-rearrangement products (Scheme 21).

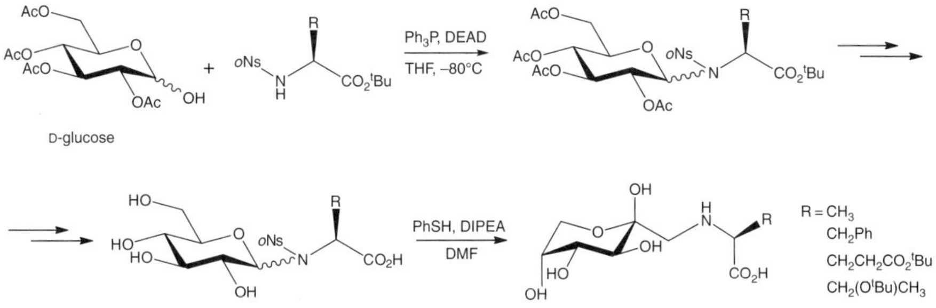

SCHEME 21

2. Enzymatic

The Amadori rearrangement has traditionally been considered as a typical non-enzymatic carbohydrate modification, and the *in vivo* formation of Amadori rearrangement products as an undesired and uncontrolled event. However, a few cases are known of the enzymatically catalyzed Amadori rearrangement, suggesting a physiological importance of this reaction and its products.

One notable example of an enzymatically synthesized 1-amino-1-deoxy-D-fructosamine derivative is the production of D-fructose–L-glutamine by roots of dicotyledonous plants infected by *Agrobacterium* sp. During the infection, the bacterium inserts T-region DNA into the nuclear genome of the plant cells. The T-DNA causes growth of tumorous crown galls on the hosts' hairy roots, but also encodes production of so-called opines by the transformed cells. Opines are secreted by the neoplastic cells and are used as a sole carbon, nitrogen, and energy source by the agrobacteria that induced the tumor.[113] Of over 20 currently known opines, N^{α}-(1-deoxy-D-mannitol-1-yl)-L-glutamine (mannopine, MOP, **48**) and N^{α}-(1-deoxy-D-fructos-1-yl)-L-glutamine (santhopine, SOP, **49**) are among those relatively well studied. These nutrients are metabolically related through the reversible redox transformation, as shown in Scheme 22. It is considered that, in crown galls, **49** is the metabolic precursor of **48**, while in *Agrobacteria* MOP is catabolized through oxidation mediated by the same enzyme, mannopine oxidoreductase, to SOP.[114] The biosynthetic pathway for D-fructose–L-glutamine in the crown gall is not certain, but probably involves an enzyme-assisted Amadori rearrangement of the glucoside formed by L-glutamine and D-glucose.[115] D-Fructose–L-glutamine is also a likely precursor of three other D-fructosamine-based opines from crown galls, which comprise the chrysopine family:[116]

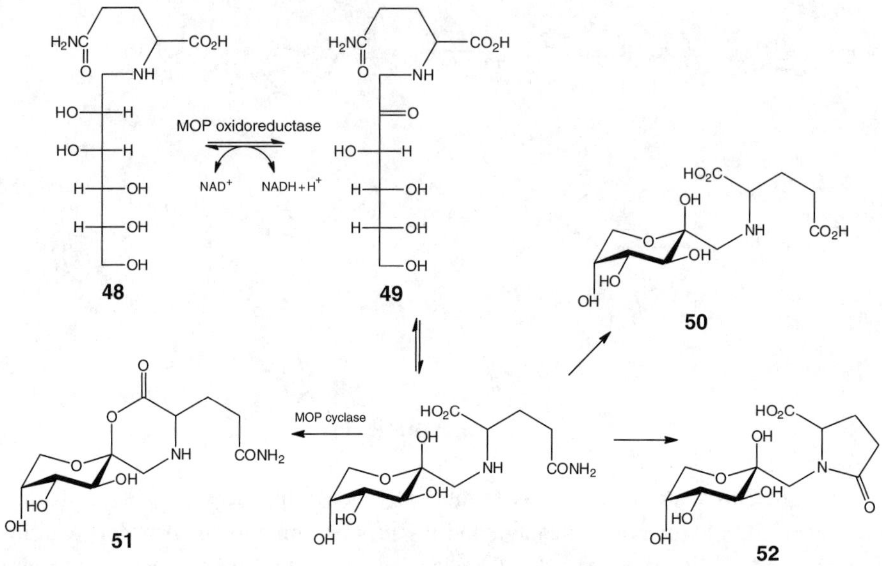

SCHEME 22

N^{α}-(1-deoxy-D-fructos-1-yl)-L-glutamic acid **50**, N^{α}-(1-deoxy-D-fructos-1-yl)-L-oxoproline (isochrysopine, **52**), and N^{α}-(1-deoxy-β-D-fructopyranos-1-yl)-L-glutamine *spiro*-α-lactone (chrysopine, **51**).

As early as 1964, Smith and Ames identified the main steps in histidine biosynthesis and concluded, to their surprise, that one of the key steps involved the Amadori rearrangement.[117] In the biosynthesis of histidine, N'-[(5′-phosphoribosyl)formimino]-5-aminoimidazole-4-carboxamide ribonucleotide (ProFAR) isomerase (HisA; EC 5.3.1.16) catalyzes rearrangement of the aminoaldose ProFAR to the aminoketose N'-[(5′-phosphoribulosyl)formimino]-5-aminoimidazole-4-carboxamide ribonucleotide (PRFAR). A similar reaction in the biosynthesis of tryptophan from chorismic acid, the isomerization of phosphoribosyl anthranilate to l-[(2-carboxyphenyl)amino]-l-deoxyribulose 5-phosphate (CdRP), is catalyzed by phosphoribosyl anthranilate (PRA) isomerase (TrpF; EC 5.3.1.24).[118] Hommel *et al.*[119] investigated the reaction kinetics and concluded that the enzyme catalyzes the 1,2-enolization of the substrate, followed by spontaneous keto–enol tautomerization. Despite the lack of detectable amino acid sequence similarity, HisA and TrpF belong to the same structural family of (βα)8-barrels[120] and can be made interchangeable after minor modifications,[121] thus revealing a likelihood of their common ancestry.

In 1970, Brown proposed that Amadori rearrangement was involved in the chemistry of pteridine metabolism.[122] Later studies identified GTP cyclohydrolase I (CYH; EC 3.5.4.16) as a catalyst for the ring expansion conducive to the formation of 7,8-dehydroneopterin 3′-triphosphate from GTP. The exact mechanism of the reaction has not yet been firmly established; it is thought to proceed through Zn-induced opening of the imidazole and α-D-ribofuranoside rings.[123,124] In the biosynthesis of the thiazole phosphate moiety of thiamine (vitamin, B^1) thiazole synthase (ThiG) forms a Schiff-base adduct with 1-deoxy-D-*threo*-pentulose-5-phosphate, which undergoes an autocatalyzed Amadori-type rearrangement to the 2-amino-1,2-dideoxy-D-pentos-3-ulose 5-phosphate intermediate.[125]

D-Glucose 6-phosphate is the second most abundant reducing sugar in animal tissues, especially in skeletal and heart muscle, and its protein glycating potential exceeds that displayed by D-glucose. Accumulation of its reactive Amadori rearrangement product, D-fructosamine 6-phosphate, may be more damaging to the muscle tissue than that of D-fructosamine; it is no coincidence that a phosphatase that catalyzes dephosphorylation of protein-bound D-fructosamine 6-phosphate (Scheme 23) has been isolated recently from skeletal muscle of rodents and identified as magnesium-dependent phosphatase-1 (MDP-1, EC 3.1.3.48).[126] In conjunction with D-fructosamine-3-kinase (see Section III.4.b), MDP-1 may be a part of the *in vivo* protein-deglycating system.

A combination of glycerol phosphate oxidase (GPO), catalase, and L-rhamnulose-1-phosphate aldolase (RhuA) allowed for an efficient conversion of aminoglycerol into 1-deoxy-1-phosphoramido-L-fructose[127] (Scheme 24).

Ribozyme-catalyzed reactions that might have led to nucleotide synthesis in the ancient “RNA world” possibly involved Amadori rearrangement products, as suggested on the basis of a pR1 ribozyme-assisted ribose 6-phosphate interaction with 6-thioguanosine in one recent study.[128]

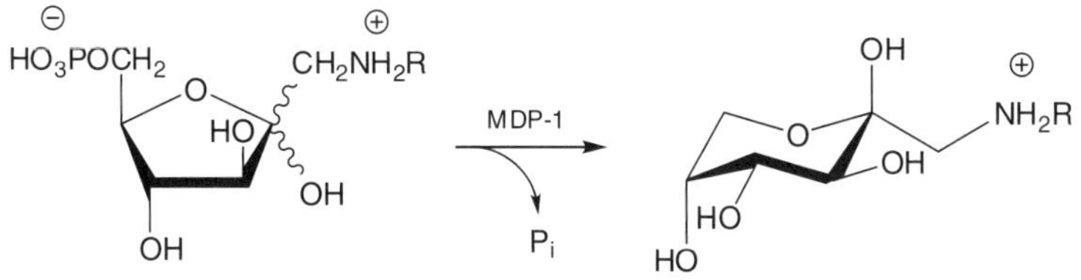

SCHEME 23

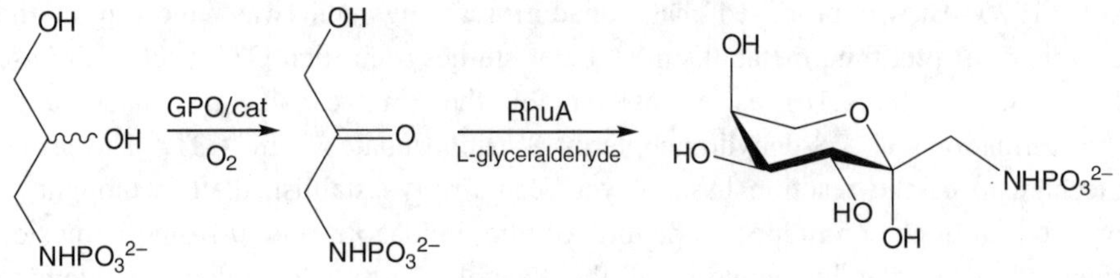

SCHEME 24

III. Structure and Reactivity

Interest in the structure and reactivity of D-fructosamine and its derivatives is determined by their practical significance for the food industry and for control/treatment of diabetes-related pathologies, but also by a notable structural flexibility of this carbohydrate derivative, which may provide for some models that are unique for the field of carbohydrate chemistry. As an important and a major class of intermediates in the Maillard reaction, the fructosamine structure serves as a precursor of a broad variety of both low- and high-molecular-weight Maillard reaction products, hundreds of which have so far been identified. Here we consider only the most immediate transformation steps of D-fructosamine, while some of the more important Maillard reaction products whose origin can be traced to the D-fructosamine intermediates will be considered in Section IV.

1. Tautomers in the Solid State and Solutions

It is convenient to separate all fructosamine derivatives into several structural subgroups which display similarities in their substituent type and reactive behavior. The most commonly considered fructosamine structures can be described by a general tautomerization scheme (Scheme 25), and include the most thermodynamically stable β-pyranose anomer and the most reactive enamine species.

a. D-Fructosamine and Its *N*-Alkyl- or -Carboxyalkyl Derivatives.—This subgroup includes most of the biologically significant fructosamines, such as derivatives of amino acids and proteins. The tautomeric equilibria of D-fructosamine,[129] D-fructose–amino acids,[17,58,130] -oligo-,[131] or -polypeptides[132–134] in aqueous solutions have been studied extensively by ^{13}C NMR. Chemical shifts of

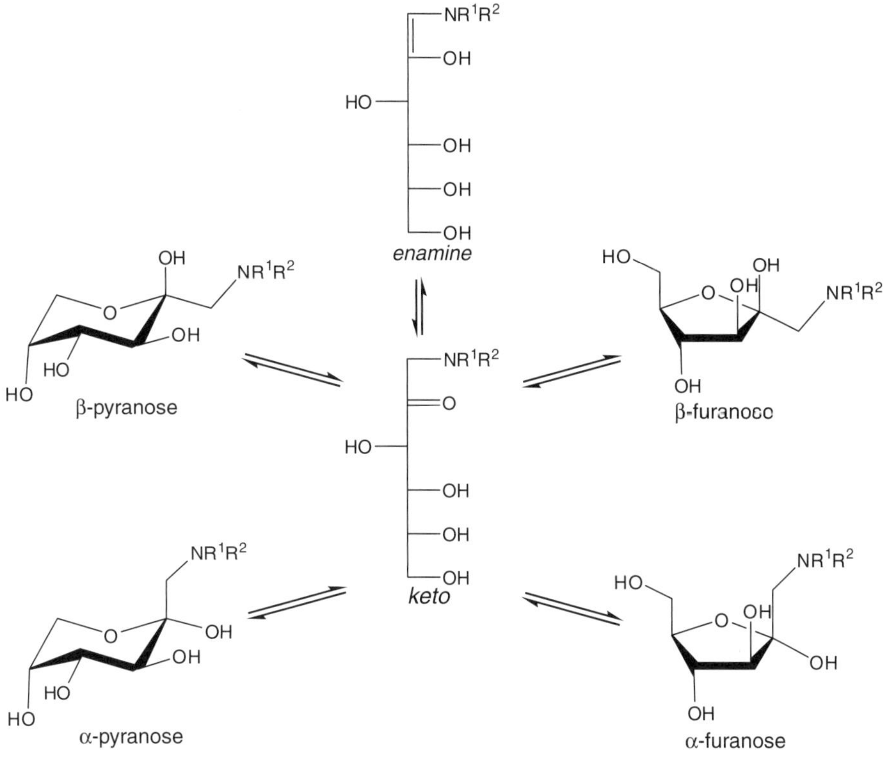

SCHEME 25

anomeric/carbonyl (carbon-2) signals in the spectra generally fall into the following narrow intervals: 210–206 ppm for the acyclic keto tautomer, 105–104 ppm for the α-furanose, 102–101 ppm for the β-furanose, 99–98 ppm for the β-pyranose, and 99–97 ppm for the α-pyranose. The enamine tautomer has not been reliably detected by the NMR technique, on account of its very low equilibrium population. Proportions of the tautomeric forms estimated by this method are given in Table II. Fructosamine does not contain an anomeric hydrogen atom and, for this reason, the utility of proton NMR spectra for estimations of the anomeric populations is limited. However, proton NMR spectra provided for estimates of the D-fructosamine ring conformations[129] by comparing the experimental first-order coupling constants $J_{3,4}$, $J_{4,5}$, and $J_{5,6A/B}$ to the values calculated for the respective D-fructose anomers.[135] Thus, it was established that the most populous β-pyranose tautomer of D-fructosamine, as well

TABLE II
Distribution of the Tautomeric Forms (%) in Aqueous Solutions of Representative Fructosamine Derivatives

Compound	β-pyr	α-pyr	β-fur	α-fur	keto	Reference
D-Fructosamine HCl	71	5	12	11	0.8	129
D-Fructose–glycine	69	5	12	13	0.8	58
N^{ε}-D-Fructose-N^{α}-formyl–L-lysine	70	4	13	13	n.r.	58
N^{α},N^{ε}-Di-D-fructose-L-lysine	68	6	11	12	1.5	38
D-Fructose–TyrGlyGlyPheLeu	69	n.r.	13	18	n.r.	131
D-Fructose–RNase A	69	n.r.	13	18	n.r.	132
D-Fructose–aniline	59	3	25	9.5	3.5	143
D-Fructose–*N*-methylaniline	51	2	35	5	10	143
D-Fructose–*N*-butylaniline	45	2	32	4	17	143
D-Fructose–*N*-cyclohexylaniline	25	1	13	3.5	57	143
Lactulosamine HCl	66	2	16	16	n.r.	28
Lactulose–L-leucine	65	3	15	17	n.r.	144
Maltulose–glycine	64	2	16	18	n.r.	145
Cellobiulose–glycine	67	2	15	16	n.r.	145
	βp-βp	**βp-βf**	**βp-αf**	**βp-ac**[a1]	**βf-ac**	
Di-D-fructose–glycine	17	10	7.5	54	3.5	38
N^{ε},N^{ε}-Di-D-fructose-N^{α}-formyl-L-lysine	4	3	2	84	5	38

[a] See Scheme 26 for the tautomer structure.
n.r. Not resolved

as of its derivatives,[17,58] exists in the solutions exclusively in the $^2C_5$ conformation. The experimental coupling constants found for the less stable D-fructosamine anomers were the most consistent with the $^2C_5$ chair for the α-pyranose, the $E_3/^4T_3$ conformation for the β-furanose, and the $^3E/^4T_5$ equilibrium for the α-furanose.[129]

The molecular and crystal structure of D-fructosamine hydrochloride and hydroacetate salts were solved by X-ray diffraction.[129] As expected, the anomeric form of the D-fructosammonium ion in these salts is β-pyranose in the $^2C_5$ conformation. The same conformation was reported for a number of other solved crystal structures of D-fructosamine with alkyl- or carboxyalkyl *N*-substituents, including D-fructose–morpholine,[136] D-fructose–dibenzylamine,[137] D-fructose–*N*-benzylmethylamine,[138] D-fructose–glycine,[139] D-fructose–L-proline,[140,141] and D-fructose–L-histidine.[142] One common feature found in crystal structures of the D-fructose–amino acids is an intramolecular hydrogen-bonding between the donor amino group and the neighboring anomeric O–2 or O–3 acceptor hydroxyl groups. The H-bonding is apparently stabilized by conjugation to the amino acid carboxylate and survives in aqueous solutions. This is manifested by the observed inequality of protons at methylene C-1, because of a lack of free rotation around the C-2—C-1 bond.[58]

b. *N*-Aryl Derivatives.—Although this subgroup lacks biological relevance, its synthetic representatives have been extensively used in earlier studies of the Amadori rearrangement reaction, because of their ease of crystallization. The tautomeric equilibria of D-fructose–aniline and its derivatives substituted at the benzene ring show the pattern similar to D-fructosamine and D-fructose–amino acids, with dominating β-pyranose, followed by β- and α-furanoses, α-pyranose, and the acyclic keto tautomer (Table II). The proportion of the keto form, however, is more than 1% and is significantly higher than that observed in D-fructosamine and its *N*-alkyl/carboxyalkyl derivatives. Upon introduction of an *N*-alkyl substituent, the proportion of the acyclic keto tautomer increases dramatically in a series of D-fructose–*N*-alkylanilines, reaching an unprecedented 57% (Table II) for some representatives. Interestingly, the relative proportions of pyranose and furanose tautomers remain the same in these compounds, which prompted a suggestion that the increase in the acyclic tautomer is facilitated by a hydrophobic, rather than a steric, effect around the carbohydrate carbonyl group. Indeed, a comparative examination of available crystal structures of D-fructose–anilines[146] and D-fructose–*N*-alkylanilines[143,147] revealed no obvious spatial hindrances for formation of the fructopyranose ring. On the other hand, the hydration rates of the fructosamine carbonyl group correlated inversely with the proportions of the acyclic keto tautomer for D-fructosamine and a number of its derivatives (Fig. 1).

The keto tautomer has been isolated in crystalline form for a number of D-fructose–*N*-alkylanilines and the structure was unequivocally confirmed by IR [148] and solid-state NMR spectroscopic data,[143] as well as by X-ray diffraction studies.[143] The ability of D-fructose–*N*-alkylanilines to crystallize spontaneously in the acyclic keto form is unprecedented among six-carbon reducing carbohydrates, and is matched only by ribulos- and xylulos-amines,[149] which themselves are the Amadori rearrangement products. This phenomenon may be explained on the basis of an interplay of the hydrophobic effect and crystal-packing forces, given that many D-fructose–*N*-alkylanilines still crystallize in the β-pyranose form, and for the majority of those which crystallize in the keto form, the acyclic tautomer constitutes a minor constituent in equilibrated solutions.[143]

Interestingly, the crystal structure of *N*-allyl-*N*-(1-deoxy-β-D-fructopyranos-1-yl)aniline contains an unusual "flip-flop" hydrogen-bonding pattern, with O-3, O-4, and O-5 hydroxyl groups rapidly changing their orientation, which results

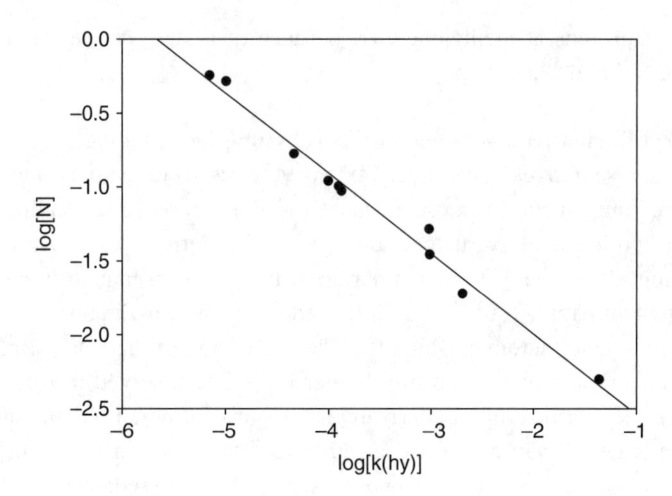

FIG. 1. A correlation between the rate constants k_{hy} for carbonyl hydration and the equilibrium molar fraction N of the acyclic keto tautomers of fructosamines in pyridine/$H_2^{18}O$ at 25 °C.[143]

in a statistical one-half occupancy of the hydrogen-atom positions.[147] This type of H-bonding is unique for a low-molecular-weight, tightly packed crystalline carbohydrate; it has previously been found only in large, loose cyclodextrin complexes.[150]

c. Disaccharides.—1-Amino-1-deoxy-4-*O*-glycopyranosyl-D-fructose analogues, such as cellobiulose-, maltulose-, and lactulose–amines tautomerize in a fashion (Table II) similar to that of D-fructose–amino acids, with a somewhat higher proportion of α- and β-furanoses, at the expense of the population of pyranoses. Other disaccharide derivatives, such as isomaltulose-, gentiobiulose-, or melibiulose–amines, have been described in the literature, but their tautomeric distribution in aqueous solutions was not reported. The presence of a glycosyl substituent at O-6 in these disaccharide derivatives forces the tautomeric equilibrium toward furanose forms only, in addition to the acyclic forms.

d. *N,N*-Di-D-fructosamines.—Although di-D-fructose–glycine was synthesized over fifty years ago, it was considered that the fructosyl residues in the molecule tautomerize independently, following the monofructose–amines tautomerization pattern (Scheme 25). However, the unexplained propensity of *N,N*-difructosamines to release one degradation product, 3-deoxy-D-*erythro*-hexos-2-ulose,[2,151] prompted a revision of their structure. In 1995, Mossine *et al.*[152] published an X-ray diffraction

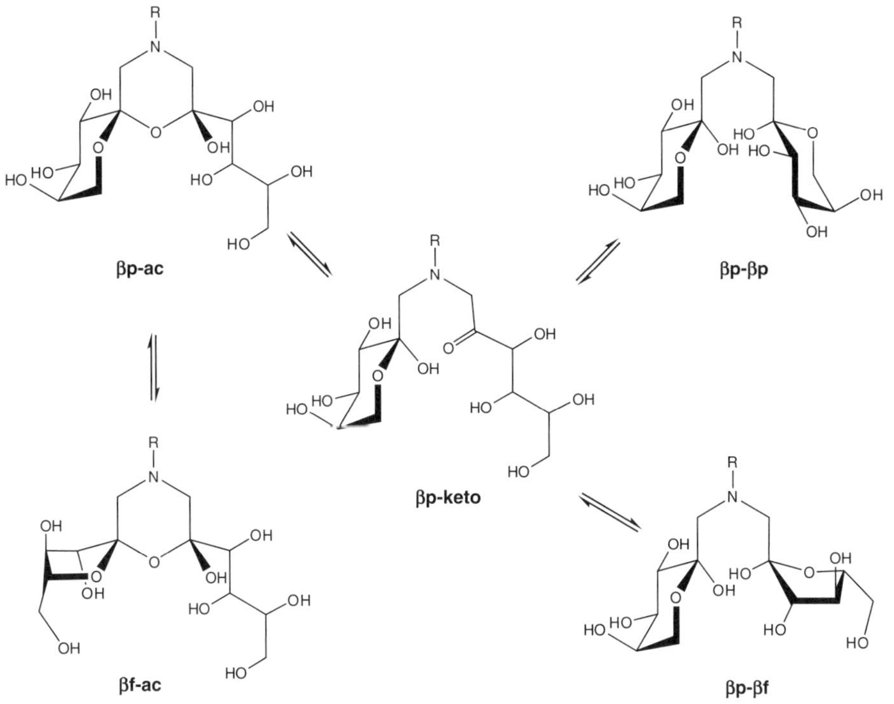

SCHEME 26

study of di-D-fructose–glycine, which showed the unexpected structure of a *spiro*-bicyclic tautomer with one of the fructosyl residues in the β-pyranose form and the second one in the acyclic hemiketal form (βp-ac in Scheme 26).

This structure appears to dominate the tautomeric equilibria in *N,N*-difructosamines (Table II), as evidenced by the NMR data.[38] Only one out of two possible stereoisomers of β*p*-ac was detected in the spectra, which suggests that the stereoelectronic anomeric effects (three in β*p*-ac versus two in β*p*-β*p*) may be the driving force behind the stabilization of such an unusual structure. *N,N*-Di-D-fructose–amino acids undergo 1,2-enolization at rates significantly higher than those demonstrated by the respective D-fructose–amino acids (Section III.3.a). The higher enolization rates explain accelerated production of 3-deoxy-*erythro*-hexos-2-ulose[153] and autoxidation products[38] by this subgroup of fructosamines.

e. Fixed Structures of Fructosamine Tautomers.—Glycosidation or carbamoylation of the anomeric hydroxyl group can be employed for the stabilization of fructosamine tautomeric forms. Thus, ethyl 1-arylamino-1-deoxy-β-D-fructopyranosides, as well as related α- and β-D-fructofuranosides, were readily isolated after Fischer glycosidation of the parent fructosamine molecule.[154] A number of *N*-alkenyl derivatives of D-fructosamine, upon glycosidation by methyl, ethyl, benzyl, or allyl alcohols in the presence of HCl, were converted into a mixture of the alkyl fructofuranosides, with about a 5:1 ratio between yields of the α and β anomers,[23] which could be separated by crystallization. The exact conformation for one of the 1-amino-1-deoxy-α-D-fructofuranosides, **53**, was determined in an X-ray diffraction study.[155]

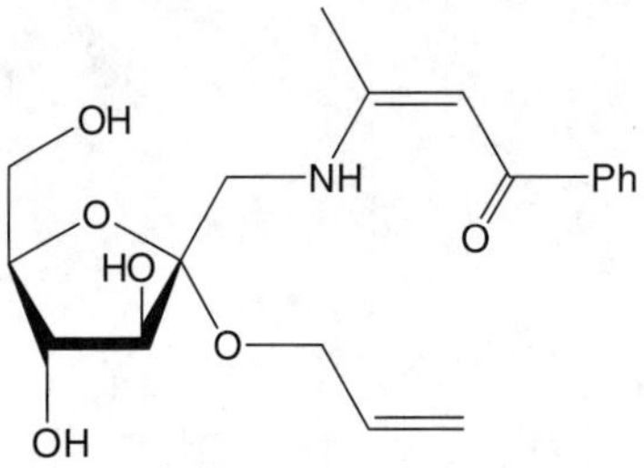

OH NH Ph O HO O OH R N O O O OBn HO BnO

53 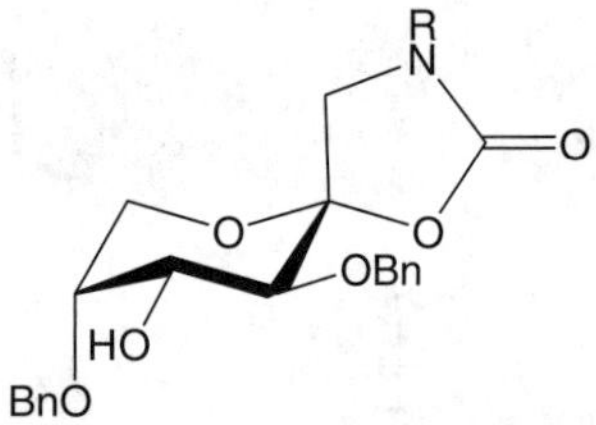 **54**

Treatment of D-fructosamines with triphosgene transformed the former into the 1-*N*,2-*O*- cyclic carbamates (**54**).[156] Notably, this method yielded unusually high α/β ratios for pyranose anomers, and may provide a convenient approach to rare α-fructopyranosyl structures.

2. Analytical Methods

Despite the proven occurrence of relatively large quantities of D-fructosamine derivatives in some common dehydrated foods, such as tomato or milk powders, no official or even recommended methods for their routine analysis in foods have thus far been offered. The situation is somewhat better with the "fructosamine assay" introduced into clinical practice; however, the assay is not particularly selective and evaluates total fructosamine modification on circulating proteins, peptides, and other biomolecules combined. Difficulties with development of universal, fast, and reliable analytical methods for fructosamine derivatives originate from the chemical nature of this carbohydrate structure, which is unstable at elevated temperatures and in the presence of nucleophilic and oxidation agents. It does not absorb appreciably in the UV-Vis range, is highly hydrophilic, and is not readily modified into analytically useful derivatives.

Historically, paper chromatography was the first method employed for direct evaluation of D-fructose–amino acids in dehydrated fruits.[157] Modern TLC methods are still very useful in the synthesis protocols,[58] as well as in food analysis, for example, of D-lactulose–phosphatidylethanolamine in dairy products by two-dimensional TLC/^{31}P-NMR.[158] Application of gas chromatography to the analysis of D-fructose–amino acids by Eichner *et al.*[159] provided a viable protocol, which requires initial separation of this fraction on ion-exchange columns, and their derivatization through oximation and trimethylsilylation reactions. The authors succeeded in assaying most D-fructose–amino acids in a number of dehydrated fruits and vegetables, including tomato, bell pepper, carrots, coconuts, and so on. Attempts to apply an amino acid analysis protocol or HPLC by the same[159] and other investigators[160] were less successful from a practical viewpoint, because of a heavy interference from free amino acids, sugars, and other components of the foods. On the other hand, HPLC separation of D-fructose–peptides and –amino acids from other peptides in enzymic hydrolyzates have been successfully optimized on boronate affinity,[161] ion-pairing reverse phase,[162] or ion-exchanger media.[163] Treatment of the hydrolyzates with the fluorescence-labeling agent 6-aminoquinolyl-*N*-hydroxysuccinimidyl-carbamate allowed for simultaneous determination of both N^{ε}-D-fructose-L-lysine and products of its degradation in glycated proteins.[164]

Mass-spectrometric methods have since the 1990s been widely considered for analytical determination of fructosamine derivatives in both food and tissue samples. Tautomerization of fructosamine derivatives in the gas phase likely favors the furanose anomers, as suggested by the fragmentation patterns that are dominated by neutral losses of 1–4 water molecules, as well as the loss of [$3H_2O$ + CH_2O].[165] Regioselective isotope labeling of the carbohydrate portion helped to formulate mechanisms of its gas-phase fragmentation.[166] For example, in N^{ε}-D-fructose–L-lysine peptides, loss of the first water molecule invariably involves elimination of the anomeric hydroxyl, followed by subsequent losses of O-3, O-4, and O-6 (Scheme 27).

Mass spectrometry proved to be a valuable method for determination of the glycation sites in peptides and proteins. A variety of ionization methods have been used, in historical order: fast-atom bombardment (FAB),[167,168] matrix-assisted laser desorption/ionization (MALDI),[69,71,75,169] and electrospray ionization (ESI).[74,170] In order to simplify the sequencing of D-fructose–peptides, it is desirable to minimize fragmentation of the fructosamine moiety, while exciting the peptide chain. Electron-transfer dissociation (ETD) or electron-capture dissociation (ECD) techniques, now well established tools for protein sequencing, are, to a large extent, fit for such a task.[166,171]

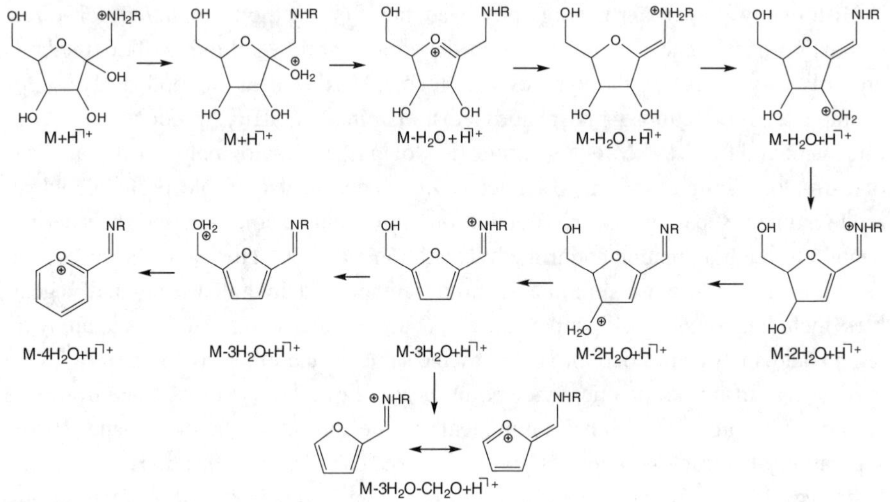

SCHEME 27

The utility of liquid chromatography–mass spectrometry (LC/MS) for determination of food-related D-fructose–amino acids has been evaluated in several studies.[172] In order to improve the resolution of D-fructose–amino acids, the use of pentafluorenylphenyl-bonded silica for the LC column has been recommended.[173]

Boronate affinity chromatography has been employed extensively for separation of glycated proteins,[174] and is a useful technique for pre-concentration of the analyte before MS analysis.[175] Phenylboronate binds strongly and selectively to fructosamine residues in the β-furanose form, and formation of this complex is not interrupted by the respective glucosylamines.[133] Incorporation of methacrylamidophenylboronic acid in polyacrylamide gels significantly improved the SDS-PAGE separation of glycated proteins from the unglycated ones.[176] Formation of stable complexes between D-fructosamine and the boronate anion may also be used for simplification of the fragmentation patterns in mass-spectrometric determination of glycated peptides, as this decreases the extent of carbohydrate dehydration in the complex.[177]

A combination of capillary electrophoresis and tandem MS is considered as a complementary technique to LC/MS, with the potential for higher resolving power in some cases. It was successfully applied to model mixtures of D-fructose–amino acids,[178] but its practical utility for determination of fructosamines in food/biosamples needs to be established.

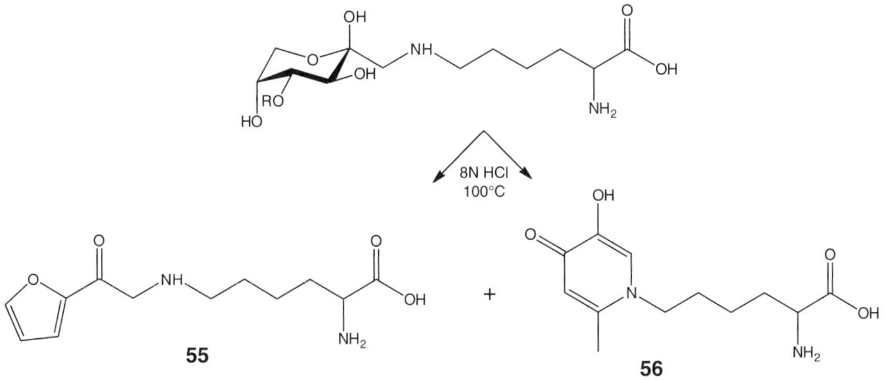

SCHEME 28

The most practical and popular method of indirect evaluation of D-fructosamine modification in food proteins is the "furosine method."[179] Upon acid hydrolysis of analyte proteins containing D-fructose–, maltulose–, or lactulose–lysine, the carbohydrate portion undergoes a dehydration reaction (Scheme 28), with the formation of N^{ε}-(2-furoylmethyl)-L-lysine (furosine, **55**). The reaction yield is about 30–50%,[180,181] but is reasonably reproducible under controlled conditions. Furosine absorbs at 280 nm and is well suited for HPLC analysis. Over the years, analysis of furosine became one of the most important methods for assessment of protein quality in processed foods, since it allows for estimates of the nutritionally unavailable ketose–lysine.[182–184]

N^{α}-D-Fructose- and -lactulose–amino acids also form respective N^{α}-2-furoylmethyl derivatives under similar conditions, although in much lower yields[181] and these were also used for HPLC assessment of D-fructose–amino acid content in dehydrated fruits,[185] vegetables,[184,186] and dairy products.[181] Along with furosine, a product second in abundance from the acidic hydrolysis of glycated proteins, pyridosine (**56**) has been proposed as an alternative indirect tracer of Amadori compounds in foods,[187] but it did not achieve as much popularity as furosine, because of lower (15–18%, based on εFruLys) yields that were less consistent.

In the clinical laboratory, determination of hemoglobin A_{1c} is one of the most widely used and important methods for mid-term glucose control in diabetic patients.[188] Attachment of a D-fructosyl residue to an amino acid nitrogen causes a decrease in its basicity (acid dissociation constant K_a) by about 1.5 orders of magnitude.[58] Although modification of hemoglobin by D-fructosyl residues also occurs at a number of lysine side-chains, these modifications do not significantly alter the protein total charge at pH 7.4, because of the relatively high basicity of lysine N^{ε} amino groups. In contrast, a single D-fructose group attached to the less basic *N*-terminal valine in the hemoglobin β-subunit causes a decrease in the level of protein protonation, significant enough for HbA_{1c} to be separated by ion-exchange HPLC and electrophoresis techniques.[189] In recent years, development of simpler techniques for determination of HbA_{1c}, which are based on enzymatic recognition of fructosamine, have attracted considerable attention by clinical analysts. The enzymatic approach utilizes the ability of fructosamine oxidases (see Section III.4) to specifically and sensitively recognize D-fructose–amino acids and –peptides, such as FruVal, after proteolytic digestion of a clinical sample.[190]

An alternative to the HbA_{1c} method for control of blood glucose was developed with the assessment of the modification by D-fructosamine of total serum protein. The serum concentration of protein-bound fructosamine is also proportional to an average glucose concentration in blood, and may serve as a mid-term glucose level marker.[191] The so-called "fructosamine assay" relies on the higher reducing potential of D-fructosamine with Nitroblue Tetrazolium dye (NBT), as compared to D-glucose and other molecules present in serum or plasma (except for ascorbic acid). Thiol groups may interfere and thus need to be blocked by iodoacetamide before addition of NBT.[192] Reduction of NBT by the analyte leads to the formation of a highly absorbing formazan dye, which can be assayed colorimetrically. As postulated by Baker *et al.*,[193] D-fructosamine reduces NBT by a one-electron mechanism, with an enamine and its enaminol radical as the principal reducing species, and "D-glucosone" (**32**) and free amine (protein) as the main products from oxidation of D-fructosamine (Scheme 29). In basic media, one-electron autoxidation of the enamine form generates superoxide free radicals, which can be quantitated by measuring the chemiluminescence produced in the superoxide–lucigenin reaction.[194] The fluorescence of fluorescein–boronic acid conjugates is quenched, dose-dependently, upon binding to fructosamine in plasma samples; this phenomenon can be used for alternative assay protocols for fructosamine with reagent selectivity higher than that shown by NBT.[195] The enzymatic

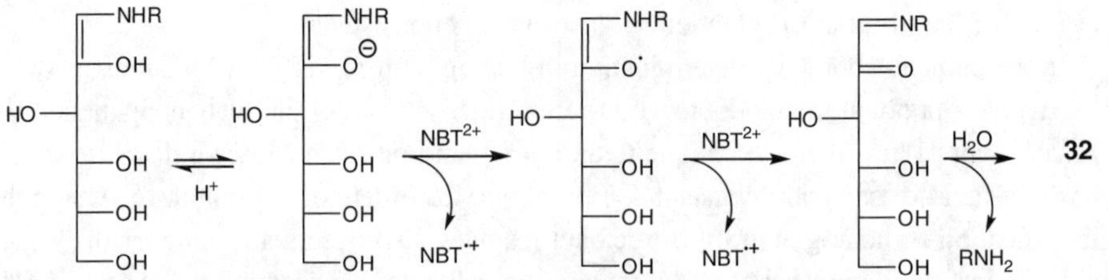

NHR OH HO OH OH OH ⇌ H^+ NHR O^- HO OH OH OH NBT^{2+} → $NBT^{\cdot+}$ NHR $O^{\cdot}$ HO OH OH OH NBT^{2+} → $NBT^{\cdot+}$ =NR =O HO OH OH OH H_2O → RNH_2 **32**

SCHEME 29

determination of fructosamine by εFruLys-specific oxidoreductases (see Section III.4) is also gaining popularity.

Although the hemoglobin A_{1c} assay is by far the more established in clinics than the fructosamine assay, the latter may be a good alternative in cases when accuracy of the former is compromised, as in anemia.[196] The content of D-fructosated serum proteins, such as albumin, transferrin, or IgG, correlates with the overall fructosamine content and their separate analysis may also be used for diagnostic purposes, in both clinical and veterinary practices.[197–200]

Immunological methods for analysis of D-fructosamine–protein conjugates are also being developed. A number of poly- and mono-clonal antibodies against lactosylated milk proteins[201,202] or glucosylated plasma proteins[203] and hemoglobin[204,205] have been reported in the literature since the 1980s.

The design of sensors for fructosamine analysis in biological samples is a rapidly expanding area of research and development. A significant effort is focused on sensors employing enzymatic detection of D-fructose–protein conjugates by fructosamine oxidases (see Section III.4.a) or other fructosamine-binding proteins. For example, a fructosyl amino acid-binding protein (FABP) from *Agrobacterium tumefaciens* was modified and conjugated with a fluorophore.[206] The construct is selective for FruVal in the presence of εFruLys, and provides for 100-fold higher sensitivity as compared to the enzyme-based detection systems. Another approach for developing fructosamine-sensing materials employs molecularly imprinted polymers (MIPs).[207] In one recent study, Rajkumar *et al.*[208] used a complex of D-fructose–L-valine with 4-vinylphenylboronate to copolymerize with ethylene glycol methacrylate (Scheme 30). The polymers were hydrolyzed to make template cavities, and such fructosamine-binding polymers were then successfully used to build amperometric[207] or thermometric[209] sensors that showed selectivity to FruVal in the presence of εFruLys.

3. Acid/Base- and Metal-Promoted Reactions

As already noted, the main interest of fructosamine derivatives in food and health sciences stems from their role as intermediates in the Maillard reaction in foods and *in vivo*. It is thought that the vast majority of D-fructosamine transformation reactions, such as dehydration, oxidation, and fragmentation, are effected by acid/base and transition metal-ion catalysts. A relatively limited effort has been spent on proper documentation of proton- and metal-binding characteristics of D-fructosamine derivatives, however. Röper *et al.*[17] evaluated acidity constants for the β-pyranose tautomer of several D-fructose–amino acids using ^{13}C-NMR. More precise pK_a values were obtained from

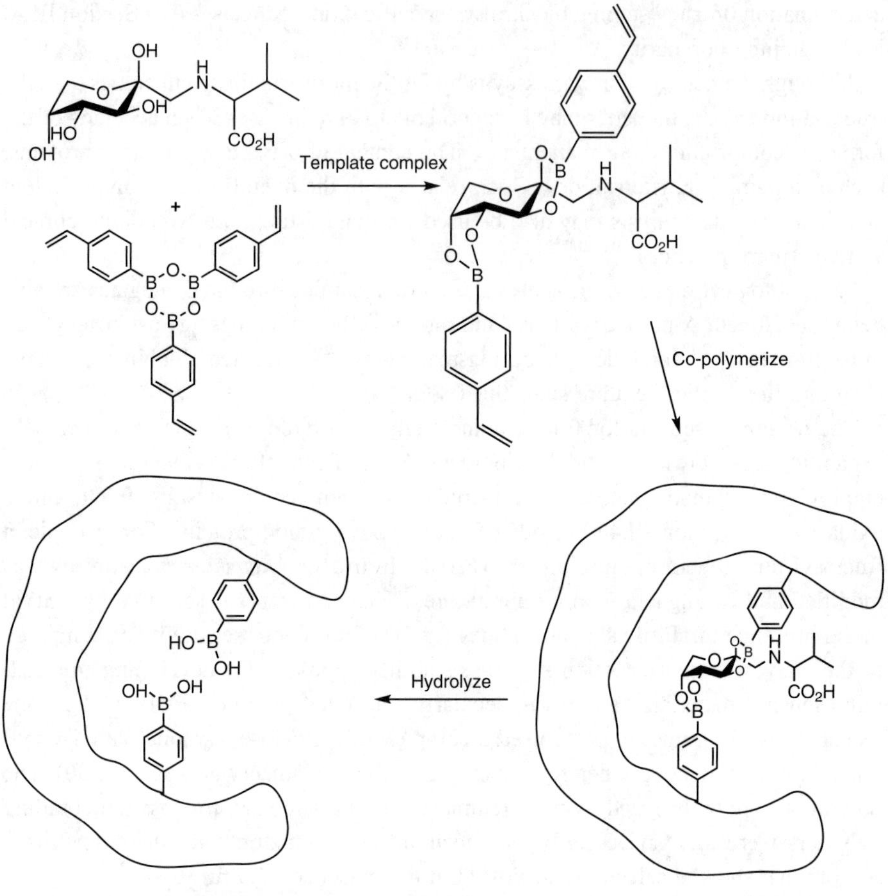

SCHEME 30

pH-potentiometric data (Table III). The increase in acid dissociation constants as a result of the amine derivatization by the 1-deoxy-fructos-1-yl residue is evident from Table III, and is more prominent for the values of amino K_{a2} (in the order of ~1.5) than to those of the neighboring α-carboxylate K_{a1} (in the order of ~0.2). Attachment of the second D-fructose group, such as in *N,N*-di-D-fructose–glycine, causes an even more dramatic drop in the basicities of both the amino and carboxylate groups. This effect has an important impact on the properties of glycated proteins, whose isoelectric point and total charge change as a result of D-fructosamine modification and the effect is widely used for analysis of fructosylated hemoglobin A_{1c} in clinical chemistry, as just mentioned.

TABLE III

Acid Dissociation Constants of Selected D-Fructose–Amino Acids at $T = 298$ K, $I = 0.2$ M

Compound	pK_{a1}[a]	pK_{a2}	Reference
D-Fructose–glycine	2.20 (2.33)[b]	8.18 (9.60)	58
D-Fructose–L-arginine	1.67 (1.98)	7.36 (9.02)	210
D-Fructose–L-histidine[c]	1.43 (1.71)	7.60 (9.09)	210
D-Fructose–L-phenylalanine	1.80 (2.18)	7.28 (9.11)	211
D-Fructose–L-valine	1.90 (2.26)	7.71 (9.42)	211
D-Fructose–β-alanine	3.35 (3.57)	8.84 (10.10)	58
D-Fructose–γ-aminobutanoic acid	4.03 (4.07)	8.93 (10.37)	58
N^{ε}-D-Fructose–N^{α}-formyl-L-lysine	3.08 (3.09)	9.02 (10.60)	58
N,N-Di-D-fructose–glycine	1.76 (2.20)	5.18 (8.18)	153

[a] K_{a1} and K_{a2} refer to the concentration acid dissociation constants of carboxyl and amino groups, respectively. [b] Respective values for parent amino acids are given in parentheses. [c] For the imidazole group, $pK_{aim} = 6.68$ (6.05).

One important omission in the foregoing acid/base studies is lack of data on the acidity of the anomeric hydroxyl group, whose pK_a is ~12.4 in fructose. It is the most acidic of the hydroxyl groups in fructose, it plays a crucial role in an unusual stabilization of metal complexes with D-fructose–α-amino acids, and its deprotonation is probably involved in the degradation mechanisms of fructosamine, as discussed next.

Affinity of D-fructosamine derivatives toward redox-active metal ions may be related to their mechanisms of oxidative degradation, but only a few studies have evaluated thermodynamic stabilities of metal ions with D-fructose–amino acids. Gyurcsik *et al.*[211] used potentiometric titration, EPR, and circular-dichroism methods to ascertain composition, structure, and stability of complexes of six D-fructose–amino acids with copper(II) and nickel(II). The study established that both metals readily form ML and ML_2 species at pH 3–5, but these complexes undergo deprotonation at pH 6–7. The authors suggested that the deprotonation may occur at the copper(II)-coordinated carbohydrate 3-hydroxyl group. In our opinion, however, the anomeric O-2 atom is better suited as a donor in the complexes, for two reasons. First, the authors' data suggest a square-planar arrangement of D-fructose–amino acid donor atoms around the metal ion, which would be better achieved with O-2 forming two conjugated 5-membered chelate rings in **57**, as compared to the 5- and 6-membered rings proposed by the authors. Second, deprotonation of the more acidic anomeric 2-hydroxyl groups (pK_a ~12.4) would be far more favored over the 3-hydroxyl group (pK_a ~14–15); however, no further deprotonation of the complexes was reported by the authors. On the other hand, the 3-hydroxyl groups may coordinate to copper(II) along with the deprotonated O-2, as was suggested for a Cu(II)/D-fructose–L-histidine complex species (**58**).[210]

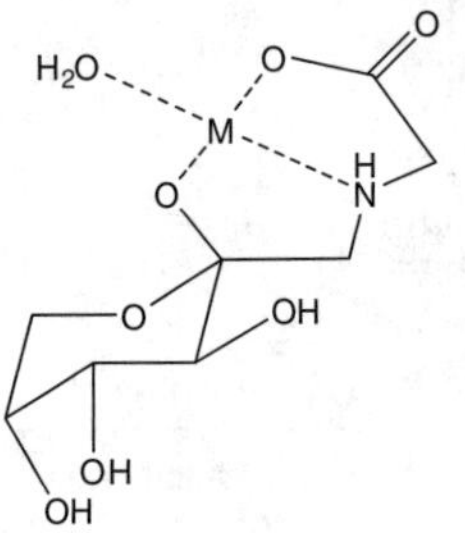

57 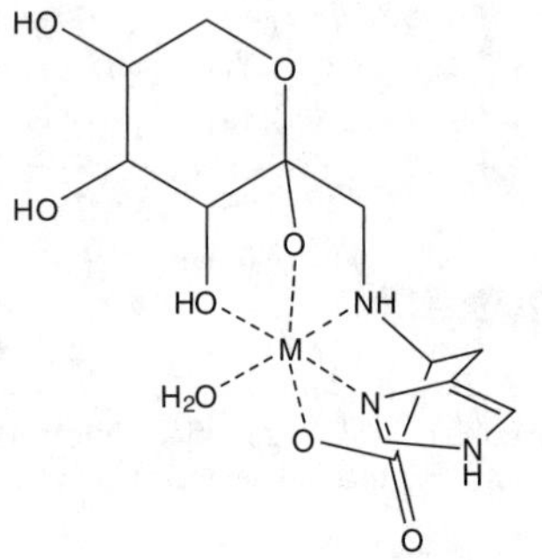 **58** 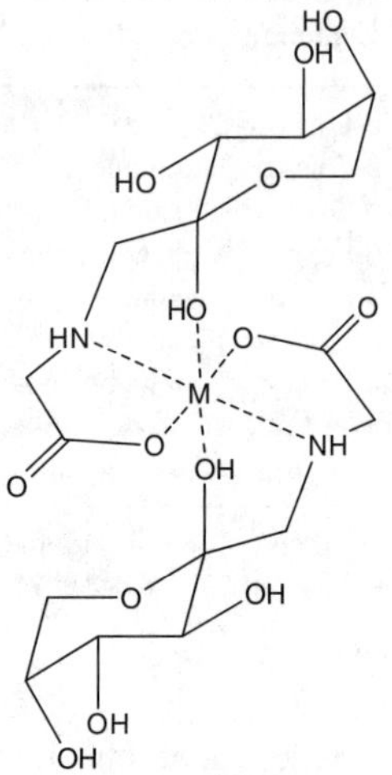 **59**

Comparison of the complex-formation constants for both 1:1 (**57** and **58**) and 1:2 (such as **59**) species[210,211] with those obtained for the respective copper(II) complexes with parent amino acids revealed that the fructosyl moiety provides for an additional chelate effect in D-fructose–α-amino acids and as a consequence, a significant increase in the complex stability. In the absence of an "anchoring" chelating group, such as α-carboxylate, the D-fructosamine structure is not a good copper(II) chelator, and Cu(II) expectably does not form stable complexes with the carbohydrate in N^{ε}-D-fructose–L-lysine peptides.[212] Although it would be expected that iron(III) complexes with D-fructose–amino acids in aqueous solutions,[213] no related thermodynamic equilibrium studies have been done so far for this important redox-active metal.

Solid complexes of D-fructose–glycine, –β-alanine, and –*p*-toluidine with palladium (II) and platinum(II) have been prepared and characterized by IR spectroscopy.[214] Using the spectral data, the authors proposed that Amadori compounds coordinate to the metals through the amine/amino acid portion, involving the amino N and carboxylate oxygen atoms. The carbonyl band at 1700–1720 cm^{-1} in the spectra indicates that the carbohydrate portion is present in the open-chain form, although the cyclic forms predominate in the complexes. Similar conclusions were drawn for the complex formation between FruGly and Pt(II) in aqueous solution on the basis of ^{1}H and ^{195}Pt NMR data.[215] When sulfur is present, as in D-fructose–L-methionine and –cystine, the coordination mode of the Amadori compounds to Pt(II) switches to that including the S and N atoms of the amino acid portion.[216]

a. Enolization.—The enolization reaction is central for understanding the mechanisms of the most fructosamine transformation reactions, leading to a multitude of Maillard reaction products, both in foods and *in vivo*. As shown in Scheme 25, formation of the enaminol, a product of 1,2-enolization, is a part of the tautomeric equilibrium displayed by D-fructosamines in solution. On the other hand, 2,3-enolization, as discussed next, probably prompts irreversible transformations of fructosamine. Enolization of D-fructosamines is catalyzed by nucleophiles in a manner similar to that already described for D-glucosylamines. Hydroxide ion and weak acids/anions, most notably phosphate and acetate, promote enolization of fructosamine, which may also occur in an autocatalytic manner, as in D-fructose–amino acids and –proteins, or in a non-catalyzed fashion at elevated temperatures. Smith and Thornalley[217] employed iodine uptake to determine kinetics of non-specific enolization of N^{ε}-D-fructose–N^{α}-hippuryl-L-lysine in phosphate-buffered solutions. The iodine uptake increased with increase of pH and phosphate/pyrophosphate concentration, but was unaffected by the protonation state of the fructosylamino group, prompting the authors to conclude the catalytic role of phosphate and hydroxide, rather than proton, in the enolization kinetics of the Amadori rearrangement product, according to the equation:

$$-d[I_2]_0/dt = [\text{fructosamine}]\cdot(k_A[H_2PO_4^-] + k_B[HPO_4^{2-}] + k_{OH}[OH^-]),$$

where $k_A = 3 \times 10^{-4}$, $k_B = 4 \times 10^{-3}$, and $k_{OH} = 37\ M^{-1}\ s^1$. The stronger the base, the stronger is its catalytic effect on fructosamine enolization.

The rate of the reversible 1,2-enolization can also be readily measured by NMR monitoring the exchange of methylene hydrogen atoms at C-1 with deuterium atoms from solvent D_2O.[38] Within a series of D-fructose–amino acids, the enolization rates were affected by the structure of Amadori compounds, the temperature, and the presence of phosphate nucleophilic catalyst (Table IV).

TABLE IV

First-Order Rate Constants of Methylene H/D Exchange at C-1 in Some D-Fructose–Amino Acids[38]

Compound	k (s^{-1})		
	Unbuffered,37 °C	**50 mM P_i,25 °C**	**50 mM P_i,37 °C**
D-Fructose–glycine	1.8×10^{-5}	9×10^{-5}	2.8×10^{-4}
N^{ε}-D-Fructose–N^{α}-formyl–L-lysine	4.7×10^{-6}	8.5×10^{-5}	
$N^{\alpha},N^{\varepsilon}$-Di-D-fructose–L-lysine	1.5×10^{-5}	1.5×10^{-4}	
N,N-Di-D-fructose–glycine	1.4×10^{-4}	1.8×10^{-3}	
$N^{\varepsilon},N^{\varepsilon}$-Di-D-fructose–$N^{\alpha}$-formyl–L-lysine	1.3×10^{-3}		

1,2-Enolization of D-fructosamine should, in principle, lead to a reversal of the Amadori rearrangement, which indeed proceeds to some extent. Davídek *et al.*[218] detected formation of D-glucose/D-mannose at 7:3 ratio in phosphate-buffered solutions of FruGly kept at 90 °C and at various pH values. We observed a slow (over a 5 year span) decomposition of lactulose–dibenzylamine solution at 8 °C, with the accompanying separation of crystalline lactose and the original amine.[28] Most of the time, however, the carbohydrate dehydration and degradation occur before any significant amounts of the parent D-glucose (and some D-mannose) can accumulate, making the reverse Amadori rearrangement an insignificant reaction.

Dehydration of the carbohydrate portion in fructosamines probably proceeds through a series of enolic intermediates (Scheme 31), which eventually tautomerize into α-dicarbonyl structures such as **60, 62**, and **63**, or cyclize into 2-furoylmethylamines (**61**) and many other heterocyclic derivatives. One of the major products of D-fructosamine degradation that occurs through the 1,2-enolization pathway is 3-deoxy-D-*erythro*-hexos-2-ulose (**60**), which is regarded as one of the most important intermediates in the Maillard reaction both in foods and *in vivo*.[219]

A number of authors have considered 2,3-enolization (Scheme 31) in order to explain the formation of many specific D-fructosamine degradation products. For example, it was proposed as a pathway to 1-deoxy-D-*erythro*-hexos-2,3-diulose (**63**), one of the major D-fructosamine degradation products captured in the form of adducts with α,β-diamines[153] or cysteine.[220] However, no isotopic exchange of hydrogen at C-3, that would be driven by 2,3-enolization, has been observed in D-fructose–amino acid (Table IV) solutions in D_2O at 25–37 °C, in the presence or absence of phosphate buffer.[38] This circumstance suggests low reversal rate of the 2,3-enol back to the original fructosamine, as compared to its further transformation reactions.

The migration of enolic double bonds along the carbohydrate chain in protein-linked D-fructosamine leads to 6-amino-3,6-dideoxy-6-*erythro*-hexos-2-ulose (**62**) and possibly represents a prevailing mechanism[221] of formation of major non-enzymatic protein cross-links, such as "glucosepane" (**81**) (also see Section IV.1.c).

b. Dehydration/Deamination/Deglycosylation.—Dehydration of D-fructosamine readily occurs upon application of minimal excitation energy, such as in the ionization chamber of a mass spectrometer with a soft ionization source (FAB, ESI), and was observed in the mass spectra of D-fructose–amino acids[58,222] or –aromatic amines. Such mass spectra normally contain the protonated molecular ion $[M + H]^+$ as a major peak, followed by intense peaks corresponding to dehydrated ions $[M - H_2O + H]^+$ and $[M - 2H_2O + H]^+$ (Scheme 27). Investigations on the fructosamine dehydration products have revealed a complex array of reactions that include dehydration,

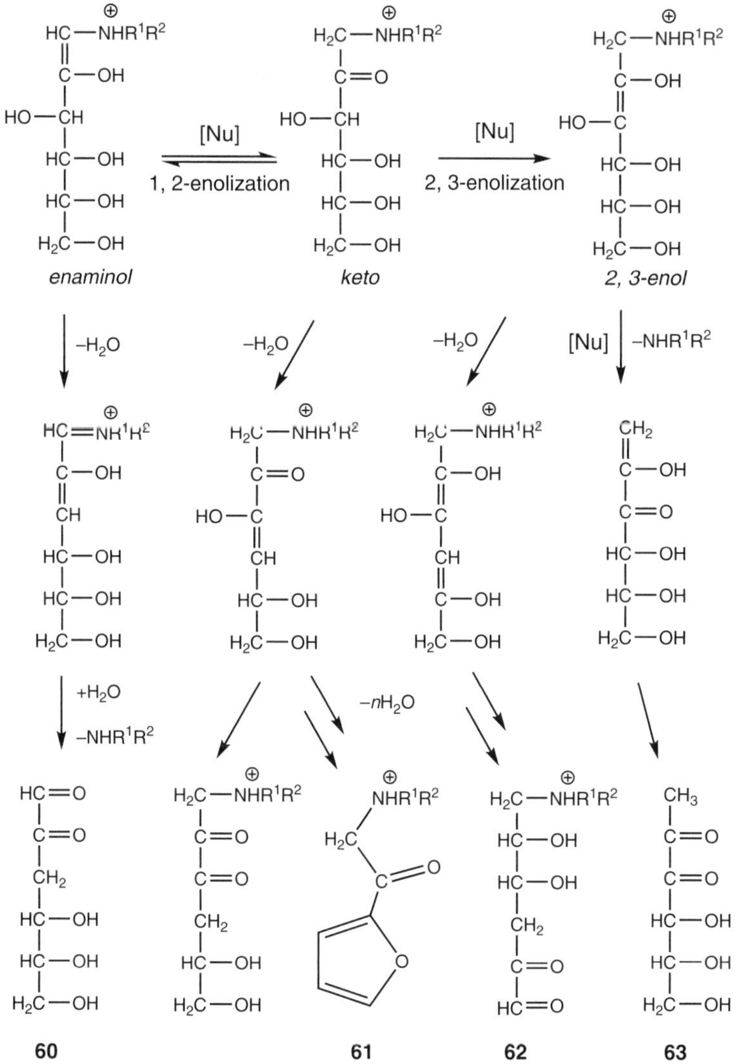

SCHEME 31

enolization, carbonyl migration, amine β-elimination, and others.[5,223] Scheme 31 provides a brief glimpse on some pathways that have been suggested, based on major reaction product structures identified so far that retained the original carbohydrate carbon skeleton.[224] A number of D-fructosamine dehydration products that retain the six-carbon skeleton and the same formal oxidation state have been identified as important intermediates and markers of the Maillard reaction. The list includes 1- and 3-deoxy-dicarbonyl products (**60** and **63**), 4,6-dideoxycarbonyl structures (**62**), furosine (**55** and **61**), pyridosine (**56**), HMF, and others.

Unsubstituted D-fructosamine is relatively unstable in comparison with its *N*-substituted derivatives and readily undergoes condensation and subsequent dehydration reactions, with "fructosazines" (**43**, **44**) as major products.[110,225] Such instability makes the preparation of D-fructosamine directly from D-glucose and ammonia impractical.

In addition to enolization pathways (Scheme 31), deamination of D-fructose–α-amino acids can proceed via the Strecker degradation reaction, with dual Strecker aldehyde formation if the reacting amine is also an α-amino acid (Scheme 32).[226,227] This mechanism widens the array of thermally generated pathways of food aroma formation in the Maillard reaction since, according to classic views, the Strecker aldehyde formation is necessarily preceded by α-dicarbonyl intermediates (see Section IV.1.a).

Di- and oligo-saccharide derivatives of D-fructosamines can undergo degradative deglycosylation following different mechanisms. Acid-catalyzed hydrolysis of the glycoside bond in lactulosamines proceeds readily and is the first step in formation of furosine (**55**) during the analysis of lactosylated milk proteins in dairy products.

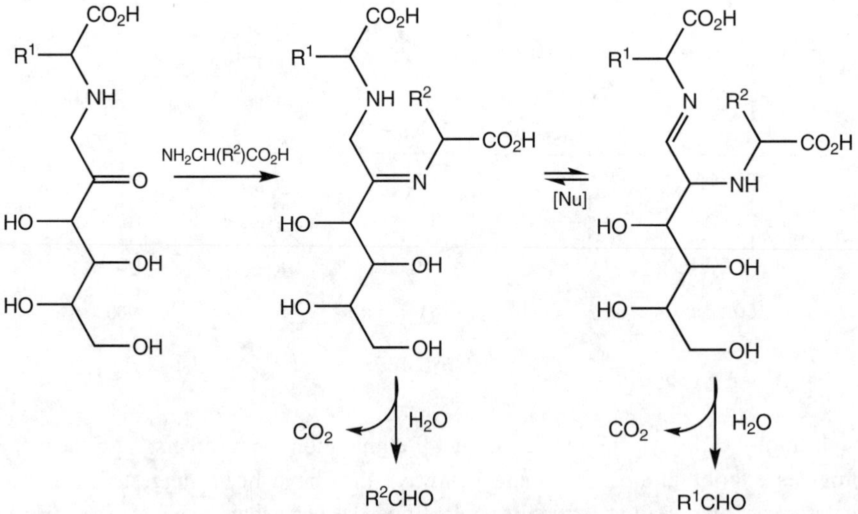

SCHEME 32

Alternatively, the Amadori compounds formed by maltose or maltotriose and lysine may degrade through the "peeling off" mechanism, which includes initial 2,3-enolization in the fructosamine moiety, followed by deaminations and deglycosylations through vinylogous β-eliminations at C-1 and C-4, with a 1,4-dideoxy-D-hexosulose as a major degradation intermediate.[228]

c. Oxidation–Reduction.—D-Fructosamines possess a reducing potential higher than that of free D-fructose and are readily oxidized by a large variety of oxidants, from redox metals, to ROS, to many organic molecules. Among these, autoxidation of D-fructosamines is by far the most important mechanism in their role as intermediates in the Maillard reaction in foods and *in vivo*. Enolization was suggested as the initial step of the oxidation of fructosamine by the one-electron mode. Indeed, enolization is a well established pathway for oxidative degradation of carbohydrates. As a matter of fact, enolization is the rate-limiting step in most of such reactions.[223] Most of the proposed pathways for oxidation of D-fructosamine, both enzymatic and non-enzymatic, suggest 1,2-enolic forms as the primary substrate species.[38,229]

(i) By Molecular Oxygen. Evidence for direct, "catalyst-free" oxidation of D-fructosamine by O_2 came from experiments where significant formation of superoxide radicals was detected in systems containing D-fructose–amino acids in Chelex-treated, DTPA-containing phosphate buffer, pH 7.[38] The rate of superoxide release correlated positively with proportions of the acyclic forms and 1,2-enolization rates measured for a number of the substrates.

(ii) By Metal Ions. Ions of copper and iron can participate in numerous redox reactions in biological systems because of their relatively low-energy one-electron transition between two oxidative states, such as in the Cu^{2+}/Cu^{+} and Fe^{3+}/Fe^{2+} pairs of aqueous ions. As discussed in Section III.3, copper(II/I) readily binds to fructosamines through anchoring amino and carboxylic groups, thus making the formation of metal complexes with the 1,2-enolized carbohydrate more likely. Iron(III/II) binding preferences are not as specific, and the redox potential does not make it a good catalyst for autoxidation of fructosamine, unless there is a good inflow of energy (temperature, irradiation), or lowering of the energy of the Fe(III)↔Fe(II) transition is attainable. Indeed, in the presence of many organic molecules, including D-fructosamine derivatives, the redox potentials of copper and iron may change dramatically, as a result of complexation and stabilization of one or both oxidation states, according to the Nernst relationship.

Addition of catalytic amounts of free copper ions, but not those of iron, to phosphate-buffered solutions of D-fructose–amino acids in the presence of molecular

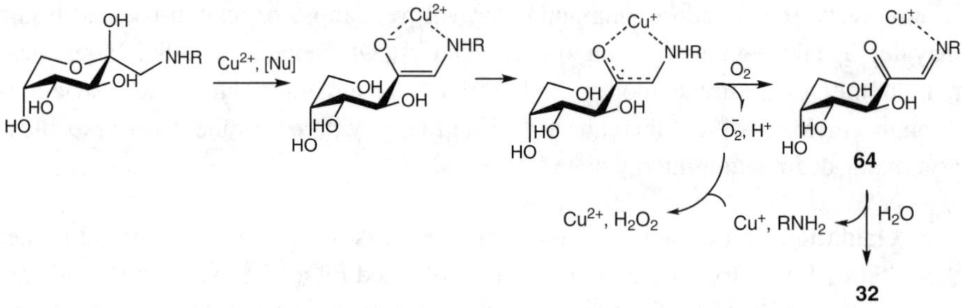

SCHEME 33

oxygen caused marked acceleration of superoxide ion-radical formation, which was attributed to direct oxidation of the Amadori compounds, presumably in the 1,2-enolic form (Scheme 33), by the metal.[38] Degradation of the copper(II)–D-fructosamine complexes at 37 °C in phosphate buffer was accompanied with the Fenton processes and generation of the hydroxyl radical.[230] D-*Arabino*-hexos-2-ulose (**32**) was established as the main carbohydrate product in copper-catalyzed autoxidation of D-fructosamines.[231]

Copper-promoted autoxidation of D-fructose–amino acids may be accompanied by the Strecker-like degradation pathway, which starts with decarboxylation of the D-*arabino*-hexos-2-ulose–amino acid adduct **64** and results in formation of the respective Strecker aldehydes.[232]

Autoxidation of D-fructosamine residues in glycated polylysine in the presence of iron(III) can be promoted by sunlight, as shown by monitoring formation of the main oxidation product, N^{ε}-carboxymethyl-L-lysine (CML, **65**).[233] The proposed oxidation mechanism includes 2,3-enolization of the carbohydrate and formation of the enolate complex with Fe^{3+} (Scheme 34). Ferricyanide anion, $Fe(CN)_6^{3-}$, is readily reduced by D-fructosamines and has traditionally been employed as a selective indicator for Amadori compounds in the presence of parent D-glucose or lactose, and amines.[234,235]

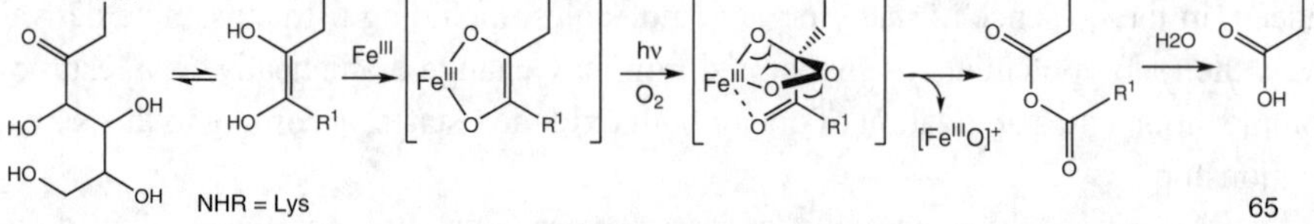

SCHEME 34

Sequestration of redox-active catalytic metal ions *in vivo* is considered as a possible mechanism of modulation of the Maillard reaction-promoted oxidative stress in diabetes by such drugs as pyridoxamine, a form of vitamin B_6.[134] On the other hand, fructosamines may themselves act as strong metal chelators and thus suppress the redox activity of copper[210] and, possibly, iron ions.

(iii) By Physiological Oxidants. A number of enzymes catalyze oxidation of D-fructosamines, with FAD or copper as electron-accepting species at the active site (see Section III.4.a). Peroxynitrite interacts with N^{ε}-D-fructose–L-lysine residues in glycated proteins, yielding CML (**65**) as a major oxidation product.[236] Hypochlorous acid also oxidatively degrades εFruLys residues in glycated proteins to CML under physiological conditions (Scheme 35). When Amadori-glycated human serum albumin was incubated in presence of activated neutrophiles, formation of CML was also detected, but could be completely abrogated by methionine, a selective HOCl scavenger.[237]

An interesting example of the biological utilization of fructosamine autoxidation reaction has been reported by Tanaka *et al.*[238] A wood-degrading fungus *Phanerochaete chrysosporum* secretes small glycoproteins that catalyze production of the hydroxyl radical needed for lignin degradation. The glycoproteins contain at least three D-fructosamine residues per molecule, oxidation of which is probably a source of H_2O_2, which is later converted into OH· in the Fenton reaction.

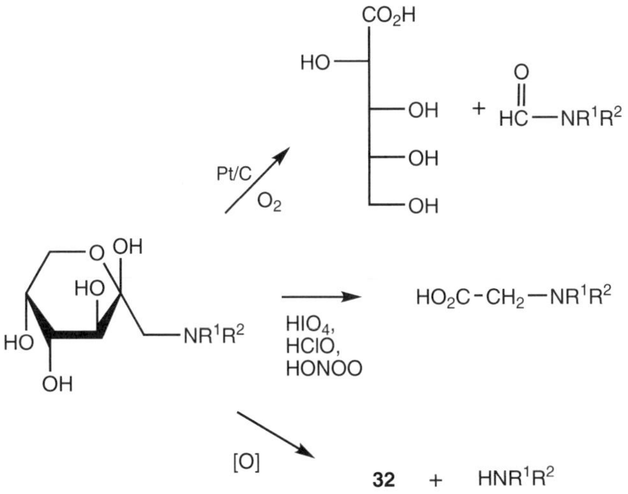

SCHEME 35

(iv) By Other Oxidants. Weygand and Bergmann[239] were the first to identify oxidation products of 1-arylamino-1-deoxy-D-fructose. Oxidation of D-fructose–*p*-toluidine by O_2 in alkaline medium in the presence of Pt/C targeted the 1,2-bond and afforded D-arabinonate and *p*-toluidine in high yield (Scheme 35). Periodate cleaves the 2,3-bond in the D-fructosamine portion to leave a carboxymethyl residue (Scheme 35). Anet[240] employed periodate oxidation of D-fructose–glycine and di-D-fructose–glycine to prove the structure of the latter. This reaction was successfully used later for identification of εFruLys formation sites in glycated proteins.[241] Oxidation of D-fructosamine by the redox dye NBT, which is widely used for the "fructosamine assay" in such biological samples as serum, severs the fructosamine C–N bond (Scheme 29) and yields D-*arabino*-hexos-2-ulose (**32**) and free amine. Other mild organic oxidants, such as benzoquinones,[242] can oxidatively deaminate the fructosamine molecule, releasing also **32**. D-Fructose–α-amino acids can probably participate in alternative Strecker degradation pathways,[243,244] which necessarily involve redox transformations of the carbohydrate. However, apart from the respective Strecker aldehydes, no other products of such reactions have been experimentally elucidated.

(v) By Reducing Agents. Reduction of D-fructosamines is of marginal interest as compared to the oxidation reactions. The carbonyl group of D-fructosamine can be reduced to an enantiomeric mixture of D-glucitol- and D-mannitol-amines by a number of suitable reagents, such as borohydrides. Formation of the enantiomeric mixture of D-hexitol-amines following reductive amination of D-glucose by substituted pyridylamines in the presence of $BH_3 \cdot Me_2NH$ was interpreted as a consequence of the intermediary Amadori rearrangement during the synthesis.[245]

4. Enzyme-Catalyzed Reactions

For years, there has been ample evidence for chemical transformations of D-fructosamine derivatives upon their exposure to bacterial,[246] fungal,[247] plant,[248] and animal[249] cells. These observations imply a number of enzymatic reactions that provide for metabolism/catabolism of the Amadori compounds *in vivo*. Several distinct groups of enzymes that recognize D-fructosamines as the substrates have been so far identified.

a. Oxidoreductases.—The largest group of known D-fructosamine-processing enzymes has been under extensive investigation for the past two decades, as part of an effort to devise approaches to lower the modification of plasma protein by D-fructosyl groups in patients on dialysis.

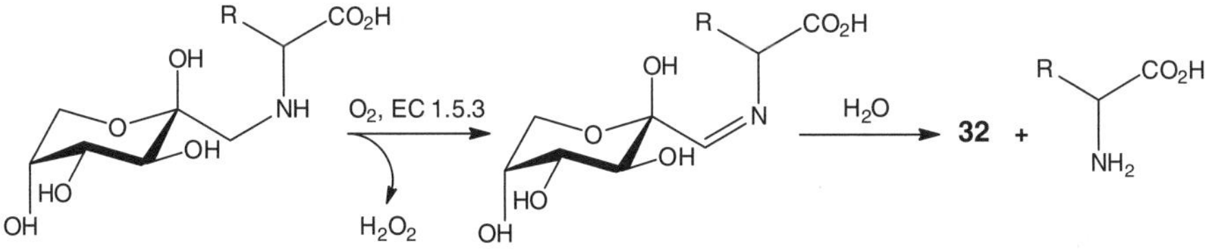

SCHEME 36

(i) Amadoriases Type I. In 1989, Horiuchi and Kurokawa reported[250] the isolation of a fructosamine oxidase from a soil *Corynebacterium* grown on D-fructose–glycine as the only carbon source. The enzyme catalyzes oxidation of D-fructose–α-amino acids by molecular oxygen, with the parent amino acid, D-*arabino*-hexos-2-ulose and hydrogen peroxide as the products of the reaction (Scheme 36). Screening the enzyme with a variety of Amadori compounds revealed that it requires a substrate as a derivative of any 1-amino-1-deoxy-ketose and an L-α-amino acid with a primary amino group, with the highest relative activity obtained for D-Fru–L-Leu, among other substrates tested.

For clinical application as a protein-deglycating enzyme, however, the carbohydrate cleavage in N^{ε}-D-fructose–L-lysine on glycated proteins is needed, and a search for enzymes capable of such activity has been ongoing for the past two decades (Table V). Most of the characterized fungal fructose–amino acid oxidases carry a relatively high

TABLE V
Amadoriases and Their Substrate Specificities

Source	**Substrate**	**Reference**
Type I[a]		
Corynebacterium sp.	Ketose–L-α-amino acid (primary amino)	250
Agrobacterium sp.	D-Fructose–L-α-amino acid	255, 256
Arthrobacter sp.	D-Fructose–α-amino acid	257
Aspergillus sp.	D-Fructose–amino acid	18, 258
Aspergillus sp.	D-Fructose–amine	259
Penicillium sp.	D-Fructose–amino acid	258
Eupenicillium terrenum	D-Fructose–dipeptide/α-amino acid	260
Fusarium oxysporum S-1F4	$N^{\alpha/\varepsilon}$-D-Fructose–$N^{\varepsilon/\alpha}$-Z-L-lysine	261
Pichia sp. N1-1	D-Fructose–L-amino acid	262
Achaetomiella virescens	D-Fructose–peptide/amino acid	252
Type II		
Pseudomonas aeruginosa	Ketose/polyol–amino acid	263

[a] To avoid any terminological confusion, the reported amadoriase I and amadoriase II from *Aspergillus* sp.[259] both operate by the type I mechanism.

level of homology with monomeric sarcosine oxidase and contain a covalently bound FAD unit as a cofactor.[251,252] The bacterial oxidases have a distinct amino acid sequence and non-covalently bound FAD. Unified under the "amadoriase" collective name, these enzymes catalyze the same D-fructosamine oxidation chemistry as depicted in Scheme 36. Substrate specificities of these two enzyme groups differ in general, with a preference to D-fructose–α-amino acids for the bacterial and to N^{ε}-D-fructose–L-lysine or *N*-alkyl-D-fructosamines for the eukaryotic amadoriases. The specificity seems to be at least partially determined by electrostatic and hydrophobic interactions in the active site, and thus could be manipulated by site-specific mutations.[253] Kinetic and structural studies with amadoriases I and II from *Aspergillus* showed[15,254] that the enzymes recognize their substrates in the D-fructopyranose form(s) and promote hydride transfer from fructosyl C-1 to the flavin coenzyme.

(ii) Amadoriase Type II. A distinct mode of the enzymatic oxidation reaction of Amadori compounds was discovered by Monnier and colleagues,[263,264] who reported isolation of an amadoriase from a *Pseudomonas* sp from soil. The enzyme catalyzes oxidation of D-fructose–amino acids at the amino acid residue, thus producing D-fructosamine and the respective carbonyl carboxylic acid, as shown in Scheme 37. In contrast to amadoriases type I, the type II enzyme has Cu(II) as a cofactor and is a cellular membrane protein.

In spite of the efforts, no amadoriase recognizing any D-fructose–protein has yet been identified. The active site in amadoriase is located too deep within the enzyme to allow access for such bulky substrates as D-fructose–proteins.[15] Unexpectedly, amadoriase I from *A. fumigatus* inhibited formation of D-fructose–proteins when it was present in a system containing D-glucose and the polypeptides.[265] The effect remains unexplained, since the amadoriase could not deglycate the same D-fructose–proteins after their formation, whether in native or denatured conformations. Nevertheless, practical utility of D-fructose–peptide/amino acid oxidases in clinics may be used in the enzymatic analysis of glycated proteins. Combinations of FruVal or εFruLys-specific amadoriases[266] and proteases/peptidases have been successfully tested for electrochemical[267–269] or spectrophotometric[270] detection of hemoglobin A_{1c} and other D-fructose–proteins in plasma. Other practical applications of amadoriases include washing and cleaning formulations.[271]

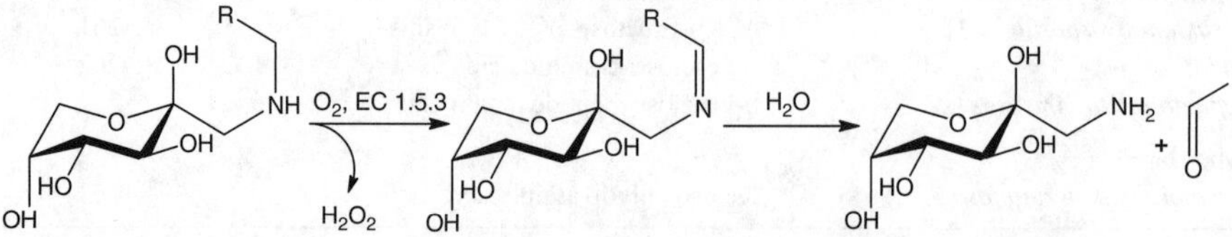

SCHEME 37

(iii) Mannopine Oxidoreductase. The enzyme from agrobacteria-induced crown galls is involved in the production of so-called opines (see Section II.2). Specifically, MOP oxidoreductase catalyzes reduction of D-fructose–L-glutamine to D-mannitol–L-glutamine (Scheme 22), which is released by the galls and utilized by the *Agrobacterium* sp. as a source of carbon and nitrogen.

b. Kinases and Phosphatases.—Discovery of the mammalian (de)phosphorylating enzymes that recognize D-fructosamine derivatives has opened a rapidly expanding area of research, because of their potential in diagnostics and the development of practical deglycating systems in diabetes and renal disease. The first report on phosphorylation of D-fructosamine by an algal fructokinase from *Anacystis montana* appeared in 1978,[272] but no further investigation of the reaction has followed.

(i) Fructosamine-3-Kinase. The enzyme (FN3K, EC 2.7.1) has been isolated from human erythrocytes,[273,274] and is particularly active in brain, heart, kidney, spinal cord, and skeletal muscle.[275] It catalyzes ATP-dependent phosphorylation of D-fructosamine derivatives, most importantly D-fructose–proteins, into respective D-fructosamine-3-phosphate derivatives (**66**). The kinase is specific to the D-fructosamine epitope and tolerates a variety of aglycon structures; however, its activity correlates positively with the substrate positive charge.[274] The proximity of the catalytic phosphate group to the anomeric and amino groups makes D-fructosamine 3-phosphate a rather unstable molecule which readily undergoes 1,2-enolization and subsequent degradation to 3-deoxy-D-*erythro*-hexos-2-ulose (**60**), inorganic phosphate, and free amine (Scheme 38). The initial excitement concerning the discovered enzyme was related to hopes of its physiological significance and clinical use as a protein deglycating tool.[276] However, it was realized that not all protein-bound D-fructose groups can be removed with FN3K;[277] presumably, those fructosamine residues that are located in more-rigid regions of a glycated protein would be less available to the catalytic site of the kinase.

The exact conformation of the D-fructosamine substrate in the active site of FN3K is not established yet; however, the best FN3K inhibitor structures, such as

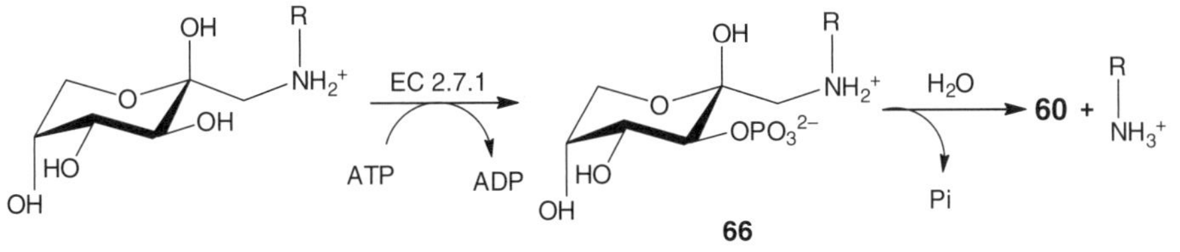

SCHEME 38

1-alkylamino-1-deoxy-3-*O*-methyl-D-glucitol, suggest that the carbohydrate likely assumes the acyclic tautomer form in the enzyme–substrate complex.[278]

Along with FN3K, mammalian genomes encode a FN3K-related protein, which shares with FN3K a significant sequence identity, but does not phosphorylate D-fructosamine derivatives.[279] FN3K-RP, however, acts on ribulos-, erythrulos-, and psicos-amines, and its possible physiological significance as a deglycating enzyme has been linked to phosphorylation of D-fructosamine-6-phosphates.[280]

(ii) Fructosamine-6-kinases. Wiame *et al.*[281,282] have identified and characterized this type of enzyme (FN6K) in bacterial extracts from *Bacillus subtilis* and *Escherichia coli.* Both kinases catalyze ATP-dependent 6-phosphorylation of D-fructose–amino acids (Scheme 39), but with essentially differing specificities. The FN6K from *E. coli* is substrate specific to N^{ε}-D-fructose–L-lysine, while the FN6K from *B. subtilis* favors D-fructose–α-amino acids.

Since FN6K from *E. coli* recognizes D-fructosamine residues on proteins, the enzyme has been proposed for use in colorimetric and electrosensor assays for glycated protein.

(iii) Phosphatases. Fortpied *et al.*[126] have reported the isolation of a magnesium-dependent phosphatase-1 (MDP-1, EC 3.1.3.48) that dephosphorylates protein-bound D-fructosamine 6-phosphate residues (see Scheme 23). The enzyme specificity is rather broad and includes, besides D-fructosamine 6-phosphate, other monosaccharides phosphorylated at the terminal carbon, while D-fructosamine 3-phosphate is not recognized by MDP-1.

c. Isomerases and Epimerases.—There is another group of bacterial enzymes that were reported to recognize D-fructosamine 6-phosphate residues and may act in concert with fructosamine-6-kinases to "deglycate" D-fructose–L-lysine and -α-amino acids.[281,282] The enzymes are encoded on the same frl (fructoselysine) operon in *E. coli* and *B. subtilis* as FN6K and were hypothesized to catalyze conversion of D-fructosamine 6-phosphate (**67**) into D-glucose 6-phosphate (**68**) and free amine (Scheme 39) immediately after **67** is produced with the aid of FN6K. The D-fructosamine 6-phosphate deglycases show the same substrate specificities with

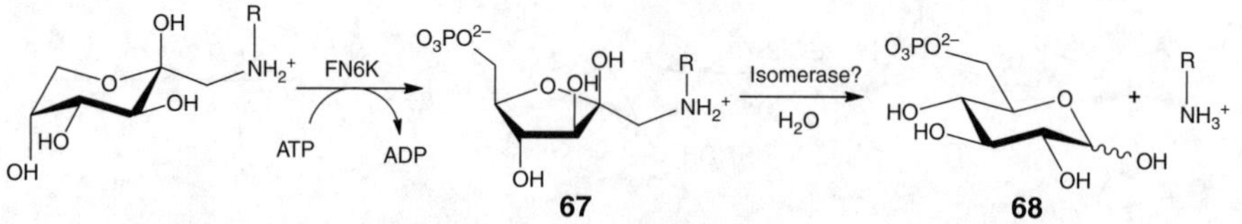

OH R NH$_2^+$ O OH HO OH FN6K ATP ADP O_3PO^{2-} O OH OH R NH$_2^+$ OH **67** Isomerase? H_2O O_3PO^{2-} HO HO O OH OH **68** + R NH$_3^+$

SCHEME 39

regard to the *N*-substituent structure as their FN6K counterparts. Although no information is available to suggest a chemical transformation mechanism served by these D-fructosamine-6-phosphate deglycases, it may be related to the mechanism provided by isomerases involved in D-glucosamine-6-phosphate synthesis and deamination,[284,285] and involve ring-opening and 1,2-enolization steps.

The frl operon in *E. coli* also contains a code for yet another D-fructosamine-modifying enzyme, fructoselysine-3-epimerase.[286] The epimerase is specific to D-fructose–L-lysine and converts it into D-psicose–L-lysine; it does not recognize D-fructose or 6-phosphorylated D-fructose–L-lysine.

d. Hydrolytic Enzymes.—Common hydrolytic proteinases readily digest D-fructose-modified proteins down to short D-fructose–peptides and –amino acids,[270] and the latter may then be subjected to the next step of the enzymatic assay by amadoriases (see Section III.4.a) or thermometric analysis on molecularly imprinted polymers.[209] Some endopeptidases, however, can also be inhibited by short D-fructose–peptides and –amino acids.[287] Efficient degradation of endogenous opioid peptides by enkephalinases requires the availability of the peptide terminal amino group for these aminopeptidases; hence, it is not surprising that the modification of the terminal amino group in D-fructose–Leu-enkephaline increased the peptide proteolytic stability in serum about 50-fold.[288]

Glycosidases can recognize, although non-specifically, D-fructosamine di- and oligo-saccharides as well. Thus, α-glucosidase from yeast catalyzes the slow hydrolysis of maltulose–glycine and maltotriulose–glycine to D-fructose–glycine and D-glucose, while glucoamylase (from *Aspergillus* sp.) and porcine alpha-amylase cleave the terminal α-glycosidic bond in maltotriulose–glycine only.[289] Human intestinal Caco-2 cells also exhibited the α-glucosidase activity, by hydrolyzing N^{ε}-maltulose– or N^{ε}-maltotriulose–N^{α}-hippuryl-L-lysine.[290] β-D-Galactosidase from intestinal bacteria, on the other hand, recognizes lactulosamine, and its activity could be readily assayed using N^{ε}-lactulose–L-lysine as a substrate.[291]

e. Miscellaneous.—Nitric oxide synthase in Korean red ginseng can use D-fructose–L-arginine as a substrate for NO production.[292] Mannopine cyclase from *Agrobacteria* converts D-fructose–L-glutamine into its *spiro*-α-lactone **51** (Scheme 22).

IV. Fructosamines as Intermediates and Scaffolds

Chemical transformations of D-fructosamine and its derivatives are of practical importance for a number of applications, especially in the food and health sciences.

1. The Maillard Reaction in Foods and *in vivo*

The Maillard reaction is defined as "an array of non-enzymatic chemical reactions between carbonyl (primarily carbohydrates) and amino compounds of biological origin". The significance and mechanistic consideration of the Maillard reaction has been reviewed in numerous publications over decades.[9,293,294] Traditionally, three main stages of the Maillard reaction are recognized. The initial stage involves the formation of glycosylamines/Schiff bases, as well as their Amadori rearrangement products, by any reducing carbohydrates and amino compounds. Section II.1.b describes reactions related to this initial stage. The intermediate stage involves dehydration, oxidation, fragmentation, and other reactions of the Schiff bases, Amadori compounds, as well as initial sugars and amines, lipids, and other biomolecules. These produce highly reactive intermediates, such as unsaturated carbonyls, dicarbonyls, furans, ROS, simple amines, and so on. Section III.3 deals with reactions related to this intermediate stage. The final stage of the Maillard reaction involves conversion of the intermediates into more stable volatile components, cross-linked products, and aromatics and colored polymers as a result of multiple recombination, degradation, and condensation reactions.

The main classes of Maillard reaction end products of practical interest include: volatile heterocyclic molecules responsible for aromas of thermally processed foods; colored oligomeric and polymeric molecules that determine non-enzymatic browning of thermally processed foods; protein and lipid cross-linking and other advanced glycation end products *in vivo*, which are implicated in pathogenesis of complications in diabetes and aging. Here we only consider the contribution of D-fructosamines implicated in the formation of these end products. A number of reviews containing detailed considerations of the formation of volatiles,[295,296] melanoidins,[297] and AGEs[298,299] can be recommended for further reading.

a. Formation of Aromas from Fructosamines.—Initial degradation reactions of D-fructosamine, as discussed in Section III.3, produce a number of deaminated carbohydrate intermediates, such as 1- and 3-deoxy-dicarbonyl hexoses, methylglyoxal, and others as well as the free amines and Strecker aldehydes, that are most commonly encountered in the Maillard reaction.[227] From this initial set, the aroma volatiles that form in foods can be conveniently separated into three groups, according to Nursten:[293]

(i) "Simple" sugar dehydration–fragmentation products, such as derivatives of furans, pyrones, cyclopentenes, carbonyl compounds, or acids (for example, structures and aroma descriptors as shown here);

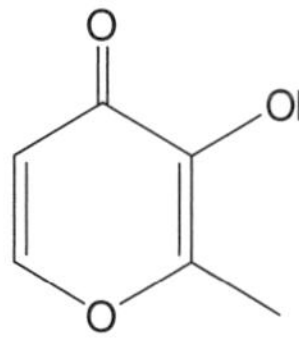

69 caramel-fruity; balsamic; 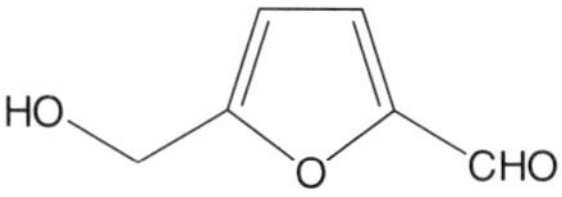 sweet-herbaceous; 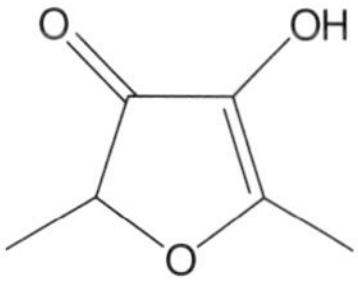 70 fruity

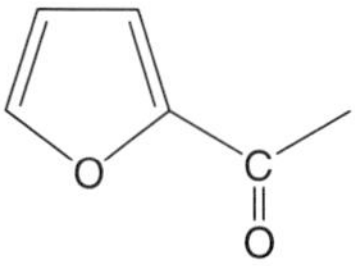

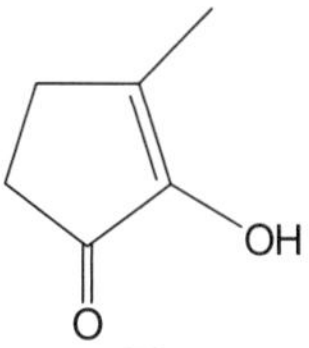

71 caramel; acidic; 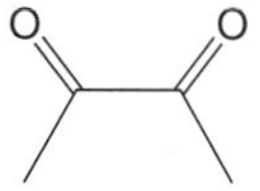 buttery

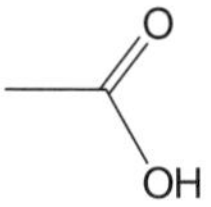

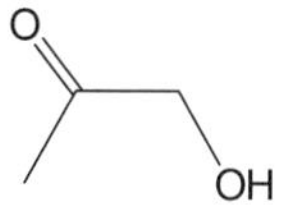

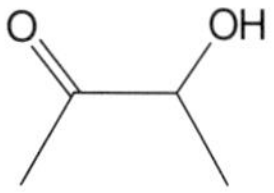

(ii) "Simple" amino acid Strecker-degradation products, such as formaldehyde, acetaldehyde, 2-methyl butanal, methional, other aldehydes, ammonia, H_2S, and other sulfides;

(iii) Volatiles produced by further interactions, such as pyrroles, pyridines, pyrazines, imidazoles, oxazoles, thiazoles, and so on (structures and aroma descriptors shown here).

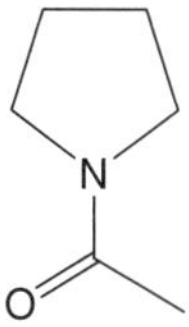

72 bread; corn; 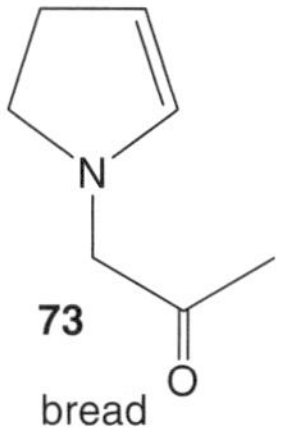 73 bread; cookie/mushroom; 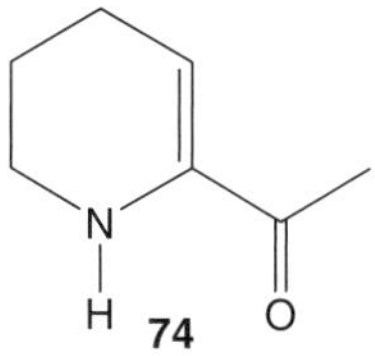 74 cracker

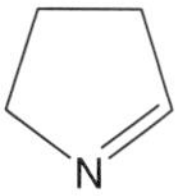

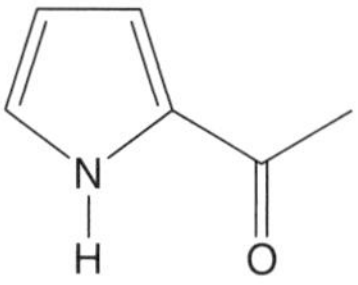

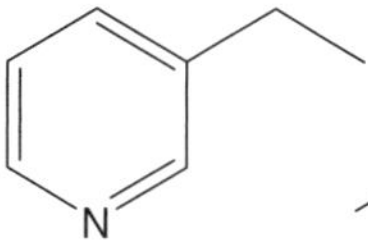

tobacco; 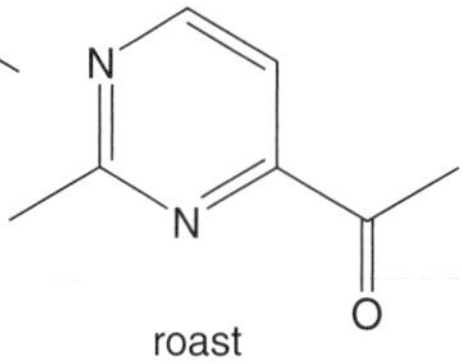 roast; 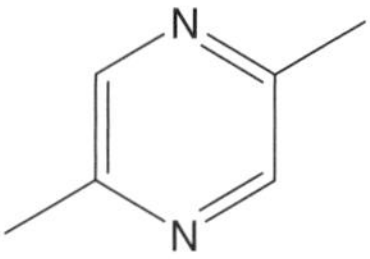 chocolate/roast; 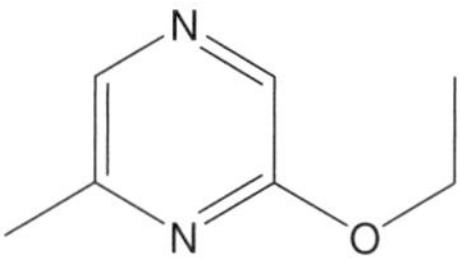 pineapple

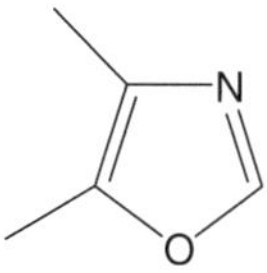

vegetable; 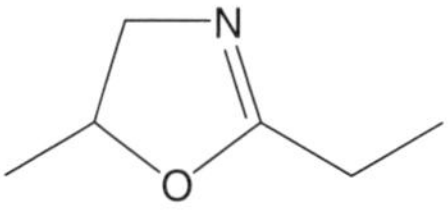 melon; 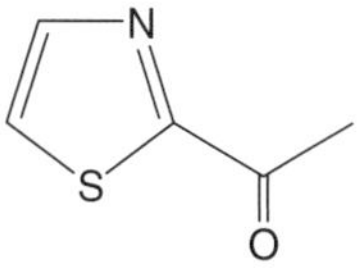 nutty; 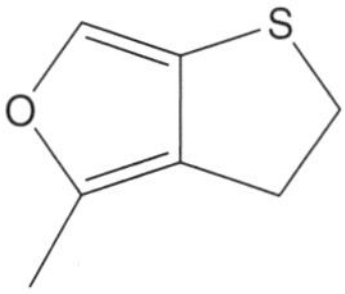 roast

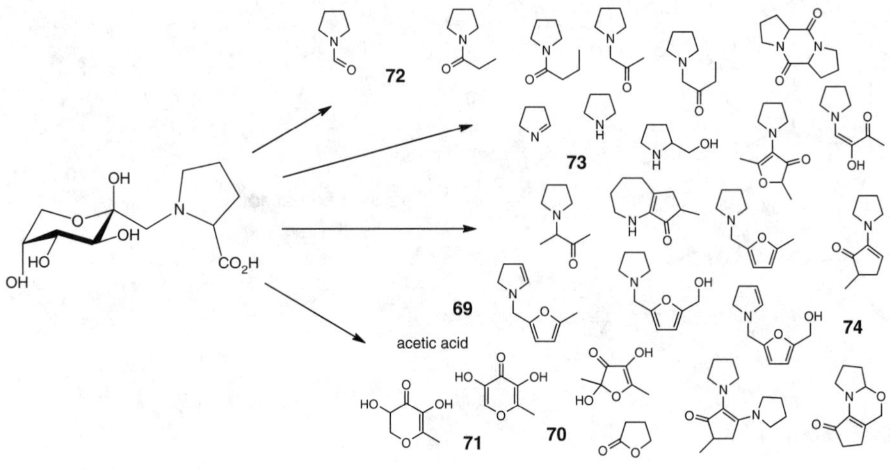

SCHEME 40

One straightforward approach to establish a casual relation between certain fructosamines and volatile products of their degradation would be the analysis of products formed during pyrolysis of individual D-fructosamine derivatives. This approach has been used to trace patterns of the heterocyclic aroma volatiles to specific D-fructose–amino acids. For example, L-proline has become one of the most studied amino acids in the Maillard reaction field, because of its proven contribution to the largely attractive character of aromas of roasted foods produced in reactions of this amino acid with monosaccharides (primarily D-glucose). D-Fructose–L-proline has been found in white vine,[300] licorice root,[301] cured tobacco,[302] dried apricots and peaches,[157,172] malts, and beer.[303] Pyrolysis or thermolysis of D-fructose–L-proline or mixtures of D-glucose and L-proline in model systems produce a variety of volatile products[304–306] (Scheme 40), with an overall caramel to nut-like aroma. These are largely similar to the volatiles detected in proline-rich products. Owing to this quality, D-fructose–L-proline has long been attractive to the food and tobacco industry as an ingredient for improving the flavor of some products.[301]

The overall aroma produced from thermally degraded D-fructosamine derivatives depends mostly on the nature of the aglycon[307] rather than on the carbohydrate, as exemplified in Table VI.

b. Melanoidins and Other Pigments.—Melanoidin is a general term for colored, polymeric structures that form as a result of the Maillard reaction in thermally processed foods. A common characteristic feature of melanoidins is strong absorption in the

TABLE VI
Thermally Generated Aroma from D-Fructosamine Derivatives or the Respective Model Systems[9]

Source	Conditions (°C)	Aroma	Reference
D-Fructose–L-alanine	100–220	Caramel	308
D-Fructose–glycine	100–220	Burnt/caramel	308
D-Fructose–L-lysine	100–220	Bread	308
D-Fructose–L-methionine	100–220	Fried potatoes	308
D-Fructose–L-phenylalanine	100–220	Chocolate	308
D-Fructose–L-proline	100–240	Nutty/caramel	304,308
D-Fructose–L-threonine	100–220	Burnt	308
Maltulose–L-alanine	130	Caramel	310
Lactulose–L-alanine	>100	Caramel	311
Lactulose–L-proline	>100	Nutty	311
L-Rhamnulose–L-methionine	>100	Fried potatoes	311

UV–blue range, which provides the familiar browning color to many baked and roasted foods. It must be noted that other types of polymers, such as those formed during caramelization of reducing sugars (with no participation from amino components) and autoxidation of phenolics, can also contribute to browning of foods on a par with melanoidins. The rates of formation of melanoidins from D-fructosamines are 1–2 orders of magnitude higher than those when D-glucose and the respective amines react under similar conditions; the presence of oxygen and catalytic metals further accelerates the rate of non-enzymatic browning by 1–2 orders of magnitude.[312] The structure of melanoidins is largely unknown; however, there is an evidence that pyrrole, pyrazine, pyridine, furan, imidazole, and other heterocyclic structures are present. The molecular weight of melanoidins is a result of many external factors, including moisture, temperature, and reaction duration, among others. Melanoidins prepared from D-fructose–amino acids generally show a high level of similarity to the materials that form in similar conditions from the corresponding mixtures of D-glucose and amino acids.

Di- and oligo-saccharide derivatives are less sensitive to dehydration reactions in the course of the Maillard reaction, and contribute less to color formation, as compared to the respective monosaccharides.[313]

c. Advanced Glycation End Products.—Another type of high-molecular-weight products of the advanced Maillard reaction forms as a result of the interactions between reducing carbohydrates, such as glucose, and such high-molecular-weight biological amines, as proteins. In these reactions, although N^{ε}-D-fructose–L-lysine forms as a major structurally defined product of protein glycation,[164] its degradation products can further rearrange and recombine with the protein amino groups of lysine and arginine side-chains and form a large variety of structures, such as **65, 75–88,**

commonly termed as "advanced glycation end products", also abbreviated as AGEs.[223] So far, several AGEs have been detected *in vivo* and in foods (structures: pyrraline,[314] pentosidine,[315] crossline,[316] argpyrimidine, glucosepane,[317] CML,[318] CEL,[319] CMA,[320] GOLA,[321] GOLD, DOLD,[164] FPPC,[322] GODIC, MODIC,[323] and DOGDIC[324] are depicted here). Formation of AGEs expectedly leads to modification of the physical and functional properties of proteins. Thus, structures **80–88** constitute AGEs that bridge parts of the same protein molecule or cross-link two separate proteins. Structures **80, 82**, and **85** are responsible for the characteristic fluorescence of glycated proteins. The immunogenic properties of structures **65** and **75** have been confirmed by generating specific antibodies to these epitopes. Some of the AGEs are recognized by specific cell-surface receptors, and many are clinically relevant as risk predictors and markers for diabetes-related diseases.[13]

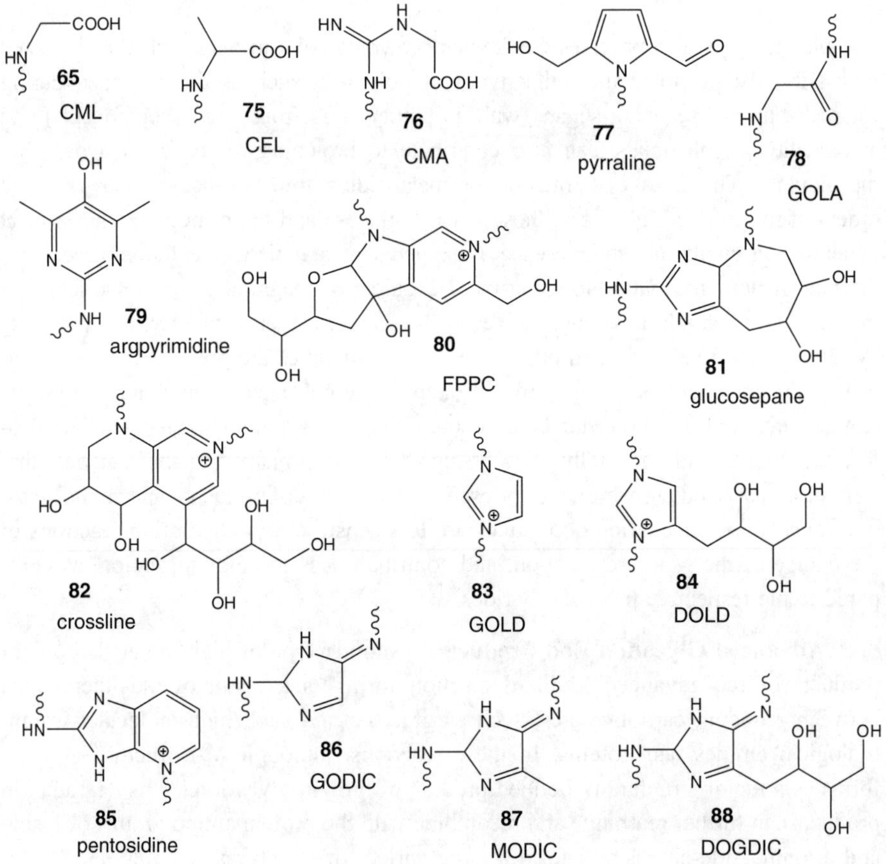

65 CML

75 CEL

76 CMA

77 pyrraline

78 GOLA

79 argpyrimidine

80 FPPC

81 glucosepane

82 crossline

83 GOLD

84 DOLD

85 pentosidine

86 GODIC

87 MODIC

88 DOGDIC

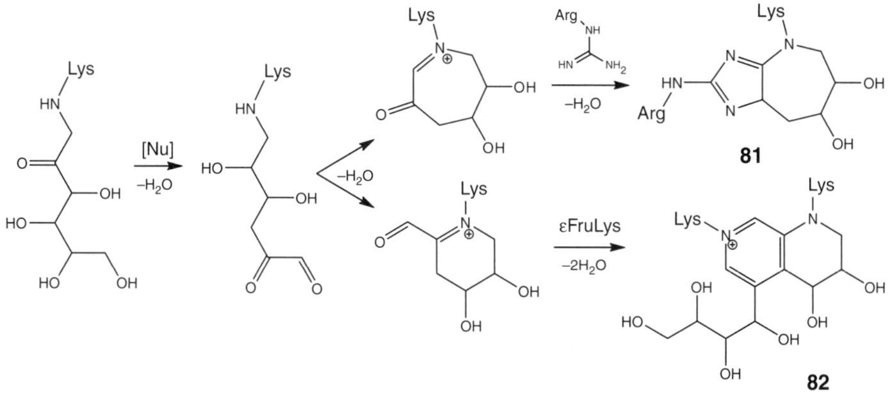

SCHEME 41

Although all of the AGE structures shown may form in systems containing D-glucose and proteins, less than half of these contain the original carbon backbone of D-glucose, suggesting that multiple sugar fragmentation and recombination reactions, which are generally characteristic for the Maillard reaction, precede formation of the AGEs. Moreover, other physiological carbonyl compounds, including pentoses, trioses and their phosphates, methylglyoxal, and lipid peroxidation products—, can lead to these and many similar AGEs found *in vivo*. For such structures as **81** (glucosepane) and **82** (crossline), N^{ε}-D-fructose–L-lysine seems to serve as a sole carbohydrate precursor[325] (Scheme 41).

2. Neoglycoconjugates

In recent years, the Amadori rearrangement reaction has attracted the attention of medicinal chemists as an alternative approach for preparation of glycoconjugate drugs. The main target properties are (i) lowered systemic or hepatic toxicity, (ii) improved bioavailability, (iii) enhanced hydrophilicity/solubility, and (iv) the multivalent presentation of a glycoepitope. In general, coupling a drug molecule to a carbohydrate can significantly influence, in many cases, the drug pharmacokinetics, for the reasons listed.

Various cytotoxic nitroso compounds have been proposed as antitumor agents over the years. Some of the D-fructosamines are included in this pool. Direct nitrosation of D-fructose–amino acids led to the respective *N*-nitroso derivatives.[326] Conjugation of

S-nitroso-*N*-acetylpenicillamine through the amide bond led to a stable and active agent, fructose–1-SNAP (**89**), with over 20-fold enhanced cytotoxicity against prostate adenocarcinoma cells.[137]

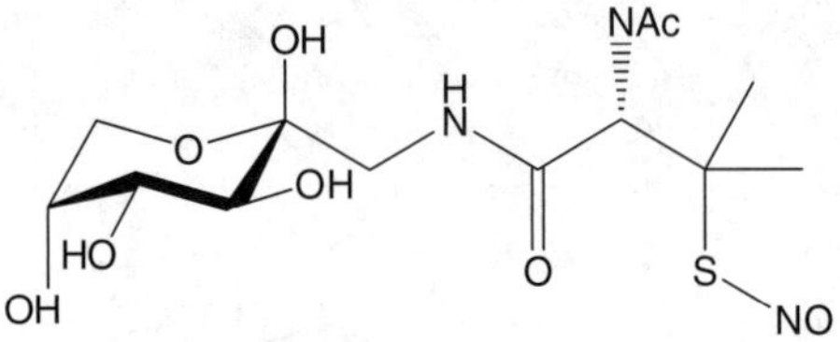

89

A few examples of therapeutically active, fructosamine-based neoglycoconjugates are also given in Section V.2.

A prospective tumor-imaging agent and radiopharmaceutical, DOTA-labeled α-melanocyte-stimulating hormone, was conjugated to D-fructosamine or maltotriulosamine in order to augment homing of the drug to melanoma tumors.[327] Accumulation of radiolabeled octreotide peptides (which target somatostatin receptors overexpressed in a variety of tumors) in medulloblastoma xenografts was tripled when the Lys-5 side-chain was conjugated with a D-fructose residue through the Amadori rearrangement.[328,329] Similar improvements in pharmacokinetics of a radioiodinated octreotide were achieved in patients who have been given a maltotriulose-[^{123}I]Tyr^{3}-octreotate.[330] It is thought that this modification diminishes the lipophilicity of the peptide and thus its hepatobiliary excretion. The lactulosamine, maltulosamine, and other Amadori-type modifications have been tested to improve the bioavailability and stability of octreotide.[331]

Radiolabeled bombesin analogues, which are also promising tumor-imaging agents, were successfully conjugated with D-fructose groups through the side-chain amino group of lysine.[332] The modification significantly lowered undesired accumulation of the agent in the liver, allowing for a better toxicity profile, as well as somewhat improved absorption and retention of the agent in experimental tumors.

Breast cancer cells overexpress the D-fructose transporter GLUT5, which also selectively takes up D-fructosamine and its conjugates. Thus, conjugation of D-fructosamine with such fluorescent aglycones, as Cy5.5, proved an important approach for development of breast-tumor imaging reagents.[25]

Insoluble globular proteins, such as casein at a pH close to its dielectric point, can be micellarized by means of graft copolymerization with dextran through the Amadori rearrangement.[333,334] Such "bio-friendly" micelles dissociate at pH values that differ from the protein pI and could be potentially used as hydrophobic or hydrophilic drug carriers, and for other biomedical applications.

3. Synthons

The Amadori rearrangement of glycosylamines has long been employed as a convenient access method to ketose derivatives. For example, Kuhn *et al.*[335] demonstrated a facile conversion of lactose into lactulose in 3 steps via diazotization of lactulosamine (Scheme 42).

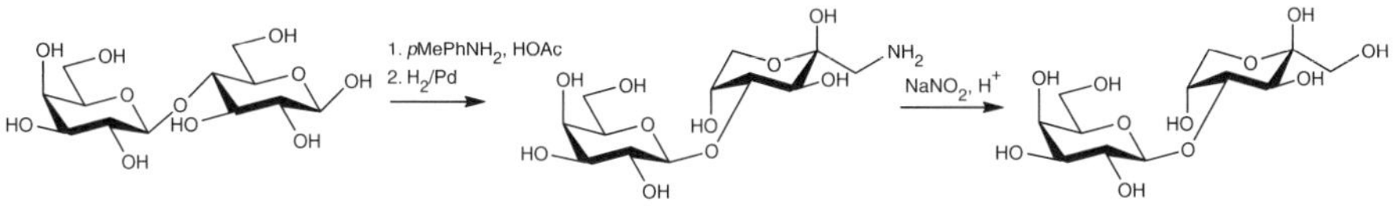

SCHEME 42

The Kiliani reaction with fructosamine and other ketosamines may provide an access to α-polyhydroxyalkyl-β-amino acids,[336] for use as chiral metal chelators or building blocks for glyco- and peptido-mimetics.

Treatment of *N,N*-disubstituted D-fructosamines with calcium hydroxide provided a simple and inexpensive one-pot access route to 2-*C*-methyl-D-ribonolactone (**90**),[337] an important synthon for branched 2′- and 4′-*C*-nucleosides, 4-*C*-methylpentuloses, and branched imino sugars, from D-glucose. 1-Deoxy-D-*erythro*-hexos-2,3-diulose (**63**) is thought to be an intermediate in this reaction (Scheme 43).

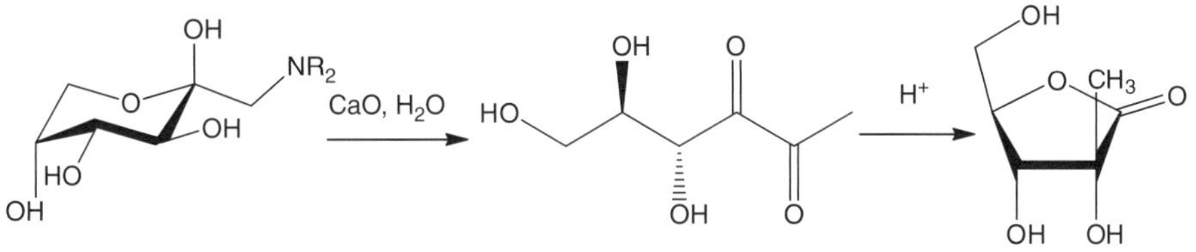

SCHEME 43

Wrodnigg *et al.*[338] developed an original synthetic approach to the glycosidase inhibitors **93**, which starts with 5-azido-5-deoxy-D-glucose (**91**) and proceeds through the intermediate 5-azido-5-deoxy-D-fructosamines (**92**) (Scheme 44).

D-Fructosamine undergoes Büchi cyclocondensation reaction with nitriles, providing access to a variety of disubstituted imidazoles (Scheme 45), some of which have been proposed as potential sphingosine-1-phosphate lyase inhibitors for the treatment of such autoimmune disorders as rheumatoid arthritis or multiple sclerosis.[339]

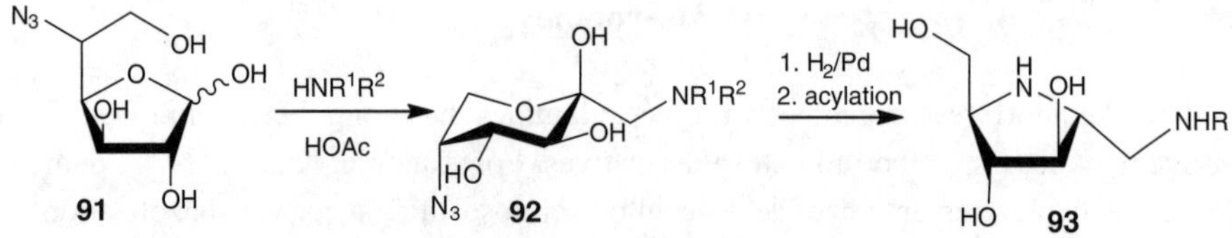

R = H, Ac, $C_3H_6NH_2$, $C_6H_{12}OH$, COC_5H_{11}, $COC_5H_{10}OH$
R^1 = CH_2Ph, C_2H_4CN, $C_6H_{12}OH$
R^2 = H, CH_2Ph

SCHEME 44

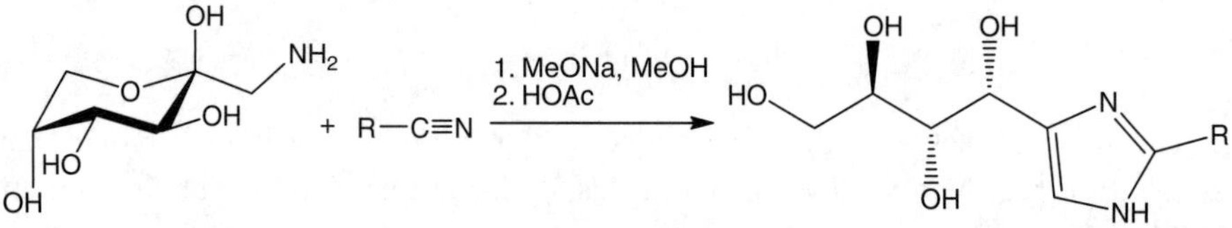

SCHEME 45

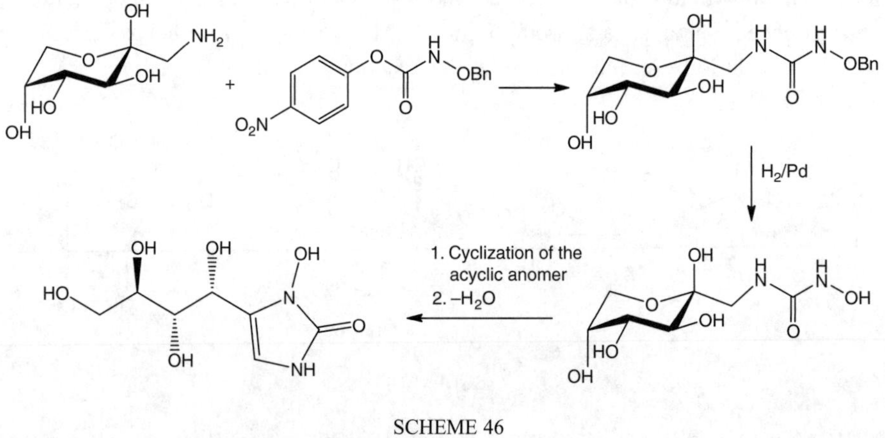

SCHEME 46

An attempt to synthesize D-fructose-*N*-hydroxyurea as a metal-chelating agent ended up with its spontaneous cyclization (Scheme 46).[24] A similar result was obtained upon employing D-glucosamine as the amine donor.

N-Protected D-fructosamine was readily converted into a fused 1,3-oxazolidine-2-thione derivative **94**, which had been prepared as a prospective inhibitor for the specific fructose transmembrane transporter GLUT5 on intestinal brush-border or breast tumor cells (Scheme 47).[108]

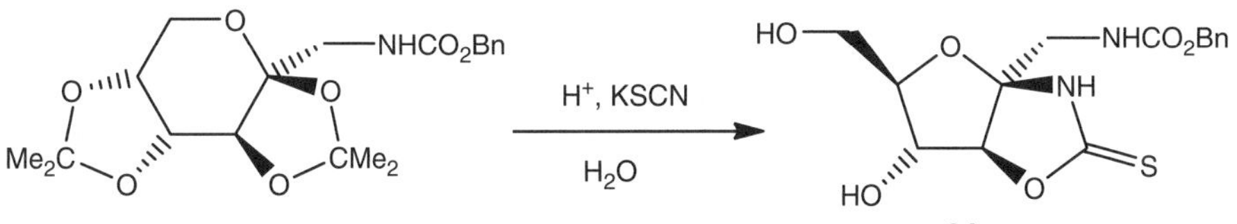

SCHEME 47

Unprotected D-fructosamines react with isothio- or isoseleno-cyanates (Scheme 48) affording respectively 4-hydroxy-4-polyhydroxyalkylimidazolidine-2-thiones or -selones **95**.[340,341] The latter were readily dehydrated to give the respective imidazoline-2-thiones and -selones **96** and **97**, which in effect are stable sulfur- and selenium-containing *C*-nucleosides.

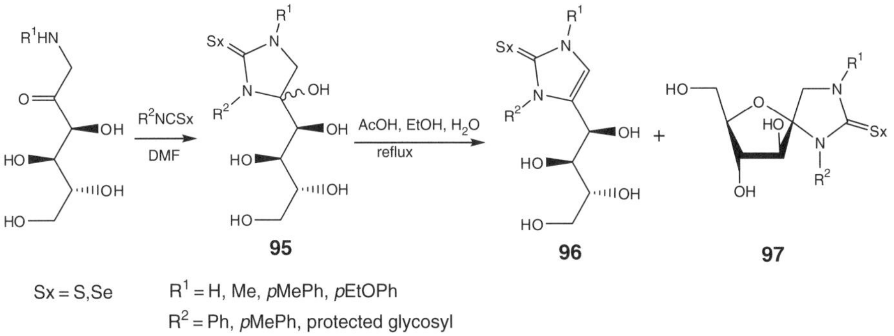

SCHEME 48

In an efficient synthesis of oxazolidinone **99** and morpholinone **100**, catalysts for the enantioselective epoxidation of asymmetric alkenes, D-fructosamines **98** have been successfully employed as starting materials (Scheme 49).[342,343] Oxazolidinone **99** (R^1=H) has also been successfully employed as the auxiliary for a chiral methacrylate reagent in an asymmetric synthesis of optically active γ-butyrolactones.[344]

Phosphoroamidites of protected D-fructosamine derivatives **101** have been prepared as chiral ligands for coordination to Ni(II) and Cu(II), and the resulting complexes were employed as catalysts for asymmetric dialkylzinc and trialkylaluminum addition reactions with carbonyls[345] and enones.[346]

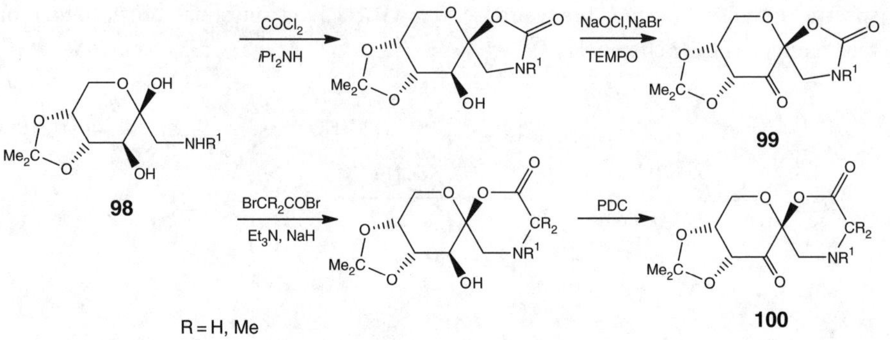

R = H, Me
R^1 = H, Ac, Boc, aryl, *n*-C_6H_{13}

SCHEME 49

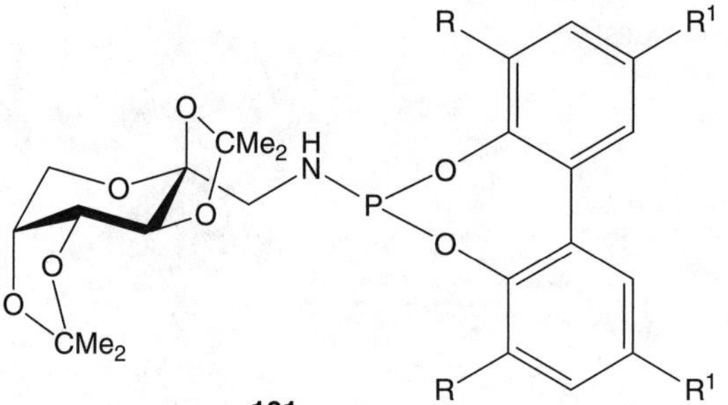

V. Biological Functions and Therapeutic Potential

Recent studies of the behavior of fructosamine in biological systems have provided solid experimental evidence that, besides being intermediates in chemical (mostly Maillard-related) transformations, D-fructosamine derivatives can regulate a number of cellular activities, from metabolism to adhesion to apoptotic signaling.

1. Endogenous Fructosamines in Plants and Animals

As free endogenous D-glucose can and does interact with proteins and other free amino compounds, it is reasonable to suggest that endogenous D-fructosamines were present in living organisms from the onset of life and that molecular mechanisms for their recognition, utilization, and/or neutralization have evolved. As attachment of the fructose residue to a protein may cause significant changes in the polypeptide conformation,[347] solubility,[348] and charge distribution, the functional properties of glycated proteins usually change as well, including decrease in enzymatic activity,[53] native ligand affinity,[72] mechanical properties,[349] and so on.

a. Opines.—Pathogenic members of the *Agrobacterium* genus highjack the root cellular machinery of the plant host by tranfection of Ti plasmid to induce uncontrolled plant-cell growth and to produce opines in the crown galls. Opines are a class of carbohydrate derivatives that serve as a nutrient source for the agrobacteria, and which incorporate several D-fructose–amino acids and related molecules. The enzymatic systems for opine synthesis and utilization were considered in Section II.2.

b. Physiological Significance of D-Fructosamines.—This question is not yet understood. Healthy humans contain about 5–7% of plasma proteins, including hemoglobin, albumin, transferrin, immunoglobulins, and fibrin, which are modified with D-fructose residues.[66,350] Low-molecular-weight D-fructosamines have also been detected in the plasma of healthy human subjects. The total fructosamine content in the blood of healthy individuals is about 0.2–0.25 mmol/L. Variations in the fructosamine content may reflect lifestyles of individuals, including the consumption of carbohydrates, fats, and alcohol, and physical activity.[351] It is possible that the physiological level of protein glycation, which exists despite presence of the FN3K deglycating system (Section III.4.b), carries certain regulatory functions. An indication to such a possibility may be served by a recently discovered phenomenon of glycation-promoted chaperon activity of human αA-crystallin.[352]

Other suggestions concerning the physiological significance of D-fructosamines may be drawn from a compilation of data on known classes of fructosamine-binding proteins as presented in Table VII.

c. Fructosamine in Diabetes and Related Complications.—D-Fructose-modified albumin was identified as the predominant form of circulating D-fructosamine in humans, both healthy and diabetic.[359] It has shown several biologic effects which were causally linked to the pathogenesis of complications in diabetes, including oxidative stress-related damage to microvascular, renal, and other cells that would be exposed to circulating D-fructose–albumin.[12]

Thus, a number of research groups have observed stimulation of the intracellular ROS production in cardiac myocytes,[360] mesangial,[361] or endothelial vascular cells[362] by D-fructose–albumin. This stimulation has been related to the activation of NADPH

TABLE VII
Types of Fructosamine-Binding Proteins

Type	Representative protein, source	Function	Substrate/ligand	Reference
Oxidoreductases	Amadoriase, bacteria, fungi	Catabolism	D-Fructose–amino acids	Table V
Kinases/phosphatases	FN3K, mammals	Catabolism	D-Fructose–proteins	Section III.4.b
Periplasmic-binding proteins	SocA, *Agrobacterium*	Transport	D-Fructose–α-amino acids	353
Hexose transporters	GLUT5, animals	Transport	Small D-fructosamines	354
	THT1, *Trypanosoma*			355
Membrane proteins	Nucleolin, myosin homologues, animals	??	D-Fructose–L-lysine	356
	Calnexin, mammal		D-Fructose–albumin	357
β-Gal-specific lectins	Galectin, animals	Signaling, cell adhesion	D-Lactulosamines	358

oxidase and protein kinase C (PKC-α), and which is also followed by overexpression of the E-selectin on the endothelial cell surface.

Modification of human serum albumin by D-glucose impaired its antiapoptotic activity for endothelial cells; a 50% loss of activity corresponded to about four fructosamine residues per protein molecule.[363] In another study, D-fructose–albumin enhanced nitric oxide synthase activity and a consequent NO-dependent apoptosis in vascular endothelial cells.[364] This effect was also attributed to a possible mechanism of hyperglycemia-induced vasculopathy in diabetic patients. In proximal tubular epithelial cells, however, the effect of co-incubated D-fructose–albumin was the opposite,[365] suggesting that the glycated protein acts as an effector of cell signaling, rather than being a direct inhibitor of NO synthase.

Vascular remodeling in diabetes may reportedly be accelerated through induction by D-fructose–albumin of the antiapoptotic protein IAP-1 or the mitogenic nerve-growth factor NGF-γ in smooth muscle cells.[366]

Binding of D-fructose–albumin to monocytes led to activation of the kinases MAPK p44/42 (ERK1/2) and p38, with subsequent translocation of a transcription factor NF-κB into the nucleus.[367] The activation of MAPK involved PKC- and MEK-1-dependent pathways, suggesting a complexity of cellular signaling induced by D-fructose–albumin, while its physiological significance remains poorly understood. As a clue for the mechanism involving D-fructose–albumin in development of atherosclerotic lesions in diabetic patients, the glycated albumin induced expression of hyaluronan receptor CD64, as well as macrophage scavenger receptors CD36 and CD68 on the surface of resting human monocytes, and thus contributed to differentiation of these cells into foam macrophages.[368]

Other stimulating activities of D-fructose–albumin include expression of the extracellular matrix proteins α1 (IV) collagen and fibronectin, the activation of cell-signaling kinases PKC-β1 and ERK, and the expression of mRNA encoding the fibrogenic Transforming Growth Factor (TGF)-β1 and its primary signaling receptor, the TGF-β type II receptor.[369] A decreasing burden of the glycated albumin in diabetic mice not only decreased overexpression of TGF-β1, but also restored renal insufficiency related to proteinuria, vascular endothelial growth factor (VEGF) production, and other pathological dysregulations.[370] Interestingly, the D-fructose–albumin-induced effects on NF-κB transcription and activation of related pro-inflammatory genes (COX-2, iNOS, TNFα, IL-1β) in human peritoneal mesothelial cells decreased with the donor's age, suggesting that the early glycated proteins may not be directly involved in the diabetes-related renal complications that accelerate with age.[371]

In human mesanglial cells, Co-incubation of human mesanglial cells with D-fructose-albumin induced transcription of the plasminogen activator inhibitor-1 (PAI-1) promoter.[372] The PAI-1 gene was probably downstream-activated by TGF-β1, whose expression is mediated by Fru–BSA. Because in diabetic nephropathy PAI-1 is involved in abnormal accumulation of extracellular matrix in the mesangium, glycated albumin is thought to be involved in promotion of such complication by upregulation of PAI-1.

Multiple cellular effects of D-fructose–albumin imply the existence of specific cellular receptors for this molecule, and a search for such receptors has identified so far a number of such sites on monocytes, peritoneal and alveolar macrophages, endothelial cells,[373] and fibroblasts. Cell-membrane proteins that may interact with D-fructose–albumin through its N^{ε}-D-fructose–L-lysine epitopes include nucleolin, myosin heavy chain,[356] and calnexin.[357]

D-Fructose–hemoglobin (HbA_{1c}), besides its diagnostic value, has recently attracted attention as a potential cause of oxidative stress in diabetes, through iron release, Fe-mediated free-radical reactions, and carbonyl formation in hemoglobin.[374] *In vitro*, it increased intracellular superoxide and activated NF-κB and activator protein-1, which are redox-regulated transcription factors, in cultured aortic smooth-muscle cells,[375] resembling similar cellular responses observed in smooth muscle cells exposed to D-fructose–albumin.[376] Such activity suggests participation of D-fructose–hemoglobin, along with D-fructose–albumin and possibly other fructosamines, in the development of diabetic vasculopathy. Transferrin glycation also leads to an impaired capacity of the Fe-binding protein to keep this redox-active metal ion in the inactive Fe^{3+} oxidation state in plasma.[377]

Glycation of glutathione may be another contributor to development of the oxidative stress, since D-fructose–L-glutathione and D-fructose–L-glutathione disulfide are poorly recognized by their processing enzymes, glutathione peroxidase, glutathione-*S*-transferase, and glutathione reductase.[88]

The *in vivo* Maillard reaction cascade (Section IV.1), which produces multiple ROS and AGEs, has been firmly identified as a major contributor to practically all complications in diabetes with poor glycemic control. Hence, inhibition of degradative transformations of endogenous or food D-fructosamine, as well as trapping of the resulting reactive intermediates, constitutes an attractive target for ameliorating such complications. The reader is directed to the relevant literature.[299,378,379]

d. Fructosamine in Other Diseases and Aging.—Elevated concentrations of circulating D-fructose–plasma proteins may be caused not only by poorly controlled D-glucose levels in blood, but also by impaired clearance of these proteins, such as in renal disease.[380] Fructosamine in plasma can be used as a non-direct indication of nematodal infections in animals.[381] In addition, accumulation of D-fructose residues occurs in proteins with low turnover rates, such as collagen. As D-fructose–protein buildup exceeds a certain threshold, it triggers, both directly (through binding to cellular receptors, Section V.1.c) and indirectly (through production of degradation and autoxidation products, Sections III.3 and IV.1.a) a host of pathophysiological responses implicated in atherosclerosis,[382] Alzheimer's disease,[383,384] and so on.

Poorly controlled hyperglycemia gives rise to glycation of not just polypeptides. D-Fructose–phosphatidylethanolamine (Fru-PE) has been identified as a major lipid-glycation product in diabetic patients.[385] This compound significantly stimulated proliferation, migration, and tube formation by human vascular endothelial cells *in vitro*.[386,387] In addition, matrix metalloproteinases-1 and -2, key enzymes in the initiation of angiogenesis, are released by the cells upon exposure to Fru-PE. The pro-angiogenic properties of Fru-PE thus may contribute to the vascular disease in diabetes, as well as to the observed positive association between plasma fructosamine level and risk of certain cancers in diabetic patients.[388,389]

2. Fructosamines as Potential Pharma- and Nutra-ceuticals

a. Fructosamine Derivatives in Foods.—Ingested food provides a constant supply of D-fructose–amino acids and –proteins for humans, estimated at about 0.5–1 g FruLys equivalent per day.[390] In 1955, Wager suggested the formation of D-fructosamines in dried carrots.[391] At about the same time, Adachi identified lactulose–L-lysine in tryptic hydrolyzates of protein from evaporated milk.[392] The contents of

TABLE VIII

Fructosamine Derivatives in Dehydrated Foods, mg/100 g Dry Weight

Compound	Carrot	Raisins	Prunes	Dates	Figs	Dried apricots	Tomato powder	Onion	Cereals	Cheese	Milk powder	Hypoallergic formula
FruγAbu	700–920[a,b]	10–76[a]	22[a]	18[a]	8[a]	4[a]	1000–1500	95–140[a,b]				
FruAla	690–1190[a,b]	0–16[a]	4.5[a]	0[a]	0[a]	0[a]						
FruArg		10–62[a]					150	940–2100[a,b]				
FruGlu							700–3800					
FruHis							40–50					
FruLys	1540–1550[a,b]	7–31[a]	21[a]	16[a]	14[a]	8[a]		580–890[a,b]	1–10[a]			
FruPro		2–8[a]	3.6[a]	2.9[a]	4.2[a]	210,5.7[a]						
FruVal	61–88[a,b]						40–50					
LctLys[c]										4–20[a,b]	50–450[a,b]	1300–2700[b]
LctAla												820–2140[c]
LctIle												730–1600[b]
LctLeu												1060–2580[b]
LctVal												480–1500[b]
References	184,400	185	185	185	185	172, 185	159,210	186	401	402	183	181

[a] Quantified as *N*-furoylmethyl-amino acids, unfactored for D-fructose– or lactulose–amino acid content.
[b] Per 100 g protein.
[c] Lct, lactulose.

some individual D-fructosamine derivatives in dehydrated foods are given in Table VIII. Such Amadori compounds as FruArg and FruLys were also found in yeast extracts.[393] In wheat gluten hydrolyzates, which are used in seasonings, both D-fructose– and maltulose–amino acids (Lys, Glu, Gln, Val) have been identified.[394] In spray-dried egg yolk and lecithin products, up to 15% of phosphatidylethanolamine is converted into the Fru-PE conjugate.[395] The total content of Amadori compounds in some dehydrated foods may be significant, reaching up to 10% per dry weight in tomato powder,[159] or 5% per protein mass in onion powder.[396]

Olfactory properties of D-fructose–amino acids have been evaluated and found to largely reflect the parent amino acid taste.[312] Most tested compounds had sweet (FruAla, FruβAla, FruGly, FruPro), bitter (FruHis, FruIle, FruLeu, FruTrp, FruTyr), or bitter-sweet (FruLys, FruPro, FruVal) taste, the acidic component was present in some Amadori compounds (FruAsp, FruGlu, FruSer) and two of these (FruAsp, FruGlu) displayed the umami taste.

Interestingly, fumonisin B1, a liver-targeting mycotoxin found in corn, is detoxified during manufacturing of tortilla chips, partially through its interaction with glucose and formation of the respective Amadori compound and products of its degradation.[397]

On the other hand, D-fructosamines may constitute a source of toxic food hazards, such as the *N*-nitroso compounds (NOC), that are suspected mutagens and carcinogens in many meat products. D-Fructose–amino acids rapidly react with nitrite to form NOC in high yield (Scheme 50), and thus constitute a major class of NOC precursors in such thermally processed, nitrite-containing foods as frankfurters.[398] Although nitrosation of 1-arylamino-1-deoxy-D-fructose caused significant cellular damage *in vitro* by the diazonium cations generated,[399] the cytotoxicity and mutagenicity of *N*-nitroso-D-fructose–α-amino acids are, however, disputed.[326]

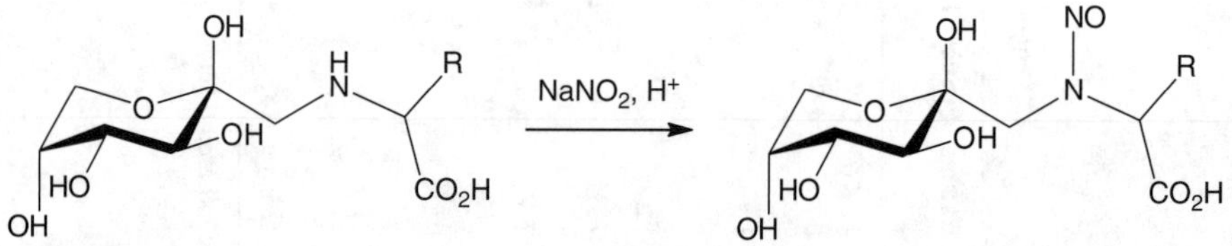

SCHEME 50

b. Bioavailability.—The nutritional value of D-fructose–amino acids has been assessed in order to explain the decreased availability of essential amino acids in thermally processed foods. Horn *et al.*[403] supplemented a methionine-deficient rat diet with D-fructose–L-methionine, but a lack of the weight gain in the animals led to the conclusion that methionine is not available from FruMet. Similar conclusions were drawn in regard to amino acid availability from FruLeu, FruTrp,[404] and FruPhe.[405] A number of studies traced the metabolic destiny of dietary N^{ε}-D-fructose–L-lysine from ingested D-fructose–proteins. Somoza *et al.*[406] established that, while fecal recovery of εFruLys from glycated casein fed to rats was around 1%, much larger amounts of unchanged εFruLys could be detected in urine (4–5% recovery) and kidneys (25–90%). Since the discovery of lactulose modifications of lysine in milk proteins, it became established that such lysine is also no longer available to mammals as a nutrient.[407] Analyses of urine samples of adults fed lactulosamine-rich milk showed,[408] however, measurable amounts of lactulose–L-lysine, demonstrating that the latter was partially absorbed into the bloodstream after digestion of milk protein, and then excreted unchanged. In experimental pigs, the maximal concentrations of LctLys/FruLys in blood were reached about 10–12 hours after a meal, amounting for about 8% of the ingested Amadori products from milk powder.[409] The level of LctLys in urine returned to its normal level within one day after its ingestion,[410] indicating rapid renal clearance of lactulosamines from blood. As only traces of the LctLys were detected in feces, most of the non-absorbed lactulose–L-lysine was probably fermented in the gut. In infants fed with heat-processed formulas, LctLys was also excreted in the urine, reaching up to 4% of the ingested amount.[411] These results suggest that D-fructose– and lactulose–amino acids are most probably absorbed passively from ingested food.

Biodistribution of N^{ε}-D-fructose–L-lysine in rats was investigated by using positron-emission tomography and -labeling in order to obtain the intravital images of the animals injected with 4-[^{18}F]-fluorobenzoyl-FruLys.[412] Half of the i.v.-injected radioactivity was excreted within 1 hour in the urine, with the rest being localized mainly in the kidneys, followed by liver, intestine, and stomach. Up to 82% of the labeled εFruLys in kidneys was phosphorylated by FN3K after 1 hour, while only small amounts of the metabolite could be found in urine or plasma.

c. Antioxidant.—Since D-fructosamines are readily oxidized by the one-electron abstraction mechanism (see Section III.3.c), the environmental context may define their antioxidant/pro-oxidant activities in the presence of ROS and other oxidants. For example, the well-documented capacity of D-fructose–amino acids to reduce iron(III) complexes[234,413] makes these compounds potent antioxidants in the FRAP and related assays. Dehydration of fruits and other plant tissues containing D-glucose and amino acids may lead to the formation of relatively large amounts of D-fructose–amino acids,[159] and at the same time to an increase in the antioxidant capacity,[413] suggesting a contribution of D-fructose–amino acids to this increase. Indeed, D-fructose– and maltulose–L-arginine have been identified as major antioxidant components of dehydrated, aged garlic,[414,415] and ginseng root.[160,416] FruArg scavenged H_2O_2, inhibited the Cu(II)-induced oxidation of low-density lipoprotein (LDL), protected endothelial cells from oxidized LDL, induced the release of lactate dehydrogenase and lipid peroxidation, and modulated release of peroxides from oxidized-LDL-treated macrophages. In addition, FruArg acted as a substrate for NO-synthase.[292] D-Fructose–amino acids and –peptides were equally potent in scavenging hydroxyl radicals produced in the Fenton reaction, while in a test of superoxide radical-scavenging activity, D-fructose–peptides showed relatively higher efficiency.[417] In tomato powder, D-fructose–L-histidine, an even more potent antioxidant, has been identified.[210] The superior ability of FruHis to protect oxidative degradation of DNA by the Fenton reaction-generated hydroxyl free radicals, as compared to FruArg, histidine, or any other water-soluble antioxidants from tomato,[413] may be explained, in part, by the high affinity of this antioxidant to redox metals such as Cu(II) that is present in the Fenton system (see Section III.3).

Naturally occurring D-fructose–L-tryptophan has been discovered[418] in tubers of nutgrass, which are widely used in the folk medicine of tropical and subtropical regions as anti-inflammatory, hypotensive, and hepatoprotective agents. However, no evaluation of the Fru-Trp as an antioxidant or any biological activities has been done so far. In another study, the ability of D-fructose–β-alanine to scavenge ROS from cigarette smoke was established.[419]

d. Antitumor.—The ability of D-fructosamine to inhibit proliferation of human tumor cells *in vitro* was mentioned as early as in 1956.[420]

Subsequently, epidemiological studies have established an inverse relation between the consumption of dehydrated fruits and vegetables and the risk of urological cancers in males.[421] One of the most notable examples is the protective effect of dietary tomato products, rather than fresh tomato, against prostate cancer. Thermal processing and dehydration of tomato brings about formation of large amounts of D-fructose–amino acids and other Maillard reaction products,[159] which may be responsible for the increased antioxidant capacity of the processed tomatoes and may contribute to the cancer-protective effects. Indeed, D-fructose–L-histidine, which is present in tomato powder at 50 mg/100 g concentration, showed the most efficient protection of ROS-induced DNA fragmentation *in vitro* as compared with other antioxidants from tomato.[210,413] In combination with lycopene, FruHis strongly inhibited the proliferation of prostate cancer cells both *in vitro* and *in vivo*, while supplementation of experimental diets with tomato paste and FruHis decreased carcinogenesis in the prostate of carcinogen-induced rats by six fold.[413] Several other D-fructose–amino acids, most notably D-fructose–L-leucine and D-fructose–D-leucine, also showed promising *in vivo* antitumor activity in experimental models of melanoma[422] and breast cancer.[144,423]

A number of lactulosamine derivatives have demonstrated the ability to bind to mammalian β-galactosyl-binding lectins (galectins).[358] Galectins are implicated in cellular adhesion and signaling, are overexpressed in many solid tumors, and have been identified as novel tumor markers and targets for cancer therapies.[424] Lactulose–amino acids bind to galectin-1 with enhanced affinity, as compared to a galectin standard ligand, lactose (K_D 70–380 μM vs 800 μM, respectively).[425] The divalent lactulosamine **102** shows even higher affinity, at $K_D = 2$ μM. Not surprisingly, lactulosamines inhibited the binding of galectins-1 and -3 to their natural glycoprotein ligands, such as laminin, 90K glycoprotein, or IgE,[358,426] and have been employed as blockers of galectin-dependent cancer-cell adhesion and proliferation *in vitro* and *ex vivo*.[427] *In vitro*, synthetic divalent lactulosamines inhibited the galectin-dependent morphogenesis of vascular endothelial cells that is an essential element of tumor-induced angiogenesis.[426] In animal prostate and breast-cancer models, lactulose–L-leucine inhibited the spread of metastatic tumors into lungs and lymph nodes,[423,428] while **102** enhanced the efficiency of dendritic-cell immunotherapy in a mouse model of breast cancer.[425] LctLeu demonstrated the ability to induce apoptosis and dose-dependent inhibition of clonogenic growth of an endothelial cancer-derived murine hemangiosarcoma SVR cell line, with IC_{50} about 350 μM.[429] Notably, LctLeu synergized with the chemotherapeutic drug doxorubicin in a dose-dependent manner. This interaction suggested that lactulose–amino acids may modulate the drug resistance of cancer cells by blocking a galectin-3-dependent, anti-apoptotic signaling pathway. Indeed, in a recent study,[428] LctLeu inhibited the galectin-3-induced phosphorylation of the proapoptotic Bad protein in human breast carcinoma MDA-MB-435 cells treated with Taxol. At the same time, LctLeu did not affect the Taxol-induced phosphorylation of antiapoptotic Bcl-2. The sensitization of MDA-MB-435 cells to Taxol by LctLeu led to their pronounced synergistic interaction, both in cell culture and in an *in vivo* xenograft model.

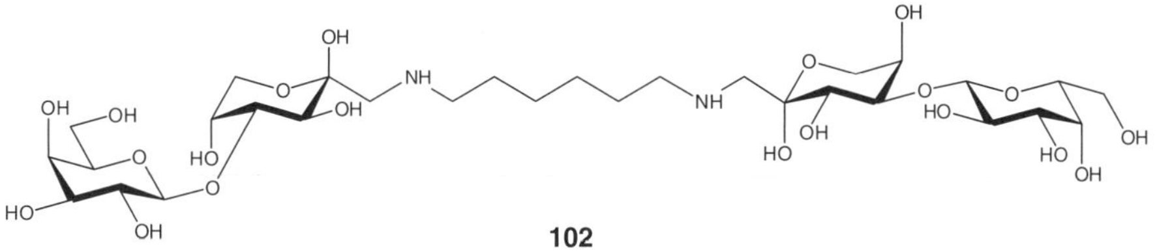

102

e. Antibacterial/Antifungal.—Mester and colleagues[430] reported the bacteriostatic effect of D-fructose–serotonin (**103**) on *Mycobacterium leprae*. The effect was linked to the inhibition of DOPA uptake by *M. leprae*, a factor essential for growth of the microorganism. Despite its poor availability (10% orally) and pharmacokinetics, **103** showed clinical and bacteriological improvements in pilot clinical studies, is non-toxic, and was proposed as an active anti-leprosy metabolite of serotonin in subjects receiving tryptophan-enriched diets. A metabolic precursor of **103**, D-fructose–5-hydroxy-L-tryptophan is equally active.

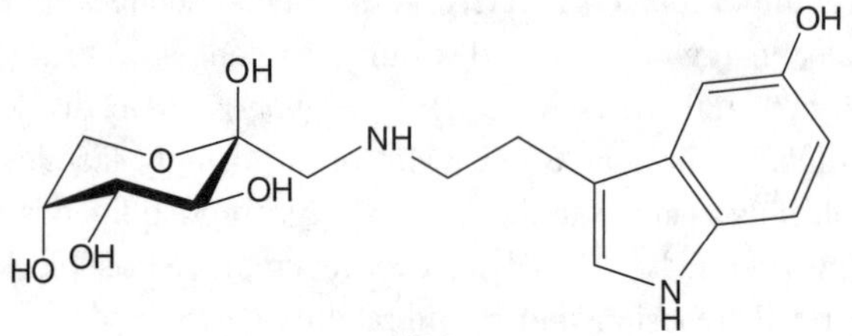

103

Trypanosoma brucei, a blood-borne parasite causing African sleeping disease, can be treated by drugs internalizing through its principal hexose transporter, THT1. A series of 1-arylamino-1-deoxy-D-fructose derivatives showed affinities to THT1 that were similar or exceeded those for D-Glc and D-Fru.[355] Importantly, these derivatives were toxic to the parasite *in vitro*, with LD_{50} values as low as 2.6 μM, thus validating the transporter itself as a potential therapeutic target.

Lactulose–memantine, an adamantylamine derivative, showed significant growth inhibition of *Staphylococcus aureus*.[431] Several amphotericin and nystatin analogues have been modified by D-glucose or lactose through the Amadori rearrangement reaction, and the respective D-fructosamine– and –lactulosamine product derivatives, such as **104**, demonstrated antifungal activities comparable with amphotericin B, while their hemolytic activity against human erythrocytes was significantly lower.[432,433]

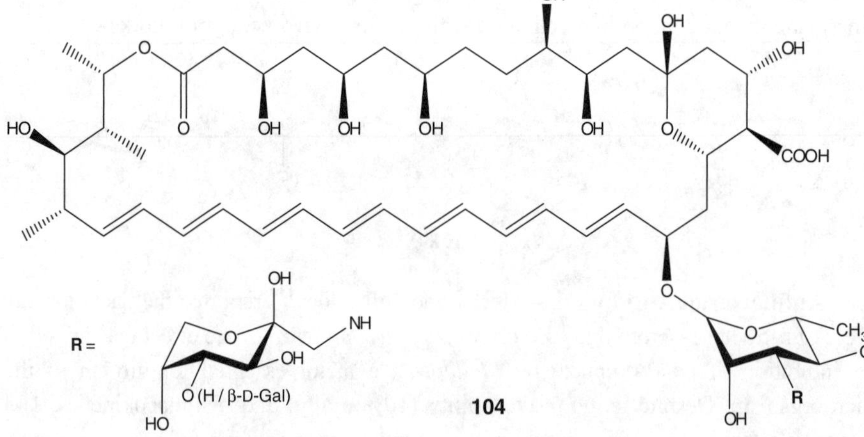

104

f. Immunostimulatory and Other Activities.—D-Fructose–L-lysine epitopes present in glycated plasma proteins possess immunogenicity, as evidenced by the detection of the respective human autoantibodies in diabetic patients.[434] Antibodies against both D-fructose- and lactulose-modified proteins have also been produced in experimental animals for analytical and diagnostic purposes (see Section III.2).

D-Fructose–L-proline was suggested as an active ingredient from an immunostimulatory peat extract.[435] When tested in immunized Balb/c mice, this Amadori compound stimulated the production of serum antibodies, and the number and activity of spleen lymphocytes. In addition, FruPro protected the splenocytes from the cytotoxic effect of hydrocortisone *in vitro*.

The lactulosamine structure may also act in an immunoprotective fashion, by blocking the immunosuppressive galectin-1. Thus, a divalent lactulosamine (**102**) restored galectin-1-impaired activity and proliferation of activated cytotoxic T-lymphocytes *in vitro*, and enhanced the efficiency of dendritic-cell immunotherapy in a rodent model of breast cancer.[425]

An analgesia-enhancing effect was suggested for a number of D-fructose–dipeptides, such as FruIleAsp or FruValGlu. It was found that these Amadori compounds, dubbed enkastines, may modulate the activity of endopeptidases, thus prolonging the analgesic activity of such endogenous opioid peptides as enkephalines.[287]

A lactulose conjugate (**105**) with the anticonvulsant drug pregabaline forms spontaneously upon storage of pharmaceutical formulations containing the drug and a lactose carrier.[436] The conjugate has been proposed for treatment of CNS disorders, such as depression or Parkinson disease.[437]

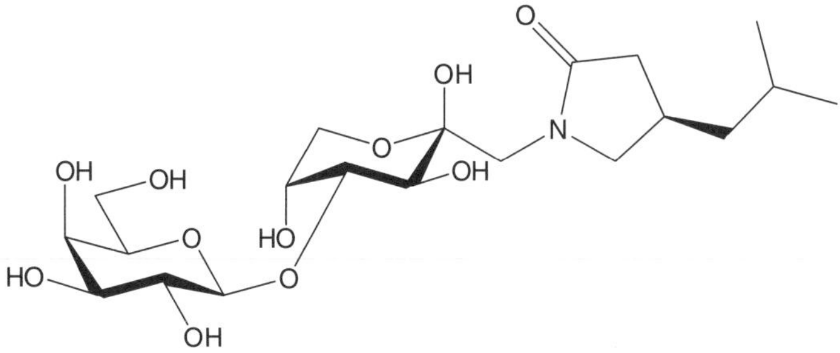

105

Thioureido derivatives of 1-deoxy-2,3:4,5-di-*O*-isopropylidene-β-D-fructopyranose were obtained in the reaction of the diacetal of D-fructosamine with substituted isothiocyanates[107] (Scheme 51). The compounds were successfully tested as potential

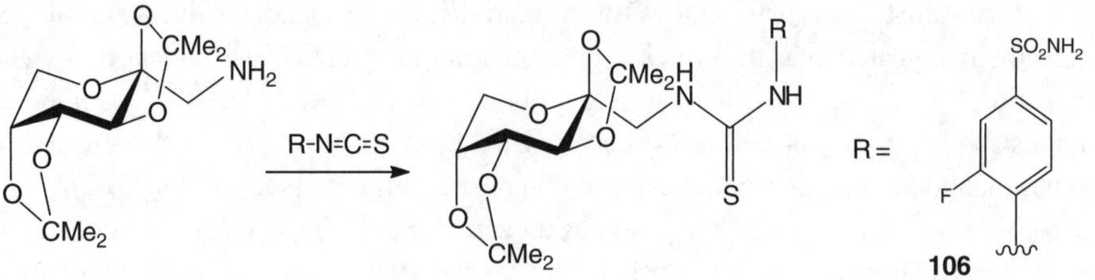

SCHEME 51

inhibitors of several isoforms of carbonic anhydrase (hCA), which play critical physiological functions in the central nervous system of vertebrates. The most efficient inhibitors of hCA II, such as **106**, showed a pronounced anticonvulsant effect in an animal model, with the efficiency exceeding that of antiepileptic drug topiramate.

D-Fructose–peptides, such as FruValTyr isolated from ginger root, were proposed as active ingredients in anti-wrinkle and hair growth-deterring cosmetics.[438,439] D-Fructose conjugates with cephaeline alkaloids have been isolated from the dried roots of *Cephaelis acuminate*, long used as an emetic and expectorant.[440]

VI. Concluding Remarks

As simple and trivial as it may look to a carbohydrate chemist, fructosamine is a truly unique carbohydrate structure, as exemplified in this chapter, on account of its outstanding chemical properties and significance as a key intermediate in a great many reactions encountered whenever D-glucose, the most common sugar, is present. Although research of its role in the Maillard reaction in foods has currently reached a steady state, there is now an explosion of interest in fructosamine in biomedical sciences, which shows signs of expansion from diabetes research into many other areas of medicinal chemistry and cancer.

It is thus fair to expect that, in the near future, considerable new effort in the area will be focused on discovery and development of novel approaches to analysis of fructosamine and its control *in vivo*. In particular, highly specific sensors recognizing such D-fructose–proteins as HbA_{1c} are in great need, and these may be developed on the basis of genetically engineered fructosamine-processing enzymes. The same kind of enzymes may be employed for efficient deglycation of circulating proteins in peritoneal dialysis and diabetes. With continuously growing interest among the general public to the functional properties of foods, assessment of the nutraceutical potential of dietary fructosamines will continue to attract the attention of investigators. Current trends strongly suggest that fructosamine, within the next decade, will attain celebrity status among many carbohydrate and food chemists, biochemists, and nutritionists, who represent a diverse Maillard-reaction community.[441]

References

1. J. E. Hodge, Amadori rearrangement, *Adv. Carbohydr. Chem.*, 10 (1955) 169–205.
2. T. M. Reynolds, Chemistry of nonenzymic browning. I. The reaction between aldoses and amines, *Adv. Food Res.*, 12 (1963) 1–52.
3. H. Paulsen and K. W. Pflughaupt, Glycosylamines, in W. Pigman, and D. Horton, (Eds.), *The Carbohydrates: Chemistry and Biochemistry*, Academic Press, N.Y., 1980, Vol. 1B, pp. 881–927.
4. T. M. Wrodnigg, and B. Eder, The Amadori and Heyns rearrangements: Landmarks in the history of carbohydrate chemistry or unrecognized synthetic opportunities?, *Top. Curr. Chem.*, 215 (2001) 115–152.
5. V. A. Yaylayan and A. Huyghues-Despointes, Chemistry of Amadori rearrangement products: Analysis, synthesis, kinetics, reactions, and spectroscopic properties, *Crit. Rev. Food Sci. Nutr.*, 34 (1994) 321–369.
6. E. Fischer, On isoglucosamine, *Ber. Dtsch. Chem. Ges.*, 19 (1886) 1920–1924.
7. M. Amadori, Products of condensation between glucose and *p*-phenetidine. I, *Atti Accad. Naz. Lincei, Cl. Sci. Fis., Mat. Nat., Rend.*, 2 (1925) 337–342.
8. R. Kuhn and F. Weygand, The Amadori rearrangement, *Ber. Dtsch. Chem. Ges.*, 70B (1937) 769–772.
9. J. E. Hodge, Browning reactions in model systems, *J. Agric. Food Chem.*, 1 (1953) 928–943.
10. J. Mauron, The Maillard reaction in food: A critical review from the nutritional standpoint, *Prog. Food Nutr. Sci.*, 5 (1981) 5–35.
11. H. F. Bunn, D. N. Haney, K. H. Gabbay, and P. M. Gallop, Further identification of the nature and linkage of the carbohydrate in hemoglobin A_{1c}, *Biochem. Biophys. Res. Commun.*, 67 (1975) 103–109.
12. M. P. Cohen, F. N. Ziyadeh, and S. Chen, Amadori-modified glycated serum proteins and accelerated atherosclerosis in diabetes: Pathogenic and therapeutic implications, *J. Lab. Clin. Med.*, 147 (2006) 211–219.

13. V. M. Monnier, D. R. Sell, Z. Dai, I. Nemet, F. Collard, and J. Zhang, The role of the Amadori product in the complications of diabetes, *Ann. N. Y. Acad. Sci.*, 1126 (2008) 81–88.
14. R. Flückiger, T. Woodtli, and W. Berger, Evaluation of the fructosamine test for the measurement of plasma protein glycation, *Diabetologia*, 30 (1987) 648–652.
15. F. Collard, J. Zhang, I. Nemet, K. R. Qanungo, V. M. Monnier, and V. C. Yee, Crystal structure of the deglycating enzyme fructosamine oxidase (amadoriase II), *J. Biol. Chem.*, 283 (2008) 27007–27016.
16. Nomenclature of carbohydrates, *Pure Appl. Chem.*, 68 (1996) 1919–2008.
17. H. Röper, S. Röper, K. Heyns, and B. Meyer, NMR spectroscopy of *N*-(1-deoxy-D-fructos-1-yl)-L-amino acids ("fructose-amino acids"), *Carbohydr. Res.*, 116 (1983) 183–195.
18. T. Horiuchi and T. Kurokawa, Purification and properties of fructosylamine oxidase from *Aspergillus* sp. 1005, *Agric. Biol. Chem.*, 55 (1991) 333–338.
19. K. Maurer and B. Schiedt, A fruitful preparation of glucosamine and a contribution to the catalytic hydrogenation of osazones, *Ber. Dtsch. Chem. Ges.*, 68B (1935) 2187–2191.
20. H. Hopff, H. Spanig, and A. Steimmig, *N*-Acylated amino sugars, *Ger. Pat.*, (1954) DE 915566.
21. A. Gómez-Sánchez, F. Javier Hidalgo, and J. L. Chiara, Studies on nitroenamines. II. Syntheses of (2-nitrovinyl)amino sugars and 2- and 3-(alditol-1-yl)-4-nitropyrroles, *Carbohydr. Res.*, 167 (1987) 55–66.
22. A. Gómez-Sánchez, M. D. G. García Martín, and C. Pascual, Protection of the amino group of amino sugars by the acylvinyl group. Part II. Synthesis and glycosidation of 1-deoxy-1-[(2,2-diacylvinyl)amino]-D-fructoses, *Carbohydr. Res.*, 149 (1986) 329–345.
23. M. D. G. García Martín, C. Gasch, and A. Gómez-Sánchez, Glycosides of 1-amino-1-deoxy-D-fructose, *Carbohydr. Res.*, 199 (1990) 139–151.
24. D. A. Parrish, Z. Zou, C. L. Allen, C. S. Day, and S. B. King, A convenient method for the synthesis of *N*-hydroxyureas, *Tetrahedron Lett.*, 46 (2005) 8841–8843.
25. J. Levi, Z. Cheng, O. Gheysens, M. Patel, C. T. Chan, Y. Wang, M. Namavari, and S. S. Gambhir, Fluorescent fructose derivatives for imaging breast cancer cells, *Bioconjug. Chem.*, 18 (2007) 628–634.
26. R. Kuhn and W. Kirschenlohr, Synthesis of 2-acetamidolactose, *Chem. Ber.*, 87 (1954) 1547–1552.
27. J. Druey and G. Huber, Amino sugars. I. The preparation of 1-amino-1-deoxyketoses, *Helv. Chim. Acta*, 40 (1957) 342–349.
28. V. V. Mossine, Personal communication.

29. G. P. Ellis and J. Honeyman, Glycosylamines, *Adv. Carbohydr. Chem.*, 10 (1955) 95–168.
30. R. Kuhn and L. Birkofer, *N*-Glucosides and the Amadori rearrangement, *Ber. Dtsch. Chem. Ges.*, 71B (1938) 621–635.
31. B. Capon and B. E. Connett, Structure of some *N*-aryl-D-glucosylamines, *J. Chem. Soc.*, (1965) 4492–4497.
32. C. L. Perrin and K. B. Armstrong, Conformational analysis of glucopyranosylammonium ions: Does the reverse anomeric effect exist?, *J. Am. Chem. Soc.*, 115 (1993) 6825–6834.
33. G. Westphal and L. Kroh, Mechanism of the "early phase" of the Maillard reaction. Part 1. Influence of the structure of the carbohydrate and the amino acid on the formation of the *N*-glycoside, *Nahrung*, 29 (1985) 757–764.
34. G. Weitzel, H. U. Geyer, and A. M. Fretzdorff, Preparation and stability of salts of amino acid *N*-glycosides, *Chem. Ber.*, 90 (1957) 1153–1161.
35. F. Micheel and I. Dijong, Mechanism of the Amadori rearrangement, *Justus Liebigs Ann. Chem.*, 658 (1962) 120–127.
36. S. V. Metlitskikh, A. M. Koroteev, M. P. Koroteev, A. S. Shashkov, A. A. Korlyukov, M. Y. Antipin, A. I. Stash, and E. E. Nifantiev, Synthesis of bis(glycosylamino) alkanes and bis(glycosylamino)arenes, *Russ. Chem. Bull.*, 54 (2005) 2890–2898.
37. K. Linek, J. Alföldi, J. Defaye, and P. I. Glycosylamines., Structure and rearrangement reactions of some diglycosylamines, *Carbohydr. Res.*, 164 (1987) 195–205.
38. V. V. Mossine, M. Linetsky, G. V. Glinsky, B. J. Ortwerth, and M. S. Feather, Superoxide free radical generation by Amadori compounds: The role of acyclic forms and metal ions, *Chem. Res. Toxicol.*, 12 (1999) 230–236.
39. G. Westphal and L. Kroh, Mechanism of the "early phase" of the Maillard reaction. Part 2. Consecutive reactions of *N*-glycosides, *Nahrung*, 29 (1985) 765–775.
40. N. Bridiau, M. Benmansour, M. D. Legoy, and T. Maugard, One-pot stereoselective synthesis of β-*N*-aryl-glycosides by N-glycosylation of aromatic amines: Application to the synthesis of tumor-associated carbohydrate antigen building blocks, *Tetrahedron*, 63 (2007) 4178–4183.
41. K. Heyns and W. Beilfuß, Aldosylamine and ketosylamine rearrangement of weakly basic *N*-glycosides, *Chem. Ber.*, 106 (1973) 2693–2709.
42. D. Palm and H. Simon, Hydride transfer in the Amadori rearrangement, *Z. Naturforsch*, 18b (1963) 419–420.
43. S. Kolka and J. Sokolowski, Kinetics of the Amadori rearrangement of *N*-D-glucosylamines, *Rocz. Chem.*, 48 (1974) 439–444.
44. S. Kolka, Kinetics of Amadori rearrangement of *N*-D-glucopyranosylamines in the presence of Lewis acids, *Pol. J. Chem.*, 58 (1984) 689–695.

45. L. Rosen, J. W. Woods, and W. Pigman, Reactions of carbohydrates with nitrogenous substances. VI. The Amadori rearrangement in methanol, *J. Am. Chem. Soc.*, 80 (1958) 4697–4702.
46. J. R. Ramos and C. Gonzalez, Kinetic study of the Amadori rearrangement of *N*-glucosylglycinate to *N*-fructosylglycinate. Part I, *Rev. Latinoam. Quím.*, 8 (1977) 94–97.
47. S. Yano, Coordination compounds containing sugars and their derivatives, *Coord. Chem. Rev.*, 92 (1988) 113–156.
48. W. Zhang, T. Jiang, S. Ren, Z. Zhang, H. Guan, and J. Yu, Metal-*N*-saccharide chemistry: Synthesis and structure determination of two Cu(II) complexes containing glycosylamines, *Carbohydr. Res.*, 339 (2004) 2139–2143.
49. H. Gil, J. F. Mata-Segreda, and R. L. Schowen, Proton transfer is not rate-limiting in buffer-induced non-enzymic glucation of hemoglobin, *J. Am. Chem. Soc.*, 110 (1988) 8265–8266.
50. S. J. Angyal, The Lobry de Bruyn-Alberda van Ekenstein transformation and related reactions, *Top. Curr. Chem.*, 215 (2001) 1–14.
51. H. Gil, D. Salcedo, and R. Romero, Effect of phosphate buffer on the kinetics of glycation of proteins, *J. Phys. Org. Chem.*, 18 (2005) 183–186.
52. H. Gil, B. Vásquez, M. Peña, and J. Uzcategui, Effect of buffer carbonate and arsenate on the kinetics of glycation of human hemoglobin, *J. Phys. Org. Chem.*, 17 (2004) 537–540.
53. N. G. Watkins, C. I. Neglia-Fisher, D. G. Dyer, S. R. Thorpe, and J. W. Baynes, Effect of phosphate on the kinetics and specificity of glycation of protein, *J. Biol. Chem.*, 262 (1987) 7207–7212.
54. H. Gil, M. Peña, B. Vásquez, and J. Uzcategui, Catalysis by organic phosphates of the glycation of human hemoglobin, *J. Phys. Org. Chem.*, 15 (2002) 820–825.
55. V. V. Mossine, C. L. Barnes, G. V. Glinsky, and M. S. Feather, Molecular and crystal structure of *N*-(2-deoxy-D-aldohexos-2-yl)-glycines (Heyns compounds), *Carbohydr. Res.*, 284 (1996) 11–24.
56. A. Jakas, A. Katić, N. Bionda, and Š. Horvat, Glycation of a lysine-containing tetrapeptide by D-glucose and D-fructose—influence of different reaction conditions on the formation of Amadori/Heyns products, *Carbohydr. Res.*, 343 (2008) 2475–2480.
57. K. Heyns and H. Breuer, Preparation and properties of additional N-substituted 2-amino-2-deoxyaldoses from D-fructose and amino acids, *Chem. Ber.*, 91 (1958) 2750–2762.

58. V. V. Mossine, G. V. Glinsky, and M. S. Feather, The preparation and characterization of some Amadori compounds (1-amino-1-deoxy-D-fructose derivatives) derived from a series of aliphatic ω-amino acids, *Carbohydr. Res.*, 262 (1994) 257–270.
59. S. M. Poling, R. D. Plattner, and D. Weisleder, *N*-(1-Deoxy-D-fructos-1-yl) fumonisin B1, the initial reaction product of fumonisin B1 and D-glucose, *J. Agric. Food Chem.*, 50 (2002) 1318–1324.
60. J. Z. Mioduszewski, K. Witkiewicz, and A. Inglot, Amadori reaction products, process for their manufacture, and their use in pharmaceuticals and cosmetics, *Int. Pat. Appl.*, (1993) WO 9316087.
61. L. Jiménez-Castaño, M. Villamiel, P. J. Martín-Álvarez, A. Olano, and R. López-Fandiño, Effect of the dry-heating conditions on the glycosylation of β-lactoglobulin with dextran through the Maillard reaction, *Food Hydrocoll.*, 19 (2005) 831–837.
62. M. Corzo-Martinez, F. J. Moreno, M. Villamiel, and F. M. Harte, Characterization and improvement of rheological properties of sodium caseinate glycated with galactose, lactose and dextran, *Food Hydrocoll.*, 24 (2010) 88–97.
63. R. H. Stadler, F. Robert, S. Riediker, N. Varga, T. Davídek, S. Devaud, T. Goldmann, J. Hau, and I. Blank, In-depth mechanistic study on the formation of acrylamide and other vinylogous compounds by the Maillard reaction, *J. Agric. Food Chem.*, 52 (2004) 5550–5558.
64. V. A. Yaylayan and R. H. Stadler, Acrylamide formation in food: A mechanistic perspective, *J. AOAC Int.*, 88 (2005) 262–267.
65. R. Shapiro, M. J. McManus, C. Zalut, and H. F. Bunn, Sites of nonenzymic glycosylation of human hemoglobin A, *J. Biol. Chem.*, 255 (1980) 3120–3127.
66. I. Penndorf, C. Li, U. Schwarzenbolz, and T. Henle, N-terminal glycation of proteins and peptides in foods and *in vivo*: Evaluation of *N*-(2-furoylmethyl)valine in acid hydrolyzates in human hemoglobin, *Ann. N. Y. Acad. Sci.*, 1126 (2008) 118–123.
67. J. W. C. Brock, D. J. S. Hinton, W. E. Cotham, T. O. Metz, S. R. Thorpe, J. W. Baynes, and J. M. Ames, Proteomic analysis of the site specificity of glycation and carboxymethylation of ribonuclease, *J. Proteome Res.*, 2 (2003) 506–513.
68. B. H. Shilton and D. J. Walton, Sites of glycation of human and horse liver alcohol dehydrogenase *in vivo*, *J. Biol. Chem.*, 266 (1991) 5587–5592.
69. C. Wa, R. Cerny, and D. S. Hage, Obtaining high sequence coverage in matrix-assisted laser desorption time-of-flight mass spectrometry for studies of protein modification: Analysis of human serum albumin as a model, *Anal. Biochem.*, 349 (2006) 229–241.

70. R. L. Garlick and J. S. Mazer, The principal site of nonenzymic glycosylation of human serum albumin *in vivo*, *J. Biol. Chem.*, 258 (1983) 6142–6146.
71. C. Wa, R. L. Cerny, W. A. Clarke, and D. S. Hage, Characterization of glycation adducts on human serum albumin by matrix-assisted laser desorption/ionization time-of-flight mass spectrometry, *Clin. Chim. Acta*, 385 (2007) 48–60.
72. B. Zhang, Y. Yang, I. Yuk, R. Pai, P. McKay, C. Eigenbrot, M. Dennis, V. Katta, and K. C. Francissen, Unveiling a glycation hot spot in a recombinant humanized monoclonal antibody, *Anal. Chem.*, 80 (2008) 2379–2390.
73. A. Scaloni, V. Perillo, P. Franco, E. Fedele, R. Froio, L. Ferrara, and P. Bergamo, Characterization of heat-induced lactosylation products in caseins by immunoenzymatic and mass spectrometric methodologies, *Biochim. Biophys. Acta*, 1598 (2002) 30–39.
74. M. Lima, C. Moloney, and J. M. Ames, Ultra performance liquid chromatography-mass spectrometric determination of the site specificity of modification of β-casein by glucose and methylglyoxal, *Amino Acids*, 36 (2009) 475–481.
75. J. Meltretter, C. M. Becker, and M. Pischetsrieder, Identification and site-specific relative quantification of β-lactoglobulin modifications in heated milk and dairy products, *J. Agric. Food Chem.*, 56 (2008) 5165–5171.
76. P. Nacharaju and A. S. Acharya, Amadori rearrangement potential of hemoglobin at its glycation sites is dependent on the three-dimensional structure of protein, *Biochemistry*, 31 (1992) 12673–12679.
77. N. G. Watkins, S. R. Thorpe, and J. W. Baynes, Glycation of amino groups in protein. Studies on the specificity of modification of RNase by glucose, *J. Biol. Chem.*, 260 (1985) 10629–10636.
78. J. Venkatraman, K. Aggarwal, and P. Balaram, Helical peptide models for protein glycation: Proximity effects in catalysis of the Amadori rearrangement, *Chem. Biol.*, 8 (2001) 611–625.
79. C. Mennella, M. Visciano, A. Napolitano, M. D. Del Castillo, and V. Fogliano, Glycation of lysine-containing dipeptides, *J. Pept. Sci.*, 12 (2006) 291–296.
80. R. Tressi, C. T. Piechotta, D. Rewicki, and E. Krause, Modification of peptide lysine during Maillard reaction of D-glucose and D-lactose, *Int. Congr. Ser.*, 1245 (2002) 203–209.
81. Š. Horvat, M. Roščić, L. Varga-Defterdarović, and J. Horvat, Intramolecular rearrangement of the monosaccharide esters of an opioid pentapeptide: Formation and identification of novel Amadori compounds related to fructose and tagatose, *J. Chem. Soc. Perkin Trans. 1*, (1998) 909–914.

82. I. Jerić, L. Šimičić, M. Stipetić, and Š. Horvat, Synthesis and reactivity of the monosaccharide esters of amino acids as models of teichoic acid fragment, *Glycoconjug. J.*, 17 (2000) 273–282.
83. I. Jerić, P. Novak, M. Vinković, and Š. Horvat, Conformational analysis of sugar-peptide adducts in the solution state by NMR spectroscopy and molecular modeling, *J. Chem. Soc. Perkin Trans.* 2 (2001) 1944–1950.
84. L. Varga-Defterdarović and G. Hrlec, Synthesis and intramolecular reactions of Tyr-Gly and Tyr-Gly-Gly related 6-*O*-glucopyranose esters, *Carbohydr. Res.*, 339 (2004) 67–75.
85. R. Kuhn and H. J. Haas, Amino sugar syntheses. VI. Preparation of D-isoglucosamine by the catalytic hydrogenation of Amadori compounds, *Justus Liebigs Ann. Chem.*, 600 (1956) 1948–1955.
86. R. Bognár, L. Somogyi, and Z. Györgydeák, Heterocyclic compounds from sugars. III. Substituted thiazolidines by reactions of L-cysteine with monosaccharides, *Justus Liebigs Ann. Chem.*, 738 (1970) 68–78.
87. K. B. de Roos, K. Wolswinkel, and G. Sipma, Amadori compounds of cysteine and their role in the development of meat flavor, *ACS Symp. Ser.*, 905 (2005) 117–129.
88. M. D. Linetsky, E. V. Shipova, R. D. Legrand, and O. O. Argirov, Glucose-derived Amadori compounds of glutathione, *Biochim. Biophys. Acta*, 1724 (2005) 181–193.
89. B. Rönner, H. Lerche, W. Bergmüller, C. Freilinger, T. Severin, and M. Pischetsrieder, Formation of tetrahydro-β-carbolines and β-carbolines during the reaction of L-tryptophan with D-glucose, *J. Agric. Food Chem.*, 48 (2000) 2111–2116.
90. P. Manini, A. Napolitano, and M. d'Ischia, Reactions of D-glucose with phenolic amino acids: Further insights into the competition between Maillard and Pictet-Spengler condensation pathways, *Carbohydr. Res.*, 340 (2005) 2719–2727.
91. P. Manini, M. d'Ischia, and G. Prota, An unusual decarboxylative Maillard reaction between L-DOPA and D-glucose under biomimetic conditions: Factors governing competition with Pictet-Spengler condensation, *J. Org. Chem.*, 66 (2001) 5048–5053.
92. M. Roščić and Š. Horvat, Transformations of bioactive peptides in the presence of sugars—characterization and stability studies of the adducts generated via the Maillard reaction, *Bioorg. Med. Chem.*, 14 (2006) 4933–4943.
93. Š. Horvat, M. Roščić, and J. Horvat, Synthesis of hexose-related imidazolidinones: Novel glycation products in the Maillard reaction, *Glycoconjug. J.*, 16 (1999) 391–398.

94. D. J. Walton, J. D. McPherson, T. Hvidt, and W. A. Szarek, Synthetic routes to *N*-(1-deoxy-D-fructos-1-yl)amino acids by way of reductive amination of hexos-2-uloses, *Carbohydr. Res.*, 167 (1987) 123–130.
95. A. Frolov, D. Singer, and R. Hoffmann, Solid-phase synthesis of glucose-derived Amadori peptides, *J. Pept. Sci.*, 13 (2007) 862–867.
96. P. Stefanowicz, K. Kapczyńska, A. Kluczyk, and Z. Szewczuk, A new procedure for the synthesis of peptide-derived Amadori products on a solid support, *Tetrahedron Lett.*, 48 (2007) 967–969.
97. S. Carganico, P. Rovero, J. A. Halperin, A. M. Papini, and M. Chorev, Building blocks for the synthesis of post-translationally modified glycated peptides and proteins, *J. Pept. Sci.*, 15 (2009) 67–71.
98. M. P. Vitek, A. Cerami, R. J. Bucala, P. C. Ulrich, H. Vlassara, and X. Zhang, Compositions and methods for advanced glycosylation endproduct-mediated modulation of amyloidosis, *U.S. Pat.*, (2002) 6,410,598
99. Š. Horvat, A. Jakas, E. Vass, J. Samu, and M. Hollósi, CD and FTIR spectroscopic studies of Amadori compounds related to the opioid peptides, *J. Chem. Soc. Perkin Trans.* 2 (1997) 1523–1528.
100. J. M. J. Tronchet, F. Habashi, O. R. Martin, A. P. Bonenfant, B. Baehler, and J. B. Zumwald, Diglycosyl derivatives, *Helv. Chim. Acta*, 62 (1979) 894–898.
101. T. Hofmann, P. Münch, and P. Schieberle, Quantitative model studies on the formation of aroma-active aldehydes and acids by Strecker-type reactions, *J. Agric. Food Chem.*, 48 (2000) 434–440.
102. L. Vargha, L. Toldy, and E. Kasztreiner, Synthesis of new sugar derivatives of potential antitumor activity. III. 2-Haloethylamino- and ethylenimino derivatives of sugar alcohols, *Acta Chim. Acad. Sci. Hung.*, 19 (1959) 295–306.
103. W. M. Corbett and D. Winters, Cross-linking of cellulose and its derivatives. IV. Action of hydrazine on the *p*-toluenesulfonates of di-*O*-isopropylidene-D-galactose and -D-fructose, *J. Chem. Soc.*, (1961) 4823–4826.
104. M. G. López and D. W. Gruenwedel, Synthesis of aromatic Amadori compounds, *Carbohydr. Res.*, 212 (1991) 37–45.
105. S. N. Noomen, G. J. Breel, and C. Winkel, The synthesis of pure Amadori rearrangement products, *Recl. Trav. Chim. Pays Bas*, 114 (1995) 321–324.
106. C. M. Utzmann and M. O. Lederer, Independent synthesis of aminophospholipid-linked Maillard products, *Carbohydr. Res.*, 325 (2000) 157–168.
107. J. Y. Winum, A. Thiry, K. El Cheikh, J. M. Dogné, J. L. Montero, D. Vullo, A. Scozzafava, B. Masereel, and C. T. Supuran, Carbonic anhydrase inhibitors. Inhibition of isoforms I, II, IV, VA, VII, IX, and XIV with sulfonamides

incorporating fructopyranose-thioureido tails, *Bioorg. Med. Chem. Lett.*, 17 (2007) 2685–2691.

108. A. Tatibouët, A. C. Simao, and P. Rollin, Fused 1,3-oxazolidine-2-thiones on keto-hexose backbones: Functional modulation processes, *Lett. Org. Chem.*, 2 (2005) 47–50.
109. K. Sumoto, M. Irie, N. Mibu, S. Miyano, Y. Nakashima, K. Watanabe, and T. Yamaguchi, Formation of pyrazine derivatives from D-glucosamine and their deoxyribonucleic acid (DNA) strand breakage activity, *Chem. Pharm. Bull.*, 39 (1991) 792–794.
110. R. Kuhn, G. Krüger, J. Haas, and A. Seeliger, Amino sugar synthesis. XXXIV. Pyrazine formation from amino sugars, *Justus Liebigs Ann.Chem.*, 644 (1961) 122–127.
111. S. Fujii and Y. Kosaka, Reaction between 2-amino-2-deoxy-D-glucose derivatives and sulfite. 2. Synthesis of 2-(D-*arabino*-tetrahydroxybutyl)-5-(3′,4′-dihydroxy-2′-sulfobutyl)pyrazine, *J. Org. Chem.*, 47 (1982) 4772–4774.
112. J. J. Turner, N. Wilschut, H. S. Overkleeft, W. Klaffke, G. A. van der Marel, and J. H. van Boom, Mitsunobu glycosylation of nitrobenzenesulfonamides: Novel route to Amadori rearrangement products, *Tetrahedron Lett.*, 40 (1999) 7039–7042.
113. C. E. White and S. C. Winans, Cell-cell communication in the plant pathogen *Agrobacterium tumefaciens*, *Philos. Trans. R. Soc. B*, 362 (2007) 1135–1148.
114. K. S. Kim, W. S. Chilton, and S. K. Farrand, A Ti plasmid-encoded enzyme required for degradation of mannopine is functionally homologous to the T-region-encoded enzyme required for synthesis of this opine in crown gall tumors, *J. Bacteriol.*, 178 (1996) 3285–3292.
115. J. G. Ellis, M. H. Ryder, and M. E. Tate, *Agrobacterium tumefaciens* TR-DNA encodes a pathway for agropine biosynthesis, *Mol. Gen. Genet.*, 195 (1984) 466–473.
116. W. S. Chilton, A. M. Stomp, V. Beringue, H. Bouzar, V. Vaudequin-Dransart, A. Petit, and Y. Dessaux, The chrysopine family of Amadori-type crown gall opines, *Phytochemistry*, 40 (1995) 619–628.
117. D. W. E. Smith and B. N. Ames, Intermediates in the early steps of histidine biosynthesis, *J. Biol. Chem.*, 239 (1964) 1848–1855.
118. R. Sterner, A. Merz, R. Thoma, and K. Kirschner, Phosphoribosylanthranilate isomerase and indoleglycerol-phosphate synthase: Tryptophan biosynthetic enzymes from *Thermotoga maritima, Meth. Enzymol.*, 331 (2001) 270–280.
119. U. Hommel, M. Eberhard, and K. Kirschner, Phosphoribosyl anthranilate isomerase catalyzes a reversible Amadori reaction, *Biochemistry*, 34 (1995) 5429–5439.

120. M. Henn-Sax, R. Thoma, S. Schmidt, M. Hennig, K. Kirschner, and R. Sterner, Two (βα)8-barrel enzymes of histidine and tryptophan biosynthesis have similar reaction mechanisms and common strategies for protecting their labile substrates, *Biochemistry*, 41 (2002) 12032–12042.
121. C. Jurgens, A. Strom, D. Wegener, S. Hettwer, M. Wilmanns, and R. Sterner, Directed evolution of a (βα)8-barrel enzyme to catalyze related reactions in two different metabolic pathways, *Proc. Natl. Acad. Sci. U. S. A.*, 97 (2000) 9925–9930.
122. G. M. Brown, Enzymic synthesis of pterins and dihydropteroic acid, *Chem. Biol. Pteridines, Proc. 4th Int. Symp.*, 1970, pp. 243–264.
123. G. Auerbach, A. Herrmann, A. Bracher, G. Bader, M. Gutlich, M. Fischer, M. Neukamm, M. Garrido-Franco, J. Richardson, H. Nar, R. Huber, and A. Bacher, Zinc plays a key role in human and bacterial GTP cyclohydrolase I, *Proc. Natl. Acad. Sci. U.S.A.*, 97 (2000) 13567–13572.
124. N. Schramek, A. Bracher, M. Fischer, G. Auerbach, H. Nar, R. Huber, and A. Bacher, Reaction mechanism of GTP cyclohydrolase I: Single turnover experiments using a kinetically competent reaction intermediate, *J. Mol. Biol.*, 316 (2002) 829–837.
125. P. C. Dorrestein, H. Zhai, S. V. Taylor, F. W. McLafferty, and T. P. Begley, The biosynthesis of the thiazole phosphate moiety of thiamin (vitamin B1): The early steps catalyzed by thiazole synthase, *J. Am. Chem. Soc.*, 126 (2004) 3091–3096.
126. J. Fortpied, P. Maliekal, D. Vertommen, and E. Van Schaftingen, Magnesium-dependent phosphatase-1 is a protein-fructosamine-6-phosphatase potentially involved in glycation repair, *J. Biol. Chem.*, 281 (2006) 18378–18385.
127. W. D. Fessner and G. Sinerius, Enzymes in organic synthesis. 7. Synthesis of dihydroxyacetone phosphate (and isosteric analogs) by enzymic oxidation: Sugars from glycerol, *Angew. Chem.*, 106 (1994) 217–220.
128. M. W. L. Lau and P. J. Unrau, A promiscuous ribozyme promotes nucleotide synthesis in addition to ribose chemistry, *Chem. Biol.*, 16 (2009) 815–825.
129. V. V. Mossine, C. L. Barnes, and T. P. Mawhinney, Structure of D-fructosamine hydrochloride and D-fructosamine hydroacetate, *J. Carbohydr. Chem.*, 28 (2009) 245–263.
130. J. H. Altena, G. A. M. Van den Ouweland, C. J. Teunis, and S. B. Tjan, Analysis of the 220-MHz, PMR spectra of some products of the Amadori- and Heyns-rearrangements, *Carbohydr. Res.*, 92 (1981) 43–55.
131. A. Jakas and Š. Horvat, Synthesis and ^{13}C NMR investigation of novel Amadori compounds (1-amino-1-deoxy-d-fructose derivatives) related to the opioid peptide, leucine-enkephalin, *J. Chem. Soc. Perkin Trans.* 2 (1996) 789–794.

132. C. I. Neglia, H. J. Cohen, A. R. Garber, P. D. Ellis, S. R. Thorpe, and J. W. Baynes, Carbon-13 NMR investigation of nonenzymatic glucosylation of protein. Model studies using RNase A, *J. Biol. Chem.*, 258 (1983) 14279–14283.
133. J. Rohovec, T. Maschmeyer, S. Aime, and J. A. Peters, The structure of the sugar residue in glycated human serum albumin and its molecular recognition by phenylboronate, *Chem. Eur. J.*, 9 (2003) 2193–2199.
134. P. A. Voziyan, R. G. Khalifah, C. Thibaudeau, A. Yildiz, J. Jacob, A. S. Serianni, and B. G. Hudson, Modification of proteins *in vitro* by physiological levels of glucose: Pyridoxamine inhibits conversion of Amadori intermediate to advanced glycation end-products through binding of redox metal ions, *J. Biol. Chem.*, 278 (2003) 46616–46624.
135. A. D. French, M. K. Dowd, and P. J. Reilly, MM3 modeling of fructose ring shapes and hydrogen bonding, *THEOCHEM*, 395–396 (1997) 271–287.
136. R. Rodríguez, J. Cobo, M. Nogueras, J. N. Low, and C. Glidewell, Hydrogen-bonded sheets of $R_2^2(10)$ and $R_4^4(24)$ rings in 1-deoxy-1-morpholino-D-fructopyranose, *Acta Crystallogr.*, C63 (2007) o507–o509.
137. Y. Hou, X. Wu, W. Xie, P. G. Braunschweiger, and P. G. Wang, The synthesis and cytotoxicity of fructose-1-SNAP, a novel fructose conjugated *S*-nitroso nitric oxide donor, *Tetrahedron Lett.*, 42 (2001) 825–829.
138. E. Moreno, S. Perez-Garrido, P. Villares, and R. Jimenez-Garay, *N*-Benzyl-1-methylamino-1-deoxy-β-D-*arabino*-2-hexulopyranose, ($C_{14}H_{21}NO_5$), *Cryst. Struct. Commun.*, 7 (1978) 547–551.
139. V. V. Mossine, G. V. Glinsky, C. L. Barnes, and M. S. Feather, Crystal structure of an Amadori compound, *N*-(1-deoxy-β-D-fructopyranos-1-yl)-glycine ("D-fructose-glycine"), *Carbohydr. Res.*, 266 (1995) 5–14.
140. M. Tarnawski, K. Ślepokura, T. Lis, R. Kuliś-Orzechowska, and B. Szelepin, Crystal structure of *N*-(1-deoxy-β-D-fructopyranos-1-yl)-l-proline—an Amadori compound, *Carbohydr. Res.*, 342 (2007) 1264–1270.
141. V. V. Mossine, C. L. Barnes, and T. P. Mawhinney, The structure of *N*-(1-deoxy-β-D-fructopyranos-1-yl)-L-proline monohydrate ("D-fructose-L-proline") and *N*-(1,6-dideoxy-α-L-fructofuranos-1-yl)-L-proline ("L-rhamnulose-L-proline"), *J. Carbohydr. Chem.*, 26 (2007) 249–266.
142. V. V. Mossine, C. L. Barnes, and T. P. Mawhinney, Solubility and crystal structure of *N*-(1-deoxy-β-D-fructopyranos-1-yl)-L-histidine monohydrate ("D-fructose-L-histidine"), *Carbohydr. Res.*, 342 (2007) 131–138.
143. V. V. Mossine, C. L. Barnes, D. L. Chance, and T. P. Mawhinney, Stabilization of the acyclic tautomer in reducing carbohydrates, *Angew. Chem. Int. Ed.*, 48 (2009) 5517–5520.

144. G. V. Glinsky, V. V. Mossine, J. E. Price, D. Bielenberg, V. V. Glinsky, H. N. Ananthaswamy, and M. S. Feather, Inhibition of colony formation in agarose of metastatic human breast carcinoma and melanoma cells by synthetic glycoamine analogs, *Clin. Exp. Metastasis*, 14 (1996) 253–267.
145. M. S. Feather and V. V. Mossine, Correlations between structure and reactivity of Amadori compounds: The reactivity of acyclic forms, *Spec. Publ. R. Soc. Chem.*, 223 (1998) 37–42.
146. D. Gomez de Anderez, H. Gil, M. Helliwell, and J. M. Segreda, *N*-(*p*-Tolyl)-amine-1-D-fructose from a small crystal, *Acta Crystallogr.*, C52 (1996) 252–254.
147. V. V. Mossine, C. L. Barnes, and T. P. Mawhinney, Disordered hydrogen bonding in *N*-(1-deoxy-β-D-fructopyranos-1-yl)-*N*-allylaniline, *Carbohydr. Res.*, 344 (2009) 948–951.
148. F. Weygand, H. Simon, and R. Ardenne, 1-Deoxy-1-(methylarylamino)-D-fructoses by Amadori rearrangement, *Chem. Ber.*, 92 (1959) 3117–3121.
149. V. V. Mossine, C. L. Barnes, M. S. Feather, and T. P. Mawhinney, Acyclic tautomers in crystalline carbohydrates: The keto forms of 1-deoxy-1-carboxy-methylamino-D-2-pentuloses (pentulose-glycines), *J. Am. Chem. Soc.*, 124 (2002) 15178–15179.
150. W. Saenger and T. Steiner, Cyclodextrin inclusion complexes: Host-guest interactions and hydrogen-bonding networks, *Acta Crystallogr.*, A54 (1998) 798–805.
151. M. S. Feather, V. Mossine, and J. Hirsch, The use of aminoguanidine to trap and measure dicarbonyl intermediates produced during the Maillard reaction, *ACS Symp. Ser.*, 631 (1996) 24–31.
152. V. V. Mossine, C. L. Barnes, G. V. Glinsky, and M. S. Feather, Interaction between two C1-X-C1′ branched ketoses: Observation of an unprecedented crystalline *spiro*-bicyclic hemiketal tautomer of *N*,*N*-di(1-deoxy-D-fructos-1-yl)-glycine ("difructose glycine") having open chain carbohydrate, *Carbohydr. Lett.*, 1 (1995) 355–362.
153. J. Hirsch, V. V. Mossine, and M. S. Feather, The detection of some dicarbonyl intermediates arising from the degradation of Amadori compounds (the Maillard reaction), *Carbohydr. Res.*, 273 (1995) 171–177.
154. W. Baltes and K. Franke, Model studies on Maillard reaction. I. Nonvolatile reaction products from the reaction of D-glucose with *p*-chloroaniline, *Z. Lebensm.-Unters. -Forsch.*, 167 (1978) 403–409.
155. M. J. Diánez, A. López-Castro, and R. Márquez, Structure of allyl 1-deoxy-1-[(1-methyl-2-benzoylvinyl)amino]-α-D-fructofuranoside, *Acta Crystallogr.*, C44 (1988) 657–660.

156. T. M. Wrodnigg, C. Kartusch, and C. Illaszewicz, The Amadori rearrangement as key reaction for the synthesis of neoglycoconjugates, *Carbohydr. Res.*, 343 (2008) 2057–2066.
157. E. F. L. J. Anet and T. M. Reynolds, Chemistry of nonenzymic browning. I. Reactions between amino acids, organic acids, and sugars in freeze-dried apricots and peaches, *Aust. J. Chem.*, 10 (1957) 182–192.
158. A. MacKenzie, M. Vyssotski, and E. Nekrasov, Quantitative analysis of dairy phospholipids by ^{31}P NMR, *J. Am. Oil Chem. Soc.*, 86 (2009) 757–763.
159. K. Eichner, M. Reutter, and R. Wittmann, Detection of Maillard reaction intermediates by high-pressure liquid chromatography (HPLC) and gas chromatography, in P. A. Finot, H. U. Aeschbacher, R. F. Hurrell, and R. Liardon, (Eds.), *The Maillard Reaction in Food Processing, Human Nutrition and Physiology*, Birkhäuser Verlag, Basel, 1990, pp. 63–77.
160. K.-M. Joo, C.-W. Park, H.-J. Jeong, S. J. Lee, and I. S. Chang, Simultaneous determination of two Amadori compounds in Korean red ginseng (*Panax ginseng*) extracts and rat plasma by high-performance anion-exchange chromatography with pulsed amperometric detection, *J. Chromatogr.*, B865 (2008) 159–166.
161. A. Frolov and R. Hoffmann, Analysis of Amadori peptides enriched by boronic acid affinity chromatography, *Ann. N. Y. Acad. Sci.*, 1126 (2008) 253–256.
162. A. Frolov and R. Hoffmann, Separation of Amadori peptides from their unmodified analogs by ion-pairing RP-HPLC with heptafluorobutyric acid as ion-pair reagent, *Anal. Bioanal. Chem.*, 392 (2008) 1209–1214.
163. T. Henle, A. W. Walter, and H. Klostermeyer, Simultaneous determination of protein-bound Maillard products by ion exchange chromatography and photodiode array detection, *Spec. Publ. R. Soc. Chem.*, 151 (1994) 195–200.
164. N. Ahmed and P. J. Thornalley, Chromatographic assay of glycation adducts in human serum albumin glycated *in vitro* by derivatization with 6-aminoquinolyl-*N*-hydroxysuccinimidyl-carbamate and intrinsic fluorescence, *Biochem. J.*, 364 (2002) 15–24.
165. Q. Zhang, V. A. Petyuk, A. A. Schepmoes, D. J. Orton, M. E. Monroe, F. Yang, R. D. Smith, and T. O. Metz, Analysis of non-enzymatically glycated peptides: Neutral-loss-triggered MS^3 versus multi-stage activation tandem mass spectrometry, *Rapid Commun. Mass Spectrom.*, 22 (2008) 3027–3034.
166. P. Stefanowicz, K. Kapczyńska, M. Jaremko, L. Jaremko, and Z. Szewczuk, A mechanistic study on the fragmentation of peptide-derived Amadori products, *J. Mass Spectrom.*, 44 (2009) 1500–1508.

167. Y. Wada, Primary sequence and glycation at lysine-548 of bovine serum albumin, *J. Mass Spectrom.*, 31 (1996) 263–266.
168. T. Miyata, R. Inagi, Y. Wada, Y. Ueda, Y. Iida, M. Takahashi, N. Taniguchi, and K. Maeda, Glycation of human β_2-microglobulin in patients with hemodialysis-associated amyloidosis: Identification of the glycated sites, *Biochemistry*, 33 (1994) 12215–12221.
169. A. Lapolla, L. Baldo, R. Aronica, C. Gerhardinger, D. Fedele, G. Elli, R. Seraglia, S. Catinella, and P. Traldi, Matrix-assisted laser desorption/ionization mass spectrometric studies on protein glycation. 2. The reaction of ribonuclease with hexoses, *Biol. Mass Spectrom.*, 23 (1994) 241–248.
170. V. Fogliano, S. M. Monti, A. Visconti, G. Randazzo, A. M. Facchiano, G. Colonna, and A. Ritieni, Identification of a β-lactoglobulin lactosylation site, *Biochim. Biophys. Acta*, 1388 (1998) 295–304.
171. Q. Zhang, A. Frolov, N. Tang, R. Hoffmann, T. van de Goor, T. O. Metz, and R. D. Smith, Application of electron transfer dissociation mass spectrometry in analyses of non-enzymatically glycated peptides, *Rapid Commun. Mass Spectrom.*, 21 (2007) 661–666.
172. T. Davidek, K. Kraehenbuehl, S. Devaud, F. Robert, and I. Blank, Analysis of Amadori compounds by high-performance cation exchange chromatography coupled to tandem mass spectrometry, *Anal. Chem.*, 77 (2005) 140–147.
173. H. Yoshida, J. Yamazaki, S. Ozawa, T. Mizukoshi, and H. Miyano, Advantage of LC-MS metabolomics methodology targeting hydrophilic compounds in the studies of fermented food samples, *J. Agric. Food Chem.*, 57 (2009) 1119–1126.
174. B. Zhang, S. Mathewson, and H. Chen, Two-dimensional liquid chromatographic methods to examine phenylboronate interactions with recombinant antibodies, *J. Chromatogr.*, A1216 (2009) 5676–5686.
175. A. Takátsy, K. Böddi, L. Nagy, G. Nagy, S. Szabó, L. Markó, I. Wittmann, R. Ohmacht, T. Ringer, G. K. Bonn, D. Gjerde, and Z. Szabó, Enrichment of Amadori products derived from the nonenzymatic glycation of proteins using microscale boronate affinity chromatography, *Anal. Biochem.*, 393 (2009) 8–22.
176. M. P. Pereira Morais, J. D. Mackay, S. K. Bhamra, J. G. Buchanan, T. D. James, J. S. Fossey, and J. M. H. van den Elsen, Analysis of protein glycation using phenylboronate acrylamide gel electrophoresis, *Proteomics*, 10 (2010) 48–58.
177. M. Kijewska, A. Kluczyk, P. Stefanowicz, and Z. Szewczuk, Electrospray ionization mass spectrometric analysis of complexes between peptide-derived Amadori products and borate ions, *Rapid Commun. Mass Spectrom.*, 23 (2009) 4038–4046.

178. J. Hau, S. Devaud, and I. Blank, Detection of Amadori compounds by capillary electrophoresis coupled to tandem mass spectrometry, *Electrophoresis*, 25 (2004) 2077–2083.
179. H. F. Erbersdobler and V. Somoza, Forty years of furosine—forty years of using Maillard reaction products as indicators of the nutritional quality of foods, *Mol. Nutr. Food Res.*, 51 (2007) 423–430.
180. R. Krause, K. Knoll, and T. Henle, Studies on the formation of furosine and pyridosine during acid hydrolysis of different Amadori products of lysine, *Eur. Food Res. Technol.*, 216 (2003) 277–283.
181. I. Penndorf, D. Biedermann, S. V. Maurer, and T. Henle, Studies on *N*-terminal glycation of peptides in hypoallergenic infant formulas: Quantification of α-*N*-(2-furoylmethyl) amino acids, *J. Agric. Food Chem.*, 55 (2007) 723–727.
182. H. Erbersdobler, Loss of lysine during manufacture and storage of milk powder, *Milchwissenschaft*, 25 (1970) 280–284.
183. J. A. B. Baptista and R. C. B. Carvalho, Indirect determination of Amadori compounds in milk-based products by HPLC/ELSD/UV as an index of protein deterioration, *Food Res. Int.*, 37 (2004) 739–747.
184. A. Wellner, C. Hüttl, and T. Henle, Influence of heat treatment on the formation of Amadori compounds in carrots, *Czech J. Food Sci.*, 27 (2009) S143–S145.
185. M. L. Sanz, M. D. Del Castillo, N. Corzo, and A. Olano, Formation of Amadori compounds in dehydrated fruits, *J. Agric. Food Chem.*, 49 (2001) 5228–5231.
186. A. Cardelle-Cobas, F. J. Moreno, N. Corzo, A. Olano, and M. Villamiel, Assessment of initial stages of Maillard reaction in dehydrated onion and garlic samples, *J. Agric. Food Chem.*, 53 (2005) 9078–9082.
187. P. A. Finot, J. Bricout, R. Viani, and J. Mauron, Identification of a new lysine derivative obtained upon acid hydrolysis of heated milk, *Experientia*, 24 (1968) 1097–1099.
188. D. B. Sacks, Translating hemoglobin A_{1c} into average blood glucose: Implications for clinical chemistry, *Clin. Chem.*, 54 (2008) 1756–1758.
189. W. G. John, Haemoglobin A_{1c}: Analysis and standardisation, *Clin. Chem. Lab. Med.*, 41 (2003) 1199–1212.
190. Y. Nanjo, R. Hayashi, and T. Yao, Determination of fructosyl amino acids and fructosyl peptides in protease-digested blood sample by a flow-injection system with an enzyme reactor, *Anal. Sci.*, 22 (2006) 1139–1143.
191. R. N. Johnson, P. A. Metcalf, and J. R. Baker, Fructosamine: A new approach to the estimation of serum glycosylprotein. An index of diabetic control, *Clin. Chim. Acta*, 127 (1983) 87–95.

192. Y. Xu, X. Wu, W. Liu, X. Lin, J. Chen, and R. He, A convenient assay of glycoserum by nitroblue tetrazolium with iodoacetamide, *Clin. Chim. Acta*, 325 (2002) 127–131.
193. J. R. Baker, D. V. Zyzak, S. R. Thorpe, and J. W. Baynes, Chemistry of the fructosamine assay: D-glucosone is the product of oxidation of Amadori compounds, *Clin. Chem.*, 40 (1994) 1950–1955.
194. T. Hanai, M. Uchida, M. Amao, C. Ikeda, K. Koizumi, and T. Kinoshita, Selective chemiluminescence analysis of Amadori form of glycated human serum albumin, *J. Liq. Chromatogr. Relat. Technol.*, 23 (2000) 3119–3131.
195. S. Blincko, D. Colbert, W. G. John, and R. Edwards, A reliable non-separation fluorescence quenching assay for total glycated serum protein: A simple alternative to nitroblue tetrazolium reduction, *Ann. Clin. Biochem.*, 37 (2000) 372–379.
196. T. Tharavanij, T. Froud, C. B. Leitao, D. A. Baidal, C. N. Paz-Pabon, M. Shari, P. Cure, K. Bernetti, C. Ricordi, and R. Alejandro, Clinical use of fructosamine in islet transplantation, *Cell Transplant*, 18 (2009) 453–458.
197. A. Van Campenhout, C. Van Campenhout, Y. S. Olyslager, O. Van Damme, A. R. Lagrou, and B. Keenoy, A novel method to quantify *in vivo* transferrin glycation: Applications in diabetes mellitus, *Clin. Chim. Acta*, 370 (2006) 115–123.
198. R. Paroni, F. Ceriotti, R. Galanello, G. Battista Leoni, A. Panico, E. Scurati, R. Paleari, L. Chemello, V. Quaino, L. Scaldaferri, A. Lapolla, and A. Mosca, Performance characteristics and clinical utility of an enzymatic method for the measurement of glycated albumin in plasma, *Clin. Biochem.*, 40 (2007) 1398–1405.
199. A. Mori, P. Lee, H. Mizutani, T. Takahashi, D. Azakami, M. Mizukoshi, H. Fukuta, N. Sakusabe, A. Sakusabe, Y. Kiyosawa, T. Arai, and T. Sako, Serum glycated albumin as a glycemic control marker in diabetic cats, *J. Vet. Diagn. Invest.*, 21 (2009) 112–116.
200. K. Mistry and K. Kalia, Non enzymatic glycosylation of IgG and their urinary excretion in patients with diabetic nephropathy, *Indian J. Clin. Biochem.*, 24 (2009) 159–165.
201. T. Matsuda, Y. Kato, K. Watanabe, and R. Nakamura, Direct evaluation of β-lactoglobulin lactosylation in early Maillard reaction using an antibody specific to protein-bound lactose, *J. Agric. Food Chem.*, 33 (1985) 1193–1196.
202. V. Fogliano, S. M. Monti, A. Ritieni, C. Marchisano, G. Peluso, and G. Randazzo, An immunological approach to monitor protein lactosylation of heated food model systems, *Food Chem.*, 58 (1996) 53–58.
203. M. P. Cohen and E. Hud, Production and characterization of monoclonal antibodies against human glycoalbumin, *J. Immunol. Methods*, 117 (1989) 121–129.

204. W. G. John, M. R. Gray, D. L. Bates, and J. L. Beacham, Enzyme immunoassay—a new technique for estimating hemoglobin A_{1c}, *Clin. Chem.*, 39 (1993) 663–666.
205. M. Plebani, R. Venturini, L. Marchioro, and A. Burlina, Measurement of hemoglobin A_{1c} by enzyme immunoassay, *Eur. J. Lab. Med.*, 1 (1993) 155–159.
206. A. Sakaguchi, S. Ferri, W. Tsugawa, and K. Sode, Novel fluorescent sensing system for α-fructosyl amino acids based on engineered fructosyl amino acid binding protein, *Biosens. Bioelectron.*, 22 (2007) 1933–1938.
207. T. Yamazaki, S. Ohta, Y. Yanai, and K. Sode, Molecular imprinting catalyst based artificial enzyme sensor for fructosylamines, *Anal. Lett.*, 36 (2003) 75–89.
208. R. Rajkumar, A. Warsinke, H. Möhwald, F. W. Scheller, and M. Katterle, Development of fructosyl valine binding polymers by covalent imprinting, *Biosens. Bioelectron.*, 22 (2007) 3318–3325.
209. R. Rajkumar, M. Katterle, A. Warsinke, H. Möhwald, and F. W. Scheller, Thermometric MIP sensor for fructosyl valine, *Biosens. Bioelectron.*, 23 (2008) 1195–1199.
210. V. V. Mossine and T. P. Mawhinney, N^{α}-(1-Deoxy-D-fructos-1-yl)-L-histidine ("D-fructose-L-histidine"): A potent copper chelator from tomato powder, *J. Agric. Food Chem.*, 55 (2007) 10373–10381.
211. B. Gyurcsik, T. Gajda, L. Nagy, K. Burger, A. Rockenbauer, and L. Korecz, Proton, copper(II) and nickel(II) complexes of some Amadori rearrangement products of D-glucose and amino acids, *Inorg. Chim. Acta*, 214 (1993) 57–66.
212. S. T. Seifert, R. Krause, K. Gloe, and T. Henle, Metal complexation by the peptide-bound Maillard reaction products N^{ε}-fructoselysine and N^{ε}-carboxymethyllysine, *J. Agric. Food Chem.*, 52 (2004) 2347–2350.
213. M. Tonković, A. Jakas, and Š. Horvat, Preparation and properties of an Fe(III)-complex with an Amadori compound derived from L-tyrosine, *Biometals*, 10 (1997) 55–59.
214. J. Chen, T. Pill, and W. Beck, Metal complexes of biologically important ligands. L. Palladium(II), platinum(II) and copper(II) complexes of α-amino acid N-glycosides and of fructose amino acids (Amadori compounds), *Z. Naturforsch.*, B44 (1989) 459–464.
215. E. Ferrari, R. Grandi, S. Lazzari, and M. Saladini, NMR study on Pt(II) interaction with Amadori compounds, *Inorg. Chim. Acta*, 360 (2007) 3119–3122.
216. E. Ferrari, R. Grandi, S. Lazzari, G. Marverti, M. C. Rossi, and M. Saladini, ^{1}H, ^{13}C, ^{195}Pt NMR study on platinum(II) interaction with sulphur containing Amadori compounds, *Polyhedron*, 26 (2007) 4045–4052.

217. P. R. Smith and P. J. Thornalley, Influence of pH and phosphate ions on the kinetics of enolization and degradation of fructosamines. Studies with the model fructosamine, N^{ε}-1-deoxy-D-fructos-1-yl-hippuryl-lysine, *Biochem. Int.*, 28 (1992) 429–439.
218. T. Davídek, N. Clety, S. Aubin, and I. Blank, Degradation of the Amadori compound *N*-(1-deoxy-D-fructos-1-yl)glycine in aqueous model systems, *J. Agric. Food Chem.*, 50 (2002) 5472–5479.
219. F. Hayase, Recent development of 3-deoxyosone related Maillard reaction products, *Food Sci. Technol. Res.*, 6 (2000) 79–86.
220. M. Ota, M. Kohmura, and H. Kawaguchi, Characterization of a new Maillard type reaction product generated by heating 1-deoxymaltulosyl-glycine in the presence of cysteine, *J. Agric. Food Chem.*, 54 (2006) 5127–5131.
221. K. M. Biemel and M. O. Lederer, Site-specific quantitative evaluation of the protein glycation product N^6-(2,3-dihydroxy-5,6-dioxohexyl)-L-lysinate by LC-(ESI)MS peptide mapping: Evidence for its key role in AGE formation, *Bioconjug. Chem.*, 14 (2003) 619–628.
222. A. A. Staempfli, I. Blank, R. Fumeaux, and L. B. Fay, Study on the decomposition of the Amadori compound *N*-(1-deoxy-D-fructos-1-yl)-glycine in model systems: Quantification by fast atom bombardment tandem mass spectrometry, *Biol. Mass Spectrom.*, 23 (1994) 642–646.
223. S. J. Cho, G. Roman, F. Yeboah, and Y. Konishi, The road to advanced glycation end products: A mechanistic perspective, *Curr. Med. Chem.*, 14 (2007) 1653–1671.
224. J. Gobert and M. A. Glomb, Degradation of glucose: Reinvestigation of reactive α-dicarbonyl compounds, *J. Agric. Food Chem.*, 57 (2009) 8591–8597.
225. H. Tsuchida, M. Komoto, H. Kato, and M. Fujimaki, Formation of deoxyfructosazine and its 6-isomer on the browning reaction between glucose and ammonia in weak acidic medium, *Agric. Biol. Chem.*, 37 (1973) 2571–2578.
226. D. R. Cremer, M. Vollenbroeker, and K. Eichner, Investigation of the formation of Strecker aldehydes from the reaction of Amadori rearrangement products with α-amino acids in low moisture model systems, *Eur. Food Res. Technol.*, 211 (2000) 400–403.
227. G. P. Rizzi, The Strecker degradation of amino acids: Newer avenues for flavor formation, *Food Rev. Int.*, 24 (2008) 416–435.
228. A. Hollnagel and L. W. Kroh, Degradation of oligosaccharides in nonenzymatic browning by formation of α-dicarbonyl compounds via a "peeling off" mechanism, *J. Agric. Food Chem.*, 48 (2000) 6219–6226.
229. S. M. Culbertson, E. I. Vassilenko, L. D. Morrison, and K. Ingold, Paradoxical impact of antioxidants on post-Amadori glycoxidation: Counterintuitive

increase in the yields of pentosidine and N^{ε}-carboxymethyllysine using a novel multifunctional pyridoxamine derivative, *J. Biol. Chem.*, 278 (2003) 38384–38394.
230. H. Horikawa, M. Okada, Y. Nakamura, A. Sato, and N. Iwamoto, Production of hydroxyl radicals and α-dicarbonyl compounds associated with Amadori compound-Cu^{2+} complex degradation, *Free Radic. Res.*, 36 (2002) 1059–1065.
231. R. Liedke and K. Eichner, Radical induced formation of D-glucosone from Amadori compounds, ACS Symp. Ser., 807 (2002) 69–82.
232. T. Hofmann and P. Schieberle, Formation of aroma-active Strecker-aldehydes by a direct oxidative degradation of Amadori compounds, *J. Agric. Food Chem.*, 48 (2000) 4301–4305.
233. T. Sakurai, K. Fujimori, T. Ueda, H. Shindo, Y. Shibusawa, and M. Nakano, Sunlight induces N^{ε}-(carboxymethyl)lysine formation from glycated polylysine-iron(III) complex, *Photochem. Photobiol.*, 74 (2001) 407–411.
234. A. Abrams, P. H. Lowy, and H. Borsook, Preparation of 1-amino-1-deoxy-2-ketohexoses from aldohexoses and α-amino acids, *J. Am. Chem. Soc.*, 77 (1955) 4794–4796.
235. E. L. Richards, A quantitative study of changes in dried skim milk and lactose-casein in the dry state during storage, *J. Dairy Res.*, 30 (1963) 223–234.
236. R. Nagai, Y. Unno, M. C. Hayashi, S. Masuda, F. Hayase, N. Kinae, and S. Horiuchi, Peroxynitrite induces formation of N^{ε}-(carboxymethyl)lysine by the cleavage of Amadori product and generation of glucosone and glyoxal from glucose: Novel pathways for protein modification by peroxynitrite, *Diabetes*, 51 (2002) 2833–2839.
237. K. Mera, R. Nagai, N. Haraguchi, Y. Fujiwara, T. Araki, N. Sakata, and M. Otagiri, Hypochlorous acid generates N^{ε}-(carboxymethyl)lysine from Amadori products, *Free Radic. Res.*, 41 (2007) 713–718.
238. H. Tanaka, G. Yoshida, Y. Baba, K. Matsumura, H. Wasada, J. Murata, M. Agawa, S. Itakura, and A. Enoki, Characterization of a hydroxyl-radical-producing glycoprotein and its presumptive genes from the white-rot basidiomycete *Phanerochaete chrysosporium, J. Biotechnol.*, 128 (2007) 500–511.
239. F. Weygand and A. Bergmann, N-Glycosides. VI. Catalytic oxidation of arylisoglycosamines, *Chem. Ber.*, 80 (1947) 261–263.
240. E. F. L. J. Anet, Chemistry of nonenzymic browning. VII. Crystalline di-D-fructoseglycine and some related compounds, *Aust. J. Chem.*, 12 (1959) 280–287.
241. R. Badoud, L. Fay, F. Hunston, and G. Pratz, Periodate oxidative degradation of Amadori compounds. Formation of N^{ε}-carboxymethyllysine and

N-carboxymethylamino acids as markers of the early Maillard reaction, *ACS Symp. Ser.*, 631 (1996) 208–220.

242. H. J. Haas and P. Schlimmer, Pure D-arabino hexosulose, *Justus Liebigs Ann. Chem.*, 759 (1972) 208–210.
243. H. Weenen and J. G. M. Van der Ven, The formation of Strecker aldehydes, *ACS Symp. Ser.*, 794 (2001) 183–195.
244. V. A. Yaylayan, Recent advances in the chemistry of Strecker degradation and Amadori rearrangement: Implications to aroma and color formation, *Food Sci. Technol. Res.*, 9 (2003) 1–6.
245. K. Fukase, H. Nakayama, M. Kurosawa, T. Ikegaki, T. Kanoh, S. Hase, and S. Kusumoto, Functional fluorescence labeling of carbohydrates and its use for preparation of neoglycoconjugates, *J. Carbohydr. Chem.*, 13 (1994) 715–736.
246. D. R. Griffiths and J. B. Pridham, The metabolism of 1-(ε-N-lysyl)-1-deoxy-D-fructose by *Escherichia coli*, *J. Sci. Food Agric.*, 31 (1980) 1214–1220.
247. K. G. Akamanchi, R. Meenakshi, and K. J. Ranbhan, A novel ability of bakers' yeast in hydrolysis of Amadori compounds (1-deoxy-1-amino-2-ketoses), *Indian J. Chem.*, 31B (1992) 908–910.
248. N. Yoshida, Fructosyl amino acid kinase germinating seeds of Japanese radish and use for assaying Amadori compounds, *Jpn. Pat. Appl.*, (2005) JP 2005058157.
249. L. Mester, L. Szabados, M. Mester, and C. Dautheville, Enzymic cleavage of Maillard-type sugar-amine compounds, *Acta Chim. Hung.*, 113 (1983) 437–442.
250. T. Horiuchi, T. Kurokawa, and N. Saito, Purification and properties of fructosyl-amino acid oxidase from *Corynebacterium* sp. 2-4-1, *Agric. Biol. Chem.*, 53 (1989) 103–110.
251. X. Wu, M. Takahashi, S. G. Chen, and V. M. Monnier, Cloning of amadoriase I isoenzyme from *Aspergillus* sp.: Evidence of FAD covalently linked to Cys342, *Biochemistry*, 39 (2000) 1515–1521.
252. K. Hirokawa, K. Gomi, M. Bakke, and N. Kajiyama, Distribution and properties of novel deglycating enzymes for fructosyl peptide in fungi, *Arch. Microbiol.*, 180 (2003) 227–231.
253. X. Wu, S. G. Chen, J. M. Petrash, and V. M. Monnier, Alteration of substrate selectivity through mutation of two arginine residues in the binding site of amadoriase II from *Aspergillus* sp., *Biochemistry*, 41 (2002) 4453–4458.
254. X. Wu, B. A. Palfey, V. V. Mossine, and V. M. Monnier, Kinetic studies, mechanism, and substrate specificity of amadoriase I from *Aspergillus* sp., *Biochemistry*, 40 (2001) 12886–12895.

255. K. Hirokawa and N. Kajiyama, Recombinant *Agrobacterium* AgaE-like protein with fructosyl amino acid oxidase activity, *Biosci. Biotechnol. Biochem.*, 66 (2002) 2323–2329.
256. C. H. Baek, S. K. Farrand, K. E. Lee, D. K. Park, J. K. Lee, and K. S. Kim, Convergent evolution of Amadori opine catabolic systems in plasmids of *Agrobacterium tumefaciens, J. Bacteriol.*, 185 (2003) 513–524.
257. S. Ferri, A. Sakaguchi, H. Goto, W. Tsugawa, and K. Sode, Isolation and characterization of a fructosyl-amine oxidase from an *Arthrobacter* sp., *Biotechnol. Lett.*, 27 (2005) 27–32.
258. N. Yoshida, Y. Sakai, A. Isogai, H. Fukuya, M. Yagi, Y. Tani, and N. Kato, Primary structures of fungal fructosyl amino acid oxidases and their application to the measurement of glycated proteins, *Eur. J. Biochem.*, 242 (1996) 499–505.
259. M. Takahashi, M. Pischetsrieder, and V. M. Monnier, Isolation, purification, and characterization of amadoriase isoenzymes (fructosyl amine-oxygen oxidoreductase EC 1.5.3) from *Aspergillus* sp., *J. Biol. Chem.*, 272 (1997) 3437–3443.
260. K. Hirokawa, K. Nakamura, and N. Kajiyama, Enzymes used for the determination of HbA_{1C}, *FEMS Microbiol. Lett.*, 235 (2004) 157–162.
261. Y. Sakai, N. Yoshida, A. Isogai, Y. Tani, and N. Kato, Purification and properties of fructosyl lysine oxidase from *Fusarium oxysporum* S-1F4, *Biosci. Biotechnol. Biochem.*, 59 (1995) 487–491.
262. K. Sode, F. Ishimura, and W. Tsugawa, Screening and characterization of fructosyl-valine-utilizing marine microorganisms, *Mar. Biotechnol.*, 3 (2001) 126–132.
263. A. K. Saxena, P. Saxena, and V. M. Monnier, Purification and characterization of a membrane-bound deglycating enzyme (1-deoxyfructosyl alkyl amino acid oxidase, EC 15.3) from a *Pseudomonas* sp. soil strain, *J. Biol. Chem.*, 271 (1996) 32803–32809.
264. C. Gerhardinger, M. S. Marion, A. Rovner, M. Glomb, and V. M. Monnier, Novel degradation pathway of glycated amino acids into free fructosamine by a *Pseudomonas* sp. soil strain extract, *J. Biol. Chem.*, 270 (1995) 218–224.
265. E. Capuano, F. Fedele, C. Mennella, M. Visciano, A. Napolitano, S. Lanzuise, M. Ruocco, M. Lorito, M. D. Del Castillo, and V. Fogliano, Studies on the effect of amadoriase from *Aspergillus fumigatus* on peptide and protein glycation in vitro, *J. Agric. Food Chem.*, 55 (2007) 4189–4195.
266. S. Kim, S. Miura, S. Ferri, W. Tsugawa, and K. Sode, Cumulative effect of amino acid substitution for the development of fructosyl valine-specific fructosyl amine oxidase, *Enzyme Microb. Technol.*, 44 (2009) 52–56.

267. T. Yamazaki, S. Ohta, and K. Sode, Operational condition of a molecular imprinting catalyst-based fructosyl-valine sensor, *Electrochemistry*, 76 (2008) 590–593.
268. A. Sakaguchi, W. Tsugawa, S. Ferri, and K. Sode, Development of highly-sensitive fructosyl-valine enzyme sensor employing recombinant fructosyl amine oxidase, *Electrochemistry*, 71 (2003) 442–445.
269. L. Fang, W. Li, Y. Zhou, and C. -C. Liu, A single-use, disposable iridium-modified electrochemical biosensor for fructosyl valine for the glycosylated hemoglobin detection, *Sens. Actuators*, B137 (2009) 235–238.
270. I. Sakurabayashi, T. Watano, S. Yonehara, K. Ishimaru, K. Hirai, T. Komori, and M. Yagi, New enzymatic assay for glycohemoglobin, *Clin. Chem.*, 49 (2003) 269–274.
271. T. O'Connell, N. Hoven, P. Siegert, and K. -H. Maurer, Amadoriases for use in washing and cleaning products, *Int. Pat. Appl.*, (2007) WO 2007128542.
272. J. A. Delvalle and C. Asensio, Distribution of adenosine 5′-triphosphate (ATP)-dependent hexose kinases in microorganisms, *BioSystems*, 10 (1978) 265–282.
273. G. Delpierre, M. H. Rider, F. Collard, V. Stroobant, F. Vanstapel, H. Santos, and E. Van Schaftingen, Identification, cloning, and heterologous expression of a mammalian fructosamine-3-kinase, *Diabetes*, 49 (2000) 1627–1634.
274. B. S. Szwergold, S. Howell, and P. J. Beisswenger, Human fructosamine-3-kinase: Purification, sequencing, substrate specificity, and evidence of activity in vivo, *Diabetes*, 50 (2001) 2139–2147.
275. J. R. Conner, P. J. Beisswenger, and B. S. Szwergold, The expression of the genes for fructosamine-3-kinase and fructosamine-3-kinase-related protein appears to be constitutive and unaffected by environmental signals, *Biochem. Biophys. Res. Commun.*, 323 (2004) 932–936.
276. M. Veiga-da-Cunha, P. Jacquemin, G. Delpierre, C. Godfraind, I. Théate, D. Vertommen, F. Clotman, F. Lemaigre, O. Devuyst, and E. Van Schaftingen, Increased protein glycation in fructosamine 3-kinase-deficient mice, *Biochem. J.*, 399 (2006) 257–264.
277. G. Delpierre, D. Vertommen, D. Communi, M. H. Rider, and E. Van Schaftingen, Identification of fructosamine residues deglycated by fructosamine-3-kinase in human hemoglobin, *J. Biol. Chem.*, 279 (2004) 27613–27620.
278. H. J. Tsai, S. Y. Chou, F. Kappler, M. L. Schwartz, and A. M. Tobia, A new inhibitor for fructosamine 3-kinase (amadorase), *Drug Dev. Res.*, 67 (2006) 448–455.
279. F. Collard, G. Delpierre, V. Stroobant, G. Matthijs, and E. Van Schaftingen, A mammalian protein homologous to fructosamine-3-kinase is a ketosamine-3-

kinase acting on psicosamines and ribulosamines but not on fructosamines, *Diabetes*, 52 (2003) 2888–2895.

280. B. S. Szwergold, Fructosamine-6-phosphates are deglycated by phosphorylation to fructosamine-3,6-bisphosphates catalyzed by fructosamine-3-kinase (FN3K) and/or fructosamine-3-kinase-related-protein (FN3KRP), *Med. Hypotheses*, 68 (2007) 37–45.
281. E. Wiame, G. Delpierre, F. Collard, and E. Van Schaftingen, Identification of a pathway for the utilization of the Amadori product fructoselysine in *Escherichia coli*, *J. Biol. Chem.*, 277 (2002) 42523–42529.
282. E. Wiame, A. Duquenne, G. Delpierre, and E. Van Schaftingen, Identification of enzymes acting on α-glycated amino acids in *Bacillus subtilis*, *FEBS Lett.*, 577 (2004) 469–472.
283. M. Yamada, W. Tsugawa, and K. Sode, Novel enzyme sensor for glycated protein biosensing without the proteolytic processes, *215th Meeting of the Electrochemical Society*, San Francisco, CA, 2009.
284. S. Milewski, Glucosamine-6-phosphate synthase—the multi-facets enzyme, *Biochim. Biophys. Acta*, 1597 (2002) 173–192.
285. C. Liu, D. Li, Y. -H. Liang, L. -F. Li, and X. -D. Su, Ring-opening mechanism revealed by crystal structures of NagB and its ES intermediate complex, *J. Mol. Biol.*, 379 (2008) 73–81.
286. E. Wiame and E. Van Schaftingen, Fructoselysine 3-epimerase, an enzyme involved in the metabolism of the unusual Amadori compound psicoselysine in *Escherichia coli*, *Biochem. J.*, 378 (2004) 1047–1052.
287. L. Vértesy, H. W. Fehlhaber, H. Kogler, and P. W. Schindler, Enkastines: Amadori products with a specific inhibiting action against endopeptidase—24.11—from *Streptomyces albus* and by synthesis, *Liebigs Ann.*, (1996) 121–126.
288. A. Jakas and Š. Horvat, The effect of glycation on the chemical and enzymatic stability of the endogenous opioid peptide, leucine-enkephalin, and related fragments, *Bioorg. Chem.*, 32 (2004) 516–526.
289. D. Schumacher and L. W. Kroh, Studies on the degradation of Maillard reaction products by amylolytic enzymes. Part 1. Reversible inhibition of α- and glucoamylase as well as α-glucosidase by oligosaccharide-Amadori-compounds, *Z. Lebensm.-Unters. -Forsch.*, 199 (1994) 270–274.
290. A. Seidowski, D. Lunow, and T. Henle, Amadori products—substrates and inhibitors for intestinal brush border glycosidases, *Czech J. Food Sci.*, 27 (2009) S146–S148.
291. H. A. Schreuder and G. W. Welling, Determination of β-galactosidase activity in the intestinal tract of mice by ion-exchange high-performance liquid

chromatography using ε-N-1-(1-deoxylactulosyl)-L-lysine as substrate, *J. Chromatogr.*, 278 (1983) 275–282.
292. H. Yoneyama, T. Ohnishi, T. Takaku, H. Okuda, H. Hori, and Y. Ichikawa, Arginyl-fructose in Korean red ginseng (*Panax ginseng* C. A. Meyer) is a substrate of nitric oxide synthase (cytochrome P-450 arg), *Wakan Iyakugaku Zasshi*, 13 (1996) 346–347.
293. H. Nursten, The Maillard Reaction: Chemistry, Biochemistry and Implications, The Royal Society of Chemistry, Cambridge, UK, 2005.
294. F. Ledl and E. Schleicher, The Maillard reaction in food and in the human body—new results in chemistry, biochemistry and medicine, *Angew. Chem.*, 102 (1990) 597–626.
295. S. Fors, Sensory properties of volatile Maillard reaction products and related compounds: A literature review, *ACS Symp. Ser.*, 215 (1983) 185–286.
296. D. S. Mottram, Flavor compounds formed during the Maillard reaction, *ACS Symp. Ser.*, 543 (1994) 104–126.
297. T. Obretenov and G. Vernin, Melanoidins in the Maillard reaction, *Dev. Food Sci.*, 40 (1998) 455–482.
298. S. R. Thorpe and J. W. Baynes, Maillard reaction products in tissue proteins: New products and new perspectives, *Amino Acids*, 25 (2003) 275–281.
299. V. M. Monnier, Intervention against the Maillard reaction in vivo, *Arch. Biochem. Biophys.*, 419 (2003) 1–15.
300. H. Hashiba, Isolation and identification of Amadori compounds from miso, white wine and sake, *Agric. Biol. Chem.*, 42 (1978) 1727–1731.
301. H. Nishi and I. Morishita, Components of Licorice root used for tobacco flavoring. I. Fractionation of the substances in Licorice root effective in improving the tobacco smoking quality, *Nippon Nogei Kagaku Kaishi*, 45 (1971) 507–512.
302. H. Tomita, M. Noguchi, and E. Tamaki, Chemical studies on ninhydrin-positive compounds in cured tobacco leaves. II. Isolation of 1-deoxy-1-L-prolino-D-fructose, *Agric. Biol. Chem.*, 29 (1965) 515–519.
303. R. Wittmann and K. Eichner, Detection of Maillard products in malts, beers, and brewing colorants, *Z. Lebensm.-Unters. -Forsch.*, 188 (1989) 212–220.
304. F. D. Mills and J. E. Hodge, Amadori compounds: Vacuum thermolysis of 1-deoxy-1-L-prolino-D-fructose, *Carbohydr. Res.*, 51 (1976) 9–21.
305. A. Huyghues-Despointes, V. A. Yaylayan, and A. Keyhani, Pyrolysis/GC/MS analysis of 1-[(2′-carboxyl)pyrrolidinyl]-1-deoxy-D-fructose (proline Amadori compound), *J. Agric. Food Chem.*, 42 (1994) 2519–2524.

306. I. Blank, S. Devaud, W. Matthey-Doret, and F. Robert, Formation of odorants in Maillard model systems based on L-proline as affected by pH, *J. Agric. Food Chem.*, 51 (2003) 3643–3650.
307. V. A. Yaylayan, N. G. Forage, and S. Mandeville, Microwave and thermally induced Maillard reactions, *ACS Symp. Ser.*, 543 (1994) 449–456.
308. M. J. Lane and H. E. Nursten, The variety of odors produced in Maillard model systems and how they are influenced by reaction conditions, *ACS Symp. Ser.*, 215 (1983) 141–158.
309. K. H. Wong, S. A. Aziz, and S. Mohamed, Sensory aroma from Maillard reaction of individual and combinations of amino acids with glucose in acidic conditions, *Int. J. Food Sci. Technol.*, 43 (2008) 1512–1519.
310. H. H. M. Fadel and A. Farouk, Caramelization of maltose solution in presence of alanine, *Amino Acids*, 22 (2002) 199–213.
311. V. Pravdova, C. Boucon, S. de Jong, B. Walczak, and D. L. Massart, Three-way principal component analysis applied to food analysis: An example, *Anal. Chim. Acta*, 462 (2002) 133–148.
312. H. Hashiba, A. Okuhara, and N. Iguchi, Oxygen-dependent browning of soy sauce and some brewed products, *Prog. Food Nutr. Sci.*, 5 (1981) 93–113.
313. O. Frank and T. Hofmann, On the influence of the carbohydrate moiety on chromophore formation during food-related Maillard reactions of pentoses, hexoses, and disaccharides, *Helv. Chim. Acta*, 83 (2000) 3246–3261.
314. M. Portero-Otin, R. H. Nagaraj, and V. M. Monnier, Chromatographic evidence for pyrraline formation during protein glycation in vitro and in vivo, *Biochim. Biophys. Acta*, 1247 (1995) 74–80.
315. S. K. Grandhee and V. M. Monnier, Mechanism of formation of the Maillard protein cross-link pentosidine. Glucose, fructose, and ascorbate as pentosidine precursors, *J. Biol. Chem.*, 266 (1991) 11649–11653.
316. H. Obayashi, K. Nakano, H. Shigeta, M. Yamaguchi, K. Yoshimori, M. Fukui, M. Fujii, Y. Kitagawa, N. Nakamura et al., Formation of crossline as a fluorescent advanced glycation end product in vitro and in vivo, *Biochem. Biophys. Res. Commun.*, 226 (1996) 37–41.
317. K. M. Biemel, D. A. Friedl, and M. O. Lederer, Identification and quantification of major Maillard cross-links in human serum albumin and lens protein. Evidence for glucosepane as the dominant compound, *J. Biol. Chem.*, 277 (2002) 24907–24915.
318. S. Reddy, J. Bichler, K. J. Wells-Knecht, S. R. Thorpe, and J. W. Baynes, N^{ε}-(Carboxymethyl)lysine is a dominant advanced glycation end product (AGE) antigen in tissue proteins, *Biochemistry*, 34 (1995) 10872–10878.

319. M. Ahmed, E. Brinkmann-Frye, T. P. Degenhardt, S. R. Thorpe, and J. W. Baynes, N^{ε}-(Carboxyethyl)lysine, a product of the chemical modification of proteins by methylglyoxal, increases with age in human lens proteins, *Biochem. J.*, 324 (1997) 565–570.
320. K. Iijima, M. Murata, H. Takahara, S. Irie, and D. Fujimoto, Identification of N^{ω}-carboxymethylarginine as a novel acid-labile advanced glycation end product in collagen, *Biochem. J.*, 347 (2000) 23–27.
321. M. A. Glomb and C. Pfahler, Amides are novel protein modifications formed by physiological sugars, *J. Biol. Chem.*, 276 (2001) 41638–41647.
322. M. Prabhakaram and V. V. Mossine, Characterization of a blue fluorophore isolated from in vitro reaction of N-α-acetyllysine and 3-deoxyglucosone, *Prep. Biochem. Biotechnol.*, 28 (1998) 319–338.
323. M. O. Lederer and R. G. Klaiber, Cross-linking of proteins by Maillard processes: Characterization and detection of lysine-arginine cross-links derived from glyoxal and methylglyoxal, *Bioorg. Med. Chem.*, 7 (1999) 2499–2507.
324. K. M. Biemel, O. Reihl, J. Conrad, and M. O. Lederer, Formation pathways for lysine-arginine cross-links derived from hexoses and pentoses by Maillard processes. Unraveling the structure of a pentosidine precursor, *J. Biol. Chem.*, 276 (2001) 23405–23412.
325. D. R. Sell, K. M. Biemel, O. Reihl, M. O. Lederer, C. M. Strauch, and V. M. Monnier, Glucosepane is a major protein cross-link of the senescent human extracellular matrix: Relationship with diabetes, *J. Biol. Chem.*, 280 (2005) 12310–12315.
326. G. Sosnovsky, C. T. Gnewuch, and E. S. Ryoo, In the search for new anticancer drugs. XXV: Role of N-nitrosated Amadori compounds derived from glucose-amino acid conjugates in cancer promotion or inhibition, *J. Pharm. Sci.*, 82 (1993) 649–656.
327. J. -P. Bapst, M. Calame, H. Tanner, and A. N. Eberle, Glycosylated DOTA-α-melanocyte-stimulating hormone analogues for melanoma targeting: Influence of the site of glycosylation on in vivo biodistribution, *Bioconjug. Chem.*, 20 (2009) 984–993.
328. G. Vaidyanathan, H. S. Friedman, D. J. Affleck, M. Schottelius, H.-J. Wester, and M. R. Zalutsky, Specific and high-level targeting of radiolabeled octreotide analogues to human medulloblastoma xenografts, *Clin. Cancer Res.*, 9 (2003) 1868–1876.
329. G. Vaidyanathan, D. J. Affleck, M. Schottelius, H. Wester, H. S. Friedman, and M. R. Zalutsky, Synthesis and evaluation of glycosylated octreotate analogues labeled with radioiodine and ^{211}At via a tin precursor, *Bioconjug. Chem.*, 17 (2006) 195–203.

330. A. Stahl, G. Meisetschläger, M. Schottelius, K. Bruus-Jensen, I. Wolf, K. Scheidhauer, and M. Schwaiger, [^{123}I]Mtr-TOCA, a radioiodinated and carbohydrated analogue of octreotide: Scintigraphic comparison with [^{111}In] octreotide, *Eur. J. Nucl. Med. Mol. Imaging*, 33 (2006) 45–52.
331. R. Albert, P. Marbach, W. Bauer, U. Briner, G. Fricker, C. Bruns, and J. Pless, SDZ CO 611: A highly potent glycated analog of somatostatin with improved oral activity, *Life Sci.*, 53 (1993) 517–525.
332. C. Schweinsberg, V. Maes, L. Brans, P. Blauenstein, D. A. Tourwe, P. A. Schubiger, R. Schibli, and E. G. Garayoa, Novel glycated [^{99m}Tc(CO)$_3$]-labeled bombesin analogues for improved targeting of gastrin-releasing peptide receptor-positive tumors, *Bioconjug. Chem.*, 19 (2008) 2432–2439.
333. R. Shepherd, A. Robertson, and D. Ofman, Dairy glycoconjugate emulsifiers: Casein-maltodextrins, *Food Hydrocoll.*, 14 (2000) 281–286.
334. X. Pan, M. Mu, B. Hu, P. Yao, and M. Jiang, Micellization of casein-graft-dextran copolymer prepared through Maillard reaction, *Biopolymers*, 81 (2006) 29–38.
335. R. Kuhn, G. Krüger, and A. Seeliger, Coupling Amadori compounds with diazonium salts; preparation of lactulose from lactose, *Justus Liebigs Ann. Chem.*, 628 (1959) 240–255.
336. S. F. Jenkinson, D. J. Hotchkiss, A. R. Cowley, G. W. J. Fleet, and D. J. Watkin, (*S*)-3-Dimethylamino-2-{(4*S*,5*R*)-5-[(*R*)-2,2-dimethyl-1,3-dioxolan-4-yl]-2,2-dimethyl-1,3-dioxolan-4-yl}-2-hydroxypropanoic acid, *Acta Crystallogr.*, E64 (2008) o294–o295.
337. K. V. Booth, F. P. da Cruz, D. J. Hotchkiss, S. F. Jenkinson, N. A. Jones, A. C. Weymouth-Wilson, R. Clarkson, T. Heinz, and G. W. J. Fleet, Carbon-branched carbohydrate chirons: Practical access to both enantiomers of 2-*C*-methyl-ribono-1,4-lactone and 2-*C*-methyl-arabinonolactone, *Tetrahedron Asymmetry*, 19 (2008) 2417–2424.
338. T. M. Wrodnigg, W. Gaderbauer, P. Greimel, H. Häusler, F. K. Sprenger, A. E. Stütz, C. Virgona, and S. G. Withers, Biologically active 1-aminodeoxy and 1-*O*-alkyl derivatives of the powerful D-glucosidase inhibitor 2,5-dideoxy-2,5-imino-D-mannitol, *J. Carbohydr. Chem.*, 19 (2000) 975–990.
339. J. T. Bagdanoff, M. S. Donoviel, A. Nouraldeen, J. Tarver, Q. Fu, M. Carlsen, T. Jessop, H. Zhang, J. Hazelwood, H. Nguyen, S. D. P. Baugh, M. Gardyan, K. M. Terranova, J. Barbosa, J. Yan, M. Bednarz, S. Layek, J. Taylor, A. M. Digeorge-Foushee, S. Gopinathan, D. Bruce, T. Smith, L. Moran, E. O'Neill, J. Kramer, Z. Lai, S. D. Kimball, Q. Liu, W. Sun, S. Yu, J. Swaffield, A. Wilson, A. Main, K. G. Carson, T. Oravecz, and D. J. Augeri, Inhibition of

sphingosine-1-phosphate lyase for the treatment of autoimmune disorders, *J. Med. Chem.*, 52 (2009) 3941–3953.

340. S. Maza, O. López, S. Martos, I. Maya, and J. G. Fernández-Bolaños, Synthesis of the first selenium-containing acyclic nucleosides and anomeric spironucleosides from carbohydrate precursors, *Eur. J. Org. Chem.*, (2009) 5239–5246.
341. C. Gasch, M. A. Pradera, B. A. B. Salameh, J. L. Molina, and J. Fuentes, Chiral thioxohydroimidazoles with two sugar moieties. *N*-, *C*-, and spiro-nucleosides, *Tetrahedron Asymmetry*, 11 (2000) 435–452.
342. M. X. Zhao, D. Goeddel, K. Li, and Y. Shi, A practical synthesis of *N*-aryl-substituted oxazolidinone-containing ketone catalysts for asymmetric epoxidation, *Tetrahedron*, 62 (2006) 8064–8068.
343. O. A. Wong, B. Wang, M. -X. Zhao, and Y. Shi, Asymmetric epoxidation catalyzed by α,α-dimethylmorpholinone ketone. Methyl group effect on spiro and planar transition states, *J. Org. Chem.*, 74 (2009) 6335–6338.
344. L. L. Huang, M. H. Xu, and G. Q. Lin, A new entry to asymmetric synthesis of optically active α,γ-substituted γ-butyrolactones, using a carbohydrate derived amide as both a chiral auxiliary and a proton source, *J. Org. Chem.*, 70 (2005) 529–532.
345. E. Raluy, M. Diéguez, and O. Pàmies, Screening of a modular sugar-based phosphoramidite ligand library in the asymmetric nickel-catalyzed trialkylaluminum addition to aromatic aldehydes, *Tetrahedron Asymmetry*, 20 (2009) 1575–1579.
346. E. Raluy, O. Pàmies, M. Diéguez, S. Rosset, and A. Alexakis, Sugar-based phosphite and phosphoroamidite ligands for the Cu-catalyzed asymmetric 1,4-addition to enones, *Tetrahedron Asymmetry*, 20 (2009) 2167–2172.
347. M. J. Howard and C. M. Smales, NMR analysis of synthetic human serum albumin α-helix 28 identifies structural distortion upon Amadori modification, *J. Biol. Chem.*, 280 (2005) 22582–22589.
348. M. Meli, R. Granouillet, E. Reynaud, A. Chamson, J. Frey, and C. Perier, Changes in glycation of fibrous type I collagen during long-term *in vitro* incubation with glucose, *J. Protein Chem.*, 22 (2003) 521–525.
349. E. J. Menzel and R. Reihsner, Alterations of biochemical and biomechanical properties of rat tail tendons caused by nonenzymic glycation and their inhibition by dibasic amino acids arginine and lysine, *Diabetologia*, 34 (1991) 12–16.
350. A. Jaleel, P. Halvatsiotis, B. Williamson, P. Juhasz, S. Martin, and K. S. Nair, Identification of Amadori-modified plasma proteins in type 2 diabetes and the effect of short-term intensive insulin treatment, *Diabetes Care*, 28 (2005) 645–652.

351. G. Misciagna, G. De Michele, A. M. Cisternino, V. Guerra, G. Logroscino, and J. L. Freudenheim, Dietary carbohydrates and glycated proteins in the blood in non diabetic subjects, *J. Am. Coll. Nutr.*, 24 (2005) 22–29.
352. A. Biswas, S. Lewis, B. Wang, M. Miyagi, P. Santoshkumar, M. H. Gangadhariah, and R. H. Nagaraj, Chemical modulation of the chaperone function of human αA-crystallin, *J. Biochem.*, 144 (2008) 21–32.
353. A. Sakaguchi, S. Ferri, and K. Sode, SocA is a novel periplasmic binding protein for fructosyl amino acid, *Biochem. Biophys. Res. Commun.*, 336 (2005) 1074–1080.
354. J. Yang, J. Dowden, A. Tatibouët, Y. Hatanaka, and G. D. Holman, Development of high-affinity ligands and photoaffinity labels for the D-fructose transporter GLUT5, *Biochem. J.*, 367 (2002) 533–539.
355. L. Azema, S. Claustre, I. Alric, C. Blonski, M. Willson, J. Perié, T. Baltz, E. Tetaud, F. Bringaud, D. Cottem, F. R. Opperdoes, and M. P. Barrett, Interaction of substituted hexose analogues with the *Trypanosoma brucei* hexose transporter, *Biochem. Pharmacol.*, 67 (2004) 459–467.
356. R. Salazar, R. Brandt, J. Kellermann, and S. Krantz, Purification and characterization of a 200 kDa fructosyllysine-specific binding protein from cell membranes of U937 cells, *Glycoconjug. J.*, 17 (2000) 713–716.
357. V. Y. Wu, C. W. Shearman, and M. P. Cohen, Identification of calnexin as a binding protein for Amadori-modified glycated albumin, *Biochem. Biophys. Res. Commun.*, 284 (2001) 602–606.
358. V. V. Mossine, V. V. Glinsky, and T. P. Mawhinney, Food-related carbohydrate ligands for galectins, in A. Klyosov, D. Platt, and Z. J. Witczak, (Eds.), Galectins, Wiley, Hoboken, NJ, 2008, pp. 235–270.
359. C. G. Schalkwijk, N. Ligtvoet, H. Twaalfhoven, A. Jager, H. G. T. Blaauwgeers, R. O. Schlingemann, L. Tarnow, H. H. Parving, C. D. A. Stehouwer, and V. W. M. van Hinsbergh, Amadori albumin in type 1 diabetic patients correlation with markers of endothelial function, association with diabetic nephropathy, and localization in retinal capillaries, *Diabetes*, 48 (1999) 2446–2453.
360. M. Zhang, A. L. Kho, N. Anilkumar, R. Chibber, P. J. Pagano, A. M. Shah, and A. C. Cave, Glycated proteins stimulate reactive oxygen species production in cardiac myocytes, *Circulation*, 113 (2006) 1235–1243.
361. C. W. Yoo, C. Y. Song, B. C. Kim, H. K. Hong, and H. S. Lee, Glycated albumin induces superoxide generation in mesangial cells, *Cell Physiol. Biochem.*, 14 (2004) 361–368.
362. K. Higai, A. Shimamura, and K. Matsumoto, Amadori-modified glycated albumin predominantly induces E-selectin expression on human umbilical vein

endothelial cells through NADPH oxidase activation, *Clin. Chim. Acta*, 367 (2006) 137–143.

363. H. Zoellner, S. Siddiqui, E. Kelly, and H. Medbury, The anti-apoptotic activity of albumin for endothelium is inhibited by advanced glycation end products restricting intramolecular movement, *Cell. Mol. Biol. Lett.*, 14 (2009) 575–586.
364. A. Amore, P. Cirina, G. Conti, F. Cerutti, N. Bagheri, S. N. Emancipator, and R. Coppo, Amadori-configurated albumin induces nitric oxide-dependent apoptosis of endothelial cells: A possible mechanism of diabetic vasculopathy, *Nephrol. Dial. Transplant.*, 19 (2004) 53–60.
365. P. Verbeke, M. Perichon, B. Friguet, and H. Bakala, Inhibition of nitric oxide synthase activity by early and advanced glycation end products in cultured rabbit proximal tubular epithelial cells, *Biochim. Biophys. Acta*, 1502 (2000) 481–494.
366. B. W. Lee, J. Ihm, J. G. Kang, M. G. Choi, H. J. Yoo, and S. H. Ihm, Amadori-glycated albumin-induced vascular smooth muscle cell proliferation and expression of inhibitor of apoptosis protein-1 and nerve growth factor-γ, *Biofactors*, 31 (2007) 145–153.
367. R. Brandt and S. Krantz, Glycated albumin (Amadori product) induces activation of MAP kinases in monocyte-like MonoMac 6 cells, *Biochim. Biophys. Acta*, 1760 (2006) 1749–1753.
368. H. Kishikawa, S. Mine, C. Kawahara, T. Tabata, A. Hirose, Y. Okada, and Y. Tanaka, Glycated albumin and cross-linking of CD44 induce scavenger receptor expression and uptake of oxidized LDL in human monocytes, *Biochem. Biophys. Res. Commun.*, 339 (2006) 846–851.
369. M. P. Cohen, Intervention strategies to prevent pathogenetic effects of glycated albumin, *Arch. Biochem. Biophys.*, 419 (2003) 25–30.
370. M. P. Cohen, G. T. Lautenslager, E. Hud, E. Shea, A. Wang, S. Chen, and C. W. Shearman, Inhibiting albumin glycation attenuates dysregulation of VEGFR-1 and collagen IV subchain production and the development of renal insufficiency, *Am. J. Physiol.*, 292 (2007) F789–F795.
371. L. Rodríguez-Mañas, C. Sánchez-Rodríguez, S. Vallejo, M. El-Assar, C. Peiró, V. Azcutia, N. Matesanz, C. F. Sánchez-Ferrer, and J. Nevado, Pro-inflammatory effects of early non-enzymatic glycated proteins in human mesothelial cells vary with cell donor's age, *Br. J. Pharmacol.*, 149 (2006) 979–987.
372. H. S. Lee, K. C. Moon, C. Y. Song, B. C. Kim, S. Wang, and H. K. Hong, Glycated albumin activates PAI-1 transcription through Smad DNA binding sites in mesangial cells, *Am. J. Physiol.*, 287 (2004) F665–F672.

373. V. Y. Wu and M. P. Cohen, Identification of aortic endothelial cell binding proteins for Amadori adducts in glycated albumin, *Biochem. Biophys. Res. Commun.*, 193 (1993) 1131–1136.
374. M. Roy, S. Sen, and A. S. Chakraborti, Action of pelargonidin on hyperglycemia and oxidative damage in diabetic rats: Implication for glycation-induced hemoglobin modification, *Life Sci.*, 82 (2008) 1102–1110.
375. C. Peiró, N. Matesanz, J. Nevado, N. Lafuente, E. Cercas, V. Azcutia, S. Vallejo, L. Rodríguez-Mañas, and C. F. Sánchez-Ferrer, Glycosylated human oxyhemoglobin activates nuclear factor-κB and activator protein-1 in cultured human aortic smooth muscle, *Br. J. Pharmacol.*, 140 (2003) 681–690.
376. Y. Hattori, M. Suzuki, S. Hattori, and K. Kasai, Vascular smooth muscle cell activation by glycated albumin (Amadori adducts), *Hypertension*, 39 (2002) 22–28.
377. A. Van Campenhout, C. Van Campenhout, A. R. Lagrou, and B. Manuel-y-Keenoy, Effects of in vitro glycation on Fe^{3+} binding and Fe^{3+} isoforms of transferrin, *Clin. Chem.*, 50 (2004) 1640–1649.
378. T. R. Brown, B. Su, K. A. Brown, M. A. Schwartz, A. M. Tobia, and F. Kappler, Modulation of *in vivo* 3-deoxyglucosone levels, *Biochem. Soc. Trans.*, 31 (2003) 1433–1437.
379. S. Rahbar and J. L. Figarola, Inhibitors and breakers of advanced glycation endproducts (AGEs): A review, *Curr. Med. Chem. Immunol. Endocr. Metab. Agents*, 2 (2002) 135–161.
380. N. Selvaraj, Z. Bobby, A. K. Das, R. Ramesh, and B. C. Koner, An evaluation of level of oxidative stress and protein glycation in nondiabetic undialyzed chronic renal failure patients, *Clin. Chim. Acta*, 324 (2002) 45–50.
381. M. J. Stear, P. D. Eckersall, P. A. Graham, Q. A. McKellar, S. Mitchell, and S. C. Bishop, Fructosamine concentration and resistance to natural, predominantly *Teladorsagia circumcincta* infection, *Parasitology*, 123 (2001) 211–218.
382. C. A. L. S. Colaco (Ed.), The Glycation Hypothesis of Atherosclerosis, Landes Bioscience, Austin, TX, 1997.
383. V. V. Shuvaev, I. Laffont, J. M. Serot, J. Fujii, N. Taniguchi, and G. Siest, Increased protein glycation in cerebrospinal fluid of Alzheimer's disease, *Neurobiol. Aging*, 22 (2001) 397–402.
384. V. P. Reddy, M. E. Obrenovich, C. S. Atwood, G. Perry, and M. A. Smith, Involvement of Maillard reactions in Alzheimer disease, *Neurotox. Res.*, 4 (2002) 191–209.
385. K. Nakagawa, J. H. Oak, O. Higuchi, T. Tsuzuki, S. Oikawa, H. Otani, M. Mune, H. Cai, and T. Miyazawa, Ion-trap tandem mass spectrometric

analysis of Amadori-glycated phosphatidylethanolamine in human plasma with or without diabetes, *J. Lipid Res.*, 46 (2005) 2514–2524.

386. K. Nakagawa, J. H. Oak, and T. Miyazawa, Angiogenic potency of Amadori-glycated phosphatidylethanolamine, *Ann. N. Y. Acad. Sci.*, 1043 (2005) 413–416.
387. J. H. Oak, K. Nakagawa, S. Oikawa, and T. Miyazawa, Amadori-glycated phosphatidylethanolamine induces angiogenic differentiations in cultured human umbilical vein endothelial cells, *FEBS Lett.*, 555 (2003) 419–423.
388. G. Misciagna, G. De Michele, V. Guerra, A. M. Cisternino, A. Di Leo, and J. L. Freudenheim, Serum fructosamine and colorectal adenomas, *Eur. J. Epidemiol.*, 19 (2004) 425–432.
389. M. Platek, V. Krogh, A. Micheli, R. Browne, E. Meneghini, S. Sieri, H. J. Schünemann, V. Pala, M. Barba, G. E. Wilding, F. Berrino, and P. Muti, Serum fructosamine and subsequent breast cancer risk: A nested case-control study in the ORDET prospective cohort study, *Cancer Epidemiol. Biomarkers Prev.*, 14 (2005) 271–274.
390. T. Henle, AGEs in foods: Do they play a role in uremia?, *Kidney Int.*, 63 (2003) S145–S147.
391. H. G. Wager, Browning reaction in dehydrated carrot and potato: Its initiation and the separation and partial characterization of an intermediate from dehydrated carrot, *J. Sci. Food Agric.*, 6 (1955) 57–64.
392. S. Adachi, New amino-acid glycosides isolated from tryptic hydrolyzates of milk products, *Nature*, 177 (1956) 936–937.
393. N. Yoshida, K. Takatsuka, T. Katsuragi, and Y. Tani, Occurrence of fructosyl-amino acid oxidase-reactive compounds in fungal cells, *Biosci. Biotechnol. Biochem.*, 69 (2005) 258–260.
394. H. Schlichtherle-Cerny, M. Affolter, and C. Cerny, Hydrophilic interaction liquid chromatography coupled to electrospray mass spectrometry of small polar compounds in food analysis, *Anal. Chem.*, 75 (2003) 2349–2354.
395. C. M. Utzmann and M. O. Lederer, Identification and quantification of amino-phospholipid-linked Maillard compounds in model systems and egg yolk products, *J. Agric. Food Chem.*, 48 (2000) 1000–1008.
396. F. J. Moreno, M. Corzo-Martinez, M. Dolores del Castillo, and M. Villamiel, Changes in antioxidant activity of dehydrated onion and garlic during storage, *Food Res. Int.*, 39 (2006) 891–897.
397. K. A. Voss, S. M. Poling, F. I. Meredith, C. W. Bacon, and D. S. Saunders, Fate of fumonisins during the production of fried tortilla chips, *J. Agric. Food Chem.*, 49 (2001) 3120–3126.

398. J. Haorah, L. Zhou, X. Wang, G. Xu, and S. S. Mirvish, Determination of total N-nitroso compounds and their precursors in frankfurters, fresh meat, dried salted fish, sauces, tobacco, and tobacco smoke particulates, *J. Agric. Food Chem.*, 49 (2001) 6068–6078.
399. B. Pignatelli, C. Malaveille, M. Friesen, A. Hautefeuille, H.Bartsch, D. Piskorska, and G. Descotes, Synthesis, analysis and mutagenic activity of *N*-nitroso derivatives of glycosylamines and Amadori compounds: Nitrosated model substances for the early Maillard reaction products, *IARC Sci. Publ.*, 84 (1987) 277–283.
400. A. C. Soria, A. Olano, J. Frías, E. Peñas, and M. Villamiel, 2-Furoylmethyl amino acids, hydroxymethylfurfural, carbohydrates and β-carotene as quality markers of dehydrated carrots, *J. Sci. Food Agric.*, 89 (2009) 267–273.
401. J. A. Rufián-Henares and C. Delgado-Andrade, Effect of digestive process on Maillard reaction indexes and antioxidant properties of breakfast cereals, *Food Res. Int.*, 42 (2009) 394–400.
402. U. Schwietzke, U. Schwarzenbolz, and T. Henle, Influence of cheese type and maturation time on the early Maillard reaction in cheese, *Czech J. Food Sci.*, 27 (2009) S140–S142.
403. M. J. Horn, H. Lichtenstein, and M. Womack, Availability of amino acids. A methionine-fructose compound and its availability to microorganisms and rats, *J. Agr. Food Chem.*, 16 (1968) 741–745.
404. V. C. Sgarbieri, J. Amaya, M. Tanaka, and C. O. Chichester, Nutritional consequences of the Maillard reaction. Amino acid availability from fructose-leucine and fructose-tryptophan in the rat, *J. Nutr.*, 103 (1973) 657–663.
405. G. H. Johnson, D. H. Baker, and E. G. Perkins, Nutritional implications of the Maillard reaction: The availability of fructose-phenylalanine to the chick, *J. Nutr.*, 107 (1977) 1659–1664.
406. V. Somoza, E. Wenzel, C. Weiß, I. Clawin-Rädecker, N. Grübel, and H. F. Erbersdobler, Dose-dependent utilisation of casein-linked lysinoalanine, *N*(epsilon)-fructoselysine and *N*(epsilon)-carboxymethyllysine in rats, *Mol. Nutr. Food Res.*, 50 (2006) 833–841.
407. R. F. Hurrell, Influence of the Maillard reaction on the nutritional value of foods, in P. A. Finot, H. U. Aeschbacher, R. F. Hurrell, and R. Liardon, (Eds.), *The Maillard Reaction in Food Processing, Human Nutrition and Physiology*, Birkhäuser Verlag, Basel, 1990, pp. 245–258.
408. T. Henle, V. Schwenger, and E. Ritz, Preliminary studies on the renal handling of lactuloselysine from milk products, *Czech J. Food Sci.*, 18 (2000) 101–102.

409. A. Rérat, R. Calmes, P. Vaissade, and P.-A. Finot, Nutritional and metabolic consequences of the early Maillard reaction of heat treated milk in the pig. Significance for man, *Eur. J. Nutr.*, 41 (2002) 1–11.
410. V. Schwenger, C. Morath, K. Schönfelder, W. Klein, K. Weigel, R. Deppisch, T. Henle, E. Ritz, and M. Zeier, An oral load of the early glycation compound lactuloselysine fails to accumulate in the serum of uraemic patients, *Nephrol. Dial. Transplant.*, 21 (2006) 383–388.
411. J. -P. Langhendries, R. F. Hurrell, D. E. Furniss, C. Hischenhuber, P. A. Finot, A. Bernard, O. Battisti, J. M. Bertrand, and J. Senterre, Maillard reaction products and lysinoalanine: Urinary excretion and the effects on kidney function of preterm infants fed heat-processed milk formula, *J. Pediatr. Gastroenterol. Nutr.*, 14 (1992) 62–70.
412. C. Hultsch, M. Hellwig, B. Pawelke, R. Bergmann, K. Rode, J. Pietzsch, R. Krause, and T. Henle, Biodistribution and catabolism of ^{18}F-labeled N-ε-fructoselysine as a model of Amadori products, *Nucl. Med. Biol.*, 33 (2006) 865–873.
413. V. V. Mossine, P. Chopra, and T. P. Mawhinney, Interaction of tomato lycopene and ketosamine against rat prostate tumorigenesis, *Cancer Res.*, 68 (2008) 4384–4391.
414. N. Ide, B. H. S. Lau, K. Ryu, H. Matsuura, and Y. Itakura, Antioxidant effects of fructosyl arginine, a Maillard reaction product in aged garlic extract, *J. Nutr. Biochem.*, 10 (1999) 372–376.
415. K. Ryu, N. Ide, H. Matsuura, and Y. Itakura, *N*α-(1-deoxy-D-fructos-1-yl)-L-arginine, an antioxidant compound identified in aged garlic extract, *J. Nutr.*, 131 (2001) 972S–976S.
416. Y. Suzuki, K. -J. Choi, K. Uchida, S. -R. Ko, H. -J. Sohn, and J. -D. Park, Arginyl-fructosyl-glucose and arginyl-fructose, compounds related to browning reaction in the model system of steaming and heat-drying processes for the preparation of red ginseng, *J. Ginseng Res.*, 28 (2004) 143–148.
417. N. V. Chuyen, K. Ijichi, and K. Moteki, Antioxidative properties of products from amino acids or peptides in the reaction with glucose, *Adv. Exp. Med. Biol.*, 434 (1998) 201–212.
418. H. M. Sayed, M. H. Mohamed, S. F. Farag, G. A. Mohamed, O. R. M. Omobuwajo, and P. Proksch, Fructose-amino acid conjugate and other constituents from *Cyperus rotundus* L., *Nat. Prod. Res.*, 22 (2008) 1487–1497.
419. L. Rong, F. Q. Yin, and J. W. He, Method for scavenging active oxygen free radical with 1-carboxyethylamino-1-deoxy-D-fructose, *Yancao Keji*, (2005) 17–19.
420. E. Sorkin and A. Fjelde, The effect of D-glucosamine and related products on human cancer cells in tissue culture, *G. Ital. Chemioter.*, 3 (1956) 355–361.

421. V. V. Mossine and T. P. Mawhinney, Significance of processing for the chemopreventive potential of tomato-based products, in R. R. Watson and V. R. Preedy, (Eds.), *Bioactive Foods and Extracts: Cancer Treatment and Prevention*, Taylor & Francis, Boca Raton, FL, (2010) 279–300.
422. T. V. Denisevitch, R. A. Semyonova-Kobzar, V. V. Glinsky, and V. V. Mosin, The influence of synthetic aminoglycoconjugates on the aggregation ability and metastatic potential of tumor cells, *Exp. Oncol.*, 17 (1995) 111–117.
423. G. V. Glinsky, J. E. Price, V. V. Glinsky, V. V. Mossine, G. Kiriakova, and J. B. Metcalf, Inhibition of human breast cancer metastasis in nude mice by synthetic glycoamines, *Cancer Res.*, 56 (1996) 5319–5324.
424. F.-T. Liu and G. A. Rabinovich, Galectins as modulators of tumor progression, *Nat. Rev. Cancer*, 5 (2005) 29–41.
425. M. E. Huflejt, V. V. Mossine, and M. Croft, Galectin-1 and galectin-4 in tumor development and use in prognosis and treatment of breast cancer, *Int. Pat. Appl.*, (2003) WO 2003026494
426. G. A. Rabinovich, A. Cumashi, G. A. Bianco, D. Ciavardelli, I. Iurisci, M. D'Egidio, E. Piccolo, N. Tinari, N. Nifantiev, and S. Iacobelli, Synthetic lactulose amines: Novel class of anticancer agents that induce tumor-cell apoptosis and inhibit galectin-mediated homotypic cell aggregation and endothelial cell morphogenesis, *Glycobiology*, 16 (2006) 210–220.
427. V. V. Glinsky, G. V. Glinsky, O. V. Glinskii, V. H. Huxley, J. R. Turk, V. V. Mossine, S. L. Deutscher, K. J. Pienta, and T. P. Quinn, Intravascular metastatic cancer cell homotypic aggregation at the sites of primary attachment to the endothelium, *Cancer Res.*, 63 (2003) 3805–3811.
428. V. V. Glinsky, G. Kiriakova, O. V. Glinskii, V. V. Mossine, T. P. Mawhinney, J. R. Turk, A. B. Glinskii, V. H. Huxley, J. E. Price, and G. V. Glinsky, Synthetic galectin-3 inhibitor increases metastatic cancer cell sensitivity to Taxol-induced apoptosis *in vitro* and *in vivo*, *Neoplasia*, 11 (2009) 901–909.
429. K. D. Johnson, O. V. Glinskii, V. V. Mossine, J. R. Turk, T. P. Mawhinney, D. C. Anthony, C. J. Henry, V. H. Huxley, G. V. Glinsky, K. J. Pienta, A. Raz, and V. V. Glinsky, Galectin-3 as a potential therapeutic target in tumors arising from malignant endothelia, *Neoplasia*, 9 (2007) 662–670.
430. L. Mester de Parajd, B. Kraska, and M. Mester de Parajd, Specific diets to increase the formation *in vivo* of Maillard-type compounds with antibacterial activity, *Dev. Food Sci.*, 13 (1986) 509–522.
431. S. P. Assenza, N. Shen, S. Iacono, Y. B. Kalyan, P. Viski, and M. Ronsheim, Synthesis of reducing carbohydrate adamantane amines for use in treating Gram pos. or Gram neg. bacteria infections, *Int. Pat. Appl.*, (2006) WO 2006066006.

432. J. Grzybowska, P. Sowiński, J. Gumieniak, T. Zieniawa, and E. Borowski, *N*-Methyl-*N*-D-fructopyranosylamphotericin B methyl ester, new amphotericin B derivative of low toxicity, *J. Antibiot.*, 50 (1997) 709–711.
433. M. N. Preobrazhenskaya, E. N. Olsufyeva, S. E. Solovieva, A. N. Tevyashova, M. I. Reznikova, Y. N. Luzikov, L. P. Terekhova, A. S. Trenin, O. A. Galatenko, I. D. Treshalin, E. P. Mirchink, V. M. Bukhman, H. Sletta, and S. B. Zotchev, Chemical modification and biological evaluation of new semisynthetic derivatives of 28,29-didehydronystatin A_1 (S44HP), a genetically engineered antifungal polyene macrolide antibiotic, *J. Med. Chem.*, 52 (2009) 189–196.
434. N. A. Ansari, Moinuddin, K. Alam, and A. Ali, Preferential recognition of Amadori-rich lysine residues by serum antibodies in diabetes mellitus: Role of protein glycation in the disease process, *Hum. Immunol.*, 70 (2009) 417–424.
435. M. Tarnawski, R. Kuliś-Orzechowska, and B. Szelepin, Stimulatory effect of *N*-(1-deoxy-β-D-fructopyranos-1-yl)-L-proline on antibody production in mice, *Int. Immunopharmacol.*, 7 (2007) 1577–1581.
436. M. J. Lovdahl, T. R. Hurley, B. Tobias, and S. R. Priebe, Synthesis and characterization of pregabalin lactose conjugate degradation products, *J. Pharm. Biomed. Anal.*, 28 (2002) 917–924.
437. T. R. Hurley, M. J. Lovdahl, and B. Tobias, Pregabalin-lactose conjugates for pharmaceuticals, *Int. Pat. Appl.*, (2002) WO 2002078747
438. M. Sugiyama, M. Sakai, Y. Shibuya, and Y. Nishizawa, Fructosyldipeptides and skin preparations, hair growth inhibitors, and antiwrinkle agents containing them, *Jpn. Pat. Appl.*, (2005) JP 2005170930
439. M. Nagasawa, Antiwrinkle cosmetics containing fructosyl dipeptides and antioxidants and/or antiinflammatory agents, *Jpn. Pat. Appl.*, (2007) JP 2007008847.
440. A. Itoh, Y. Ikuta, Y. Baba, T. Tanahashi, and N. Nagakura, Ipecac alkaloids from *Cephaelis acuminata, Phytochemistry*, 52 (1999) 1169–1176.
441. www.imars.org.

SIALIDASES IN VERTEBRATES: A FAMILY OF ENZYMES TAILORED FOR SEVERAL CELL FUNCTIONS*

BY EUGENIO MONTI[a], ERIK BONTEN[b], ALESSANDRA D'AZZO[b], ROBERTO BRESCIANI[a], BRUNO VENERANDO[c], GIUSEPPE BORSANI[a], ROLAND SCHAUER[d] and GUIDO TETTAMANTI[e]

[a] Department of Biomedical Science and Biotechnology, University of Brescia, 25123, Italy
[b] Department of Genetics, St. Jude Children's Research Hospital, Memphis, TN, 28105-2794, USA
[c] Department of Medical Chemistry, Biochemistry and Biotechnology, University of Milan, Segrate, 20090, Italy
[d] Institute of Biochemistry, University of Kiel, Kiel, D-24098, Germany
[e] Laboratory of Stem Cells for Tissue Engineering, IRCCS Policlinico San Donato, San Donato Milanese, 20097, Italy

I. Introduction 405
II. The Lysosomal Sialidase NEU1 406
1. Background 406
2. Lysosomal Routing and Formation of the Multienzyme Complex 409
3. Human NEU1 Deficiency 411
4. New Functions of NEU1 in Tissue Remodeling and Homeostasis 413
5. Mouse Model of NEU1 Deficiency and Study of Disease Pathogenesis 418
III. The Cytosolic Sialidase NEU2 422
1. General Properties of NEU2 422
2. Crystal Structure of Human Sialidase NEU2 423
3. Functional Implication of NEU2 425
IV. The Plasma Membrane-Associated Sialidase NEU3 429
1. General Properties of NEU3 429
2. Functional Implication of NEU3 431
V. The Particulate Sialidase NEU4 435
1. General Properties of NEU4 435
2. Functional Implication of NEU4 436
VI. Sialidases and Cancer 438
VII. Sialidases and Immunity 439
VIII. Further Evidence for Possible Functional Implications of Sialidases 442
IX. *In silico* Analysis of Sialidase Gene Expression Patterns 444
X. Amino Acid Sequence Variants in Human Sialidases 448
XI. Sialidases in Teleosts 449

ISBN: 978-0-12-380854-7
DOI: 10.1016/S0065-2318(10)64007-3

Abbreviations

9-OAcGD3, 9-*O*-acetylated disialo-ganglioside GD3; ACF, aberrant crypt foci; AKT, serine/theonine-specific protein kinase family; ALL, acute lymphoblastic leukemia; ARIs, axon regeneration inhibitors; ASMC, aortic smooth muscle cells; BCL-2, B-cell leukemia/lymphoma 2 integral membrane protein; BM, bone marrow; BMT, bone marrow transplantation; BMEF, bone marrow extracellular fluid; CD, cluster of differentiation; CFU-E, colony-forming unit-erythroid cells; CG, cathepsin G; CHO, Chinese Hamster Ovary (cells); CSER, cell-surface elastin receptor; DRMs, detergent-resistant microdomains; EBP, Elastin-Binding Protein; *EBP*, human gene encoding the Elastin-Binding Protein; EGF, epidermal growth factor; EGFR, epidermal growth-factor receptor; ELISA, enzyme-linked immunosorbent assay; EMH, extramedullary hematopoiesis; ER, endoplasmic reticulum; ERK, extracellular signal-regulated kinase; EST, Expressed Sequence Tag; *β-GAL*, human gene encoding the β-galactosidase; β-GAL, β-galactosidase; GEMs, glycosphingolipid-enriched microdomains; GS, galactosialidosis; HeLa, cell line was derived from cervical cancer cells taken from a patient named Henrietta Lacks; HLA, human leukocyte antigen; HPCs, hematopoietic progenitor cells; ICAM-1, Intercellular Adhesion Molecule 1 (also known as cluster of differentiation 54 (CD54)); IFN, interferon; IGF-1R, receptor of insulin-like growth factor-1; IL, interleukin; JNK, protein kinase member of the MAP kinase; LacCer, lactosyl-ceramide; LAMP, lysosome-associated membrane protein; Le^x, $Lewis^x$ antigens; LIMP, lysosome integral membrane protein; M6P, mannose 6-phosphate; MAG, myelin-associated glycoprotein; MAPK, mitogen-activated protein kinase; MEK, mitogen-activated or extracellular signal-regulated protein kinase; MHC, major histocompatibility complex; *MMP9*, gene encoding the matrix metalloproteinase-9; MMP9, matrix metalloproteinase-9; MU, methylumbelliferone; NCAM, neural cell adhesion molecule; NCBI, National Center for Biotechnology Information; NE, neutrophil elastase; *NEU1*, gene encoding the lysosomal sialidase; NEU1, human lysosomal sialidase; *Neu1*, mouse gene encoding the mouse lysosomal

---

Author contributions: Erik Bonten and Alessandra D'Azzo contributed Section II. Giuseppe Borsani contributed Sections IX–X. Roland Schauer contributed Section XI. The remaining authors, together with Giuseppe Borsani, contributed the other sections.

sialidase; Neu1, mouse lysosomal sialidase; Neu5Ac, *N*-acetylneuraminic acid; NGF, nerve growth factor; NMR, nuclear magnetic resonance; nsSNP, non-synonymous single nucleotide polymorphism; PAG, phosphoprotein associated with GEMs; PDGF, platelet-derived growth factor; PDGFR, platelet-derived growth-factor receptor; PI3K, phosphoinositide-3 kinase; PKC, protein kinase C; PLC, phosphoinositide-specific phospholipase C; PMA, phorbol 12-tetradecanoate 13-acetate; PPCA, protective protein/cathepsin A; Rac, subfamily of the Rho family of GTPases; RD, cell line derived from a human rhabdomyosarcoma; RMS, rhabdomyosarcoma; RSV, respiratory syncytial virus; RT-PCR, reverse transcription-polymerase chain reaction; SFK, Src family protein tyrosine kinases; shRNA, short hairpin RNA; sLe^x, sialyl Lewis^x antigens; SNP, single nucleotide polymorphism; TERMs, tetraspanin-enriched microdomains; THP-1, human acute monocytic leukemia cell line; TLR, toll-like receptor; TNF, tumor necrosis factor; TrKA, high-affinity catalytic receptor for nerve growth factor; TS, trans-sialidase; Tsk, tight-skin (mouse); VCAM, vascular cell adhesion molecule; VSMC, vascular smooth muscle cells.

Note

Gene and protein symbols are reported according to:

H. M. Wain, E. A. Bruford, R. C. Lovering, M. J. Lush, M. W. Wright, and S. Povey, Guidelines for human gene nomenclature, *Genomics* 79 (2002) 464–470;

Refer Guidelines for Nomenclature of Genes, Genetic Markers, Alleles, and Mutations in Mouse and Rat at the Mouse Genome Informatics (MGI) website (http://www.informatics.jax.org/mgihome/nomen/gene.shtml); and Zebrafish Nomenclature Guidelines at the ZFIN website (http://zfin.org/zf_info/nomen.html).

Ganglioside nomenclature is according to:

L. Svennerholm, Ganglioside designation, *Adv. Exp. Med. Biol.* 125 (1980) 11.

I. Introduction

Sialidases or neuraminidases (EC 3.2.1.18; *N*-acylneuraminyl glycohydrolases) are a family of exo-glycosidases that catalyze the hydrolytic cleavage of non-reducing sialic acid residues[1] ketosidically linked to the saccharide chains of glycoproteins and glycolipids (gangliosides) as well as to oligo- and poly-saccharides. They are widely distributed in nature, from viruses and microorganisms (such as bacteria, protozoa, and fungi) to vertebrates, but are absent in plants, insects, and yeast.[2] The molecular cloning of several mammalian sialidases since 1993 and the great development that followed early afterward has been summarized in two reviews.[3,4]

In vertebrates, mammalian sialidases and their target substrates have been implicated in crucial biological processes, including the regulation of cell proliferation/-differentiation, clearance of plasma proteins, control of cell adhesion, metabolism of gangliosides and glycoproteins, immunocyte function, modification of receptors, and the developmental modeling of myelin. The pivotal and diverse functions of these enzymes, many of which have not yet been discovered, probably account for the existence of four mammalian sialidases, encoded by different genes and defined as lysosomal (NEU1), cytosolic (NEU2), plasma-membrane (NEU3), and mitochondrial/-lysosomal/-intracellular membranes (NEU4). These enzymes differ in their subcellular localizations, pH optima, kinetic properties, responses to ions and detergents, and substrate specificities. There appears to be little overlap in function of the individual sialidases, even though they share a common mechanism of action.

Here we survey the data published since 2002 on vertebrate (mainly mammalian) sialidase biology. The subject is organized as follows: (i) a description of the four different sialidase forms and their functional implication; (ii) an introductory section on some *in silico* analysis of the sialidase gene family; (iii) a description of recent findings in lower vertebrates (teleosts); and (iv) a comparison between vertebrate sialidases and trans-sialidases, a peculiar group of sialidases that are able to catalyze the transfer of sialic acid from a donor to an acceptor substrate. These enzymes are present in specific microorganisms, and have been implicated in some mammalian infectious diseases. Finally, based on recent discoveries, we attempt to provide a comprehensive picture of the increasing complexity of the biological roles performed by this enzyme family.

II. The Lysosomal Sialidase NEU1

1. Background

Lysosomal sialidase (NEU1) initiates the hydrolysis of sialyl-glycoconjugates by removing their terminal sialic acid residues. The enzyme is present in almost all vertebrate tissues and cell types, and functions in a multienzyme complex containing at least two other hydrolases: the glycosidase β-galactosidase (β-GAL) and the serine carboxypeptidase protective protein/cathepsin A (PPCA). Both the human and murine neuraminidase genes (*NEU1* and *Neu1*) map within the major histocompatibility locus. The corresponding proteins share primary structure characteristics with other mammalian and microbial sialidases. However, unlike other members of this family,

NEU1 reaches the lysosome and becomes catalytically active by interacting with the auxiliary protein PPCA. NEU1 is also the only sialidase that is linked to two neurodegenerative disorders of glycoprotein metabolism: sialidosis, caused by structural lesions in the *NEU1* gene, and galactosialidosis (GS), a combined deficiency of NEU1 and β-GAL that is secondary to the absence of functional PPCA. Currently, there is no effective therapy for these disorders. The cloning of the cDNAs and genes for both human and mouse *NEU1* and the generation of a mouse model for sialidosis have enabled studies of disease pathogenesis as well as the implementation of therapeutic modalities that could eventually become clinically useful. Here we summarize the latest findings on this pivotal enzyme that have uncovered new functions of NEU1 in tissue homeostasis.

a. Primary Structure of NEU1.—The human *NEU1* gene is localized within the HLA histocompatibility locus on chromosome band 6p21.3, while the murine gene maps to the H-2 locus on chromosome 17, a genomic region syntenic to the human chromosome 6p region.[5–8] Both genes have similar genomic organization. The human 1.9 kb *NEU1* mRNA encodes a single-chain polypeptide of 415 amino acids that shares 90% sequence identity with murine NEU1 and is highly conserved in other mammalian species. Nevertheless, NEU1 shows higher sequence identity to bacterial sialidases (such as enzymes from *Salmonella typhimurium, and Micromonospora viridifaciens*, *Vibrio cholerae)* than to the NEU2, NEU3, and NEU4 polypeptides. The latter three enzymes are in fact more similar to each other (42–44%) than to NEU1 (<28%), hence their genes must have evolved through "recent" gene duplication events within the vertebrate lineage.

The primary structures of human and murine NEU1 start with a conventional signal sequence, which promotes translocation of the nascent polypeptide into the lumen of the endoplasmic reticulum (ER), and include three and four Asn-linked glycosylation sites, respectively.[5,7] All glycosylation sites are utilized, and differentially affect the proteins' stability and folding.[9] As with all non-viral sialidases, NEU1 contains a short sequence motif of eight amino acids (Ser/Thr-X-Asp-X-Gly-X-Thr-Trp/Phe), designated as "Asp-box" because of the presence of a conserved Asp residue.[10] This motif usually recurs multiple times within the sequence: in NEU1 there are four conserved and one degenerated Asp boxes. Although the precise function of these motifs is still a subject of debate, the absence of Asp boxes in most viral neuraminidases argues against their involvement in the catalysis of the substrates. In addition, Asp box motifs have also been identified in proteins that do not belong to the sialidase superfamily, but exert other functions.[11,12] Most of these proteins share the ability to bind carbohydrates and are in fact carbohydrate-processing enzymes,[11] which suggests a role for Asp-boxes in substrate recognition and binding (see also Section III.2). A second highly conserved domain in non-viral sialidases is the (F/Y)RIP motif, which is located near the N-terminus of NEU1 and includes the Arg residue, which is part of the catalytic pocket. The distance between the (F/Y)RIP domain and the first Asp-box (35–41 residues) and the spacing between the remaining Asp boxes are also conserved among the non-viral sialidases. Besides the Asp box motifs and the (F/Y)RIP domain, nine other amino acids are strictly conserved among sialidases, six of which are also found in the viral enzymes.[5]

b. Structural Model of NEU1.—The crystal structure of the influenza virus neuraminidase was the first 3D structure of a neuraminidase to be determined; it revealed a tetrameric association of identical monomers, whose fold was described as a superbarrel or β-propeller, also named the "β-propeller fold."[13,14] Each monomer consists of six four-stranded, antiparallel β-sheets that are arranged as the blades of a propeller around a pseudo-sixfold axis. Three-dimensional structure determination of bacterial neuraminidases identified a superfamily of multidomain enzymes arranged around the canonical β-propeller fold, which is conserved across species.[15–17] Like the active site of the viral enzymes, that of bacterial sialidases consists of an arginine triad, a hydrophobic pocket, and a key tyrosine residue and glutamic acid residue. The Asp-boxes are located at topologically identical positions in the β-propeller, so that the Asp residues are exposed to the solvent.

Structural models of NEU1, NEU3, and NEU4 were generated by molecular replacement[18] on the basis of the atomic coordinates of the crystal structure of NEU2.[19] As predicted, all four mammalian sialidases share the canonical β-propeller fold. However, two important variations were identified among these sialidases: the first is in the twist of the β-strands and their relative arrangement to the propeller axis; the second is in the conformation of the loops that connect the β-strands. The variable loops on the catalytic surface appear to adopt various conformations according to the substrates present in the environment.[18] Thus the length and conformation of these loops may dictate the substrate specificity of the different enzymes. The four common functional groups of sialic acid or 5-*N*-acetyl-2-deoxy-2,3-didehydro-neuraminic acid (Neu5Ac2en, also termed DANA), a synthetic sialic acid analogue, that bind to the active site of NEU2 are the 2-carboxylate, 4-hydroxyl, 5-*N*-acetyl, and 6-glycerol moieties.[19] Most of the Neu5Ac2en-binding residues are preserved in the active site crevice of the other three modeled human sialidases.[18] However, in agreement with their primary sequence homology, the active site crevices of NEU2, NEU3, and NEU4 are more similar to each other than to that of NEU1. The predicted differences in the tertiary structures of the four mammalian sialidases may be useful for the design of specific inhibitors for each enzyme, and help in dissecting their unique functions.

2. Lysosomal Routing and Formation of the Multienzyme Complex

a. NEU1 Expression and Lysosomal Routing.—NEU1 is ubiquitously but differentially expressed in tissues and cell types, and has been purified from many sources, including placenta, mammary gland, brain, kidney, liver, testis, thyroid, salivary gland, leukocytes, lymphocytes, macrophages, and fibroblast.[2] The enzyme has an acidic pH optimum and is active toward sialylated oligosaccharides and glycoproteins with a preference for α-(2→3) and α-(2→6) sialyl linkages. Being poorly mannose 6-phophorylated,[20] NEU1 is inefficiently routed to the endosome/-lysosome via interaction with the mannose 6-phosphate receptor (MPR), the canonical mode of lysosomal transport for soluble lysosomal enzymes.[21] Instead, two distinctive and alternative mechanisms of lysosomal targeting of NEU1 have been proposed in the literature. The first is based on the assumption that NEU1 is not a soluble enzyme, but instead behaves as an integral membrane protein that is compartmentalized in lysosomes via the interaction of its C-terminal tetrapeptide (^{412}YGTL^{415}) with the adaptor protein complex-3 (AP-3).[22] The latter is a common mechanism whereby integral membrane proteins, including the lysosomal-associated membrane proteins (LAMPs), lysosomal integral membrane proteins (LIMPs), and acid phosphatase, are transported to the lysosomal compartment.[23,24] It has been proposed that NEU1 uses a mode of compartmentalization similar to that of acid phosphatase, and is transported to the lysosomes linked to membrane vesicles and subsequently released as soluble enzyme into the lysosomal lumen by proteolytic cleavage.[22] However, in this model it is difficult to envisage how NEU1 that misses a stretch of hydrophobic residues characteristic of transmembrane proteins could expose its C-terminal tetrapeptide to the cytosolic face of the membrane and interact with the AP-3 complex. In addition, unlike what has been shown for acid phosphatase,[25] there is no clear evidence of intralysosomal proteolytic processing of NEU1.

The second and more widely accepted mechanism of intracellular routing of NEU1 derives from purification studies. The fact that under different purification conditions, NEU1 consistently segregates in the particulate membrane fractions of cell and tissue lysates can be explained by changes in the biochemical and structural properties of the enzyme when in complex or not in complex with its auxiliary protein PPCA.[26] Subcellular fractionation and co-immunoprecipitation experiments with cultured cells have demonstrated that a tight interaction of NEU1 with the PPCA zymogen, equipped with a functional M6P recognition marker, occurs already in an early biosynthetic compartment.[20] Through this interaction, NEU1 is compartmentalized in lysosomes via the MPR pathway. In absence of functional PPCA, an inactive NEU1 is either retained in the ER and eventually degraded, or it is targeted to the secretory pathway or to an early endosomal compartment, but remains catalytically inactive.[20] Thus, PPCA functions as a molecular chaperone for NEU1. Interestingly, the β-GAL precursor also co-immunoprecipitates with the NEU1 polypeptide. Although there is no genetic or biochemical evidence that the latter interaction is required for the transport or functioning of either enzyme, it may provide a means of regulation of their activities, or may enhance their stability.

b. Catalytic Activation and the Multienzyme Complex.—Once in the acidic lysosomal environment, PPCA, NEU1, and β-GAL form a multienzyme complex of high molecular weight in which NEU1 acquires full enzymatic activity.[20] The PPCA zymogen undergoes proteolytic processing into a 32/20-kDa two-chain mature and active enzyme, which remains in part assembled with NEU1 and β-GAL. The molecular nature of the catalytic activation of NEU1 is not yet fully understood. However, the carboxypeptidase activity of PPCA itself does not seem to play a role in the activation of NEU1, because a catalytically inert mutant of PPCA retains the capacity to activate NEU1 and β-GAL.[27] It is likely that the multienzyme complex is a rather dynamic module, whose exact molecular composition is dictated by the nature and characteristics of the available substrates. Within this module, NEU1 is the only enzyme that is strictly dependent on association with PPCA for catalytic activity. The extent of available PPCA is rate limiting, because increasing the intralysosomal concentration of PPCA by addition of incremental amounts of exogenous enzyme to wild-type cultured cells results in a proportional increase of endogenous NEU1 activity.[28]

In vitro experiments with purified enzymes have also shown that, at acidic pH, PPCA can promote the formation of NEU1 oligomers, which are catalytically active.[26] In contrast, a change from acidic to neutral pH causes disassembly of NEU1 and subsequent loss of enzyme activity. Thus, within the multienzyme complex, NEU1 tends to form oligomers whose composition is controlled by the levels of PPCA. Binding assays carried out by surface plasmon resonance with purified NEU1 and PPCA proteins and peptides have shown that PPCA and NEU1 compete for the same binding sites on other NEU1 molecules.[28] Therefore, in the absence of sufficient amounts of PPCA, as is the case in GS patients, NEU1 prematurely self-associates in an early biosynthetic compartment, forming oligomers and aggregates that vary in length. On the other hand, in the presence of PPCA, NEU1 preferentially binds to this enzyme, forming a NEU1-PPCA heterodimer, thereby preventing its premature oligomerization. It is likely that the PPCA–NEU1 heterodimer allows for efficient trafficking of NEU1 to the lysosome, and, that once in the lysosome, the enzymes dissociate and reassemble in the multienzyme complex, together with the other known

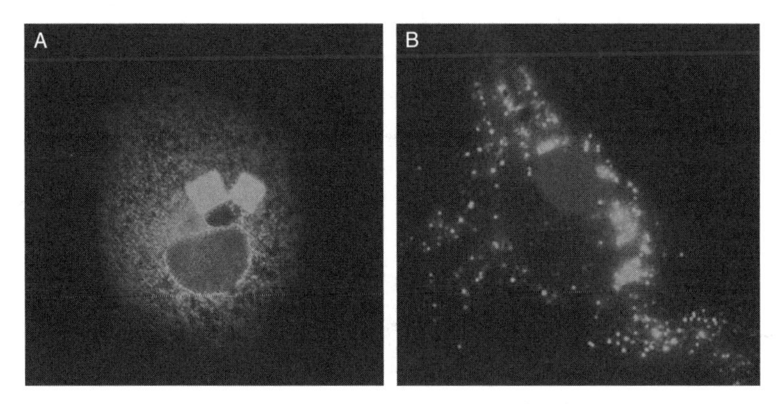

FIG. 1. Crystal formation of overexpressed NEU1 in absence of PPCA. (A) Localization of NEU1 in large crystal-like bodies in galactosialidosis fibroblasts that were transfected with a NEU1 cDNA expression construct. (B) A typical punctated lysosomal localization of NEU1 in galactosialidosis fibroblasts that were transfected with both NEU1 and PPCA cDNAs. Cells were incubated with affinity-purified anti-NEU1 antibody and FITC-conjugated secondary antibody. Nuclei were stained with DAPI. Magnification 400×. (See Color Plate 45.)

enzyme components. Together these studies have provided compelling evidence of the crucial role of PPCA in the transport, maturation, and activation of NEU1. They also define a novel mechanism of intracellular routing of a lysosomal enzyme that has evolved to depend on the interaction with an auxiliary protein for its correct compartmentalization (Fig. 1).

3. Human NEU1 Deficiency

a. Sialidosis and Galactosialidosis (GS).—NEU1 is the only mammalian sialidase that is clinically relevant: defective or deficient enzyme activity is associated with two neurodegenerative lysosomal storage disorders, sialidosis, which is caused by structural lesions in the *NEU1* gene (OMIM 256550), and galactosialidosis (GS), a combined deficiency of NEU1 and β-GAL that is caused by the absence of functional PPCA (OMIM 256540). Patients with sialidosis and those with GS share clinical and biochemical features that have been attributed at least in part to the loss of NEU1 function in both diseases.[29,30] Different clinical phenotypes of sialidosis are distinguished according to the age of onset and severity of the symptoms. Type I (normomorphic) sialidosis is an attenuated, non-neuropathic form of the disease that corresponds to the cherry-red spot-myoclonus syndrome. Symptoms typically appear in the second decade of life, and are mostly restricted to myoclonus and progressive impaired vision, albeit nystagmus, ataxia, and grand mal seizures have also been reported in some of these patients. Type II (dysmorphic) sialidosis is a severe, neuropathic form of the disease, classified in three subtypes: congenital or hydropic with onset *in utero*, infantile with onset between birth and 12 months, and juvenile with onset past 2 years of age. All patients with type II disease develop a progressive mucopolysaccharidosis-like phenotype, including macular cherry-red spots and myoclonus, coarse facies, hepatosplenomegaly, dysostosis multiplex, vertebral deformities, and severe mental retardation. Patients with the most acute congenital disease experience hydrops fetalis, neonatal ascites, or both; they are stillborn or die shortly after birth. Their clinical presentation includes facial edema, inguinal hernias, hepatosplenomegaly, stippling of the epiphyses, and periosteal cloaking. These phenotypic aberrations are also common in patients with the early infantile, severe, neurodegenerative form of GS, which is fatal in early childhood. The severity of the disease correlates closely with the type of gene mutation(s), which determines whether a patient has a complete or partial loss of functional NEU1.

b. Lysosomal Storage.—Residual lysosomal neuraminidase activity in fibroblasts or leukocytes from sialidosis patients ranges from 0 to 10% of control values. Obligate carriers have 50% of normal NEU1 activity, but they are phenotypically normal. Vacuolated lymphocytes in peripheral blood smears and foam cells in bone marrow smears are prominent in type II, but absent in type I sialidosis. Chemical and NMR-spectroscopic analyses have shown that the concentrations of linear and branched sialyloligosaccharides are increased in patients' urine and fibroblasts, and correlate with the severity of the clinical phenotype. The majority of the excreted sialoglycoconjugates contain *N*-acetylglucosamine at the reducing ends and *N*-acetyl neuraminic acid at the non-reducing termini. The sialic acid residues are linked to galactose via either a α-(2→6) (70%) or α-(2→3) linkage (30%), and probably originate from carbohydrate side-chains that have been cleaved from glycoproteins by an endo-*N*-acetyl-glucosaminidase.

c. NEU1 Mutations.—Important information about the biochemical properties of NEU1 came from studies of the mutant enzymes found in patients with sialidosis. To date more than 40 disease-causing mutations have been identified, the majority of which are missense mutations, resulting in single amino acid substitutions. Interestingly, more than 70% of all mutations cluster within exons 4-6 of the *NEU1* gene.[31–33] Several common mutations have also been found in seemingly unrelated patients, who nonetheless may have originated from the same geographic region or have similar ethnicity, indicating the existence of ancestral "founder mutations." However, the majority of *NEU1* mutations are *de novo* mutations. Based on their biochemical properties, the mutant NEU1 enzymes can be grouped as follows: (i) enzymes that are catalytically inactive and do not reach the lysosomal compartment; (ii) enzymes that localize to lysosomes but are catalytically inactive; (iii) enzymes that localize to lysosomes and do have residual catalytic activity.[31] Mutations that fall in the first two categories are mostly found in type II sialidosis patients, whereas mutations that belong to the third group are generally found in type I patients. Many of the single amino acid substitutions involve conserved residues. Other gene alterations include nonsense mutations, deletions, insertions, duplications, and splice-site variants. Such mutations are mostly associated with severe type II sialidosis patients. One exception is a 6-nucleotide in-frame duplication found in exon 6, which results in the duplication of amino acids H399/Y400. This NEU1 variant has nonetheless relatively high residual enzyme activity. This mutation was identified in the *NEU1* gene of a relatively mild type I sialidosis patient.[31] This illustrates that the careful evaluation of mutant NEU1 enzymes is critical for the assignment or confirmation of clinical phenotypes and predictions about the progression of the disease.

The structural effects of several mutations have been analyzed by using a 3D model of NEU1 that was built based on the atomic coordinates of the crystal structures of three different bacterial sialidases that share ~30% identity with NEU1.[32] This study validated the assignment of NEU1 mutations to 3 groups: (1) mutations that either directly or indirectly involve active site residues were found in type II sialidosis patients; (2) mutations on the periphery of the enzyme that have no obvious structural implications and mostly allow for correct lysosomal targeting and residual catalytic activity are mostly present in type I patients; (3) mutations on the surface of the enzyme that are near one of the seven β-strands mostly lead to the complete inactivation of the enzyme and are present in type II sialidosis patients. Three of these mutations (F260Y, L270F, A298V) were predicted to interfere with the capacity of NEU1 to assemble with PPCA. In contrast, more than 100 reported PPCA mutations do not seem to directly influence binding to NEU1.

4. New Functions of NEU1 in Tissue Remodeling and Homeostasis

a. The Role of NEU1 and Elastin-Binding Protein (EBP) in Elastogenesis.— The process of elastic fiber assembly[34] is mediated by a non-integrin cell surface complex, named the "cell-surface elastin receptor" (CSER),[35] which can be captured on elastin-affinity columns and eluted with a solution of galactose. It consists of at least three components. The main component is a 67-kDa elastin-binding protein (EBP) that belongs to a family of S-type, Ca^{2+}-independent, β-galactoside-specific lectins.[36] EBP is identical to a catalytically inactive, non-lysosomal, splice variant of β-GAL.[37] It shares with β-GAL most of its amino acid sequence, but embeds a unique stretch of 32-residue peptide with high affinity for a VGVAPG hydrophobic domain of elastin.[38] EBP co-localizes with tropoelastin, both intracellularly and at the cell surface, indicating that the two proteins already associate in an early biosynthetic compartment and as such are routed to the cell surface.[38] Binding of galactose to the lectin site of EBP promotes its dissociation from elastin and from the plasma membrane. EBP functions as a chaperone to tropoelastin, protecting it from premature self-aggregation and degradation by elastases, and facilitating its extracellular deposition on the microfibrillar scaffold.

As is the case for β-GAL, the EBP has the capacity to associate at the cell surface with NEU1 and PPCA, the other two components identified in the CSER, which suggests an active role for these two proteins in the assembly process.[38] These authors showed that in a 3-day culture of aortic smooth muscle cells (ASMCs), a stage in which newly produced microfibrils are not yet fully embedded in the elastic core, sialic acid residues and galactose co-localize on the extracellular microfibrillar network.[38] Addition of an exogenous sialidase to the culture results in the hydrolysis of terminal sialic acid residues from the microfibrillar scaffold, and in the exposure of galactose for which EBP has a high affinity. Binding of the lectin-site of EBP to these galactose moieties likely causes a conformational change that promotes its dissociation from elastin after it is deposited onto the microfibrillar network. Dissociated EBP is then endocytosed and recycled back to the cell surface for another round of extracellular elastin deposition.[39] These findings point to a direct involvement of NEU1 at the cell surface in the process of elastic fiber assembly. In line with this is the observation that ASMCs, cultured in the presence of sialidase inhibitors or NEU1-specific antibodies, have significantly reduced deposition of cross-linked extracellular elastin, and higher levels of unassembled intracellular tropoelastin. In addition, treatment of full-term chicken embryos with sialidase inhibitors results in a marked inhibition of elastogenesis in lungs and aortas. Thus, NEU1 activity appears to be required for the deposition of extracellular elastin, although these experiments do not rule out the involvement of another sialidase, besides NEU1, in this process.

Based on these observations, the proposed model of elastic fiber assembly is that the EBP–tropoelastin complex first traverses the plasma membrane near a growing elastic fiber and then, by virtue of its binding to the PPCA/NEU1, releases tropoelastin that is then incorporated into the elastic fiber.

b. Elastogenesis in Fibroblasts from Sialidosis and GS Patients.—A role of NEU1 in elastogenesis has been further suggested by experiments with NEU1-deficient fibroblasts from sialidosis and GS patients. These cells deposit significantly less insoluble elastin as compared to control fibroblasts, but have increased levels of soluble tropoelastin.[38] This phenotype was reversed by the addition of exogenous sialidase to the culture medium. In contrast, exogenous sialidase did not improve the elastin deposition in fibroblasts from a GM1-gangliosidosis patient with deficiency of both β-GAL and EBP, but with normal NEU1 activity. Thus, although NEU1 and EBP are in a complex at the cell surface, a defect or absence of the latter protein cannot be compensated for by normal endogenous NEU1 activity or by adding exogenous sialidase. Another intriguing observation is that, in cells from a juvenile GM1-gangliosidosis patient, EBP depletion or mislocalization and impaired elastin deposition was the result of excessive accumulation of keratan sulfate arising from the β-GAL deficiency.[39] These authors postulated that EBP is quenched by the galactose residues of keratan sulfate prior to reaching the cell surface, and this either prevents it from binding to soluble elastin or provokes the premature intracellular release of elastin from EBP.[39] Combined, these studies imply that EBP and NEU1 assert their action on different components of the elastogenesis pathway.

c. EBP/NEU1/PPCA and Signal Transduction.—The CSER has also been involved in the transduction of intracellular signals. In human skin fibroblasts, hydrolysis of elastin and elastin peptides (namely κ-elastin) that are derived from alkaline degradation of insoluble elastin can activate the ERK1/2 pathway which, in turn, induces pro-matrix metalloproteinase-1 (pro-MMP1).[40] In contrast to cathepsin A, NEU1 activity apparently takes part directly in this process because treatment of cells with a neuraminidase inhibitor reduced pro-MMP-1 accumulation as well as hyper-phosphorylation of ERK1/2, while a serine protease inhibitor did not.[40] Moreover, fibroblasts cultured in the presence of κ-elastin have increased sialidase activity at the cell surface and increased phosphorylated ERK1/2 that could both be inhibited by siRNA silencing of NEU1.[40] A further interesting observation is that addition of sialic acid to the medium of fibroblasts also increased pro-MMP1 levels and ERK1/2 phosphorylation. These studies suggest the following sequence of events: binding of the EBP tropoelastin complex to NEU1/PPCA at the cell surface activates NEU1, which in turn desialylates components of the microfibrillar scaffold, exposing their galactose moieties that can then bind to the lectin-site of EBP. The latter process leads to the dissociation of EBP from elastin and the inactivation of NEU1. The sialic acids that are released during this process can then induce ERK1/2 phosphorylation, resulting in the transcriptional activation and extracellular accumulation of MMP1.[40]

d. Elastogenesis in NEU1-Deficient Mice.—The function of the EBP/NEU1/-PPCA-CSER in the process of elastogenesis has thus far been studied only in cultured human fibroblasts. It is noteworthy that the splicing mechanism of human *β-GAL* mRNA that gives rise to the *EBP* transcript is not conserved in the mouse.[41] Thus, it is still unclear whether in mice the chaperone function of EBP toward elastin is controlled by another EBP-like protein, or whether mice have an EBP-independent mechanism of elastogenesis. For the purpose of ascertaining whether the other components of the CSER, NEU1, and PPCA are involved in the extracellular deposition of elastin, the process of elastogenesis was investigated in *Neu1* null mice, a faithful model of Type II sialidosis (a detailed description of these mice is given in paragraph 5 of this section).[42] The first indication that Neu1-deficient mice could have defective elastic fiber deposition came from the observation that they have a tight skin,[43] a phenotype similar to that of homozygous tight-skin (Tsk) mice. The latter is a model of fibrosis that is characterized by the production of mutant fibrillin-1 and abnormal accumulation of connective tissue in the skin and visceral organs, including the heart.[44] As illustrated in Fig. 2, this phenotypic abnormality is detected in *Neu1*-null mice already at young age. In addition, as compared to wild-type mice, the lungs of Neu1-deficient mice have an emphysematous appearance, with areas of enlarged air spaces, reduced elastic fibers in the crest region, and disordered and dispersed myofibroblasts around the alveolar walls. Concomitantly, fibulin-5 (Fib-5), an extracellular basement membrane protein required for the polymerization of elastin, and fibrillin-2 (Fbn-2), an extracellular matrix protein involved in elastic fiber assembly, are present in reduced amount and their localization appear diffuse and disorganized in *Neu1*-null lungs.[43]

Elastin concentration is also severely reduced in the aortic wall of *Neu1*-null mice.[43] Elastin lamellae are thinner and straighter than in the wild-type tissue and are separated by large spaces abnormally filled with sialic acid-containing material. At the ultrastructural level, the skin, lungs, and thoracic aortas from Neu1-deficient mice show overt differences in the organization of the cells and the extracellular matrix; in deficient mice vacuolated fibroblasts or smooth muscle cells are surrounded by loose extracellular matrix, and abnormally high number of collagen fibers, but scarce, short, and highly immature elastic fibers.[43] Together these findings point to an essential role played by Neu1 in the deposition and assembly of the elastic fibers both in human and mouse tissues, although it remains unclear whether in mice a CSER similar to that in humans is present at the cell surface.

e. Role of NEU1 in Proliferation and Cancer.—Recent studies have focused on the potential function of NEU1 in cell proliferation and cancer. ASMCs exposed to a bacterial sialidase have decreased proliferation capacity, while the same cells treated with

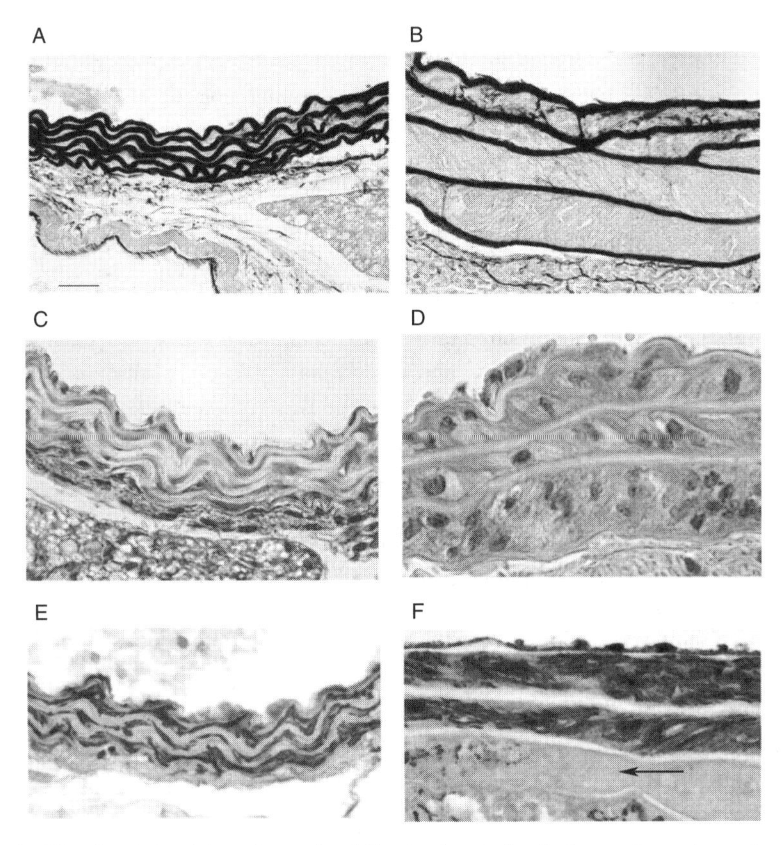

FIG. 2. Histochemical staining of control and *Neu1*-null aorta for elastin, mucins, and smooth muscle. (A) Control aorta stained for elastin with Hart's stain showing 5–6 elastic lamellae and (B) *Neu1*-null aorta showing the abnormal separation between the elastic lamellae. (C) Control aorta stained for sialomucins with Alcian blue (blue) with elastic lamellae appearing opaque and (D) *Neu1*-null aorta stained for sialomucins indicating the large mass of positive material between elastic lamellae. (E) Control aorta immunostained for smooth muscle with an antibody to alpha-smooth muscle actin (red) showing the thin area between elastic lamellae and (F) *Neu1*-null aorta showing the increased mass of smooth muscle between elastic lamellae, whereas some areas were devoid of smooth muscle (arrow). Scale bar, 22 μm. From Starcher *et al.*[43] (See Color Plate 46.)

either a sialidase inhibitor or a NEU1-specific antibody have an increased proliferation rate.[35] In support of NEU1 being a negative modulator of proliferation is the observation that fibroblasts from sialidosis patients have higher basal proliferation rates than normal fibroblasts, and respond more strongly than normal fibroblasts to the addition of recombinant growth factors, such as platelet-derived growth factor (PDGF) and insulin-like growth factor-2 (IGF-2), both potent stimulators of ASMC proliferation.[35]

This study proposes that the extent of sialylation of the cell-surface receptors PDGF-R and IGF-1R activate pro-mitogenic signals in ASMCs via the PI3K/Akt, PLCγ, and Ras-Raf1-MEKs-ERK1/2 pathways upon interaction with their cognate ligands. In addition, ASMCs pre-treated with exogenous sialidase no longer proliferate when exposed to recombinant PDGF and IGF-2, probably because of desialylation of the PDGF-R and IGF-1R and in turn reduced AKT and ERK1/2 phosphorylation. Thus, NEU1 inhibits the proliferation of ASMCs and fibroblasts as well as regulating pro-mitogenic signals by desialyation of the cell-surface receptors for PDGF and IGF-2 growth factors.[35]

There is evidence that cancer cells undergo alterations in glycosylation and sialylation of cell-surface molecules, which seems to correlate with malignant transformation, increasing their metastatic potential and invasiveness.[45] These phenotypic alterations are mirrored by changes in the expression profiles of endogenous mammalian sialidases.[46] NEU3 is up-regulated 3- to 100-fold in colon carcinomas and other cancer cell lines,[46] and overexpression of NEU3 inhibits apoptosis by increasing the level of anti-apoptotic Bcl-X_L and decreasing the amount of pro-apoptotic caspase-3.[47] In contrast, the expression of NEU1 appears to be down-regulated in highly metastatic colon adenocarcinoma cells as compared to tumor cells with low metastatic potential.[46] In line with this observation, overexpression of NEU1 in human colon adenocarcinoma HT-29 cells suppressed cell migration and invasion, while knockdown of NEU1 had the opposite effect.[46] Mice that had been transsplenically injected with NEU1-overexpressing HT-29 cells show decreased *in vivo* liver metastasis as compared with mice injected with the parental HT-29 cells.[48] These authors demonstrated that β4-integrin becomes hyposialylated in the tumor cells as a result of NEU1 overexpression. This is accompanied by decreased phosphorylation of β4-integrin and attenuation of the focal adhesion kinase, and the ERK1/2 pathways. In addition, overexpression of NEU1 also results in downregulation of such matrix metalloproteinases as MMP7, which are known to be induced in cancer cells having high metastatic potential.

In the future, therapeutic agents may be identified or synthesized that can enhance the expression or catalytic activity of NEU1. Such drugs may be targeted to tumor cells that have low or null NEU1 expression, and potentially could reduce or prevent metastasis.

5. Mouse Model of NEU1 Deficiency and Study of Disease Pathogenesis

a. Mouse Model of NEU1 Deficiency.—Little attention has been given to the potential role played by NEU1, beside its primary hydrolytic activity, in other biological processes. A careful dissection of the multifaceted phenotypes that develop progressively in a faithful mouse model of sialidosis has led to the discovery of unexpected functions of this pivotal enzyme. *Neu1*$^{-/-}$ mice develop a multisystemic phenotype closely resembling that of type II sialidosis patients.[42] Homozygous null mice have severely attenuated sialidase activity in most tissues tested as compared to wild-type littermates, while heterozygous mice have intermediate enzyme values. A relatively high residual sialidase activity is measured in muscle, thymus, and brain, probably due to the overlapping catalytic activities of Neu2 and Neu3, which are unaffected in these mice, and well expressed in these tissues.[49] Newborn *Neu1*$^{-/-}$ mice weigh about 25% less than their heterozygous or wild-type littermates. However, a percentage of the *Neu1*$^{-/-}$ pups, depending on the genetic background, die around the time of weaning; hyperkeratosis of the stomach suggests that the cause of death is inanition after 1–2 days of anorexia. Shortly after birth, mutant mice exhibit profound oligosacchariduria, which is diagnostic of the disease in children with either sialidosis or GS. Edema of the subcutaneous tissues, limbs, penis, forehead, and eyelids develops already at 4–6 weeks of age, and progressively worsens. Kyphosis of the lumbar spine and lordosis of the cervical and thoracic spine becomes prominent by the age of 6 months. A shock-like twitch indicative of peripheral nervous system involvement is also a consistent phenotype. At the end of their lifespan (6–9 months) *Neu1*$^{-/-}$ mice appear weak and debilitated, and suffer from dyspnea, loss of weight, gait abnormalities, and tremor.[42]

b. Extramedullary Hematopoiesis and Impaired Bone Marrow Retention in Neu1$^{-/-}$ Mice.—An overt phenotype of *Neu1* null mice that is nearly fully penetrant and is characteristic of the severe form of sialidosis in children is splenomegaly, which becomes apparent already in 6-week-old mice. The spleen size increases steadily to reach a maximum at 3–5 months of age, but normalizes later in life.[42,50] This time-dependent change in size is accompanied by a progressive increase in the number of hematopoietic progenitor cells (HPCs) in the spleen and peripheral blood, and by a corresponding decrease in the number of these cells in the bone marrow (BM). At the histopathological level both the BM and the spleen appear filled with vacuolated cells, particularly macrophages. These findings were consistent with extramedullary hematopoiesis (EMH) as the primary cause of splenic hypertrophy in *Neu1*$^{-/-}$ mice. In addition, BM transplantation (BMT) of null mice with wild-type or transgenic BM over-expressing NEU1 around the time of occurrence of EMH failed to engraft. As summarized in Fig. 3, five months after BMT the majority of transplanted *Neu1*$^{-/-}$ mice had no detectable Neu1 activity in their hematopoietic tissues, which indicated a failure in the long-term engraftment of donor BM. Combined, these results suggest a serious defect in bone marrow homing and/or retention in these mice.[42,50]

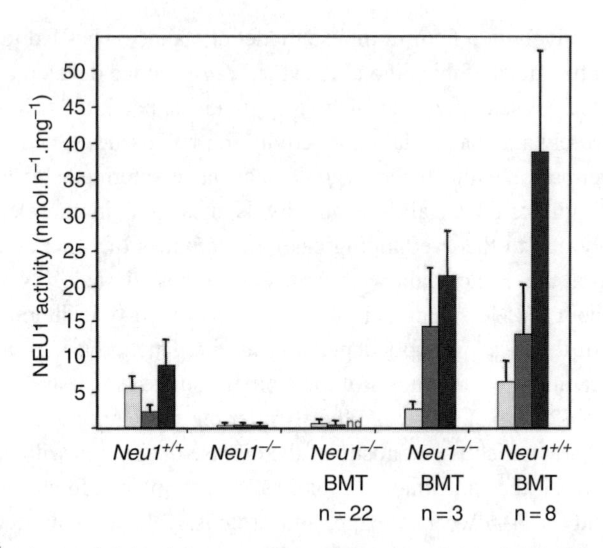

FIG. 3. *Neu1*$^{-/-}$ mice are deficient in long-term bone marrow engraftment. Sub-lethally irradiated *Neu1*-null and wild-type mice were transplanted with transgenic NEU1-expressing bone marrow (BM) and analyzed 5 months after bone-marrow transplantation. To assess engraftment of the donor bone marrow, NEU1 activity was measured in hematopoietic tissues from both bone marrow-transplanted and untreated wild-type and *Neu1*-null mice. While all transplanted wild-type mice had successfully engrafted the donor bone marrow, 22 of the 25 transplanted *Neu1*-null mice failed to engraft. Liver, yellow bars; spleen, red bars; BM, green bars. (See Color Plate 47.)

A series of experiments using BM and BM-derived macrophages were performed to assess whether in *Neu1*-deficient mice the lack of long-term BM engraftment and the EMH phenotype shared a common determinant. Proteomic analyses of the BM extracellular fluid (BMEF) isolated from *Neu1*$^{-/-}$ and wild-type mice revealed that the *Neu1*$^{-/-}$ BMEF contains severely reduced levels of two clusters of uncleaved clade A serine protease inhibitors, serpina1 and serpina3. Serpins are normally present extracellularly as an uncleaved inert form and a cleaved activated form.[51] They are targeted by specific serine proteases that cleave the reactive center loop, resulting in the formation of a serpin–protease covalent complex and inactivation of the protease activity.[51] Serpina1 and serpina3 are responsible for the regulation of the levels of the serine proteases neutrophil elastase (NE) and Cathepsin G (CG) in the BM microenvironment. These proteases play a central role in the regulated degradation of cell-adhesion molecules, such as VCAM-1, which are expressed on BM stromal cells (BMSC) and are responsible for the homing/retention of HPCs in the bone niche.[52] NE and CG are released into the BM microenvironment by the degranulation of secondary granules (secretory lysosomes). This process occurs via lysosomal exocytosis, a multistep mechanism that entails the attachment of secretory lysosomes to the cytoskeleton, their migration to and docking at the plasma membrane, the Ca^{2+}-dependent fusion of the lysosomal membrane with the plasma membrane and finally the release of the luminal content into the extracellular milieu.[53]

The presence of the lysosomal-associated membrane protein-1 (LAMP-1) at the plasma membrane has been used as a parameter to identify cell populations engaged in lysosomal exocytosis. An increased percentage of cell surface-LAMP-1-positive cells was detected in the total BM of *Neu1*$^{-/-}$ mice. This was accompanied by elevated levels of NE and CG in the BMEF of these mice.[50] Under normal conditions, an excess of NE and CG in the BMEF would be inactivated by serpina1 and serpina3. However, in the *Neu1*$^{-/-}$ BMEF, the natural balance between the levels of these proteases and their inhibitors has been severely compromised, as evidenced by the low amounts of serpins detected in the proteomic analysis of BMEF samples of deficient mice. These combined features indicate increased lysosomal exocytosis of *Neu1*$^{-/-}$ BM cells. The resulting elevated proteolytic load in the bone niche of *Neu1*$^{-/-}$ mice leads to premature degradation of VCAM-1 on BMSCs and reduced retention of HPCs. These findings offer an explanation for the time-dependent loss of BM retention in the bone niche of *Neu1*$^{-/-}$ mice, and provide a rationale for the occurrence of EMH in this mouse model and probably in children with sialidosis.

c. NEU1 Is a Negative Regulator of Lysosomal Exocytosis.—To reconcile the loss of NEU1 activity with the excessive lysosomal exocytosis of BM cells, it was postulated that NEU1 plays a role in the processing of glycans on lysosomal proteins implicated in the exocytic process, thereby changing their properties.[50] It was found that proteins that participate in the membrane-fusion process are not affected by the loss of NEU1. However, in *Neu1*$^{-/-}$ cells, LAMP-1 is increased in quantity and molecular weight because of abnormal processing of its sialic acid moieties and decreased turnover rate in absence of Neu1. These results prove that LAMP-1 is a natural substrate of NEU1. In addition, oversialylated LAMP-1 accumulates at the cell surface of *Neu1*$^{-/-}$ macrophages, indicating that a larger population of lysosomes fuses with the plasma membrane.[50] A direct involvement of LAMP-1 in the process of lysosomal exocytosis is substantiated by the observation that silencing of LAMP-1 expression in *Neu1*$^{-/-}$ macrophages normalizes LAMP-1 intracellular distribution and reduced the level of exocytosis.

Unlike macrophages, which naturally have a high level of lysosomal exocytosis, this process occurs marginally in normal fibroblasts, and LAMP-1 is usually not detectable at their cell surface. In contrast, fibroblasts from Type II sialidosis patients with the congenital, early-onset form of the disease and complete lack of NEU1 activity have significantly increased levels of LAMP-1 localized at the cell surface, which is accompanied by highly increased lysosomal exocytosis of active hydrolases. Fibroblasts from a patient with the late-onset, attenuated form of disease and with residual NEU1 activity have near-normal levels of LAMP-1 at the cell surface, and low levels of lysosomal exocytosis. Thus, the residual NEU1 activity in sialidosis patients with different phenotypes correlates inversely with the levels of cell-surface LAMP-1 and lysosomal exocytosis. Based on these results it is predicted that also in non-secretory cells, a complete lack of NEU1 activity increases the pool of lysosomes that are docked at or near the plasma membrane, and become engaged in lysosomal exocytosis.

Evidence that excessive lysosomal exocytosis also plays a role in other pathologic manifestations in the *Neu1*-null mice is presented in a recent publication.[54] As is the case for sialidosis patients, *Neu1*-null mice display profound hearing loss involving both conductive and sensorineural components.[54] In cochleae of adult wild-type mice, Neu1 is expressed in several cell types in the stria vascularis, the organ of Corti, and spiral ganglion. As early as age P9, progressive morphological abnormalities and vacuolization are detected in the cochleae of *Neu1*-null mice. Similarly to BM cells, these morphologic changes are associated with oversialylation of lysosomal-associated membrane proteins (LAMPs) in the stria vascularis. The increased expression and apical localization of LAMP-1 in marginal cells of the stria vascularis predicts exacerbation of lysosomal exocytosis into the endolymph, and consequently, a reduction in the endolymphatic potential, which, in turn, may cause dysfunction of transduction in sensory hair cells, resulting in a loss of hearing.

A recent study by Seyrantepe and co-workers also implicate NEU1 in the mechanism of phagocytosis. They show that mouse macrophages and immature dendritic cells with a deficiency of NEU1 secondary to a defective PPCA have a reduced capacity to engulf bacteria and synthetic polymer beads. The cells have increased sialylation of cell-surface proteins, probably affecting several yet undetermined phagocytosis receptors.[55]

Together these studies have proven crucial for understanding the physiological functions of NEU1.

III. The Cytosolic Sialidase NEU2

1. General Properties of NEU2

NEU2 was the first vertebrate sialidase characterized using the recombinant DNA techniques. The enzyme was first cloned[56] from rat skeletal muscle in 1993, and subsequently a highly similar protein was cloned from CHO cells.[57] Overexpression of the human cytosolic sialidase NEU2 homologue in *E. coli*[58] and the purification of the enzyme to homogeneity permitted its detailed kinetic characterization.[59] The pH optimum was 5.6, V/E and V/t had the expected features, and the $V/[S]$ relationships obeyed Michaelis–Menten kinetics. The highest catalytic activity and affinity was exhibited on α–(2→3) linkages present in different sialoconjugates (gangliosides, glycoproteins, sialyl-lactose), whereas the α–(2→8) linkage in colominic acid and the α–(2→6) linkage carried by α–(2→6) sialyl-lactose were resistant to hydrolysis. The enzyme exhibited the highest activity on gangliosides [GM3, GD1a, GD1b, GT1b, and α–(2→3) sialylparagloboside] in their micellar (vesicular in the case of GM3) supramolecular organization. Instead, micellar-form monosialoganglioside GM2 and GM1, both carrying a galactose residue linked to sialic acid and *N*-acetylgalactosamine were not affected by NEU2. Remarkably, NEU2 was active on a monomeric dispersion of gangliosides, including GM2 and GM1, with K_m values as low as 10 nM. The ability of the enzyme to recognize not only sialosyl linkages, but also saccharide units close to sialic acid and the supramolecular organization of such amphipathic substrates as gangliosides, presumably reflects peculiar substrate conformations and interactions within and around the active site of the enzyme necessary to approach the sialosyl linkage.

The use of molecular-modeling techniques and site-directed mutagenesis at the level of the active site and in its boundary regions may help to clarify the complex enzyme–substrate interactions of NEU2.

2. Crystal Structure of Human Sialidase NEU2

Human sialidase NEU2 was the first, and so far the only mammalian sialidase characterized at the structural level.[19] The atomic coordinates and structural features of NEU2 in its free (*apo*) form, or in the forms induced by its interaction with maltose or with the competitive inhibitor Neu5Ac2en are available in the Protein Data Bank with the codes 1SNT, 1SO7, and 1VCU, respectively. As shown in Fig. 4A, the NEU2 structure has the six-bladed β-propeller typical of bacterial and viral sialidases,[60] as well as that of trypanosomal trans-sialidases (Section XI of this article). In the NEU2-Neu5Ac2en complex structure, the competitive inhibitor lies in the shallow crevice of the active site, surrounded by ten amino acids (Fig. 4B). The arginine triad (R21, R237, and R304), highly conserved among sialidases, positions the carboxylate group of Neu5Ac2en within the active site. In addition, the tyrosine residue at position 334 that lies below the Neu5Ac2en molecule presents its hydroxyl group close to the carbon atom 2 of the inhibitor ring and possibly stabilizes a putative carbonium ion

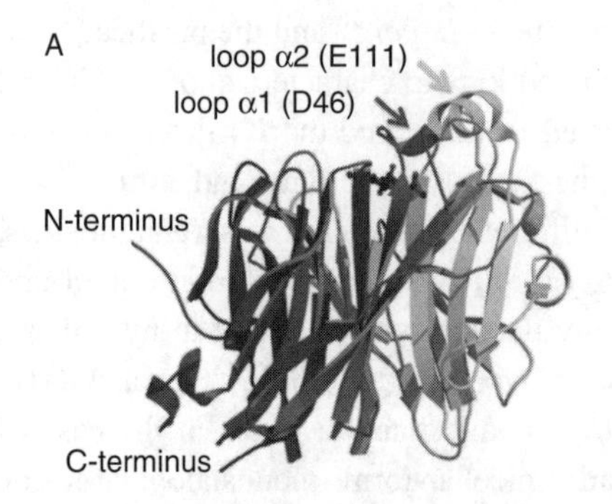

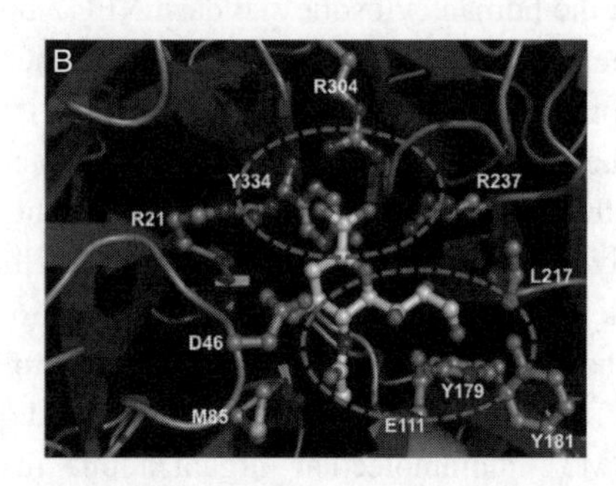

FIG. 4. Structural features of the cytosolic sialidase NEU2. (A) Ribbon diagram of NEU2–Neu5Ac2en (DANA) complex viewed from the side. Neu5Ac2en is represented using ball and stick model. The six blades of the β-propeller are colored differently. The colored arrows indicate loops α1 and α2 that become structured upon the competitive-inhibitor binding to the NEU2 active site and contain D46 and E111, respectively. (B) The catalytic crevice of NEU2–Neu5Ac2en complex. Neu5Ac2en and R-chains of the residues involved in the competitive-inhibitor coordination are represented as ball and stick models. Blue arrows represent β-strands and a red ribbon represents a portion of α2 helix bearing the E111 residue. Conserved residues involved in the catalytic process and in the Neu5Ac2en coordination, namely, R21, D46, R237, R304, and Y334, are highlighted by a green dotted oval. Other residues, namely E111, Y179, or Y181, highlighted by a red dotted oval, are present only in the NEU2–Neu5Ac2en complex and interact with the glycerol terminal oxygen atoms of the inhibitor. Modified from Chavas *et al.*[19] (See Color Plate 48.)

formed in the transition state of the enzyme-catalyzed reaction.[61] Interestingly, the *N*-acetyl and glycerol portions of Neu5Ac2en interact with NEU2 amino acid residues in a manner different from those of both bacterial and viral sialidases. These structural differences could be useful for design of inhibitory drugs specifically targeted toward the viral/microbial or mammalian enzymes. Actually, using an homology-modeling approach based on the crystal structure of NEU2, the predicted structures of the other members of the family confirmed such peculiar structural differences[18] (see also Section II). On this basis, some C-9 amide-linked hydrophobic derivatives of Neu5Ac2en have been synthesized showing a selective inhibitory activity on the human lysosomal sialidase NEU1,[62] whereas a series of fluorobenzoic acid derivatives did not.[63]

Similarly to NEU1, NEU2 contains multiple Asp-box motifs of which one is less conserved (residues 247–254) and two are canonical Asp-boxes (residues 129–136 and 199–206). Structurally, Asp boxes show a β-hairpin structure stabilized by a water molecule at the center of three conserved residues of serine, aspartate, and tryptophan, and are found in topologically equivalent positions in all members of the sialidase gene family.[10,64] Interestingly, Asp-boxes are found in protein families having different sequences and three-dimensional structures, such as bacterial ribonucleases, reelin, netrins, sulfite oxidases, some lipoprotein receptors, and a variety of glycol-hydrolases.[64] Thus far the function of Asp boxes is still unknown. The presence of these structural motifs in glyco-hydrolases (namely sialidases), mostly exposed at the surface of these proteins, has suggested a role in the binding of saccharide structures, although in some proteins, such as in barnase-like ribonucleases, the Asp box is located in the active site, containing the conserved His and Tyr residues involved in the binding of guanosine 3′-monophosphate.[64] It is noteworthy that, in NEU2 the Asp boxes are located on the opposite site of the active site, and no hydrophobic or extended surface of contact is detectable between the motifs.[19] The characterization of the three-dimensional structure of NEU2 has represented a pivotal step in sialidase biology, and efforts are ongoing to obtain structural data regarding the other members of the mammalian sialidase family.

3. Functional Implication of NEU2

a. NEU2 and Muscle Differentiation.—The involvement of NEU2 in myoblast differentiation *in vitro* was first demonstrated in rat L6 myogenic cells.[65] Subsequent studies showed that in murine C2C12 myoblasts, NEU2 appears to undergo up-regulation along with the process of differentiation.[66] The authors reported that C2C12 stable transfectants overexpressing NEU2 are characterized by a marked decrease of their proliferation rate with a marked decrease of cyclin D1 protein. In addition, these transfectants undergo spontaneous differentiation, giving rise to myogenin-positive myotubes under standard growth conditions. Overall, these results provide evidences that NEU2 up-regulation constitutes an important event to trigger *in vitro* myoblast differentiation. The possible involvement of NEU2 in muscle regeneration *in vivo* has been suggested in the SLJ mouse model.[67] In fact, in a mouse model for human dysferlinopathy (OMIM 603009, 253601),[68] the down-regulation of the *Neu2* gene is associated with impairment of muscle regeneration.[67]

In addition, a link between the insulin-like growth factor-1 (IGF1) signaling pathways and NEU2 expression has been shown.[69] In IGF1-induced hypertrophy of myoblasts, the activation of the phosphatidylinositol-3-kinase (PI3)/serine–threonine protein kinase AKT1/mammalian target of rapamycin (mTOR) pathway leads to an increase of NEU2 levels, both in terms of transcript level and enzyme activity. This up-regulation can be obtained by using a constitutively activated form of AKT1, whereas transfection of a kinase-inactive mutant of AKT1 inhibits myotube formation and causes *NEU2* down-regulation. Although *NEU2* transfected myoblasts differentiate more easily, the enzyme overexpression does not overrule the block of differentiation induced by PI3 and mTOR inhibitors. Based on these results, the increase of NEU2 appears to represent a downstream event in myoblast differentiation and hypertrophy induced by IGF1. This *in vitro* evidence was further confirmed using different myoblast cell models in relation to myofiber hypertrophy and atrophy.[70] As expected, hypertrophy was found to be coupled to high levels of NEU2, whereas atrophy is characterized by a down-regulation of the enzyme. In addition, in experimentally induced *in vitro* myoblast atrophy, the enzyme appeared to undergo degradation through autophagy with the involvement of lysosomal proteases but independently of proteasome activation.[71]

Finally, a study in rhabdomyosarcoma (RMS), a pediatric sarcoma originating from mesenchymal precursor cells that fail to undergo proper differentiation,[72] revealed that NEU2 expression is absent in these cells.[73] The embryonal RMS RD cell line that is pharmacologically or genetically forced toward myogenic differentiation shows delayed growth in size of the myotubes together with low NEU2 content, both in terms of transcript levels and enzyme activity. Although further experiments (namely transfection of RD cells with *NEU2*) are needed in order to support a correlation between the lack of NEU2 and the reduced size of the myotubes observed, this study may constitute a stimulating line of research.

In summary, these results demonstrate the involvement of *NEU2* in the complex series of events leading to myoblast differentiation, although more information is needed in order to define the exact role played by this enzyme. For example, one important and open question is the identification of the possible substrate(s) and product(s) of NEU2 in myofibers.

b. NEU2 and Cancer Cell Differentiation.—Some information about the involvement of NEU2 in differentiation of cells different from myoblasts has been obtained by using PC12 cells, a cell line widely used as a model to study neuronal differentiation. The administration of growth factors to these cells promotes the process of neuritogenesis and this effect, as well as proliferation induced by epidermal growth factor (EGF), is coupled with a transient overexpression of NEU2.[74] The implication of NEU2 in the differentiation of colony-forming unit-erythroid (CFU-E) cells and K562 erythroleukemic cells was subsequently ascertained.[75] CFU-E cells, the second stage of erythroid precursors, correspond to the differentiation stage that precedes hemoglobin production in erythropoiesis. Erythroid differentiation can also be studied in human leukemic K562 cells upon treatment with sodium butanoate.[76] During differentiation, CFU-E cells become progressively sensitive to erythropoietin, and *NEU2* mRNA is detectable only in a narrow temporal window at the end of process, although at lower levels of expression when compared to *NEU1* and *NEU3* transcripts.[75] In the case of K562 cells treated with sodium butanoate, erythroid differentiation is characterized by higher expression levels of *NEU1* and *NEU3* as compared to CFU-E cells, whereas *NEU2* transcript appears to be absent. Based on these results, the possible contribution of NEU2 in the pathogenesis of such blood cancer forms as leukemia was investigated in K562 clones expressing the enzyme.[77] NEU2 expression impairs Bcr-Abl activity, modifying the signaling cascades stimulated by the oncoprotein and leading to a decrease of BCL-2 and Bcl-XL gene expression. As a result, K562 cells become more sensitive to apoptotic stimuli and do not progress toward malignancy. The molecular link between NEU2 activity and these signaling pathways does not appear to depend on the desialylation of gangliosides, which are not affected, but rather on the desialylation of some yet unidentified glycoproteins in the mass range of 45–66 kDa.

The possibility that the sialylation level of glycoproteins is crucial for the shift from apoptosis to malignant proliferation is an intriguing working hypothesis, and opens new perspectives in therapeutical approaches for chronic myeloid leukemia based on the cytosolic sialidase NEU2 expression.

c. Inhibitors of Influenza Virus Sialidase and NEU2.—The influenza virus sialidase, also known as neuraminidase (NA), plays a key role in the life cycle of viruses and is a primary drug target against influenza, especially in the event of an avian and swine influenza pandemic.[78,79] At present, two inhibitors of the viral sialidase, zanamivir (Relenza) and oseltamivir-phosphate (Tamiflu), are the first choice drugs to fight influenza,[80,81] on the basis that these molecules inhibit only the viral enzymes and not the human sialidases. Regarding Tamiflu, the seasonal treatment is associated with generally mild side effects. However, some rare but severe adverse events have been reported in Asia, particularly in Japan.[82] Chinese researchers have proposed a possible explanation for these adverse side effects, based on the identification of a non-synonymous single nucleotide polymorphism (nsSNP) in the SNP database available at NCBI, resulting in the substitution of Arg 41 with Glu (R41Q) of human NEU2.[83] This nsSNP segregates only in the Asian population and, surprisingly, *in vitro* sialidase assays using the wild-type enzyme or the R41Q mutant indicates a modest increase of the binding affinity of the mutant NEU2 form to oseltamivir carboxylate, the active form of the drug, and a decrease of enzyme activity. In this perspective, the Tamiflu treatment in people carrying this nsSNP could result in the decrease of NEU2 activity leading, as suggested by the authors, to the severe adverse complications in patients treated with this drug.

However, it should be noted that *NEU2* is generally expressed in humans at only very low levels and predominantly in muscle,[58] which is difficult to reconcile with the amplitude and severity of these complications. Curiously, these complications include the same neuropsychiatric symptoms caused by mutations in lysosomal *NEU1*.[84,85]

In order to better clarify this issue, the effect of zanamivir and Tamiflu on recombinant human sialidases has been investigated.[86] Intriguingly, the active form of Tamiflu hardly affects the activity of all the four sialidases, even at a very high concentration (1 mM), while zanamivir is a remarkably good inhibitor of NEU2, NEU3, and NEU4 in the micromolar concentration range and a modest inhibitor of NEU1 in the millimolar range of concentration. Thus, the potential involvement of NEU2 in the reported harmful effects of Tamiflu treatment seems improbable.

A possible inhibitory effect on endogenous sialidase(s) exerted *in vivo* by Tamiflu has been observed using a mouse model of respiratory syncytial virus (RSV) infection.[87] RSV infection increased sialidase activity in lung mononuclear cells as well as the content of ganglioside GM1 (GM1 is the product of sialidase activity on polysialylated gangliosides of the ganglio-series) in activated T cells. The administration of Tamiflu to RSV-infected mice decreased the GM1 content on the cell surface of T cells and inhibited viral clearance. Overall these results suggest a direct effect of the antiviral drug on endogenous sialidase(s), NEU1, NEU3, or both, because these are the only endogenous sialidases expressed in immune cells, thus indicating an involvement of these glycohydrolases in the regulation of GM1 expression in T cells and antiviral immunity.

d. Perspective on the Biotechnological Application of NEU2.—Terminal sialic acid is a biological mask and, for example, the removal of this acidic sugar from the glycans of serum glycoproteins significantly increases their clearance from plasma as compared to the sialylated counterparts.[88] For this reason, a challenge in production of therapeutic recombinant protein by mammalian cells is to avoid or significantly decrease the cleavage of terminal sialic acids by glycosidases released by lysed cells into the culture medium. The production of recombinant IFN-γ by CHO cells has been studied, using RNA interference to silence *NEU2* expression.[89] Cell lines characterized by a reduction in cytosolic sialidase activity of more than 60% as compared to control CHO cells allow the recovery of fully sialylated recombinant IFN-γ, thus demonstrating that RNA interference of sialidase(s) can be a useful biotechnological approach to produce fully sialylated recombinant glycoproteins.

e. New Developments in NEU2 Biology.—The notion that NEU2 is a soluble, cytosolic enzyme is well established. However, a later study showed the occurrence in mice of five types (A, B, C, D, and N) of alternative spliced *Neu2* mRNA, encoding enzyme isoforms that have short additional stretches of amino acids at the N-terminus that depend on the first exon transcribed.[90] The expression of these enzyme variants is different in various tissues; in particular type B is the only one expressed in the thymus. Two transcript variants, encoding "long" (type B) and "short" (type C) isoforms of NEU2 (the latter corresponding to the "canonical" cytosolic sialidase from skeletal muscle) were expressed in COS7 cells and were analyzed for their subcellular localization and substrate specificity. Interestingly, both isoforms of NEU2 appear to be only partially soluble, with a significant amount (30–40%) associated to membranous structures surrounding the nucleus and in a few peripheral ruffled spots. Moreover, the substrate specificities and the K_m values appeared different for the soluble and membrane-associated forms. Although, as commented by the authors, it cannot be excluded that the membrane-associated forms are a result of an artifact derived from the overexpression of the enzyme isoforms, the finding may contribute to a better understanding of the physiological role(s) of sialidases.

IV. The Plasma Membrane-Associated Sialidase NEU3

1. General Properties of NEU3

The membrane-associated sialidase NEU3 was first cloned from bovine brain in 1999, as a plasma membrane-associated sialidase specific for gangliosides.[49] NEU3 homologues subsequently characterized at the molecular level from various animal species were found to be ubiquitously expressed and were confirmed, at least by *in vitro* enzyme assays, to be highly specific for gangliosides.[4,91,92] The enzyme shows a high level of sequence identity with NEU2 and, together with NEU4 and NEU1, shares the same β-barrel structure of the cytosolic enzyme,[19] as demonstrated by homology modeling.[18] The puzzling mechanism of NEU3 anchorage to the membrane has been studied in COS7 and HeLa cells, overexpressing murine NEU3. In these cells the enzyme behaves as a peripherally associated membrane protein, present in both the plasma membrane and the membranous structures corresponding to the recycling endosomal compartment, from which it can be released by treatment with carbonate.[93] The movement of NEU3 between these two subcellular compartments has been convincingly demonstrated,[93] confirming a previous observation carried out in

A431 squamous carcinoma cells in response to EGF.[94] Under these conditions, NEU3 redistributes from the leading edges of the plasma membrane to the ruffling cell membranes. Within these particular areas of the plasma membrane, the enzyme not only co-localizes with Rac-1,[95] but overexpressed NEU3 actually binds to and activates Rac-1, leading to increased cell migration in both A431 and HeLa cells.[94]

The activity of NEU3 toward gangliosides inserted into the membranes of intact, living cells has been demonstrated by transient transfection of COS7 cells with the murine enzyme.[96] Metabolic labeling of the sphingolipids with tritiated sphingosine demonstrated that NEU3 overexpression led to a decrease of about one third of GM3 and GD1a ganglioside cell content, with a parallel 35% increase in ganglioside GM1. Moreover, a mixed culture of NEU3-overexpressing cells with sphingosine-labeled cells (the latter acting as substrate carriers) showed that the enzyme present at the cell surface was able to remove sialic acid residues from the oligosaccharide moieties of ganglioside exposed on the membrane of neighboring cells (Fig. 5). Overall these results demonstrated that the enzyme on the cell surface modifies the ganglioside pattern not only of the cells where it resides (*cis*-activity), but also of adjacent cells (*trans*-activity), indicating a possible involvement of the enzyme in cell-to-cell interactions. These results are further supported by the enrichment of the NEU3 at the plasma membrane sites corresponding to cellular contact regions (Fig. 6).

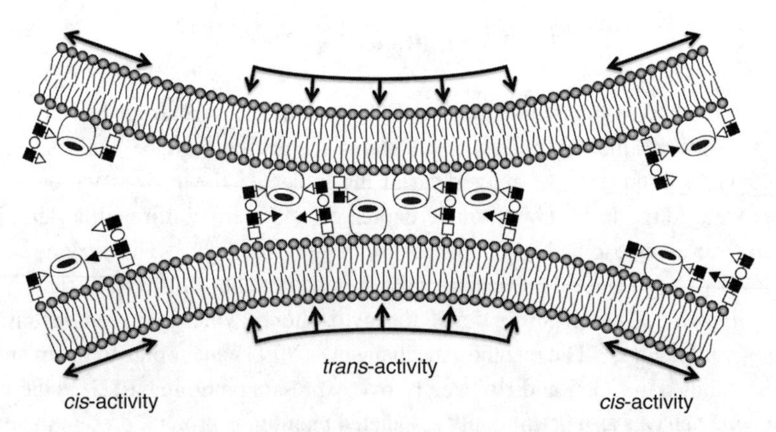

FIG. 5. *Cis-* and *trans*-activity of sialidase NEU3 toward gangliosides. The membrane-associated sialidase NEU3 recognizes the saccharide portions of ganglioside substrates on the same membrane where it resides, *cis*-activity, as well as those belonging to gangliosides inserted into the membranes of neighboring cells, *trans*-activity. The hydrolyzing activity of NEU3 toward the sialyl linkages of gangliosides of neighboring cells, *trans*-activity, has been demonstrated in COS7 cells,[96] in C2C12 myoblasts,[118] and in neutrophil.[152]

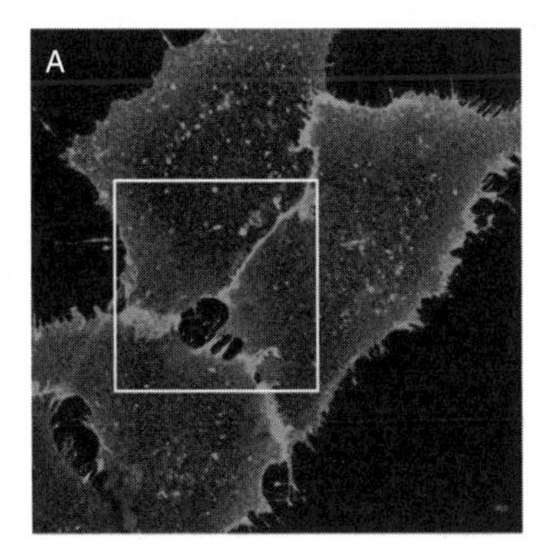

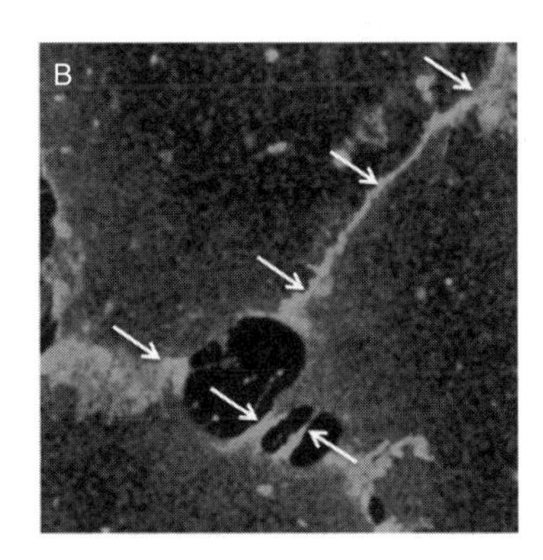

FIG. 6. Localization of NEU3 at the cell-to-cell contact areas. HeLa cells were transfected with pcDNA 3.1-NEU3-HA, a chimeric form of the murine protein that has been demonstrated to be fully active[96,112] and without significant variations in the subcellular localization compared with the wild-type protein of human origin.[101] Laser confocal analysis of transfectants at the cellular adhesion plane (panel A, magnification 250×) shows that the protein is highly enriched at the cellular boundaries as expected. Higher magnification (panel B showing inset in A, magnification 600×) demonstrates a significant enrichment of the protein at cell-to-cell contact areas (arrows in B). (See Color Plate 49.)

2. Functional Implication of NEU3

a. NEU3 and Membrane Microdomains.—Lipid rafts (or detergent-resistant membranes, DRMs, or glycosphingolipid-enriched microdomains, GEMs) are defined as small (10–200 nm), heterogeneous, highly dynamic platforms or aggregates resulting from the preferential packing of some proteins, cholesterol, and sphingolipids (namely ceramide, sphingomyelin, and gangliosides) that float within the liquid disordered bilayer of cellular membranes.[97] Lipid rafts function in several pivotal cellular processes, such as membrane trafficking, cell polarization and signaling, as well as in the process of pathogen invasion, lipid homeostasis, angiogenesis, and neurodegenerative diseases.[98,99] Because of the presence of NEU3 at the cell surface and its specificity toward gangliosidic substrates, the enzyme could be involved in the regulation, maintenance, and even formation of these specialized areas of the membranes. The first data supporting an association of NEU3 with lipid raft markers were produced in 2001.[100] Later on, a co-fractionation of NEU3 with caveolin-1 in low-density Triton-X-100-insoluble membrane fractions was demonstrated for both the endogenous and overexpressing enzyme, using HeLa cells and COS1 cells, respectively.[101] The tight association between NEU3 and caveolin-1 has been demonstrated by co-purification of both polypeptides using affinity chromatography and a His-tagged form of NEU3, as well as by co-immunoprecipitation experiments using an anti-caveolin-1 antibody.[101] The enzyme was found not only in membrane microdomains associated with caveolin but also in tetraspanin-enriched microdomains (TERMs).[102] Tetraspanins are a family of membrane proteins involved in membrane compartmentalization and dynamics;[103,104] these proteins provide, through self-interactions, platforms for recruiting other proteins such as integrins, tyrosine kinases, and G-protein-coupled receptors into TERMs.[105,106] Overexpression of NEU3 in HB2 mammary epithelial cells affected the stability of CD82-containing TERMs. Indeed, the association between tetraspanin CD82 and CD151 was diminished, and, this in turn, affected the EGF-induced signaling. Overall, these results demonstrated the relevance of gangliosides, as NEU3 substrates, in maintaining the integrity and functionality of CD82-enriched microdomains and suggested a role of NEU3 as possible physiological regulator of these membrane structures.

Furthermore, the involvement of NEU3 in the regulation of PDGF-induced Src mitogenic signaling has been demonstrated.[107] Caveolin-enriched membrane microdomains regulate the association of Src family protein tyrosine kinases (SFK) with platelet-derived growth-factor receptor (PDGFR),[108] which, in turn, represents a pivotal site for kinase activation and cell-cycle progression. Experimentally induced variations of intracellular levels of PAG (phosphoprotein associated with GEMs), a member of the transmembrane adaptor protein family,[109] enhance NEU3 activity, leading to GM1 accumulations and PDGFR exclusion from *caveolae*. This, in turn, hampers Src mitogenic signaling and the induction of DNA synthesis. Again, the biological action of NEU3 is exerted through its activity on gangliosides, which are essential structural and functional components of membrane microdomains.

b. NEU3 and Cell Differentiation.—The first evidence concerning a possible role of NEU3 in cell differentiation was obtained in human neuroblastoma cells several years before the cloning of the corresponding cDNA.[110] The molecular characterization of NEU3 allowed more-detailed studies, as reported by Hasegawa *et al.* using Neuro2a mouse neuroblastoma cells.[111] Experimentally induced differentiation of these cells increased NEU3 expression, whereas the stable overexpression of the enzyme led to accelerated neurite arborization. Important details on the role played by NEU3 in axonal growth/regeneration and neuronal polarization were provided by studies on cultured rat hippocampal neurons.[112,113] The enrichment of NEU3 at the tip of one growth cone of unpolarized neurons was shown to influence the spatially restricted regulation of proteins involved in axon outgrowth that, in turn, led to depolymerization of actin defining axonal fate. The cascade of events start with the local activation of TrkA, which triggers PI3K- and

Rac1-dependent inhibition of RhoA signaling, which finally results in reduced actin stability. Interestingly, this NEU3-mediated regulatory mechanism appears to be dependent on the formation of GM1, a potent modulator of intracellular signaling, which is promoted by the enzyme activity on polysialylated gangliosides,[96,111] possibly generating GM1-enriched membrane domains.[114,115] Support for this notion has been provided by experiments showing that the overexpression of NEU3 in neuroblastoma Neuro2a cells elicits neuritogenesis, whereas NEU3 silencing inhibits neurite formation.[116] In contrast to this evidence are the results obtained from experiments carried out on the same cells, which show that a reduction of NEU3 by mRNA interference is followed by a change of the ganglioside pattern with marked decrease of GM1 and with concomitant enhancement of neurite elongation.[117] A reason for this apparent discrepancy has not yet been found. However, since GM1 is present at basal levels in the plasma membrane of virtually all cells, we hypothesize that the activation of neuritogenesis is induced by a modulation in the level of GM1, either up or down, specifically at the site of neurite formation. A better understanding of this mechanism and the role played by NEU3 may be obtained when potential interactions between GM1, NEU3, and other neuritogenic effectors in the plasma membrane have been identified.

New evidence about the role played by NEU3 in cell differentiation has been obtained in muscle cells.[118] The constitutive silencing of the enzyme with shRNA in murine C2C12 myoblasts induced: (i) complete inhibition of the differentiation processes; (ii) massive apoptosis under differentiation conditions; and (iii) EGFR inhibition and down-regulation as a consequence of the markedly increased levels of endogenous ganglioside GM3, a substrate for NEU3. The direct involvement of this ganglioside in the effects observed was demonstrated by the reduction of the differentiation capability of wild-type C2C12 cells cultured in a medium supplemented with GM3. Furthermore, when NEU3-silenced myoblasts were co-cultured with wild-type C2C12 cells, they re-acquired the ability to fuse differentiating into MHC-expressing myotubes. Since the activity of NEU3 toward ganglioside substrates inserted into the membranes of neighboring cells has been already demonstrated in COS7 cells,[96] these results could be explained by the decrease of GM3 ganglioside content in NEU3-silenced myoblasts as a consequence of the trans-hydrolyzing activity of the enzyme present on the cell surface of co-cultured wild-type C2C12 cells.

NEU3-silencing experiments have also been carried out on chronic myeloid leukemic K562 cells.[119] In this cell line, the decrease in NEU3 levels caused: (i) alterations in the molecular machinery involved in proliferation and apoptosis control, leading to a decrease in cell growth and resistance to apoptotic stimuli, and (ii) establishment of a new ganglioside pattern with a remarkable increment of GM3. This ganglioside is likely to be responsible for the activation of PKC, MAPK, and JNK signaling cascades, which are known to be involved in megakaryocytic differentiation of K562 cells.[120] Similar results were obtained when megakaryocytic differentiation of K562 cells was induced by treatment with phorbol 12-tetradecanoate 13-acetate (PMA).[121] In this case, PMA treatment led to an increase of GM3 synthase (hST3Gal V) and a decrease in NEU3 transcripts. The resulting increase in cellular GM3 cell content[122] activated the PCK/ERK/p38 MAPK signaling cascade, resulting in formation of the CD41b (GPIIb) surface antigen and megakaryocytic differentiation. In contrast, the overexpression of NEU3 led to a decrease of ganglioside sialic acid content, and the cellular levels of GM3 became inadequate to activate the aforementioned signaling cascade. This, in turn, resulted in the inhibition of megakaryocytic differentiation of K562 cells.[121]

Another interesting effect of NEU3 on ganglioside substrates, in particular GM3, was observed in human fibroblasts.[123] When these cells overexpress the murine form of NEU3, they produce roughly six times more ceramide as compared to mock-transfected cells. Presumably, the increase in ceramide level is derived from NEU3-mediated conversion of GM3 into LacCer, which is subsequently further degraded to GlcCer and, finally, to Cer by β-galactosidase and β-glucosidase, respectively. Thus, in human fibroblasts the overexpression of NEU3 results, through the sequential action of downstream hydrolases, in the formation of ceramide, a well known bioactive sphingolipid,[124] which in turn leads to decreased cell proliferation and increased sensitivity to apoptosis.

c. NEU3 and Virus Infection.—The well-established presence of sialidase(s) or neuraminidase(s) on the envelope of influenza viruses, and its role in the virus penetration into cells, led to the discovery of specific inhibitors of the viral enzyme(s), which constitute, at the moment, the forefront tool to fight influenza pandemics.[125] On the other hand, much less is known about a possible involvement in viral biology of the endogenous sialidase(s) present at the surface of mammalian cells that are a target of the virus particles. For example, Newcastle Disease Virus (NDV) belongs to the *Paramyxoviridae* family, and is the causative agent of a contagious bird disease that causes severe economic losses in industrial poultry breeding.[126] *Paramyxoviridae* membranes contain two transmembrane glycoproteins,[126] the fusion (F) protein directly responsible for the fusion of viral envelope with the target cell membrane, and the multifunctional hemagglutinin-neuraminidase (HN) that binds to and cleaves the sialic acid receptor molecules exposed on the host cell surface.[127] Among the sialic acid-containing molecules acting as receptors, gangliosides have been reported to interact with NDV.[128] Therefore, it may be reasoned that an extensive modification of the cell-surface ganglioside pattern induced by the overexpression of NEU3[96] may interfere with the virus–host interaction. In this perspective, COS7 cells overexpressing murine NEU3 have been used as a model system to study the behavior of NDV.[129] The overexpression of the enzyme had no significant effect on NDV binding to cells itself, whereas it caused a significant decrease of NVD infection, as well as viral propagation through cell–cell fusion. In addition, it was shown that HN acted directly on ganglioside GD1a of host cells, converting it to GM1, and that the viral particles were preferentially associated with the detergent-resistant microdomains (DRMs) of COS7 cells. These results shed some light on the complex series of events governing virus–host cell interactions, and suggest a role in this process of the endogenous enzymes on the host's plasma membrane, which are involved in the fine regulation of ganglioside levels.

V. The Particulate Sialidase NEU4

1. General Properties of NEU4

The fourth member of the mammalian sialidase family, NEU4, has been identified by searching sequence databases for entries showing homologies to human cytosolic sialidase NEU2.[4,130,131] One peculiar feature of this enzyme is the presence of a long stretch of about 80 amino acids (aa 294–373) that appears unique among mammalian sialidases,[131] and teleosts.[132] Noteworthy is the fact that human NEU4 is present in two isoforms, differing in the presence or absence of a 12 amino acid stretch at the N-terminus.[133] Surprisingly, the long isoform, named NEU4L, which was also described by Seyrantepe *et al.*,[134] showed a mitochondrial localization,[133] whereas the short isoform, named NEU4S, was associated with intracellular membranes, as previously reported.[131] Thus, the N-terminal portion of NEU4L functions as a mitochondrial targeting sequence, as demonstrated by the import into mitochondria of an engineered fusion product containing the Enhanced Green Fluorescent Protein (EGFP), which served as a reporter, fused to the N-terminus region (aa 1–33) of NEU4L.[133] Subsequent experiments showed that NEU4L is bound to the outer mitochondrial membrane and NEU4S to the endoplasmic reticulum, and that both enzymes are extrinsic membrane proteins, anchored via protein–protein interactions.[135] Both isoforms show broad substrate specificity toward sialoglycoconjugates, including gangliosides.[94] Interestingly, the mitochondrial localization of NEU4L suggests that the protein could be involved in the pathways of mitochondrial apoptosis through the regulation of the level of apoptogenic ganglioside GD3 within the organelle.[136] The link between NEU4L, ganglioside GD3, and apoptosis has been further supported by a study on SH-SY5Y neuronal cell-lines expressing human tyrosinase under the control of exogenous inducers.[137] Indeed, catechol-oxidized metabolites that are generated by tyrosinase cause apoptotic neurodegeneration characterized by release of cytochrome into the cytosol. The mitochondrial damage is coupled to the trafficking of GD3 to mitochondria and a dramatic decrease of NEU4L expression that starts at the beginning of the apoptotic process. Although these results suggest that NEU4L and one of its substrates, GD3 ganglioside, are involved in apoptosis, it is surprising that this mitochondrial isoform has been identified only in humans[133] and not in rodents[116,130] and teleosts.[132]

2. Functional Implication of NEU4

a. NEU4 and Lysosomes.—In addition to its intracellular membrane and mitochondrial membrane localization, human NEU4 has also been described as an enzyme of the lysosomal lumen, with wide substrate specificity, from glycoproteins to gangliosides and oligosaccharides.[134] Moreover, the same authors reported that overexpression of NEU4 in cultured fibroblasts of patients with sialidosis and galactosialidosis led to clearance of the stored materials in lysosomes, indicating that the enzyme is somehow able to complement the defect of the lysosomal sialidase NEU1.[134] This rescue activity exerted by NEU4 on NEU1-deficient cells has been further sustained by another paper from the same research group.[138] First of all, neuroglia cells from Tay-Sachs patients were transfected with *NEU4*, resulting in the clearance of accumulated GM2 ganglioside. This would imply that NEU4 has the ability to hydrolyze GM2: in this respect it would be interesting to ascertain whether NEU4, as demonstrated for NEU2,[59] is capable of hydrolyzing monomeric dispersions of such monosialogangliosides as GM2 and GM1. Seyrantepe and colleagues also generated *in vitro* and *in vivo* loss-of-NEU4 models in HeLa cells using RNA interference, and in mice by the targeted disruption of *Neu4* gene, respectively. As expected, NEU4-silenced HeLa cells are characterized by diminished enzyme activity, measured both on the artificial fluorescent substrate 4MU-NeuAc and on a mixture of gangliosides. In addition, cell loading with a ganglioside mixture from pig brain led to extensive vacuolization, but only in the silenced cells. Electron microscopy of these cells revealed the presence of large, heterogeneous lysosomes filled with lamellar structures, thus supporting the notion of an impaired lysosomal degradation due to lowered *Neu4* expression. Surprisingly, *Neu4* knockout mice appeared to be healthy, fertile, and without consistent gross abnormalities. Detailed tissue analysis revealed the presence of vacuolization and lysosomal storage only in cells of the lung and spleen cells, which was accompanied by increased levels of polyunsaturated fatty acids, ceramide, and cholesterol. In addition, analysis of brain gangliosides of *Neu4* knockout mice revealed an increased level of GD1a and a decrease of GM1, suggesting a possible involvement of the enzyme in the ganglioside desialylation processes within the central nervous system.

b. NEU4 and Cancer.—The involvement of NEU4 in the complex series of phenomena leading to cancerous transformation has been studied in colon cancer.[139] The main isoform found in these cells is NEU4S, and its corresponding transcript is decreased in tumors as compared to adjacent non-tumor mucosa. However, in cultured human colon cancer cells, the enzyme is up-regulated during the early stages of apoptosis and its overexpression leads to accelerated apoptosis, and to a decreased invasion and motility. As expected, the use of siRNA specific for *NEU4* transcript caused a remarkable inhibition of apoptosis and promoted cell motility and invasion. Intriguingly, no significant variations in the glycolipid pattern were detectable upon experimental induction of *NEU4*, albeit lectin-blot analysis revealed the presence of desialylated forms of glycoproteins having a molecular weight of about 100 kDa. Thus, in colon cells, NEU4 seems to be involved in the maintenance of normal mucosa acting on glycoprotein substrate(s) and, conversely, its down-regulation could contribute to invasive properties typical of colon cancerous cells.

c. NEU4 and Neuronal Cell Differentiation.—A recent paper has discovered the potential involvement of NEU4 in the regulation of neuronal cell differentiation.[116] The murine *Neu4* gene is also transcribed into two alternatively spliced variants, named NEU4a and NEU4b, corresponding to the human NEU4L and NEU4S, respectively.[133] However, contrary to what has been described for the two human isoforms, the two corresponding murine NEU4 isoforms are both partially localized in the calnexin-positive endoplasmic reticulum membranes, without any apparent distribution in other membranous structures and/or organelles, namely mitochondria and lysosomes. The analysis of the primary structure of human and murine NEU4 isoforms revealed a high degree of amino acid identity, with significant variations in the long stretch of about 80 amino acids first described in human NEU4S (residues 284–373).[131] In addition, in the case of human NEU4L, the short stretch of N-terminal residues contains the consensus sequence responsible for the mitochondrial targeting of the enzyme,[133] whereas in the mouse counterpart the additional 23 amino acid stretch does not appear to contain any canonical targeting signals. The two murine isoforms differ in their expression levels and tissue distribution, as well as in their expression pattern during development. Moreover, the substrate specificities of the two isoforms are different, with the more abundant short NEU4b isoform being more active toward ganglioside and glycoprotein substrates as compared to the long NEU4a isoform. Interestingly, NEU4a seems to act on polysialylated NCAM, one of the molecules involved in neural development and plasticity,[140] whereas the short isoform, NEU4b, when overexpressed in Neuro2a cells, suppresses neurite formation upon treatment with retinoid acid.[116] Conversely, NEU4b knockdown accelerates neurite formation, a phenomenon that seems to be related to prevention of desialylation of a yet unidentified 95 kDa glycoprotein.

VI. Sialidases and Cancer

As discussed in previous sections of this survey, during the past few years many of the scientific papers published on mammalian sialidases pointed out the potential involvement of sialidases in the occurrence of various kinds of tumors. A comprehensive picture of the results obtained thus far has already been presented in a series of reviews.[46,141–144] Therefore, here we give only an overview of the most recent contributions. Briefly, human NEU3 is up-regulated in human colon cancer[145] and *in vitro* experiments demonstrate that, depending on the composition of the extracellular matrix, the enzyme differentially regulates cell proliferation through integrin-mediated signaling.[146] Interestingly, when compared to normal mucosa, colon-cancer tissues are characterized by a significant increase of LacCer.[145] LacCer is produced by the action of NEU3 on GM3, and plays important roles as a signaling molecule.[147] In addition, NEU3 interacts directly with signaling molecules such as the EGF receptor (EGFR), as demonstrated by co-immunoprecipitation experiments. In cancer cells, NEU3 suppresses apoptosis by increasing phosphorylation of EGFR and activation of the Ras/ERK pathway.[47] The *in vivo* tumorigenic potential of NEU3 has been tested in *Neu3* transgenic mice as the model.148,[149] The induction of aberrant crypt foci (ACF) by injection of azoxymethane [MeN=(N→O)Me] in these mice demonstrates that up-regulation of NEU3 is an important event in the promotion of colon carcinogenesis. As already observed in human colon cancer,[145] the ganglioside pattern of transgenic mucosa shows a decrease in GM3 and an increase in LacCer, which may contribute to the increased occurrence of ACF in azoxymethane-treated *Neu3*-transgenic mice. Upon azoxymethane treatment, NEU3-induced variations in the levels of GM3 and

LacCer enhance EGFR activation, which in turn, leads to the phosphorylation of Akt and ERK and the up-regulation of the pro-survival protein Bcl-xL. Overall, this molecular machinery results in the reduction of apoptosis in colonocytes, so that these surviving damaged cells develop ACF that may eventually progress into tumors.

Another interesting report concerns about the role of NEU3 in acute lymphoblastic leukemia (ALL).[150] These authors describe that, in ALL lymphoblasts, NEU3 decreases about 25–40%, both in terms of gene expression and activity toward ganglioside GD1a, as compared to healthy control cells. Overexpression of NEU3 in the MOLT4 ALL cell-line led to a decrease in the percentage of cells that are positive for the ALL marker 9-OAcGD3, and guided the lymphoblast cells to apoptosis. These phenomena are possibly linked to a decrease in the level of GD3, concurrently with an increase of ceramide content, as a direct consequence of NEU3-degrading activity toward cellular gangliosides. Thus, it appears that in ALL, NEU3 exerts a similar role to that observed in human NEU3-overexpressing fibroblasts,[123] but a clearly opposite role to what was observed in certain carcinomas[46] and chronic myeloid leukemic cells,[119] which show increased expression of the membrane-associated sialidase. The correlation between decreased NEU3 expression and leukemic disease progression was confirmed in ALL patients. Actually, NEU3 activity increases in clinical remission after chemotherapy, but decreases again in relapsed patients.[150] Whether the expression and activity of NEU3 is similarly affected in other types of leukemia is yet unknown; however, NEU3 may ultimately be considered as another biomarker in the diagnosis of ALL.

VII. Sialidases and Immunity

There is evidence for an involvement of sialidase(s) in the complex cellular network of immunity in higher organisms. It has been shown that, during bone marrow differentiation, myeloid precursors highly express sialoglycoconjugates. The latter are remodeled during maturation of the precursors with the concurrent release of sialic acid and gain in plasticity and mobility required to reach the peripheral circulation. Once in the peripheral circulation, the activation of neutrophils is associated with a further decrease of exposed sialic acid, resulting in an additional loss of negative charge on the cell surface.[151] *In vivo*, the recruitment of murine neutrophils into inflammatory sites is inhibited by treatment with a polyclonal antibody raised against *Clostridium perfringens* sialidase.[152] This treatment was shown to: (i) decrease the sialidase activity detectable in neutrophil cell lysates; (ii) increase cell surface expression of a target protein, possibly a membrane-bound form of sialidase, upon IL-8 *in vitro* and *in vivo* stimulation of human and murine neutrophils; (iii) inhibit both pulmonary leukostasis and transendothelial migration of neutrophils in mice. Western-blot analysis of polymorphonuclear lymphocytes and granules with anti-clostridial sialidase antibody revealed the presence of five protein bands, with molecular weights ranging from approximately 46 to 89 kDa. Based on the molecular weight of the four different mammalian sialidases,[3,4] the 46-kDa protein present at the cell surface (as demonstrated by the flow cytometry experiments), this protein might represent either a short variant of the plasma membrane-associated sialidase NEU3 or the lysosomal sialidase NEU1. Further experiments are needed in order to verify this hypothesis.

The relevance of cell sialioglycoconjugates on the surface of neutrophils has been further substantiated by an *in vitro* study in which their adhesion to, and migration across, a monolayer of pulmonary vascular endothelial cells were tested upon treatment with *Clostridium perfringens* sialidase.[153] Trimming of sialic acids from the cell surface of neutrophils and endothelial cells by the exogenous sialidase increased their adhesion and migration capability of the cells without significant variations in the expression of adhesion molecules like ICAM-1 (CD54) and E-selectin (CD62E) or the release of IL-8 and IL-6. Accordingly, the inhibition of endogenous sialidase activity on the cell surface of neutrophils obtained by the addition of either a polyclonal antiserum against the *C. perfringens* enzyme or by the sialidase competitive inhibitor Neu5Ac2en led to a significant inhibition of the adhesion of activated neutrophils to endothelial cells. In addition, the cell-surface sialidase recruited on neutrophils upon their activation[152] acts directly on endothelial cells, desialylating their cell-surface sialoglycoconjugates. This sialidase *trans*-activity* first demonstrated in COS7 cells overexpressing NEU3[96] and later in C2C12 myoblasts,[118] produces a condition of hyperadhesivity for neutrophils and their eventual extravasation.

The possible role of sialidase(s) has been studied in other immune cells, such as monocytes and T cells. Peripheral blood monocytes can differentiate into either dendritic cells or macrophages.[154] Desialylation of monocyte cell-surface glycoconjugates obtained by treatment with bacterial sialidase leads to the activation of the ERK 1/2 pathway, resulting in the production of specific cytokines and enhancement of the monocyte response to bacterial lipopolysaccharide.[155] Real-time PCR analysis of the expression levels of endogenous sialidases during monocyte differentiation into macrophages revealed a consistent increase of *NEU1* and *NEU3* transcripts, but curiously a decrease of *NEU4* mRNA[156] and a complete absence of *NEU2* transcripts.[156]

---

* In this context the term "*trans*" means the ability of the enzyme, exposed on the surface of a cell, to affect sialoglycoconjugates present on the surface of a vicinal cell, in contrast with the term "*cis*" indicating the ability to affect the sialoglycoconjugates present on the membrane of the same cell.

Interestingly, most of the sialidase activity detectable after differentiation is abolished upon immunoprecipitation with a polyclonal rabbit antiserum raised against recombinant NEU1,[157] suggesting a predominant role of this sialidase in the regulation of cell-surface sialoglycoconjugates during monocyte differentiation. A detailed study on NEU1 during differentiation of both human monocytes and the acute monocytic leukemia cell line THP-1 into macrophages demonstrates that most of the NEU1 migrates from the lysosomes to the cell surface.[158] These authors show that the *NEU1* promoter is activated during monocyte differentiation and the newly synthesized NEU1 translocates, together with cathepsin A, from lysosomes to the plasma membrane via major histocompatibility complex II-positive vesicles. In addition, the inhibition of NEU1, either by using NEU1-specific siRNA or anti-NEU1 antiserum results in a reduction of the cell's ability to phagocytose bacteria or to produce such cytokines as IL-1b, IL-6, and TNF-α.

In the case of T-lymphocytes, stimulation via the T-cell receptor leads to the induction of both NEU1 and NEU3.[159] Overexpression experiments of the two enzymes suggest their involvement of NEU1 in the production of various cytokines by Th1 and Th2 cells, and overexpression of NEU3 was shown to stimulate cytokine gene expression (except for IL-4). Similar experiments have been carried out by a different research group, which obtained slightly different results.[160] As just mentioned, only *NEU1* and *NEU3* transcripts are detectable in lymphocytes, with the lysosomal sialidase being the predominant form. Activation of lymphocytes using anti-CD3 and anti-CD28 immunoglobulins leads to a significant increase of *NEU1* transcript and specific activity of the enzyme, but has no effect on *NEU3* expression.

In 2008, the involvement of sialidases in human myeloid cell differentiation and neutrophil adhesion to dendritic cells was reported.[161] The expression of Lewis X (Le^x or CD15) is known to be associated with these processes and its trisaccharide unit exposed on cell-surface glycoconjugates is a useful marker for detection of granulocytic cells along the myeloid differentiation lineage.[162] The authors provide evidence that increased expression of Le^x depends on the removal of a sialic acid group linked through an α–(2→3) linkage to the galactose residue of the trisaccharide (sLe^x or sialyl-CD15) and not to an increase of its biosynthetic rate. This unmasking activity correlates with an increase in the expression of NEU1 rather than of NEU3, whose transcript levels remain constant. These results highlight the effect of glycosidases, and specifically sialidase(s), in the fine regulation of cell-surface carbohydrate composition, and emphasize the intriguing biological relevance of sialidase(s) intracellular trafficking.

NEU1 was also found to be a key regulator of toll-like receptors (TLRs) activation. These receptors are located at the cell surface of immune cells and play a key role in the innate immune system. It was shown that NEU1 forms a complex with TLR-2, -3, and -4 receptors on the cell surface of naive and activated macrophages and its activity appears to regulate the initial mechanism of TLR activation and subsequent cell function.[163] Ligand binding to TLR activates Neu1 by a yet unknown mechanism. These authors' hypothesis is that Neu1 may hydrolyze sialic acid residues of TLRs, which, in turn, unmasks its ectodomain and allows TLR dimerization and activation.

Overall these data strongly support the involvement of NEU1 in cellular immune responses, and identify novel function(s) that are gained because of migration of the enzyme from its original intracellular lysosomal localization to the cell surface, where a variety of biological processes take place that are based on specific molecular recognition. A question that remains unanswered is whether, besides NEU1, the NEU1-specific accessory protein cathepsin A plays a direct or supporting role in cellular immune responses, and whether NEU3, already present on the cell surface, and NEU1 can complement each other in these processes.

VIII. Further Evidence for Possible Functional Implications of Sialidases

Several biological processes are known to involve sialidase activity, but attribution of this activity to any of the four mammalian enzymes is still lacking. A direct link between activation of the TrkA tyrosine protein kinase receptor for nerve growth factor (NGF) and sialidase(s) has been described in PC12 rat pheochromocytoma cell-lines and primary cortical neurons.[164] Upon NGF stimulation, a sialidase activity is induced that specifically removes sialic acid residues that are α–(2→3) linked to an α-galactose residue of the receptor. This trimming action constitutes an early step for receptor dimerization, internalization, and final activation, resulting in neurite outgrowth in both cell types. Several points need further clarification, including which mammalian sialidase is involved in the desialylation of TrkA. However, studies on the mechanism of TrkA increased expression on the cell surface, and the mode of recognition/activation of a sialidase toward this particular membrane-bound receptor upon NGF stimulation suggest a new function for sialidase(s) as a key regulator of receptor activation by its ligand.

The importance of sialidase(s) in axon outgrowth has been shown in an *in vivo* animal model.[165] It is well known that, because of the presence of axon regeneration inhibitors (ARIs) that accumulate at the sites of injury in the central nervous system, axon outgrowth is in part inhibited, thus severely limiting the recovery from traumatic injury.[166] Conversely, peripheral nerve sheaths support axon outgrowth,[167] thus peripheral nerve graft implantation into the central nervous system offers a potential therapeutic application in the treatment of brachial plexus avulsion.[168] To improve functional recovery after brachial plexus injury, a combination of surgical reconnection and molecular therapies based on ARI blockers has been explored and, in this perspective, the effect of sialidase activity has been tested. Among ARIs, myelin-associated glycoprotein (MAG), belonging to the Siglec family of sialic acid-binding lectins,[169] is supposed to inhibit axon regeneration by binding to axonal gangliosides GD1a and GT1b,[170,171] both potential substrates for sialidase. Using a rat model of brachial plexus avulsion, infusion of *C. perfringens* sialidase to the injury site leads to a 2.6-fold increase of the spinal axon growth into the graft. These results suggest that sialidase treatment elicits two effects: (i) desialylation of GD1a and GT1b causes release of MAG, thus removing its blocking action, and (ii) GM1 formed by desialylation of GD1a and GT1b could by itself exert locally protective and neuritogenesis effects.[172] It is not known whether an increase of endogenous sialidase activity may lead to the same results.

A study has analyzed sialidase activity in normal and atherosclerotic human aortic intima.[173] Although no information is provided on the expression of single sialidase genes in the intima, measurements of sialidase activity by ELISA assays[174] demonstrated an increased sialidase level in the intima areas with atherosclerotic lesions as compared with unaffected ones. It is tempting to speculate on the biological significance of these data, considering the ganglioside and glycoprotein modification associated to the deposition of the atherosclerotic plaque,[175] and the demonstration of a significant decrease of elastin concentration in *Neu1*-null mice.[43] In this respect, an increased level of sialidase, probably NEU1, may represent an attempt to bypass the vessel alteration induced by the atherosclerotic plaque.

In general, the altered extracellular matrix proteins and the abnormal growth of vascular smooth muscle cells (VSMC) constitute important features of such vascular diseases as atherosclerosis and restenosis after angioplasty. For example, TNF-α is released by macrophages in atherosclerotic lesions and by VSMC after balloon injury of the vessel.[176] Based on the possible involvement of sialidase(s) in intima thickening[173] and in the binding of fibrinogen and low-density lipoproteins to the vessel wall,[177] the effect of NEU3 overexpression on mouse aortic VSMC has been investigated.[178] These studies indicated that NEU3 overexpression has no effect on the rate of cell proliferation, but instead leads to inhibition of the TNF-α induction of the matrix metalloproteinase-9 (MMP9). This inhibitory effect seems to derive from a down-regulation of the corresponding gene, as a result of the binding of NF-κB and activation of protein-1 to the *MMP9* gene promoter. Thus, *NEU3* expression may play a role in modulating VSMC responses to TNF-α and, through inhibition of *MMP9*, in favoring atherosclerotic plaque instability.

The presence of sialidases on membranes of the nuclear envelope was demonstrated in 2009 in two tumor cell-lines of neuronal origin.[179] The first evidence of a sialidase active toward various ganglioside substrates and *N*-acetylneuraminyllactitol in this particular membranous structure was obtained in a purified nuclear-membrane preparation from adult rat brain.[180] Using a procedure for separation of the nuclear membranes[181,182] and a combination of Western-blot and immunocytochemistry techniques, Wang and colleagues showed that NEU1 is present on the outer membrane and NEU3 on the inner membrane of the nuclear envelope.[179] Both enzymes exhibited the ability to hydrolyze *in vitro* endogenous and exogenous ganglioside GD1a into the corresponding GM1 at pH 5.5. The authors hypothesized that the functional role of this nuclear fraction of NEU3 is to maintain an optimal level of the GM1 for regulating nuclear Ca^{2+} homeostasis and cell viability. The Na^{+}/Ca^{2+} exchanger (NCX) is actually localized on the inner membrane of the nuclear envelope and its activity is potentiated through the formation of a NCX/GM1 complex.[182] Although several questions are still open, such as the possible role of NEU1 on the external membrane of the nuclear envelope and the sorting mechanism responsible for this peculiar topology of the two proteins, these data further support the notion of sialidases as multipurpose enzymes, depending on their sub-cellular localization and substrate specificity/availability in a particular area of the cell.

IX. *In Silico* Analysis of Sialidase Gene Expression Patterns

A few studies have been performed on sialidases that describe the expression of these genes at the transcriptional level. Reports presenting the identification of *NEU* genes in man or in the mouse mostly included RNA expression data based either on Northern Blot or RT-PCR analyses (as reviewed in Refs. 3 and 4, as well as in the original papers cited in the different sections of this article). More recent papers provide detailed expression studies of murine sialidases carried out by quantitative RT-PCR in adult tissues and in the developing brain, together with RNA *in situ* hybridization in adult brain.[116,138] These studies can now be complemented by the wealth of public data generated by a number of systematic studies on the transcriptome of *Homo sapiens* and other vertebrates, which allow researchers to infer the expression pattern of the gene(s) of interest, using *in silico* strategies.[183]

For instance, a survey of tissue expression levels of *NEUs* can be performed using UniGene's EST Expression Profile Viewer that is hosted by the NCBI. This analysis is based on *in silico* gene-expression profiling of Expressed Sequence Tag (EST) organized in discrete gene clusters in the UniGene database (http://www.ncbi.nlm.nih.gov/sites/entrez?db=unigene).[183] Each UniGene cluster is assessed for the total number of ESTs from a specified tissue or cell type and the total number of ESTs in the cluster. The ratio of the two gives a measure of the level of expression (in transcripts per million) of the gene in that biological sample. Despite some limitations, the availability of more than 8 million human and 4.8 million mouse EST in GenBank (dbEST division) makes the analysis of EST profiles one of the most important strategies for the study of gene expression.

As shown in Table I, *NEU1* is by far the highest expressed sialidase gene in both man and mouse. The EST Expression Profile Viewer analysis reported in Figs. 7 and 8 indicates that the gene is mostly ubiquitously but differentially expressed, both in normal and pathological samples. The three other *NEU* genes appear to be expressed at similar levels in the mouse, based on total EST counts. Remarkably, in man there is a complete absence of *NEU2* transcripts in dbEST, which indicates that the gene is expressed at extremely low levels or, alternatively, it is transcribed mainly in tissues that have not been sampled by the EST generation efforts. On the other hand, the presence of 24 *Neu2* ESTs in mouse tissues indicates that this enzyme is more expressed in rodents than in man. The expression profiling data presented in Fig. 7 suggest that, in contrast to *NEU1*, human *NEU3* and *NEU4* are both transcribed at low levels and in a tissue-specific manner. The two genes also appear to be transcribed in a number of cancer tissues (Fig. 8). In particular, about one-third of the total number of *NEU4* ESTs in dbEST have been obtained from gliomas, indicating that this gene is unusually highly expressed in tumors of glial origin. These data have been experimentally confirmed by quantitative Real-Time PCR on a panel of brain tumors and the significance of these intriguing findings is currently being investigated (G. Borsani and E. Monti, unpublished results).

Using an alternative approach, the tissue-specific RNA expression pattern of a gene of interest can be evaluated by using the GeneAtlas Gene Expression Database developed by the Genomics Institute of the Novartis Research Foundation.

TABLE I

Total Number of Sialidase Gene ESTs in the UNIGENE Database as of December 2008

	Homo sapiens	*Mus musculus*	*Rattus norvegicus*
NEU1	672	416	29
NEU2	0	24	13
NEU3	31	16	5
NEU4	79	19	1

EST Pool Name	*NEU1* TPM	*NEU1* SI	*NEU1* GE/TE	*NEU3* TPM	*NEU3* SI	*NEU3* GE/TE	*NEU4* TPM	*NEU4* SI	*NEU4* GE/TE
adipose tissue	0		0 / 13159	0		0 / 13159	0		0 / 13159
adrenal gland	270		9 / 33324	0		0 / 33324	0		0 / 33324
ascites	0		0 / 40067	0		0 / 40067	0		0 / 40067
bladder	132		4 / 30121	0		0 / 30121	0		0 / 30121
blood	64		8 / 123959	0		0 / 123959	8		1 / 123959
bone	111		8 / 71802	0		0 / 71802	0		0 / 71802
bone marrow	20		1 / 48949	40		2 / 48949	0		0 / 48949
brain	75		83 / 1104257	1		2 / 1104257	46		51 / 1104257
cervix	164		8 / 48506	0		0 / 48506	0		0 / 48506
connective tissue	60		9 / 149531	0		0 / 149531	0		0 / 149531
ear	0		0 / 16345	0		0 / 16345	0		0 / 16345
embryonic tissue	27		6 / 215831	13		3 / 215831	0		0 / 215831
esophagus	247		5 / 20212	98		2 / 20212	0		0 / 20212
eye	47		10 / 211771	0		0 / 211771	4		1 / 211771
heart	66		6 / 90351	0		0 / 90351	0		0 / 90351
intestine	161		38 / 235458	0		0 / 235458	29		7 / 235458
kidney	221		47 / 212588	0		0 / 212588	9		2 / 212588
larynx	0		0 / 24438	0		0 / 24438	0		0 / 24438
liver	110		23 / 208370	0		0 / 208370	33		7 / 208370
lung	85		29 / 338117	2		1 / 338117	5		2 / 338117
lymph	22		1 / 44401	0		0 / 44401	0		0 / 44401
lymph node	32		3 / 91923	10		1 / 91923	0		0 / 91923
mammary gland	278		43 / 154363	32		5 / 154363	19		3 / 154363
mouth	29		2 / 67225	0		0 / 67225	0		0 / 67225
muscle	27		3 / 108174	0		0 / 108174	0		0 / 108174
nerve	63		1 / 15820	0		0 / 15820	0		0 / 15820
ovary	224		23 / 102623	0		0 / 102623	0		0 / 102623
pancreas	69		15 / 215311	0		0 / 215311	0		0 / 215311
parathyroid	145		3 / 20643	0		0 / 20643	0		0 / 20643
pharynx	0		0 / 41507	0		0 / 41507	0		0 / 41507
pituitary gland	60		1 / 16651	0		0 / 16651	0		0 / 16651
placenta	214		61 / 283990	0		0 / 283990	0		0 / 283990
prostate	99		19 / 190802	0		0 / 190802	0		0 / 190802
salivary gland	0		0 / 20279	0		0 / 20279	0		0 / 20279
skin	264		56 / 211591	0		0 / 211591	0		0 / 211591
spleen	110		6 / 54068	0		0 / 54068	0		0 / 54068
stomach	51		5 / 97185	0		0 / 97185	0		0 / 97185
testis	21		7 / 331358	9		3 / 331358	0		0 / 331358
thymus	24		2 / 81242	36		3 / 81242	0		0 / 81242
thyroid	166		8 / 47936	0		0 / 47936	0		0 / 47936
tonsil	176		3 / 17024	0		0 / 17024	0		0 / 17024
trachea	38		2 / 52435	19		1 / 52435	0		0 / 52435
umbilical cord	0		0 / 13767	0		0 / 13767	0		0 / 13767
uterus	68		16 / 233959	8		2 / 233959	4		1 / 233959
vascular	38		2 / 51954	0		0 / 51954	0		0 / 51954

FIG. 7. Expression profile suggested by analysis of EST counts: breakdown by body sites. EST profiles show approximate gene expression patterns as inferred from EST counts and the cDNA library sources (as reported by sequence submitters). Libraries known to be normalized, subtracted, or otherwise biased have been removed. Since no ESTs for the *NEU2* gene are found in dbEST, the gene is not represented in the figure. TPM: transcripts per million; SI: spot intensity based on TPM; GE/TE: gene EST/total EST in pool ratio. Data were obtained from the UniGene EST Profile Viewer web site of the National Center for Biotechnology Information on December 2008.

	NEU1			*NEU3*			*NEU4*		
EST Pool Name	TPM	SI	GE/TE	TPM	SI	GE/TE	TPM	SI	GE/TE
adrenal tumor	77		1 / 12851	0		0 / 12851	0		0 / 12851
bladder carcinoma	56		1 / 17751	0		0 / 17751	0		0 / 17751
breast (mammary gland) tumor	359		34 / 94595	31		3 / 94595	0		0 / 94595
cervical tumor	115		4 / 34598	0		0 / 34598	0		0 / 34598
chondrosarcoma	96		8 / 82863	0		0 / 82863	0		0 / 82863
colorectal tumor	147		17 / 114933	0		0 / 114933	0		0 / 114933
esophageal tumor	231		4 / 17294	115		2 / 17294	0		0 / 17294
gastrointestinal tumor	100		12 / 119863	0		0 / 119863	0		0 / 119863
germ cell tumor	109		29 / 264709	0		0 / 264709	0		0 / 264709
glioma	111		12 / 107374	0		0 / 107374	214		23 / 107374
head and neck tumor	43		6 / 137337	0		0 / 137337	0		0 / 137337
kidney tumor	158		11 / 69390	0		0 / 69390	28		2 / 69390
leukemia	10		1 / 96400	10		1 / 96400	0		0 / 96400
liver tumor	165		16 / 96641	0		0 / 96641	20		2 / 96641
lung tumor	57		6 / 103455	0		0 / 103455	0		0 / 103455
lymphoma	97		7 / 72061	0		0 / 72061	0		0 / 72061
non-neoplasia	102		10 / 97520	0		0 / 97520	0		0 / 97520
normal	80		272 / 3375491	4		14 / 3375491	13		46 / 3375491
ovarian tumor	233		18 / 77188	0		0 / 77188	0		0 / 77188
pancreatic tumor	85		9 / 104967	0		0 / 104967	0		0 / 104967
primitive neuroectodermal tumor...	300		38 / 126470	0		0 / 126470	0		0 / 126470
prostate cancer	115		12 / 103845	0		0 / 103845	0		0 / 103845
retinoblastoma	64		3 / 46531	0		0 / 46531	0		0 / 46531
skin tumor	326		41 / 125536	0		0 / 125536	0		0 / 125536
soft tissue/muscle tissue tumor	7		1 / 125769	0		0 / 125769	0		0 / 125769
uterine tumor	121		11 / 90813	22		2 / 90813	11		1 / 90813

FIG. 8. Expression profile suggested by analysis of EST counts: breakdown by Health State. For details see the legend of Fig. 7.

The GeneAtlas data are based on gene-expression measurements originated by using Affymetrix high-density oligonucleotide arrays, and provide expression information for most protein-encoding transcripts across more than 100 human tissues, and over 60 murine tissues.[184] Expression data are also available on the website for over 80 cell lines, including the NCI60 dataset. This wealth of data available at https://biogps.gnf.org/ suggests a ubiquitous transcription of *NEU2* and a broader pattern of expression for human *NEU3* as compared to the one obtained from EST profiles (data not shown). Data for *NEU4* are only available for the mouse gene, which is expressed at comparable levels in most tissues tested.

Although *in silico* analysis of large-scale gene expression data offers some interesting clues about the biological relevance of sialidases in different tissues and cell types, there remains of course the need of more specific and deeper studies, both at the RNA and at the protein level, to better define the role of these enzymes in health and disease.

X. Amino Acid Sequence Variants in Human Sialidases

Almost all (99.9%) nucleotide bases are identical in all humans, with the remaining 0.1% accounting for about 3 millions single-base DNA differences/polymorphisms (SNPs). While the vast majority of SNPs fall within DNA stretches between genes or in non-coding regions of genes, a more limited number of SNPs are found within coding sequences. These SNPs may fall into two categories: (i) synonymous SNPs (sSNPs) that, due to degeneracy of the genetic code, do not result in an amino acid change in the encoded protein; and (ii) non-synonymous SNPs (nsSNPs) that result in a change of the amino acid sequences of the corresponding proteins, potentially affecting protein functions and interactions.

A recent study by Li *et al.*[83] identified a nsSNP (R41Q) in the human cytosolic sialidase NEU2 occurring in 9.29% of Asian populations that could not be detected in European and African American populations. As already discussed in the NEU2 section of this article, structural analyses and *in vitro* sialidase assays indicated that this SNP could increase the binding affinity of NEU2 to oseltamivir carboxylate (the active form of Tamiflu) thus decreasing sialidase activity. Theoretically, the administration of Tamiflu to people having the R41Q NEU2 variant might further diminish their cytosolic sialidase activity; this according to the authors of the study may be associated with certain severe adverse reactions to this drug that have been observed, mainly in Japan. Although further studies are needed to support this hypothesis, it is probable that amino acid sequence-variations in sialidases may affect their biochemical properties and in some instances have consequences for individuals expressing these protein variants.

Analysis of the SNP database (dbSNP) allowed the identification of nsSNPs in all human sialidase genes, resulting in the amino acid changes listed in Table II. It is

Table II

Amino Acid Changes in Human Sialidase Genes Due to Non-Synonymous Single Nucleotide Polymorphisms (nsSNP) as Reported in dbSNP as of April 2009

Gene	nsSNP1	nsSNP2	nsSNP3	nsSNP4	nsSNP5
NEU1	V317L rs1802188	–	–	–	–
NEU2	S11R rs2233384	R41Q	A145T rs2233390	H168N rs2233391	R182Q rs2233393
NEU3	R48Q rs7115499	P344L rs7936236	G355R rs7935409	P357L rs7936249	–
NEU4	G313R rs11545301	R470H rs2293762	–	–	–

Below the change is indicated the dbSNP rs# cluster id.

anticipated that, with the advent of personal genomics, an increasing number of nsSNP will be identified within the coding region of genes.

Prediction of the possible impact of the amino acid substitutions on the structure and function of a human protein has been evaluated with the bioinformatic tool PolyPhen (http://genetics.bwh.harvard.edu/pph/). The PolyPhen predictions are based on straightforward empirical rules that are applied to the sequence, phylogenetic and structural information characterizing the substitution.[185] Besides the rare but disease-causing amino acid substitutions in sialidosis patients (see Section II), all of the amino acid substitutions in NEU enzymes resulting from nsSNPs appear to be benign, with the exception of R41Q and of H168N in the cytosolic sialidase NEU2, which are predicted to be "damaging" and "possibly damaging", respectively. Overall, the biological relevance of nsSNPs awaits experimental validation by a detailed biochemical characterization of the protein isoforms present in the human population.

XI. Sialidases in Teleosts

Until recently, little was known about sialidases in vertebrates other than mammals. A 2007 study describes the identification and characterization of the sialidase gene family in the zebrafish *Danio rerio*.[132] While in mammals four genes are known to encode sialidases, in zebrafish the picture is more complex, with seven genes homologous with human sialidases (Fig. 9). In the fish a single orthologue exists for the mammalian *NEU1* gene, which was named *neu1*. This finding was supported by the presence of partial synteny between the genomic sequences containing the orthologous genes. Similarly, *neu4* from *Danio rerio* appears to be closely related to human NEU4, although no synteny can be detected between the zebrafish and mammal genomic sequences. The remaining five genes (*neu3.1–neu3.5*) are arranged in a cluster on chromosome 21 and they all encode polypeptides related to human NEU3. A conserved synteny is detectable between the *neu3* gene cluster of the zebrafish chromosome 21, and the chromosome locus 11q13.4 region harboring *NEU3* in humans (Fig. 9). Surprisingly, no clear orthologue of mammalian *NEU2* has been identified in *Danio rerio.* A bioinformatic analysis performed on *Takifugu rubripes* (fugu) and *Tetraodon nigroviridis* genomes revealed that only one *NEU3*-like gene is present in both fish species, suggesting that the redundancy observed in zebrafish might be the result of an independent gene duplication event that occurred in this teleost. It is noteworthy that, in fugu and *Tetraodon nigroviridis*, only three sialidase genes exist, representing the putative orthologues of *NEU1, NEU3*, and

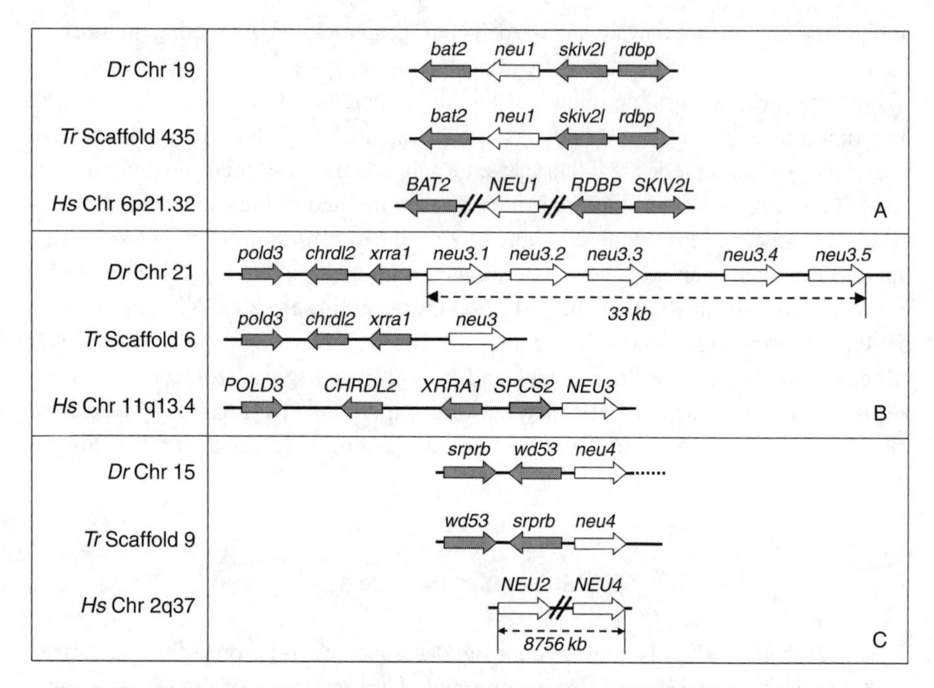

FIG. 9. Organization of sialidase genes in man and teleosts. A, B, and C indicate, respectively, the *NEU1, NEU3*, and *NEU4* chromosomal regions in zebrafish (*Dr*), fugu (*Tr*), and man (*Hs*). The arrows indicate the direction of gene transcription within each chromosome. Sialidase genes are depicted with white arrows, while gray arrows represent the adjacent genes in syntenic regions. The size of the arrows is not proportional to that of the genes, and the distances between genes are arbitrary. The gray arrow indicates the human *SPCS2* gene that appears to be absent both in zebrafish and fugu genomes. Modified from Manzoni *et al.*[132]

NEU4, while in chicken (E. Giacopuzzi, R. Bresciani, G. Borsani, and E. Monti, manuscript in preparation) and mammals, a fourth gene, *NEU2*, is also present. In teleosts, a *NEU2* orthologue may not be necessary or, alternatively, one (or more) of the NEU3-like proteins could play the biological role of a cytosolic sialidase.

Functional studies were carried out on NEU3- and NEU4-like zebrafish sialidases.[132] Transient expression of *neu3.1, neu3.2, neu3.3*, and *neu4* zebrafish genes in COS7 cells demonstrates that they encode enzymatically active sialidases. All but Neu3.2 show an extremely acidic pH optimum (<3.0), which is lower than the pH range of mammalian sialidases. Based on these studies, Neu3.1, Neu3.3, and Neu4 behave as higher vertebrate NEU3 and NEU4 proteins, all characterized by very low pH optima, whereas Neu3.2, with a less acidic pH optimum of 5.6, behaves as the cytosolic sialidase NEU2. Cell-fractionation experiments further confirmed this hypothesis, with the group of the extremely acidic sialidases segregating with the particulate membranous fraction, while Neu3.2 behaving as a soluble enzyme. Immunofluorescence localizations of the zebrafish *myc*-tagged enzymes are in agreement with the subcellular fractionation data: Neu3.1 and Neu4 show a substantial co-localization with the calnexin-positive cell membranes, whereas Neu3.3 localizes both at the cell surface and inner membranous structures, with only a partial co-localization with LAMP1-positive vesicles. Finally, Neu3.2 shows a distribution largely superimposable with that of the human cytosolic sialidase NEU2.

The expression of zebrafish sialidases in COS7 cells metabolically labeled with tritiated sphingosine allowed testing of the activity of these enzymes toward ganglioside substrates inserted in the lipid bilayer of living cells. All zebrafish Neus studied except Neu3.2 induce a significant decrease of GM3 and an increase of GM1 content when compared with mock-transfected cells and, in the case of cells expressing Neu4, a significant decrease of the relative content of GD1 was also detected. Similar results were obtained with the mammalian plasma membrane-associated sialidase NEU3,[96] suggesting that the overall substrate specificity, as well as the action of the enzymes in living cells, is roughly similar, despite the evolutionary distance between teleosts and mammals. These biochemical data further support the classification inferred by bioinformatic analysis. Neu3.1 and Neu3.3 belong to the group of the membrane-associated NEU3-like enzymes, whereas Neu4 belongs to the NEU4-like enzymes. Neu3.2, despite its high sequence identity to mammalian NEU3 and its localization in the Neu3-like cluster on chromosome 21, is a soluble protein having a pH optimum typical of the mammalian cytosolic NEU2 sialidases thus far cloned. Neu3.2 is more probably a NEU3 paralogue that has lost its capability to interact with the lipid bilayer, thus becoming a soluble enzyme.

Other proteins involved in sialoglyconjugate biology have already been described in zebrafish. Among them, the sialic acid-binding protein siglec-4 shows binding features very similar to the human orthologue.[186] Genes encoding putative sialyltransferases have likewise been identified in zebrafish, as well as in the other vertebrates and in some invertebrates.[187] Furthermore, the fact that overexpression of GM3 synthase in zebrafish embryos resulted in neuronal cell death in the central nervous system[188] indicates the relevance of gangliosides in the biology of fishes. These findings are consistent with a published glycomic survey map of zebrafish which reveals unique sialylation features, as well as variations during embryogenesis.[189]

Overall, the studies in zebrafish suggest the presence of a complex regulation pattern for expression of the enzymes involved in sialoglycoconjugate synthesis and degradation. Study of the different members of the sialidase gene family in this model organism offers a novel and attractive field of glycobiology. In particular, the developmental roles of sialidases in vertebrates can now be tested in zebrafish, using overexpression and morpholino loss-of-function approaches.

Data obtained from the study of sialidases in *Danio rerio* suggest that other unexpected results may arise from the characterization of members of this gene family in other vertebrates, such as amphibians, reptiles, and avian species.

XII. Trans-sialidases: What Distinguishes Them from Sialidases?

A remarkable group of sialidases are the trans-sialidases (TS), found mainly in microorganisms. The groups best investigated are those from such pathogenic trypanosomes as the African *Trypanosoma brucei*,[190] *T. congolense*,[191] and the American *T. cruzi*.[192] The TS are glycosylphosphatidylinositol (GPI)-anchored in the cell surface of the parasites. In contrast to the "classical" sialidases, trypanosomal TS (EC 3.2.1.18) catalyze the transfer of preferably α-(2→3)-linked sialic acids from donor glycans directly to terminal β-Gal- or β-GalNAc-containing acceptor molecules, giving rise to new α-(2→3) linkages[193,194] (Fig. 10). In the absence of an appropriate acceptor, this enzyme acts as a sialidase, although this activity is slower than the transfer reaction, by transfer of the glycosidically linked sialic acid to a water molecule instead of Gal. The TSs constitute a kind of sialyltransferase that can form new sialic acid–glycan linkages without the need for prior activation of sialic acid by CTP.

TS activity has also been described in *Endotrypanum* species.[195] These microorganisms, together with the trypanosomes, are protozoic, that is, they are eukaryotic cells. TS activity has also been detected in human serum,[196] although this activity was not further investigated. TS activity was also reported in such prokaryotes as

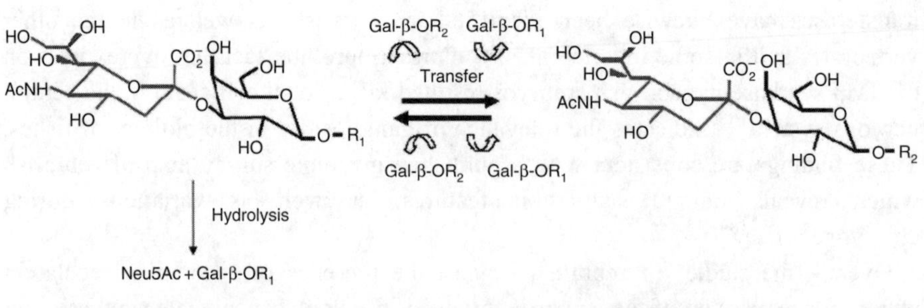

Fig. 10. Trans-sialidase reaction. Sialic acid is transferred from one galactose with R_1 as penultimate glycan to another galactose bound to R_2 in a reversible reaction. From Schrader *et al.*,[194] with the permission of the publishers.

Corynebacterium diphtheria,[197] *Campylobacter jejuni*,[198] and *Pasteurella multocida*.[199] Remarkably, the *C. jejuni* TS exemplifies an oligosaccharide trans-sialidase that specifically transfers (or hydrolyzes) α-(2→8)-linkages of sialic acid. The *P. multocida* enzyme exhibits α-(2→3) TS activity. Even more striking is the fact that the latter two bacterial enzymes possess, besides their TS and sialidase activities, additional classical sialyltransferase activities with CMP-Neu5Ac as donor.

In contrast to these transmolecular TSs, the group of Yu-Teh Li described an intramolecular TS in the leech that forms 2,7-anhydro-Neu5Ac upon release of sialic acid from the glycosidic linkage.[200] Such an enzyme ("neuraminidase B") was found in *Streptococcus pneumoniae* by Gut *et al.*[201] This enzyme has a strict specificity for α-(2→3)-linked sialic acid substrates.

Various radioactive and non-radioactive assays exist for the detection of TS activities.[194,202] A fluorimetric, 96-well plate assay enables the specific, sensitive, and relatively rapid detection of these enzyme activities on a larger scale. 4-Methylumbelliferyl ß-D-galactopyranoside (MUGal) is used as acceptor substrate and sialyllactose as the sialic acid donor. The MUGalNeu5Ac formed is separated from the substrate MUGal by ion-exchange chromatography on 96-well filter plates, hydrolyzed, and the MU liberated measured in a fluorimeter, thus giving the TS activity.

The trans-sialidases of the African and American trypanosomes are potent virulence factors, being involved in the onset of sleeping sickness and of Chagas disease, respectively. Both diseases affect millions of people in Africa and South America. *T. brucei* expresses this enzyme only in the procyclic vector stage in the gut of the transmitting tsetse fly (*Glossina* species).[203,204] *T. cruzi*, which is transmitted by the blood-sucking bug *Triatoma infestans*, exhibits TS activity during its life in the human host.[204–206] In the midgut of the tsetse fly, *T. brucei* acquires sialic acid from blood components sucked from the host for sialylation of the cell surface of the parasites in order to protect them from the digestive and trypanocidal environment.[203,207] In blood circulation, the TS of *T. cruzi* trypanomastigotes is used to shield the trypanosomes from the immune system in different ways. Much research has been performed on the pathophysiological role of *T. cruzi* TS, and only a few examples can be given here. General features of the effects of TS are cell invasion by trypanosomes, and disturbance of the cytokine network, leading to inflammation, trans-membrane signaling, and compromise of the immune system. For example, many different copies of active and inactive TS molecules, products of a multigene family, are produced in order to evade the host's immune system.[208] The virulence of *T. cruzi* is enhanced because TS abnormally activates polyclonal B cells.[209] TS induce endothelial cell signaling, and can block apoptosis of these cells.[210] Furthermore, stimulation of T lymphocytes, disorganization of thymus histoarchitecture, and thymus aplasia were observed in mice and humans.[209,211,212] All of these, and other events, lead to the chronic stages of Chagas disease that is often lethal.

TS may be used in biotechnology, because they enable α-(2→3)-sialylation of oligosaccharides and glycoproteins in high yield when using α-(2→3)-sialyllactose as the most suitable donor.[213,214]

The next topic compares some properties of the vertebrate sialidases,[3,4] including the recently crystallized NEU2,[19] with those of trypanosomal TS to emphasize the difference between a sialic acid hydrolase and a sialic acid transferase (trans-glycosidase), with special reference to structural and enzyme-mechanistic aspects. Among the classical prokaryotic and eukaryotic sialidases studied, Neu5Ac is the most rapidly hydrolyzed or transferred, followed by Neu5Gc and, significantly less rapidly, by *O*-acetylated sialic acid.[3,4,46,204,215] Sialidases and trans-sialidases can act on a great variety of sialylated complex carbohydrates, as detailed in the references cited. In most instances the best substrates are oligosaccharides and glycoproteins, whereas gangliosides are often less readily desialylated. Studies on human sialidases revealed great differences in substrate specificity. While the lysosomal NEU1 has a narrow substrate-specificity, showing best activity with sialylated oligosaccharides and glycopeptides,[3] the cytosolic NEU2 and NEU4 localized at various subcellular membrane sites can hydrolyze both glycoproteins and gangliosides.[59,116,134] The plasma membrane NEU3 almost specifically hydrolyzes gangliosides.[46,91] Trypanosomal trans-sialidases also have limited sialyl donor or acceptor specificity, and favor oligosaccharides and glycoproteins, as demonstrated with *T. brucei*[216] and *T. congolense*.[203] In *T. brucei*, the procyclic acidic repetitive membrane (glyco-)protein (PARP) is the physiological trans-sialidase substrate.[216,217] Likewise, in *T. cruzi*, mucin-like glycoproteins on the cell surface are excellent sialic acid acceptors for the trans-sialidase from this trypanosome.[218] Gangliosides and mucins have almost no donor activity with the African trans-sialidases.[191,204,216] In contrast, for *T. cruzi*, glycolipids were also described as sialic acid donors.[219]

A marked difference between "classical" sialidases and trans-sialidases is their behavior toward potential inhibitors, such as Neu5Ac2en, 4-amino-Neu5Ac2en or 4-guanidino-Neu5Ac2en, and *N*-(4-nitrophenyl)oxamic acid.[191,204,216] These molecules inhibit trans-sialidases from both the African and American pathogenic trypanosomes by 50%, but only in the higher millimolar range. In particular Neu5Ac2en, which is in wider use as sialidase inhibitor than the oxamic acid derivative, can inhibit vertebrate, bacterial, and viral sialidases at micromolar concentrations to various extents. Reports from M. Kiso's group have shown inhibition of the four human sialidases at various degrees by Neu5Ac2en derivatized at C-9,[62] and the low inhibitory effect of 4-acetamido-5-acylamido-2-fluorobenzoic acid derivatives.[63] Free

Neu5Ac is a weak inhibitor of "classical" sialidases[204] but does not act at all on the trans-sialidases. Correspondingly, the 2,3-difluoro derivative of Neu5Ac, forming a covalent intermediate with sialidases and trans-sialidases (see later) was required at high concentrations (20 mM) to inactivate the enzyme completely.[220] Since trans-sialidases possess a lactose-binding site in their active center (as described next), lactose and its derivatives, especially lactitol, inhibit the sialic acid transfer of the enzyme from *T. cruzi.*[221] Oligosaccharides from the mucins of this trypanosome can also inhibit the transfer of sialic acid to, for example, the substrate *N*-acetyllactosamine.[218] It should be noted that heavy metal ions, especially mercury, are potent inhibitors of sialidases and trans-sialidases. The latter enzymes do not require calcium ions, in contrast to some sialidases.[3,4,46,204]

These observations point to the existence of characteristic differences in the structural and catalytic mechanisms between the microbial and animal sialidases on the one hand and the trypanosomal trans-sialidases on the other, although they share similar protein structures and conserved amino acids involved in enzyme catalysis. It was first observed by Pereira *et al.*[222] that the *T. cruzi* trans-sialidase consists of an N-terminal half, containing the enzymatic function, and a C-terminal portion, mainly composed of a tandem series of twelve amino acid repeats. They are called Shed Acute Phase Antigens (SAPA) repeats, and are highly antigenic. The repeats were found not to be involved in enzyme catalysis, because a recombinant *T. cruzi* TS lacking the repeats was shown to retain enzyme activity.[223] Four Asp-boxes exist within the N-terminus of the enzyme, and while three of these conserved motifs occur in positions of the amino acid sequence similar to those found in bacterial sialidases, the distance between the third and the fourth Asp box is much larger in the parasite enzyme than in the bacterial proteins. Three Asp-boxes were also found in the human sialidase NEU2.[19] As in the "pure" sialidases, the Asp boxes are not directly involved in enzyme catalysis, but they may preserve the protein structure. The American parasite *T. rangeli* secretes an enzyme having only sialidase activity, but shows 70% amino acid identity with the amino acid sequence with *T. cruzi* TS. The secreted protein is composed of an enzymatic and a lectin-like domain connected by an α-helix. Moreover, the SAPA repeats are missing.[224] The crystal structure of *T. rangeli* sialidase shows that the active site of the enzyme has a canonical ß-propeller topology similar to that of viral, bacterial, and human NEU2, the latter being the first vertebrate sialidase to be crystallized.[18,19] The C-terminal domain of *T. rangeli* sialidase shows the ß-barrel structure of plant lectins. This X-ray crystallographic study enabled modeling of the structure of *T. cruzi* trans-sialidase, and together with mutagenesis experiments allowed the identification of the amino acids of the active site that transformed this sialidase into an efficient sialyltransferase.[224] Most important was the discovery of a binding site for the acceptor carbohydrate that is distinct from the donor (sialic acid) binding site. At this site a tyrosine residue (Tyr 120) was found by mutagenesis to be crucial for binding of the acceptor substrate. This structural feature, not present in sialidases known so far, contributed to our understanding of how a glycosidase scaffold can achieve glycosyltransferase activity. The authors[224] provided a model of both hydrolysis and transfer reactions as catalyzed by *T. cruzi* TS. As in the activity clefts of all sialidases, the carboxylic group of sialic acid interacts with an arginine triad, a glutamic acid residue stabilizes one of these arginines, an aspartic acid (Asp 59) is essential for catalysis by proton transfer, and a tyrosine (Tyr 342) is in contact with the transient oxacarbenium ion at the C-2 carbon of sialic acid.[224,225] The reaction is completed after the nucleophilic attack of either water or, preferentially, the hydroxyl group at C-3 of galactose or a galactoside.

Investigation of *T. cruzi* TS crystals revealed the structural differences of the active-site cleft between the *T. cruzi* TS and the *T. rangeli* hydrolase, which could be responsible for the different types of enzyme reaction. The substrate-binding pocket of *T. cruzi* TS appears narrower and more hydrophobic, and may favor trans-sialylation by the exclusion of water.[225] A proline residue (Pro 283), which influences the position of a conserved tryptophan near the active center, is also essential for *T. cruzi* TS activity. The most important differences between classical sialidases and trans-sialidases are marked structural changes occurring on the soaking of substrates or Neu5Ac2en into the crystals. Even the crystals become instable due to disturbance of the molecular packing. This is because of conformational changes caused by sialic acid of donor substrates. Most important is the steric translocation of Tyr 342 by about 2 Å upon substrate binding, which leads to a shortening of the distance between the hydroxyl group of the tyrosine residue and the C-2 atom of sialic acid.[225,226] This plasticity of the catalytic pocket induced by binding of sialic acid is strictly conserved, is essential for catalysis by TS, and was not observed in all classical prokaryotic and eukaryotic sialidases having known 3D structures, including the enzyme from *T. rangeli*. Furthermore, the latter sialidase and *T. cruzi* TS interact in a different way with the sialic acid side-chain. Only a single interaction was seen in *T. rangeli* sialidase, but multiple ones occur with *T. cruzi* TS.[227]

Most information on how the trans-sialidase reaction proceeds was obtained from binding studies with either active *T. cruzi* TS[225,228,229] or with inactive natural mutants[230] using surface plasmon resonance technology and NMR spectroscopy. While, for example, sialyllactose was readily bound to the enzyme protein, the acceptor substrate lactose or other asialo glycoconjugates did not interact with the enzyme at all unless it had been pre-incubated with the sialylated donor substrate. All of these studies clearly show that sialic acid modulates the affinity for the asialo acceptor substrate. The association of the asialo receptor with the active site is an absolute requirement for the transfer reaction. That this occurs by a structural change in the sequential binding of the enzyme was confirmed by investigation of crystals soaked with donor or acceptor substrates. The amino acids Trp 312 and the Tyr 119 were found to make stacking interactions with lactose, and to enable positioning of the 3-OH group of the galactose moiety in the ternary enzyme complex, allowing nucleophilic attack on the anomeric carbon in the sialylated transition species.[225] Point mutations of these amino acids confirmed their role in catalysis and that Tyr 119 is part of the second binding-site. They furthermore abolished TS activity but enabled the protein to hydrolyze both α-(2→3)- and α-(2→6)-linked sialic acid.[231] This observation shows the necessity of a precise orientation of the substrates within the catalytic center for the transfer reaction, in contrast to hydrolysis, and the presence of distinct binding-sites for acceptor and donor substrates. These structural features are unique to trans-sialidases. Interestingly, sialic acid α-(2→6)-glycosidically linked to lactose cannot trigger the conformational switch required for the trans-glycosylation reaction, in contrast to α-(2→3)-linked sialic acid.[230] Fig. 11 shows the sequential binding of donor α-(2→3)-linked sialic acid and acceptor galactose, including a change of protein conformation involved in *T. cruzi* TS reaction.

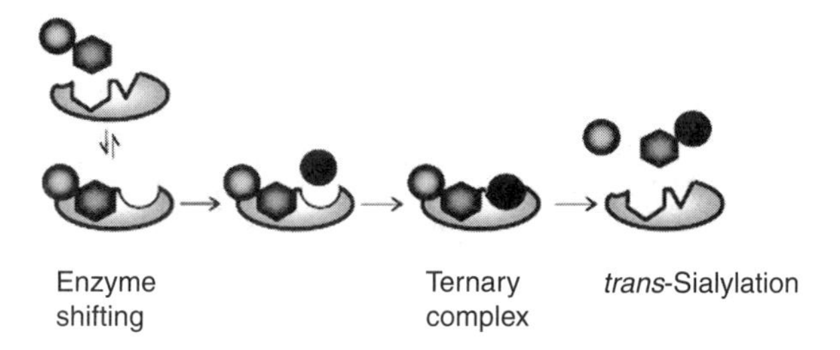

Fig. 11. Conformational switch in the active center of trans-sialidase. This occurs after the binding of α-(2→3)-linked sialic acid, for example, sialyllactose, and enables binding of galactose or galactosides and the trans-glycosylation to proceed. Hexagon, sialic acid; black circle, accepting galactoside; empty circle, leaving galactoside, for example, lactose. Modified from Todeschini *et al.*[230]

The binding of donor substrate to *T. cruzi* TS was also demonstrated by STD NMR experiments using *p*-nitrophenyl glycosides of Neu5Ac.[229] These experiments also allowed comparison of the rates of substrate hydrolysis with those of sialic acid transfer. Interestingly, shortening of the Neu5Ac side-chain favors hydrolysis over transfer, whereas 9-*O*-acetylation of Neu5Ac has the opposite effect.

Mutants of *T. rangeli* sialidase have been obtained that exhibit some TS activity.[232] This was shown to be due to the formation of a TS-like binding-site for the acceptor sugar. This binding is a prerequisite for TS activity, together with fine-tuning of protein–substrate interactions and the flexibility of crucial active-site residues.

There is reason to assume that the catalytic mechanisms of the TS from both the American and African trypanosomes (*T. cruzi* and *T. brucei*, respectively, which can catalyze the same chemical reaction) are similar. Evidence is presented for a common ancestor of both trypanosome types around 100 million years ago, carrying a primitive TS gene,[233] and only later did the African and American continents separate, together with the protozoa. Although a crystal structure of *T. brucei* TS is still missing, its protein sequence from gene coding experiments is available.[234] It shows 45% identity in its catalytic domain with the corresponding region of *T. cruzi* TS, and it conserves most of the amino acids relevant for the catalytic site of the American TS. Seven positions are invariant in the two TS. Remarkably, and similar to the *T. cruzi* TS W312A mutant,[231] exchange of the corresponding Trp (W400) against Ala in the TS of *T. brucei* abolished trans-sialylation activity and enabled the mutant to additionally hydrolyze α-(2→6)-sialic acid linkages.

Two trans-sialidases isolated from the animal-pathogenic African trypanosome *T. congolense* show pronounced differences in their capacity for sialic acid transfer as compared with the hydrolytic activity.[191] Partial sequences were obtained from these two enzymes by a PCR-based approach, showing 50% identity with each other, but are similar to viral, bacterial, animal sialidases and other trans-sialidases.[235] Most of the critical active-site residues common to other trypanosomal sialidases and TS are conserved. This similarity, together with the amino acid difference at the active site, between American and African (trans-)sialidases is depicted in the publication.[235]

Structural, kinetic, and mechanistic analyses of *T. cruzi* TS revealed the transfer reaction to proceed by a classical ping-pong bi-bi kinetic mechanism, in which the donor and acceptor substrates must separately bind at the same site. In the first step, an intermediate with sialic acid covalently linked to Tyr 342 as the active-site nucleophile is formed, as shown with fluorinated Neu5Ac derivatives.[220,226,236] Asp 59 serves as the acid–base catalyst in this reaction. Corresponding studies with *T. rangeli* sialidase, which is a "true" sialidase, has shown a corresponding mechanism,[227] leading to the assumption that probably all sialidases, as retaining glycosidases, operate through a similar mechanism, with the transient formation of a covalently sialylated enzyme. For the trans-sialidases, however, sialic acid is transferred in a second step to galactose by essentially the reverse of the foregoing sialylation process, again involving an electrophilic migration of the anomeric center onto the 3-hydroxyl group of galactose.[226] This trans-glycosylation process may be facilitated by the longer lifetime of the sialylated tyrosine intermediate observed in *T. cruzi* TS together with other intriguing differences in the reaction mechanism that have made trans-sialidases unique.

XIII. Final Remarks

The cells and tissues of vertebrates, particularly mammals, contain a variety of sialic acid-carrying compounds (sialoglycoproteins, sialoglycolipids, sialo-oligosaccharides) that are mainly associated with the plasma and intracellular membranes but are also present in the cytosol and body fluids. The removal of sialic residues from sialoglycoconjugates is catalyzed by sialidases. Four different sialidases occur in vertebrates (NEU1, NEU2, NEU3, and NEU4), and each sialidase is encoded by a specific gene. The fact that all of these genes have been conserved during evolution and that the different sialidases have a preferential subcellular localization and substrate specificity underscore their involvement in the multitude of cellular processes that are regulated by the extent of sialylation of glycoconjugates. Moreover, the specific role for each sialidase is further supported by the fact that a deficiency of NEU1, despite the presence of the other sialidases, is the cause of two severe inherited disorders in humans.

Remarkably, mammalian sialidases do not appear to contain amino acid sequences for incorporation into membranes as integral proteins, nor for prenylation, acylation, or attachment to glycanphosphoinositide anchors. Nevertheless, all of them, including the "canonical" cytosolic sialidase NEU2, may adhere to membranes, sometimes very firmly, as in the case of NEU3 and NEU4, presumably by ionic interactions or through a bond with a specific protein component(s) of the membrane. This ability may be instrumental for regulating their intracellular trafficking and substrate recognition.

As reviewed here, the process of desialylation governed by sialidases generates less sialylated or completely desialylated compounds with consequent changes in their negative charge and conformation that, in turn, may influence their specific interactions with partner molecules. Moreover, the removal of sialic acid may alter the concentration of potentially bioactive molecules, with dramatic downstream consequences. For example, among glycolipids, sialylated compounds such as gangliosides (GM3, GM2, GM1, GD3, GD2, GD1a, GD1b, GT1b, and GQ1b) or desialylated compounds such as lactosyl-ceramide, glucosyl-ceramide, and ceramide may function as bioactive molecules, playing distinct roles in the regulation of cell proliferation, apoptosis, differentiation, "social behavior," and even conversion to malignancy, depending on their concentrations and subcellular localization. On the basis of recent findings it can be predicted that individual sialoglycoproteins may also have similar implications.

It is expected that the huge progress in glycomics technology applied to investigations in single cells at specific subcellular sites will provide leads toward a better understanding of the subtle role of sialylation/desialylation in cellular functions.

References

1. R. Schauer, Sialic acids: Fascinating sugars in higher animals and man, *Zoology (Jena)*, 107 (2004) 49–64.
2. M. Saito and R. K. Yu, Biochemistry and function of sialidases, in A. Rosenberg, (Ed.), *Biology of the Sialic Acids*, Plenum Press, New York, 1995, pp. 261–313.
3. K. E. Achyuthan and A. M. Achyuthan, Comparative enzymology, biochemistry and pathophysiology of human exo- alpha-sialidases (neuraminidases), *Comp. Biochem. Physiol. B Biochem. Mol. Biol.*, 129 (2001) 29–64.
4. E. Monti, A. Preti, B. Venerando, and G. Borsani, Recent development in mammalian sialidase molecular biology, *Neurochem. Res.*, 27 (2002) 649–663.
5. E. Bonten, A. van der Spoel, M. Fornerod, G. Grosveld, and A. d'Azzo, Characterization of human lysosomal neuraminidase defines the molecular basis of the metabolic storage disorder sialidosis, *Genes Dev.*, 10 (1996) 3156–3169.
6. A. V. Pshezhetsky, C. Richard, L. Michaud, S. Igdoura, S. Wang, M. A. Elsliger, J. Qu, D. Leclerc, R. Gravel, L. Dallaire, and M. Potier, Cloning, expression and chromosomal mapping of human lysosomal sialidase and characterization of mutations in sialidosis, *Nat. Genet.*, 15 (1997) 316–320.
7. R. J. Rottier, E. Bonten, and A. d'Azzo, A point mutation in the neu-1 locus causes the neuraminidase defect in the SM/J mouse, *Hum. Mol. Genet.*, 7 (1998) 313–321.
8. S. A. Igdoura, C. Gafuik, C. Mertineit, F. Saberi, A. V. Pshezhetsky, M. Potier, J. M. Trasler, and R. A. Gravel, Cloning of the cDNA and gene encoding mouse lysosomal sialidase and correction of sialidase deficiency in human sialidosis and mouse SM/J fibroblasts, *Hum. Mol. Genet.*, 7 (1998) 115–121.
9. D. Wang, S. Zaitsev, G. Taylor, A. d'Azzo, and E. Bonten, Protective protein/cathepsin A rescues N-glycosylation defects in neuraminidase-1, *Biochim. Biophys. Acta*, 1790 (2009) 275–282.
10. P. Roggentin, R. Schauer, L. L. Hoyer, and E. R. Vimr, MicroReview: The sialidase superfamily and its spread by horizontal gene transfer, *Mol. Microbiol.*, 9 (1994) 915–921.
11. B. Rothe, P. Roggentin, and R. Schauer, The sialidase gene from *Clostridium septicum*: Cloning, sequencing, expression in *Escherichia coli* and identification of conserved sequences in sialidases and other proteins, *Mol. Gen. Genet.*, 226 (1991) 190–197.
12. I. Chaudhuri, J. Soding, and A. N. Lupas, Evolution of the beta-propeller fold, *Proteins*, 71 (2008) 795–803.

13. J.N. Varghese, W. G. Laver, and P. M. Colman, Structure of the influenza virus glycoprotein antigen neuraminidase at 2.9 Å resolution, *Nature*, 303 (1983) 35–40.
14. P. M. Colman, Influenza virus neuraminidase: Structure, antibodies, and inhibitors, *Protein Sci.*, 3 (1994) 1687–1696.
15. S. Crennell, E. Garman, G. Laver, E. Vimr, and G. Taylor, Crystal structure of *Vibrio cholerae* neuraminidase reveals dual lectin-like domains in addition to the catalytic domain, *Structure*, 2 (1994) 535–544.
16. A. Gaskell, S. Crennell, and G. Taylor, The three domains of a bacterial sialidase: A beta-propeller, an immunoglobulin module and a galactose-binding jelly-roll, *Structure*, 3 (1995) 1197–1205.
17. S. J. Crennell, E. F. Garman, C. Philippon, A. Vasella, W. G. Laver, E. R. Vimr, and G. L. Taylor, The structures of *Salmonella typhimurium* LT2 neuraminidase and its complexes with three inhibitors at high resolution, *J. Mol. Biol.*, 259 (1996) 264–280.
18. S. Magesh, T. Suzuki, T. Miyagi, H. Ishida, and M. Kiso, Homology modeling of human sialidase enzymes NEU1, NEU3 and NEU4 based on the crystal structure of NEU2: Hints for the design of selective NEU3 inhibitors, *J. Mol. Graph. Model.*, 25 (2006) 196–207.
19. L. M. Chavas, C. Tringali, P. Fusi, B. Venerando, G. Tettamanti, R. Kato, E. Monti, S.and Wakatsuki, Crystal structure of the human cytosolic sialidase Neu2. Evidence for the dynamic nature of substrate recognition, *J. Biol. Chem*, 280 (2005) 469–475.
20. A. van der Spoel, E. Bonten, and A. d'Azzo, Transport of human lysosomal neuraminidase to mature lysosomes requires protective protein/cathepsin A, *EMBO J.*, 17 (1998) 1588–1597.
21. A. Hille-Rehfeld, Mannose 6-phosphate receptors in sorting and transport of lysosomal enzymes, *Biochim. Biophys. Acta*, 1241 (1995) 177–194.
22. K. E. Lukong, V. Seyrantepe, K. Landry, S. Trudel, A. Ahmad, W. A. Gahl, S. Lefrancois, C. R. Morales, and A. V. Pshezhetsky, Intracellular distribution of lysosomal sialidase is controlled by the internalization signal in its cytoplasmic tail, *J. Biol. Chem.*, 276 (2001) 46172–46181.
23. R. Le Borgne, A. Alconada, U. Bauer, and B. Hoflack, The mammalian AP-3 adaptor-like complex mediates the intracellular transport of lysosomal membrane glycoproteins, *J. Biol. Chem.*, 273 (1998) 29451–29461.
24. A. A. Peden, V. Oorschot, B. A. Hesser, C. D. Austin, R. H. Scheller, and J. Klumperman, Localization of the AP-3 adaptor complex defines a novel endosomal exit site for lysosomal membrane proteins, *J. Cell Biol.*, 164 (2004) 1065–1076.

25. S. Gottschalk, A. Waheed, B. Schmidt, P. Laidler, and K. von Figura, Sequential processing of lysosomal acid phosphatase by a cytoplasmic thiol proteinase and a lysosomal aspartyl proteinase, *EMBO J.*, 8 (1989) 3215–3219.
26. E. J. Bonten and A. d'Azzo, Lysosomal neuraminidase. Catalytic activation in insect cells is controlled by the protective protein/cathepsin A, *J. Biol. Chem.*, 275 (2000) 37657–37663.
27. N. J. Galjart, H. Morreau, R. Willemsen, N. Gillemans, E. J. Bonten, and A. d'Azzo, Human lysosomal protective protein has cathepsin A-like activity distinct from its protective function, *J. Biol. Chem.*, 266 (1991) 14754–14762.
28. E. J. Bonten, Y. Campos, V. Zaitsev, A. Nourse, B. Waddell, W. Lewis, G. Taylor, and A. d'Azzo, Heterodimerization of the sialidase NEU1 with the chaperone protective protein/cathepsin A prevents its premature oligomerization, *J. Biol. Chem.*, 284 (2009) 28430–28441.
29. A. d'Azzo, G. Andria, P. Strisciuglio, and H. Galjaard, Galactosialidosis, in, C.R. Scriver, A. L. Beaudet, W. S. Sly, and D. Velle, (Eds.), *The Metabolic and Molecular Bases of Inherited Disease*, McGraw-Hill Inc, New York, 2001, pp. 3811–3826.
30. G.H. Thomas, Disorders of glycoprotein degradation and structure: A-mannosidosis, b-mannosidosis, fucosidosis, and sialidosis, in C. R. Scriver, A. L. Beaudet, W. S. Sly, and D. Valle (Eds.), *The Metabolic and Molecular Bases of Inherited Disease*, McGraw-Hill Inc, New York, 2001, pp. 3507–3534.
31. E.J. Bonten, W. F. Arts, M. Beck, A. Covanis, M. A. Donati, R. Parini, E. Zammarchi, and A. d'Azzo, Novel mutations in lysosomal neuraminidase identify functional domains and determine clinical severity in sialidosis, *Hum. Mol. Genet.*, 9 (2000) 2715–2725.
32. V. Seyrantepe, H. Poupetova, R. Froissart, M. T. Zabot, I. Maire, and A.V. Pshezhetsky, Molecular pathology of NEU1 gene in sialidosis, *Hum. Mutat.*, 22 (2003) 343–352.
33. S. Pattison, M. Pankarican, C. A. Rupar, F. L. Graham, and S. A. Igdoura, Five novel mutations in the lysosomal sialidase gene (NEU1) in type II sialidosis patients and assessment of their impact on enzyme activity and intracellular targeting using adenovirus-mediated expression, *Hum. Mutat.*, 23 (2004) 32–39.
34. S. M. Mithieux and A. S. Weiss, Elastin, *Adv. Protein Chem.*, 70 (2005) 437–461.
35. A. Hinek, T. D. Bodnaruk, S. Bunda, Y. Wang, and K. Liu, Neuraminidase-1, a subunit of the cell surface elastin receptor, desialylates and functionally inactivates adjacent receptors interacting with the mitogenic growth factors PDGF-BB and IGF-2, *Am. J. Pathol.*, 173 (2008) 1042–1056.

36. A. Hinek, M. Rabinovitch, F. Keeley, Y. Okamura-Oho, and J. Callahan, The 67-kD elastin/laminin-binding protein is related to an enzymatically inactive, alternatively spliced form of beta-galactosidase, *J. Clin. Invest.*, 91 (1993) 1198–1205.
37. H. Morreau, N. J. Galjart, N. Gillemans, R. Willemsen, G. T. van der Horst, and A. d'Azzo, Alternative splicing of beta-galactosidase mRNA generates the classic lysosomal enzyme and a beta-galactosidase-related protein, *J. Biol. Chem.*, 264 (1989) 20655–20663.
38. A. Hinek, A. V. Pshezhetsky, M. von Itzstein, and B. Starcher, Lysosomal sialidase (neuraminidase-1) is targeted to the cell surface in a multiprotein complex that facilitates elastic fiber assembly, *J. Biol. Chem.*, 281 (2006) s3698–3710.
39. A. Caciotti, M. A. Donati, T. Bardelli, A. d'Azzo, G. Massai, L. Luciani, E. Zammarchi,, and A. Morrone, Primary and secondary elastin-binding protein defect leads to impaired elastogenesis in fibroblasts from GM1-gangliosidosis patients, *Am. J. Pathol.*, 167 (2005) 1689–1698.
40. L. Duca, C. Blanchevoye, B. Cantarelli, C. Ghoneim, S. Dedieu, F. Delacoux, W. Hornebeck, A. Hinek, L. Martiny, and L. Debelle, The elastin receptor complex transduces signals through the catalytic activity of its Neu-1 subunit, *J. Biol. Chem.*, 282 (2007) 12484–12491.
41. H. Morreau, E. Bonten, X. Y. Zhou, and A. D'Azzo, Organization of the gene encoding human lysosomal beta-galactosidase, *DNA Cell Biol.*, 10 (1991) 495–504.
42. N. de Geest, E. Bonten, L. Mann, J. de Sousa-Hitzler, C. Hahn, and A. d'Azzo, Systemic and neurologic abnormalities distinguish the lysosomal disorders sialidosis and galactosialidosis in mice, *Hum. Mol. Genet.*, 11 (2002) 1455–1464.
43. B. Starcher, A. d'Azzo, P. W. Keller, G. K. Rao, D. Nadarajah,, and A. Hinek, Neuraminidase-1 is required for the normal assembly of elastic fibers, *Am. J. Physiol. Lung Cell Mol. Physiol.*, 295 (2008) L637–L647.
44. C. M. Kielty, M. Raghunath, L. D. Siracusa, M. J. Sherratt, R. Peters, C. A. Shuttleworth, and S. A. Jimenez, The tight skin mouse: Demonstration of mutant fibrillin-1 production and assembly into abnormal microfibrils, *J. Cell Biol.*, 140 (1998) 1159–1166.
45. J. W. Dennis, M. Granovsky, and C. E. Warren, Glycoprotein glycosylation and cancer progression, *Biochim. Biophys. Acta*, 1473 (1999) 21–34.
46. T. Miyagi, T. Wada, K. Yamaguchi, K. Shiozaki, I. Sato, Y. Kakugawa, H. Yamanami, and T. Fujiya, Human sialidase as a cancer marker, *Proteomics*, 8 (2008) 3303–3311.
47. T. Wada, K. Hata, K. Yamaguchi, K. Shiozaki, K. Koseki, S. Moriya, and T. Miyagi, A crucial role of plasma membrane-associated sialidase in the survival of human cancer cells, *Oncogene*, 26 (2007) 2483–2490.

48. T. Uemura, K. Shiozaki, K. Yamaguchi, S. Miyazaki, S. Satomi, K. Kato, H. Sakuraba, and T. Miyagi, Contribution of sialidase NEU1 to suppression of metastasis of human colon cancer cells through desialylation of integrin beta4, *Oncogene*, 28 (2009) 1218–1229.
49. T. Miyagi, T. Wada, A. Iwamatsu, K. Hata, Y. Yoshikawa, S. Tokuyama, and M. Sawada, Molecular cloning and characterization of a plasma membrane-associated sialidase specific for gangliosides, *J. Biol. Chem.*, 274 (1999) 5004–5011.
50. G. Yogalingam, E. J. Bonten, D. van de Vlekkert, H. Hu, S. Moshiach, S.A. Connell, and A. d'Azzo, Neuraminidase 1 is a negative regulator of lysosomal exocytosis, *Dev. Cell*, 15 (2008) 74–86.
51. I.G. Winkler, J. Hendy, P. Coughlin, A. Horvath, and J. P. Levesque, Serine protease inhibitors serpina1 and serpina3 are down-regulated in bone marrow during hematopoietic progenitor mobilization, *J. Exp. Med.*, 201 (2005) 1077–1088.
52. J. P. Levesque, Y. Takamatsu, S. K. Nilsson, D. N. Haylock, and P. J. Simmons, Vascular cell adhesion molecule-1 (CD106) is cleaved by neutrophil proteases in the bone marrow following hematopoietic progenitor cell mobilization by granulocyte colony-stimulating factor, *Blood*, 98 (2001) 1289–1297.
53. C. Dahlgren, S. R. Carlsson, A. Karlsson, H. Lundqvist, and C. Sjölin, The lysosomal membrane glycoproteins LAMP-1 and LAMP-2 are present in mobilizable organelles, but are absent from the azurophil granules of human neutrophils, *Biochem. J.*, 311 (1995) 667–674.
54. X. Wu, K. A. Steigelman, E. Bonten, H. Hu, W. He, T. Ren, J. Zuo, and A. d'Azzo, Vacuolization and alterations of lysosomal membrane proteins in cochlear marginal cells contribute to hearing loss in neuraminidase 1-deficient mice, *Biochim. Biophys. Acta*, 1802 (2010) 259–268.
55. V. Seyrantepe, A. Iannello, F. Liang, E. Kanshin, P. Jayanth, S. Samarani, M.R. Szewczuk, A. Ahmad, and A. V. Pshezhetsky, Regulation of phagocytosis in macrophages by the neuraminidase 1, *J. Biol. Chem.*, 285 (2010) 206–215.
56. T. Miyagi, K. Konno, Y. Emori, H. Kawasaki, K. Suzuki, A. Yasui, and S. Tsuik, Molecular cloning and expression of cDNA encoding rat skeletal muscle cytosolic sialidase, *J. Biol. Chem.*, 268 (1993) 26435–26440.
57. J. Ferrari, R. Harris, and T. G. Warner, Cloning and expression of a soluble sialidase from Chinese hamster ovary cells: Sequence alignment similarities to bacterial sialidases, *Glycobiology*, 4 (1994) 367–373.
58. E. Monti, A. Preti, C. Nesti, A. Ballabio, and G. Borsani, Expression of a novel human sialidase encoded by the NEU2 gene, *Glycobiology*, 9 (1999) 1313–1321.
59. C. Tringali, N. Papini, P. Fusi, G. Croci, G. Borsani, A. Preti, P. Tortora, G. Tettamanti, B. Venerando, and E. Monti, Properties of recombinant human

cytosolic sialidase HsNEU2. The enzyme hydrolyzes monomerically dispersed GM1 ganglioside molecules, *J. Biol. Chem.*, 279 (2004) 3169–3179.

60. P. Roggentin, B. Rothe, J. B. Kaper, J. Galen, L. Lawrisuk, E. R. Vimr, and R. Schauer, Conserved sequences in bacterial and viral sialidases, *Glycoconj. J.*, 6 (1989) 349–353.
61. W.P. Burmeister, B. Henrissat, C. Bosso, S. Cusack, and R. W. Ruigrok, Influenza B virus neuraminidase can synthesize its own inhibitor, *Structure*, 1 (1993) 19–26.
62. S. Magesh, S. Moriya, T. Suzuki, T. Miyagi, H. Ishida, and M. Kiso, Design, synthesis, and biological evaluation of human sialidase inhibitors. Part 1: Selective inhibitors of lysosomal sialidase (NEU1), *Bioorg. Med. Chem. Lett.*, 18 (2008) 532–537.
63. S. Magesh, V. Savita, S. Moriya, T. Suzuki, T. Miyagi, H. Ishida, and M. Kiso, Human sialidase inhibitors: Design, synthesis, and biological evaluation of 4-acetamido-5-acylamido-2-fluoro benzoic acids, *Bioorg. Med. Chem.*, 17 (2009) 4595–4603.
64. R. R. Copley, R. B. Russell, and C. P. Ponting, Sialidase-like Asp-boxes: Sequence-similar structures within different protein folds, *Protein Sci.*, 10 (2001) 285–292.
65. K. Sato and T. Miyagi, Involvement of an endogenous sialidase in skeletal muscle cell differentiation, *Biochem. Biophys. Res. Commun.*, 221 (1996) 826–830.
66. A. Fanzani, R. Giuliani, F. Colombo, D. Zizioli, M. Presta, A. Preti, and S. Marchesini, Overexpression of cytosolic sialidase Neu2 induces myoblast differentiation in C2C12 cells, *FEBS Lett.*, 547 (2003) 183–188.
67. N. Suzuki, M. Aoki, Y. Hinuma, T. Takahashi, Y. Onodera, A. Ishigaki, M. Kato, H. Warita, M. Tateyama, and Y. Itoyama, Expression profiling with progression of dystrophic change in dysferlin-deficient mice (SJL), *Neurosci. Res.*, 52 (2005) 47–60.
68. R. E. Bittner, L. V. Anderson, E. Burkhardt, R. Bashir, E. Vafiadaki, S. Ivanova, T. Raffelsberger, I. Maerk, H. Hoger, M. Jung, M. Karbasiyan, M. Storch, H. Lassmann, J. A. Moss, K. Davison, R. Harrison, K. M. Bushby, and A. Reis, Dysferlin deletion in SJL mice (SJL-Dysf) defines a natural model for limb girdle muscular dystrophy 2B, *Nat. Genet.*, 23 (1999) 141–142.
69. A. Fanzani, F. Colombo, R. Giuliani, A. Preti, and S. Marchesini, Insulin-like growth factor 1 signaling regulates cytosolic sialidase Neu2 expression during myoblast differentiation and hypertrophy, *FEBS J.*, 273 (2006) 3709–3721.
70. A. Fanzani, R. Giuliani, F. Colombo, S. Rossi, E. Stoppani, W. Martinet, A. Preti, and S. Marchesini, The enzymatic activity of sialidase Neu2 is inversely regulated during *in vitro* myoblast hypertrophy and atrophy, *Biochem. Biophys. Res. Commun.*, 370 (2008) 376–381.

71. S. Rossi, E. Stoppani, W. Martinet, A. Bonetto, P. Costelli, R. Giuliani, F. Colombo, A. Preti, S. Marchesini, and A. Fanzani, The cytosolic sialidase Neu2 is degraded by autophagy during myoblast atrophy, *Biochim. Biophys. Acta*, 1790 (2009) 817–828.
72. K. M. Skubitz and D. R. D'Adamo, Sarcoma, *Mayo Clin. Proc.*, 82 (2007) 1409–1432.
73. E. Stoppani, S. Rossi, S. Marchesini, A. Preti, and A. Fanzani, Defective myogenic differentiation of human rhabdomyosarcoma cells is characterized by sialidase Neu2 loss of expression, *Cell Biol. Int.*, 33 (2009) 1020–1025.
74. A. Fanzani, F. Colombo, R. Giuliani, A. Preti, and S. Marchesini, Cytosolic sialidase Neu2 upregulation during PC12 cells differentiation, *FEBS Lett.*, 566 (2004) 178–182.
75. C. Tringali, L. Anastasia, N. Papini, A. Bianchi, L. Ronzoni, M. D. Cappellini, E. Monti, G. Tettamanti, and B. Venerando, Modification of sialidase levels and sialoglycoconjugate pattern during erythroid and erythroleukemic cell differentiation, *Glycoconj. J.*, 24 (2007) 67–79.
76. C.B. Lozzio, B. B. Lozzio, E. A. Machado, J. E. Fuhr, S. V. Lair, and E. G. Bamberger, Effects of sodium butyrate on human chronic myelogenous leukaemia cell line K562, *Nature*, 281 (1979) 709–710.
77. C. Tringali, B. Lupo, L. Anastasia, N. Papini, E. Monti, R. Bresciani, G. Tettamanti, and B. Venerando, Expression of sialidase Neu2 in leukemic K562 cells induces apoptosis by impairing Bcr-Abl/Src kinases signaling, *J. Biol. Chem.*, 282 (2007) 14364–14372.
78. A. W. Hampson and J. S. Mackenzie, The influenza viruses, *Med. J. Aust.*, 185 (2006) S39–S43.
79. G. Neumann, T. Noda, and Y. Kawaoka, Emergence and pandemic potential of swine-origin H1N1 influenza virus, *Nature*, 459 (2009) 931–939.
80. M. von Itzstein, The war against influenza: Discovery and development of sialidase inhibitors, *Nat. Rev. Drug Discov.*, 6 (2007) 967–974.
81. T. Islam, and M. von Itzstein, Anti-influenza drug discovery: Are we ready for the next pandemic?, *Adv. Carbohydr. Chem. Biochem.*, 61 (2007) 293–352.
82. I. Fuyuno, Tamiflu side effects come under scrutiny, *Nature*, 446 (2007) 358–359.
83. C. Y. Li, Q. Yu, Z. Q. Ye, Y. Sun, Q. He, X. M. Li, W. Zhang, J. Luo, X. Gu, X. Zheng, and L. Wei, A nonsynonymous SNP in human cytosolic sialidase in a small Asian population results in reduced enzyme activity: Potential link with severe adverse reactions to oseltamivir, *Cell Res.*, 17 (2007) 357–362.
84. S.R. Maxwell, Tamiflu and neuropsychiatric disturbance in adolescents, *Br. Med. J.*, 334 (2007) 1232–1233.

85. M. Long, Side effects of Tamiflu: Clues from an Asian single nucleotide polymorphism, *Cell Res.*, 17 (2007) 309–310.
86. K. Hata, K. Koseki, K. Yamaguchi, S. Moriya, Y. Suzuki, S. Yingsakmongkon, G. Hirai, M. Sodeoka, M. von Itzstein, and T. Miyagi, Limited inhibitory effects of oseltamivir and zanamivir on human sialidases, *Antimicrob. Agents Chemother.*, 52 (2008) 3484–3491.
87. M.L. Moore, M. H. Chi, W. Zhou, K. Goleniewska, J. F. O'Neal, J. N. Higginbotham, and R. S. PeeblesJr., Cutting Edge: Oseltamivir decreases T cell GM1 expression and inhibits clearance of respiratory syncytial virus: Potential role of endogenous sialidase in antiviral immunity, *J. Immunol.*, 178 (2007) 2651–2654.
88. G. Ashwell and J. Hartford, Carbohydrate-specific receptors of the liver, *Annu. Rev. Biochem.*, 51 (1982) 531–554.
89. F. A. Ngantung, P. G. Miller, F. R. Brushett, G. L. Tang, and D. I. Wang, RNA interference of sialidase improves glycoprotein sialic acid content consistency, *Biotechnol. Bioeng.*, 95 (2006) 106–119.
90. T. Koda, S. Kijimoto-Ochiai, S. Uemura, and J. Inokuchi, Specific expression of Neu2 type B in mouse thymus and the existence of a membrane-bound form in COS cells, *Biochem. Biophys. Res. Commun.*, 387 (2009) 729–735.
91. C. Oehler, J. Kopitz, and M. Cantz, Substrate specificity and inhibitor studies of a membrane-bound ganglioside sialidase isolated from human brain tissue, *J. Biol. Chem.*, 383 (2002) 1735–1742.
92. K. T. Ha, Y. C. Lee, S. H. Cho, J. K. Kim, and C. H. Kim, Molecular characterization of membrane type and ganglioside-specific sialidase (Neu3) expressed in *E. coli, Mol. Cells*, 17 (2004) 267–273.
93. G. Zanchetti, P. Colombi, M. Manzoni, L. Anastasia, L. Caimi, G. Borsani, B. Venerando, G. Tettamanti, A. Preti, E. Monti, and R. Bresciani, Sialidase NEU3 is a peripheral membrane protein localized on the cell surface and in endosomal structures, *Biochem. J.*, 408 (2007) 211–219.
94. K. Yamaguchi, K. Hata, T. Wada, S. Moriya, and T. Miyagi, Epidermal growth factor-induced mobilization of a ganglioside-specific sialidase (NEU3) to membrane ruffles, *Biochem. Biophys. Res. Commun.*, 346 (2006) 484–490.
95. A. B. Jaffe and A. Hall, Rho GTPases: Biochemistry and biology, *Annu. Rev. Cell Dev. Biol.*, 21 (2005) 247–269.
96. N. Papini, L. Anastasia, C. Tringali, G. Croci, R. Bresciani, K. Yamaguchi, T. Miyagi, A. Preti, A. Prinetti, S. Prioni, S. Sonnino, G. Tettamanti, B. Venerando, and E. Monti, The plasma membrane-associated sialidase MmNEU3 modifies the ganglioside pattern of adjacent cells supporting its involvement in cell-to-cell interactions, *J. Biol. Chem.*, 279 (2004) 16989–16995.

97. G. Schmitz and M. Grandl, Update on lipid membrane microdomains, *Curr. Opin. Clin. Nutr. Metab. Care*, 11 (2008) 106–112.
98. K. Simons and D. Toomre, Lipid rafts and signal transduction, *Nat. Rev. Mol. Cell Biol.*, 1 (2000) 31–39.
99. K. Simons and R. Ehehalt, Cholesterol, lipid rafts, and disease, *J. Clin. Invest.*, 110 (2002) 597–603.
100. D. Kalka, C. von Reitzenstein, J. Kopitz, and M. Cantz, The plasma membrane ganglioside sialidase cofractionates with markers of lipid rafts, *Biochem. Biophys. Res. Commun.*, 283 (2001) 989–993.
101. Y. Wang, K. Yamaguchi, T. Wada, K. Hata, X. Zhao, T. Fujimoto, and T. Miyagi, A close association of the ganglioside-specific sialidase Neu3 with caveolin in membrane microdomains, *J. Biol. Chem.*, 277 (2002) 26252–26259.
102. E. Odintsova, T. D. Butters, E. Monti, H. Sprong, G. van Meer, and F. Berditchevski, Gangliosides play an important role in the organization of CD82-enriched microdomains, *Biochem. J.*, 400 (2006) 315–325.
103. M. Yunta and P. A. Lazo, Tetraspanin proteins as organisers of membrane microdomains and signalling complexes, *Cell. Signal.*, 15 (2003) 559–564.
104. S. Levy and T. Shoham, The tetraspanin web modulates immune-signalling complexes, *Nat. Rev. Immunol.*, 5 (2005) 136–148.
105. F. Berditchevski, Complexes of tetraspanins with integrins: More than meets the eye, *J. Cell. Sci.*, 114 (2001) 4143–4151.
106. M. E. Hemler, Tetraspanin functions and associated microdomains, *Nat. Rev. Mol. Cell Biol.*, 6 (2005) 801–811.
107. L. Veracini, V. Simon, V. Richard, B. Schraven, V. Horejsi, S. Roche, and C. Benistant, The Csk-binding protein PAG regulates PDGF-induced Src mitogenic signaling via GM1, *J. Cell Biol.*, 182 (2008) 603–614.
108. P. A. Bromann, H. Korkaya, and S. A. Courtneidge, The interplay between Src family kinases and receptor tyrosine kinases, *Oncogene*, 23 (2004) 7957–7968.
109. T. Shima, S. Nada, and M. Okada, Transmembrane phosphoprotein Cbp senses cell adhesion signaling mediated by Src family kinase in lipid rafts, *Proc. Natl. Acad. Sci. U.S.A.*, 100 (2003) 14897–14902.
110. J. Kopitz, C. von Reitzenstein, C. Muhl, and M. Cantz, Role of plasma membrane ganglioside sialidase of human neuroblastoma cells in growth control and differentiation, *Biochem. Biophys. Res. Commun.*, 199 (1994) 1188–1193.
111. T. Hasegawa, K. Yamaguchi, T. Wada, A. Takeda, Y. Itoyama, and T. Miyagi, Molecular cloning of mouse ganglioside sialidase and its increased expression in Neuro2a cell differentiation, *J. Biol. Chem.*, 275 (2000) 8007–8015.

112. J. A. Rodriguez, E. Piddini, T. Hasegawa, T. Miyagi, and C. G. Dotti, Plasma membrane ganglioside sialidase regulates axonal growth and regeneration in hippocampal neurons in culture, *J. Neurosci.*, 21 (2001) 8387–8395.
113. J. S. Da Silva, T. Hasegawa, T. Miyagi, C. G. Dotti, and J. Abad-Rodriguez, Asymmetric membrane ganglioside sialidase activity specifies axonal fate, *Nat. Neurosci.*, 8 (2005) 606–615.
114. R. W. Ledeen, G. Wu, Z. H. Lu, D. Kozireski-Chuback, and Y. Fang, The role of GM1 and other gangliosides in neuronal differentiation. Overview and new finding, *Ann. N. Y. Acad. Sci.*, 845 (1998) 161–175.
115. A. M. Duchemin, N. H. Neff, and M. Hadjiconstantinou, Induction of Trk phosphorylation in rat brain by GM1 ganglioside, *Ann. N. Y. Acad. Sci.*, 845 (1998) 406.
116. K. Shiozaki, K. Koseki, K. Yamaguchi, M. Shiozaki, H. Narimatsu, and T. Miyagi, Developmental change of sialidase neu4 expression in murine brain and its involvement in the regulation of neuronal cell differentiation, *J. Biol. Chem.*, 284 (2009) 21157–21164.
117. R. Valaperta, M. Valsecchi, F. Rocchetta, M. Aureli, S. Prioni, A. Prinetti, V. Chigorno, and S. Sonnino, Induction of axonal differentiation by silencing plasma membrane-associated sialidase Neu3 in neuroblastoma cells, *J. Neurochem.*, 100 (2007) 708–719.
118. L. Anastasia, N. Papini, F. Colazzo, G. Palazzolo, C. Tringali, L. Dileo, M. Piccoli, E. Conforti, C. Sitzia, E. Monti, M. Sampaolesi, G. Tettamanti, and B. Venerando, NEU3 sialidase strictly modulates GM3 levels in skeletal myoblasts C2C12 thus favoring their differentiation and protecting them from apoptosis, *J. Biol. Chem.*, 283 (2008) 36265–36271.
119. C. Tringali, B. Lupo, F. Cirillo, N. Papini, L. Anastasia, G. Lamorte, P. Colombi, R. Bresciani, E. Monti, G. Tettamanti, and B. Venerando, Silencing of membrane-associated sialidase Neu3 diminishes apoptosis resistance and triggers megakaryocytic differentiation of chronic myeloid leukemic cells K562 through the increase of ganglioside GM3, *Cell Death Differ.*, 16 (2009) 164–174.
120. A. Jacquel, M. Herrant, V. Defamie, N. Belhacene, P. Colosetti, S. Marchetti, L. Legros, M. Deckert, B. Mari, J. P. Cassuto, P. Hofman, and P. Auberger, A survey of the signaling pathways involved in megakaryocytic differentiation of the human K562 leukemia cell line by molecular and c-DNA array analysis, *Oncogene*, 25 (2006) 781–794.
121. U. H. Jin, K. T. Ha, K. W. Kim, Y. C. Chang, Y. C. Lee, J. H. Ko, and C. H. Kim, Membrane type sialidase inhibits the megakaryocytic differentiation of human leukemia K562 cells, *Biochim. Biophys. Acta*, 1780 (2008) 757–763.

122. M. Nakamura, K. Kirito, J. Yamanoi, T. Wainai, H. Nojiri, and M. Saito, Ganglioside GM3 can induce megakaryocytoid differentiation of human leukemia cell line K562 cells, *Cancer Res.*, 51 (1991) 1940–1945.
123. R. Valaperta, V. Chigorno, L. Basso, A. Prinetti, R. Bresciani, A. Preti, T. Miyagi, and S. Sonnino, Plasma membrane production of ceramide from ganglioside GM3 in human fibroblasts, *FASEB J.*, 20 (2006) 1227–1229.
124. M. Schenck, A. Carpinteiro, H. Grassme, F. Lang, and E. Gulbins, Ceramide: Physiological and pathophysiological aspects, *Arch. Biochem. Biophys.*, 462 (2007) 171–175.
125. M. Itzstein and R. Thomson, Anti-influenza drugs: The development of sialidase inhibitors, *Handb. Exp. Pharmacol.*, 189 (2009) 111–154.
126. R. A. Lamb and D. Kolakofsky, Paramyxoviridae: The viruses and their replication, in D. M. Knippe, and P. M. Howley (Eds.), *Fields Virology*, Lippincot-Williams and Wilkins, New York, 2001.
127. S. Crennell, T. Takimoto, A. Portner, and G. Taylor, Crystal structure of the multifunctional paramyxovirus hemagglutinin-neuraminidase, *Nat. Struct. Biol.*, 7 (2000) 1068–1074.
128. Y. Suzuki, M. Suzuki, M. Matsunaga, and M. Matsumoto, Gangliosides as paramyxovirus receptor. Structural requirement of sialo-oligosaccharides in receptors for hemagglutinating virus of Japan (Sendai virus) and Newcastle disease virus, *J. Biochem. (Tokyo)*, 97 (1985) 1189–1199.
129. L. Anastasia, J. Holguera, A. Bianchi, F. D'Avila, N. Papini, C. Tringali, E. Monti, E. Villar, B. Venerando, I. Munoz-Barroso, and G. Tettamanti, Overexpression of mammalian sialidase NEU3 reduces Newcastle disease virus entry and propagation in COS7 cells, *Biochim. Biophys. Acta*, 1780 (2008) 504–512.
130. E. M. Comelli, M. Amado, S. R. Lustig, and J. C. Paulson, Identification and expression of Neu4, a novel murine sialidase, *Gene*, 321 (2003) 155–161.
131. E. Monti, M. T. Bassi, R. Bresciani, S. Civini, G. L. Croci, N. Papini, M. Riboni, G. Zanchetti, A. Ballabio, A. Preti, G. Tettamanti, B. Venerando, and G. Borsani, Molecular cloning and characterization of NEU4, the fourth member of the human sialidase gene family, *Genomics*, 83 (2004) 445–453.
132. M. Manzoni, P. Colombi, N. Papini, L. Rubaga, N. Tiso, A. Preti, B. Venerando, G. Tettamanti, R. Bresciani, F. Argenton, G. Borsani, and E. Monti, Molecular cloning and biochemical characterization of sialidases from zebrafish (*Danio rerio*), *Biochem. J.*, 408 (2007) 395–406.
133. K. Yamaguchi, K. Hata, K. Koseki, K. Shiozaki, H. Akita, T. Wada, S. Moriya, and T. Miyagi, Evidence for mitochondrial localization of a novel human sialidase (NEU4), *Biochem. J.*, 390 (2005) 85–93.

134. V. Seyrantepe, K. Landry, S. Trudel, J. A. Hassan, C. R. Morales, and A. V. Pshezhetsky, Neu4, a novel human lysosomal lumen sialidase, confers normal phenotype to sialidosis and galactosialidosis cells, *J. Biol. Chem.*, 279 (2004) 37021–37029.
135. A. Bigi, L. Morosi, C. Pozzi, M. Forcella, G. Tettamanti, B. Venerando, E. Monti, and P. Fusi, Human sialidase NEU4 long and short are extrinsic proteins bound to outer mitochondrial membrane and the endoplasmic reticulum, respectively, *Glycobiology*, 20 (2009) 148–157.
136. F. Malisan and R. Testi, GD3 ganglioside and apoptosis, *Biochim. Biophys. Acta*, 1585 (2002) 179–187.
137. T. Hasegawa, N. Sugeno, A. Takeda, M. Matsuzaki-Kobayashi, A. Kikuchi, K. Furukawa, T. Miyagi, and Y. Itoyama, Role of Neu4L sialidase and its substrate ganglioside GD3 in neuronal apoptosis induced by catechol metabolites, *FEBS Lett.*, 581 (2007) 406–412.
138. V. Seyrantepe, M. Canuel, S. Carpentier, K. Landry, S. Durand, L. Feng, J. Zeng, A. Caqueret, R. A. Gravel, S. Marchesini, C. Zwingmann, J. Michaud, C.R. Morales, T. Levade, and A. V. Pshezhetsky, Mice deficient in Neu4 sialidase exhibit abnormal ganglioside catabolism and lysosomal storage, *Hum. Mol. Genet.*, 17 (2008) 1556–1568.
139. H. Yamanami, K. Shiozaki, T. Wada, K. Yamaguchi, T. Uemura, Y. Kakugawa, T. Hujiya, and T. Miyagi, Down-regulation of sialidase NEU4 may contribute to invasive properties of human colon cancers, *Cancer Sci.*, 98 (2007) 299–307.
140. H. Hildebrandt, M. Muhlenhoff, B. Weinhold, and R. Gerardy-Schahn, Dissecting polysialic acid and NCAM functions in brain development, *J. Neurochem.*, 103 Suppl 1 (2007) 56–64.
141. T. Miyagi, T. Wada, K. Yamaguchi, and K. Hata, Sialidase and malignancy: A minireview, *Glycoconj. J.*, 20 (2004) 189–198.
142. T. Miyagi, Aberrant expression of sialidase and cancer progression, *Proc. Jpn. Acad., Ser. B, Phys. Biol. Sci.*, 84 (2008) 407–418.
143. T. Miyagi, T. Wada, and K. Yamaguchi, Roles of plasma membrane-associated sialidase NEU3 in human cancers, *Biochim. Biophys. Acta*, 1780 (2008) 532–537.
144. T. Miyagi, T. Wada, K. Yamaguchi, K. Hata, and K. Shiozaki, Plasma membrane-associated sialidase as a crucial regulator of transmembrane signalling, *J. Biochem.*, 144 (2008) 279–285.
145. Y. Kakugawa, T. Wada, K. Yamaguchi, H. Yamanami, K. Ouchi, I. Sato, and T. Miyagi, Up-regulation of plasma membrane-associated ganglioside sialidase (Neu3) in human colon cancer and its involvement in apoptosis suppression, *Proc. Natl. Acad. Sci. U.S.A.*, 99 (2002) 10718–10723.

146. K. Kato, K. Shiga, K. Yamaguchi, K. Hata, T. Kobayashi, K. Miyazaki, S. Saijo, and T. Miyagi, Plasma-membrane-associated sialidase (NEU3) differentially regulates integrin-mediated cell proliferation through laminin- and fibronectin-derived signalling, *Biochem. J.*, 394 (2006) 647–656.
147. S. Chatterjee and A. Pandey, The Yin and Yang of lactosylceramide metabolism: Implications in cell function, *Biochim. Biophys. Acta*, 1780 (2008) 370–382.
148. K. Shiozaki, K. Yamaguchi, I. Sato, and T. Miyagi, Plasma membrane-associated sialidase (NEU3) promotes formation of colonic aberrant crypt foci in azoxymethane-treated transgenic mice, *Cancer Sci.*, 100 (2009) 588–594.
149. A. Sasaki, K. Hata, S. Suzuki, M. Sawada, T. Wada, K. Yamaguchi, M. Obinata, H. Tateno, H. Suzuki, and T. Miyagi, Overexpression of plasma membrane-associated sialidase attenuates insulin signaling in transgenic mice, *J. Biol. Chem.*, 278 (2003) 27896–27902.
150. C. Mandal, C. Tringali, S. Mondal, L. Anastasia, S. Chandra, and B. Venerando, Down regulation of membrane-bound Neu3 constitutes a new potential marker for childhood lymphoblastic leukemia and induces apoptosis suppression of neoplastic cells, *Int. J. Cancer*, 126 (2010) 337–349.
151. A. S. Cross and D. G. Wright, Mobilization of sialidase from intracellular stores to the surface of human neutrophils and its role in stimulated adhesion responses of these cells, *J. Clin. Invest.*, 88 (1991) 2067–2076.
152. A. S. Cross, S. Sakarya, S. Rifat, T. K. Held, B. E. Drysdale, P. A. Grange, F.J. Cassels, L. X. Wang, N. Stamatos, A. Farese, D. Casey, J. Powell, A.K. Bhattacharjee, M. Kleinberg, and S. E. Goldblum, Recruitment of murine neutrophils *in vivo* through endogenous sialidase activity, *J. Biol. Chem.*, 278 (2003) 4112–4120.
153. S. Sakarya, S. Rifat, J. Zhou, D. D. Bannerman, N. M. Stamatos, A. S. Cross, and S. E. Goldblum, Mobilization of neutrophil sialidase activity desialylates the pulmonary vascular endothelial surface and increases resting neutrophil adhesion to and migration across the endothelium, *Glycobiology*, 14 (2004) 481–494.
154. C. Auffray, M. H. Sieweke, and F. Geissmann, Blood monocytes: Development, heterogeneity, and relationship with dendritic cells, *Annu. Rev. Immunol.*, 27 (2009) 669–692.
155. N. M. Stamatos, S. Curreli, D. Zella, and A. S. Cross, Desialylation of glycoconjugates on the surface of monocytes activates the extracellular signal-related kinases ERK 1/2 and results in enhanced production of specific cytokines, *J. Leukoc. Biol.*, 75 (2004) 307–313.
156. N. M. Stamatos, F. Liang, X. Nan, K. Landry, A. S. Cross, L. X. Wang, and A.V. Pshezhetsky, Differential expression of endogenous sialidases of human

monocytes during cellular differentiation into macrophages, *FEBS J.*, 272 (2005) 2545–2556.

157. M. V. Vinogradova, L. Michaud, A. V. Mezentsev, K. E. Lukong, M. El-Alfy, C. R. Morales, M. Potier, and A. V. Pshezhetsky, Molecular mechanism of lysosomal sialidase deficiency in galactosialidosis involves its rapid degradation, *Biochem. J.*, 330 Pt 2 (1998) 641–650.
158. F. Liang, V. Seyrantepe, K. Landry, R. Ahmad, A. Ahmad, N. M. Stamatos, and A. V. Pshezhetsky, Monocyte differentiation up-regulates the expression of the lysosomal sialidase, neu1, and triggers its targeting to the plasma membrane via major histocompatibility complex class II-positive compartments, *J. Biol. Chem.*, 281 (2006) 27526–27538.
159. P. Wang, J. Zhang, H. Bian, P. Wu, R. Kuvelkar, T. T. Kung, Y. Crawley, R.W. Egan, and M. M. Billah, Induction of lysosomal and plasma membrane-bound sialidases in human T-cells via T-cell receptor, *Biochem. J.*, 380 (2004) 425–433.
160. X. Nan, I. Carubelli, and N. M. Stamatos, Sialidase expression in activated human T lymphocytes influences production of IFN-gamma, *J. Leukoc. Biol.*, 81 (2007) 284–296.
161. S. Z. Gadhoum and R. Sackstein, CD15 expression in human myeloid cell differentiation is regulated by sialidase activity, *Nat. Chem. Biol.*, 4 (2008) 751–757.
162. W. Tao, M. Wang, E. D. Voss, R. R. Cocklin, J. A. Smith, S. H. Cooper, and H.E. Broxmeyer, Comparative proteomic analysis of human CD34+ stem/progenitor cells and mature CD15+ myeloid cells, *Stem Cells*, 22 (2004) 1003–1014.
163. S. R. Amith, P. Jayanth, S. Franchuk, S. Siddiqui, V. Seyrantepe, K. Gee, S. Basta, R. Beyaert, A. V. Pshezhetsky, and M. R. Szewczuk, Dependence of pathogen molecule-induced Toll-like receptor activation and cell function on Neu1 sialidase, *Glycoconj. J.*, 26 (2009) 1197–1212.
164. A. Woronowicz, S. R. Amith, K. De Vusser, W. Laroy, R. Contreras, S. Basta, and M. R. Szewczuk, Dependence of neurotrophic factor activation of Trk tyrosine kinase receptors on cellular sialidase, *Glycobiology*, 17 (2007) 10–24.
165. L. J. Yang, I. Lorenzini, K. Vajn, A. Mountney, L. P. Schramm, and R. L. Schnaar, Sialidase enhances spinal axon outgrowth *in vivo, Proc. Natl. Acad. Sci. U.S.A.*, 103 (2006) 11057–11062.
166. T. Spencer, M. Domeniconi, Z. Cao, and M. T. Filbin, New roles for old proteins in adult CNS axonal regeneration, *Curr. Opin. Neurobiol.*, 13 (2003) 133–139.
167. S. Mears, M. Schachner, and T. M. Brushart, Antibodies to myelin-associated glycoprotein accelerate preferential motor reinnervation, *J. Peripher. Nerv. Syst.*, 8 (2003) 91–99.

168. J. A. Bertelli and M. F. Ghizoni, Brachial plexus avulsion injury repairs with nerve transfers and nerve grafts directly implanted into the spinal cord yield partial recovery of shoulder and elbow movements, *Neurosurgery*, 52 (2003) 1385–1389, discussion 1389–1390.
169. A. Varki and T. Angata, Siglecs—the major subfamily of I-type lectins, *Glycobiology*, 16 (2006) 1R–27R.
170. A. A. Vyas, H. V. Patel, S. E. Fromholt, M. Heffer-Lauc, K. A. Vyas, J. Dang, M. Schachner, and R. L. Schnaar, Gangliosides are functional nerve cell ligands for myelin-associated glycoprotein (MAG), an inhibitor of nerve regeneration, *Proc. Natl. Acad. Sci. U.S.A.*, 99 (2002) 8412–8417.
171. A. A. Vyas, O. Blixt, J. C. Paulson, and R. L. Schnaar, Potent glycan inhibitors of myelin-associated glycoprotein enhance axon outgrowth *in vitro*, *J. Biol. Chem.*, 280 (2005) 16305–16310.
172. G. Wu, Z. H. Lu, J. Wang, Y. Wang, X. Xie, M. F. Meyenhofer, and R. W. Ledeen, Enhanced susceptibility to kainate-induced seizures, neuronal apoptosis, and death in mice lacking gangliotetraose gangliosides: Protection with LIGA 20, a membrane-permeant analog of GM1, *J. Neurosci.*, 25 (2005) 11014–11022.
173. N. K. Golovanova, E. V. Gracheva, O. P. Il'inskaya, E. M. Tararak, and N.V. Prokazova, Sialidase activity in normal and atherosclerotic human aortic intima, *Biochemistry (Mosc)*, 67 (2002) 1230–1234.
174. K. Ogura, M. Ogura, R. L. Anderson, and C. C. Sweeley, Peroxidase-amplified assay of sialidase activity toward gangliosides, *Anal. Biochem.*, 200 (1992) 52–57.
175. M. I. Noble, A. J. Drake-Holland, and H. Vink, Hypothesis: Arterial glycocalyx dysfunction is the first step in the atherothrombotic process, *QJM*, 101 (2008) 513–518.
176. N. Clausell, V. C. de Lima, S. Molossi, P. Liu, E. Turley, A. I. Gotlieb, A.G. Adelman, and M. Rabinovitch, Expression of tumour necrosis factor alpha and accumulation of fibronectin in coronary artery restenotic lesions retrieved by atherectomy, *Br. Heart J.*, 73 (1995) 534–539.
177. L. A. Cuniberti, V. Martinez, J. Schachter, G. Magarinos, P. C. Meckert, R.P. Laguens, J. Levenson, and J. P. Werba, Sialic acid as a protective barrier against neointima development, *Atherosclerosis*, 181 (2005) 225–231.
178. S. K. Moon, S. H. Cho, K. W. Kim, J. H. Jeon, J. H. Ko, B. Y. Kim, and C. H. Kim, Overexpression of membrane sialic acid-specific sialidase Neu3 inhibits matrix metalloproteinase-9 expression in vascular smooth muscle cells, *Biochem. Biophys. Res. Commun.*, 356 (2007) 542–547.

179. J. Wang, G. Wu, T. Miyagi, Z. H. Lu, and R. W. Ledeen, Sialidase occurs in both membranes of the nuclear envelope and hydrolyzes endogenous GD1a, *J. Neurochem.*, 111 (2009) 547–554.
180. M. Saito, C. L. Fronda, and R. K. Yu, Sialidase activity in nuclear membranes of rat brain, *J. Neurochem.*, 66 (1996) 2205–2208.
181. J. S. Gilchrist and G. N. Pierce, Identification and purification of a calcium-binding protein in hepatic nuclear membranes, *J. Biol. Chem.*, 268 (1993) 4291–4299.
182. X. Xie, G. Wu, Z. H. Lu, and R. W. Ledeen, Potentiation of a sodium-calcium exchanger in the nuclear envelope by nuclear GM1 ganglioside, *J. Neurochem.*, 81 (2002) 1185–1195.
183. D. Murray, P. Doran, P. MacMathuna, and A. C. Moss, *In silico* gene expression analysis—an overview, *Mol. Cancer*, 6 (2007) 50.
184. A. I. Su, T. Wiltshire, S. Batalov, H. Lapp, K. A. Ching, D. Block, J. Zhang, R. Soden, M. Hayakawa, G. Kreiman, M. P. Cooke, J. R. Walker, and J.B. Hogenesch, A gene atlas of the mouse and human protein-encoding transcriptomes, *Proc. Natl. Acad. Sci. U.S.A.*, 101 (2004) 6062–6067.
185. V. Ramensky, P. Bork, and S. Sunyaev, Human non-synonymous SNPs: Server and survey, *Nucleic Acids Res.*, 30 (2002) 3894–3900.
186. F. Lehmann, H. Gathje, S. Kelm, and F. Dietz, Evolution of sialic acid-binding proteins: Molecular cloning and expression of fish siglec-4, *Glycobiology*, 14 (2004) 959–968.
187. A. Harduin-Lepers, R. Mollicone, P. Delannoy, and R. Oriol, The animal sialyltransferases and sialyltransferase-related genes: A phylogenetic approach, *Glycobiology*, 15 (2005) 805–817.
188. H. Sohn, Y. S. Kim, H. T. Kim, C. H. Kim, E. W. Cho, H. Y. Kang, N. S. Kim, C. H. Kim, S. E. Ryu, J. H. Lee, and J. H. Ko, Ganglioside GM3 is involved in neuronal cell death, *FASEB J.*, 20 (2006) 1248–1250.
189. Y. Guerardel, L. Y. Chang, E. Maes, C. J. Huang, and K. H. Khoo, Glycomic survey mapping of zebrafish identifies unique sialylation pattern, *Glycobiology*, 16 (2006) 244–257.
190. M. Engstler, G. Reuter, and R. Schauer, Purification and characterization of a novel sialidase found in procyclic culture forms of *Trypanosoma brucei*, *Mol. Biochem. Parasitol.*, 54 (1992) 21–30.
191. E. Tiralongo, S. Schrader, H. Lange, H. Lemke, J. Tiralongo, and R. Schauer, Two trans-sialidase forms with different sialic acid transfer and sialidase activities from *Trypanosoma congolense*, *J. Biol. Chem.*, 278 (2003) 23301–23310.

192. S. Schenkman, M. S. Jiang, G. W. Hart, and V. Nussenzweig, A novel cell surface trans-sialidase of *Trypanosoma cruzi* generates a stage-specific epitope required for invasion of mammalian cells, *Cell*, 65 (1991) 1117–1125.
193. F. Vandekerckhove, S. Schenkman, L. Pontes, S. de Carvalho, Tomlinson, M. Kiso, M. Yoshida, A. Hasegawa, and V. Nussenzweig, Substrate specificity of the *Trypanosoma cruzi* trans-sialidase, *Glycobiology*, 2 (1992) 541–548.
194. S. Schrader and R. Schauer, Nonradioactive trans-sialidase screening assay, in I. Brockhausen (Ed.), *Glycobiology Protocols*, Humana Press Inc, Totowa, 2006, pp. 93–107.
195. E. Medina-Acosta, S. Paul, S. Tomlinson, and L. C. Pontes-de-Carvalho, Combined occurrence of trypanosomal sialidase/trans-sialidase activities and leishmanial metalloproteinase gene homologues in *Endotrypanum* sp, *Mol. Biochem. Parasitol.*, 64 (1994) 273–282.
196. V. V. Tertov, V. V. Kaplun, I. A. Sobenin, E. Y. Boytsova, N. V. Bovin, and A.N. Orekhov, Human plasma trans-sialidase causes atherogenic modification of low density lipoprotein, *Atherosclerosis*, 159 (2001) 103–115.
197. A. L. Mattos-Guaraldi, L. C. Formiga, and A. F. Andrade, Trans-sialidase activity for sialic acid incorporation on *Corynebacterium diphtheriae*, *FEMS Microbiol. Lett.*, 168 (1998) 167–172.
198. J. Cheng, H. Yu, K. Lau, S. Huang, H. A. Chokhawala, Y. Li, V. K. Tiwari, and X. Chen, Multifunctionality of *Campylobacter jejuni* sialyltransferase CstII: Characterization of GD3/GT3 oligosaccharide synthase, GD3 oligosaccharide sialidase, and trans-sialidase activities, *Glycobiology*, 18 (2008) 686–697.
199. H. Yu, H. Chokhawala, R. Karpel, B. Wu, J. Zhang, Y. Zhang, Q. Jia, and X. Chen, A multifunctional *Pasteurella multocida* sialyltransferase: A powerful tool for the synthesis of sialoside libraries, *J. Am. Chem. Soc.*, 127 (2005) 17618–17619.
200. Y. Luo, S. C. Li, Y. T. Li, and M. Luo, The 1.8 Å structures of leech intramolecular trans-sialidase complexes: Evidence of its enzymatic mechanism, *J. Mol. Biol.*, 285 (1999) 323–332.
201. H. Gut, S. J. King, and M. A. Walsh, Structural and functional studies of *Streptococcus pneumoniae* neuraminidase B: An intramolecular trans-sialidase, *FEBS Lett.*, 582 (2008) 3348–3352.
202. S. Schrader, E. Tiralongo, G. Paris, T. Yoshino, and R. Schauer, A nonradioactive 96-well plate assay for screening of trans-sialidase activity, *Anal. Biochem.*, 322 (2003) 139–147.
203. M. Engstler, R. Schauer, and R. Brun, Distribution of developmentally regulated trans-sialidases in the Kinetoplastida and characterization of a shed trans-

sialidase activity from procyclic *Trypanosoma congolense, Acta Trop.*, 59 (1995) 117–129.

204. R. Schauer and J. P. Kamerling, Chemistry, biochemistry and biology of sialic acids, in J. Montreuil, J. F. Vliegenthart, and H. Schachte (Eds.), *Glycoproteins II*, Elsevier Science B.V., Amsterdam, 1997, pp. 243–402.
205. S. Schenkman, L. Pontes de Carvalho, and V. Nussenzweig, *Trypanosoma cruzi* trans-sialidase and neuraminidase activities can be mediated by the same enzymes, *J. Exp. Med.*, 175 (1992) 567–575.
206. M. J. Alves and W. Colli, *Trypanosoma cruzi*: Adhesion to the host cell and intracellular survival, *IUBMB Life*, 59 (2007) 274–279.
207. K. Nagamune, A. Acosta-Serrano, H. Uemura, R. Brun, C. Kunz-Renggli, Y. Maeda, M. A. Ferguson, and T. Kinoshita, Surface sialic acids taken from the host allow trypanosome survival in tsetse fly vectors, *J. Exp. Med.*, 199 (2004) 1445–1450.
208. L. Ratier, M. Urrutia, G. Paris, L. Zarebski, A. C. Frasch, and F. A. Goldbaum, Relevance of the diversity among members of the *Trypanosoma cruzi* trans-sialidase family analyzed with camelids single-domain antibodies, *PLoS ONE*, 3 (2008) e3524.
209. W. Gao, H. H. Wortis, and M. A. Pereira, The *Trypanosoma cruzi* trans-sialidase is a T cell-independent B cell mitogen and an inducer of non-specific Ig secretion, *Int. Immunol.*, 14 (2002) 299–308.
210. W. B. Dias, F. D. Fajardo, A. V. Graca-Souza, L. Freire-de-Lima, F. Vieira, M.F. Girard, B. Bouteille, J. O. Previato, L. Mendonca-Previato, and A.R. Todeschini, Endothelial cell signalling induced by trans-sialidase from *Trypanosoma cruzi*, *Cell Microbiol.*, 10 (2008) 88–99.
211. M. G. Risso, G. B. Garbarino, E. Mocetti, O. Campetella, S. M. Gonzalez Cappa, C. A. Buscaglia, and M. S. Leguizamon, Differential expression of a virulence factor, the trans-sialidase, by the main *Trypanosoma cruzi* phylogenetic lineages, *J. Infect. Dis.*, 189 (2004) 2250–2259.
212. A.R. Todeschini, M. P. Nunes, R. S. Pires, M. F. Lopes, J. O. Previato, L. Mendonca-Previato, and G. A. DosReis, Costimulation of host T lymphocytes by a trypanosomal trans-sialidase: Involvement of CD43 signaling, *J. Immunol.*, 168 (2002) 5192–5198.
213. B. Neubacher, D. Schmidt, P. Ziegelmuller, and J. Thiem, Preparation of sialylated oligosaccharides employing recombinant trans-sialidase from *Trypanosoma cruzi*, *Org. Biomol. Chem.*, 3 (2005) 1551–1556.
214. V. M. Mendoza, R. Agusti, C. Gallo-Rodriguez, and R. M. de Lederkremer, Synthesis of the O-linked pentasaccharide in glycoproteins of *Trypanosoma cruzi* and selective sialylation by recombinant trans-sialidase, *Carbohydr. Res.*, 341 (2006) 1488–1497.

215. R. Agusti, M. E. Giorgi, and R. M. de Lederkremer, The trans-sialidase from *Trypanosoma cruzi* efficiently transfers alpha-(2→3)-linked *N*-glycolylneuraminic acid to terminal beta-galactosyl units, *Carbohydr. Res.*, 342 (2007) 2465–2469.
216. M. Engstler, G. Reuter, and R. Schauer, The developmentally regulated trans-sialidase from *Trypanosoma brucei* sialylates the procyclic acidic repetitive protein, *Mol. Biochem. Parasitol.*, 61 (1993) 1–13.
217. L. C. Pontes, S. de Carvalho, Tomlinson, F. Vandekerckhove, E. J. Bienen, A.B. Clarkson, M. S. Jiang, G. W. Hart, and V. Nussenzweig, Characterization of a novel trans-sialidase of *Trypanosoma brucei* procyclic trypomastigotes and identification of procyclin as the main sialic acid acceptor, *J. Exp. Med.*, 177 (1993) 465–474.
218. R. Agusti, M. E. Giorgi, V. M. Mendoza, C. Gallo-Rodriguez, and R. M. de Lederkremer, Comparative rates of sialylation by recombinant trans-sialidase and inhibitor properties of synthetic oligosaccharides from *Trypanosoma cruzi* mucins-containing galactofuranose and galactopyranose, *Bioorg. Med. Chem.*, 15 (2007) 2611–2616.
219. S. Schenkman, D. Eichinger, M. E. Pereira, and V. Nussenzweig, Structural and functional properties of *Trypanosoma* trans-sialidase, *Annu. Rev. Microbiol.*, 48 (1994) 499–523.
220. A. G. Watts, I. Damager, M. L. Amaya, A. Buschiazzo, P. Alzari, A. C. Frasch, and S. G. Withers, *Trypanosoma cruzi* trans-sialidase operates through a covalent sialyl-enzyme intermediate: Tyrosine is the catalytic nucleophile, *J. Am. Chem. Soc.*, 125 (2003) 7532–7533.
221. R. Agusti, G. Paris, L. Ratier, A. C. Frasch, and R. M. de Lederkremer, Lactose derivatives are inhibitors of *Trypanosoma cruzi* trans-sialidase activity toward conventional substrates *in vitro* and *in vivo*, *Glycobiology*, 14 (2004) 659–670.
222. M. E. Pereira, J. S. Mejia, E. Ortega-Barria, D. Matzilevich, and R. P. Prioli, The *Trypanosoma cruzi* neuraminidase contains sequences similar to bacterial neuraminidases, YWTD repeats of the low density lipoprotein receptor, and type III modules of fibronectin, *J. Exp. Med.*, 174 (1991) 179–191.
223. O. E. Campetella, A. D. Uttaro, A. J. Parodi, and A. C. Frasch, A recombinant *Trypanosoma cruzi* trans-sialidase lacking the amino acid repeats retains the enzymatic activity, *Mol. Biochem. Parasitol.*, 64 (1994) 337–340.
224. A. Buschiazzo, G. A. Tavares, O. Campetella, S. Spinelli, M. L. Cremona, G. Paris, M. F. Amaya, A. C. Frasch, and P. M. Alzari, Structural basis of sialyltransferase activity in trypanosomal sialidases, *EMBO J.*, 19 (2000) 16–24.

225. A. Buschiazzo, M. F. Amaya, M. L. Cremona, A. C. Frasch, and P. M. Alzari, The crystal structure and mode of action of trans-sialidase, a key enzyme in *Trypanosoma cruzi* pathogenesis, *Mol. Cell*, 10 (2002) 757–768.
226. M. F. Amaya, A. G. Watts, I. Damager, A. Wehenkel, T. Nguyen, A. Buschiazzo, G. Paris, A. C. Frasch, S. G. Withers, and P. M. Alzari, Structural insights into the catalytic mechanism of *Trypanosoma cruzi* trans-sialidase, *Structure (Camb)*, 12 (2004) 775–784.
227. A. G. Watts, P. Oppezzo, S. G. Withers, P. M. Alzari, and A. Buschiazzo, Structural and kinetic analysis of two covalent sialosyl-enzyme intermediates on *Trypanosoma rangeli* sialidase, *J. Biol. Chem.*, 281 (2006) 4149–4155.
228. T. Haselhorst, J. C. Wilson, A. Liakatos, M. J. Kiefel, J. C. Dyason, and M. von Itzstein, NMR spectroscopic and molecular modeling investigations of the trans-sialidase from *Trypanosoma cruzi*, *Glycobiology*, 14 (2004) 895–907.
229. A. Blume, B. Neubacher, J. Thiem, and T. Peters, Donor substrate binding to trans-sialidase of *Trypanosoma cruzi* as studied by STD NMR, *Carbohydr. Res.*, 342 (2007) 1904–1909.
230. A. R. Todeschini, W. B. Dias, M. F. Girard, J. M. Wieruszeski, L. Mendonca-Previato, and J. O. Previato, Enzymatically inactive trans-sialidase from *Trypanosoma cruzi* binds sialyl and beta-galactopyranosyl residues in a sequential ordered mechanism, *J. Biol. Chem.*, 279 (2004) 5323–5328.
231. G. Paris, M. L. Cremona, M. F. Amaya, A. Buschiazzo, S. Giambiagi, A.C. Frasch, and P. M. Alzari, Probing molecular function of trypanosomal sialidases: Single point mutations can change substrate specificity and increase hydrolytic activity, *Glycobiology*, 11 (2001) 305–311.
232. G. Paris, L. Ratier, M. F. Amaya, T. Nguyen, P. M. Alzari, and A. C. Frasch, A sialidase mutant displaying trans-sialidase activity, *J. Mol. Biol.*, 345 (2005) 923–934.
233. W. Gibson, Sex and evolution in trypanosomes, *Int. J. Parasitol.*, 31 (2001) 643–647.
234. G. Montagna, M. L. Cremona, G. Paris, M. F. Amaya, A. Buschiazzo, P. M. Alzari, and A. C. Frasch, The trans-sialidase from the African trypanosome *Trypanosoma brucei*, *Eur. J. Biochem.*, 269 (2002) 2941–2950.
235. E. Tiralongo, I. Martensen, J. Grötzinger, J. Tiralongo, and R. Schauer, Trans-sialidase-like sequences from *Trypanosoma congolense* conserve most of the critical active site residues found in other trans-sialidases, *Biol. Chem.*, 384 (2003) 1203–1213.
236. I. Damager, S. Buchini, M. F. Amaya, A. Buschiazzo, P. Alzari, A. C. Frasch, A. Watts, and S. G. Withers, Kinetic and mechanistic analysis of *Trypanosoma cruzi* trans-sialidase reveals a classical ping-pong mechanism with acid/base catalysis, *Biochemistry*, 47 (2008) 3507–3512.

AUTHOR INDEX

Page numbers in roman type indicate that the listed author is cited on that page; page numbers in italic denote the page where the literature citation is given.

A

Abad, J., 432, *469*
Abd El-Thalouth, T., 132, *195*
Abdel-Malik, M. M., 129, *193*
Abdelmouleh, M., 73, *112*
Abe, M., 229, 269, *277*
Abrams, A., 336, 362, *385*
Acharya, A. S., 304, *372*
Achyuthan, A. M., 406, *460*
Achyuthan, K. E., 406, *460*
Ackerson, C. J., 215, 238, *273*
Ackman, R. G., 151, 157, *200*
Acosta-Serrano, A., 453, *477*
Adachi, S., 358, *398*
Adak, A. K., 220, 223, 231, 233, 255, 269, *276*
Adams, P., 128, *192*
Adams, Y., 124, *189*
Adden, A., 161, *204*
Adden, R., 120, 139–140, 142, 157–158, 163–164, 165–166, 173–175, 178–179, 181–182, *188*, *196–197*, *203*, *205–207*, *209*
Addison, R. F., 151, *200*
Adelman, G., 443, *474*
Adibekian, A., 250, 267, *286*
Aeschbacher, H. U., 323, 360–363, *379*, *399*
Affleck, D. J., 350, *392–393*
Affolter, M., 360, *398*
Agawa, M., 337, *385*
Aggarwal, K., 304, *372*
Aggarwal, R. J., 245, *282*
Agoston, K., 157, *203*
Agusti, R., 454–455, *477–478*
Ahmad, A., 409, 422, 441, *461*, *464*, *473*
Ahmad, R., 441, *473*
Ahmed, A. U., 49, *106*
Ahmed, M., 248, 250, *286*, *392*
Ahmed, N., 323, 347–348, *379*
Aime, S., 324, *377*
Akamanchi, K. G., 338, *386*
Akita, H., 249, 259, 265, *285*, *288*, 435–438, *470*
Akiyama, Y., 229–230, 235–236, 254, 269, *277–278*, *287*
Al-Afaleq, E., 72, 74, *112*
Alam, K., 365, *402*
Albersheim, P., 151, 157, *200*, *203*
Albert, R., 350, *393*
Albertin, L., 220, *276*
Albrecht, G., 130, 137, 147, 151, *194*, *197*, *201*
Albrecht, W., 118, *185*
Alcamí, J., 263, *289*
Alcántara, D., 225–226, 240–242, 270, *277*, *281*, *290*
Alconada, A., 409, *461*
Alejandro, R., 149, *382*
Alexakis, A., 353, *394*
Alföldi, J., 298, *369*
Ali, A., 365, *402*
Alivisatos, A. P., 245–248, *282–285*
Allen, C. L., 295, *368*

Alloin, F., 81, *113*
Almin, K. E., 177, *209*
Alric, I., 364, *395*
Altaner, C., 124, 175, 178, *189*, *208*
Altena, J. H., 317, *376*
Alvarez-Lorenzo, C., 118, *185*
Alves, M. J., 453, *477*
Alzari, P. M., 455–458, *478–479*
Alzari, P., 455, 458, *478–479*
Amado, M., 435–436, *470*
Amao, M., 326, *382*
Amass, A., 90, *115*
Amass, W., 90, *115*
Amaya, J., 361, *399*
Amaya, M. F., 455–458, *478–479*
Amaya, M. L., 455, 458, *478*
Ambronn, H., 41, *104*
Ames, B. N., 314, *375*
Ames, J. M., 304, 323, *371–372*
Amith, S. R., 442, *473*
Amore, A., 356, *396*
An, J. S., 97, *116*
Anas, A., 246, *283*
Anastasia, L., 426–427, 429–431, 433, 435, 439–440, 451–452, *466–467*, *469–470*, *472*
Anderegg, J., 216, *274*
Anderson, L. V., 425, *465*
Anderson, R. L., 443, *474*
Anderson, T., 175, 178, 181, *208*
Andrade, A. F., 453, *476*
Andress, K. R., 48, *106*
Andrews, K. T., 124, *189*
Andria, G., 411, *462*
Anet, E. F. L. J., 323, 338, 346, *379*, *386*
Angata, T., 443, *474*
Angyal, S. J., 9, 10, 131, 155, *195*, *202*, 301, *370*
Anilkumar, N., 355, *395*
Ansari, N. A., 365, *402*
Anthony, D. C., 243, 268, *281*, 363, *401*
Antipin, M. Y., 298, *369*
Aoki, M., 425, *465*
Aoki, N., 75, *112*
Aono, H., 120, *188*
Aoyama, Y., 247, 251, 264, *284, 286*
Appelhans, D., 237, *279*
Arai, T., 327, *382*
Araki, T., 337, *385*
Ardenne, R., 319, *378*
Argenton, F., 435–436, 449–450, *470*
Argirov, O. O., 435–436, *470*
Arimoto, N., 97, *116*
Arisz, P. W., 138–139, 160, 162–163, *197*
Armstrong, K. B., 297, 300, *369*
Arndt, D. J., 248, 266, *285*
Arndt-Jovin, D. J., 248, 266, *285*
Aronica, R., 323, *380*
Arts, W. F., 412–413, *462*
Asensio, C., 341, *388*
Ashwell, G., 428, *467*
Aspinall, G. O., 27, *101*
Assenza, S. P., 364, *401*
Assunção, R. M. N., 92, *115*
Astheimer, A. -J., 118, *185*
Astruc, D., 213, 270, *271*, *290*
Atalla, R. H., 40, 42–43, *104–105*
Atwood, C. S., 358, *397*
Auberger, P., 433, *469*
Aubin, S., 332, *384*
Auerbach, G., 315, *376*
Auffray, C., 440, *472*
Augeri, D. J., 351, *394*
Aureli, M., 433, *469*
Austin, C. D., 409, *461*
Auzély-Velty, R., 258, *288*
Aveyard, J., 238, *279*
Azakami, D., 327, *382*
Azcutia, V., 357, *396–397*
Azema, L., 364, *395*
Aziz, S. A., 347, *391*
Azizi, M. A. S., 81, 92, *113*, *116*

B

Baar, A., 129, *193*
Baba, Y., 337, 366, *385*, *402*
Babic, M., 245, *282*
Babu, P., 249, 257, *285*

Bacher, A., 315, 350, *376*
Backson, S. C. E., 63, *110*
Bacon, C. W., 360, *398*
Bader, G., 315, 350, *376*
Badoud, R., 338, *386*
Badyal, J. P. S., 63, *110*
Baehler, B., 309, *374*
Baek, C. H., 339, *387*
Bagdanoff, J. T., 351, *393*
Bagheri, N., 356, *396*
Bahrke, S., 163, *204*
Baidal, D. A., 327, *382*
Bakala, H., 356, *396*
Baker, A. A., 63–64, *109–110*
Baker, D. H., 361, *399*
Baker, J. R., 326, *382*
Bakke, M., 339–340, *386*
Balaram, P., 304, *372*
Baldo, L., 323, *380*
Baldwin, P. M., 125–126, *191*
Ballabio, A., 423, 428, 435, 437, *464*, *470*
Balser, K., 118, *185*
Baltes, W., 322, *378*
Baltz, T., 356, 364, *395*
Bamrungsap, S., 214, *272*
Bannerman, D. D., 440, *472*
Bantjes, A., 130, *194*
Bapst, P., 350, *392*
Baptista, J. A. B., 325, *381*
Barba, M., 358, *398*
Barbosa, J., 351, *394*
Barchi, J. J., 220, 222, 249, *275–276*, *285*
Bardelli, T., 414–415, *463*
Barnes, C. L., 302, 317–320, *370*, *376–378*
Barnoud, F., 41, 56, *104*
Barrett, M. P., 356, 364, *395*
Barrientos, A. G., 217, 219–222, 224, 226, 251–253, 254, 256, 260, 269–270, , *275–277*, *286–288*, *290*
Bartsch, H., 360, *399*
Bashir, R., 425, *465*
Bassi, M. T., 435, 437, *470*
Basso, L., 434, 439, *470*
Basta, S., 442, *473*
Basu, A., 251, *286*
Batalov, S., 447, *475*
Batelaan, J. G., 151, *201*
Bates, D. L., 327, *383*
Battista, G., 327, *382*
Battisti, O., 361, *400*
Bauer, U., 409, *461*
Bauer, W., 119–120, *187*, 350–351, *393*
Baugh, S. D. P., 351, *393*
Baum, M., 132, *195*
Baumann, H., 124, *189*
Baumann, M., 84, *113*
Bawendi, M. G., 213, 246–247, *271*, *283–284*
Bayer, E. A., 82, *113*
Baynes, J. W., 301–302, 304, 317–318, 326, 344, 348, *370*
Beacham, J. L., 327, *383*
Beaudet, A. L., 411, *462*
Beck, M., 412–413, *462*
Beck, W., 330, *383*
Becker, C. M., 304, 323, *372*
Becker, U., 150, *199*
Becker, W., 124, 130, *191*
Bednarz, M., 351, *394*
Bedoya, L. M., 263, *289*
Begley, T. P., 315, *376*
Behrendt, M., 249, 266, *286*
Beilfuß, W., 298, *370*
Beisswenger, P. J., 341, *388*
Belaich, J.-P., 82, *113*
Belgacem, M. N., 73, 85–86, 93, *112*, *114*, *116*, 123–124, *187*
Belhacene, N., 433, *469*
Belton, P. B., 43, *105*
Belton, P. S., 43, *105*
BeMiller, J. N., 3, 126, 143, 161, 182, *191*, *203*
ben Salah, A., 73, *112*
Bengele, H. H., 239, *280*
Benistant, C., 432, *468*
Benkovic, S. J., 214, *272*
Benmansour, M., 298, *369*
Bentolila, L. A., 246, *283*
Benziman, M., 28, *101*
Berditchevski, F., 432, *468*
Bergamo, P., 304, *372*

Bergen, J. M., 258, *288*
Berger, W., 294, *368*
Bergeron, M. G., 244, *282*
Berglund, L., 92, *116*
Bergmann, A., 338, *386*
Bergmann, R., 361, *400*
Bergmüller, W., 306, *373*
Bergquist, K. -E., 143, *198*
Bergsma, J., 143, 182–183, *198*, *210*
Beringue, V., 314, *375*
Berlin, P., 112, *143*
Berlioz, S., 86, *115*
Bernade, A., 260, *288*
Bernetti, K., 327, *382*
Berrino, F., 358, *398*
Berry, C. C., 239, *279*
Berry, J. W., 126, *192*
Bertelli, J. A., 443, *474*
Bertoft, E., 175–176, *198*
Bertoniere, N. R., 123, 150, *189*, *199*
Bertozzi, C. R., 217, *274–275*
Bertrand, J. M., 361, *400*
Besemer, A. C., 71, *111*
Beshay, A. D., 85, *113*
Bethell, D., 215–216, 235, *273*, *278*
Betley, T. A., 239, *280*
Betzel, C., 39, 50, *103*
Beyaert, R., 442, *473*
Bhamra, S. K., 324, *380*
Bhaskaran, S., 237, *279*
Bhattacharjee, K., 439–440, *472*
Bhattacharjee, S. S., 119, 177, *187*
Bhumkar, D. R., 237, *279*
Bian, H., 441, *473*
Bianchi, A., 426–427, 435, *466*, *470*
Bianco, G. A., 363, *401*
Bichler, J., 348, *392*
Biedermann, D., 325, *381*
Bielenberg, D., 318, 363, *378*
Biely, P., 124, *189*
Biemel, K. M., 332, 348–349, *384*, *391–392*
Bienen, E. J., 454, *478*
Biermann, C. J., 155, *202*
Bigi, A., 435, *471*
Biju, V., 246, *283*
Bikales, N. M., 73, *112*
Bikalesand, N. M., 73, *112*
Billah, M. M., 441, *473*
Bindoli, C., 234–236, 262, *278*
Bionda, N., 309, *371*
Birkofer, L., 297, *369*
Birsan, C., 208, *348*
Biswas, A., 355, *395*
Bittner, R. E., 425, *465*
Björndal, H., 119, *186*
Blaauwgeers, H. G. T., 355, *395*
Blackwell, J., 42, 44, 50, 53–54, 79, *104*, *106*, *108*
Blakeney, B., 157, *202*
Blanchevoye, C., 415, *463*
Blank, I., 303, 324, 332, 346, *371*, *380–381*, *384*, *391*
Blanton, R. L., 28, *101*
Blauenstein, P., 350, *393*
Bledzki, A. K., 350, *393*
Bleeker, I. P., 143, 182–183, *198*, *210*
Blincko, S., 326, 330, *382*
Blixt, O., 443, *474*
Block, D., 447, *475*
Blonski, C., 356, 364, *395*
Blume, A., 120, 187, *456–457*, *479*
Bobby, Z., 358, *397*
Bociek, S. M., 143, *198*
Böddi, K., 324, *380*
Bodnaruk, T. D., 413, 417–418, *462*
Bogdanov, A., 240, *281*
Boggs, J. M., 251, *286*
Bognár, R., 10, 306, *373*
Bohn, A., 121, *187*
Bohnert, H. J., 128, *192*
Bohrn, A., 133, *196*
Boisselier, E., 270, *290*
Boisset, C., 82, *113*
Boissinot, M., 244, *282*
Bokranz, W., 28, *101*
Bols, M., 161, *204*
Bonenfant, A. P., 309, *374*

Bonetto, A., 426, *466*
Bonn, G. K., 324, *380*
Bonn, R., 129, *193*
Bonomi, R., 234–236, *278*
Bonten, E. J., 403–458, *460–462*, *463–464*
Boon, J. J., 138–139, 160, 162–163, *197*
Boote, E. J., 237, *279*
Booth, K. V., 351, *393*
Borgards, A., 181, *196*
Bork, P., 449, *475*
Borkman, R. F., 215, *273*
Borowski, E., 364, *402*
Borredon, M. E., 86, *114*
Borsani, G., 403–458, *460*, *464*, *467*, *470*
Borsook, H., 336, 362, *385*
Bösch, A., 120, *188*
Bosso, C., 424, 441, *465*
Botter, H., 151, *201*
Bouchet, B., 125–126, *191*
Boucon, C., 347, *391*
Boudeulle, M., 55–56, *108*
Boufi, S., 73, *112*
Boullanger, P., 220, *276*
Bouma, B., 44, *106*
Bouteille, B., 453, *477*
Bouzar, H., 314, *375*
Bovin, N. V., 452, *476*
Bowen, J. E., 247, *284*
Bowers, M. T., 128, *192*
Boylan, J. R., 90, *115*
Boytsova, E. Y., 452, *476*
Bracher, A., 315, *376*
Braconnot, H., 33, *102*
Bragd, P. L., 71, *111*
Brandt, R., 356–357, *395–396*
Brans, L., 350, *393*
Braun, E., 34, *102*
Braun, P. J., 175–176, *207*
Braunschweiger, P. G., 318, 350, *377*
Breel, G. J., 310, *374*
Brendler, E., 69, *111*
Bresciani, R., 403–458, *466–467*, *469*, *470*
Breuer, H., 302, *371*
Bricout, J., 325, *381*
Bridiau, N., 298, *369*
Briner, U., 350, *393*
Bringaud, F., 356, 364, *395*
Brinkmalm, G., 139, 164–166, 173–176, 178–179, *197*, *205–209*
Brinkmann-Frye, E., 348, *392*
Briones, F., 241–242, *281*
Brock, J. W. C., 304, *371*
Brockhausen, I., 452–453, *476*
Brogniart, A., 118, *184*
Bromann, P. A., 432, *468*
Brome, V. A., 131, *195*
Brook, M. A., 254, *287*
Brown, E. B., 246, *283*
Brown, G. M., 315, *376*
Brown, K. A., 358, *397*
Brown, M. R. J., 28, *101*
Brown, M. R., 32, *101*
Brown, R. M. J., 28, 50, 57, *101*, *107*, *109*
Brown, S., 264, *289*
Brown, T. R., 358, *397*
Browne, R., 358, *398*
Broxmeyer, E., 441, *473*
Bruce, D., 351, *394*
Bruchez, M. P., 171, 246, *283*
Bruchez, M., 245–246, *283*
Brumer, H., 84, *113*
Brun, R., 453, *476–477*
Bruns, C., 350, *393*
Brus, L. E., 245, *282*
Brushart, T. M., 442, *473*
Brushett, F. R., 428, *467*
Brust, M., 214–216, *272–273*
Bruus-Jensen, K., 350, *393*
Bucala, R. J., 309, *374*
Buchanan, J. G., 324, *380*
Buchini, S., 458, *479*
Buijtenhuijs, F. A., 129, 181, *193*, *209*
Bukhman, V. M., 364, *402*
Buleon, A., 81, 86, *113–114*, 175, *198*
Bunda, S., 413, *462*
Bunn, H. F., 293, 304, *368*, *371*
Burchard, W., 120, 124, 150, 162, *188*, *190*, *200*

Burda, C., 213, *271*
Burger, K., 329–330, *383*
Burger, M. M., 251, *286*
Burkhardt, E., 425, *465*
Burlina, A., 327, *383*
Burmeister, W. P., 424, *465*
Burnfield, K. E., 54, *108*
Buscaglia, C. A., 454, *477*
Buschiazzo, A., 455–458, *478–479*
Bushby, K. M., 425, *465*
Bushnell, D. A., 215–216, 238, *273*
Butters, T. D., 432, *468*

C

Caciotti, A., 414–415, *463*
Cai, H. J., 220, *276*
Cai, H., 358, *397*
Caimi, L., 429, *467*
Calame, M., 350, *392*
Calero, G., 215–216, 238, *273*
Callahan, J., 414, *463*
Calmes, R., 361, *400*
Cambi, A., 248, 266, *285*
Cameron, N. R., 220, *276*
Cammarata, R. C., 212, *271*
Campbell, F. T., 118, *185*
Campbell, S. J., 243, 268, *281*
Campetella, O. E., 454–456, *477–478*
Campion, C., 225–226, *276*
Campos, Y., 410, *462*
Cañada, F. J., 253, *287*
Cancer, J., 217, 219–221, 251, 256, *275–276*
Cantaert, T., 170, *206*
Cantarelli, B., 415, *463*
Cantz, M., 429, 431–432, *467–468*
Canuel, M., 436, 444, *471*
Cao, Z., 442, *473*
Capon, B., 161, *206*, 297, *369*
Cappellini, M. D., 426–427, *466*
Capuano, E., 340, *387*
Caqueret, A., 436, 444, *471*
Cardelle, A., 325, *381*
Carganico, S., 309, *374*
Carlsen, M., 351, *393*
Carlsson, K., 84, *113*
Carlsson, S. R., 421, *464*
Carpentier, S., 436, 444, *471*
Carpinteiro, A., 434, *470*
Carpita, N. C., 155, *202*
Carson, K. G., 351, *394*
Cartier, N., 43, *105*
Carubelli, I., 441, *473*
Caruthers, J. M., 86, *114*
Carvalho de Souza, A., 220, 222, 253–255, 269, *276*, *287–288*
Carvalho, R. C. B., 325, *381*
Casey, D., 439–440, *472*
Cassels, F. J., 439–440, *472*
Cassuto, J. P., 433, *469*
Casteel, S. W., 237, *279*
Casti, T. E., 137, 147, *197*
Catinella, S., 323, *380*
Cavaillé, J. Y., 81, 92, *113*, *115*
Cave, A. C., 355, *395*
Cerami, A., 309, *374*
Cercas, E., 357, *397*
Cerdán, S., 225–226, 242–243, 268, *277*
Ceriotti, F., 327, *382*
Cerny, C., 360, *398*
Černý, M., 155, *202*
Cerny, R. L., 304, 323, *372*
Cerny, R., 304, 323, *372*
Cerqueira, D. A., 92, *115*
Cerutti, F., 356, *396*
Chabre, Y. M., 217, *275*
Chaikof, E. L., 247, 257, *285*
Chakraborti, A. S., 357, *397*
Chamson, A., 355, *394*
Chan, C. T., 295, *369*
Chan, W. C. W., 245–246, *283*
Chan, Y. C., 175–176, *207*
Chance, D. L., 318, 319–320, *378*
Chandra, R., 90, *115*
Chandra, S., 439, *472*
Chandrasekhar, M., 237, *279*
Chang, C.–Y., 269, *289*

Chang, E., 247, *284*
Chang, H.-T., 234–235, 242, *278*
Chang, I. S., 323, 362, *379*
Chang, L. Y., 451, *475*
Chang, P. S., 71, *111*
Chang, Y. C., 434, *469*
Chang, Y.-W., 269, *289*
Chanzy, H., 27, 31, 39, 41, 43–45, 47–48, 50, 52–57, 61, 63, 68, 76, 79–82, 83, 92, *100*, *103–110*, *112–113*, *115*, 121, *189*
Charlton, W., 34, *102*
Charreyre, M.–T., 220, *276*
Chatterjee, D. K., 247, *284*
Chatterjee, S., 438, *472*
Chaudhuri, I., 407, *460*
Chaussy, D. A. D., 93, *116*
Chauvelon, G., 86, *114*
Chavas, L. M., 408, 423–425, 429, *461*
Chemello, L., 327, *382*
Chen, A. C., 220–221, 255, *275*
Chen, C.–C., 220–221, 223, 231, 233, 249, 255, 257, *275–276, 285*
Chen, C.–H., 214, *272*
Chen, C.–L., 220–221, 255, *274*
Chen, C.–T., 220, 223, 231, 233–235, 242, 249, 255, 257, 269, *276, 278, 285, 288*
Chen, G.-F., 220–221, 255, *275*
Chen, H., 324, *380*
Chen, J., 326, 330, *382–383*
Chen, J.-H., 266, *289*
Chen, P., 247, 266, *284*
Chen, R., 85, *113*
Chen, S. G., 340, *386–387*
Chen, S. H., 215, *273*
Chen, S. W., 215, *273*
Chen, S., 293, 357, *368*, *396*
Chen, S.-H., 269, *289*
Chen, X., 213, *271*, 453, *476*
Chen, Y., 247, 257, *284*
Chen, Y.-C., 220–221, 255, 269, *275*
Chen, Y.-J., 220, 223, 231, 233, 255, 269, *276*, *289*
Cheng, F., 127, *192*
Cheng, F.-Y., 266, *289*
Cheng, J., 453, *476*
Cheng, Z., 295, *369*
Cheon, J., 239, 264, *280*, *289*
Cherniak, R., 150, *200*
Chi, M. H., 428, *467*
Chiara, J. L., 295, *368*
Chiba, T., 149, *199*
Chibber, R., 355, *395*
Chien, Y.-Y., 220, 223, 231, 233, 255, 269, *276*, *289*
Chigorno, V., 433–434, 439, *469–470*
Chikae, M., 229, 257, *277*
Chilton, W. S., 314, *374*
Ching, K. A., 447, *475*
Chizov, O. S., 119, *186*
Cho, E. W., 451, *475*
Cho, S. H., 429, 443, *467*, *474*
Cho, S. -J., 220, 223, 262, *276*, 333, 335, 348, *384*
Choi, K. -J., 362, *400*
Choi, M. G., 356, *396*
Choi, S. H., 231–232, *277*
Choi, S. W., 120, *187*
Choi, S.-K., 217, 254, *275*
Chokhawala, H. A., 453, *476*
Chompoosor, A., 212, 214, *271*
Chopra, P., 362–363, *400*
Chorev, M., 309, *374*
Chou, P.-T., 249, 257, 265, *285*
Chou, S. Y., 342, *389*
Chu, R.-M., 249, 257, 265, *285*
Chu, S. S. C., 35, 39, *102*
Chun, S. W., 97, *116*
Chuyen, N. V., 362, *400*
Ciavardelli, D., 363, *401*
Cirillo, F., 433, 439, *469*
Cirina, P., 356, *396*
Cisternino, A. M., 355, 358, *395*, *398*
Citterio, D., 243, *281*
Ciucanu, I., 142, 157, *197*, *203*
Civini, S., 435, *470*
Claeyssens, M., 170, *207*
Claffey, W., 42, *104*
Clarke, W. A., 304, 323, *372*

Clarkson, A. B., 454, *478*
Clarkson, R., 351, *393*
Clausell, N., 443, *474*
Claustre, S., 356, 364, *395*
Clavel, C., 220, 222, 225–226, 258, 262–263, 269, *276*, *288–290*
Clawin, I., 361, *399*
Clety, N., 332, *384*
Cliffel, D. E., 215, *273*
Clode, D. M., 124, *190*
Clotman, F., 341, *388*
Cobler, J. G., 118, *185*
Cobo, J., 318, *377*
Cocklin, R. R., 441, *473*
Cohen, A., 173, 176, *207*
Cohen, H. J., 317–318, *377*
Cohen, K. S., 246, *283*
Cohen, M. P., 293, 327, 356–357, *368*, *383*, *395–397*
Colaco, C. A. L. S., 358, *397*
Colazzo, F., 430, 433, 440, *469*
Colbert, D., 326, 330, *382*
Collard, F., 293–294, 341–342, *368*, *388–389*
Colli, W., 453, *477*
Collier, C. P., 216, *274*
Colman, P. M., 408, *461*
Colombi, P., 429, 433, 435–436, 439, 449–450, *467*, *469–470*
Colombo, F., 425–426, *465–466*
Colonna, G., 323, *380*
Colosetti, P., 433, *469*
Colvin, J. R., 57, *109*
Colvin, V. L., 247, *284*
Comelli, E. M., 435–436, *470*
Communi, D., 341, *388*
Concheiro, A., 118, *185*
Conforti, E., 430, 433, 440, *469*
Connell, A., 419, 421, *464*
Conner, H. T., 129, 135, 137–138, 147–148, *193*
Conner, J. R., 341, *388*
Connett, B. E., 297, *369*
Connolly, J., 239, *280*
Conrad, H. E., 157, *202*
Conrad, J., 348, *392*
Conti, G., 356, *396*
Contreras, R., 442, *473*
Controll, J., 442, *473*
Cooke, M. P., 447, *475*
Cooper, S. H., 441, *473*
Copley, R. R., 441, *473*
Corbett, W. M., 310, *374*
Cordoret, J.-S., 86, *115*
Cornelissen, I., 262, *289*
Corzo, N., 325, *381*
Corzo-Martinez, M., 303, 360, *371*, *398*
Cosgrove, D. J., 30, 32, *101*
Costelli, P., 426, *466*
Costello, C. E., 158, *203*
Costello, E. C., 157, *203*
Cotham, W. E., 304, *371*
Cottem, D., 356, 364, *395*
Coughlin, P., 420, *464*
Coupas, A.-C., 114, *150*
Courtneidge, S. A., 432, *468*
Cousin, S. K., 33, *102*
Covanis, A., 412–413, *462*
Cowley, A. R., 351, *393*
Crawley, Y., 441, *473*
Creighton, J. A., 254, *287*
Cremer, D. R., 334, *384*
Cremona, M. L., 455–458, *478–479*
Crennell, S. J., 408, 434, *461*, *470*
Crespo, P., 240–242, *281*
Croci, G. L., 423, 430–431, 435–437, *464*, *467*, *470*
Croft, M., 363, 365, *401*
Croon, I., 129, *194*
Cross, A. S., 439–440, *472*
Crowell, E. F., 33, *102*
Cui, W., 247, 257, *285*
Culbertson, S. M., 335, *385*
Cumashi, A., 363, *401*
Cunha, A. G., 86, *114*
Cuniberti, L. A., 443, *474*
Cure, P., 327, *382*
Curreli, S., 440, *472*
Curtis, A. S. G., 239, *279*

Curvelo, A. A. D., 85, *114*
Cusack, S., 424, *465*

D

da Cruz, F. P., 351, *393*
D'Adamo, D. R., 426, *466*
Dahlgren, C., 421, *464*
Dai, Z., 293, *368*
Dallaire, L., 407, *460*
Damager, I., 455–456, 458, *478–479*
D'Ambra, A. J., 153, *202*
Daneault, C., 85, *113*
Dang, J., 443, *474*
Daniel, J. R., 125–126, 129–130, 143, *191*
Daniel, M.-C., 213, *271*
Darvill, A. G., 157, *203*
Das, A. K., 358, *397*
Da Silva, J. S., 432, *469*
Daub, J., 120, *187*
Dautheville, C., 338, *386*
Davidek, T., 303, 324, 332, *371*, *380*, *384*
Davies, G. J., 170, 175, *207–208*
D'Avila, F., 435, *470*
Davis, B. G., 243, 268, *281*
Davison, K., 425, *465*
Dawes, K., 155, *202*
Day, C. S., 295, *368*
d'Azzo, A., 403–479
de Carvalho, S., 452, 454, *476*, *478*
de Geest, N., 416, 419, *463*
D'Egidio, M., 363, *401*
d'Ischia, M., 306, *373*
de Jong, S., 347, *391*
de la Fuente, J. D. M., 249, *285*
de la Fuente, J. M., 214, 217, 219–221, 224–226, 240–243, 251–252, 254, 256, 260–261, 268–270, *271–272*, *277*, *281*, *286–288, 290*
de Lederkremer, R. M., 454–455, *477–478*
de Lima, V. C., 443, *474*
De Michele, G., 355, 358, *395*, *398*
de Paz, J.-L., 220, 222, 224, 226, 228, *275, 277*
de Roos, K. B., 306, *373*
de Sousa-Hitzler, J., 416, 419, *463*
De Vusser, K., 442, *473*
de Weijs, L. G. R., 126, 175, 177, *191*
De, K. K., 161, *203*
De, M., 270, *290*
Dean, B., 251, *286*
Debelle, L., 415, *463*
DeBolt, S., 33, *102*
Decan, M. R., 216, *274*
Deckert, M., 434, *469*
Dedier, J., 63, *110*
Dedieu, S., 415, *463*
Defamic, V., 434, *469*
Defaye, J., 120, *188*, 298, *369*
Degenhardt, T. P., 348, *392*
Del Castillo, M. D., 305, 325, 340, *372*, *381*, *387*
Delacoux, F., 415, *463*
Delair, T., 220, 258, *276*, *288*
Delannoy, P., 451, *475*
Delgado-Andrade, C., 359, *399*
Delpierre, G., 341–342, *388–389*
Delvalle, J. A., 341, *388*
Demchenko, A. V., 249, *285*
Demeter, J., 130, *194*
Deng, Z., 250, *286*
Denisevitch, T. V., 363, *401*
Dennis, J. W., 418, *463*
Dennis, M., 304, *372*
Denooy, A. E. J., 71, *111*
Deppisch, R., 361, *400*
Descotes, G., 360, *399*
Desmet, T., 170, *206*
Desprez, T., 33, *102*
Dessaux, Y., 314, *375*
Deus, C., 131, 148–150, 162, *195*
Deutscher, S. L., 363, *401*
Deutschman, A. J., 126, 157, *192*
Devaud, S., 303, 324, 346, *371*, *380*, *381*, *391*
Devuyst, O., 341, *388*

Dewhirst, M. W., 239, *279*
Dhar, S., 237, *279*
Di Leo, A., 358, *398*
Diánez, M. J., 322, *379*
Dias, W. B., 453, 456–457, *477*, *479*
Díaz, V., 260, *288*
Dicke, R., 68, 74–75, *111*, 131, *195*
Diéguez, M., 353, *394*
Dietz, F., 451, *475*
Dijong, I., 298, 300, *369*
Dileo, L., 430, 433, 440, *469*
Dimitriu, S., 66, *110*
Dinand, E., 54, 61, 64–65, 79–80, *108–109*, *113*
Ding, B., 124, *190*
Diotallevi, F., 33, *102*
Dobberpuhl, D., 151, *201*
Dobo, A., 157, *203*
Dobson, J., 239, 270, *280*, *290*
Dogné, J. M., 311, *375*
Dole, M., 163, *204*
Dolores del Castillo, M., 360, *398*
Domeniconi, M., 442, *473*
Domon, B., 158, *203*
Donald, A. M., 125, *191*
Donati, M. A., 412–415, *462–463*
Dönges, R., 123, 129, *189*
Donoviel, M. S., 351, *393*
Doose, S., 246, *283*
Doran, P., 444–445, *475*
Dorn, S., 44, 48, *190*
Dorrestein, P. C., 315, *376*
DosReis, G. A., 454, *477*
Dotti, C. G., 431–432, *469*
Doty, R. C., 214, *272*
Dowd, M. K., *103*, 318, *377*
Dowden, J., 356, *395*
Doyle, H., 239, *280*
Drake-Holland, A. J., 443, *474*
Drechsler, U., 212, *270*
Drezek, R., 247, *284*
Druey, J., 296, *369*
Drysdale, B. E., 440, *472*
Du, Y.-M., 247, 266, *284*
Duan, H. W., 264, *283*
Duarte, J., 92, *115*
Duca, L., 415, *463*
Duchemin, A. M., 433, *469*
Duda, D. G., 246, *283*
Dufour, M., 85, *114*
Dufresne, A., 59, 81, 92, *109*, *113*, *115*
Duft, D., 163, *204*
Duguet, E., 239–240, *279*
Dulmage, W. J., 55, *108*
Dumas, R., 118, *184*
Dumitriu, S., 70, *111*
Duquenne, A., 342, *389*
Durachko, D. M., 30, 32, *101*
Durán, R. V., 258–259, *288*
Durand, S., 436, 444, *471*
Durugkar, K. A., 237, *279*
Dutton, G. C. S., 119, *186*
Dwek, R. A., 166, *206*, 217, *274*
Dyason, J. C., 456, *479*
Dyer, D. G., 301–302, 304, 355, *370*

E

Eadon, D. G., 254, *287*
Earhart, C., 229–230, 244, 250, 257, *277*
Eaton, P., 240–242, 252, 258–259, *281*, *286*, *288*
Eberhard, M., 315, *376*
Eberle, A. N., 350, *392*
Ebert, A., 130, 183, *194*
Ebner, G., 181, *196*
Eckersall, P. D., 358, *397*
Edelstein, A. S., 212, *271*
Eder, B., 292, *367*
Edgar, K., 132, *196*
Edward, J. T., 161, *204*
Edwards, R., 326, 330, *382*
Egan, W., 441, *473*
Eggens, I., 251, *286*
Ehehalt, R., 431, *468*
Ehrhardt, D. W., 33, *102*
Ehrhardt, L., 120, *187*
Ehrler, R., 139, *197*
Eichinger, D., 454, *478*
Eichler, T., 118, *185*

Eichner, K., 323, 334, 336, 346, 360, *379*, *384–385*, *390*
Eigenbrot, C., 304, *372*
Einfeldt, L., 126, 130, *192*, *194*
El Cheikh, K., 311, 366, *375*
El Rassi, Z., 151, *201*
El Seoud, O. A., 124, *190–191*
El-Alfy, M., 441, *473*
El-Assar, M., 357, *396*
El-Boubbou, K., 244, 261–262, *282*
Elizondo, G., 239, *280*
El-Kashouti, M. A., 132, *195*
Elli, G., 323, *380*
Ellis, G. P., 297, *369*
Ellis, J. G., 314, *375*
Ellis, P. D., 317–318, *377*
El-Nokaly, M. A., 69, *111*
El-Sayed, M. A., 213, *271*
Elsliger, M. A., 407, *460*
Emancipator, S. N., 356, *396*
Emmons, A. M. C., 34, *102*
Emori, Y., 422, *464*
Enebro, J., 164, 175, 178–179, *205*, *207–209*
Engstler, M., 452–454, *475–476*, *478*
Enoki, A., 337, *385*
Erathodiyil, N., 229–230, 244, *277*
Erbersdobler, H. F., 325, 361, *381*, *399*
Erdogan, B., 212, *270*
Erhradt, D. W., 33, *102*
Eriksson, K. E., 177, *209*
Erler, U., 151–152, *200*
Ernst, B., 39, *104*
Esnault, J., 251, *286*
Estroff, L. A., 213, *271*
Evans, J. P., 126, 157, *192*
Evans, N. D., 216, *274*
Evthushenko, E. V., 131, *195*
Eyzaguirre, J., 124, *189*

F

Facchiano, A. M., 323, *380*
Fadel, H. H. M., 347, *391*
Fajardo, F. D., 453, *477*
Fallarini, S., 234–236, 262, *278*
Fang, J.-M., 249, *285*
Fang, L., 340, *388*
Fang, Y., 433, *469*
Fang, Z., 244, *282*
Fannon, J. E., 126, 143, *191*
Fanzani, A., 425–426, *465–466*
Faraday, M., 213, *271*
Farag, S. F., 362, *400*
Farag, S., 72, 74, *112*
Fardim, P., 86, *114*
Farese, A., 439–440, *472*
FaroKhzad, O. C., 269, *290*
Farouk, A., 347, *391*
Farrand, S. K., 314, 339, *375*, *387*
Faulds, C. B., 124, *189*
Faust, V., 124, *189*
Favier, V., 92, *115*
Fay, L. B., 332, 338, *384, 386*
Feather, M. S., 298, 302, 317–320, 323, 325, 329, 332, 363, *369–371*, *377–378*
Fedele, D., 323, *380*
Fedele, E., 304, *372*
Fedele, F., 340, *387*
Fehlhaber, W., 343, 365, *389*
Fenderson, B., 251, *286*
Feng, L., 436, 444, *471*
Fengel, D., 40, 57, *104*, *109*
Fenn, D., 119, 123–124, 143, *187*, *190*
Fenn, J. B., 163, *204*
Fent, G. M., 237, *279*
Ferguson, L. D., 163, *204*
Ferguson, M. A., 453, *477*
Fernández, A., 217, 219–221, 224, 226, 240–242, 251–252, 254, 256, *275, 281*
Fernandez, J. M., 216, *273*
Fernández-Bolaños, J. G., 353, *394*
Fernig, D. G., 214, 238, *272*, *279*
Ferrara, L., 304, *372*
Ferrari, E., 330, *384*
Ferrari, J., 423, *464*
Ferri, S., 327, 339–340, 356, *383*, *387–388*, *395*
Fessner, W. D., 316, *376*

Fidale, L. C., 124, *190–191*
Field, R. A., 231–232, 236–237, 249, 255, *277–278*, *285*, *287*
Figarola, J. L., 358, *397*
Figdor, C. G., 248, 262, 266, *285*, *289*
Filbin, M. T., 442, *473*
Fink, H. -P., 121, 142–143, 147, 150, *187*, *197–198*
Fink, J., 215–216, 235, *273*, *278*
Finkendstadt, V. L., 44, *106*
Finne, J., 251, *286*
Finot, P. A., 323, 325, 361, *379*, *381*, *399–400*
Fischer, D. G., 323, 325, 361, *379*, *381*, *399–400*
Fischer, E., 292, 294–295, *367*
Fischer, K., 69, *111*
Fischer, M., 315, 350, *376*
Fischer, S., 69, *111*, 132, *190*
Fitzpatrick, F., 175, 178, 181, *208*
Fjelde, A., 362, *400*
Fleer, G. J., 142, *198*
Fleet, G. W. J., 351, *393*
Flückiger, R., 294, *368*
Fogelström, L., 92, *116*
Fogliano, V., 305, 323, 327, 340, *372*, *380*, *383*, *387*
Forage, N. G., 346, *391*
Forbes, N. S., 216, *273*
Forcella, M., 435, *471*
Formiga, L. C., 453, *476*
Fornerod, M., 407–408, *460*
Fors, S., 344, *390*
Fortpied, J., 315, *342*
Fossey, J. S., 324, *380*
Fraizy, J., 93, 95, *116*
Franchuk, S., 442, *473*
Francissen, K. C., 304, *372*
Franco, P., 304, *372*
Frank, O., 350, *391*
Franke, K., 322, *378*
Fransson, L., 82, *113*
Frasch, A. C., 453, 455–458, *477–479*
Fraschini, C., 82, *113*
Fratzl, P., 65, *110*
Fréchet, J. M. J., 217, *275*
Freichels, H., 258, *288*
Freilinger, C., 306, *373*
Freire, C. S. R., 86, *114*
Freire-de-Lima, L., 453, *477*
French, A. D., 27, 39, *100*, *103*, 318, *377*
French, D., 126, 175, *191*, *207*
Frens, G., 213–214, *271*
Fretzdorff, A. M., 298, 300, *369*
Freudenberg, K., 34, 102, *118*, *185*
Freudenheim, J. L., 355, 360, *395*, *398*
Frey, J., 355, *394*
Frey-Wyssling, A., 56, *108*
Frías, J., 359, *399*
Fricker, G., 350, *393*
Friebolin, H., 131, 148–150, 162, *195*
Friedl, D. A., 348, *391*
Friedman, H. S., 350, *392–393*
Friedrich, M., 151, *200*
Friesen, M., 360, *399*
Friguet, B., 356, *396*
Froio, R., 304, *372*
Froissart, R., 413, *462*
Frolov, A., 309, 323, *374*, *379–380*
Fromholt, S. E., 443, *474*
Fronda, C. L., 444, *475*
Froud, T., 327, *382*
Fu, Q., 351, *393*
Fuentes, J., 353, *394*
Fuhr, J. E., 426, *466*
Fujii, J., 358, *397*
Fujii, M., 348, *391*
Fujii, S., 311, *375*
Fujimaki, M., 334, *384*
Fujimori, K., 336, *385*
Fujimoto, D., 348, *392*
Fujimoto, T., 431–432, *468*
Fujiwara, Y., 337, *385*
Fujiya, T., 418, 438–439, 454–455, *463*
Fujiyoshi, Y., 56, 58, *108*
Fukase, K., 338, *386*
Fukuda, T., 229, 257, *277*
Fukui, M., 348, *391*
Fukumura, D., 246, *283*

Fukuta, H., 327, *382*
Fukuya, H., 339, *387*
Fuller, D., 28, *101*
Fumeaux, R., 332, *384*
Furniss, D. E., 361, *400*
Furuhata, K., 75, *112*
Furukawa, K., 436, *471*
Fusi, P., 408, 423, 435–436, *461*, *464*, *471*
Fuss, M., 241–242, *281*
Fuyuno, I., 427, *466*

G

Gabbay, K. H., 293, *368*
Gaderbauer, W., 351, *393*
Gadhoum, S. Z., 441, *473*
Gafuik, C., 407, *460*
Gahl, W. A., 409, *461*
Gajda, T., 329–330, *383*
Galanello, R., 327, *382*
Galatenko, O. A., 364, *402*
Galen, J., 423, *465*
Galjaard, H., 411, *462*
Galjart, N. J., 410, 414, *462–463*
Gallant, D. J., 125–126, *191*
Gallop, P. M., 293, *368*
Gallo-Rodriguez, C., 454–455, *477–478*
Gambhir, S. S., 246, *283*, 295, *369*
Ganchev, D. N., 253, *287*
Gandini, A., 73, 85–86, *112*, *114*, 123–124, 143, *187*
Gangadhariah, M. H., 355, *395*
Gao, W., 453–454, *477*
Gao, X., 246, *283*
Garayoa, E. G., 350, *393*
Garbarino, G. B., 454, *477*
Garber, A. R., 317–318, *377*
Garcia Fernandez, J. M., 120, *188*
García, I., 258, *288*
García, M. A., 241, *281*
García, M. D. G., 295, *368*
García, R., 252–253, *287*
García-Martin, M. L., 225–226, 242–243, 268, *277*
Gardiner, E. S., 53, *107*
Gardner, K. H., 42, 44, *104*
Gardyan, M., 351, *393*
Garidel, P., 120, *187*
Garlick, R. L., 304, *372*
Garman, E. F., 408, *461*
Garman, E., 408, *461*
Garrido-Franco, M., 315, 350, *376*
Gasch, C., 295, 353, *368*, *394*
Gaschler, W., 239, *280*
Gaskell, A., 408, *461*
Gassan, J., 73, *112*
Gathje, H., 451, *475*
Gauthier, C., 85, *114*
Gauthier, G., 92, *115*
Gauthier, H., 86, *114*
Gauthier, R., 86, *114*
Gautier, C., 39, *104*
Ge, N., 246, *283*
Geddes, C. D., 229–230, *277*
Gee, K., 442, *473*
Geijtenbeek, T. B. H., 256, 262, *288–289*
Geissmann, F., 440, *472*
Gelbart, W. M., 215, *273*
Gelman, R. A., 177, *209*
Gerardy-Schahn, R., 438, *471*
Gergaud, N., 86, *114*
Gerhardinger, C., 323, 340, *380*, *387*
Gerhartz, W., 118, *185*
Gericke, M., 124, 132, *190*
Germgård, U., 127, 129, 143, *192–193*, *198*
Gervay-Hague, J., 220, 223, 262, *276*, *289*
Gerwig, G. J., 220, 223, 255, 269, *276*, *288*
Gessler, K., 39, 50, *103*
Gestwicki, J. E., 217, *275*
Geyer, H. U., 298, 300, *369*
Gheysens, O., 295, *369*
Ghizoni, M. F., 443, *474*
Ghoneim, C., 415, *463*
Ghosh, I., 246, *284*
Ghosh, P., 270, *290*
Giambiagi, S., 457–458, *479*
Giasson, J., 63, *110*
Gibson, W., 458, *479*
Gidley, M. J., 143, *198*

Giersig, M., 215, 246, 270, *272*, *284*, *290*
Gil, G. -C., 166, *206*
Gil, H., 300–302, *319*, *370*, *378*
Gilchrist, J. S., 444, *475*
Gillemans, N., 410, 414, *462–463*
Gin, P., 245, *283*
Giorgi, M. E., 454–455, *478*
Girard, M. F., 453, 456–457, *477*, *479*
Girardeau, S., 86, *114*
Giuliani, R., 425–426, *465–466*
Gjerde, D., 324, *380*
Glasser, W. G., 74, 86, *112*, *114*, 124, 137, 142, *189*, *196*
Glasser, W., 123, 150, *189*, *199*
Gleiter, H., 245, *282*
Glidewell, C., 318, *377*
Glinskii, A. B., 363, *401*
Glinskii, O. V., 363, *401*
Glinsky, G. V., 298, 302, 317–320, 323, 363, *369*, *370–371*, *377–378*, *401*
Glinsky, V. V., 320, 356, 363, *378*, *395*, *401*
Gloe, K., 330, *383*
Glomb, M. A., 333, 340, 348, *384*, *387*, *392*
Gnewuch, C. T., 349, 360, *392*
Goatley, J. L., 118, *186*
Gobert, J., 333, *384*
Goclik, V., 130, 132, 153, *194–195*
Godfraind, C., 341, *388*
Gody, G., 220, *276*
Goeddel, D., 353, *394*
Gohdes, M., 119, 130, 132, 153, 157, 161, *187*, *195*, *201*
Goldbaum, F. A., 453, *477*
Goldblum, S. E., 439–440, *472*
Goldman, E. R., 247, *283*
Goldmann, T., 303, *371*
Goleniewska, K., 428, *467*
Golovanova, N. K., 443, *474*
Gomez de Anderez, D., 319, *378*
Gomez, A., 163, 204, *295*, *368*
Gómez, D. E., 246, *284*
Gomez-Amoza, J. L., 118, *185*
Gómez-Sánchez, A., 295, *368*
Gomi, K., 339–340, *386*
Gonera, A., 132, *195*
Gonneau, M., 33, *102*
Gonzalez, C., 300, *370*
Gonzalez, S. M., 454, *477*
Gonzalez-Carreno, T., 239, *280*
Goodman, T. T., 258, *288*
Gopinathan, S., 351, *394*
Gore, J. L., 153, *201*
Gorin, P. A. J., 119, *186*
Goring, D. A. I., 54, 79, *108*
Gorton, L., 119, 143, 166, 173–176, 178–179, *187*, *198*, *206–207*
Gotlieb, A. I., 443, *474*
Goto, H., 339, *387*
Goto, T., 43, *105*
Gottschalk, S., 409, *462*
Graca-Souza, A. V., 453, *477*
Gracheva, E. V., 443, *474*
Grafflin, M. W., 27, 118, 136, *100*, *186*, *196*
Graham, F. L., 412, *462*
Graham, P. A., 358, *397*
Graham, P., 258, *288*
Grandhee, S. K., 348, *391*
Grandi, R., 330, *384*
Grandl, M., 431, *468*
Grange, P. A., 439, *472*
Granouillet, R., 355, *394*
Granovsky, M., 418, *463*
Grasset, F., 239–240, *279*
Grassme, H., 434, *470*
Gravel, R. A., 407, 436, 444, *460*, *471*
Gray, D. G., 63, *110*
Gray, G. R., 129, 150, 153, 153, 156, *193*, *200–202*
Gray, J. A., 126, 143, *191*
Gray, M. R., 327, *383*
Green, S. J., 215, *273*
Greimel, P., 351, *393*
Griesgraber, G. W., 129, 153, 156, *193*
Griffiths, D. R., 338, *386*
Grimson, M. J., 28, *101*
Groff, J. L., 150, *200*
Gross, L., 170, *206*
Gross, S. M., 215–216, *273–274*
Grosveld, G., 407, *460*

Grötzinger, J., 458, *479*
Grübel, N., 361, *399*
Gruber, P. R., 153, 156, *202*
Gruden, C., 244, 261, *282*
Gruenwedel, D. W., 310, *374*
Grülc, D., 130, 183, *194*
Grzybowska, J., 364, *402*
Gu, H. W., 234–235, 244, *278*
Gu, H., 244, *282*
Gu, X., 427, *466*
Gualfetti, P., 170, *206*
Guan, H., 300, *370*
Guerardel, Y., 451, *475*
Guerra, V., 355, 358, *395*, *398*
Guifeng, L., 127, *192*
Guillaumie, F., 270, *290*
Gulbins, E., 434, *470*
Gumaa, K., 226, *277*
Gumieniak, J., 364, *402*
Gunawan, N., 126, 143, *191*
Günther, W., 126, *192*
Gupta, A. K., 239, *280*
Gupta, K. C., 72, *112*
Gupta, M., 239, *280*
Gustavsson, M., 82, *113*
Gut, H., 453, *476*
Gutierrez, R., 33, *102*
Gutlich, M., 315, 350, *376*
Györgydeák, Z., 306, *373*
Gyurcsik, B., 329–330, *383*

H

Ha, K. T., 429, 434, *467*, *469*
Haag, R., 237, *278*
Haas, H. J., 306, 338, *373*, *386*
Haas, J., 311, 334, *375*
Habashi, F., 309, *374*
Habibi, Y., 81, *113*
Hadjiconstantinou, M., 433, *469*
Haebel, S., 163, *204*
Hage, D. S., 304, 323, *372*
Hahn, C., 416, 419, *463*
Haines, A. H., 231–232, 253, 255, *277*, *287*
Haiss, W., 238, *279*
Hajek, M., 245, *282*
Haji Begli, A., 130, 183, *194*
Hajji, P., 92, *115*
Hakomori, S.(-I.), 157, *186*, 217, 251, *274*, *286*
Haley, K. N., 246, *282*
Halkes, K. M., 220, 222–223, 253–255, 269, *276*, *287–288*
Hall, A., 430, *467*
Hall, L. D., 119, *186*
Haller, C., 247, 257, *285*
Halperin, J. A., 309, *374*
Halvatsiotis, P., 355, *395*
Ham, J. T., 39, *103*
Hamed, O. A., 123, *189*
Hampson, A. W., 427, *466*
Han, G., 216, 270, *273*, *290*
Han, J. -A., 143, *198*
Hanai, T., 326, *382*
Handley, D. A., 215, *272*
Haney, D. N., 293, *368*
Hang, H. C., 217, *275*
Hanley, S. J., 63, *110*
Haorah, J., 360, *399*
Harada, H., 56, 58, *108*
Haraguchi, N., 337, *385*
Harashima, H., 249, 259, 265, *285*, *288*
Harduin-Lepers, A., 451, *475*
Hardy, S. M., 236–237, 255, *278*
Harris, R., 423, *464*
Harrison, R., 425, *465*
Hart, G. W., 452, 454, *476*, *478*
Hart, G., 39, *104*
Harte, F. M., 303, *371*
Hartford, J., 428, *467*
Hartman, K. B., 270, *290*
Harvey, D. J., 163, 166, *204–206*
Harwood, L. M., 131, *195*
Hase, S., 338, *386*
Hasegawa, A., 432–433, 436, *468–469*, *471*
Hasegawa, T., 452, *476*
Haselhorst, T., 456, *479*
Hashiba, H., 346–347, 360, *390–391*

Hashida, M., 250, 266, *286*
Hassan, J. A., 435–436, 454, *471*
Hata, K., 418–419, 428–432, 435–438, *463–464*, *467–468*, *470–472*
Hatanaka, Y., 356, *395*
Hatch, D. M., 244, *282*
Hattori, S., 357, *397*
Hattori, Y., 357, *397*
Hau, J., 303, 324, *371*, *381*
Häusler, H., 351, *393*
Hautefeuille, A., 360, *399*
Haworth, W. N., 34, 102, *118*, *185*
Hayakawa, M., 447, *475*
Hayase, F., 332, 337, *384–385*
Hayashi, J., 48, 52–54, 79, *106–107*
Hayashi, M. C., 337, *385*
Hayashi, R., 326, *381*
Hayat, M. A., 213, 215, *271–272*
Haylock, D. N., 420, *464*
Hazelwood, J., 351, *393*
He, L., 214, *272*
He, Q., 427, 448, *466*
He, R., 326, *382*
He, W., 362, *400*, 422, *464*
Heath, J. R., 215–216, *273–274*
Hebeish, A., 132, *195*
Hedlund, A., 129, 143, *193*
Heffer-Lauc, M., 443, *474*
Heiner, A. P., 44, *106*
Heinrich, J., 119–120, 131, 151, *187*, *195*, *201*
Heins, D., 131, *195*
Heinz, T., 351, *393*
Heinze, J., 137, 142, *196*
Heinze, T., 68, *111*, 124, 150, 155, 175, 181, *190*, *198–199*, *202*, *208*
Heinze, Th, 119–120, 123–124, 127, 132, 137, 142, 147–151, 155, *187–192*, *195–200*, *202*, *208*
Heinze, U., 119, 147, *187*, *198*
Helbert, W., 44, 46, 57, 63–64, 81, *106*, *109–110*, *113*
Held, T. K., 439, *472*
Hellerqvist, C. G., 119, *186*
Helliwell, M., 319, *378*
Hellwig, M., 361, *400*
Hemler, M. E., 432, *468*
Hemmingson, J. A., 64–65, *110*
Hendy, J., 420, *464*
Henle, T., 304, 323, 325, 330, 343, 358, 361, *371*, *379*, *381*, *383*, *390*, *398–400*
Henn, M., 315, *376*
Hennig, C., 168, *206*
Hennig, M., 315, *376*
Henrichs, S. E., 216, *274*
Henrissat, B., 39, 43–44, 79, 82, *103*, *105–106*, *109*, *112–113*, 170, *207*, 424, *465*
Henry, C. J., 363, *401*
Herault, A., 120, *188*
Hermans, P. E., 40, *104*
Hermans, R. H., 41, 55, *104*
Hernáiz, M. J., 252, 269, *287*
Hernando, A., 240–242, *281*
Herrant, M., 434, *469*
Herrero-Martinez, J. M., 183, *210*
Herrmann, A., 315, 350, *376*
Herynek, V., 245, *282*
Hess, K., 55, *108*
Hesse-Ertelt, S., 150, *199*
Hesser, B. A., 409, *461*
Hettrich, K., 132, *190*
Hettwer, S., 315, *376*
Heublein, B., 120–121, 130, 150, *187–188*, *200*
Heux, L., 61, 81, *109*, *113*
Heyns, K., 294, 298, 302, *368*, *370–371*
Heyraud, A., 50, *107*
Hieta, K., 57, *109*
Hifumi, H., 243, *281*
Higai, K., 355, *396*
Higginbotham, J. N., 428, *467*
Higuchi, O., 358, *397*
Higuchi, T., 43, *105*
Higuchi, Y., 250, *286*
Hijazi, K., 225–227, 256, *276*
Hildebrandt, H., 438, *471*
Hillenkamp, F., 119, *187*

Hille-Rehfeld, A., 409, *461*
Hillier, J., 213, *271*
Hinek, A., 413–417, *462–463*
Hines, R. L., 163, *204*
Hinton, D. J. S., 304, *371*
Hinuma, Y., 425, *465*
Hirai, A., 43, 52, *105*, *107*
Hirai, G., 428, *467*
Hirai, K., 340, 343, *388*
Hirokawa, K., 339–340, *386–387*
Hirose, A., 357, *396*
Hirsch, J., 320–321, *378*
Hirst, E. L., 34, *102*
Hischenhuber, C., 361, *400*
Ho, F. F. L., 129, *193*
Ho, P.-L., 234–235, 244, *278*, *282*
Hodge, J. E., 292–293, 346–347, *367*, *390*
Hoffmann, R., 309, 323, *374*, *379–380*
Hoflack, B., 409, *461*
Hofman, P., 434, *469*
Hofmann, T., 309, 336, 347, *374*, *385*, *391*
Hofmeister, G., 163–164, *205*
Hofte, H., 33–34, *102*
Hogenesch, B., 447, *475*
Hoger, H., 425, *465*
Holbrey, J. D., 69, *111*
Holguera, J., 435, *470*
Hollmark, B. H., 177, *209*
Hollnagel, A., 335, *385*
Hollósi, M., 309, *374*
Holman, G. D., 356, *395*
Holmbom, B., 86, 114, *129*, *193*
Holzer, W., 120, *187*
Hommel, U., 315, *376*
Hon, D. N. S., 27, *101*
Hone, D. C., 231–232, 255, *277*
Honeyman, J., 297–298, *369*
Hong, H. K., 355, *395*
Hong, R., 216, *273*
Hong, S., 270, *290*
Honjo, G., 43, *105*
Hood, L. F., 176, *209*
Höök, J. E., 161, *204*
Hopf, H., 126, *192*
Hopff, H., 295, *368*
Hoppe, L., 118, *185*
Horak, D., 245, *282*
Horejsi, V., 432, *468*
Hori, H., 343, 362, *390*
Horii, F., 43–44, 52, *105*, *107*
Horikawa, H., 336, *385*
Horiuchi, S., 337, *385*
Horiuchi, T., 294, 339, *368*, *386*
Horn, M. J., 360, *399*
Hornebeck, W., 415, *463*
Horner, S., 147, 151, 155, 173, 175, 177–178, 181, *198*, *201–202*, *207–208*
Horton, D., 124, 146, *190*, *198*, 292, *367*
Horvat, J., 305, 307, *373–374*
Horvat, Š, 305, 307, 309, 317–318, 330, 343, *371*, *373–374*, *377*, *383*, *389*
Horvath, A., 420, *464*
Hostetler, M. J., 215–216, *273–274*
Hotchkiss, D. J., 351, *393*
Hou, Y., 318, *377*
Houseman, B. T., 217, *275*
Housni, A., 220, *276*
Hoven, N., 340, *388*
Howard, M. J., 355, *394*
Howarth, M., 246, *283*
Howell, S., 341, *388*
Howley, P. M., 434, *470*
Howley, P. S., 63, 109, 161, *204*
Howsmon, J. A., 434, *470*
Hoyer, L. L., 407, *460*
Hrlec, G., 305, 307, *373*
Hsieh, P. C.–H., 266, *289*
Hu, B., 350, *393*
Hu, H., 419, 422, *464*
Hu, Q., 244, *282*
Huang, C. J., 451, *475*
Huang, C.–C., 234–235, 242, *278*
Huang, H., 237, *279*
Huang, L. L., 353, *394*
Huang, N., 247, *283*
Huang, S., 453, *476*
Huang, X., 244, 261, *281*
Huber, B. A., 163, *204*

Hüber, D. L., 239, *279*
Huber, G., 296, *369*
Huber, K. C., 126, 143, *191*, *198*
Huber, R., 315, *376*
Hud, E., 327, 357, *383*, *396*
Hudson, B. G., 317, 337, *377*
Huflejt, M. E., 363, 365, *401*
Hujiya, T., 437, *471*
Huletsky, A., 244, *282*
Hult, A., 33, *116*
Hultsch, C., 361, *400*
Hunston, F., 338, *386*
Hurley, T. R., *436–437*
Hurrell, R. F., 379, *399–400*
Husemann, E., 118–119, *185–186*
Hussain, I., 214, *272*
Hussain, M. A., 150, *199*
Hussain, S. M., 240, *280*
Hüttl, C., 325, *381*
Huxley, V. H., 363, *401*
Huyghues-Despointes, A., 292, 346, *367*, *391*
Hvidt, T., 308, *374*
Hyeon, T., 239, *280*

I

Iacobelli, S., 363, *401*
Iacono, S., 364, *401*
Iannello, A., 422, *464*
Ichikawa, Y., 343, 362, *390*
Ide, M., 237, *278*
Ide, N., 362, *400*
Idegami, K., 229, 257, *277*
Igdoura, S. A., 407, 412, *460*, *462*
Igdoura, S., 407, *460*
Iguchi, N., 347, *391*
Ihm, J., 356, *396*
Ihm, S. H., 356, *396*
Iida, Y., 323, *380*
Iijima, K., 348, *392*
Ijichi, K., 362, *400*
Ijiro, K., 249, 259, 265, *285*, *288*
Ikeda, C., 326, *382*
Ikegaki, T., 338, *386*
Ikuta, Y., 366, *402*
Il'inskaya, O. P., 443, *474*
Illaszewicz, C., 322, *379*
Illy, N., 119, *187*
Imada, K., 41, 56, *104*
Imai, K., 149, *199*
Imai, T. T., 44, 46, 57, *105–106*
Imberty, A., 39, *103–104*, 258, *288*
Inagaki, H., 149–150, *199*
Inagi, R., 323, *380*
Inglot, A., 303, *371*
Ingold, K., 335, *385*
Inokuchi, J., 429, *467*
Iranfar, N., 28, *101*
Iribarne, J. V., 163, *204*
Irie, M., 311, *375*
Irie, S., 348, *392*
Irvine, J. C., 34, *102*
Ishida, H., 408, 424, 454, *461*, *465*
Ishigaki, A., 425, *465*
Ishikawa, M., 246, *283*
Ishimaru, K., 340, 343, *388*
Ishimura, F., 339, *387*
Islam, T., 427, *466*
Isogai, A., 44, 50, 70–71, *106*, *111*, 124, *190*, 339, *387*
Isogai, O., 44, *106*
Isogai, T., 124, *190*
Itakura, S., 337, *385*
Itakura, Y., 362, *400*
Ito, K. -I., 149, *199*
Ito, T., 59, *109*
Itoh, A., 366, *402*
Itoh, T., 44, 105, *246*, 283
Itoyama, Y., 425, 432–433, 436, *465*, *468*, *471*
Itzstein, M., 434, *470*
Iurisci, I., 363, *401*
Ivanova, S., 425, *465*
Iwamatsu, A., 419, 429, *464*
Iwamoto, N., 336, *385*
Iwata, S., 161, *204*
Iyer, S. S., 244, *282*

J

Jacob, J., 317, 337, *377*
Jacobs, P., 240, *280*
Jacquel, A., 434, *469*
Jacquemin, P., 341, *388*
Jadzinsky, P. D., 215–216, *273*
Jaffe, A. B., 430, *467*
Jager, A., 355, *395*
Jäger, C., 39, *103*, 150, *200*
Jahngier, K., 122, *189*
Jain, R. K., 246, *283*
Jaiswal, J. K., 246, *283*
Jakas, A., 309, 317–318, 330, 343, *371*, *374*, *377*, *383*, *389*
James, T. D., 324, *380*
Jan, M.–D., 220, 223, 231, 233, 255, 269, *276*, *289*
Jana, N. R., 229–230, 244, 257, *277*
Jandura, P., 114, *150*
Jane, J. L., 183, *210*
Jang, S., 166, *206*
Jardeby, K., 143, *198*
Jaremko, L., 323, *380*
Jaremko, M., 323, *380*
Jarvis, M. C., 30–31, *101*
Jasieniak, J., 246, *284*
Javier, F., 295, *368*
Jayanth, P., 422, 442, *464*, *473*
Jeffrey, G. A., 35, 39, *102*
Jendelova, P., 245, *282*
Jenkins, P. J., 125, *191*
Jenkinson, S. F., 351, *393*
Jensen, H. H., 161, *204*
Jensen, K. J., 231, 233, 256, 270, *278*, *290*
Jensen, R. G., 128, *192*
Jeon, J. H., 443, *474*
Jeong, H.-J., 323, *379*
Jerić, I., 305, *373*
Jérôme, C., 258, *288*
Jespersen, H. M., 170, *207*
Jessop, T., 351, *393*
Ji, T., 247, 257, *284*
Jia, Q., 453, *476*
Jiang, B., 124, *190*
Jiang, M. S., 452, 454, *476*, *478*
Jiang, M., 350, *393*
Jiang, S., 265, *283*
Jiang, T., 300, *370*
Jiang, X., 250, 270, *286*, *290*
Jianxin, F., 127, *191*
Jiménez, M., 220, 256, *276*
Jimenez, S. A., 416, *463*
Jiménez-Barbero, J., 253, *287*
Jiménez-Castaño, L., 303, *371*
Jimenez-Garay, R., 318, *377*
Jin, U. H., 434, *469*
Jingwu, Z., 127, *192*
Johansson, G., 169, *206*
Johansson, M., 178, *209*
John, W. G., 326–327, 330, *381–383*
Johnson, D. C., 151, *201*
Johnson, D. W., 165, *206*
Johnson, G. H., 361, *399*
Johnson, G. P., 39, *103*
Johnson, K. D., 363, *401*
Johnson, R. N., 191, *326*
Jones, N. A., 351, *393*
Jones, P., 153, *202*
Jones, R. B., 90, *115*
Jones, S. K., 239, *280*
Joo, K.-M., 323, 362, *379*
Jordan, R., 216, *274*
Jörntén-Karlsson, M., 165, 175, 179, *205*
Josephson, L., 239–240, *280–281*
Joshi, H. M., 237, *279*
Jouy, H., 33, *102*
Jovin, T. M., 248, 266, *285*
Joy, P. A., 245, *282*
Juhasz, P., 355, *395*
Jun, Y.-W., 264, *289*
Jung, C. W., 240, *280*
Jung, M., 425, *465*
Juraniec, M., 33, *102*
Jurgens, C., 315, *376*
Jyothi, A. N., 129, *194*

K

Kadavanich, A. V., 248, *285*
Kafka, M., 118, *185*

Kagan, C. R., 213, 239, *271*, *280*
Kajiyama, N., 339–340, *386–387*
Kakugawa, Y., 418, 437–439, 454–455, *463*, *471*
Kale, R. R., 244, *282*
Kalia, K., 327, *382*
Kalka, D., 431, *468*
Kalyan, Y. B., 364, *401*
Kamel, S., 122, *189*
Kamerling, J. P., 129, 143, 151, 182–183, *193*, *198*, *201*, *210*, 220, 222, 253–255, 269, *276*, *287–288*, 453–455, *477*
Kaminaga, J., 71, *111*
Kamitakahara, H., 120, *188*
Kamitani, R., 249, *285*
Kammer, H. W., 68, *110*
Kanamori, T., 247, 251, 264, *284, 286*
Kanayama, N., 237, *278*
Kang, H. W., 240, *281*
Kang, H. Y., 451, *475*
Kang, J. G., 356, *396*
Kannan, R., 237, *279*
Kanoh, T., 338, *386*
Kanshin, E., 422, *464*
Kapczyńska, K., 309, 323, *374*, *380*
Kaper, J. B., 423, *465*
Kaplun, V. V., 452, *476*
Kappler, F., 342, 358, *389*, *397*
Karakawa, M., 150, *200*
Karamanska, R., 231–232, 249, *277*, *285*
Karas, M., 163, *204*
Karbasiyan, M., 425, *465*
Karlsson, A., 421, *464*
Karlsson, J., 164, 170, 173, 178, *205–206*, *209*
Karlsson, K.-E., 165, 175, 178–179, *205*
Karlsson, S., 164, 175, 178–179, *205*, *207–208*
Karnani, R., 73, *112*
Karp, J. M., 270, *290*
Karpel, R., 453, *476*
Karrasch, A., 150, *200*
Kartusch, C., 322, *379*
Kasai, K., 357, *397*
Kasture, M., 245, *282*
Kasztreiner, E., 310, *374*
Kataoka, K., 229, 235–236, 254, 269, *277–278*, *287*
Katić, A., 309, *371*
Kato, H., 334, *384*
Kato, K., 418, 438, *464*, *472*
Kato, M., 425, *465*
Kato, N., 339, *387*
Kato, R., 408, *461*
Kato, Y., 71, 111, *327*, *382*
Katsuragi, T., 360, *398*
Katta, V., 304, *372*
Katterle, M., 327, 343, *383*
Katti, K. V., 237, *279*
Katti, K., 237, *279*
Kattumuri, V., 237, *279*
Katz, E., 212, *270*
Käuper, P., 131, 147, *195*, *198*
Kauw, H. J. J., 138, 160, 162–163, *197*
Kauzlarich, S., 220, 223, *276*
Kawaguchi, H., 332, *384*
Kawahara, C., 357, *396*
Kawakami, S., 250, 266, *286*
Kawaoka, Y., 427, *466*
Kawasaki, H., 422, *464*
Keating, C. D., 214, *272*
Keeley, F., 414, *463*
Keely, C. M., 150, *199*
Keenoy, B., 327, 357, *382*, *397*
Kell, A. J., 244, *282*
Keller, P. W., 416–417, *463*
Kellerhals, M. B., 59, *109*
Kellermann, J., 356–357, *395*
Kelly, C., 225–226, 256–257, 262, , *276*
Kelly, E., 356, *396*
Kelly, K., 240, *281*
Kelm, S., 451, *475*
Kennedy, J. F., 27, *100–101*, 123, 129, *189*, *193*
Kerek, F., 142, 157, *197*
Kerekgyarto, J., 157, *203*
Kerman, K., 229, 257, *277*
Kern, H., 120, *187*
KewalRamani, V. N., 262, *289*

Keyhani, A., 346, *391*
Khalifah, R. G., 317, 337, *377*
Khandekar, K., 72, *112*
Kho, A. L., 355, *395*
Khomutov, G. B., 270, *290*
Khoo, K. H., 451, *475*
Kieboom, A. P. G., 151, *201*
Kiefel, M. J., 456, *479*
Kiefer, L. L., 157, *203*
Kielty, C. M., 416, *463*
Kiely, C. J., 215–216, 235, *273, 278*
Kiesewetter, R., 129, *193*
Kiessling, L. L., 217, *274–275*
Kijewska, M., 324, *380*
Kijimoto, S., 429, *467*
Kikkeri, R., 250, 267, *286*
Kikuchi, A., 436, *471*
Kim, B. C., 355, 357, *395–396*
Kim, B. -G., 166, *206*
Kim, B. H., 97, *116*
Kim, B. J., 216, *273*
Kim, B. Y., 443, *474*
Kim, C. H., 429, 443, 451, *467*, *474–475*
Kim, C. K., 269, *290*
Kim, H. T., 451, *475*
Kim, I. K., 231, 233, *278*
Kim, I. S., 97, *116*
Kim, J. K., 429, *467*
Kim, J. P., 97, *116*
Kim, J., 239, *280*
Kim, K. S., 255, *288*, 314, 339, *375*, *387*
Kim, K. W., 434, 443, *469*, *474*
Kim, N. S., 451, *475*
Kim, S. S., 97, *116*
Kim, S., 340, *388*
Kim, U. J., 72, *111*
Kim, Y. -G., 166, *206*
Kim, Y. S., 451, *475*
Kimball, S. D., 351, *394*
Kimura, K., 215, *273*
Kinae, N., 337, *385*
King, A., 216, *274*
King, S. B., 295, *368*
King, S. J., 453, *476*
Kinoshita, T., 326, *382*, 453, *477*
Kircher, M. F., 240, *281*
Kiriakova, G., 363, *401*
Kirito, K., 434, *470*
Kirschenlohr, W., 296, *369*
Kirschner, K., 315, *375–376*
Kishikawa, H., 357, *396*
Kishimoto, Y., 269, *289*
Kiso, M., 408, 424, 452, *461*, *465*, *476*
Kitagawa, Y., 348, *391*
Kitamaru, R., 52, *107*
Kitano, H., 237, *278*
Kitaoka, T., 223, 226, 229, *277*
Kiyosawa, Y., 327, *382*
Klaffke, W., 311, *375*
Klaiber, R. G., 348, *392*
Klatt, N., 124, *189*
Klein, W., 361, *400*
Kleinberg, M., 439–440, *472*
Klemm, D., 75, *112*, 119–121, 123–124, 126, 130–131, 150–152, *187–190*, *192*, *194*, *200*
Klohr, E. A., 68, 74–75, *111*, 147, 151, *198*, *201*
Kloow, G., 127, *192*
Klosiewicz, D. W., 129, *193*
Klostermeyer, H., 323, *379*
Kluczyk, A., 309, 324, *374*, *380*
Klüfers, P., 127, *192*
Klumperman, J., 409, *461*
Klyosov, A., 356, 363, *395*
Knight, G., 215, *273*
Knippe, D. M., 434, *470*
Knobler, C. M., 216, *274*
Knochenmuss, R., 163, *204*
Knoll, K., 325, *381*
Knop, S., 130, *194*
Ko, J. H., 434, 443, 451, *469*, *474–475*
Ko, R., 362, *400*
Kobayashi, T., 436, 438, *471–472*
Koch, W., 68, 74–75, *111*, 147, 155, *198*, *202*
Kochetkov, N. K., 119, *186*
Koda, T., 429, *467*
Koetz, J., 237, *279*

Kogler, H., 343, *389*
Kogure, K., 249, 259, 265, *285*, *288*
Koh, J., 97, *116*
Köhler, S., 124, *190–191*
Kohmura, M., 332, *384*
Koizumi, K., 326, *382*
Kokta, B. V., 85–86, *113–114*
Kolakofsky, D., 434, *470*
Kolka, S., 299, *370*
Kolpak, F. J., 50, *106*
Komori, T., 340, *388*
Komoto, M., 334, *384*
Kondo, T., 150, 172, 177, *199–200*, *207*
Koner, B. C., 358, *397*
König, W. A., 157, *203*
Konishi, Y., 333, 335, 348, *384*
Konno, K., 422, *464*
Kopitz, J., 429, 431–432, *467–468*
Korecz, L., 329–330, *383*
Korkaya, H., 432, *468*
Korlyukov, A. A., 298, *369*
Kornberg, R. D., 215–216, 238, *273*
Koroteev, A. M., 298, *369*
Koroteev, M. P., 298, *369*
Kosaka, Y., 311, *375*
Koschella, A., 68, 74–75, *111*, 119–120, 123–124, 130, 143, 150, 155, *187–190*, *200*, *202*
Koseki, K., 418, 428, 433, 435–438, 444, 454, *463*, *467*, *469–470*
Kosma, P., 39, *103*, 133, *196*
Köth, A., 237, *279*
Kothe, O., 119–120, *187*
Koyama, M., 44, 46, 57, *106*
Kozireski-Chuback, D., 433, *469*
Kraehenbuehl, K., 324, *380*
Kragten, E. A., 129, 151, *193*, *201*
Kramer, J., 351, *394*
Krämer, M., 237, *278*
Krantz, S., 356–357, *395–396*
Kraska, B., 364, *401*
Krässig, H., 118, *185*
Kratky, O., 56, *108*
Krause, E., 305, *372*
Krause, R., 325, 330, 381, *383*, *400*, *409*
Krauss, N., 39, 50, *103*
Krausz, P., 86, *114*
Kreiman, G., 447, *475*
Krestin, G. P., 240, *280*
Kriebel, J. K., 213, *271*
Krishnan, M., 73, *112*
Krogh, V., 358, *398*
Kroh, L. W., 335, 343, *385*, *389*
Kroh, L., 297–298, *369*
Kroon-Batenburg, L. M. J., 44, *106*, 121, *188–189*
Krüger, G., 311, 334, *351*, 375, *393*
Krüger, R., 163, *204*
Kruse, T., 175, 178, 181, *208*
Kubo, T., 42, *104*
Kuga, S., 50, 57, 72, 86, *107*, *109*, *111*, *115*
Kühn, G., 137, 141–142, 160–161, 163, *197*
Kuhn, R., 292, 296–297, 306, 311, 351, *367*, *369*, *373*, *375*, *393*
Kulicke, W. M., 129–131, 147, 193–195, *198*
Kuliś-Orzechowska, R., 318, 365, *377*, *402*
Kulkarni, M. P., 71–72, *111*
Kull, A. H., 68, 74–75, *111*
Kumar, C. S. S. R., 212, *270*
Kung, S.-D., 27, *100*
Kung, T. T., 441, *473*
Kunze, J., 142–143, 147, 150, *197–198*
Kunz-Renggli, C., 453, *477*
Kurokawa, T., 294, 339, *368*, *386*
Kurosawa, M., 338, *386*
Küster, B., 163, *204*
Kuutti, L., 63, *110*
Kuvelkar, R., 441, *473*
Kvick, A., 39, *103*
Kvick, Å, 50, *107*
Kvien, I., 92, *115*
Kwak, S. Y., 97, *116*
Kwatra, H. S., 86, *114*
Kwon, D. S., 262, *289*
Kwon, D., 247, *285*
Kwon, Y. K., 97, *116*

L

Laffont, I., 358, *397*
Lafuente, N., 357, *397*
Lagrou, A. R., 327, 357, *382*, *397*
Laguens, P., 443, *474*
Lai, Z., 351, *394*
Laidler, P., 409, *462*
Lair, S. V., 426, *466*
Lakin, M., 44, 48, *106*
Lakowicz, J. R., 229–230, *277*
Lamb, R. A., 434, *470*
Lamed, R., 82, *113*
Lamorte, G., 433, 439, *469*
Landry, K., 409, 435, 436, 440–441, 444, 454, *461, 471–473*
Lane, M. J., 347, *391*
Lang, F., 434, *470*
Lang, H., 133, *196*
Langan, P., 39, 44–45, 47, 50–51, 53, 63, *103*, *106–107*, 121, 131, *189*
Lange, H., 452, 454, 458, *475*
Lange, M., 130, 153, 157, *195*
Lange, T., 181, *196*
Langer, R., 270, *290*
Langhendries, P., 361, *400*
Lanzuise, S., 340, *387*
Lapolla, A., 323, 326, *380*, *382*
Lapp, H., 447, *475*
Laroy, W., 442, *473*
Larsen, K., 270, *290*
Larson, P., 246, *283*
Lasseuguette, E., 71, *111*
Lassmann, H., 425, *465*
Lau, B. H. S., 362, *400*
Lau, K., 453, *476*
Lau, M. W. L., 316, *376*
Laurino, P., 250, 267, *286*
Lautenslager, G. T., 357, *396*
Lauth-de Viguerie, N., 237, *278*
Laver, W. G., 408, *461*
Lawrisuk, L., 423, *465*
Lay, L., 234–236, 262, *278*
Layek, S., 351, *394*
Lazik, W., 137, 147–148, 151, *197*
Lazo, P. A., 132, *468*
Lazzari, S., 330, *384*
Le borgne, R., 409, *461*
Le bouar, T., 251, *286*
Leary, J. A., 163–164, *205*
Lebioda, S., 178, *208*
Leclerc, D., 407, *460*
Lecourt, T., 120, *188*
Ledeen, R. W., 433, 443–444, *469*, *474–475*
Lederer, M. O., 311, 332, 348–349, 360, *374*, *384*, *391–392*, *398*
Ledl, F., 344, *390*
Lee, B. W., 356, *396*
Lee, B.-Y., 255, *288*
Lee, C. C., 217, *275*
Lee, C. K., 156, *201*
Lee, D. M., 54, *108*
Lee, H. S., 355, 357, *395*, *396*
Lee, H., 231–233, *277–278*
Lee, J. H., 451, *475*
Lee, J. K., 339, *387*
Lee, J.-H., 239, 264, *280*, *289*
Lee, K. E., 339, *387*
Lee, K., 231, 233, 266, *278*
Lee, P., 327, *382*
Lee, R. T., 217, 251, 254, *274*
Lee, S. J., 323, 362, *379*
Lee, S.-J., 249, 257, 265, *285*
Lee, W.-Y., 255, *288*
Lee, Y. C., 217, 251, 254, *274*, 429, 434, *452*, *467*, *469*
Leff, D. V., 216, *273–274*
Lefrancois, S., 409, *461*
Legoy, M. D., 298, *369*
Legrand, R. D., 306, 358, *373*
Legros, L., 434, *469*
Leguizamon, M. S., 454, *477*
Lehmann, F., 451, *475*
Lehtio, J., 169, *206*
Lehtlö, J., 82, *113*
Leipner, H., 69, *111*
Leisner, T., 163, *204*
Leitao, C. B., 327, *382*
Lemaigre, F., 341, *388*

Lemesle, L., 170, *207*
Lemke, H., 452, 454, 458, *475*
Lennholm, H., 143, *198*
Lepenies, B., 250, 267, *286*
Lerche, H., 306, *373*
Letsinger, R. L., 214, *272*
Levade, T., 436, 444, *471*
Levchenko, T. S., 246, *283*
Levenson, J., 443, *474*
Levesque, J. P., 420, *464*
Levi, J., 295, *369*
Levy, R., 214, *272*
Levy, S., 432, *468*
Lewis, S., 355, *395*
Lewis, W., 410, *462*
Li, C. Y., 427, 448, *466*
Li, C., 304, 355, *371*
Li, D., 343, *389*
Li, H., 270, *290*
Li, J. J., 246, *283*
Li, K., 353, *394*
Li, L.–F., 343, *389*
Li, S. C., 453, *476*
Li, W., 340, *388*
Li, X. M., 427, 448, *466*
Li, Y. T., 453, *476*
Li, Y., 453, *476*
Liakatos, A., 456, *479*
Liang, C. Y., 61, *109*
Liang, F., 422, 440–441, *464*, *472–473*
Liang, Y.-H., 343, *389*
Liao, H.-K., 269, *289*
Liao, K.-W., 249, 257, 265, *285*
Liardon, R., 323, 360–363, *379*, *399*
Lichtenegger, H., 65, *110*
Lichtenstein, H., 360, *399*
Lidke, D. S., 248, 266, *285*
Liebert, T. F., 120, 123–124, 127, 130, 132, 137, 142, 149–150, 155, 175, 178, 181, *187–192*, *195–196*, *199*, *202*, *208*
Liebig, J., 118, *186*
Liedke, R., 336, *385*
Ligtvoet, N., 355, *395*
Lim, K. H., 247, *285*
Lim, K.-R., 255, *288*
Lima, M., 304, 323, *372*
Lin, C.–C., 220–221, 223, 231, 233, 249, 255, 266, 269, *275–276, 285, 289*
Lin, G. Q., 353, *394*
Lin, R.-K., 249, 266, *285*
Lin, S.-Y., 214, *272*
Lin, X., 129, 137, 143, *193*, 326, *382*
Lin, Y.-P., 220, 223, 231, 233, 255, 269, *276*
Lin, Z.-H., 234–235, 242, *278*
Lincke, T., 124, 143, *190*
Lindberg, B., 138, *197*
Linder, M., 82, *113*, 169–170, *206–207*
Lindhorst, T. K., 217, *275*
Lindquist, U., 138, *197*
Linek, K., 298, *369*
Linetsky, M. D., 298, 306, 358, *369, 373*
Linow, K.-J., 35, *103*
Lis, T., 318, *377*
Litrán, R., 241, *281*
Littman, D. R., 262, *289*
Liu, C., 340, *388–389*
Liu, F.-T., 363, *401*
Liu, G.-Y., 220, 223, 262, *276*
Liu, H. Q., 68, *110*
Liu, H., 246, *283*
Liu, H.-H., 247, 266, *284*
Liu, J., 214, 246, *272*, *283*
Liu, K., 413, 417–418, *462*
Liu, P., 443, *474*
Liu, Q., 351, *394*
Liu, S. Y., 220, *276*
Liu, W., 246, *283*, 326, *382*
Liu, X., 263, *282*
Liu, Y.-H., 249, 266, *285*
Logroscino, G., 355, *395*
Lomax, J. A., 178, *197*
Lombardi, G., 234, 235–236, 262, *278*
Long, M., 428, *467*
Lönnberg, H., 92, *116*
Lönngren, J., 119, *186*
Loomis, W. F., 28, *101*
Lopes, M. F., 454, *477*
López, M. G., 310, 317, 364, *374*

López, O., 353, *394*
López-Castro, A., 322, *379*
López-Fandiño, R., 303, *371*
Lorenzini, I., 442, *473*
Lorenzo-Ferreira, R. A., 143, *185*
Lorito, M., 340, *387*
Loth, F., 142–143, 150, *197*
Lourdin, D., 86, *114*
Lovdahl, M. J., 365, *402*
Love, J. C., 213, *271*
Low, J. N., 318, *377*
Lowy, P. H., 336, 362, *385*
Lozzio, B. B., 426, *466*
Lozzio, C. B., 426, *466*
Lu, A.-H., 239, *280*
Lu, Y., 214, *272*
Lu, Z. H., 433, 443–444, *469*, *474–475*
Luciani, L., 415, *463*
Lukong, K. E., 409, 441, *461*, *473*
Luna, M., 241–242, *281*
Lundquist, J. J., 217, 251, *274*
Lundqvist, H., 421, *464*
Lunow, D., 343, *390*
Luo, J., 124, 142, *190*
Luo, M., 453, *476*
Luo, P. G., 260, *288*
Luo, Y., 453, *476*
Lupas, A. N., 407, *460*
Lupo, B., 427, 433, 439, *466*, *469*
Lustig, S. R., 435–436, *470*
Lutz, M., 120, 131, *187*
Luzikov, Y. N., 364, *402*
Ly, B., 93, *116*
Lyu, Y.-K., 255, *288*

M

Mac Gregor, E. A., 170, *207*
Machado, E. A., 426, *466*
Machemer, H., 118, *185*
Mack, L. L., 163, *204*
MacKay, J. A., 217, *275*
Mackay, J. D., 324, *380*
MacKenzie, A., 323, *379*
Mackenzie, J. S., 427, *466*
MacMathuna, P., 444–445, *475*
MacQueen-Mason, S. J., 30, 32, *101*
Madson, M. A., 126, 143, *191*
Maeda, K., 323, *380*, *453*, *477*
Maehara, Y., 237, *278*
Maerk, I., 425, *465*
Maes, E., 451, *475*
Maes, V., 350, *393*
Magarinos, G., 443, *474*
Magesh, S., 408, 424, 429–430, 455, *461*, *465*
Mahmood, U., 240, *281*
Mahot, O., 151, *201*
Mai, M., 132, *195*
Maia, E., 68, 69, *110*
Main, A., 351, *394*
Maire, I., 413, *462*
Malaveille, C., 360, *399*
Malcom Bown, R., 33, *102*
Maliekal, P., 315, 342, *376*
Malisan, F., 436, *471*
Maljaars, C. E. P., 220–223, 255, 269, *276*, *288*
Mallard, A., 27, 56, *101*
Mallet, J.-M., 120, *188*, 251, *286*
Malmström, E., 92, *116*
Mammen, M., 217, 254, *275*
Mancin, F., 234–235, *278*
Mandal, C., 439, *472*
Mandeville, S., 346, *391*
Manea, F., 234–235, , *278*
Manelius, R., 175–176, *198*
Manini, P., 306, *373*
Manley, R. S. J., 47, *106*
Mann, G., 142–143, 150, *197*
Mann, J., 42–43, *105*
Mann, L., 416, 419, *463*
Manuel, B., 357, *397*
Manzoni, M., 429, 435–436, 449–450, *467*, *470*
Marbach, P., 350, *393*
March, J. F., 155, *202*
Marchesini, S., 304, 425–426, 436, *465*, *471*
Marchessault, R. H., 27, 30, 38–39, 55–56, 61, *101*, *103*, *108–109*

Marchetti, S., 434, *469*
Marchioro, L., 327, *383*
Marchisano, C., 327, *383*
Marco, M. P., 270, *290*
Mares, P., 151, *200*
Margalit, R., 270, *290*
Margolles-Clark, E., 169, *206*
Mari, B., 434, *469*
Marion, M. S., 340, *387*
Mark, H., 56, *108*, 118, *185*
Mark, R. E., 35, *103*
Markó, L., 324, *380*
Márquez, R., 322, *379*
Marradi, M., 225–227, 242–243, 256–257, 262–263, 268, 270, *276–277, 289–290*
Marrinan, H. J., 43, *105*
Marshall, L. F., 246, *283*
Martensen, I., 458, *479*
Martin, O. R., 309, *374*
Martin, S., 355, *395*
Martín-Álvarez, P. J., 303, *371*
Martinet, W., 426, *465*
Martinez, V., 443, *474*
Martínez-Ávila, O., 225–226, 256–257, 262–263, 270, *276*, *289–290*
Martinez-Pacheco, R., 118, 145, *185*
Martín-Lomas, M., 220, 224, 226, 228, 256, *276–277*
Martins, M. B., 249, *285*
Martiny, L., 415, *463*
Martos, S., 353, *394*
Marty, J. D., 237, *278*
Marverti, G., 330, *384*
Marx-Figini, M., 35, *103*
Maschmeyer, T., 324, *377*
Masereel, B., 53, *375*
Maskos, K., 123, 150, *189*, *199*
Massai, G., 414–415, *463*
Massart, D. L., 347, *391*
Massey, A. P., 258, *288*
Masuda, S., 337, *385*
Masuda, T., 249, 259, 265, *285*, *288*
Mata-Segreda, J. F., 300–302, *370*
Matesanz, N., 357, *396–397*
Matheson, N. K., 169, *206*
Mathewson, S., 324, *380*
Mathieson, A., 50, *107*
Matsuda, T., 327, *382*
Matsumoto, K., 355, *396*
Matsumoto, M., 435, *470*
Matsumura, H., 86, *114*
Matsumura, K., 337, *385*
Matsumura, S., 90, *115*
Matsunaga, M., 435, *470*
Matsuo, R., 71, *111*
Matsuo, Y., 249, 265, *285*
Matsuura, H., 362, *400*
Matsuzaki-Kobayashi, M., 436, *471*
Matthey-Doret, W., 346, *391*
Matthijs, G., 342, *389*
Mattos-Guaraldi, A. L., 453, *476*
Mattoussi, H., 246–247, *283*
Matyi, R. J., 245, *282*
Matzilevich, D., 455, *478*
Maugard, T., 298, *369*
Maurer, A., 57, *109*
Maurer, K., 295, *368*
Maurer, K.-H., 340, *388*
Maurer, S. V., 325, *381*
Mauro, J. M., 246, *283*
Mauron, J., 293, *367*
Mawhinney, T. P., 362–363, *401*
Maxwell, R., 428, *466*
Maya, I., 353, *394*
Mayer, R., 28, *101*
Maysinger, D., 249, *286*
Maza, S., 353, *394*
Mazeau, K., 44, 48, 66, *106*, *110*
Mazer, J. S., 304, *372*
McBain, S. C., 270, *290*
McBrierty, V. J., 150, *199*
McCann, M., 30, *101*
McCarthy, J. R., 239, *280*
McCleary, B. V., 169, *206*
McDonald, M. A., 243, *281*
McDonnell, M. B., 236–237, 255, *278*
McGilvray, K. L., 216, *274*
McGinnis, G. D., 155, *202*

McKay, P., 304, *372*
McKellar, Q. A., 358, *397*
McLafferty, F. W., 315, *376*
McLean, J. A., 269, *289*
McManus, M. J., 304, *371*
McPherson, J. D., 308, *374*
McReynolds, K. D., 262, *289*
Mears, S., 442, *473*
Meckert, P. C., 443, *474*
Medbury, H., 356, *396*
Medina-Acosta, E., 452, *476*
Medintz, I. L., 247, *283*
Medley, C. D., 214, *272*
Meeldijk, J. D., 220, 222, 253–254, *276*, *287*
Meenakshi, R., 338, *386*
Meerwein, H., 157, *203*
Meireles, S., 92, *115*
Meisetschläger, G., 350, *393*
Meister, F., 127, *192*
Mejia, J. S., 455, *478*
Melander, C., 166, 173–174, 178–179, *206–207*
Meli, M., 355, *394*
Mello, K., 92, *115*
Meltretter, J., 304, 323, *372*
Mendonca-Previato, L., 453, 454–457, *477*, *479*
Mendoza, V. M., 454–455, *477–478*
Meneghini, E., 358, *398*
Menéndez, M., 252, *286*
Menikh, A., 251, *286*
Mennella, C., 305, 340, *372*, *387*
Menzel, E. J., 355, *394*
Mera, K., 337, *385*
Mercier, C., 176, *209*
Meredith, F. I., 360, *398*
Mertineit, C., 407, *460*
Merz, A., 315, *375*
Mester de Parajd, L., 364, *401*
Mester de Parajd, M., 364, *401*
Metcalf, J. B., 363, *401*
Metcalf, P. A., 326, *382*
Metlitskikh, S. V., 298, *369*
Metz, T. O., 304, 323, *371*, *379–380*
Meyenhofer, M. F., 443, *474*
Meyer, B., 294, *368*
Meyer, C. A., 343, 362, *390*
Meyer, K. H., 42–43, 57, 104, *118*, *185*
Mezentsev, A. V., 441, *473*
Mibu, N., 311, *375*
Michalet, X., 246, *283*
Michaud, F., 92, *115*
Michaud, J., 436, 444, *471*
Michaud, L., *460*, *473*
Micheel, F., 298, 300, *369*
Micheli, A., 358, *398*
Microsc, J., 63, *110*
Middel, J., 262, *289*
Miihlethaler, K., 56, *108*
Mijland, P. J. H. C., 143, 175, 177, 182–183, *191*, *198*, *210*
Mikawa, Y., 120, *188*
Milane, R. P., 44, *106*
Miles, D. T., 215, *273*
Miles, M. J., 63, *109–110*
Milewski, S., 343, *389*
Miller, D. L., 118, *185*
Miller, G. D., 90, *115*
Miller, P. G., 428, *467*
Mills, F. D., 346, *390*
Mine, S., 357, *396*
Mingotaud, C., 237, *278*
Minoda, M., 68, *110*
Mioduszewski, J. Z., 303, *371*
Mirchink, E. P., 364, *402*
Mirkhalaf, F., 216, *274*
Mirkin, C. A., 212, 214, 246, *270–272*, *283*
Mirvish, S. S., 360, *399*
Misch, L., 42–43, 57, *104*
Mischnick, P., 117–184, *187–189*, *191–192*, *194–198*, *200–207*, *209*
Misciagna, G., 355, 358, *395*, *398*
Misevic, G. N., 251, *286*, 329–330, 337, *362–363*
Mistry, K., 327, *382*
Mitchell, S., 358, *397*
Mitchinson, C., 170, *206*
Mithieux, S. M., 413, *462*
Miura, S., 340, *388*
Miura, Y., 229, 257, *277*

Miyagi, M., 355, *395*
Miyagi, T., 408, 418–419, 422, 424–425, 428–440, 444, 454–455, *461*, *463–465*, *467–472*, *475*
Miyamoto, T., 68, 110, *149–150*, *199*
Miyano, H., 324, *380*
Miyano, S., 311, *375*
Miyata, T., 323, *380*
Miyazaki, K., 438, *472*
Miyazaki, S., 418, *464*
Miyazawa, T., 358, *397–398*
Mizukoshi, M., 327, *382*
Mizukoshi, T., 324, *380*
Mizutani, H., 327, *382*
Mobley, R. C., 163, *204*
Mocetti, E., 454, *477*
Mohamed, G. A., 362, *400*
Mohamed, M. H., 362, *400*
Mohamed, S., 348, *391*
Mohs, A. M., 246, *283*
Möhwald, H., 327, 343, *383*
Molina, J. L., 353, *394*
Mollard, A., 41, 56, *104*
Mollicone, R., 451, *475*
Möllmann, E., 124, 130, *191*
Moloney, C., 304, 323, *372*
Molossi, S., 443, *474*
Momcilovic, D., 117–184, *201*, *205–209*
Mondal, S., 439, *472*
Monnier, V. M., 293–294, 327, 339, 340, 344, 348–349, 358, *368*, *386–387*, *390–392*
Monroe, M. E., 323, *379*
Montagna, G., 458, *479*
Montanari, S., 81, *113*
Montero, J. L., 311, 366, *375*
Monti, E., 403–459, *460–461*, *464*, *466–471*
Monti, S. M., 323, 327, *380*, *383*
Montreuil, J., 453–455, *477*
Moon, K. C., 357, *396*
Moon, S. K., 443, *474*
Moore, A., 240, *280*
Moore, L., 428, *467*
Moore, T. M., 245, *282*
Moorthy, S. N., 129, *194*
Morales, C. R., 409, 435–436, 441, 444, 454, *461*, *471*, *473*
Morales, J. C., 217, 251, *274*
Morales, M. D., 239, *280*
Moran, L., 351, *394*
Morath, C., 361, *400*
Moreno, E., 318, *377*
Moreno, F. J., 303, 325, 359–360, *371*, *381*, *398*
Morgenstern, B., 68, *110*
Mori, A., 327, *382*
Morishita, I., 346, *390*
Moriya, S., 418, 424, 428, 430, 435–438, 454, *463*, *465*, *467*, *470*
Mormann, W., 130, *194*
Mornet, S., 239–240, *279*
Mornon, J. P., 170, *207*
Moronne, M., 245–246, *283*
Morooka, T., 132, *195*
Morosi, L., 435, *471*
Morreau, H., 410, 414, 416, *462–463*
Morrison, L. D., 335, *385*
Morrone, A., 414–415, *463*
Mosca, A., 327, *382*
Moshiach, S., 419, 421, *464*
Mosin, V. V., 363, *401*
Moss, A. C., 444–445, *475*
Moss, J. A., 425, *465*
Mossine, V. V., 291–366, *369–371*, *376–378*, *383*, *387*, *392*, *395*, *400–401*
Moteki, K., 362, *400*
Mottram, D. S., 344, *390*
Moudgil, B., 264, *289*
Mountney, A., 442, *473*
Mrksich, M., 217, *275*
Mu, M., 350, *393*
Mucic, R. C., 214, *272*
Muetgeert, J., 143, 176, 182–183, *191*
Muggli, R., 42, *104*
Muhl, C., 432, *468*
Muhlenhoff, M., 438, *471*
Mukhopadhyay, B., 231–232, 249, 255, *277–278*, *285*
Mulder, B. M., 33–34, *102*

Müller, M., 65, *110*
Müller, R., 126, 139, 142, 163, 165–167, 181–182, *192*, *197*, *204*, *209*
Müller, T. M., 44, 50, *106*
Multigner, M., 241, *281*
Mulvaney, P., 215, 246, *272*, *284*
Münch, P., 309, *374*
Mune, M., 358, *397*
Munot, Y. S., 249, 266, *285*
Munoz-Barroso, I., 435, *470*
Murata, J., 337, *385*
Murata, M., 348, *392*
Murray, C. B., 213, 239, 247, *271*, *280*, *284*
Murray, D., 444–445, *475*
Murray, R. W., 213, 215–216, *271*, *273–274*, *280*
Musick, M. D., 214, *272*
Muti, P., 358, *398*

N

Na, H. B., 239, *280*
Nacharaju, P., 304, *372*
Nada, S., 432, *468*
Nadarajah, D., 416–417, *463*
Nagahori, N., 229, 269, *277*, *289*
Nagai, K., 54, *116*
Nagai, R., 337, *385*
Nagakura, N., 366, *402*
Nagamune, K., 453, *477*
Nagaraj, R. H., 348, 355, *391*, *395*
Nagasaki, Y., 229, 235, 254, 269, *277–278*, *287*
Nagasawa, M., 366, *402*
Nagy, G., 324, *380*
Nagy, L., 324, 329–330, *380*, *383*
Nair, K. S., 355, 363, *395*
Nakagawa, K., 358, *397–398*
Nakahara, S., 245, *282*
Nakai, T., 251, *286*
Nakamura, K., 339, *387*
Nakamura, M., 434, *470*
Nakamura, N., 348, *391*
Nakamura, R., 327, *382*
Nakamura, S., 75, *112*
Nakamura, T., 235–236, 254, *278*
Nakamura, Y., 336, *385*
Nakamura-Tsuruta, S., 269, *289*
Nakano, K., 348, *391*
Nakano, M., 336, *385*
Nakashima, Y., 311, *375*
Nakatsubo, F., 120, 150, *188*, *200*
Nakayama, H., 338, *386*
Namavari, M., 295, 350, *369*
Namchuk, M. N., 175, *208*
Nan, X., 440–441, *472–473*
Nanjo, Y., 326, *381*
Napolitano, A., 305–306, 340, *372–373*, *387*
Nar, H., 315, 350, *376*
Narain, R., 220, 250, *276*, *286*
Narayan, R., 73, *112*
Narayanan, R., 213, *271*
Narimatsu, H., 433, 436–438, 444, 454, *469*
Narui, T., 161, *204*
Narute, S. B., 237, *279*
Natan, M. J., 214, *272*
Naturforsch, Z., 298, 330, *370*, *383*
Naven, T. J. P. M., 165, *205*
Naven, T. J. P., 163, 166, *204*, *205*
Needs, P. W., 157, *203*
Neff, N. H., 433, *469*
Neglia, C. I., 301–302, 304, 317–318, 355, *370*, *377*
Nehls, I., 132, 147, 149, *190*, *199*
Nekrasov, E., 323, *379*
Nemet, I., 293–294, 317, 327, 340, 348, *368*
Nerinckx, W., 170, *206*
Nesti, C., 423, 428, *464*
Neto, C. P., 86, *114*
Neubacher, B., 454, 456–457, *477*, *479*
Neumann, G., 427, *466*
Nevado, J., 357, *396–397*
Nevell, T. P., 150, *200*
Newman, R. H., 64–65, *110*
Ngantung, F. A., 428, *467*
Nguyen, H., 351, *393*
Nguyen, T., 456–458, *479*
Nicewarner, S. R., 214, *272*

Nichols, R. J., 214, *272*
Nicolai, E., 27, *101*
Nie, Q. L., 247, *284*
Nie, S. M., 245–246, *283*
Niedner, W., 163–167, *197*, *205*
Nieduszynski, I. A., 42, *105*
Niemczura, W. P., 123, 150, *189, 199*
Niemelä, K., 129, *193*
Niemeyer, C. M., 212, 214, 246, *270–272*, *283*
Nifantiev, E. E., 298, *369*
Nifantiev, N., 363, *401*
Niikura, K., 249, 259, 265, *285*, *288*Nilsson, G., 143, 173, 176, *198*, *207*
Nilsson, S. K., 420, *464*
Nimtz, M., *101, 188*
Nishi, H., 346, *390*
Nishikawa, S., 35, *103*
Nishimura, H., 54, 79, *108*
Nishimura, S.-I., 229–230, 269, *277*, *289*
Nishimura, T., 268, *288*
Nishino, T., 97, *116*
Nishio, T., 249, 259, 265, *285*, *288*
Nishiyama, Y., 27, 39, 44, 47, 50–51, 53, 71, 86, 101, 103, *106–107*, *111*, *115*, *121*, *131*, *189*
Nishizawa, Y., 366, *402*
Nitz, M., 178, *208*
Noble, M. I., 443, *474*
Nobles, D. R., 28, *101*
Noda, T., 427, *466*
Noether, H. D., 55, *108*
Noguchi, M., 346, *390*
Nogueras, M., 318, *377*
Nojiri, H., 434, *470*
Nojiri, M., 172, 177, *207*
Nolting, B., 220, 223, 262, *276*
Nonvolatile, I., 322, *378*
Noomen, S. N., 310, *374*
Nordholt, M. G., 121–122, *188*
Norimoto, M., 132, *195*
Norris, D. J., 247, *284*
Notenboom, V., 178, *208*
Nottet, H., 262, *289*
Nouraldeen, A., 351, *393*
Nourse, A., 410, *462*
Novak, P., 305, *373*
Nucl, J., 350, *393*
Nukushina, Y., 59, *109*
Nunes, M. P., 454, *477*
Nurmi, K., 175–176, *198*
Nursten, H. E., 347, *391*
Nursten, H., 344, *390*
Nussenzweig, V., 452–454, *476–478*
Nutr, J., 361–362, *399–400*
Nuzzo, R. G., 213, *271*
Nyburg, S. C., 50, *106*

O

Oak, J. H., 358, *397–398*
Obayashi, H., 348, *391*
Obinata, M., 438, *472*
Obrenovich, M. E., 358, *397*
Obretenov, T., 344, *390*
O'Connell, T., 340, *388*
Odintsova, E., 432, *468*
Oehler, C., 429, *467*
Ofman, D., 350, *393*
Ogura, K., 443, *474*
Ogura, M., 443, *474*
Ohara, P. C., 215, *273*
Ohkita, J., 53, *107*
Ohmacht, R., 324, *380*
Ohnishi, T., 343, 362, *390*
Ohta, S., 327, 340, *383*, *388*
Oikawa, S., 358, *397–398*
Ojeda, R., 220, 222, 224, 226, 228, *275*, *277*
Oka, M., 250, 266, *286*
Okada, M., 336, *385*, 432, *468*
Okada, Y., 357, *396*
Okamura, Y., 414, *463*
Okamura-Oho, K., 15, *101*
Okano, T., 44, 48, 50, 52, 54, 79, *106–108*
Oksman, K., 92, *115*
Okuda, H., 343, 362, *390*

Okuhara, A., 347, 360, *391*
Olano, A., 303, 325, *371*, *381*, *399*
Olaru, N., 127, 129, *192–193*
Olaru, O., 127, 129, *192–193*
Olsufyeva, E. N., 364, *402*
Olyslager, Y. S., 327, *382*
O'Neal, J. F., 428, *467*
O'Neill, E., 339, *351*
Ono, S., 35, *103*
Onodera, Y., 425, *465*
Oorschot, V., 409, *461*
Opietnik, M., 223, 226, 229, *277*
Opperdoes, F. R., 356, 364, *395*
Oppezzo, P., 456, 458, *479*
Oravecz, T., 351, *393*
Orekhov, N., 452, *476*
Oriol, R., 451, *475*
Ortega-Barria, E., 455, *478*
Ortiz, C., 120, *188*
Orton, D. J., 323, *379*
Ortwerth, B. J., 298, 318, 321, 331–332, 335–336, *369*
Osaki, F., 247, 264, *284*
Osborn, H. M. I., 131, *195*
Oshima, M., 149, *199*
O'Sullivan, A., 27, *101*
Ota, M., 332, *384*
Otagiri, M., 337, *385*
Otani, H., 358, *397*
Otsuka, H., 229–230,235–236, 254, *277–278*
Ott, E., 27, 100, 118, 136–137, 141, *147*, *186*, *196*
Ouchi, K., 438, *471*
Oudhoff, K. A., 181, *209*
Overkleeft, H. S., 311, *375*
Ovodoc, Y. S., 131, *195*
Ozawa, S., 324, *380*

P

Paatero, E., 129, *193*
Pagano, P. J., 355, *395*
Pai, R., 304, *372*
Pala, V., 358, *398*
Palazzolo, G., 430, 433, 440, *469*
Paleari, R., 327, *382*
Palfey, B. A., 340, *387*
Palladium, L., 251, *383*
Palm, D., 298, *370*
Pàmies, O., 353, *394*
Pan, B.-Q., 247, 266, *284*
Pan, X., 350, *393*
Pandey, A., 438, *472*
Pang, D.–W., 247, 266, *284*
Panico, A., 327, *382*
Pankarican, M., 412, *462*
Pankhurst, Q. A., 239, *280*
Papini, A. M., 309, *374*
Papini, N., 423, 426–427, 430–431, 433, 435–436, 439–440, 449–451, 454, *464*, *466–467*, *469–470*
Paprotny, J., 216, *274*
Parcy, F., 33, *102*
Pardoe, W. D., 146, *198*
Paredez, A. R., 33, *102*
Parfondry, A., 175, *208*
Parikh, D. V., 150, *199*
Parini, R., 412, *462*
Paris, G., 453, 455–458, *476–479*
Paris, O., 65, *110*
Park, C., 247, *285*
Park, C.–W., 323, 362, *379*
Park, D. K., 339, *387*
Park, D., 362, *400*
Park, H., 166, *206*
Park, T. G., 231–233, 267, *278*
Parodi, A. J., 455, *478*
Paroni, R., 327, *382*
Parrish, D. A., 295, *368*
Parving, H. H., 355, *395*
Pasch, H., 180, *209*
Pascoal Neto, C., 86, *114*
Pasquini, D., 85–86, *114*
Patel, C. K., 150, *200*
Patel, H. V., 443, *474*
Patel, M., 295, 350, *369*
Patel, P., 245, *282*
Patel, R. D., 150, *200*

Pattison, S., 412, *462*
Paul, S., 452, *476*
Pauli, J., 39, *103*
Paulsen, H., 292, *367*
Paulson, J. C., 435–436, 443, *470*, *474*
Pavlov, V., 214, *272*
Pawelke, B., 400, *409*
Payen, A., 28, 33, *101*
Payen, M., 118, *184*
Paz-Pabon, C. N., 327, *382*
Peale, F., 246, *283*
Pearce, A. J., 120, *188*
Peat, S., 34, *102*
Peden, A. A., 409, *461*
Peer, D., 269, *290*
Peiró, C., 357, *396–397*
Pelouze, T. J., 118, *184*
Peltonen, J., 63, *110*
Peluso, G., 327, *383*
Peña, M., 301–302, *370*
Penadés, S., 211–269, *272*, *275–277*, *281*, *285*, *287–290*
Peñas, E., 359, *399*
Peng, X., 247, 248, *284–285*
Penndorf, I., 304, 325, 355, *371*, *381*
Penzkofer, A., 120, *187*
Peralta-Inga, Z., 39, *103*
Pere, J., 63, *110*
Pereira, M. A., 453–454, *477*
Pereira, M. E., 454–455, *478*
Pereira, M. P., 324, *380*
Perez, E., 251, *286*
Perez, J. M., 240, *281*
Perez, S., 25–100, *103–104*, *106*, *110*, 318, *377*
Perichon, M., 356, *396*
Perié, J., 356, 364, *395*
Perier, C., 355, *394*
Pérignon, N., 237, *278*
Perillo, V., 304, *372*
Perkins, E. G., 361, *399*
Perlin, A. S., 71, 75, 111, 119, 155, *175*, *187*, *202*, *208*
Perrin, C. L., 297, 300, *369*
Perry, G., 358, *397*
Persson, J., 48, 81, *106*
Peter, M. G., 163, *204*
Peter-Katalinic, J., 119, *187*
Peters, J. A., 132, 148, 151, 195, *201*, *324*, *377*
Peters, R., 416, *463*
Peters, T., 456–457, *479*
Petersson, L., 92, *115*
Petit, A., 314, *375*
Petrash, J. M., 340, *387*
Petrovsky, A., 240, *281*
Pettersson, G., 169, *206*
Petyuk, V. A., 323, *379*
Petzold, K., 123–124, 126, 130, 143, 150, 157, *187*, *192*, *195*, *200*
Peydecastaing, J., 86, *114*
Peytavi, R., 244, *282*
Pfahler, C., 348, *392*
Pfefferkorn, R., 118, *185*
Pfeiffer, K., 137, 147–149, 151, *197*, *199*
Pflughaupt, K. W., 292, *367*
Philipp, B., 35, 103, 119, 121, 123, 132, 143, *147–149*, *187*, *190*, *199*
Philippon, C., 408, *461*
Piao, Y., 239, 264, *280*
Piccoli, M., 430, 433, 440, *469*
Piccolo, E., 363, *401*
Pickford, R., 119, *187*
Piddini, E., 431–432, *469*
Piechotta, C. T., 305, *372*
Piens, K., 170, *206*
Pienta, K. J., 363, *401*
Pientka, Z., 245, *282*
Pierce, G. N., 444, *475*
Pieters, R. J., 254, *287*
Pietzsch, J., 361, *400*
Pigman, W., 292, 299, *367*, *370*
Pignatelli, B., 360, *399*
Pill, T., 330, *383*
Pinaud, F. F., 246, *283*
Pincet, F., 251, *286*
Pinel, E. F., 241, *281*
Pinto, B. M., 46, *106*
Pires, R. S., 454, *477*
Pischetsrieder, M., 304, 306, 323, 339, *372–373*, *387*

Piskorska, D., 360, *399*
Platek, M., 358, *398*
Platt, D., 356, 363, *395*
Plattner, R. D., 302, *371*
Plebani, M., 327, *383*
Pless, J., 350, *393*
Pochylova, Z., 33, *102*
Pokharkar, V. B., 129, 153, 237, *279*
Polak, J. M., 213, *271*
Poling, S. M., 302, 360, *371*, *398*
Polito, L., 234, *278*
Polizzi, S., 234–236, *278*
Pollert, E., 245, *282*
Pontes, L. C., 452, 454, *476*, *478*
Pontes, L., 452–453, *476–477*
Pontes-de-Carvalho, L. C., 452–453, *476–477*
Ponting, C. P., 424–425, *465*
Poppleton, B. J., 50, *107*
Porter, M., 216, *274*
Portero, M., 348, *391*
Portner, A., 434, *470*
Potier, M., 407, 441, *460*, *473*
Potthast, A., 39, 103, 133, 150, *181*, *196*, *200*
Poupetova, H., 412–413, *462*
Powell, J., 430, 439–440, *472*
Pozzi, C., 435, *471*
Prabhakaram, M., 348, *392*
Prabhune, A. A., 245, *282*
Pradera, M. A., 353, *394*
Prasad, B. L. V., 237, 245, *279*, *282*
Pratz, G., 338, *386*
Pravdova, V., 347, *391*
Preedy, V. R., 362, *401*
Prehm, P., 157, *203*
Preobrazhenskaya, M. N., 364, *402*
Presta, M., 425, *465*
Preston, R. D., 27, 42, 56, *101*, *105*, *108*
Preti, A., 406, 423, 425–426, 428–431, 433–437, 439–440, 449–451, 454, *460*, *464–467*, *470*
Previato, J. O., 453–454, 456, *477*, *479*
Price, J. E., 318, 363, *378*, *401*
Pridham, J. B., 338, *386*
Priebe, S. R., 365, *402*
Prinetti, A., 430–431, 433–435, 439–440, 451, *467*, *469–470*
Prioli, R. P., 455, *478*
Prioni, S., 430–431, 433, 435, 440, 451, *467*, *478*
Prokazova, N.V., 443, *474*
Proksch, P., 362, *400*
Prota, G., 306, *373*
Pshezhetsky, A. V., 407, 409, 413–415, 422, 435–436, 440, 441–442, 444, 454, *460–464*, *471–473*
Puls, J., 124, 147, 151, 155, 173, 175, 177–178, 181, *189*, *198*, *201–202*, *207–208*
Pun, S. H., 220, 258, *276*, *288*
Puranik, V. G., 237, *279*
Purves, C. B., 129, *194*
Putaux, J.-L., 258, *288*
Puthenveetil, S., 246, *283*

Q

Qanungo, K. R., 294, 340, *368*
Qi, S., 129, 137, 143, *193*
Qu, J., 407, *460*
Quaino, V., 327, *382*
Qui, D. T., 39, *103*
Quinn, T. P., 363, *401*

R

Rabinovich, G. A., 363, *401*
Rabinovitch, M., 414, 443, *463*, *474*
Rabito, C. A., 239, *280*
Radcliffe, C., 166, *206*
Rademacher, T. W., 226, *277*
Radosta, S., 130, 183, *194*
Rafailovich, M., 216, *274*
Raffelsberger, T., 425, *465*
Raghunath, M., 416, *463*
Ragusa, A., 258–259, *288*
Rahbar, S., 358, *397*
Rajasekharan, K. N., 129, *194*
Rajkumar, R., 327, 343, *383*
Rako, J., 157, *203*

Raluy, E., 353, *394*
Ramalingham, K. V., 118, *185*
Ramana, C. V., 237, 245, *279*, *282*
Ramensky, V., 449, *475*
Ramesh, R., 358, *397*
Ramos, J. R., 300, *370*
Ranbhan, K. J., 338, *386*
Randall, J. N., 245, *282*
Randazzo, G., 323, 327, *380*, *383*
Rangaraj, G., 251, *286*
Rao, G. K., 416–417, 443, *463*
Ratier, L., 453, 455, 457, *477–479*
Raymond, S., 39, 50, *103*, *107*
Raz, A., 363, *401*
Reddy, E. M., 237, *279*
Reddy, S., 348, *392*
Reddy, V. P., 358, *397*
Redl, F. X., 119–120, *187*
Reed, M. A., 245, *282*
Reed, R., 27, *101*
Reese, E. T., 119, *186*
Reihl, O, 348–349, *392*
Reihmane, S., 73, *112*
Reihsner, R., 355, *394*
Reilly, P. J., 318, *377*
Reinikainen, T., 169, *206*
Reis, A., 425, *465*
Ren, S., 300, *370*
Ren, T., 422, *464*
Rencurosi, A., 39, *103*
Rendelman, J. A., 39, *103*
Rérat, A., 361, *400*
Reuben, J., 129, 135–138, 147–148, *193*, *196–198*
Reuter, G., 452, 454, *475*, *478*
Reutter, M., 323, 360, 362–363, *379*
Revol, J. F., 54, 56, 58, 63, 79, *108*, *110*, *112*
Rewicki, D., 305, *372*
Reynaud, E., 355, *394*
Reynolds, A. J., 231–232, 253, *277*
Reynolds, J. C., 119, 171, 173, *187*
Reynolds, T. M., 292, 323, 346, *367*, *379*
Reznikova, M. I., 364, *402*
Riboni, M., 435, 437, *470*
Rice, M. J., 156, *202*
Richard, C., 407, *460*
Richard, V., 432, *468*
Richards, E. L., 336, *385*
Richards, R. W., 63, *110*
Richardson, J., 315, 350, *376*
Richardson, S., 119, 143, 171, 173, 175–176, 178, *187*, *198*, *207–209*
Richter, A., 124, 126, *189–190*, *192*
Ricordi, C., 327, *382*
Rider, M. H., 341, *388*
Riediker, S., 303, *371*
Riedl, B., 114, *150*
Rieger, J., 258, *288*
Riekel, C., 65, *110*
Riese, H. H., 260, *288*
Rieseler, R., 75–76, *112*
Rifat, S., 430, 439–440, *472*
Rijpkema, B., 130, *194*
Riley, J. L., 55, *108*
Ring, S. -G., 155, *202*
Ringer, T., 324, *380*
Risso, M. G., 454, *477*
Ritieni, A., 323, 327, *380*, *383*
Ritz, E., 361, *399–400*
Rizzi, G. P., 334, 344, *385*
Robert, F., 303, 324, 346, *371*, *380*, *391*
Roberts, E. J., 63, *109–110*, 128, *192–193*
Roberts, K., 30, *101*
Roberts, R., 151, *201*
Robertson, A., 350, *393*
Robertson, D. J., 237, *279*
Robinson, A., 249, 257, 265, *285*
Robyt, J. F., 71, 111, *175–176*, *207*
Rocchetta, F., 433, *469*
Roche, E., 52, 55–56, 81, *107–108*, *113*
Roche, S., 432, *468*
Rockenbauer, A., 329–330, *383*
Rockl, K., 119–120, *187*
Rode, K., 361, *400*
Rodrigues Filho, G., 92, *115*
Rodrigues, J. A., 119, *187*
Rodriguez, J. A., 431–432, *469*
Rodríguez, R., 318, *377*

Rodríguez-Mañas, L., 357, *397*
Rogers, R. D., 69, *111*
Roggentin, P., 407, 423, 460, *465*
Rogmann, N., 119, 130–131, 153, *187*, *194*, *202*
Rohde, M., 28, *101*
Rohovec, J., 324, *377*
Röhrling, J., 39, 103, *181*, *196*
Rojas, C. T., 258, *288*
Rojas, T. C., 217, 219–221, 226, 240–242, 251, 256, *275*, *281*
Rojo, J., 217, 219–221, 251–252, 254, 256, 261, *274–275*, *287–288*
Rolf, D., 150, *200*
Rollin, P., 311, 353, *375*
Roman, G., 333, 335, 348, *384*
Romanovicz, D. K., 28, *101*
Röme, D., 165, *205*
Romero, R., 301–302, *370*
Römling, U., 28, *101*
Rong, L., 362, *400*
Rönner, B., 306, *373*
Ronsheim, M., 364, *401*
Ronzoni, L., 426–427, *466*
Röper, H., 125–126, 129–130, 191, *368*
Röper, S., 294, *368*
Roščić, M., 305, 307, *373–374*
Rose, D. R., 178, *208*
Rosen, L., 299, *370*
Rosenau, T., 133, 150, 196, 200, 223, 226, 229, *277*
Rosenberg, A., 405, *460*
Rosenblum, M. G., 270, *290*
Rosenzweig, Z., 247, 257, *284*
Rosi, N. L., 214, *272*
Ross, P., 28, *101*
Rosset, S., 353, *394*
Rossetti, R., 245, *282*
Rossi, M. C., 330, *384*
Rossi, S., 426, *465–466*
Rotello, V. M., 212–213, 216, *270*, *273*, *290*
Rothe, B., 407, 423, *460*, *465*
Rottier, R. J., 407, *460*
Roumani, M., 81, *113*
Rounsaville, J. F., 118, *185*
Roux, D., 71, *111*
Rovero, P., 309, *374*
Rovner, A., 340, *387*
Rowland, S. O., 63, *109–110*
Rowland, S. P., 128, *192–193*
Rowland, S. R., 161, *204*
Roy, M., 357, *397*
Roy, N., 161, *204*
Roy, R., 217, *275*
Royle, L., 166, *206*
Rubaga, L., 435–436, 449–450, *470*
Rudd, P. M., 166, *206*
Rufián-Henares, J. A., 359, *399*
Ruigrok, R. W., 424, *465*
Ruiz, N., 124, *191*
Ruocco, M., 340, *387*
Rupar, C. A., 412, *462*
Russell, D. A., 230–232, 236–237, 249, 253–255, *277*, *285*, *287*
Russell, D. H., 269, *289*
Russell, R. B., 424–425, *465*
Rustgi, R., 115, *151–152*
Ryan, S., 244, *282*
Ryder, M. H., 314, *375*
Ryoo, E. S., 349, 360, *392*
Ryu, K., 362, *400*
Ryu, S. E., 451, *475*

S

Saake, B., 124, 147, 150–151, 155, 173, 175, 177, *189*, *198*, *200–202*, *207–208*
Saberi, F., 407, *460*
Sachinvala, N. D., 123, 150, *189*, *199*
Sacks, D. B., 325, *381*
Sackstein, R., 441, *473*
Saegusa, Y., 75, *112*
Saenger, W., 39, 50, *103*, 320, *378*
Sahoo, S., 72, *112*
Saijo, S., 438, *472*
Saito, M., 93, *116*, 405, 434, 444, *460*, *470*, *475*
Saito, N., 339, *386*

Sakaguchi, A., 327, 339–340, 356, *383*, *387–388*, *395*
Sakai, M., 366, *402*
Sakai, Y., 339, *387*
Sakamoto, M., 75, *112*
Sakarya, S., 430, 439, 440, *472*
Sakata, N., 337, *385*
Sako, T., 327, *382*
Sakuraba, H., 418, *464*
Sakurabayashi, I., 340, 343, *388*
Sakurada, L., 59, *109*
Sakurai, T., 336, *385*
Sakusabe, A., 327, *382*
Sakusabe, N., 327, *382*
Salabas, E. L., 239, *280*
Saladini, M., 330, *384*
Salameh, B. A. B., 353, *394*
Salazar, R., 356–357, *395*
Salcedo, D., 301–302, *370*
Salinas, F. G., 214, *272*
Salmi, T., 129, *193*
Salunke, S. B., 249, 266, *285*
Samain, D., 25–100, *115–116*
Samaranayake, G., 74, *112*
Samarani, S., 422, *464*
Sampaolesi, M., 430, 433, 440, *469*
Samsel, E. P., 118, *185*
Samu, J., 309, *374*
Sánchez, J. C., 241, *281*
Sánchez-Ferrer, C. F., 357, *396–397*
Sánchez-Rodríguez, C., 357, *396*
Sanders, P., 126, *191*
Sandford, P. A., 157, *202*
Sando, S., 247, 264, *284, 286*
Sandros, M. G., 249, 266, *286*
Santacroce, P. V., 251, *286*
Santos, H., 341, *388*
Santos, J. I., 253, *287*
Santoshkumar, P., 355, *395*
Santra, S., 264, *289*
Sanvicens, N., 256, *290*
Sanz, M. L., 325, *381*
Sarko, A., 25, 39, 42, 44, 50, 52–55, 79, *101*, *103–104*, *107–108*
Sartori, J., 133, *196*
Sasaki, A., 438, *472*
Sassi, J. F., 61, 76, 92, *109*, *112*
Sastry, M., 237, *279*
Sato, A., 336, *385*
Sato, I., 418, 438–439, 454–455, *463*, *471–472*
Sato, K., 425, *465*
Sato, Y., 149–150, *199*
Satomi, S., 418, *464*
Sauer, J., 231–233, 256, *278*
Saulnier, L., 86, *114*
Saunders, D. S., 360, *398*
Savin, G., 150, *200*
Savita, V., 424, 454, *465*
Sawada, M., 419, 438, *464*, *472*
Saxena, A. K., 339–340, *387*
Saxena, I. M. J., 33, *101*
Saxena, P., 339–340, *387*
Sayed, H. M., 362, *400*
Saykally, R. J., 216, *274*
Scadden, D. T., 246, *283*
Scaiano, J. C., 216, *274*
Scaldaferri, L., 327, *382*
Scaloni, A., 304, *372*
Schaaff, T. G., 215, *273*
Schachner, M., 442–443, *473–474*
Schachte, H., 453–455, *477*
Schachter, J., 443, *474*
Schagerlöf, H., 165, 175, 178–179, 181, *205*, *208–209*
Schagerlöf, U., 175, *208*
Schalkwijk, C. G., 355, *395*
Schaller, J., 150, *199*
Schauer, R., 403–459, *460*, *465*, *475–479*
Scheidhauer, K., 350, *393*
Scheller, F. W., 327, 343, *383*
Scheller, R. H., 409, *461*
Schenck, M., 434, *470*
Schenkman, S., 452–454, *476–478*
Schepmoes, A. A., 323, *379*
Scheraga, H. A., 44, *106*
Scheutjens, M. H. M., 142, *198*
Schibli, R., 350, *393*

Schieberle, P., 309, 317, 336, 353, *374*, *385*
Schiedt, B., 295, *368*
Schiehser, S., 133, *196*
Schiffrin, D. J., 215–216, 229, 235, *272–274*, *278*
Schilling, C. H., 126, *191*
Schindler, W., 343, 365, *389*
Schlamp, M. C., 248, *285*
Schleicher, E., 344, *390*
Schleicher, H., 133, *196*
Schlichtherle-Cerny, H., 360, *398*
Schliefer, K., 118, *185*
Schlimmer, P., 338, *386*
Schlingemann, R. O., 355, *395*
Schmid, G., 213, *271*
Schmidt, B., 409, *462*
Schmidt, D., 454, *477*
Schmidt, M. M., 245, *283*
Schmidt, M., 150, *200*
Schmidt, S., 315, *376*
Schmitz, G., 431, *468*
Schnaar, R. L., 442–443, *473–474*
Schoenmakers, P. J., 181, 183, *209–210*
Schofield, C. L., 236, 254–255, *278*, *287*
Schönbein, C. F., 118, *185*
Schönfelder, K., 361, *400*
Schottelius, M., 350, *392–393*
Schowen, R. L., 300–302, *370*
Schrader, S., 452–454, 458, *475–476*
Schramek, N., 315, *376*
Schramm, L. P., 442, *473*
Schraven, B., 432, *468*
Schreuder, H. A., 343, *390*
Schuerch, C., 56, *109*
Schulein, M., 82, *113*, 173, 178, *207*, *209*
Schulz, L., 124, 162, *190*
Schumacher, D., 343, *389*
Schurz, J., 118, *185*
Schüth, F., 239, *280*
Schwaiger, M., 350, *393*
Schwartz, M. A., 358, *397*
Schwartz, M. L., 342, *389*
Schwartz-Albiez, R., 124, *189*
Schwarzenbolz, U., 304, 355, *371*, *399*
Schweinsberg, C., 350, *393*
Schwenger, V., 361, *399–400*
Schwietzke, U., *399*
Scotti, M., 216, *274*
Scozzafava, A., 311, 366, *375*
Scrimin, P., 234–236, 262, *278*
Scriver, C. R., 411, *462*
Scurati, E., 327, *382*
Seeberger, P. H., 250, 267, *286*
Seeliger, A., 311, 334, 351, *375*, *393*
Segal, L., 73, *112*
Seger, B., 124, 162, *190*
Segreda, J. M., 319, *378*
Segura, I., 260, *288*
Seidel, J., 130–131, *194*
Seidowski, A., 343, *390*
Seifert, S. T., 330, *383*
Sekiguchi, S., 249, 264–265, *285*
Sell, D. R., 293, 349, *368*, *392*
Selvaraj, N., 358, *397*
Selvendran, R. R., 155, 157, *202–203*
Semyonova-Kobzar, R. A., 363, *401*
Sen, S., 357, *397*
Senterre, J., 361, *400*
Sera, T., 247, 264, *284*
Seraglia, R., 323, *380*
Serianni, A. S., 317, 337, *377*
Serna, C. J., 239, *280*
Serot, J. M., 358, *397*
Serres, S., 243, 268, *281*
Severin, T., 306, *373*
Seyrantepe, V., 409, 413, 435–436, 441–442, 444, 454, *461–462*, *464*, *471*, *473*
Sgarbieri, V. C., 361, *399*
Shafigullin, M. N., 215, *273*
Shah, A. M., 355, *395*
Shapiro, R. H., 90, *200*
Shapiro, R., 304, *371*
Shari, M., 327, *382*
Sharma, P., 264, *289*
Sharon, N., 260, *288*
Shashkov, A. S., 298, *369*
Shea, E. M., 155, *202*
Shea, E., 357, *396*

Shearman, C. W., 356–357, *396–395*
Shen, J. J., 183, *210*
Shen, N., 364, *401*
Shepherd, R., 350, *393*
Sherman, J. S., 156, *201*
Sherratt, M. J., 416, *463*
Shi, Y., 342–343, *353*
Shiang, J. J., 216, *274*
Shiang, Y.-C., 234–235, 242, *278*
Shibata, S., 161, *204*
Shibata, T., 149–150, *199*
Shibusawa, Y., 336, *385*
Shibuya, Y., 366, *402*
Shieh, D.–B., 266, *289*
Shiga, K., 438, *472*
Shigeta, H., 348, *391*
Shilton, B. H., 304, *372*
Shim, W. S., 97, *116*
Shima, T., 432, *468*
Shimamura, A., 355, *396*
Shimizu, Y.-L., 48, 54, *106*
Shindo, H., 336, *385*
Shiozaki, K., 418, 433, 435–439, 444, 454–455, *463–464*, *469–472*
Shiozaki, M., 433, 436–438, 444, 454, *469*
Shipova, E. V., 306, 358, *373*
Shiraishi, N., 27, *101*
Shiras, A., 237, *279*
Shlyahovsky, B., 214, *272*
Shoham, T., 432, *468*
Shoham, Y., 82, *113*
Shtatland, T., 240, *281*
Shuvaev, V. V., 358, *397*
Sibson, N. R., 243, 268, *281*
Siddiqui, S., 355, 396, *442*, *473*
Siefert, E., 131, 148–150, 162, *195*
Siegert, P., 340, *388*
Sieri, S., 358, *398*
Sierks, M. R., 170, *207*
Siest, G., 358, *397*
Sieweke, M. H., 440, *472*
Siika-aho, M., 170, 173, 175–176, 178–179, 182–183, 189, *206–209*
Silvestre, A. J. D., 86, *114*
Silvestre, A., 86, *114*
Simao, A. C., 311, 353, *375*
Simard, B., 244, *282*
Šimičić, L., 302, *373*
Simmons, P. J., 420, *464*
Simon, H., 298, 319, *370*, *378*
Simon, I., 47, *106*
Simon, M., 245, *283*
Simon, V., 432, *468*
Simons, K., 431, *468*
Sinaÿ, P., 39, *104*, 120, *188*, 251, *286*
Sinerius, G., 316, *376*
Singer, D., 309, *374*
Singh, S., 245, *282*
Sipma, G., 306, *373*
Siracusa, L. D., 416, *463*
Sisson, W. A., 40, *104*
Sitzia, C., 430, 433, 440, *469*
Sixta, H., 133, *196*
Sjöholm, R., 129, *193*
Sjölin, C., 421, *464*
Sjöström, E., 129, *193*
Skorepa, J., 151, *200*
Skubitz, K. M., 426, *466*
Ślepokura, K., 318, *377*
Sletta, H., 364, *402*
Sly, W. S., 411, *462*
Smales, C. M., 355, *394*
Smith, A. M., 264, *283*
Smith, D. W. E., 314, *375*
Smith, E., 143, 176, 182–183, *191*
Smith, J. A., 441, *473*
Smith, J. E., 214, *272*
Smith, M. A., 358, *397*
Smith, P. R., 331, *384*
Smith, R. D., 323, 323, *379–390*
Smith, T., 351, *394*
Smolders, C. A., 130, *194*
Snel, M. M. E., 253, *287*
Sobenin, I. A., 452, *476*
Sode, K., 317, 327, 339–340, 356, *383*, *387–389*, *395*
Soden, R., 447, *475*
Sodeoka, M., 428, *467*

Soding, J., 407, *460*
Sohn, H., 451, *475*
Sohn, H.-J., 362, *400*
Soini, H. A., 69, *111*
Sokolov, J., 216, *274*
Sokolowski, J., 299, *370*
Solís, D., 226, 256, *276*
Sollogoub, M., 120, *188*
Solovieva, S. E., 364, *402*
Somerville, C., 33, *102*
Sommerville, C. R., 33, *102*
Somogyi, L., 306, *373*
Somoza, V., 325, 361, *381*, *399*
Song, C. Y., 355, 357, *395–396*
Song, I. C., 239, *280*
Sonnino, S., 430–431, 433, 434–435, 439–440, 451, *467*, *469–470*
Soria, A. C., *399*
Sorkin, E., 362, *400*
Sosnovsky, G., 349, 360, *392*
Southwick, J., 52, *107*
Souto, C., 118, *185*
Sowiński, P., 364, *402*
Spain, S. G., 220, *276*
Spanig, H., 295, *368*
Spear, S. K., 69, *111*
Spencer, T., 442, *473*
Spencer, W. W., 118, *186*
Spinelli, S., 455–456, *478*
Sprague, B. S., 55, *108*
Sprenger, F. K., 351, *393*
Sprong, H., 432, *468*
Spurlin, H. M., 27, *100*, 118, 128–129, 133, 135–137, 141, 147, 181, *186*, *193*, *196*, *209*
Srisodsuk, M., 169, *206*
Stadler, R. H., 303, *371*
Staempfli, A. A., 332, *384*
Stahl, A., 350, *393*
Stahlberg, J., 169, *206*
Stamatos, N. M., 321–323, 329, 332, 362, 430, 439–440, *472–473*
Staňek, Jr., J., 155, *202*
Starcher, B., 414–417, *463*
Stash, A. I., 298, *369*
Staudinger, H., 34, *102*, 118, *185*
Steadman, R. G., 118, *185*
Stear, M. J., 358, *397*
Steeneken, P. A. M., 143, 176, 182–183, *191*
Stefanowicz, P., 123, 309, 324, *374*, *380*
Stehouwer, C. D. A., 355, *395*
Steigelman, K. A., 422, *464*
Steimmig, A., 295, *368*
Stein, A., 126, 130, 132, 139, 151–152, 153, 157, *190*, *192*, *194–195*, *200*
Steiner, T., 39, 50, 103, *320*, *378*
Stenberg, O., 138, *197*
Sterner, R., 315, *375–376*
Stevens, E. D., 39, *103*
Stevenson, P. C., 213–214, *271*
Stewart, G., 244, *282*
Stigsson, V., 127, *192*
Stinga, C., 86, 98, *115–116*
Stipanovich, A. J., 50, 55, *107–108*
Stipetić, M., 305, *373*
Stokes, J. J., 215, *273*
Stoleriu, A., 127, *192*
Stomp, A. M., 314, *375*
Stone, B. A., 157, *202*
Stoppani, E., 426, *465–466*
Storch, M., 425, *465*
Storhoff, J. J., 214, *271–272*
Strauch, C. M., 349, *392*
Strisciuglio, P., 411, *462*
Stroh, M., 247, *283*
Strom, A., 315, *376*
Strong, L. E., 217, *275*
Stroobant, V., 341, 342, *388–389*
Stroud, M., 251, *286*
Stscherbina, D., 147, 149, *199*
Stumpo, K. A., 269, *289*
Stutzenberger, F. J., 260, *288*
Su, A. I., 447, *475*
Su, B., 358, *397*
Su, C.-H., 266, *289*
Su, C.-L., 269, *289*
Su, X.-D., 343, *389*
Suda, Y., 269, *289*

Sufoka, A., 58, 79, *107*
Sugeno, N., 436, *471*
Sugiyama, J., 33, 39, 44, 46–48, 52, 56–58, 63–64, 81–82, 86, 103, 105–110, *113–114*, *121*, *131*, *189*
Sugiyama, M., 366, *402*
Sujith, A., 247, *283*
Sukumar, N., 39, *103*
Sumoto, K., 311, *375*
Sumpton, D. P., 119, *187*
Sun, E. Y., 240, *281*
Sun, S. H., 239, *280*
Sun, W., 351, *394*
Sun, X.–L., 247, 257, *285*
Sun, Y., 427, 448, *466*
Sundararajan, P. R., 27, *101*
Sundaresan, G., 246, *283*
Sundgren, A., 220, 222, *276*
Sunyaev, S., 449, *475*
Supuran, C. T., 311, 366, *375*
Surolia, A., 249, 257, *285*
Sutherland, A. J., 246, *283*
Suthers, W. G., 131, *195*
Suzuki, H., 438, *472*
Suzuki, K., 243, *281*, 422, *464*
Suzuki, M., 357, *397*, 435, *470*
Suzuki, N., 425, *465*
Suzuki, S., 75, *112*, 438, *472*
Suzuki, T., 408, 424, 429, 454, *461*, *465*
Suzuki, Y., 362, *400*, 428, 435, *467*, *470*
Svarovsky, S. A., 220, 222, 249, *275*, *285*
Svensson, B., 170, *207*
Svensson, S., 119, *186*
Svrivastava, H. C., 118, *185*
Swaffield, J., 351, *394*
Swatloski, R. P., 69, *111*
Sweeley, C. C., 443, *474*
Sweet, D. P., 151, 157, *200*
Sykova, E., 245, *282*
Szabados, L., 338, *386*
Szablikowski, K., 129, *193*
Szabó, S., 324, *380*
Szabó, Z., 324, *380*
Szarek, W. A., 308, *374*
Szekely, Z., 220, 222, *275*
Szelepin, B., 318, 365, *377*
Szewczuk, M. R., 422, 442, *464*, *473*
Szewczuk, Z., 309, 323–324, *374*, *380*
Szoka, F. C., 217, 255, *275*
Szurmai, Z., 157, *203*
Szwergold, B. S., 341, 342, *388–389*

T

Tabata, T., 357, *396*
Tabrizian, M., 249, 266, *286*
Taiz, L., 69, *111*
Takae, S., 234–235, 254, 269, *278*, *287*
Takagi, S., 50, *107*
Takahara, H., 348, *392*
Takahashi, K., 161, *204*
Takahashi, M., 323, 339–340, *380*, *386–387*
Takahashi, T., 327, 382, *425*, *465*
Takaku, T., 343, 362, *390*
Takamatsu, Y., 420, *464*
Takaragi, A., 68, *110*
Takatsuka, K., 360, *398*
Takátsy, A., 324, *380*
Takeda, A., 432–433, 436, *468*, *471*
Takimoto, T., 434, *470*
Tamaki, E., 346, *390*
Tamiya, E., 229–230, 257, *277*
Tan, W. B., 237, 247, 266, *279*, *284*, *289*
Tan, W., 214, *272*
Tanahashi, M., 43, *105*
Tanahashi, T., 366, *402*
Tanaka, H., 337, *385*
Tanaka, M., 361, *399*
Tanaka, Y., 357, *396*
Tang, G. L., 428, *467*
Tang, K., 163, *204*
Tang, N., 323, *380*
Tang, Y., 240, *281*
Tang, Z., 214, *272*
Tani, Y., 339, 360, *387*, *398*
Taniguchi, N., 323, 358, *380*, *397*
Tanimoto, A., 243, *281*
Tankam, P. F., 126, *192*

Tankanen, M., 124, *189*
Tanner, H., 350, *392*
Tanner, S. F., 43, *105*
Tanner, S., 52, *105*
Tao, B. Y., 85, *114*
Tao, W., 441, *473*
Tararak, E. M., 443, *474*
Tarnawski, M., 318, 365, *377*, *402*
Tarnow, L., 355, *395*
Tartaj, P., 239, *280*
Tarver, J., 351, *393*
Tas, A. C., 126, *191*
Tasker, S., 63, *110*
Tate, M. E., 314, *375*
Tateno, H., 438, *472*
Tateyama, M., 425, *465*
Tatibouët, A., 311, 353, 356, *375*, *395*
Tavares, G. A., 455–456, *478*
Taylor, A. M., 119, *187*
Taylor, G. L., 407–408, 410, 434, *460–462*, *470*
Taylor, J., 351, *394*
Taylor, S. V., 315, *376*
Teeri, T. T., 82, 84, 113, *169–170*, *206–207*
Tekely, P., 76, *112*
Teleman, O., 42, 53, *106*, *110*
Templeton, A. C., 213, 215–216, *271*, *273–274*
Tenkanen, M., 170, 173, *206*
Terekhova, L. P., 364, *402*
Terranova, K. M., 351, *394*
Tertov, V. V., 452, *476*
Testi, R., 436, *471*
Tetaud, E., 356, 364, *395*
Tettamanti, G., 403–459, *461*, *464*, *466–467*, *469–471*
Teunis, C. J., 317, *376*
Tevyashova, A. N., 364, *402*
Tezuka, Y., 149–150, *199*
Th.KoK, W., 183, *210*
Thai, H. T. T., 138–139, *197*
Thanh, N. T. K., 214, 238, *272*, *279*
Tharavanij, T., 327, *382*
Thaxton, C. S., 214, *272*
Théate, I., 341, *388*
Thibaudeau, C., 317, 337, *377*
Thibault, J.-F., 86, *114*
Thielemans, W., 93, *116*
Thielking, H., 131, 147, 150, 151, 156, 175, 177, *194–195*, *198*, *200–201*
Thiem, J., 454, 456–457, *477*, *479*
Thiry, A., 311, 366, *375*
Thoma, R., 315, *375–376*
Thomann, R., 237, *278*
Thomas, G. H., 411, *462*
Thomas-Oates, J., 119, *187*
Thomson, B. A., 163, *204*
Thomson, R., 434, *470*
Thornalley, P. J., 323, 331, 347–348, *379*, *384*
Thorpe, S. R., 302, 304, 317–318, 326, 344, 348, *370–372*, *377*, *382*, *390*, *392*
Thygesen, M. B., 231, 233, 256, *278*
Tighe, B., 90, *115*
Tiller, D., 75–76, *112*
Tiller, J., 75, *112*
Timell, T. E., 128–129, 161, *193*, *203–204*
Timpu, D., 127, *192*
Tinari, N., 363, *401*
Ting, A. Y., 246, *283*
Tiralongo, E., 452–454, 458, *475–476*, *479*
Tiralongo, J., 452, 454, 458, *475*
Tiso, N., 435–436, 449–450, *470*
Tiwari, V. K., 453, *476*
Tjan, S. B., 317, *376*
Tjerneld, F., 165, 170, 173, 175, 178–179, 181, *205–206*, *208–209*
Tobia, A. M., 342, 358, *389*, *397*
Tobias, B., 365, *402*
Todeschini, A. R., 453–454, 456–457, *477*, *479*
Tokuyama, S., 419, 429, *464*
Toldy, L., 310, *374*
Tomasik, P., 126, *191*
Tomita, H., 346, *390*
Tomlinson, S., 452, *476*
Tomme, P., 170, *207*
Tong, E., 234–235, 244, *278*

Tonković, M., 330, *383*
Toomre, D., 431, *468*
Toone, E. J., 217, 251, *274*
Torchilin, V. P., 217, 246, *275*, *283*
Torensma, R., 262, *289*
Tortora, P., 423, 436, 454, *464*
Tourwe, D. A., 360, *393*
Toyokuni, T., 251, *286*
Traldi, P., 323, *380*
Tran, Qui. D, 50, *107*
Trasler, J. M., 407, *460*
Trchova, M., 245, *282*
Treadway, J. A., 246, *283*
Tregre, G. J., 123, 150, *189*
Trenin, A. S., 364, *402*
Treshalin, I. D., 364, *402*
Tressi, R., 305, *372*
Tringali, C., 408, 423, 426–427, 430–431, 433–436, 439–440, 451, 454, *461*, *464*, *466–467*, *469–470*, *472*
Trivedi, H. C., 150, *200*
Trogus, C., 55, *108*
Tromas, C., 252–253, *287*
Tronchet, J. M. J., 309, *374*
Trudel, S., 409, 435–436, 454, *461*, *471*
Tsai, C.–S., 255, *288*
Tsai, H. J., 342, *389*
Tsai, T.-L., 266, *289*
Tsang, K. W. T., 244, *282*
Tsay, J. M., 246, *283*
Tseng, W.-L., 269, *289*
Tsuchida, H., 334, *384*
Tsuchiya, Y., 149–150, *199*
Tsugawa, W., 327, 339–340, *383*, *387–389*
Tsuik, S., 422, *464*
Tsuji, M., 43, *105*
Tsuzuki, T., 358, *397*
Tung, C.–H., 240, *280*
Turk, J. R., 363, *401*
Turkevich, J., 213–214, *271*
Turley, E., 443, *474*
Turner, J. J., 311, *375*
Tüting, W., 151, 158, 175, *198*, *201*, *203*
Twaalfhoven, H., 355, *395*
Tzeng, H.-C., 220, 223, 231, 233, 255, 269, *276*

U

Uchida, K., 362, *400*
Uchida, M., 326, *382*
Ueda, T., 336, *385*
Ueda, Y., 323, *380*
Uemura, H., 327, *477*
Uemura, H., 453, *477*
Uemura, S., 429, *467*
Uemura, T., 418, 437, *464*, *471*
Ulman, A., 216, *274*
Ulrich, A. S., 150, *199*
Ulrich, P. C., 309, *374*
Unno, Y., 337, *385*
Unrau, P. J., 316, *376*
Urrutia, M., 453, *477*
Usuda, M., 57, *109*
Uttaro, A. D., 455, *478*
Utzmann, C. M., 311, 360, *374*, *398*
Uyeda, H. T., 247, *283*
Uyeda, N., 56, 58, *108*
Uzcategui, J., 301–302, *370*

V

Vaca-Garcia, C., 86, *114*
Vafiadaki, E., 425, *465*
Vaidyanathan, G., 350, *392–393*
Vaissade, P., 361, *400*
Vajn, K., 442, *473*
Valade, J. L., 85, *113*
Valaperta, R., 433, 434, 439, *469–470*
Valle, D., 411, *462*
Vallejo, S., 357, *396–397*
Valsecchi, M., 433, *469*
Valtakari, D., 129, *193*
van Bekkum, H., 71, *111*, 132, 151, *195*, *201*
Van Campenhout, A., 327, 357, *382*, *397*
Van Campenhout, C., 327, 357, *382*, *397*
Van Damme, O., 327, *382*

van de Goor, T., 323, *380*
van de Vlekkert, D., 419, 421, *464*
Van den Ouweland, G. A. M., 317, *376*
van der Burgt, Y. E. M., 143, 182–183, *198*, *210*
van der Kerk-van Hoof, A., 143, 182–183, *198*, *210*
van der Velden, P. M., 130, *194*
Van der Ven, J. G. M., 338, *386*
van Duin, M., 151, *201*
van Embden, J., 246, *284*
van Kooyk, Y., 256, 262, *288–289*
van Meer, G., 432, *468*
Van Schaftingen, E., 315, 341–343, *376*, *388–389*
Vanbekkum, H., 71, *111*
Vandeberg, P., 151, *201*
Vandekerckhove, F., 452, 454, *476*, *478*
VanderHart, D. L. A., 40, 42–43, *104–105*
Vanstapel, F., 341, *388*
Varela, O., 75, *112*
Varga, L., 305, 307, *373*
Varga, N., 303, *371*
Vargha, L., 310, *374*
Varghese, N., 408, *461*
Varki, A., 217, *274*, 443, *474*
Varma, A. J., 71–72, *111*
Varndell, I. M., 214, *271*
Varrot, A., 175, *208*
Vasella, A., 408, *461*
Vásquez, B., 301–302, *370*
Vass, E., 309, *374*
Vasseur, S., 239–240, *279*
Vassilenko, E. I., 335, *385*
Vaudequin -Dransart, V., 314, *375*
Veiga-da-Cunha, M., 341, *388*
Veintemillas-Verdaguer, S., 239, *280*
Velle, D., 411, *462*
Venerando, B., 403–459, *460–461*, *464*, *466–467*, *469–472*
Venkatraman, J., 304, *372*
Venturini, R., 327, *383*
Veracini, L., 432, *468*
Verbeke, P., 356, *396*
Verkleij, A. J., 220, 222, 253, *276*, *287*
Verlhac, C., 63, *110*
Vernhettes, S., 33, *102*
Vernin, G., 344, *390*
Verraest, D. L., 132, 148, *195*
Vértesy, L., 343, 365, *389*
Vertommen, D., 315, 341–342, *376*, *388*
Viani, R., 325, *381*
Vieira, F., 453, *477*
Vieira, M. C., 130, *194*
Vietor, R. J., 44, 48, *106*
Vigier, G., 92, *115*
Vignon, M., 54, 61, 64–65, 79–81, *108–109*, *113*
Vigo, T. L., 150, *199*
Villamiel, M., 303, 325, 360, *371*, *381*, *398–399*
Villar, E., 435, *470*
Villares, P., 318, *377*
Villemagne, P., 86, *114*
Vimr, E. R., 407–408, 423, *460–461*, *465*
Vimr, E., 408, *461*
Vincendon, M., 43, 52, *105*
Vink, H., 443, *474*
Vinković, M., 305, *373*
Vinogradova, M. V., 441, *473*
Virgona, C., 351, *393*
Visciano, M., 305, *372*
Visconti, A., 323, *380*
Viski, P., 364, *401*
Vitek, M. P., 309, *374*
Vlassara, H., 309, *374*
Vliegenthart, J. F. G., 129, 143, 151, 182–183, *193*, *198*, *201*, *210*, 220, 222, 253, *276*, *287*, 453–455, *477*
Vogel, V., 212, *270*
Voigt, W., 69, *111*
Voit, B., 237, *279*
Volkert, B., 123, 132, 151, *187*, *190*, *195*, *201*
Vollenbroeker, M., 334, *384*
Vollmer, A., 124, 130, 153, 156, *191*, *202*
Von Helden, G., 128, *192*
von Itzstein, M., 414, 427–428, 456, *463*, *466–467*, *469*
von Recum, H. A., 258, *288*
von Reitzenstein, C., 431–432, *468*

Vorwerg, W., 130, 183, *194*
Voss, E. D., 441, *473*
Voss, K. A., 360, *398*
Voziyan, P. A., 317, 337, *377*
Vullo, D., 311, 366, *375*
Vuong, R., 41, 44, 56, 79, *104–105*, *112*
Vyas, A. A., 443, *474*
Vyas, K. A., 443, *474*
Vyssotski, M., 323, *379*

W

Wa, C., 304, 323, *372*
Wada, M., 48, 53, *106–107*
Wada, T., 418–419, 429–433, 435–439, 454–455, *463–464*, *467–468*, *470–471*
Wada, Y., 323, *380*
Waddell, B., 410, *462*
Wade, C. P., 128, *192–193*
Wagenknecht, W., 119, 123, 126, 132, 147, 149, 153, 157, 178, *187*, *190*, *192*, *195*, *199*, *201*, *208*
Wager, H. G., 358, *398*
Wagner, T., 130, *194*
Waheed, A., 409, *462*
Wahlund, K.-G., 164–165, 175, 178–179, 181, *205*, *208*
Wainai, T., 434, *470*
Walczak, B., 347, *391*
Walker, J. R., 447, *475*
Walker, M., 215, *273*
Walsh, M. A., 453, *476*
Walter, A. W., 323, *379*
Walter, G., 264, *289*
Walton, D. J., 304, 308, *372*, *374*
Wang, A., 357, *396*
Wang, B., 353, 355, *394–395*
Wang, D. I., 428, *467*
Wang, D. S., 216, *274*
Wang, D., 407, *460*
Wang, J., 212, *270*, 443–444, *474–475*
Wang, K., 124, *190*
Wang, K.-T., 220, 223, 231, 233, 269, *276*, *289*
Wang, L. X., 439–440, *472*
Wang, L., 235, 244, 244, *278*, *282*
Wang, M., 441, *473*
Wang, P. G., 318, 350, *377*
Wang, P., 441, *473*
Wang, S. P.-H., 266, *289*
Wang, S., 357, *396*, 407, *460*
Wang, X., 360, *399*
Wang, X.–H., 247, 266, *284*
Wang, Y. A., 247, *284*
Wang, Y., 295, *369*
Wang, Y.-C., 249, *285*
Wang, Y.-X. J., 240, *280*
Wariishi, H., 223, 226, 229, *277*
Warita, H., 425, *465*
Warner, T. G., 423, *464*
Warren, C. E., 418, *463*
Warren, R. A. J., 170, 178, *207–208*
Warsinke, A., 327, 343, *383*
Wasada, H., 337, *385*
Wastyn, M., 130, 183, *194*
Watanabe, K., 311, *375*, *382*
Watanabe, M., 43, *105*
Watanabe, S., 53, 79, *107*
Watano, T., 340, 343, *388*
Watkin, D. J., 351, *393*
Watkin, K. L., 243, *281*
Watkins, N. G., 301–302, 304, 355, *370*, *372*
Watson, R. R., 362, *401*
Watts, A. G., 455–456, 458, *478–479*
Watts, A., 458, *479*
Weels, B., 30, *101*
Weenen, H., 338, *386*
Wegemann, K., 175–176, 182, 183, *198*
Wegener, D., 315, *376*
Wehenkel, A., 456, 458, *479*
Wei, L., 427, 448, *466*
Weidinger, A., 41, 55, *104*
Weigel, K., 361, *400*
Weinhold, B., 438, *471*
Weisleder, D., 302, *371*
Weiss, A. A., 244, *282*
Weiss, A. S., 413, *462*
Weiß, C., 361, *399*

Weiss, S., 246, *283*
Weissleder, R., 239, 239, 240, *280–281*
Weitzel, G., 298, 300, *369*
Wellard, H. J., 42, *104*
Welling, G. W., 343, *390*
Wellner, A., 325, *381*
Wells-Knecht, K. J., 348, *392*
Wendel, M., 118, *185*
Wenz, G., 120, *187*
Wenzel, E., 361, *399*
Werba, J. P., 443, *474*
Wester, H.-J., 350, *392–393*
Westphal, G., 297–298, *369*
Wetsel, A. E., 245, *282*
Weygand, F., 292, 319, 338, *367*, *378*, *386*
Weymouth-Wilson, A. C., 351, *393*
Whetten, R. L., 215, *273*
Whistler, R. L., *186, 191*
White, C. E., 314, *375*
White, H., 216, *274*
Whitesides, G. M., 213, 217, 254, *271*, *275*
Whitmore, R. E., 43, *105*
Whyman, R., 215, *273*
Wiame, E., 342–343, *389*
Wieruszeski, J. M., 456–457, *479*
Wijnen, P. H., 181, *209*
Wilding, G. E., 358, *398*
Wiley, J., 43, *105*
Wilke, O., 119, 153, *187*, *194*
Wilks, E. S., 125–126, 129–130, *191*
Willats, W. G. T., 270, *290*
Willemsen, R., 410, 414, *462–463*
Williams, D. G., 39, *103*
Williams, P. A., 27, 100, *129*, *193*
Williamson, B., 355, *395*
Willner, I., 212, 214, *270–272*
Willson, M., 356, 364, *395*
Willstätter, R., 33, *102*
Wilmanns, M., 315, *376*
Wilschut, N., 311, *375*
Wilson, A., 351, *394*
Wilson, J. C., 456, *479*
Wilson, K. S., 170, *207*
Wilson, L. J., 270, *290*
Wiltshire, T., 447, *475*
Winans, S. C., 314, *375*
Winkel, C., 310, *374*
Winkler, I. G., 420, *464*
Winsor, D. L., 123, *189*
Winters, D., 310, *374*
Winum, J. Y., 311, 366, *375*
Wirick, M. G., 119, *186*
Witczak, Z. J., 356, 363, *395*
Withers, S. G., 175, 208, *351*, *393*
Witholt, B., 59, *109*
Witkiewicz, K., 303, *371*
Wittenberg, J., 239, *280*
Witter, R., 150, *199*
Wittgren, B., 164–165, 175, 178, 181, *205*, *208–209*
Wittmann, I., 324, *380*
Wittmann, R., 323, 346, 360, 362–363, *379*, *390*
Wittrup, K. D., 246, *283*
Wolf, I., 350, *393*
Wolswinkel, K., 306, *373*
Womack, M., 360, *399*
Wong, K. H., *391*
Wong, O. A., 353, *394*
Woods, J. W., 299, *370*
Woodtli, T., 294, *368*
Woortman, A. J. J., 126, *191*
Woronowicz, A., 442, *473*
Wortis, H. H., 453, *477*
Wright, D. G., 439, *472*
Wrodnigg, T. M., 292, 322, 351, *367*, *379*, *393*
Wu, A. M., 246, *283*
Wu, B., 453, *476*
Wu, G., 433, 444, *469*, *474–475*
Wu, P., 441, *473*
Wu, P.-C., 266, *289*
Wu, S.-H., 214, *272*
Wu, V. Y., 356–357, *395*, *397*
Wu, X., 246, *283*, 318, 326, 340, 350, *377*, *382*, *386–387*, 422, *464*
Wu, Y., 214, *272*
Wu, Y.-C., 220–221, 255, 269, *275*

Wuelfing, M. P., 213, *271*
Wuelfing, W. P., 215, *273*
Wurzburg, O. B., 118, *185*
Wyckoff, R. W. G., 56, *108*
Wyttenbach, T., 128, *192*

X

Xiao, Y., 214, *272*
Xie, M., 247, 266, *284*
Xie, W., 318, 350, *377*
Xie, X., 443–444, *474–475*
Xie, Z.-X., 247, 266, *284*
Xu, B., 234, 244, *278*, *282*
Xu, G., 360, *399*
Xu, M. H., 353, *394*
Xu, Y., 326, *382*

Y

Yagi, M., 339–340, 343, *387–388*
Yalpani, M., 129, *193*
Yamada, M., *389*
Yamada, T., 48, 54, 106, *132*, *195*
Yamaguchi, K., 418, 428, 430–433, 435–440, 444, 451, 454–455, *463–464*, *467–472*
Yamaguchi, M., 348, *391*
Yamaguchi, T., 311, *375*
Yamamoto, H., 43, *105*
Yamamoto, Y. S., 118, *185*
Yamanami, H., 418, 437–439, 454–455, *463*, *471*
Yamanoi, J., 434, *470*
Yamaoka, S., 243, *281*
Yamasaki, Y., 254, 269, *287*
Yamazaki, J., 324, *380*
Yamazaki, T., 327, 340, *383*, *388*
Yan, J., 351, *394*
Yanagisawa, M., 124, *190*
Yanai, Y., 327, *383*
Yang, C.–Y., 132, *275*
Yang, F., 323, *379*
Yang, H., 270, *290*
Yang, J., 356, *395*
Yang, L. J., 442, *473*
Yang, S.-F., 27, *100*
Yang, X., 237, *279*
Yang, Y., 304, *372*
Yano, S., 300, *370*
Yao, P., 350, *393*
Yao, T., 326, *381*
Yasui, A., 422, *464*
Yaylayan, V. A., 292, 303, 338, 346, *367*, *371*, *386*, *391*
Ye, Z. Q., 427, 448, *466*
Yeboah, F., 333, 335, 348, *384*
Yee, C. (K.), 216, *274*
Yee, V. C., 294, *368*
Yeh, C.–S., 266, *289*
Yeh, Y.-C., 220–221, 255, 269, *275*
Yen, B. K. H., 216, *274*
Yildiz, A., 317, 337, *377*
Yin, F. Q., 362, *400*
Ying, J. Y., 229–230, 244, 250, 257, *277*
Yingsakmongkon, S., 428, *467*
Yiu, H. H. P., 270, *290*
Yogalingam, G., 419, 421, *464*
Yokota, H., 122, 131, 132–133, 137, 143, *189*
Yokota, S., 224, 226, 229, *276*
Yonehara, S., 340, 343, *388*
Yoneyama, H., 343, 362, *390*
Yoo, C. W., 355, *395*
Yoo, H. J., 356, *396*
Yoon, T. H., 247, *285*
York, W. S., 157, *203*
Yoshida, G., 337, *385*
Yoshida, H., 324, *380*
Yoshida, M., 452, *476*
Yoshida, N., 338–339, 360, *386–387*, *398*
Yoshikawa, Y., 419, 429, *464*
Yoshimori, K., 348, *391*
Yoshinaga, A., 120, *188*
Yoshino, T., 453, *476*
Yoshizumi, A., 237, *278*
You, C.–C., 212, *271*
Yq, J., 124, *190*

Yu, C.–W., 244, *282*
Yu, H., 453, *476*
Yu, J., 300, *370*
Yu, J.-J., 220, 223, 262, *276*
Yu, N., 153, 156, *201*
Yu, Q., 427, 448, *466*
Yu, R. K., 405, 444, *460*, *475*
Yu, S., 351, *394*
Yu, T.-B., 255, *288*
Yu, W. W., 247, *284*
Yuan, H., 86, *115*
Yuan, Q., 237, *279*
Yuk, I., 304, *372*
Yunta, M., 432, *468*

Z

Zabot, M. T., 413, *462*
Zaitsev, S., 407, *460*
Zaitsev, V., 410, *462*
Zalut, C., 304, *371*
Zalutsky, M. R., 350, *392–393*
Zamborini, F. P., 215, *273*
Zammarchi, E., 412–415, *462–463*
Zanchetti, G., 429, 435, 437, *467*, *470*
Zarebski, L., 453, *477*
Zegeer, A., 128, *192*
Zeier, M., 361, *400*
Zeiger, E., 69, *111*
Zeisel, S., 118, *185*
Zella, D., 440, *472*
Zeller, S. G., 129, 153, 156, *193*, *202*
Zeng, J., 436, 444, *471*
Zeni, M., 92, *115*
Zenobi, R., 163, *204*
Zeronian, S. H., 150, *200*
Zhai, H., 315, *376*
Zhang, B., 244, *282*, 304, 324, *372*, *380*
Zhang, H., 351, *393*
Zhang, J., 229–230, *277*, 293–294, 368, 441, *447*, *453*, *473*, *475–476*
Zhang, L. N., 68, *110*
Zhang, M., 355, *395*
Zhang, Q., 323, *379–380*
Zhang, W., 300, *370*
Zhang, X., 150, *199*, 244, *282*, 292, 309, *367*, *374*
Zhang, Y., 237, 247, 251, 266, *279*, *284*, *286*, *289*, *453*, *476*
Zhang, Z., 300, *370*
Zhang, Z.-L., 247, 266, *284*
Zhao, J., 126, 143, *191*
Zhao, M. X., 353, *394*
Zhao, W., 254, *287*
Zhao, X., 431–432, *468*
Zheng, J., 156, *201*
Zheng, X., 427, 448, *466*
Zheng, Y., 246, *283*
Zhong, C. J., 216, *274*
Zhou, J., 440, *472*
Zhou, L., 360, *398*
Zhou, M., 246, *284*
Zhou, Q., 84, *113*
Zhou, W., 428, *467*
Zhou, X. Y., 416, *463*
Zhou, Y., 340, *388*
Zhou, Z., 163–164, *205*
Ziegelmuller, P., 454, *477*
Zieniawa, T., 402, *464*
Zimmer, J. P., 246, *283*
Zitha-Bovens, E., 132, 148, *195*
Ziyadeh, F. N., 293, *368*
Zizioli, D., 425, *465*
Zoellner, H., 356, *396*
Zogaj, X., 28, *101*
Zotchev, S. B., 364, *402*
Zou, Z., 295, *368*
Zugenmaier, P., 27, 92, *101*, *115*
Zumwald, J. B., 309, *374*
Zuo, J., 422, *464*
Zwingmann, C., 436, 444, *471*
Zyzak, D. V., 326, *382*

SUBJECT INDEX

A

Acetylated rayon fiber–cellulose acetate composites, 97
Acid/base-and metal promoted reactions, fructosamine
affinity, redox-active metal ions, 329
anomeric hydroxyl group, 329
β-pyranose tautomer, 327–328
dehydration
di-and oligosaccharide derivatives, mechanism, 334
mass spectra, 332
occurrence, 332
peeling off mechanism, 335
products, 332–333
Strecker degradation reaction, 334
structure, carbohydrate carbon skeleton, 333–334
unsubstituted D-fructosamine and *N*-substituted derivatives, 334
D-fructose–amino acids, copper(II) and nickel(II)
anchoring absence, 330
chelate rings, 329–330
complex-formation constants, 330
deprotonation, pH 6–7, 329
3-hydroxyl groups, 329–330
D-fructose–glycine and–β-alanine
carbonyl band, 330
dissociation constants, 328–329
enolization
carbohydrate portion dehydration, 332
D-glucose/D-mannose formation, 332
2,3-enolization, 331–333
1,2-enolization product, 317, 331
first-order rate constants, 331
iodine uptake, 331
reversible 1,2-enolization, 331
Maillard reaction, food and *in vivo,* 327
oxidation
autoxidation, 335
enolization, 335
by metal ions, 335–337
by molecular oxygen, 335
oxidants, 338
physiological oxidants, 337
palladium (II) and platinum(II)
carbonyl band, 330
proton- and metal-binding characteristics, 327
p-toluidine
carbonyl band, 330
reduction agents, 338
thermodynamic stabilities, 329
transformation reactions, 327
Acrylonitrile, cellulose grafting with, 73
Acute lymphoblastic leukemia (ALL)
human NEU3 role in, 439
Acylation reactions, 131–132
of cellulose
acid anhydrides and carboxylic acids, 74
DCC with, 73–74
electron diffraction diagrams, 77
N-Acyl hydrazine, 165
N-Acylneuraminyl glycohydrolases, 405

Advanced glycation end products (AGEs)
  formation, 348
  recognization, 348
  structures, 348–349
  in vivo and in foods, 348
6-Aldehydo cellulose, 124
Alkali cellulose
  cellulose chains arrangement, 53
  X-ray diffraction characterization, 54
Alkanethiols, 214
Alkylation, 127–129
Alkylene oxides, cellulose ethers from, 73
Alkyl halides
  route for preparation of cellulose ethers from, 72
Allomorphs of cellulose
  relationships among, 40–41
2-Allyl ethers, 155
Amadoriases oxidoreductase
  substrate specificities, 339
  type I
    Corynebacterium, 339
    electrostatic and hydrophobic interactions, 340
    FAD, 340
    fungal fructose–amino acid oxidases, 339–340
    kinetic and structural studies, 340
    as protein-deglycating enzyme, 339
  type II
    D-fructose–proteins, 340
    FruVal/εFruLys-specific and proteases/peptidases, 340
    glycated proteins, 340
    Pseudomonas sp., 340
    washing and cleaning formulations, 340
Amadori rearrangement
  enzymatically catalyzed (*see* Enzymatic formation method, fructosamine)
  glucosylamines (*see* Glucosylamines, Amadori rearrangement)
American Association of Cereal Chemists (AACC), 10
Amino acid sequence variants in human sialidases
  non-synonymous SNPs, 448
  by PolyPhen, bioinformatic tool, 449
  synonymous SNPs, 448
Aminocellulose, 76
6-Amino-6-deoxy-celluloses, 124
1-Amino-1-deoxy-D-fructose. *see* Fructosamine and derivatives
1-Amino-1-deoxy-D-glucitol, 308
1-Amino-1-deoxy-L-sorbose, 296
Amylopectin, 125, 170–171
Amylose, 170
Analytical methods, fructosamine derivatives
  assay, 322
  capillary electrophoresis and MS, 324
  chromatography
    boronate affinity, 324
    D-fructose–amino acids in dehydrated fruits, 323
    gas, 323
    HPLC, 323
    TLC method, 323
  D-fructose–protein conjugates enzymatic detection, 327
  fructosamine assay
    anemia, 327
    enamine one-electron autoxidation, 326
    enzymatic determination, 326–327
    fluorescein–boronic acid, 326
    nitroblue tetrazolium dye (NBT), 326
  furosine method
    analyte proteins acid hydrolysis, 325
    N^{α}-D-fructose-and-lactulose–amino acids, 325
    protein quality assessment, 325
    pyridosine, 325
    reaction yield, 325
  HbA_{1c}
    β-subunit, 326
    diabetic patients, 325
    enzymatic approach, 326
    modification, D-fructosyl residues, 325–326

immunological, 327
ionization, 323
LC/MS, 324
mass spectrometry (MS), 323
SDS-PAGE, 324
tautomerization, 323
Aortic smooth muscle cells (ASMCs), 414, 417–418
Arabinogalactans (AGP), 32
N-Aryl-fructosamine, 292
6-Azido-6-deoxy-cellulose, 124

B

Biocompatibility, 242
Bionanomaterials, 270
Bone marrow
BM extracellular fluid (BMEF), 420
BM stromal cells (BMSC), 420
BM transplantation (BMT), 419
Boronate affinity chromatography
glycated proteins separation, 324
phenylboronate, 324
Büchi cyclocondensation reaction, 351–352
Bulk material, heterogeneities in
analysis of fractions, 182
cellulose derivatives, fractionation of, 180–181
starch derivatives, topochemical differentiation of, 182–183

C

Carbohydrates
carbohydrate–carbohydrate interactions, 251
carbohydrate–protein interactions, 269
functionalized QDs, 264
uses of, 16
N,N′-Carbonyldiimidazole, 124
Carboxylic acid chlorides, cellulose acylation, 75
Carboxymethylation
of cellulose, rate constants, 129
of starches, 126
Cathepsin G (CG), 420
Cell-surface elastin receptor (CSER), 413–415
Cellulose, 26
in cell wall environment
electron microscopy, 31
microfibrils, 30
occurrence in nature, 29
plant sources, 28
plasma membrane, formed in, 32
prokaryotic and eukaryotes organisms, 28
conformations and crystalline structure, 33
$^4C_1$ chair conformation, 37
degree of polymerization (DP), 35
distances for cellobiose fragments, 39
fiber repeat, periodicity, 36
gauche and *trans* conformation, 38–39
glycosidic torsion angles, 37, 39
helical parameters, 37
homopolymer of β-(1→4)-linked D-glucopyranose residues, 34–35
molecular representation of chain, 35
structural descriptors for, 38
three-dimensional structure, 37
X-ray fiber diffraction diagram of, 35–36
crystallinity and polymorphism
crystal density, 40
crystalline dimorphism, 43
degree of crystallinity, 40
forms, 40
intra- and intermolecular hydrogen bonds, 40
morphologies, 55
polarity of crystals, 56–57
reactions conditions
under heterogeneous conditions, 69
under homogeneous conditions, 66–69
reactions sites in, 69
factors influencing characteristics of, 70
Cellulose acetate (CTA)
crystal and molecular structure, 55
postchemical enzymatic deacetylation, 124
transmission electron microscopy of, 56
Cellulose derivatives, 118, 121
applications, 177

Cellulose derivatives (*cont*)
  fringed micelle model of, 162
  reductive depolymerization of, 156
Cellulose ethers, 118, 121
Cellulose Iα and Iβ
  ^{13}C CP-MAS spectra of, 46
  crystalline arrangement representation, 58
  crystal structure and hydrogen-bonding
    system, 45–46
  molecular modeling methodology, 44
  neutron diffraction pattern, 44
  "parallel-up" chain-packing organization, 46
  real and simulated AFM images, 64
  relationships between unit cell of, 49
  relative slippage of cellulose chains
    between, 48
  solid-state NMR spectrum of, 43
  structural details, 46–47
  trans–gauche orientation, 46
Cellulose III
  conformational changes, 53
  electron diffraction study, 52
  molecular and crystal structure, 53
  reversible crystalline forms, 52
  solid-state NMR study, 52
Cellulose-selective endo-enzymes, 172
Cellulose synthase complexes
  cartoon representation of, 34
  in plasma membrane, 33
Cellulose 6-tosylates, 124
Cellulose triacetates, ^{1}H NMR, 148
Ceric ammonium nitrate (CAN)
  reaction with cellulose
    cellulose–polyacrylonitrile graft
      copolymers, 72
CFU-E cells. *see* Colony-forming
    uniterythroid (CFU-E) cells
Chemical modification of (1→4)-glucans, 123
Chinese hamster ovary (CHO), 422–423
Chiral (–)-menthyloxyacetates esters, 124
Chitosan nanoparticles, 266
Chromatogeny principle and engineering, 87
Citrate–ligand exchange, 214
Cobalt superparamagnetic nanoparticles, 245
Cobb test, 91
Collision-induced dissociation (CID), 158
Colloidal gold, 213
Colon cancer, human NEU3 role in, 438
Colony-forming uniterythroid (CFU-E)
    cells, 426–427
Concanavalin A (ConA) lectin, 255
Cross-linked iron oxide (CLIO), 240
CSER. *see* Cell-surface elastin receptor (CSER)
Cyanobacteria, cellulose biosynthesis in, 28
Cyanoethylation, 132
  cyanoethyl glucans, 146–147
Cyclodextrins, 120, 131
Cytosolic sialidase NEU2
  CHO, 422–423
  functional implication
    biotechnological application, 428
    and cancer cell differentiation, 426–427
    developments in, 429
    influenza virus sialidase, inhibitors,
      427–428
    and muscle differentiation, 425–426
    RMS, study, 426
    transfected myoblasts, 425–426
  human sialidase NEU2, crystal structure
    arginine triad, 423–424
    Asp-box, 424–425
    N-acetyl and glycerol portions, 424
    Protein Data Bank with, 423
  molecular-modeling techniques, use, 423
  pH optimum, 423
  and site-directed mutagenesis, use, 423
  structural features of, 424

D

DCC. *see* *N,N*-Dicyclohexylcarbodiimide
    (DCC)
Degree of polymerization (DP), 120
6-Deoxy-celluloses, 124
1-Deoxy-D-fructosylation reagent, 311
N-(1-Deoxy-2,3:4,5-di-*O*-isopropylidene-β-D-
    fructopyranos-1-yl)-amino acids, 308
Deoxygenation of cellulose, 74–75
6-Deoxy-6-halo-celluloses, 124

Dependence model, 168
Detergent-resistant microdomains (DRMs), 435
Dextran, 121
D-Fructose–amino acids, 308
D-Fructose–hemoglobin (HbA_{1c}), 357
D-Fructose-Leu-enkephalin, 307
Dialdehyde cellulose, 124
2,3-Dialdehyde cellulose
  cellulose oxidation by sodium periodate, 72
Diaminals
  in acid-catalyzed Amadori rearrangement, 298
  formation, 298
N,N-Dicyclohexylcarbodiimide (DCC), 73
Diethylenetriaminepentacetate (DTPA)–Gd complex, 266
N,N-Dimethylacetamide (DMAC), 68
4-Dimethylaminopyridine (DMAP), 124
DMAC–LiCl system, cellulose in, 68
DNA-functionalized gold nanoparticles, 214
DRMs. *see* Detergent-resistant microdomains (DRMs)

E

Economy based on cellulose, 84
  cellulose–cellulose composites
    macromolecular interaction involved in, 98
    scanning electron micrograph of, 97
    uses, 90–92
  cellulose–synthetic polymer composites
    uses, 85–89
  protective films
    uses, 89–90
Elastin-binding protein (EBP), 414
Elastogenesis
  in fibroblasts from sialidosis and GS patients, 415
  NEU1 and elastin-binding protein, role
    ASMCs, 414
    CSER, 413–414
    EBP, 414
    VGVAPG hydrophobic domain, 414
  NEU1 deficient mice
    concomitantly, fibulin-5 (Fib-5), 416
    elastin concentration and lamellae, 416
    Neu1-null mice, 416
    tight-skin (Tsk) mice, 416
Electron-capture dissociation (ECD) technique, 323
Electrontransfer dissociation (ETD) technique, 323
Electrospray ionization (ESI), 323
EMH. *see* Extramedullary hematopoiesis (EMH)
Endogenous fructosamines, plants and animals, 354–355
  Alzheimer's disease, 358
  D-fructose–phosphatidylethanolamine (Fru-PE), 358
  diabetes and related complications
    biological effects, 355
    cardiac myocytes, ROS production, 355
    D-fructose–albumin activities, 357
    glutathione glycation, 358
    HbA_{1c}, 357
    human serum albumin modification, 356
    Maillard reaction, 358
    MAPK p44/42 (ERK1/2) and p38, 356–357
    NADPH oxidase and protein kinase C (PKC-α), 355–356
    PAI-1 promoters, 357
    vascular remodeling, 356
  opines
    enzymatic systems, 355
    production, 355
  physiological significance
    binding proteins, 355–356
    plasma proteins, human, 355
  plasma, 358
Enzymatic alterations and modifications of cellulose fibers
  cellulolytic enzymes, 82
  digestion, 82
  hydrolysis, 81

Enzymatic alterations and modifications of cellulose fibers (*cont*)
cellulosomes, 82–83
modification by
transglycosylating enzyme, 84
xyloglucan endotransglycosylase, 84
Enzymatic formation method, fructosamine
1-amino-1-deoxy-D-fructosamine derivative
chrysopine family, 314
crown galls, 314
production, 312–313
transfer DNA (T-DNA), 313
D-fructosamine 6-phosphate, 315–316
D-glucose 6-phosphate, 315
glycerol phosphate oxidase (GPO), catalase and aldolase (RhuA) combination, 316
guanidine triphosphate (GTP) cyclohydrolase I, 315
histidine biosynthesis, 314–315
opines, 313–314
pteridine metabolism, 315
ribozyme-catalyzed reactions, 316
thiazole biosynthesis, 315
tryptophan biosynthesis, 315
Enzymatic hydrolyzates, chemical analysis, 174
Enzyme-aided structure analysis of starch/cellulose, 172–175
Enzyme-catalyzed reactions, fructosamine derivatives
hydrolytic
amadoriases, 343
endopeptidases, 343
glycosidases, 343
isomerases and epimerases
FN6K, 342–343
frl operon, *E. coli,* 343
kinases
FN3K (*see* Fructosamine-3-kinase (FN3K))
FN6K (*see* Fructosamine-6-kinases (FN6K))
mannopine cyclase, 343
nitric oxide, 343
oxidoreductases
amadoriase, 339–340
mannopine, 341
plasma protein modification, 338
phosphatases, 342
Esterases, 32
Ester hydrolysis, 124
Etherification reactions, 131
of cellulose, 72
Extensins, 32
Extramedullary hematopoiesis (EMH), 419

F

Fast-atom bombardment (FAB), 323
Fischer glycosidation, 322
Fischer's approach, 295
Fringed micelle model of cellulose derivatives, 162
Fructosamine and derivatives
acid-labile glucosylamine, 292
biological functions and therapeutic potential
endogenous, plants and animals, 354–358
as potential pharma- and nutraceuticals, 358–366
carbohydrate modifications, 294
dehydrated foods, 359
D-fructosamine
core structures, 293
in literature, 294
formation methods
Amadori rearrangement, 294–295
enzymatic (*see* Enzymatic formation method, fructosamine)
non-enzymatic (*see* Non-enzymatic formation method, fructosamine)
hemoglobin A1c (HbA1c), 293
as intermediates and scaffolds
Maillard reaction, foods and *in vivo,* 344–349
neoglycoconjugates, 349–350
synthons, 351–354
isoglucosamine, 292
N-aryl-glucosylamine rearrangement, 292–293

stable isomer, 292
structure and reactivity
acid/base-and metal-promoted reactions, 327–338
analytical methods, 322–327
enzyme-catalyzed reactions, 338–343
tautomers, solid state and solutions, 316–322
Fructosamine-3-kinase (FN3K)
ATP-dependent phosphorylation catalyzes, 341
catalytic phosphate group proximity, 341
D-fructosamine epitope and aglycon structures, 341
inhibitor structures, 341–342
isolation and activity, 341
mammalian genomes, 342
Fructosamine-6-kinases (FN6K)
ATP-dependent 6-phosphorylation, catalyze, 342
bacterial extracts, 342
E. coli, 342
Fructosazines, 334
Fünfkugelapparat method, 118
Furan-2-carboxylates esters, 124

G

Galactosialidosis (GS), 407, 411
β-Galactosidase (β-GAL), 406
Galacturonananses, 32
Garegg, P. J.
β-mannosides formation
indirect and direct methods, 23
Norblad-Ekstrand Medal, 22
oligosaccharide synthesis, 22
Ph.D. studies, 21–22
research environment in carbohydrate chemistry, 22
schooling, 21
synthetic carbohydrate chemistry, 22
Gas barrier test, 91
Glaucocystis nostochinearum
cellulose study, 43
synchrotron and neutron diffraction data, 47
Globotriose-GNPs, 255
Glucanases, 32, 169, 171
active site, 169
Glucan derivatives, structural complexity, 132
bulk material, heterogeneities in, 142–144
glucan chains, substituent distribution, 139–142
glucose residues, distribution, 133
Spurlin and Reuben models, 133
broadened substituent distribution, 136
carboxymethylation, determination of relative rate constants, 137
(1→4)-glucans, monomer patterns, 134
hydroxyethylation, 139
inter-and intramonomeric effects, 138
mol fractions of regioisomers, 135
MS/DS ratio, 139
topochemical effects, 142–144
(1→4)-glucans chemical modification, 121
cellulose
characterization of products, 123
cross-linked celluloses, 123
glucose residues, 122
hydrogen-bond patterns, 122
kinetically controlled reactions, 122
linear β-(1→4)-linked glucan structure, 121
α-(1→6)-linked glucan, 121
polymer-based reactions, 123
kinetically controlled reactions, 127
alkylation, 127–129
esterification, 131
hydroxyalkylation, 129–130
sulfopropylation, 130
Williamson-type etherifications, 130
starch
"cationic starch", 126
derivatives, 126, 175
granule structure, 125–126
thermodynamically controlled reactions
acylations, 131–132
cyanoethylation, 132
sulfation, 132
Glucitols, 152, 154
Glucose, recovery in pre-Christian era, 16

Glucosylamines, Amadori rearrangement
acid catalysis
aldose diaminal intermediate, 300
amino group protonation in
N-alkyl-D-glucosylamines, 300
carboxylic acids, 299
D-glucosylamines derivation, 300
enaminol and ketosamine, 298
glucopyranosyl ring, 298
hemoglobin glycation, 300–301
hydrochloric acid, 299
Lewis acids, 300
metal complexation, 300
N-aryl D-glycosylamines, 298–299
protein-derived D-glucosylamines, 300
water, in foods and *in vivo,* 299
base catalysis
alcohols, 302
amines, 301
2-amino-2-deoxy-D-glucosyl- and-
mannosylamines, 302
enolization and resonance
carbocation, 301
food proteins, 303
in hemoglobin, 301
Lobry de Bruyn–Alberda van Ekenstein
transformation, 301
nucleophilic enolization catalyst, 301
phosphate, pH-buffering species, 301–302
in proteins, 301
proton-abstracting, 301
solvent nature, 302
water and aqueous syrups, 303
concurrent reactions
α-amino groups of peptides, 307
imidazolidinones, 307
L-glutathione, 306
m-tyrosine/DOPA,/catecholamines, 306
oxazolidin-5-one intermediate
decarboxylation, 305–306
thiazolidines formation, 306
tryptophan, 306–307
condensation reaction, 296
diaminals and N,N-aminodiglucosides, 298
D-mannose and amines condensation
reaction, 296–297
formation of
classic method, 298
crystalline hexose–amine adducts, 297
N-aryl and N-alkyl D-glucosylamines, 297
N-aryl D-glucosylamines preparation, 298
glycosylamine–ketosamine transformation
mechanism, 297
intramolecular
amino acid esters, 305
cyclization, amino acid/peptide
conjugates, 305
D-fructosamine structures, 305
oxazolidin-5-one intermediates
acrylamide formation, 303
intramolecular cyclization, Schiff
base, 303
protein glycation by glucose
amino group and catalyst in
microenvironment, 304
catalytic power, 304
D-glucosylamine and D-fructosamine
formation comparison, 304
εFruLys, 304
hemoglobin αβ dimer, 304
milk, lactose, 304
peptides, 304–305
Glycation, 293
Glycine, 32
Glycoconjugates
used in
direct syntheses of gold GNPs, 221–223
thiolated ligand exchange on, 232–234
Glyconanoparticles (GNPs), 213–214, 217
applications of
as antiadhesion agents, 260–264
in carbohydrate–carbohydrate interaction
studies, 251–254
in carbohydrate–protein interactions,
254–258, 269
in cellular and molecular
imaging, 264–269
interaction studies, 258–260

in MALDI-MS-based structural characterization, 269
Au/Fe-containing GNPs, 240–241
$Au_{102}(p\text{-MBA})_{44}$ nanoparticle, X-ray crystal structure, 216
O-carboxymethyldextran-coated SPIO nanoparticles, 240
CdSe–ZnS-labeled carboxymethylchitosan, 266
characterization, 218
concept, 218
methods for preparation, 219
synthesis, 218
Glyconanotechnology, 217–218
Glyco-quantum dots, 245–250
Glycosidases, 32
Glycoside-cluster effect, 251
Glycosphingolipid (GSL), 229
Gold–biomolecule nanoparticles, 213
Gold GNPs
AFM images, 259
application, 267
characterization, 237–238
glycoconjugates used in direct synthesis of, 221–223
hybrid, 228–229, 236–237
iron-free, 241
as probes for analytical techniques, 269
of saccharide, 235
by *in situ* protection with, 229
strategies for converting, 241
synthesis of, 219, 226
Gold lactose-GNPs, antimetastatic effect of, 261
Gold nanoparticles, 213, 216
Grafting cell fibers, 88
Grease barrier test, 91
Gum arabic, 237

H

Halodeoxycelluloses by halogenation, 74–75
Hemagglutinin-neuraminidase (HN), 434
Hematopoietic progenitor cells (HPCs), 419
Hemicelluloses, 121
polysaccharide molecules, 31
Hemoglobin A1c (HbA1c)
nonenzymatic glycosylation, 293
Heparin, 124
Heterogeneities, in bulk material
analysis of fractions, 182
cellulose derivatives, fractionation of, 180–181
starch derivatives, topochemical differentiation of, 182–183
Heyns rearrangement reaction, 302
High-performance liquid chromatography (HPLC)
D-fructose–peptides and–amino acid separation, 323
hydrolyzates treatment, 323
HLA. *see* Human leukocyte antigen (HLA)
HPCs. *see* Hematopoietic progenitor cells (HPCs)
Human leukocyte antigen (HLA), 407
Hyaluronic acid-immobilized gold nanoprobes, 267
Hyaluronidase (HAdase), 267
Hybrid glycoclusters, 235
Hybrid gold GNPs, 227
by direct *in situ* formation, 224–225
incorporating ligands in variable ratio, 228
by ligand place exchange, 236
2-*O*-Hydroxalkyl ethers, 155
Hydroxyalkylation, 129–130
of ethers, quantitative mass spectrometry, 165
of starches, 126
Hydroxypropyl starches, 126

I

Insulin-like growth factor-1 (IGF-1)
signaling pathways, 425
Insulin-like growth factor-2 (IGF-2), 417
International Carbohydrate Symposia, 9
Iron oxide-based GNPs, 243–245
Isoglucosamine
relation with fructose, 292

K

K562 erythroleukemic cells, 426–427
Keto tautomer
acyclic
arbonyl hydration and the equilibrium molar fraction N, 319
proportion, 319
isolation and structure, 319
Kinetically controlled reactions, 127
alkylation
kinetic data, 128
kinetics of, 129
laboratory scale, 127
nucleophilic substitution, 127
rate constants, 127
rate of reaction, 128
support for 2-OH deprotonation, 128
esterification, 131
hydroxyalkylation
carboxymethylation of cellulose, 129
sodium hydroxide concentrations, 130
sulfopropylation, 130
Williamson-type etherifications, 130

L

Lactose-GNPs, 252, 254, 260–261
Lignins, 30
Liquid chromatography–mass spectrometry (LC/MS), 324
Lobry de Bruyn–Alberda van Ekenstein transformation, 301
Lysosomal-associated membrane proteins (LAMPs), 409, 421–422
Lysosomal integral membrane proteins (LIMPs), 409
Lysosomal sialidase (NEU1)
β-GAL and PPCA, 406
cDNAs and genes, cloning, 407
galactosialidosis (GS), 407
histochemical staining, 417
human NEU1 deficiency
lysosomal storage, 412
NEU1 mutations, 412–413
sialidosis and galactosialidosis, 411–412
mouse model and disease pathogenesis, study, 418
extramedullary hematopoiesis, 419–421
homozygous null mice, 419
and impaired bone marrow retention, 419–421
lysosomal exocytosis, negative regulator, 421–422
shock-like twitch, 419
and multienzyme complex
crystal formation, 411
NEU1 and β-GAL, 410
PPCA zymogen, 410
in vitro experiments, 410
primary structure of
Asp-box, 407–408
ganglioside nomenclature, 407
gene and protein symbols, 407
glycosylation sites, 407
HLA, 407
routing and formation
acid phosphatase, 409
adaptor protein complex-3 (AP-3), 409
β-GAL precursor, 410
C-terminal tetrapeptide, 409
expression and, 409–410
intracellular routing of, 409–410
LAMPs and LIMPs, 409
PPCA functions, 410
structural model of
Asp-boxes, 408
β-propeller fold, 408
Neu5Ac2en-binding, 408
sialic acid, 408
in tissue remodeling and homeostasis, functions
deficient mice, elastogenesis in, 416
EBP/PPCA and signal transduction, 415
elastin-binding protein in elastogenesis, role, 413–414
elastogenesis in fibroblasts from sialidosis and GS patients, 415
in proliferation and cancer, role, 416–418

M

Magnetic Fe–Pt nanoparticles, 244
Magnetic glyconanoparticles (MGNPs)
  based systems, 244
  iron oxide-based GNPs, 243–245
  superparamagnetic/paramagnetic, 240, 243
    iron oxide-based magnetic glyconanoparticles, 244
    maltose-coated Au and Au/Fe GNPs, 242
    paramagnetic Gd-based gold glyconanoparticles, 243
    strategies for converting gold GNPs into, 241
Magnetic nanoparticles (MNPs), 239
Maillard reaction
  aromas formation, in food
    aglycon, 346
    amino acid Strecker degradation products, 345
    carbohydrate intermediates, 344
    from D-fructosamine derivatives, 347
    L-proline, 346
    pyrolysis, 346
    sugar dehydration–fragmentation products, 344
    volatiles, 345
  definition, 344
  di-and oligosaccharide derivatives, 347
  end product classes, 344
  glycation end products
    AGEs (*see* Advanced glycation end products (AGEs))
    lysine and arginine, 347–348
    N^{ε}-D-fructose–Llysine, 347
  melanoidins
    characteristic feature, 346–347
    formation rate, 347
    molecular weight, 347
    structure, 347
  stages, 344
Major histocompatibility complex (MHC), 433
Mammalian sialidases
  cytosolic (NEU2), 406
  lysosomal (NEU1), 406
  mitochondrial/lysosomal/intracellular membranes (NEU4), 406
  pivotal and diverse functions, 406
  plasma-membrane (NEU3), 406
  in silico analysis, 406
Mammalian target of rapamycin (mTOR) pathway, 425
Manno-GNPs, 226–227
  as inhibitors of HIV transinfection, 263
Mannopine (MOP) oxidoreductase, 341
MAPK. *see* Mitogen-activated protein kinase (MAPK)
Mariotte flask technique, 91
Matrix-assisted laser desorption/ionization (MALDI), 323
Mercerization process, 40
Metal nanoparticles functionalization, 230
2,3,6-*O*-Methylated maltooligosaccharides, 158
Methylation methods, 17
N-Methylmorpholine *N*-oxide (NMNO), 68, 226
  and cellulose, interaction between, 69
O-Methyl/*O*-methyl-d_3 derivatives, 164–165
Methyl starches, 126
MHC. *see* Major histocompatibility complex (MHC)
Michael addition synthesis of cellulose ethers, 72
Microfibril of cellulose, 58
  computer representation, 62
  molecular model of, 59
  size ranges, 59
  structural organization in plants, 66
  ultrastructural organization
    asymmetry in, 65
    distribution map of orientation of axes, 65
Mitogen-activated protein kinase (MAPK), 434
Monocotyledons, 32
Monomer composition
  glucose residue, substituent distribution in, 150–151
    collision-induced dissociation of dimers, 157–159

Monomer composition (*cont*)
electrospray ionization-mass spectrometry, 157–159
sample preparation, 154–157
separation after depolymerization, 151–154
Monomer residues along polymer chain
distribution, 159
enzymes selective tools, 169
cellulose derivatives, 177–179
enzyme-aided structure analysis, 172–175
starch derivatives, 175–177
oligomer analysis after random degradation, 160
mass-spectrometric analysis, 163–166
quantitative evaluation, 167–168
sample preparation, 161–163
Monomers
analysis
of methylated glucans by ESI-MS/CID, 158–159
patterns of (1→4)-glucans, 134
Multifunctional envelope-type nanodevices (MEND), 259
Multifunctional GNPs, tools for glycobiology, 270

N

Nanobiomaterials, 212
Nanobiotechnology, 212
Nanotechnology, 264, 269
National Center for Biotechnology Information (NCBI), 427
Native cellulose
crystalline morphology of, 57–58
crystalline structures of, 41
chain orientation, 42
dimorphism in, 43–44
Iα and Iβ, crystalline phases, 42–43
synchrotron and neutron techniques, 44
two-fold helicoidal symmetry, 42
X-ray and neutron fiber diffraction patterns, 45
liquid ammonia, treatment with, 40
regeneration and mercerization
transition to crystalline form cellulose II, 40
NCBI. *see* National Center for Biotechnology Information (NCBI)
NDV. *see* Newcastle Disease Virus (NDV)
Neoglycoconjugates
breast cancer cells, 350
casein, 350
cytotoxic nitroso compounds, 349
fructose–1-SNAP, 350
melanoma tumors, 350
N-nitroso derivatives, 349
properties, 349
radioiodinated octreotide, 350
radiolabeled bombesin analogues, 350
Neuraminidase (NA), 427
NEU4 sialidase
experiments on, 435
expression, 436
feature of, 435
functional implication
and cancer, 437
and lysosomes, 436–437
neuronal cell differentiation, 437–438
ganglioside GD3 and apoptosis, 436
mammalian sialidases and teleosts, 435
mitochondrial localization of, 435–436
NCAM activity and, 438
Neutron diffraction pattern of cellulose Iβ, 44
Neutrophil elastase (NE), 420
Newcastle Disease Virus (NDV), 434–435
NMNO. *see* *N*-Methylmorpholine *N*-oxide (NMNO)
Noble metal glyconanoparticles
characterization, 237–238
direct *in situ* formation, 219–220, 226, 228
two-step formation, 228–229, 231
chemical modification, 228–229, 231
non-covalent approaches, 237
thiolated ligand exchange, 231, 235, 237

Non-enzymatic formation method, fructosamine
catalytic hydrogenation, 311–312
D-glucose phenylosazone reduction
1-deoxy-β-D-fructopyranos-1-ylamine, 295
Fischer's approach, 295
ketosamines preparation, 296
N-substituted D-fructosamines, 295–296
Pd-catalyzed hydrogenation, 295
D-glucosone reductive amination
advantages, 309
free amino acid, NaCNBH3 in water, 308
ketosamine derivative, 309
reaction with α-amino acids, Strecker degradation pathway, 309–310
secondary amines, 309
glucosylamines, Amadori rearrangement (*see* Glucosylamines, Amadori rearrangement)
Mitsunobu condensation, 312–313
nucleophilic substitution reactions
antitumor agents, 310
D-fructose–amino acid preparation, 310
2,3:4,5-di-*O*-isopropylidene-β-D-fructopyranose activation, 310–311
hydroxyl groups, 309–310
tosylate treated with hydrazine hydrate, 310
triflate by azide, 311
regioselective epoxide opening, 311
Non-synonymous single nucleotide polymorphism (nsSNP), 427

O

Oil drilling, 120
Oligosaccharides
MALDI-TOF-MS, in quantitative analysis, 165–166
mediated nuclear transport of glyco-QDs, 265
Osazones, Fischer's discovery, 17
Oseltamivir-phosphate, 427–428
Oxidation, D-fructosamine
metal ions
copper and iron, 335
copper-promoted autoxidation, 336
glycated polylysine, iron(III), 336
organic molecules, 335
redox-active catalytic, 337
superoxide ion-radical formation, 335–336
molecular oxygen
catalyst-free, 335
superoxide release rate, 335
oxidants, 338
benzoquinones, 338
εFruLys formation, 338
redox dye NBT, 338
Strecker aldehydes, 338
physiological oxidants
flavin adenine dinucleotide (FAD)/copper, 337
peroxynitrite, 337
Phanerochaete chrysosporum, 337

P

"Parallel-up" chain-packing organization, 44–45
Particulate sialidase NEU4
experiments on, 435
expression, 436
feature of, 435
functional implication
and cancer, 437
and lysosomes, 436–437
neuronal cell differentiation, 437–438
ganglioside GD3 and apoptosis, 436
mammalian sialidases and teleosts, 435
mitochondrial localization of, 435–436
NCAM activity and, 438
Pathogens, real-time detection, 270
PDGF. *see* Platelet-derived growth factor (PDGF)
PDGFR. *see* Platelet-derived growth-factor receptor (PDGFR)

Pectins
(1→4)-α-D-galacturonan, 32
(1→2)-α-L-rhamnan, 32
arabinan–arabinogalactan type, 32
of dicotyledonous higher plants, 32
Peeling off mechanism, 335
Periodate oxidation, 118
Peroxidases, 32
Phase-transfer catalyst, 215
Phenylhydrazones
Fischer's discovery of, 17
Phorbol 12-tetradecanoate 13-acetate (PMA), 434
Phosphatidylinositol-3-kinase (PI3), 425
Pictet–Spengler phenolic condensation, 306–307
Plant
cell walls
contents, 28
enzymes, function of, 32
hydrophobicity, 30
liquid-crystalline matrix, 31
location of polysaccharide components, 30
mechanical strength, 30–31
microfibril, 30
pectins and ion-exchange capacity, 32
primary, 28
proteins in, 32
representation of, 30
rigidity, 30
secondary, 29
turgor pressure on plasmalemma, 30
Plasminogen activator inhibitor-1 (PAI-1) promoter, 357
Platelet-derived growth factor (PDGF), 417
Platelet-derived growth-factor receptor (PDGFR), 432
PMA. *see* Phorbol 12-tetradecanoate 13-acetate (PMA)
Polymeric glucan derivatives, 120
Polysaccharide Chemistry, 11
Polysaccharide derivatives
as renewable resources, 119
applications, 120
enzymatic degradability, 121
modified glucans, 120
research, 118
Polysaccharides: starch, 118
Polyvinyl alcohol (PVA)
cellulose layer modification, 89
representation of, 91
waterproofing effect, 90
chromatogeny process, 90
molecular structure, 90
Potential pharma-and nutraceuticals, fructosamines
antibacterial/antifungal
D-fructosamine–and–lactulosamine product derivatives, 364–365
Mycobacterium leprae, D-fructose–serotonin, 364
Staphylococcus aureus, 364
Trypanosoma brucei, 364
antioxidant
D-fructose–L-tryptophan, 362
Fenton reaction, 362
fruits and plant tissues, dehydration, 362
antitumor
lactulosamine derivatives, 363–364
LctLeu, 363
tomato thermal processing and dehydration, 362–363
urological cancers, 362
bioavailability
amino acids, 360
D-fructose–L-methionine, 360–361
εFruLys, urine and kidneys, 361
LctLys, feces, 361
LctLys/FruLys, blood, 361
*N*ε-D-fructose-L-lysine, 361
in food
dehydrated, 358–360
dried carrots, 358
fumonisin B1, 360
ingested, 358
milk, 358
N-nitroso compounds (NOC), 360
olfactory properties, 360

immunostimulatory
analgesia-enhancing effect, 365
D-fructose–L-lysine epitopes, 365
D-fructose–L-proline, 365
D-fructose–peptides, 366
lactulosamine structure, 365
lactulose conjugate, 365–366
thioureido derivatives, 366
PPCA. *see* Serine carboxypeptidase protective protein/cathepsin A (PPCA)
Proline, 32
Pro-matrix metalloproteinase-1 (pro-MMP1), 415

Q

Quantum dots (QDs), 245
Quartz crystal microbalance (QCM)
based biosensors, 256
Quaternary aminopropyloxirane (QUAB), 126

R

Rayon fiber–cellulose acetate composites, 97
optical microscopy, 94
scanning electron micrograph of, 95
transmission infrared spectra, 94
Rayon manufacture, 97
Reactive oxygen species (ROS), 267
Regenerated cellulose (Cellulose II), 48
cellulosic oligomers in, 50
hydroxymethyl group amount, 52
liquid ammonia, treatment with, 40
monoclinic cell, 51
packing energies of cellulose chains in unit cell, 50
resolution of, 50
structural details of, 51
synchrotron X-ray data, 51
trans–gauche and *gauche–trans* positions, 51
two-fold symmetry, 50
X-ray diffraction studies, 49
Regeneration process, 40
Relenza. *see* Zanamivir
Respiratory syncytial virus (RSV), 428
Rhabdomyosarcoma (RMS)
study, 426
Rosette particles in plasma membrane, 32–33
RSV. *see* Respiratory syncytial virus (RSV)

S

Serine carboxypeptidase protective protein/cathepsin A (PPCA), 406
Serine-threonine protein kinase AKT1, 425
SFK. *see* Src family protein tyrosine kinases (SFK)
Short hairpin RNA (shRNA), 433
Sialidase NEU3, plasma membrane-associated
A431 squamous carcinoma cells, 430
β-barrel structure, 429
cis- and *trans*-activity of, 430
functional implication of
and cell differentiation, 432–434
and membrane microdomains, 431–432
and virus infection, 434–435
localization of, 431
metabolic labeling, 430
Sialidases, 405
cancer and human NEU3
in acute lymphoblastic leukemia (ALL), 439
apoptosis, 438
in colon cancer, 438
EGF receptor, 438
LacCer and, 438
overexpression, 439
Ras/ERK pathway, 438
functional implications of
abnormal growth of VSMC, 443
axon outgrowth, 442
extracellular matrix proteins and, 443
ganglioside substrates, 444
myelin associated glycoprotein (MAG), 443
N-acetylneuraminyl-lactitol, 444

Sialidases (*cont*)
NCX/GM1 complex, 444
normal and atherosclerotic human aortic intima, 443
TNF-α induction of matrix metalloproteinase-9 (MMP9), 443
TrkA tyrosine protein kinase receptor activation, 442
and immunity
Clostridium perfringens sialidase, 439
in human myeloid cell differentiation, 441
inhibition of adhesion, 440
Lewis X expression, 441
monocyte differentiation, 441
neutrophil adhesion, 441
overexpression of, 441
regulator of toll-like receptors (TLRs) activation, 441–442
role in, 440
in silico analysis, gene expression patterns
expression profile, 446–447
GeneAtlas data, 447
GeneAtlas Gene Expression Database, 445–446
tissue-specific RNA expression pattern, 445–446
UniGene cluster, 445
UniGene's EST Expression Profile Viewer, 444
in teleosts
cell-fractionation experiments, 450–451
data from, 452
expression of, 451
GM3 synthase overexpression, 451
organization of genes in man and, 450
in zebrafish *Danio rerio,* 449
Sialidosis
type I (normomorphic) sialidosis
cherry-red spot-myoclonus syndrome, 411
clinical phenotypes of, 411
symptoms, 411–412
type II (dysmorphic) sialidosis
clinical phenotypes of, 412
Sialyl Lewis X (sLeX), 243
sLe^X-GNPs, 268
SILICA 2010 Yellow, 88
Silicone oligomers polymerizing, 86
Smith degradation, 118
Sodium dodecyl sulfate polyacrylamide gel electrophoresis (SDS-PAGE), 324
Spirobicyclic tautomer, 321
Src family protein tyrosine kinases (SFK), 432
Starch, 125
chemical modification of, 126
derivatives, applications, 175
elemental analysis for, 145–146
granule structure, 125–126
starch derivatives, 126
Strecker degradation reaction, 334
Structure analysis of (1→4)-glucans derivatives
average degree of substitution, 144
elemental analysis, 145–146
NMR spectroscopy, 146–150
Zeisel method, 145
monomer composition (*see* Monomer composition)
monomer residues along polymer chain (*see* Monomer residues along polymer chain distribution)
Sulfation, 132
Sulfopropylation, 130
Surface features of cellulose
AFM images, 63
conformations, 64–65
hydroxyl group in, 63
interreticular distance, 63
intra-and intermolecular hydrogen bonds, 62
organization of, 63
periodicities, 63
Synthons
Büchi cyclocondensation reaction, 351–352
1-deoxy-D-*erythro*-hexos-2,3-diulose, 351
D-fructose-*N*-hydroxyurea, 352
glycosidase inhibitors, 351–352
isothio-/isoseleno-cyanates, 353

N-protected D-fructosamine, 352–353
oxazolidinone and morpholinone, 353–354
phosphoroamidites, 353–354

T

Tamiflu. *see* Oseltamivir-phosphate
Tautomerization scheme, 317
Tautomers, 316
D-fructosamine and N-alkyl-/-carboxyalkyl derivatives
amino acids and proteins, 317
anomeric/carbonyl (carbon-2) signals, 317
D-fructose–amino acids crystal structure feature, 319
enamine, NMR technique, 317
hydrochloride and hydroacetate salt molecular and crystal structure, 318
ring conformations, 318
disaccharides, 320
1-amino-1-deoxy-4-O-glycopyranosyl-D-fructose analogues, 320
glycosyl substituent at O-6, 320
form proportions, 317–318
fructosamine, fixed structure
anomeric hydroxyl group, glycosidation/carbamoylation, 321–322
N-alkenyl derivatives, 322
treatment, triphosgene transformed, 322
X-ray diffraction, 322
N-aryl derivatives
acyclic keto proportion, 318–319
D-fructose–aniline equilibria, 319
fructosamine carbonyl group hydration rates, 319
keto isolation and structure, 319
N-allyl-N-(1-deoxy-β-D-fructopyranos-1-yl) aniline, 320
N,N-di-D-fructosamines
1,2-enolization, 321
monofructose–amines tautomerization pattern, 317, 320
spirobicyclic structure, 321
X-ray diffraction, 320–321
tautomerization scheme, 317
T-Cell helper peptide, 226
Tetrakis (hydroxymethyl)phosphonium chloride (THPC), 235
2,2,6,6-Tetramethylpiperidine-1-oxy radical (TEMPO)
mediated oxidation of cellulose, 71
Tetraspanin-enriched microdomains (TERMs), 432
Thermodynamically controlled reactions
acylations, 131–132
cyanoethylation, 132
sulfation, 132
Thermoplastic composites of cellulose–cellulose acetate, 93–95
"grafting" of cellulose acetate chains, 95
macromolecular interaction involved in, 96
mechanical performance, 95–96
optical microscopy, 94
production of, 95–96
T Lymphocytes, 262
Topochemistry of cellulose, 75
acetylation of cellulose, 76
conversion of cellulose I into cellulose II
model of, 79
morphological modifications, 80
"shish-kebab" morphology, 80–81
nanocrystals of, 80
"cellulose whiskers," 81
chemical modifications by cross-linking, 81
solid-state conversion of cellulose I into cellulose III_1, 76–77
morphological changes, 78
Transesterification reaction, 75
Transforming growth factor (TGF)-β1, 357
Transglycosylases, 32
Trans-sialidases (TS)
apoptosis, 453
in biotechnology use, 454

Trans-sialidases (TS) (*cont*)
conformational switch in active center of, 457
detection of activity, 453
endothelial cell signaling, 453
in *Endotrypanum* species, 452
inhibitor of, 454
lactose-binding site, 455
in microorganisms, 452
polyclonal B cells activity, 453
reaction of, 452
sialyltransferase activities with CMP-Neu5Ac, 453
sleeping sickness and Chagas disease, 453
T. brucei TS
protein sequence, 458
trans-sialylation activity, 458
T. cruzi trans-sialidase
amino acid sequence with, 455
analysis, 458
binding studies with, 456
hydrolysis and transfer reactions catalyzed by, 456
protein conformation involved in, 457
Shed Acute Phase Antigens (SAPA) repeats, 455
STD NMR experiments, 457
types of enzyme reaction, 456
T. rangeli sialidase
crystal structure of, 455
Trifluoromethylsulfonylation, 310
O-(3-Trimethylammonium-2-hydroxy)propyl starch, 126
Tri-*n*-butylphosphine (TBP), 247
Tri-*n*-octylphosphine oxide (TOPO), 247
Tri-*n*-octylphosphine (TOP), 247
Tri-*O*-methylated disaccharides, 159
Tunicin microcrystals
diffraction patterns, 60
surface chains in, 61
transmission electron micrographs, 61

V

Valonia cellulose
diffraction-contrast transmission electron microscopy, 80
electron diffraction studies, 42
microcrystals, action of cellulases on, 83
triclinic organization, 63
X-ray patterns of, 44
Vascular cell adhesion molecule (VCAM)-1, 420
Vascular endothelial growth factor (VEGF), 357
Vascular smooth muscle cells (VSMC), 443

W

Water barrier test, 91
Water-soluble metallic nanoparticles, 215
Whiskers, 58
acetylation of, 92
of celluloses, 60
mechanical modulus of, 59
production and application, 93
Whistler, R. L.
American Association of Cereal Chemists, 11
American Chemical Society
Division of Carbohydrate Chemistry Executive Committee, 10–11
Claude S. Hudson Award of Division of Carbohydrate Chemistry, 13
Fred W. Tanner Lectureship of Chicago Section, 13
Carbohydrate Chemistry for Food Scientists, 12
in carbohydrate nomenclature committee, 6
consultant to
companies producing industrial gums, 11
corn wet-milling company, 11
guar gum structural analysis, 7
Industrial Gums, 12

Methods in Carbohydrate Chemistry series, 12
National Representative of United States, 11
Polysaccharide Chemistry, 6, 11
polysaccharides
calorie intake, 8–9
concept of sweetness, 8
derivatives of, 8
industrial utilization, 3
infrared (IR) spectrophotometry for analysis of, 9
structure and function, 3
satellite conferences on carbohydrates, 10
Starch, Chemistry and Technology, 12
Starch/Stärke, 12
Steering Committee for International Carbohydrate Symposia, 9–10
U.S. Carbohydrate Nomenclature Committee
glycan rule, 12
Williamson ether synthesis, 72
Williamson-type etherifications, 130
Wood
cells crosssection, 67
cellulose in
electron microscopy, 31
helical orientation, 67
fiber structure
representation, 67

X

Xanthate system, 68
Xyloglucans, 32

Z

Zanamivir, 427–428
Zeisel method, 118
for DS/MS of alkyl/hydroxyalkyl starches, 145

Erratum

The Omitted Corrections to Volume 64 of Advances in Carbohydrate Chemistry and Biochemistry (ISBN: 9780123808547)

P. 404, footnote, line 2: "Section XI" should read "Section XII"
P. 423, line 32: "Section XI" should read "Section XII"
P. 430, last line of the legend of Fig. 5: "neutrophil" should read "neutrophils"

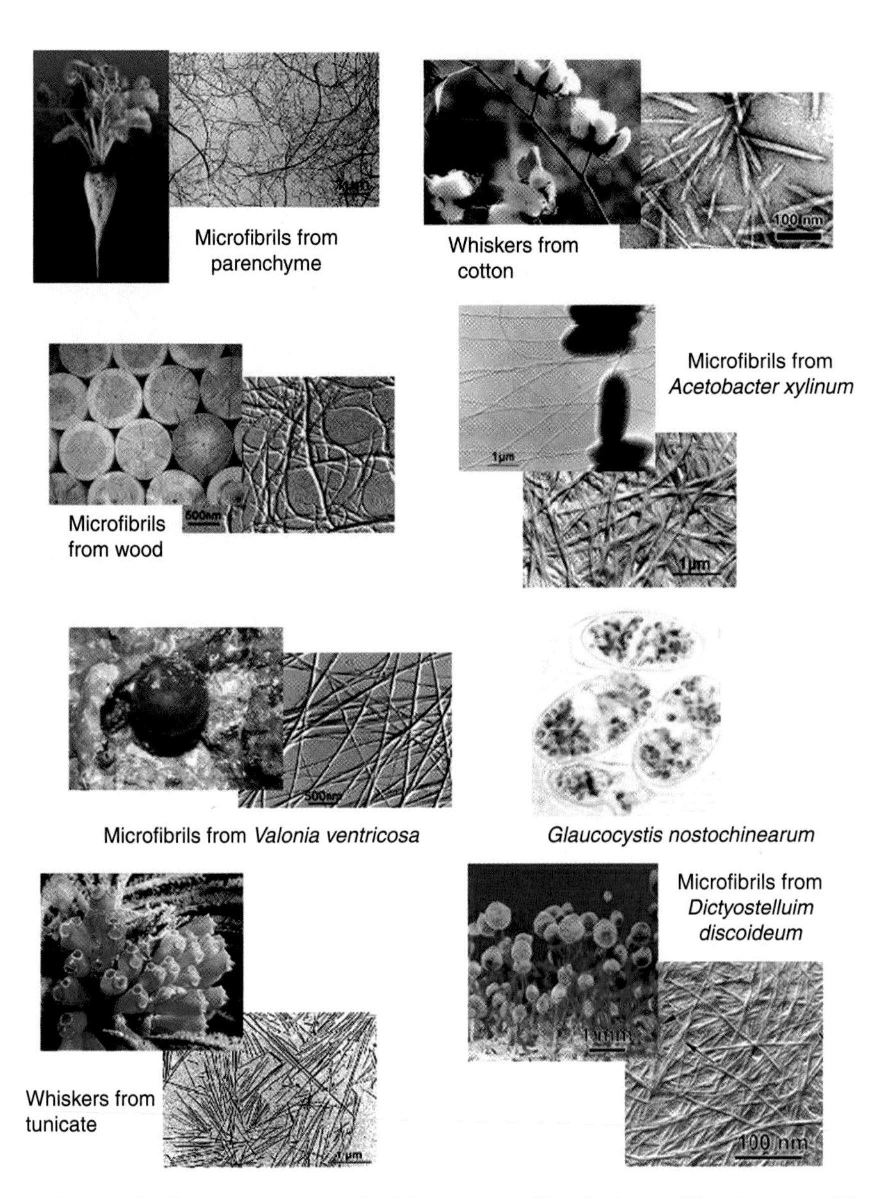

Plate 1 The ubiquitous occurrence of cellulose in nature. (See Fig. 1 in the Pérez chapter, p. 29.)

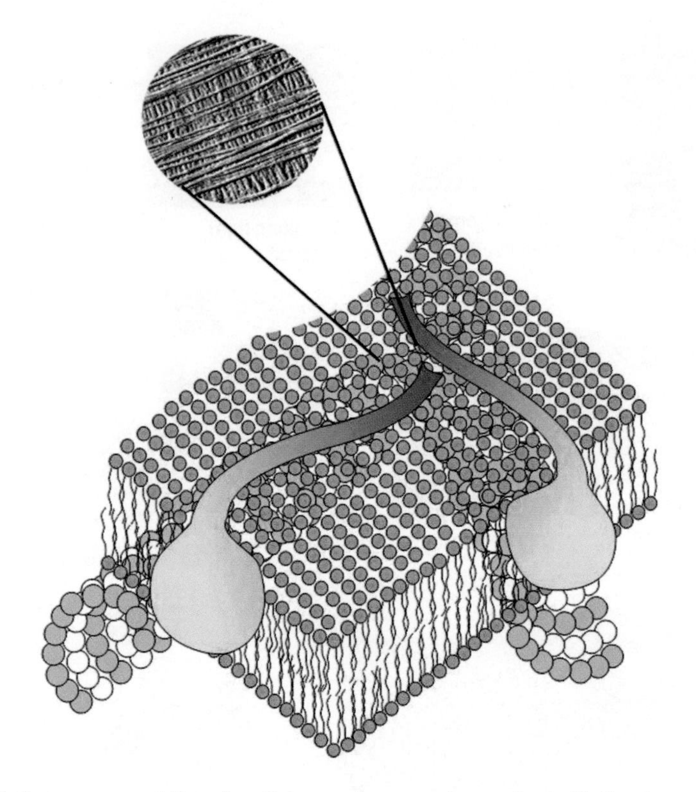

Plate 2 Cartoon representation of a cellulose synthase complex moving inside the plasma membrane, leaving a cellulose microfibril as one component of the primary cell-wall (inset: transmission electron microscopy). The cellulose synthase complex becomes active while unattached to a microtubule. (See Fig. 4 in the Pérez chapter, p. 34.)

Plate 3 Molecular representation of the chain structure of cellulose. Labeling of the cellobiose repeat unit is given in accordance with IUPAC nomenclature; *n* indicates the number of repeating disaccharide units in a given cellulose chain. The reducing end of the chain is indicated; a schematic representation of the chemical polarity of the chain is shown underneath; this representation is used throughout this chapter. (See Fig. 5 in the Pérez chapter, p. 35.)

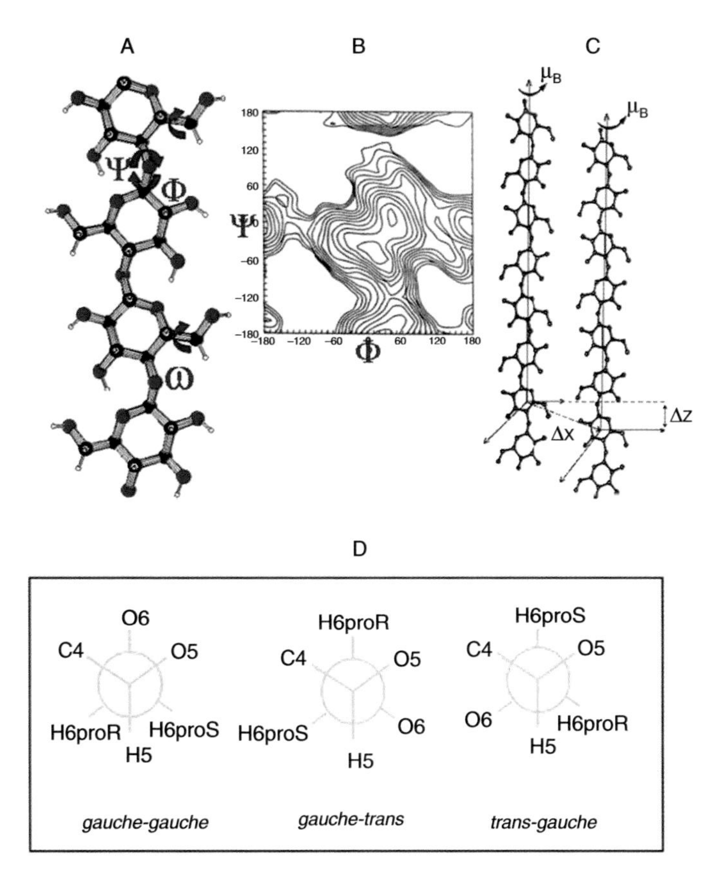

Plate 4 Structural descriptors for the cellulose conformation and 3-D structure. (A) The main conformational descriptors of the cellulose chains are the glycosidic torsional angles Φ=O-5–C-5–C-1–O-1, Ψ =C-5–C-1–O-1–C-4′, and the orientation of the primary hydroxyl group about the C-5–C-6 bond: *ω*. The low-energy conformations of the chains are derived from the potential energy surface that shows conformational energy with respect to the Φ and Ψ torsion angles (B). The primary hydroxyl group of a pyranose is locked to the C-6 carbon atom, whereas the secondary hydroxyl groups are linked to the ring carbon atoms. The orientation of the primary hydroxyl group is referred to two different torsion angles. The first is O-5–C-5–C-6–O-6 and the second is C-4–C-5–C-6–O-6. These torsion angles are said to be in a *gauche* conformation when they have a ±60° angle, and in a *trans* conformation at 180°. The three significant conformations are thus *gauche–gauche* (*gg*), *gauche–trans* (*gt*), and *trans–gauche* (*tg*). Usually, one of the three possible conformers, *gg/gt/tg*, is discarded because of the unfavorable 1-3-*syn*-axial interactions between the O-6 and O-4 atoms. The prochiral hydrogen atoms (as pro*R* and pro*S*) are noted. The relative orientation of two cellulose chains is dictated by the interhelical parameters μA, μB: rotation of chains A and B; Δ*Z*: translation of chain B with respect to chain A; Δ*X*: distance between the helical axis of chains A and B. (See Fig. 8 in the Pérez chapter, p. 38.)

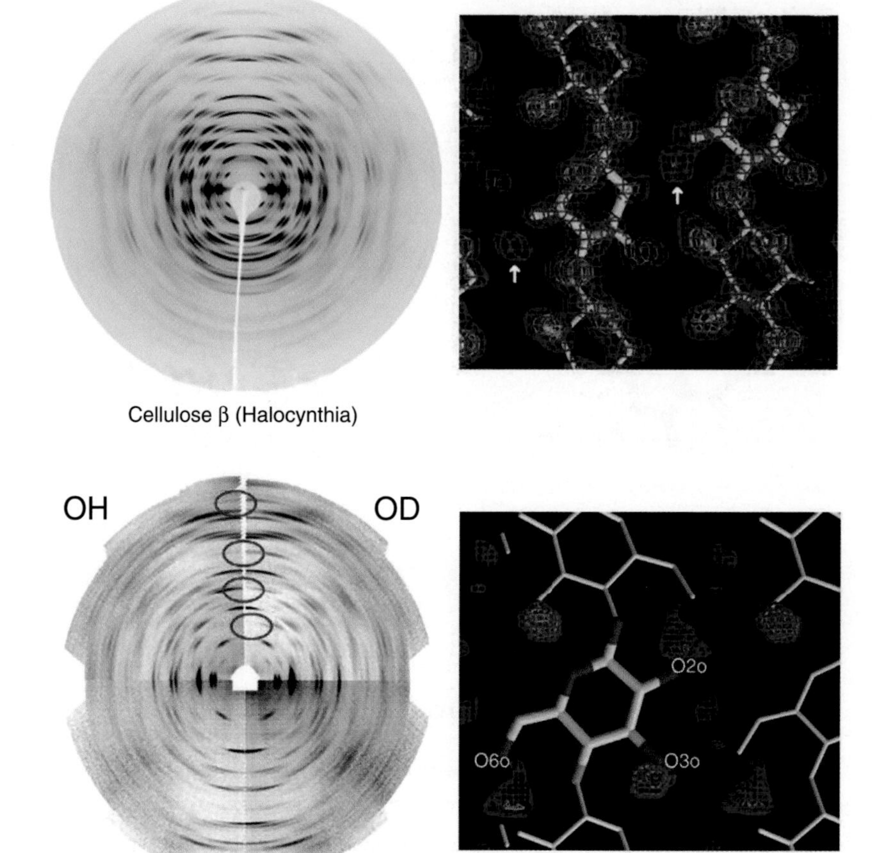

PLATE 5 X-Ray and neutron fiber-diffraction patterns recorded from native cellulose (OH) and fully deuterated (OD) cellulose samples. The difference in the scattering properties of hydrogen and deuterium (indicated by the red circles) allows the precise location of the electron density (shown in blue), which corresponds to the position of hydrogen atoms in the crystalline lattice. (See Fig. 12 in the Pérez chapter, p. 45.)

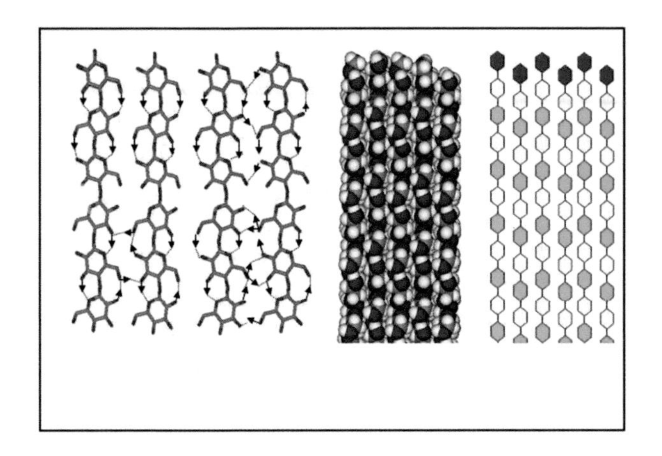

PLATE 6 Structural details of cellulose Iβ (See Fig. 13 in the Pérez chapter, p. 46.)

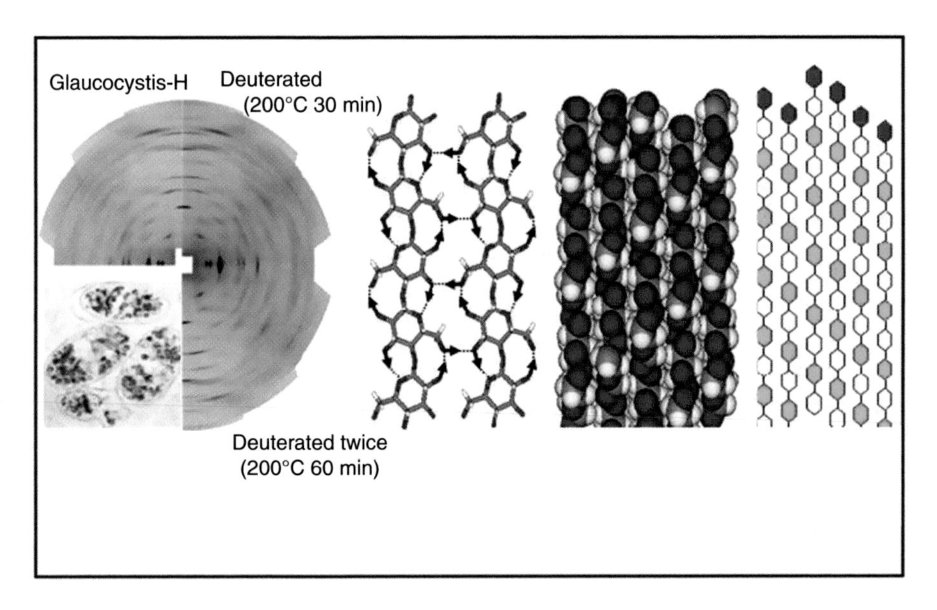

PLATE 7 Structural details of cellulose Iα. (See Fig. 14 in the Pérez chapter, p. 47.)

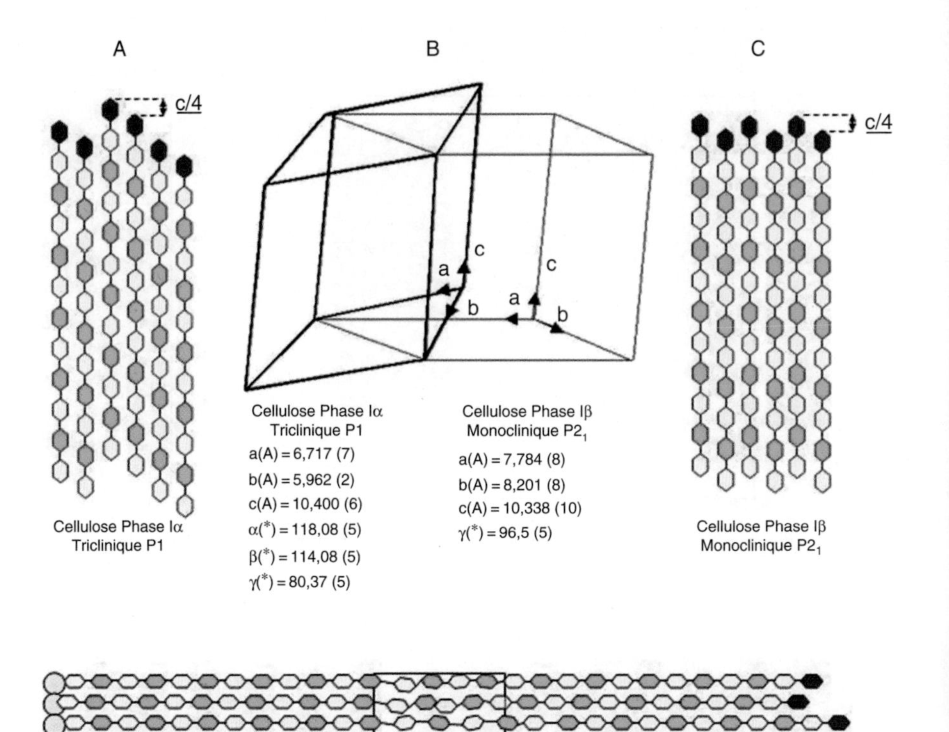

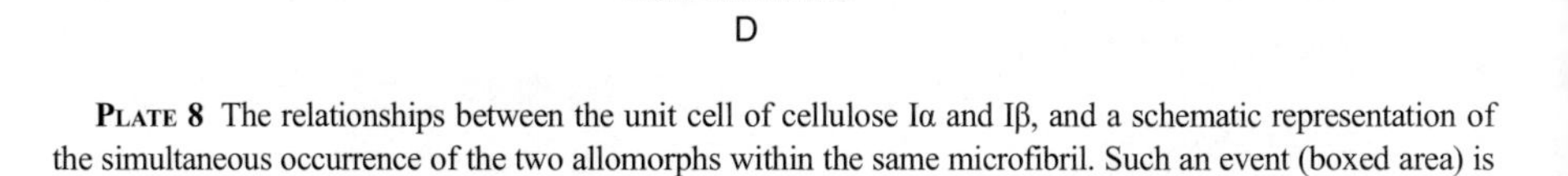

PLATE 8 The relationships between the unit cell of cellulose Iα and Iβ, and a schematic representation of the simultaneous occurrence of the two allomorphs within the same microfibril. Such an event (boxed area) is likely to be the site of an amorphous moiety within the microfibril. (See Fig. 15 in the Pérez chapter, p. 49.)

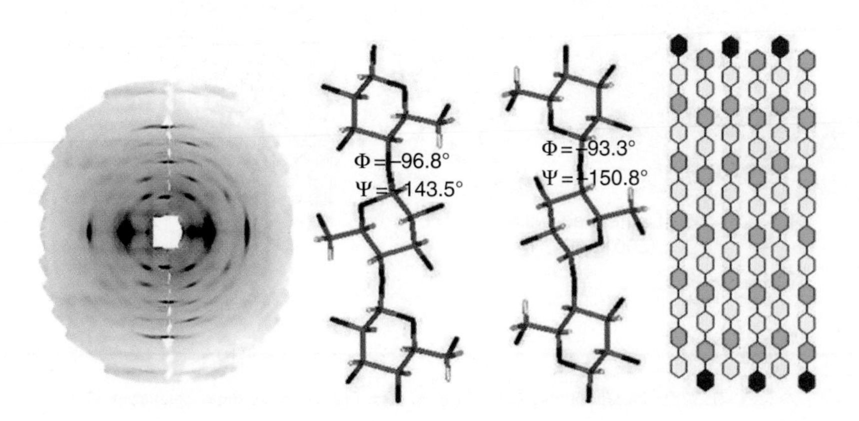

PLATE 9 Structural details of cellulose II. (See Fig. 16 in the Pérez chapter, p. 51.)

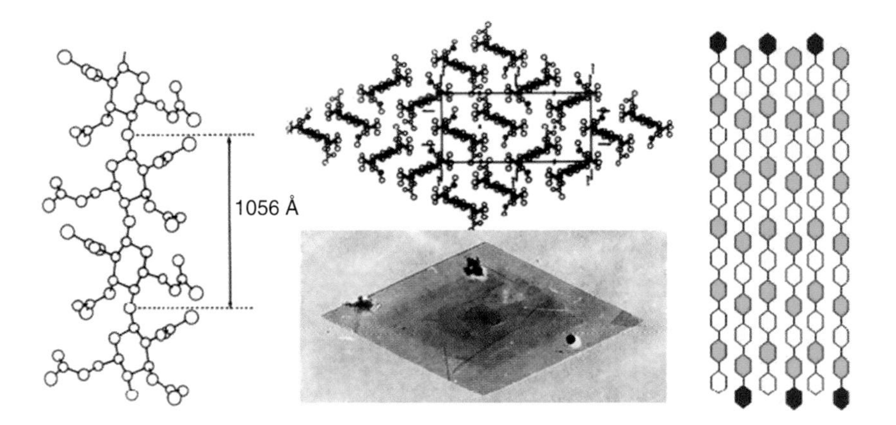

Plate 10 Cellulose triacetate (CTA) II. (A) Transmission electron microscopy of a polymeric single crystal of CTA II. (B) Projections of the chains I in the *a b* plane. (C) Electron diffraction pattern of a tip of a single crystal of CTA II in the *a–b* plane. (See Fig. 17 in the Pérez chapter, p. 56.)

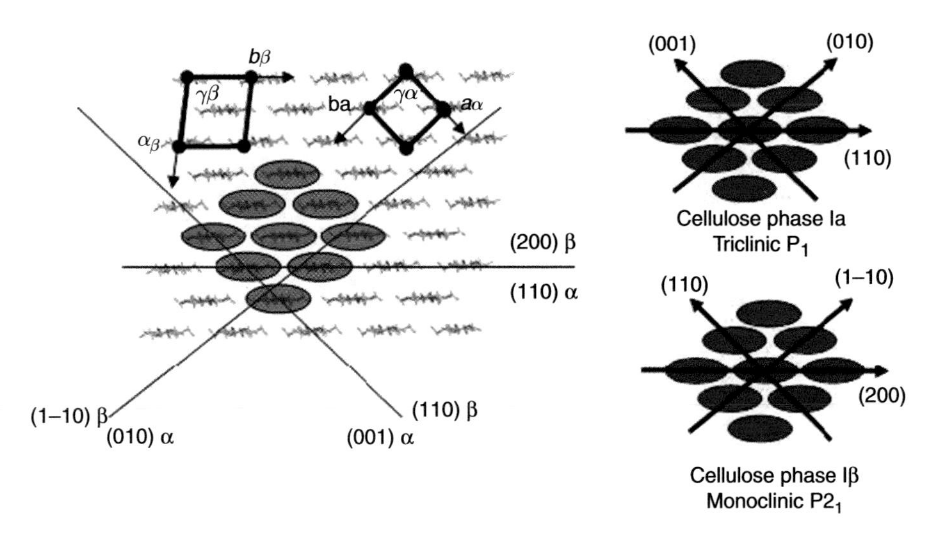

Plate 11 Schematic representation of the crystalline arrangement of cellulose Iα and Iβ in relation with their unit cells. The triclinic and monoclinic unit-cells are shown, along with the main crystallographic directions relevant for the crystalline morphologies. (See Fig. 18 in the Pérez chapter, p. 58.)

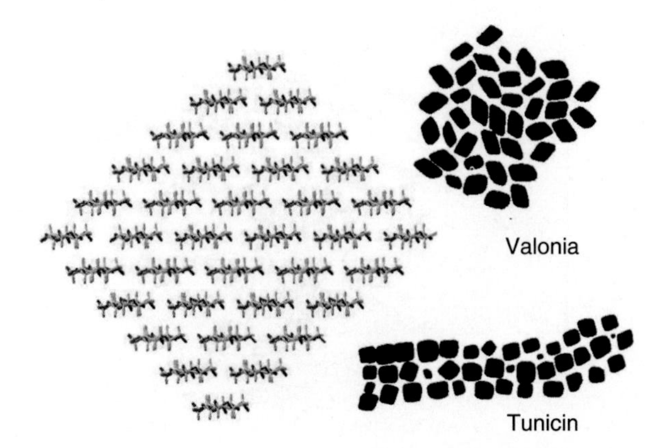

Plate 12 Molecular model of a microfibril of cellulose, projected along the fibril axes compared with the typical morphologies observed for *Valonia* cellulose and tunicin, along with the CPK (Corey–Pauling–Koltun) representation of the main crystalline faces for cellulose I. (See Fig. 19 in the Pérez chapter, p. 59.)

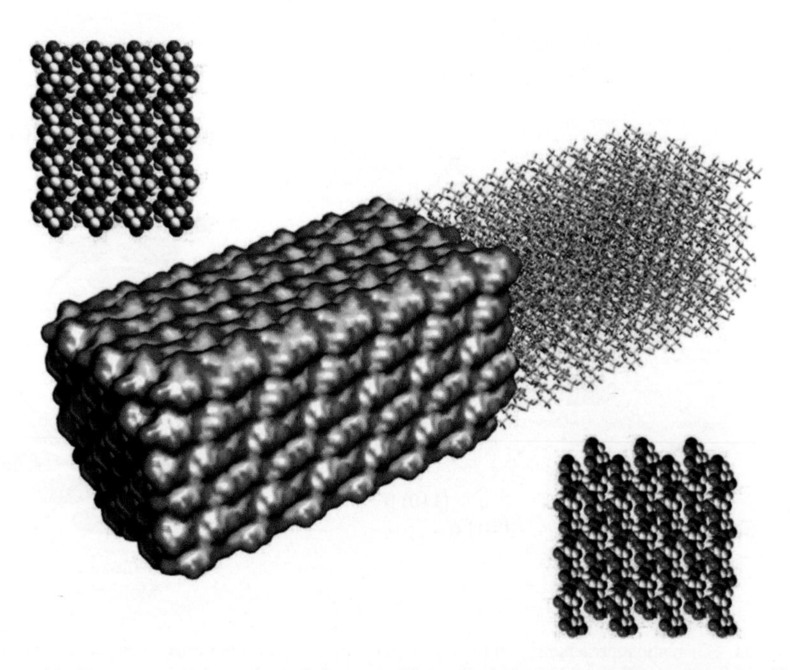

Plate 13 Computer representation of the crystalline morphology and surfaces of a microfibril of cellulose made up of 36 cellulose chains. (See Fig. 23 in the Pérez chapter, p. 62.)

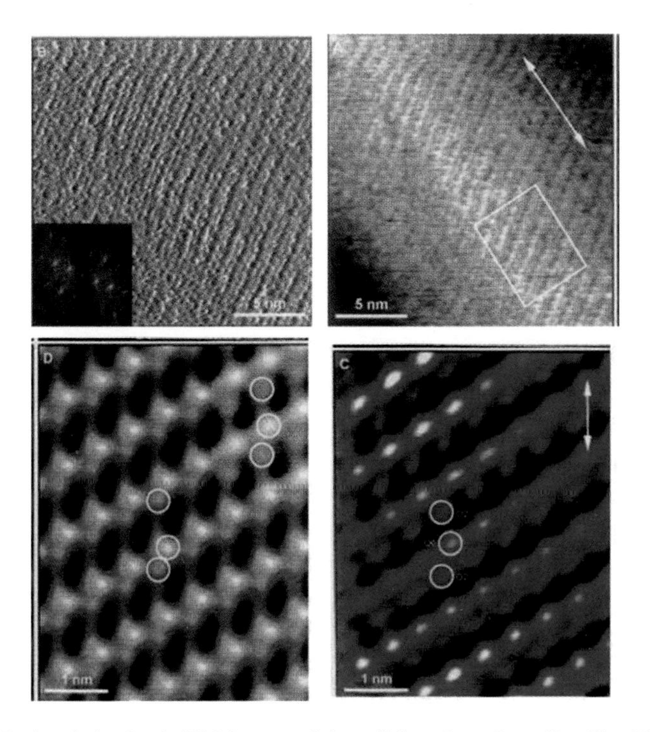

Plate 14 Real and simulated AFM images of the cellulose Iα surface. (See Fig. 24 in the Pérez chapter, p. 64.)

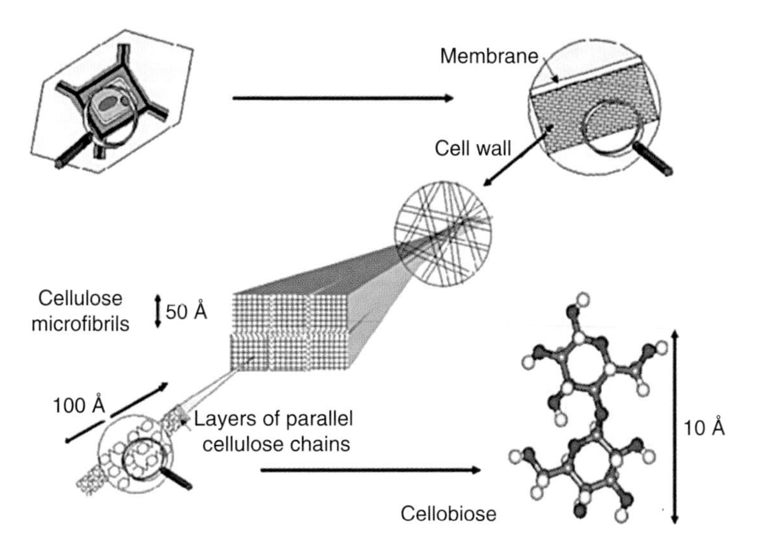

Plate 15 Structural organization in plants. (See Fig. 25 in the Pérez chapter, p. 66.)

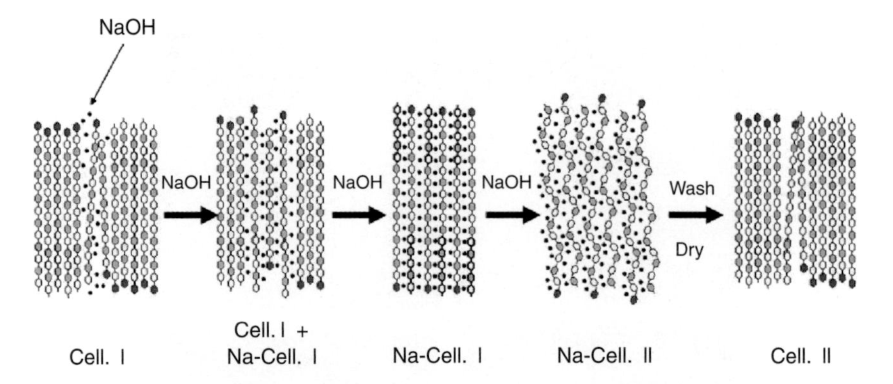

Plate 16 Model of the conversion of parallel-packed arrays of microfibrils of up and down chains of cellulose I to antiparallel-packed fibrils of cellulose II during mercerization (redrawn from Ref. 108). (See Fig. 41 in the Pérez chapter, p. 79.)

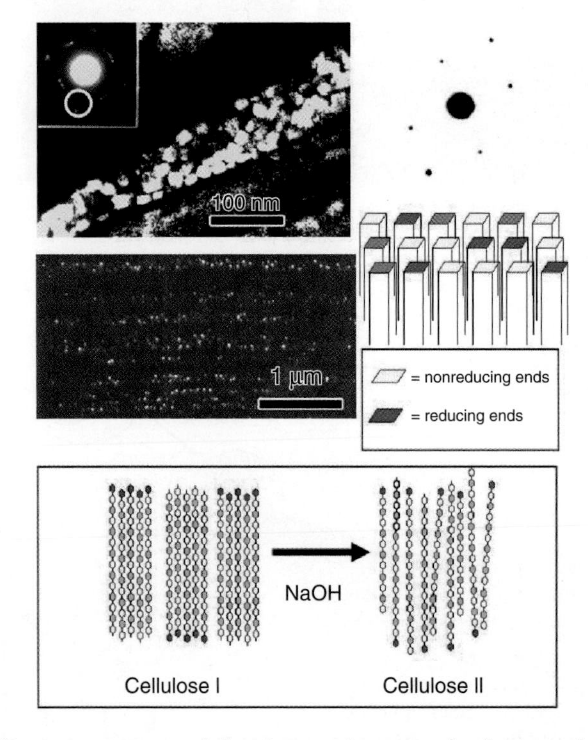

Plate 17 Diffraction-contrast transmission electron microscopy of a fragment of *Valonia ventricosa* cell wall cross-sectioned perpendicular to one of the main microfibrillar directions. The picture is printed in reverse contrast, so that the cross-sectioned microfibrils appear as white squares. (See Fig. 42 in the Pérez chapter, p. 80.)

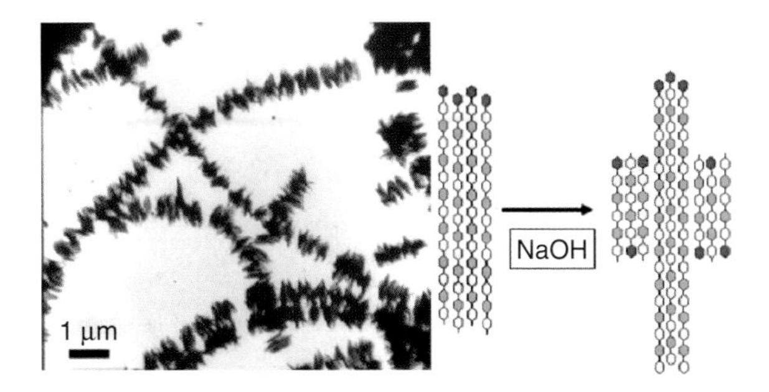

Plate 18 "Shish-kebab" morphology. (See Fig. 43 in the Pérez chapter, p. 81.)

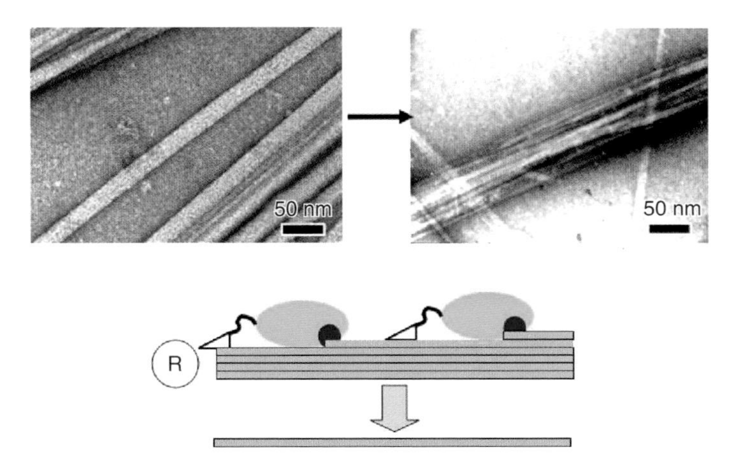

Plate 19 Microcrystals of *Valonia macrophysa* cellulose subjected to the action of cellulases (Cel7A: from *Humicola insolens*) consisting of a hydrolytic core, a cellulose-binding module, and a linker that binds the two enzymic components. The reducing end of the cellulose chains is indicated. R, transmission electron microscopy of the cellulose microcrystals before and after the enzyme action indicates that Cel7A induced a thinning of the crystals. (See Fig. 44 in the Pérez chapter, p. 83.)

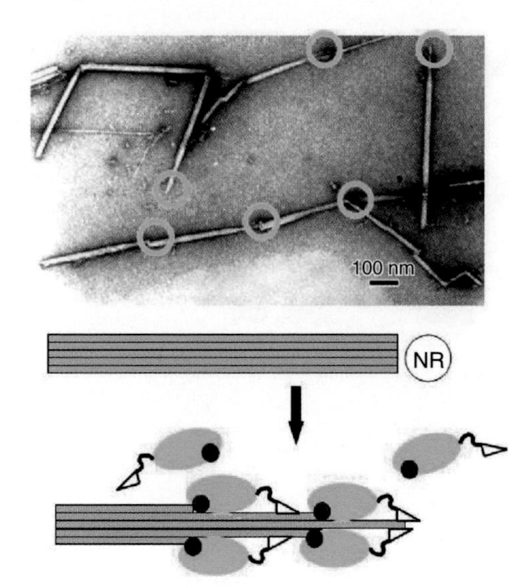

Plate 20 Microcrystals of *Valonia macrophysa* cellulose subjected to the action of cellulases (Cel6A: from *Humicola insolens*) consisting of a hydrolytic core, a cellulose-binding module, and a linker that binds the two enzymic components. The nonreducing end of the cellulose chain is indicated as NR. Transmission electron microscopy of the cellulose microcrystals before and after enzyme action indicates that the crystals are eroded only on one end of their tips, which corresponds to the nonreducing end of cellulose (marked with circles). (See Fig. 45 in the Pérez chapter, p. 83.)

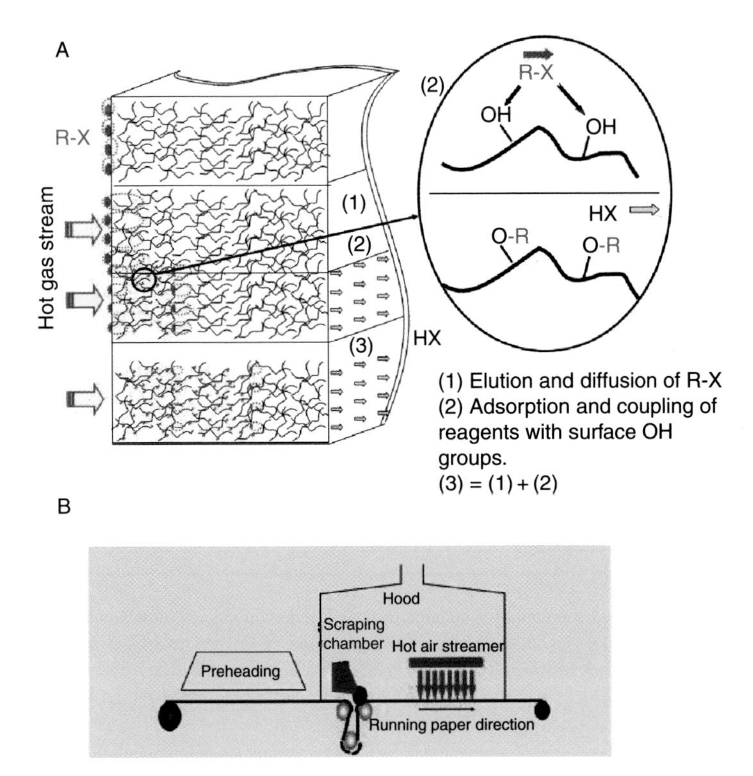

PLATE 21 Chromatogeny principle and engineering. (A) "Chromatogeny" makes use of a novel solvent-free chemical pathway for molecular grafting. It is used to achieve heterogeneous acylation of cellulose fibers through the control of the diffusion properties of acyl chloride reagents and of their by-product, HCl. This reaction occurs in the solid–gas interface, akin to the diffusion occurring in gas chromatography. This analogy is responsible for coining the expression "chromatogeny." (B) Flow sheet of the industrial implementation of "chromatogeny" for cellulose molecular grafting. (See Fig. 47 in the Pérez chapter, p. 87.)

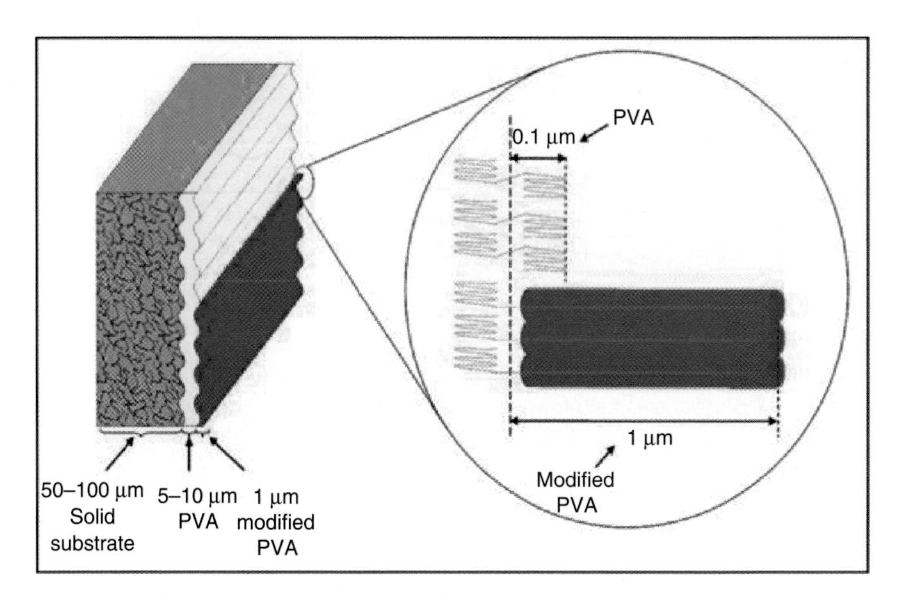

Plate 22 Schematic representation of the architecture of the cellulosic substrates coated with PVA and grafted with long-chain fatty acid molecules. This cellulosic-based material (BT3 pack TM) exhibits remarkable barrier properties to water, grease, and gases. (See Fig. 50 in the Pérez chapter, p. 91.)

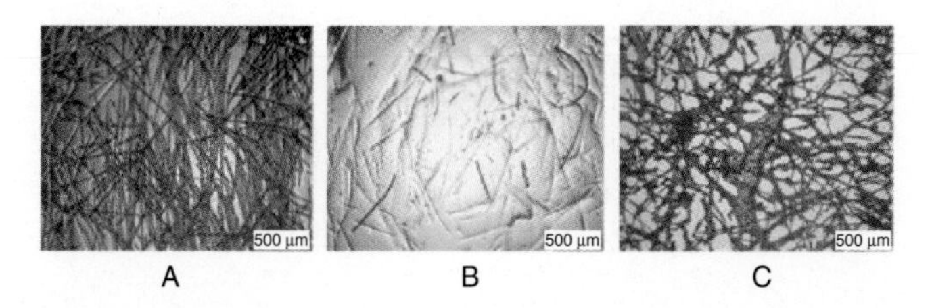

Plate 23 Optical microscopy of (A) cellulose acetate matrix and rayon fibers; (B) cellulose acetate and partially acetylated (38%) rayon fibers; and (C) partially acetylated (38%) rayon fibers after partial removal of the cellulose acetate matrix. (See Fig. 52 in the Pérez chapter, p. 94.)

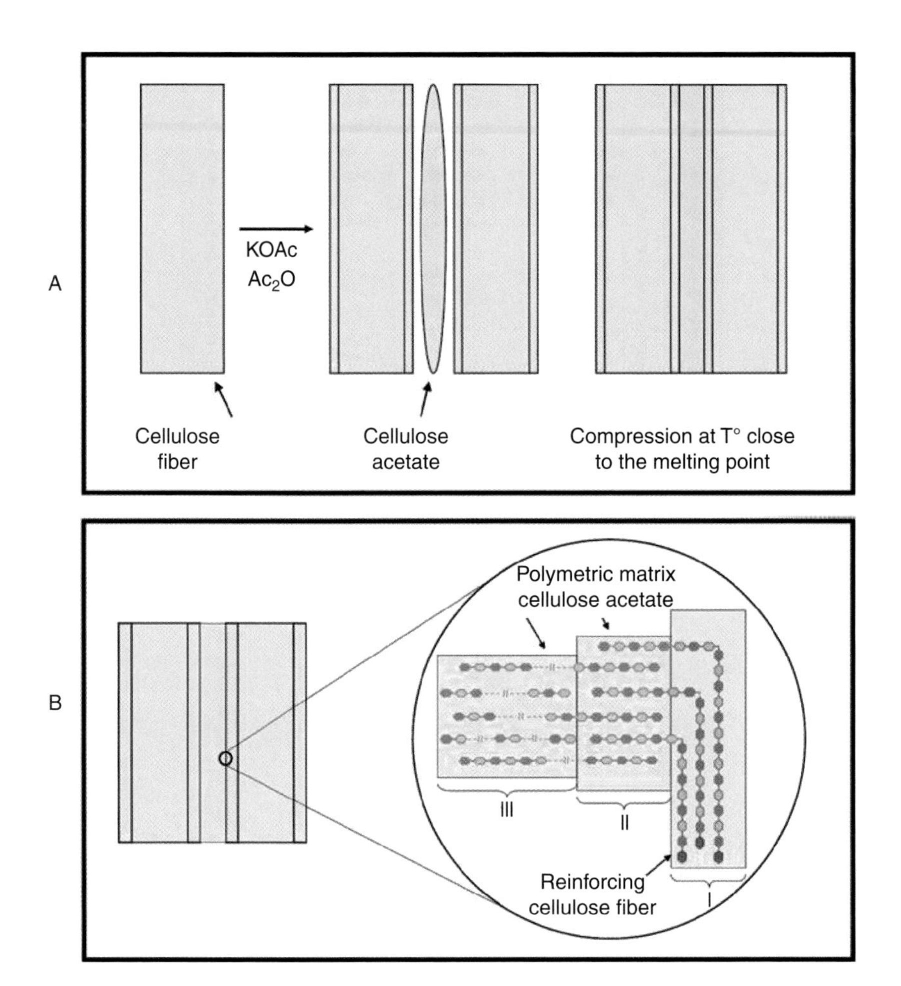

Plate 24 Principle of superficial acetylation of cellulose fibers, followed by compression at a temperature close to the melting point of cellulose acetate. Schematic representation of the intimate macromolecular interaction involved in the thermoplastic composites of cellulose–cellulose acetate. (See Fig. 54 in the Pérez chapter, p. 96.)

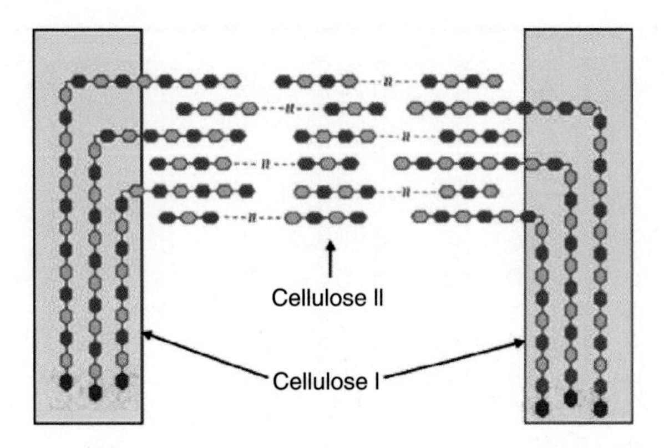

PLATE 25 Schematic representation of the macromolecular interaction involved in the cellulose–cellulose composite. (See Fig. 56 in the Pérez chapter, p. 98.)

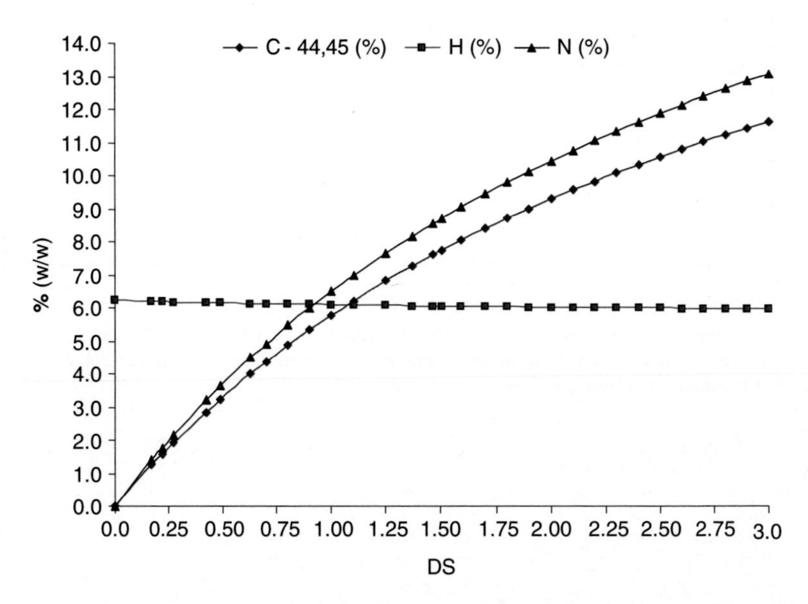

PLATE 26 C, H, and N (%) of *O*-cyanoethyl glucans, depending on DS. For better comparison the graph for C (44.45% at DS 0) is shifted to 0%. (See Fig. 13 in the Mischnick chapter, p. 146.)

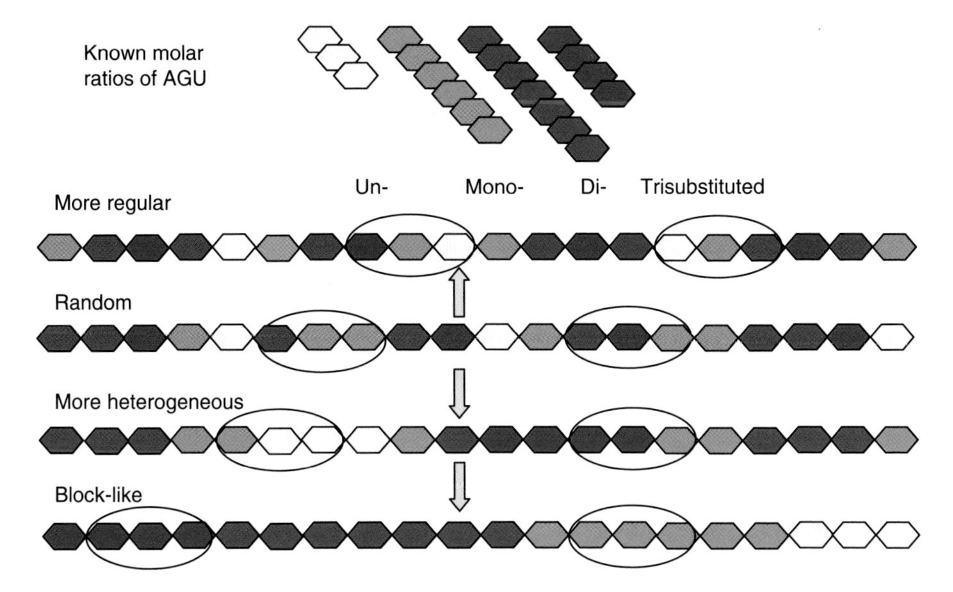

Plate 27 Schematic view of types of distribution of monomer units (AGU) in the polysaccharide chains (see also Fig. 12); "random" is used as defined reference distribution; composition of trimers from partial depolymerization are determined by MS and compared. (See Fig. 25 in the Mischnick chapter, p. 160.)

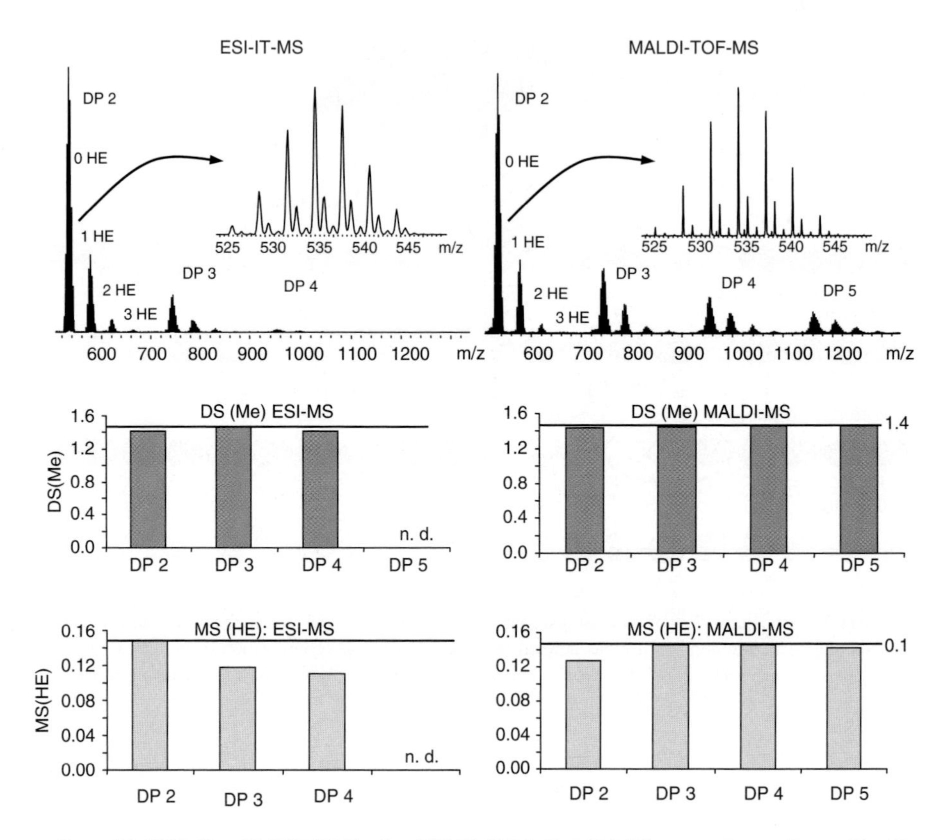

PLATE 28 ESI-MS and MALDI-MS of an HEMC (DS 1.45, MS 0.15) prepared according to Fig. 29. Reprinted from Ref. 177 with permission from the American Chemical Society, Copyright 2006. (See Fig. 30 in the Mischnick chapter, p. 166.)

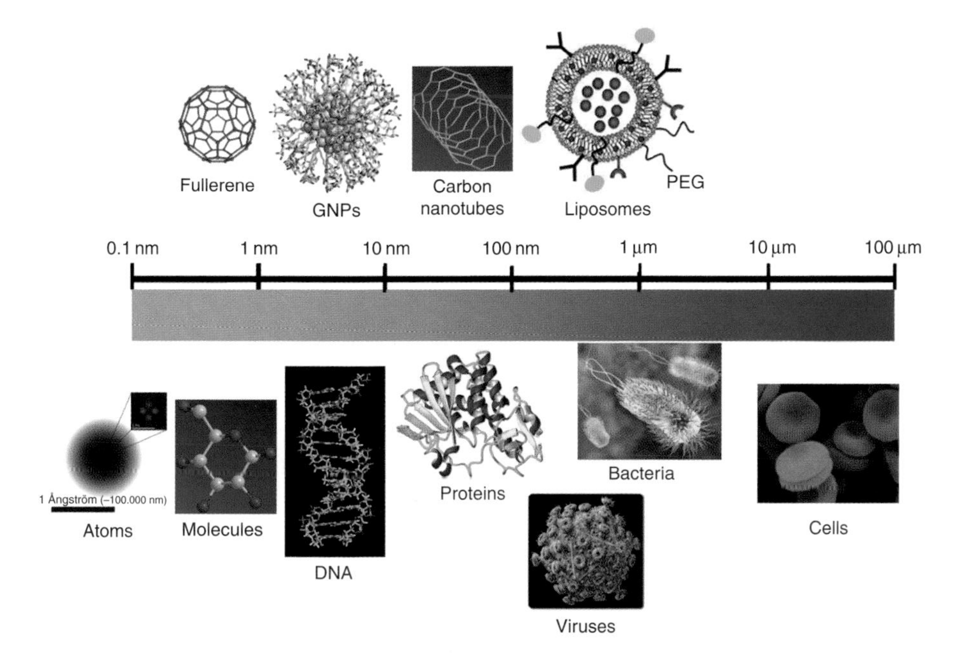

Plate 29 Sizes of glyconanoparticles (GNPs) in relation to other chemical and biological objects. (See Fig. 1 in the Penadés chapter, p. 213.)

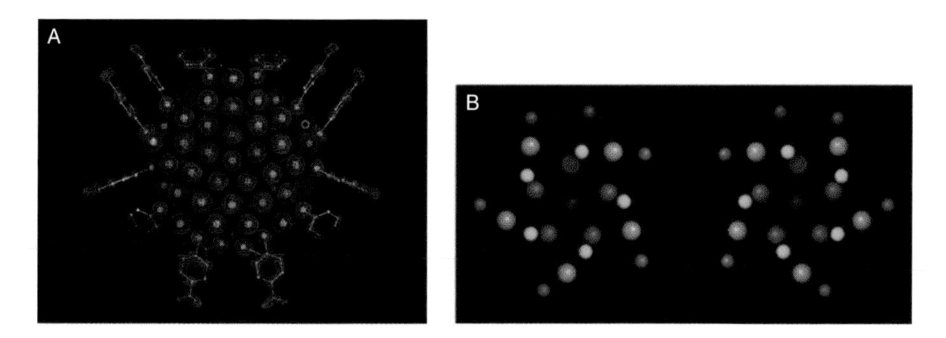

Plate 30 X-Ray crystal-structure determination of the $Au_{102}(p\text{-MBA})_{44}$ nanoparticle. (A) Electron density map (red mesh) and atomic structure (gold atoms depicted as yellow spheres, and *p*-MBA shown as framework and with small spheres [sulfur in cyan, carbon in gray, and oxygen in red]). (B) View down the cluster axis of the two enantiomeric particles. Colour scheme as in (A), except only sulfur atoms of *p*-MBA are shown. Adapted from Ref. 47. (See Fig. 2 in the Penadés chapter, p. 216.)

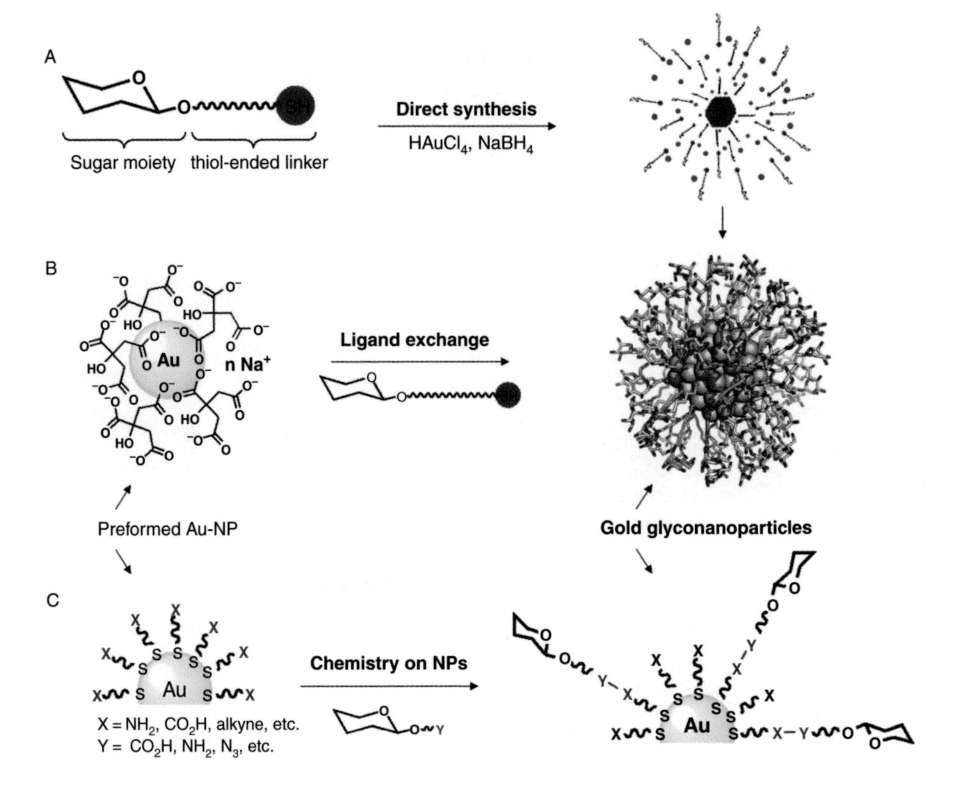

Plate 31 Main methods for the preparation of gold GNPs. (A) Direct synthesis based on the reduction of a Au(III) salt and *in situ* protection of the nascent gold nanocluster with thiol-armed glycoconjugates. (B) Ligand place exchange reactions based on the treatment of preformed gold nanoparticles with thiol-derivatized glycoconjugates. (C) Functionalization of gold nanoparticles by reaction between functional groups on gold surface and suitably derivatized carbohydrates. (See Fig. 4 in the Penadés chapter, p. 219.)

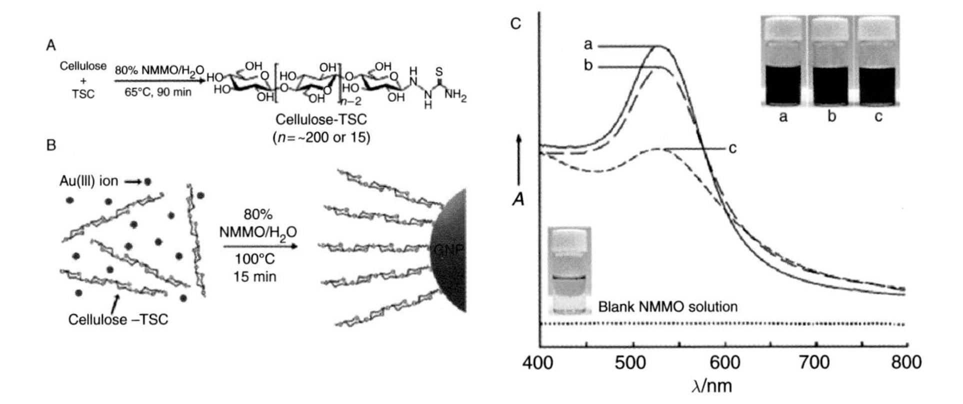

PLATE 32 Gold GNPs by *in situ* protection with thiosemicarbazones carbohydrate derivatives and their UV-Vis spectra. (A) Synthesis of cellulose thiosemicarbazone (Cellulose-TSC). (B) Preparation and possible structure of a cellulose-conjugated gold nanoparticle. (C) UV/Vis spectra and true-colour images of a GNP-free NMMO blank solution and GNPs/NMMO solutions: a) Cellulose-free GNPs, b) Cel_{200}-GNPs, and c) Cel_{15}-GNPs. Adapted from Ref. 93. (See Fig. 7 in the Penadés chapter, p. 229.)

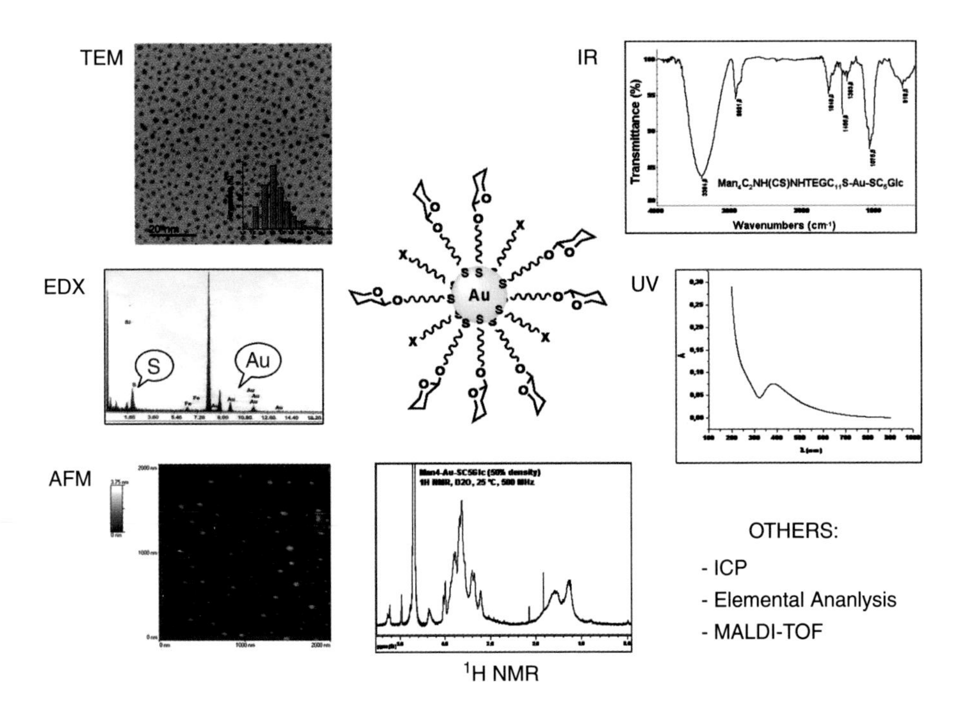

PLATE 33 Different techniques employed for the characterization of GNPs. (See Fig. 8 in the Penadés chapter, p. 238.)

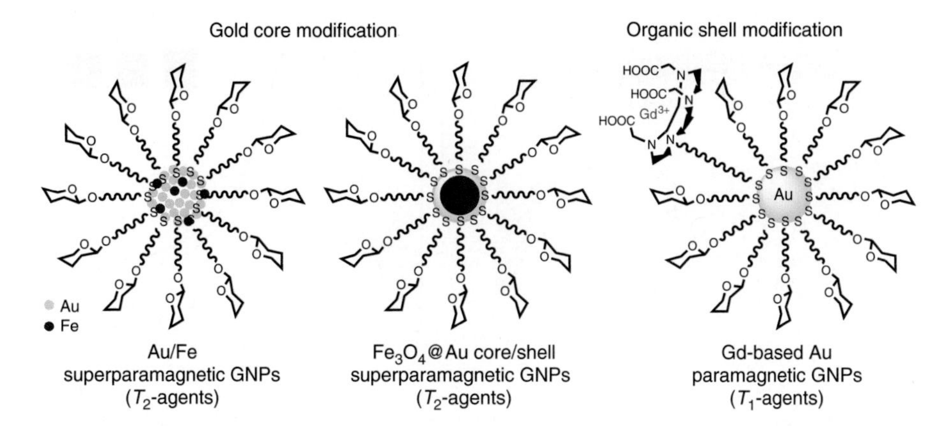

Plate 34 Strategies for converting gold GNPs into superpara- and paramagnetic GNPs. (See Fig. 9 in the Penadés chapter, p. 241.)

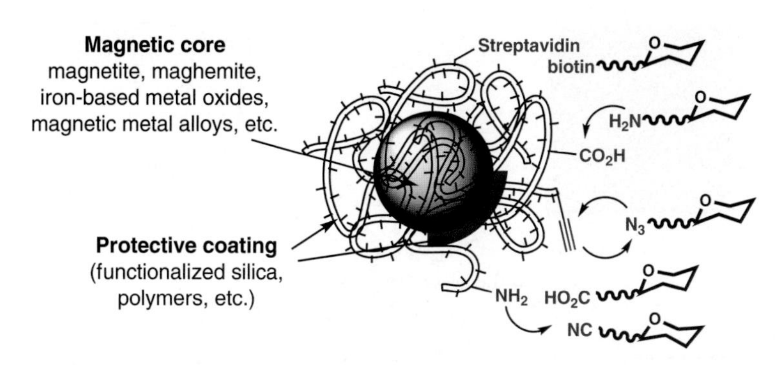

Plate 35 Strategies to prepare iron oxide-based magnetic glyconanoparticles. (See Fig. 12 in the Penadés chapter, p. 244.)

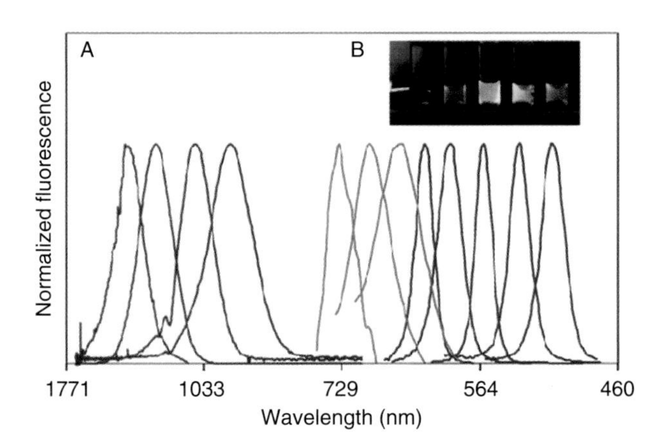

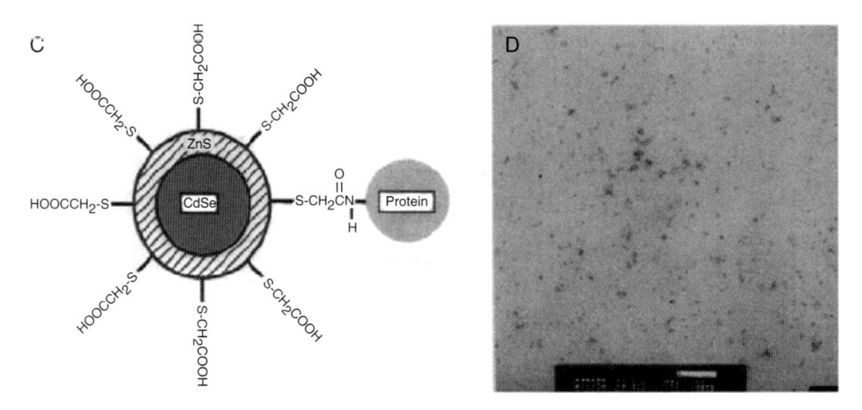

PLATE 36 Biofunctional QDs as fluorescent biological labels. (A) Size- and material-dependent emission spectra of different surfactant-coated semiconductor nanocrystals. The blue series represents CdSe nanocrystals with diameters of 2.1, 2.4, 3.1, 3.6, and 4.6 nm (from right to left). The green series represents InP nanocrystals with diameters of 3.0, 3.5, and 4.6 nm. The red series represents InAs nanocrystals with diameters of 2.8, 3.6, 4.6, and 6.0 nm. (B) True-colour images of silica coated core (CdSe)-shell (ZnS or CdS) nanocrystals in aqueous buffer, illuminated with an ultraviolet lamp. (C) Schematic representation of a core (CdSe)-shell (ZnS)quantum dot that is covalently coupled to a protein by mercaptoacetic acid. (D) TEM of QD-transferrin conjugates. Scale bar, 100 nm. Adapted from Refs. 167 and 168. (See Fig. 13 in the Penadés chapter, p. 246.)

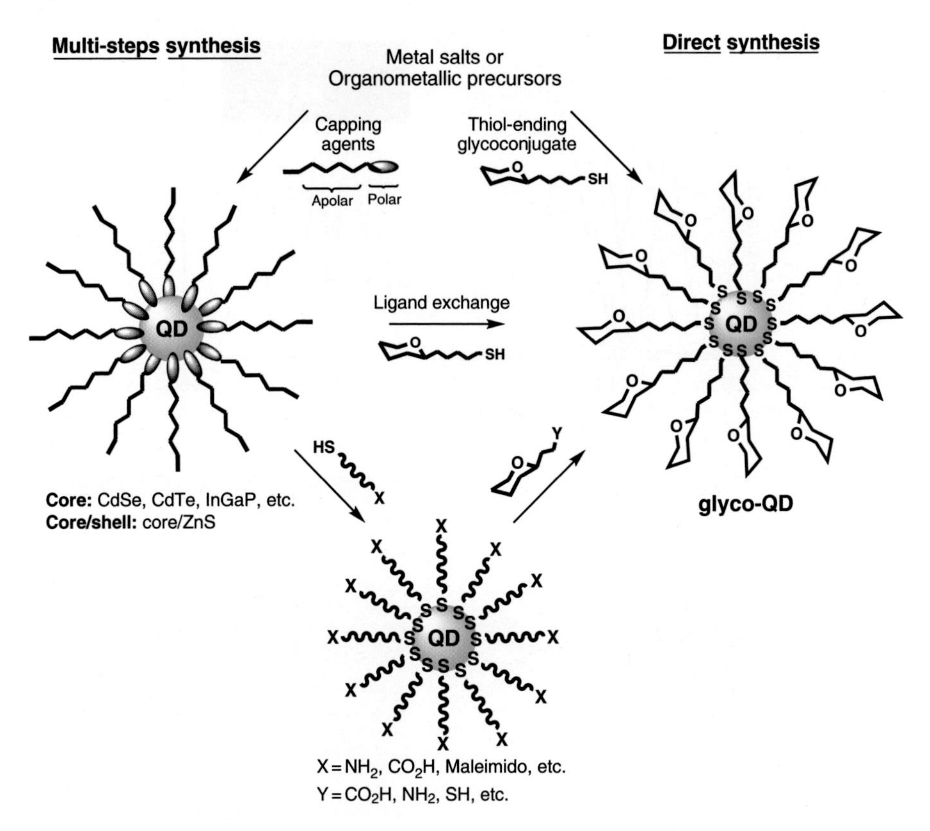

Plate 37 Main strategies for the synthesis of glyco-QDs by means of thiol chemistry. (See Fig. 14 in the Penadés chapter, p. 248.)

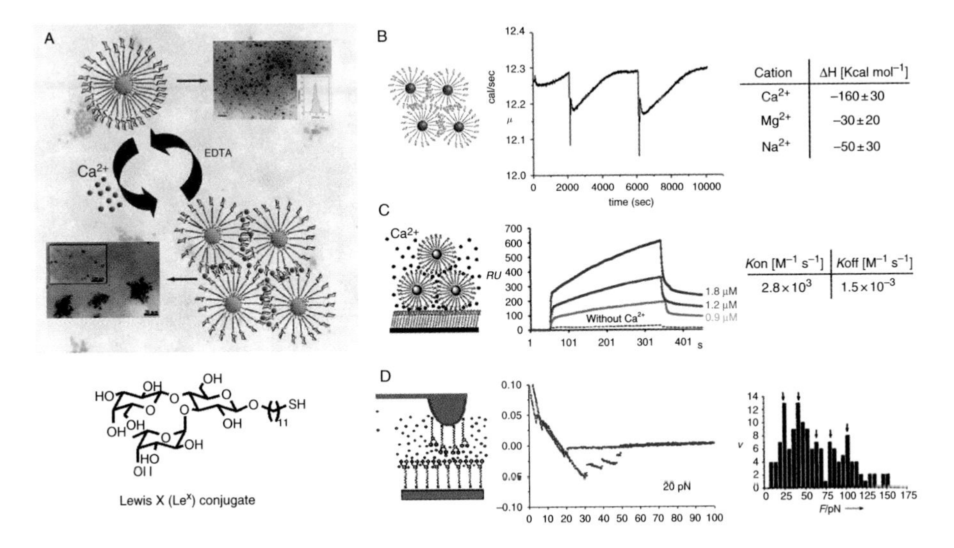

PLATE 38 Glyconanotechnology in carbohydrate–carbohydrate interactions. (A) TEM images showing that the self-aggregation of *Le*^{X}-GNPs is calcium-dependent (0.1 mg/mL of *Le*^{X}-GNPs in 10 mM $CaCl_2$ solution causes 3D *Le*^{X}-GNPs aggregates, bottom-left) and reversible (treating the previous solution with EDTA causes disgregation of the aggregates, up-right). (B) ITC curves showing thermodynamic evidence for self-aggregation of *Le*^{X}-GNPs added to 10 mM CaCl2. (C) SPR sensorgrams for the interaction of *Le*^{X}-GNPs to SAMs of *Le*^{X} conjugates on gold surfaces without (- - -) and with Ca^{2+} ions (—) (middle) and rate constants (k_{on} and k_{off}) obtained from the analysis of the sensorgrams (right). (D) AFM force-distance curves obtained between a *Le*^{X}–functionalized tip and SAMs of *Le*^{X} conjugates in 10 mM $CaCl_2$ solution (middle) and histogram of the corresponding unbinding force measured through 300 force-distance curves (right). Adapted from Refs. 70, 216, 217, and 218. (See Fig. 15 in the Penadés chapter, p. 252.)

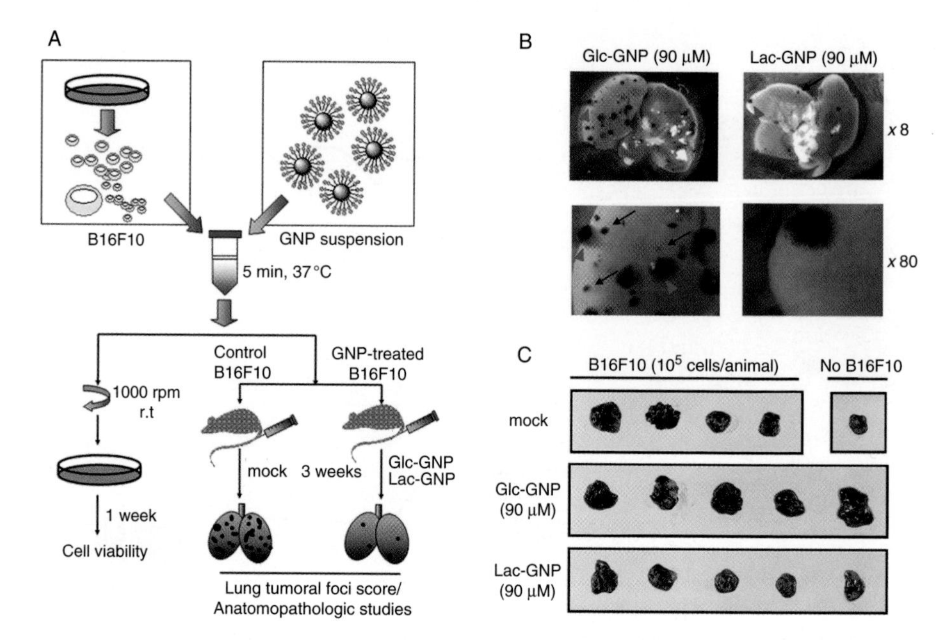

Plate 39 Anti-metastatic effect of gold *lactose*-GNPs on lung tumor development in mice. (A) Schematic representation of *ex vivo* experiments for evaluating cell viability and anti-metastatic potential of glyconanoparticles coated with lactose (Lac-GNP) or glucose (Glc-GNP). (B) Pictures of the lungs at two different magnifications (×8, ×80) corresponding to mice treated with B16F10 cells and Glc-GNP (left) or Lac-GNP (right). Black arrows indicate the small foci (<1 mm) and blue triangles denote the presence of large foci (>1 mm). (C). Lungs corresponding to mice treated with B16F10 cells (mock, up-left), Glc-GNP (middle), or Lac-GNP (bottom) in comparison with the lungs obtained from a control animal (not injected with B16F10 cells; mock, up-right). The specific anti-metastatic effect of Lac-GNPs is evident from (B) and (C). Adapted from Ref. 241. (See Fig. 17 in the Penadés chapter, p. 261.)

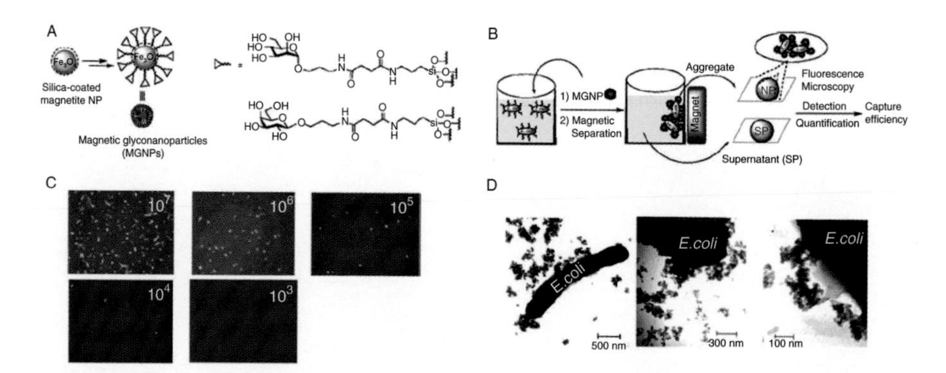

Plate 40 *Escherichia coli* detection and decontamination with Man-MGNPs. (A) Silica-coated magnetite GNPs and neoglycoconjugates used to protect the magnetic core. (B) Schematic representation of pathogen detection by MGNPs. (C) Fluorescence microscopic images of captured *E. coli*. The concentration (cells/mL) of bacteria incubated with Man-MGNPs is indicated on each image. (D) TEM images of Man-MGNPs/*E. coli* complexes. Adapted from Ref. 154. (See Fig. 18 in the Penadés chapter, p. 262.)

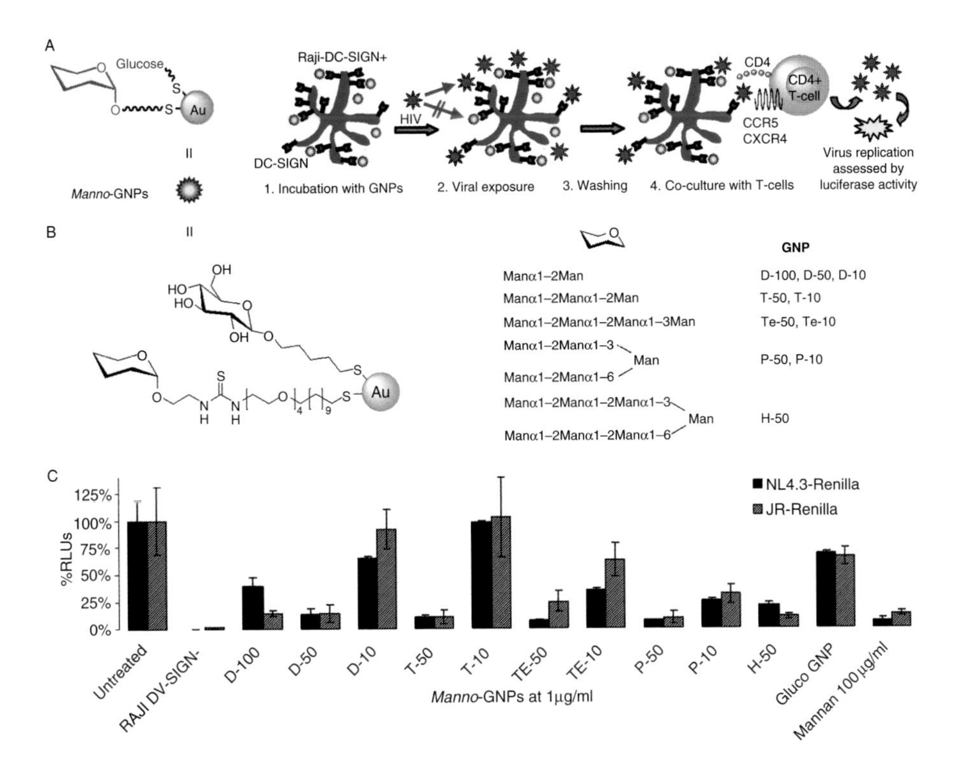

Plate 41 "*Manno*-GNPs" as inhibitors of HIV trans-infection of human lymphocites. (A) Schematic representation of cellular experiments for evaluating the potential of gold glyconanoparticles coated with oligomannosides ("*Manno*-GNPs") as inhibitors of DC-SIGN-mediated HIV-1 trans-infection of human T cells. (B) Schematic representation of "*manno*-GNPs". D, T, Te, P, and H stand for di- tri-, tetra-, penta-, and heptamannose conjugates, respectively; the numbers indicate the percentages of mannose oligosaccharides on GNP, the rest being the stealthy glucoside component. (C) Anti-HIV evaluation of "*manno*-GNPs" at 1 μg/mL in DC-SIGN-mediated trans-infection of human T cells. HIV-1 recombinant viruses NL4.3-Renilla X4 (black) or JR-Renilla R5 (striped) were used. Raji cells not expressing DC-SIGN (Raji DC-SIGN-) were used as control to allow for DC-SIGN-independent viral transfer. Mannan (100 μg/mL) was used as a positive control. Results are expressed as percentages of infection related to untreated control. Adapted from Ref. 244. (See Fig. 19 in the Penadés chapter, p. 263.)

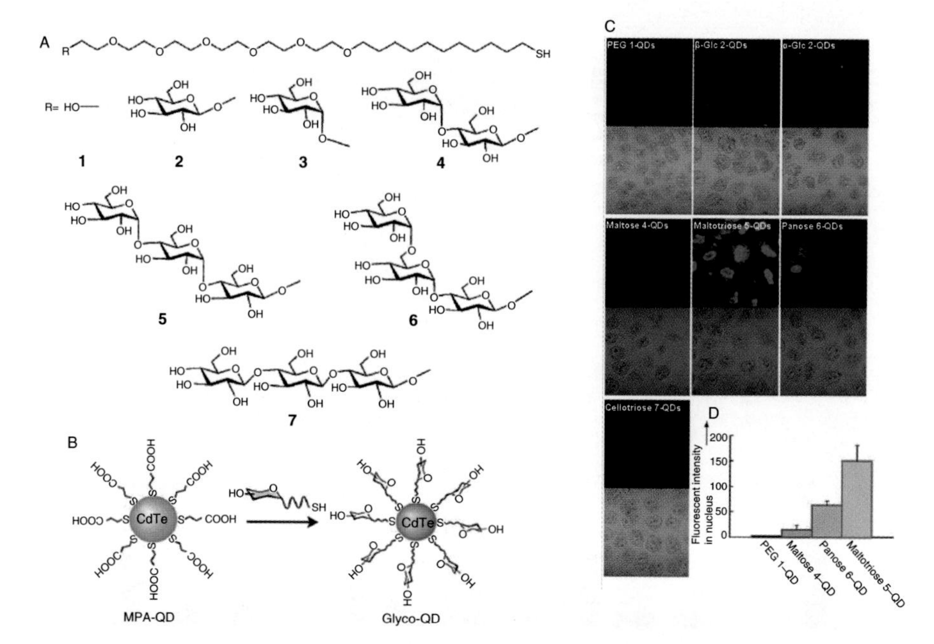

Plate 42 Oligosaccharide-mediated nuclear transport of glyco-QDs. (A) Chemical structures of neoglycoconjugates. (B) Synthesis of glyco-QDs by thiol exchange reaction. (C) Confocal fluorescence images (top panels) and differential interference contact (lower panels) of digitonin-permeabilized HeLa cells incubated with glyco-QDs. (D) Digitalized fluorescence intensity of different QDs in the nucleus. Adapted from Ref. 204. (See Fig. 20 in the Penadés chapter, p. 265.)

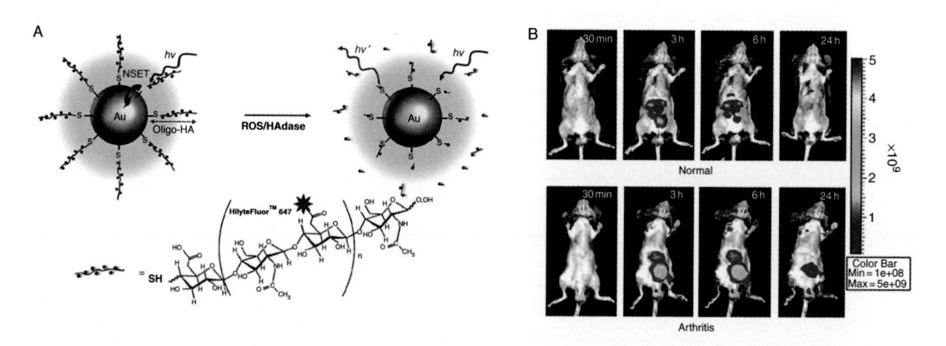

Plate 43 *In vivo* diagnostic application of hyaluronic acid immobilized gold nanoprobes. (A) The fluorescence quenching by nanoparticle surface-energy transfer between Hilyte-647 dye labelled oligo-HA and gold nanocluster (left) is followed by fluorescence recovery after addition of reactive oxygen species/HAdase which release the dye labeled oligo-HA fragments. (B) Tail vein injection of GNPs capped with HA conjugates labelled with Hilyte-647 in normal (up) and arthritis (bottom) mice. Adapted from Ref. 103. (See Fig. 21 in the Penadés chapter, p. 267.)

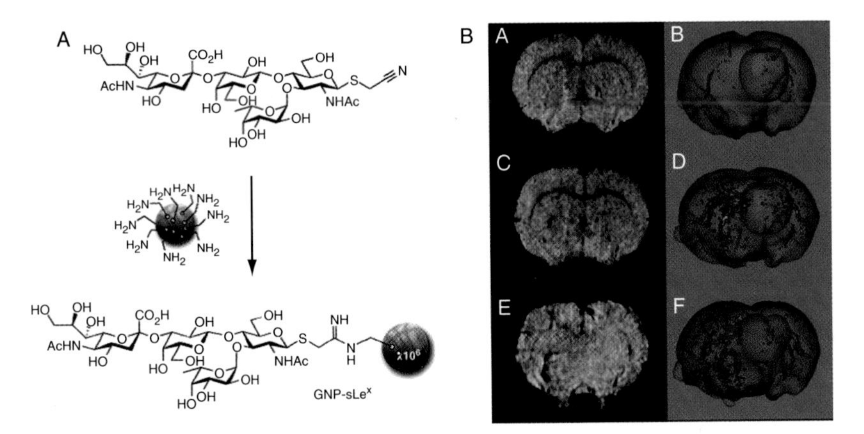

Plate 44 *In vivo* imaging of brain inflammation by means of sLe^X-GNPs. (A) Example of construction of sLe^X-GNPs. (B) T_2^*-weighted images (*A*, *C*, and *E*) and 3D reconstructions of the GNPs accumulation (*B*, *D*, and *F*) reveal that sLe^X-GNPs enables detection of lesions in models of multiple sclerosis (*C* and *D*) and stroke (*E* and *F*) in contrast to unfunctionalized control-NP (*A* and *B*). Adapted from Ref. 153. (See Fig. 22 in the Penadés chapter, p. 268.)

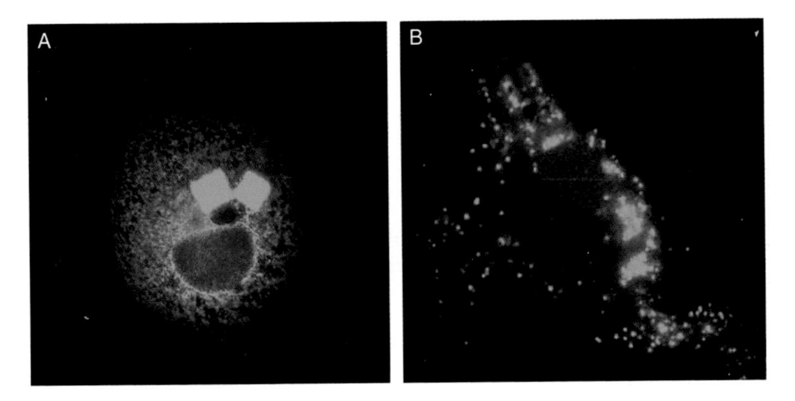

Plate 45 Crystal formation of overexpressed NEU1 in absence of PPCA. (A) Localization of NEU1 in large crystal-like bodies in galactosialidosis fibroblasts that were transfected with a NEU1 cDNA expression construct. (B) A typical punctated lysosomal localization of NEU1 in galactosialidosis fibroblasts that were transfected with both NEU1 and PPCA cDNAs. Cells were incubated with affinity-purified anti-NEU1 antibody and FITC-conjugated secondary antibody. Nuclei were stained with DAPI. Magnification 400×. (See Fig. 1 in the Monti chapter, p. 411.)

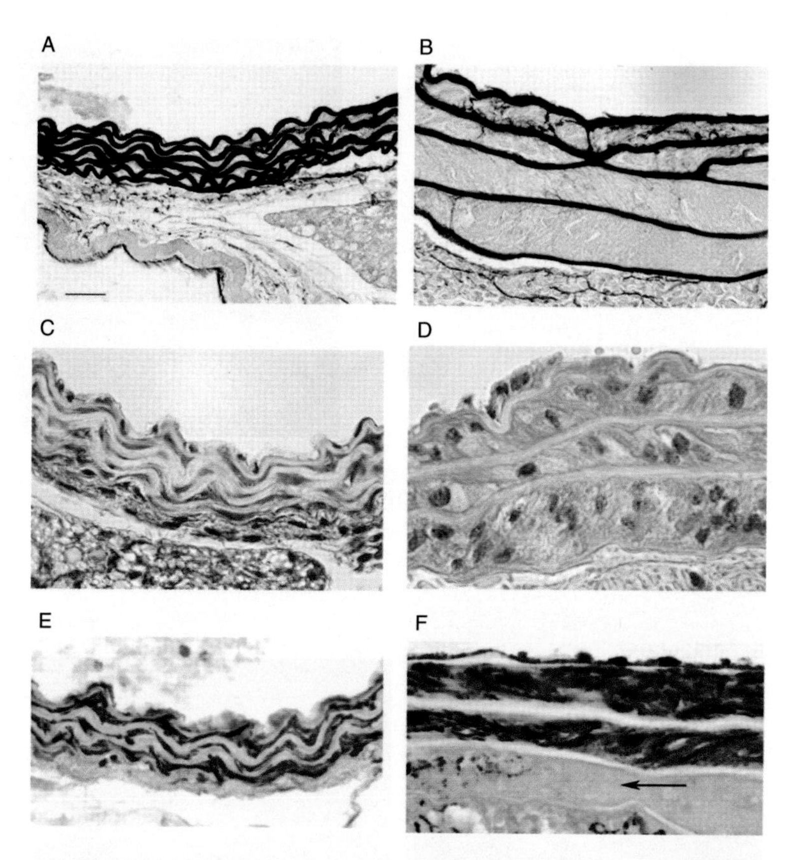

Plate 46 Histochemical staining of control and *Neu1*-null aorta for elastin, mucins, and smooth muscle. (A) Control aorta stained for elastin with Hart's stain showing 5–6 elastic lamellae and (B) *Neu1*-null aorta showing the abnormal separation between the elastic lamellae. (C) Control aorta stained for sialomucins with Alcian blue (blue) with elastic lamellae appearing opaque and (D) *Neu1*-null aorta stained for sialomucins indicating the large mass of positive material between elastic lamellae. (E) Control aorta immunostained for smooth muscle with an antibody to alpha-smooth muscle actin (red) showing the thin area between elastic lamellae and (F) *Neu1*-null aorta showing the increased mass of smooth muscle between elastic lamellae, whereas some areas were devoid of smooth muscle (arrow). Scale bar, 22 μm. From Starcher *et al.* (See Fig. 2 in the Monti chapter, p. 417.)

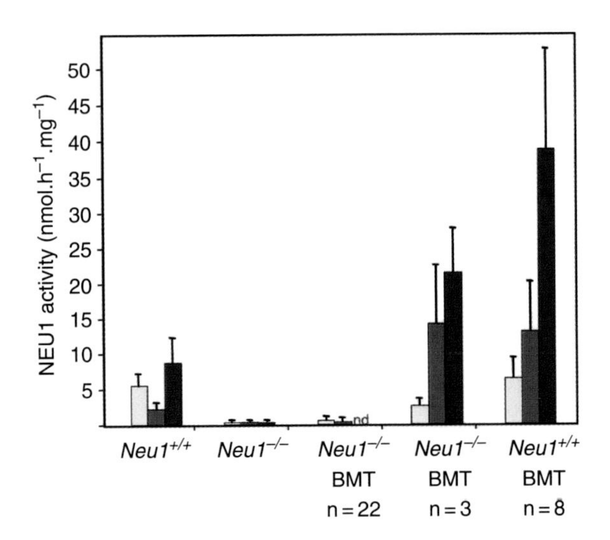

PLATE 47 *Neu1*$^{-/-}$ mice are deficient in long-term bone marrow engraftment. Sub-lethally irradiated *Neu1*-null and wild-type mice were transplanted with transgenic NEU1-expressing bone marrow (BM) and analyzed 5 months after bone-marrow transplantation. To assess engraftment of the donor bone marrow, NEU1 activity was measured in hematopoietic tissues from both bone marrow-transplanted and untreated wild-type and *Neu1*-null mice. While all transplanted wild-type mice had successfully engrafted the donor bone marrow, 22 of the 25 transplanted *Neu1*-null mice failed to engraft. Liver, yellow bars; spleen, red bars; BM, green bars. (See Fig. 3 in the Monti chapter, p. 420.)

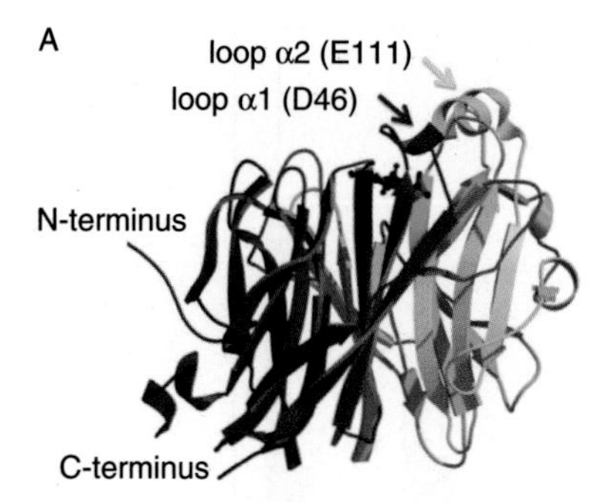

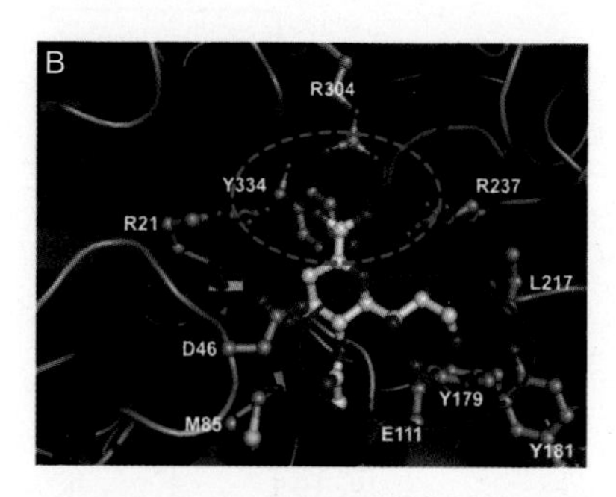

Plate 48 Structural features of the cytosolic sialidase NEU2. (A) Ribbon diagram of NEU2–Neu5Ac2en (DANA) complex viewed from the side. Neu5Ac2en is represented using ball and stick model. The six blades of the β-propeller are colored differently. The colored arrows indicate loops α1 and α2 that become structured upon the competitive-inhibitor binding to the NEU2 active site and contain D46 and E111, respectively. (B) The catalytic crevice of NEU2–Neu5Ac2en complex. Neu5Ac2en and R-chains of the residues involved in the competitive-inhibitor coordination are represented as ball and stick models. Blue arrows represent β-strands and a red ribbon represents a portion of α2 helix bearing the E111 residue. Conserved residues involved in the catalytic process and in the Neu5Ac2en coordination, namely, R21, D46, R237, R304, and Y334, are highlighted by a green dotted oval. Other residues, namely E111, Y179, or Y181, highlighted by a red dotted oval, are present only in the NEU2–Neu5Ac2en complex and interact with the glycerol terminal oxygen atoms of the inhibitor. Modified from Chavas *et al.* (See Fig. 4 in the Monti chapter, p. 424.)

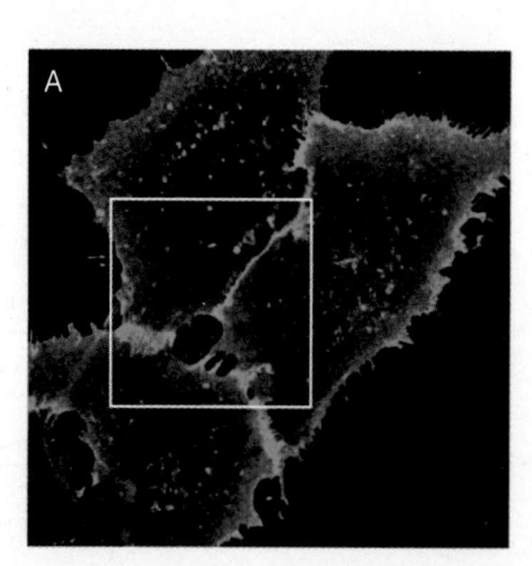

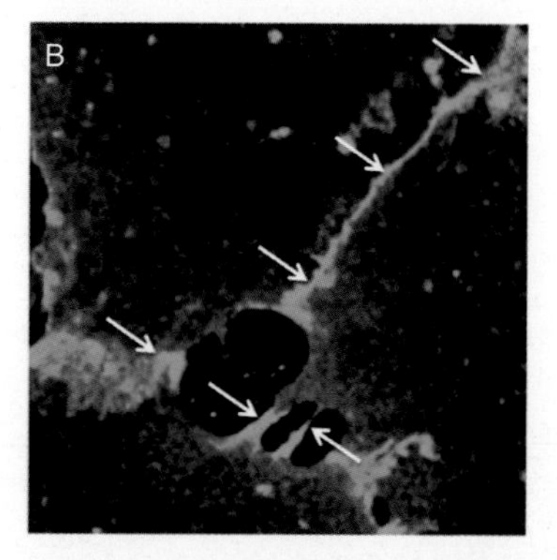

Plate 49 Localization of NEU3 at the cell-to-cell contact areas. HeLa cells were transfected with pcDNA 3.1-NEU3-HA, a chimeric form of the murine protein that has been demonstrated to be fully active and without significant variations in the subcellular localization compared with the wild-type protein of human origin. Laser confocal analysis of transfectants at the cellular adhesion plane (panel A, magnification 250×) shows that the protein is highly enriched at the cellular boundaries as expected. Higher magnification (panel B showing inset in A, magnification 600×) demonstrates a significant enrichment of the protein at cell-to-cell contact areas (arrows in B). (See Fig. 6 in the Monti chapter, p. 431.)